TOOLS FOR PROBLEM SOLVING

K. Elayn Martin-Gay

Carol Carter

Carol Ozee

Sarah Kravits

Joyce Bishop

Kathleen T. McWhorter

Maria Valeri-Gold

Frank Pintozzi

Donna Gorrell

Muriel Harris

John M. Lannon

M. Jimmie Killingsworth

Jacqueline S. Palmer

Kristin R. Woolever

Sharon J. Gerson

Steven M. Gerson

Pearson
Custom
Publishing

Prentice
Hall

Cover Art: © Jerry Blank/The Stock Illustration Source

Excerpts taken from:

Algebra: A Combined Approach;
and *Basic College Mathematics*
By K. Elayn Martin-Gay
Copyright © 1999 by Prentice-Hall, Inc.
A Pearson Education Company
Upper Saddle River, New Jersey 07458

The Writer's FAQs: A Pocket Handbook
by Muriel Harris
Copyright © 1999 by Prentice-Hall, Inc.

Technical Writing: Process and Product, Third Edition,
by Sharon J. Gerson and Steven M. Gerson
Copyright ©1999 by Prentice-Hall, Inc.

Written Communications Resource Digest,
by Pamela J. Gurman
Copyright © 1999 by Pearson Custom Publishing
A Pearson Education Company
Boston, Massachusetts 02116

Keys to Thinking and Learning: Creating Options and Opportunities,
By Carol Carter, Joyce Bishop, Sarah Lyman Kravits and Judy Block
Copyright © 1999 by Prentice-Hall, Inc.

Information in Action: A Guide to Technical Communication,
by M. Jimmie Killingworth and Jacqueline S. Palmer
Copyright © 1998 by Allyn and Bacon
A Pearson Education Company
Needham Heights, Massachusetts 02494

Pathways to Success,
by Carol Carter, Carol Ozee and Sarah Lyman Kravits
Copyright © 1998 by Prentice-Hall, Inc.

This special edition published in cooperation with
Pearson Custom Publishing

Printed in the United States of America

10   9   8   7   6   5

Please visit our web site at www.pearsoncustom.com

ISBN 0-13-028695-8

BA 992138

PEARSON CUSTOM PUBLISHING
75 Arlington Street, Boston, MA 02116
A Pearson Education Company

CONTENTS

PART I

ITT Learning Resource Center, Internet Research Tools, and Library Skills

CHAPTER **1**

The Internet, the Intranet, and the Extranet

For the first time in centuries, the written word has undergone a quantum metamorphosis, leaping from the printed page into cyberspace. This change has affected technical writing significantly. Correspondence, once limited to letters and memos, is now often online as e-mail. Corporate brochures and newsletters, once paper bound, now are online. Product and service manuals, once paper bound, now are online. Résumés are online. Research is online. Technical writing is increasingly electronic. To keep up with this communication revolution, you must go online also. By teaching you how to write successful e-mail, write online help screens, and create an effective Web site, this chapter will help you learn how to work in cyberspace.

THE INTERNET—THE "INFORMATION SUPERHIGHWAY"

Your e-mail, online help screens, and Web sites will be transmitted via the Internet, an intranet, or an extranet. But what do these words mean? Though the word "Internet" has been bandied about for years, finally on October 24, 1995, we received an official definition. The Federal Networking Council (FNC) unanimously resolved to define the word, in consultation with leadership from the Internet and intellectual property rights (IPR) communities. Here's their definition ("FNC Resolution"):

> The Federal Networking Council (FNC) agrees that the following language reflects our definition of the term "Internet."
>
> > "Internet" refers to the global information system that—
> > 1. Is logically linked together by a globally unique address . . . based on the Internet Protocol (IP) or its subsequent extensions.
> > 2. Is able to support communications using the Transmission Control Protocol/Internet Protocol (TCP/IP) suite or its subsequent extensions . . . and other IP-compatible protocols.

3. Provides, uses or makes accessible, either publicly or privately, high-level services layered on the communications and related infrastructures.

The FNC's definition, though official, is challenging. Let's try to simplify matters. The Internet is a network that electronically connects over a million computers internationally. And it is growing in importance daily. The number of Internet data packets grew from 152 million in June 1988 to 60,587 million packets in July 1994. Host growth escalated from 235 in May 1982 to over 3 million in 1994 (Pike et al., 14–15). As of January 1998, the Internet included approximately 30 million users in over 100 countries ("Internet"). Only three months later, an article in the *Kansas City Star* reported that "traffic on the Internet is doubling every 100 days" and stated that "as many as 62 million Americans are now using the world-wide network" ("Internet Traffic" B1).

Furthermore, the Internet is a decentralized medium, with very few external controls. To use the Internet, you can choose which Internet service or provider you wish, and you can access any Internet site you desire. Thus, as the synonym "information superhighway" suggests, the Internet allows the user to access data from various locations almost without restriction. The Internet lets you travel from your computer to information sites worldwide. You can speak to and hear from educators, engineers, lawyers, doctors, businesspeople, hobbyists, or anyone linked to this international communication web, day or night.

How do you find the information you need on the Internet? If you were on a highway looking for a specific site, you might consult a road atlas. To help you find information on the Internet, you can use something resembling a road atlas—the World Wide Web (WWW or Web), along with Web browsers such as *Netscape Navigator* and Microsoft's *Internet Explorer*. The World Wide Web supports documents (Web sites) formatted in a special computer language called HTML (Hyper-Text Markup Language). (See Table 1.1 for Web abbreviations and acronyms.) HTML allows the webmaster (the creator of the Web site) to include graphics, audio, video, and animation as well as hypertext links to other sites and other Web screens ("World Wide Web" 1).

THE INTRANET—A COMPANY'S INTERNAL WEB

Whereas the Internet opens the entire world of information to us, an intranet is more limited. It is a contained network. An intranet might include several, linked local area networks (LANs), and it might include connections to the Internet. Like the Internet, an intranet uses TCP/IP, HTTP and other Internet protocols. The difference, however, is that an intranet belongs to a company. Its purpose is "to share company information and computing resources among employees" ("Intranet"). Individuals outside the company cannot access the intranet unless they have corporate authorization; a "firewall" prevents access to an intranet's private network ("Firewall" 1).

How does a company benefit by using an intranet? Rather than printing hundreds of copies of a document for employees who might be officed in buildings located around a city or around the world, a company can make this information accessible to its employees through an intranet. Companies can place on their corporate intranets

- Web-based discussion forums. These allow employees to post, read, respond to, and then archive corporate e-mail messages.
- Web-based multimedia and kiosks for instructional or informational purposes.

TABLE	1.1	Common Web Site/Internet Abbreviations and Acronyms
<A HREF>		Anchor Hyper-reference
<BG>		Background
CGI		Common Gateway Interface
.com		commercial business—domain name
.edu		educational institution—domain name
ftp		file transfer protocol
.gif		graphics interchange format
.gov		governmental—domain name
HTML		HyperText Markup Language
HTTP		hypertext transfer protocol
<IMG SRC>		image source
.jpeg, jpg		joint photographic experts group
<LI>		listed item
.mil		military—domain name
moo		MUD, object oriented
MUD		multi-user dialogue
.net		network—domain name
<OL>		ordered list
.org		organization—domain name
SGML		Standard Generalized Markup Language
TCP/IP		Transmission Control Protocol/Internet Protocol
.tif		tagged image file
<UL>		unordered list
URL		uniform resource locator

- Online polls. Employees can respond to corporate polls. Then the results can be tabulated, and employees around the company can view the results.
- Company forms. These can be filled out and submitted electronically.
- Policy and procedure manuals. Whereas paper documents take up space, get dusty, and get lost, online intranet manuals save space and are available instantly.
- Employee phone directories. Company employees come and go. Online, additions and deletions can be updated easily.
- Organizational charts. Company hierarchies change rapidly. Online, they can be updated and conveyed to employees immediately.

The benefits of an intranet are numerous. An intranet is paper free. Companies can lower mailing, binding, and printing costs. Information can be updated easily. Data can be archived without taking up space. Procedures, polls, directories, forums, and forms are at an employee's fingertips rather than buried under paper or on the top shelves of inaccessible bookshelves. Most important, every corporate employee throughout a company—whether in the office or offsite—can access corporate information online at the touch of a key. Speed, efficiency, and cost savings—these are the benefits of a corporate intranet ("Intranet Applications").

THE EXTRANET—A WEB WITHIN A WEB

We've seen that the Internet gives us worldwide access to information online, and we've explained that an intranet gives employees within a company access to corporate information. What's an extranet? An extranet "is a collaborative network" that uses Internet protocols to "link businesses with their suppliers, customers, or other businesses" that have common goals. An extranet thus allows several companies to share information but also to keep this information within the confines of their collaborative unit. Extranets can include Web-based training programs of mutual benefit to the participating companies, product catalogs, private newsgroups relating to a specific industry, and management approaches for companies working on a shared project ("Extranet").

2 ITT Learning Resource Center

A LIBRARY WITHOUT WALLS

Libraries have traditionally provided service, organization, and access to resources. Technological advances now enable us take to advantage of the flexible, evolving nature of the World Wide Web to provide convenient access to digital information 7 days a week, 24 hours a day. The ITT Technical Institute Virtual Library is ITT Technical Institute's online service to provide access to periodical databases, electronic reference sources, and information services. Our online collection features two periodical databases which include millions of articles from thousands of general interest and technology-related professional journals. Users access these materials through customized interfaces directing them through search and retrieval of full-text, full-image articles. Students are guided to the myriad of electronic reference sources that are available on the Internet through subject and curricula-organized collections of links to recommended sites. Reference materials at these websites include encyclopedias, dictionaries, and directories as well as specialized resources focusing on careers, technology, science and the humanities. Textbook websites, specific to the ITT Technical Institute curricula are included in the Virtual Library to help provide an interactive learning environment for students and to supply support resources for instructors. The Virtual Library's Help Desk is an information service that provides students, faculty, and staff with the means to seek and receive library reference services electronically. Resources and services of the Virtual Library provide a system that can serve many of the information needs of the students, faculty, and staff of the ITT Technical Institutes.

HOW TO GET STARTED

Step 1: Go to the ITT Technical Institute Virtual Library webpage: **http://vl.ittesi.com/vlib**

Step 2: Click on **New User Registration** and follow the instructions. If you have already registered, click on **Enter Library.**

Step 3: Enter the user login name and password that you selected during the registration process.

Registration

ITT Technical Institute students, faculty, and staff are encouraged to become registered users of the Virtual Library. The steps of the registration process are explained in detail on the website. Users should be prepared to provide their name, student (or employee) identification number, school location, telephone number, and email address. Check with the Learning Resource Center staff at your school for more information about the registration process.

Step 4:

- Select **Online Collection** to search for magazine or journal articles
- Select **Net Resources** to connect to recommended websites
- Select **Help Desk** to ask a question
- Select **Textbook Websites** to utilize online supplemental textbook materials.

ONLINE COLLECTION

HOW TO USE A PERIODICAL DATABASE

Step 1: Choosing a Database **EbscoHost Masterfile Premier** is a good choice for general interest and cross-disciplinary information. Use this database to look for articles on art, economics, history, health, education, business, sociology, and other broad topics.

ProQuest Direct is a set of databases all relating to technology. Select from *Applied Science & Technology, Telecommunications,* or *Computing.*

Step 2: Starting a Search **EbscoHost** and **ProQuest Direct** offer very similar search screens which ask you to enter a keyword. A keyword is actually the subject of your search. For example, if you want to locate articles about physics, then physics is the keyword. You must point your mouse to the keyword entry box, click on it so that you have a flashing cursor displayed, and type in the keyword. Next you click on the "Search" button when using **ProQuest** or **EbscoHost.**

You can get help right on your screen by clicking on the "Help" button. There are also several options to help you design your search.

Step 3: Viewing the Search Results Articles that match your search request are displayed next. The listing will contain key information about the article including the title, author, and the name and date of the magazine or journal where it appeared. Additional information provided may include the number of pages, journal volume number, whether it is illustrated, etc. You will be able to determine whether the complete article (full text or full image) or an abstract of the article is available.

Step 4: Refining the Search You may find that your results are not what you wanted or expected. You may refine your search by using different keywords, adding or subtracting search terms, limiting your search to a particular publication or time period, specifying that you are only interested in full text articles, or using any of the *advanced* searching features that are available. These features are well explained on the help screens within the databases.

Step 5: Viewing / Printing Your Results You can display an article by clicking on its title from the results list. The complete citation information will be displayed at the top of the page, followed by the abstract and text content. You may print the article or email it to an account to print later.

These general steps can help you get started. The best information on how to use the databases is found within the databases themselves through the "Help" screens.

PROQUEST

COMPUTING, TELECOMMUNICATIONS, APPLIED SCIENCE & TECHNOLOGY

Keyword Search

Enter one or more words or a phrase in the text box. Phrases containing three words or more may be enclosed in quotation marks (example: "world trade organization"). You can use Boolean operators: AND, AND NOT, OR, to narrow or broaden your search. The * symbol is used as a wildcard if you're not sure about the correct spelling (example: gettysb*rg will find gettysberg and gettysburg). The ? symbol is used for truncation and finding words sharing a common root (example: educat? will find education, educated, educators)

Dropdown menus beneath the text box give you additional options. You can use the menus to select a specific date range for your search, restrict the search by publication type, or search for occurrences of your search terms in the full text of articles instead of the nine basic fields.

ProQuest displays complete MLA-required citation results in reverie chronological order (most current information first). To view an article display, click on the viewing format icon or the citation's title. To return to your list of citations, click on the **Results List** box off the main menu.

Viewing Formats

Citation / Abstract provides bibliographic information, including the article title, author, source, volume, issue, publication date, and a brief summary of the article.

Full Text Format gives bibliographic information, abstract, and article content without pictures.

Text + Graphics Format gives bibliographic information, abstract, and article content, including any photos, charts, or other graphics.

Page Image Format gives an exact scanned reproduction of the article, similar to a photocopy from the actual magazine. (Page Image requires Adobe Acrobat Reader for viewing.)

Printing, Emailing, and Saving an Article

To email articles, enter an email address in the "Email Article" textbox.

To print, copy, or save articles, click on the "Print Article" button. Instructions on saving articles to disk, copying and pasting parts of articles, and printing articles are provided.

Subject Searching

Select the **Subject List** link off the Search by Word-Basic screen, enter a search term, and you receive a list of matching topics or suggested terms that help broaden or narrow your search.

Advanced Searching

Search by Word-Advanced helps you build a customized search. You can search up to three terms simultaneously and specify operators and fields using the drop down menus. The search field drop down menu default, Basic Fields, includes searching the author, abstract, article title, company name, geographical name, personal name, product name, subject terms, and source/publication fields. Limiting by Boolean operator and date range is also available. To restrict your search further, specify an article type (such as interview) or publication type (such as reference books) or accept the default of All for both drop-downs.

Search for Publication

This feature allows you to find specific issues of magazines, journals, or newspapers. You may enter a specific title or enter just a keyword from a title. You can then select a publication, choose a specific issue, and browse the table of contents to look for articles of interest.

EBSCOHOST

KEYWORD SEARCHING

A search begins at the search screen. Select the "New Search" icon located on the main toolbar to create a new search. Use the Keyword Search Screen to create a Boolean search. The Boolean operators are AND, OR and NOT. They allow you to create a very broad or very narrow search.

SEARCH LIMITERS AND EXPANDERS

Search options allow you to adjust the focus of your search. Use Limiters to narrow your search results. For example, select "Full Text" to limit Your search results to full text articles. To limit search results to articles from a specific publication title, enter the title name in the "Magazine" field, for example, "Time". You can also limit your search by date range.

Search options also include Expanders, which you can use to broaden your search results. For example, select the option "also search, for related words" to expand a search to include synonyms and plurals of your search terms. Select the option "also search full text articles" to search for your keywords within full text articles.

VIEWING FORMATS

The Result List is the information EbscoHost retrieves from the databases that are searched. Search results can be citations, document summaries (or abstracts), or full text. The number of matching results can be found in the top left corner of the screen. Click on the right or left arrow icons (when active) to move to the next or previous page of results. The Full Record opens when you click on the citation. When available, clicking the open book icon will display the full text article, the page icon will open the result in Adobe Acrobat Reader as a PDF file, and the compound document icon indicates that the article has full text with images embedded in it.

BROWSING THE AUTHORITY FILE

The Authority File is an alternative to keyword searching that allows you to browse a list of subjects, magazine titles, or other database specific indices. Select an Authority File icon (for example: Subject Search) located on the main toolbar, to display the Authority File Screen. Enter the term(s) you want find in the text box provided. Select "Browse" to alphabetically position the Authority File List and view the term(s) entered. To browse by People, Products & Books, Companies or Subjects, click on one of the related links before entering your terms.

PRINTING SEARCH RESULTS

EbscoHost allows you to print records from the Result List or Full Text. To print records from the Result List, select the record you want to print. To print full text, select the open book icon to display an available full text article. Next, select **Print/Email/Save** to display the Delivery Options screen. On the Delivery Options screen, select the records you want to print and the format of your bibliography. When printing full text, select the option "Highlight the search term(s) in full text" to print the article with your search term(s) highlighted. Next, Select "Display to Print" delivery option and then select **Submit** to display your records without graphics or buttons.

To save your records to a floppy disk, select the "Display to Save" delivery option. To send your records via email, select the "E-mail" option and enter your email address (i.e. name@address.com) and comments (optional) in the appropriate fields. Select **Submit** to send the email.

HELP DESK

The Virtual Library's Help Desk is a service that provides ITT Technical Institute students, faculty, and staff with the means to seek and receive library reference services electronically. ITT Technical Institute librarians monitor the questions or requests for information received through the Virtual Library's webpage and provide responses to the inquiries via email.

The Help Desk Service provides online assistance to users who need help devising an effective search strategy, information about the databases and resources that comprise the Virtual Library, or help with reference questions. This service does not provide in-depth research assistance or answer questions that should be directed to subject experts. In general, the Help Desk Service is a means of providing guidance and support for the information seeker.

To access the Help Desk, users should login to the ITT Technical Institute Virtual Library, then select **Help Desk** from the Directory page. From the Help Desk page it is possible to ask a question through the form provided, view frequently asked questions and their answers, utilize a worksheet designed to help devise a search strategy, or access several popular Internet search engines.

The staff of the Help Desk will answer questions and requests for information via email within two business days. If a question needs to be answered more quickly, users are advised to ask for assistance at the school's Learning Resource Center.

INTERNET RESOURCES

The ITT Technical Institute Virtual Library provides selected collections of authoritative websites that support the curricula of the ITT Technical Institutes.

NET RESOURCES

Net Resources can be accessed from the Directory page of the Virtual Library. The website links are arranged by:

- *Type* A collection of Internet links to standard types of reference resources. Included are almanacs, encyclopedias, dictionaries, directories, and other useful resources found on the World Wide Web.

- *Subject* This collection of general-interest Internet sites are recommended for their excellent coverage of topics such as Art, Careers, Economics, Mathematics, and Science.

- *Curriculum* This collection of Internet sites, recommended by our faculty and arranged by program, support the specific curricula of the ITT Technical Institutes.

TEXTBOOK WEBSITES

Links to textbook websites are available from the Directory page of the Virtual Library. A textbook website serves as an excellent source of supplemental materials that can enhance and extend the learning experience provided by the textbook. The website provides an interactive learning environment for students as well as support resources for instructors.

ASSISTANCE

The ITT Technical Institute Virtual Library offers resources and services that can contribute to the educational experience of students of ITT Technical Institutes. For more information, contact the ITT Educational Services, Inc. Corporate Librarian at 317-594-4362 or the Learning Resource Center at any of the ITT Technical Institutes listed below:

Location	Telephone	Location	Telephone
Albany NY	518-452-9300	Mechanicsburg PA	717-691-9263
Albuquerque NM	505-828-1114	Memphis TN	901-762-0556
Anaheim CA	714-535-3700	Miami FL	305477-3080
Arlington TX	817-794-5100	Monroeville PA	412-856-5920
Arnold MO	636-464-6600	Murray UT	801-263-3313
Austin TX	512-467-6800	Nashville TN	615-889-8700
Birmingham AL	205-991-5410	Newburgh IN	812-858-1600
Boise ID	208-322-8844	Norfolk VA	757-466-1260
Bothell WA	425-485-0303	Norwood OH	513-531-8300
Buff Ridge EL	630-455-6470	Omaha NE	402-331-2900
Dayton OH	937-454-2267	Oxnard CA	805-988-0143
Earth City MO	314-298-7800	Phoenix AZ	602-231-0871
Fort Lauderdale FL	954-476-9300	Pittsburgh PA	412-937-9150
Fort Wayne IN	219-484-4107	Portland OR	503-255-6500
Framingham MA	508-879-6266	Rancho Cordova CA	916-851-3900
Greenville NY	716-689-2200	Richardson TX	972-690-9100
Grand Rapids MI	616-956-1060	Richmond VA	804-330-4992
Greenfield WI	414-282-9494	St. Rose LA	504-463-0338
Greenville SC	864-288-0777	San Antonio TX	210-694-4612
Hayward CA	510-785-8522	San Bernardino CA	909-889-3800
Henderson NV	702-558-5404	San Diego CA	858-571-8500
Hoffman Estates IL	847-519-9300	Santa Clara CA	408-496-0655
Houston West TX	713-952-2294	Seattle WA	206-244-3300
Houston North TX	281-873-0512	Spokane WA	509-926-2900
Houston South TX	281-486-2630	Strongsville OH	440-234-9091
Indianapolis IN	317-875-8640	Sylmar CA	818-364-5151
Jacksonville FL	904-573-9100	Tampa CA	813-885-2244
Knoxville TN	865-671-2800	Thornton CO	303-288-4488
Lathrop CA	209-858-0077	Torrance CA	310-380-1555
Little Rock AR	501-565-5550	Troy MI	248-524-1800
Liverpool NY	315-461-8000	Tucson AZ	520-408-7488
Louisville KY	502-327-7424	West Covina CA	626-960-8681
Maitland FL	407-660-2900	Woburn MA	781-937-8324
Matteson FL	708-747-2571	Youngstown OH	330-270-1600

3 Researching and Determining the Best Sources

An important component of technical communication precedes the actual writing and editing. Before you can write about a technical subject, you need to research it thoroughly and gather enough information to write with authority on the topic. No matter how well you write, edit, or design a document, if the information is not thorough and correct, the document will be inadequate.

Most students in college or graduate school think of research in writing as a library project in which they must find as many sources on a subject as possible and synthesize that material into a cogent academic paper. In the technical and scientific professions, the library is just one possible stop on the way to collecting sufficient information. There are many others. And in these fields the nature of the research may be different from that done by academics. For technical communicators, there are usually two kinds of research:

- Researching to solve a problem
- Researching to understand a product

RESEARCHING TO SOLVE A PROBLEM

This type of research closely resembles the traditional approach practiced in schools: You recognize a problem or an issue that needs to be resolved, and you gather the information necessary to find a feasible solution. For example, a large supermarket has gone out of business in the town where you live and you must determine the best use for the abandoned building and the lot on which it stands. Your first step is to gather information on a variety of topics: zoning laws, tax codes, accessibility issues, commercial potential, neighborhood attitudes, and so forth. You might then want to explore what has happened in other towns where similar situations have occurred. Perhaps the local real estate brokers have some relevant data for you to consider. If there are interested buyers or investors, you'll need to talk with them and evaluate their proposals for transforming the empty building into a recreation center or a shopping mall. Some land developers may want to demolish the building and sell parcels of the land for condominium development or low income housing. Your job is to sort through all these ideas and then write a report recommending the best course of action.

RESEARCHING TO UNDERSTAND A PRODUCT

Not all technical writing demands traditional research from scratch. In one assignment, you may be given data already collected by someone else and be asked to turn that material into a manual or a report. For example, an engineer might give you a completed project file and ask you to write a final report to a client. Or software developer may send you a product specification for a new product for which you must write the user's guide. In these and other cases, your first step is to gather enough information from the technical resources so that you understand the material thoroughly and thus can explain it to the intended audience. If you aren't familiar with the technology of the project, you can't write about it. Talk with the developers to make sure your preliminary outlines and descriptions are technically accurate, try to find out the exact definitions of specific terms and concepts, and learn the history of the project so you can place it in a clear context for the audience.

Both types of research—to solve a problem and to understand a product—require significant planning and sufficient time to gather material. And both require that you have an intelligent and organized approach to the research task. It's not enough to spend hours and hours collecting every imaginable piece of information about a topic unless you are doing so systematically and for a defined purpose. Too many writers waste time gathering mountains of material, only to end up overwhelmed by the sheer volume of the data they have collected, much of it unrelated to their project.

A SYSTEMATIC APPROACH TO THE RESEARCH TASK

Planning research, like planning a document, begins with the creation of a preliminary strategy. In some technical writing environments, writers move from project to project throughout the year and usually have developed an approach to researching that works well for them and their colleagues. Such an approach might specify a schedule of steps in the research process: conducting library searches, attending developers' meetings, contacting technical experts, and so forth.

When you find yourself researching something for the first time, you need to develop a strategy that will focus your information-gathering efforts. The following steps can be followed in most situations.

FIVE STEPS TO COMPLETING RESEARCH

1. Define the problem clearly. If you are researching to solve a problem, make sure you understand the issue at stake. Write it down in your own words in simple and direct language. (If you're not sure what the problem is, you may want to come back to this step after some brainstorming, a process that allows you to see issues in ways you hadn't thought of before.)

2. Brainstorm ideas. In a group meeting or alone at your desk, jot down every idea that comes to mind about the topic. Try not to eliminate any idea at first. Then, once you have generated lists of possible elements to write about, decide if there are some topics that are really off-topic and should be removed at this stage.

3. Cluster ideas into categories. Look over the lists you generated in the brainstorming session and see how you might collect the ideas into units of connected elements. You might want to try **mind-mapping** (see Figure 3.1) at this stage if you are more of a visual thinker than an abstract one. Sometimes known as "branching," "ballooning," or "clustering," mind-mapping is a visual technique that

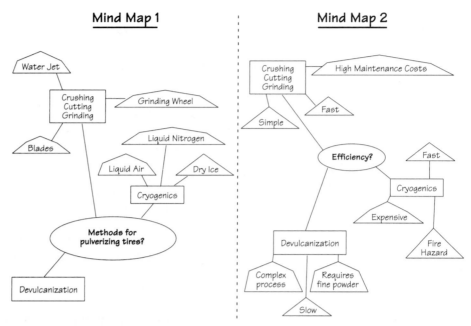

FIGURE 3.1 Two Mind Maps

allows writers to see what they are thinking and make appropriate connections between thoughts. Although there are many variations, the basic process is as follows:

a. Write the main topic you're researching in the middle of a large sheet of paper or other writing surface (a white board or a flip chart), and draw a circle around it. Make sure that the paper is large enough to allow for extensive branches of thought.

b. Think of subtopics and write these around the main topic, but put these subissues in boxes to differentiate them from the circled main idea. (Make an effort to express each idea in one or two key words rather than long phrases.)

c. Connect the boxes to the main topic with solid lines.

d. Repeat steps (b) and (c), for each subtopic, generating supporting points for

When you have finished the mind map, you will have a visual "map" of relationships among topics—your ideas will be grouped in visible clusters and levels.

4. Organize the categories. Give each category a label or name, and arrange all the categories in a logical order. Check the categories against the problem definition. Does each category make sense in light of that problem? Is each a legitimate area for research? Is anything missing? Can you combine some categories?

5. Develop research questions. Once you have clear categories, think about ways to phrase the category names into question form. For instance, instead of "Zoning Laws" phrase the research topic more actively: "How will zoning laws limit the options for using the land?" Phrasing the topics in this interrogatory way gives you more direction in your research and will preclude time wasted researching peripheral information. Make sure you investigate only the material that directly answers your research questions—unless you discover a significant element you hadn't considered before. If that happens during your research (as it likely will), see if you can fit it into one of the preexisting categories. If not, generate a new category for it and phrase it as a question as well.

When you have generated a logical series of research questions, you're ready to begin gathering information. You may want to keep a separate file for the information you find on each of the questions so you can be as organized as possible when the research phase is over and you must transform your notes into an effective report, proposal, manual, or other document.

DETERMINING THE BEST SOURCES

Even with a great strategy for researching, you can still be stalled in the process if you don't know where to look for information or how to evaluate the sources you find. When you understand the basics of where to look and how to know if the material you find is a valuable resource, you can approach any research project with confidence.

WHERE TO LOOK FOR INFORMATION

Information can be found in many sources. The following list is by no means comprehensive.

Hardcopy Resources

Traditional research takes place in the library where you can find books, periodicals, and other hardcopy material. Keep in mind that these sources are often not the most recent information on a subject—especially if the information is contained in books, which are usually in production for at least six months, if not longer. With the rapid pace of technology these days, what may have been current when an author wrote a book can be outdated by the time the book is placed on the library shelf. Unless you are researching the history of a product or problem, books will be of limited use. It's a better idea to rely on current sources such as periodicals, newspapers, and electronic databases.

Electronic Resources

Most libraries today are computerized and therefore allow users access to electronic indexes of periodicals, conference proceedings, unpublished papers, dissertations, theses, and other current information. By using these indexes, you can search for information with a few keystrokes instead of taking hours to sort through documents by hand. But you can do a lot of electronic research in your own home or at your office if you have access to the Internet—a global network of connected databases. With World Wide Web technology, you can easily conduct Internet searches on almost any topic and download printed material onto your personal computer.

Interviews

One of the best ways to obtain up-to-date information is to interview experts. If you plan your interviews well, you can gather extensive material on your topic.

Questionnaires and Surveys

If you want to know what an audience or a specified clientele thinks about a subject, send a questionnaire or survey to a sufficient sampling of people. The key to getting good information via this method is to ask the right questions and design the surveys well.

Product Reviews

Reading about other products (or earlier versions of the product you're working with) is a valuable source of information about the technology, user attitudes, and success rates of particular products. These reviews usually appear in trade magazines such as *PC WEEK, MacWorld,* and *Design News.*

Other Reports, Manuals, or Proposals

If you must write a document in a particular format or on a particular subject, it helps to see what other writers have done. You can gather especially good information about how to write your own document by looking at the competition's work or by looking at documents on file in your own company. If you are a student, check with your instructor to see if there are sample documents on file in the department.

HOW TO EVALUATE SOURCES

Once you have the information in front of you, the next step is to decide if the source is an appropriate one. After many years of working in a specialized field, most technical professionals develop a practiced eye for what is trustworthy in that field and what is not. The following are a few general guidelines for evaluating sources.

1. Is the source timely and current? Check the publication date of any written material you use. If the material is over two years old, it most likely has been superseded by new ideas or by updates on earlier concepts. In medicine, technology, or similar disciplines, older publications often give a historical perspective on new developments, but they may no longer be authoritative sources.

2. Is it grounded in previous research? A second test for the authority of written material is its reliance on previous sources. An article or a book usually combines the author's experience with his or her own research in the field. If the article offers no grounding in previous research and lists insufficient references to outside sources, the author may not be familiar enough with the work already done in the field. Perhaps the article is "reinventing the wheel," so to speak, or is not trustworthy in its assertions because they are based only on the author's personal experience. But, at the other extreme, if written material is choked with outside references, the author may have nothing original to say and may be simply rehashing what others have already said. Look for a balance of external references and personal experience. Both are valuable if used together.

3. Look for bias. Check the breadth of the argument. In weak articles, authors offer biased opinions with no discussion of the opposing views. A hallmark of poor arguments is their one-sided nature, while good arguments usually include all possibilities and suggest why the author's view is the best alternative. Be wary of articles that use emotion rather than logic to make their points. If an author depends on emotion as the primary persuasive element, it is probably because the argument's logic is weak.

4. Check the reliability of the sources. Note the credentials of the author or the interviewee, and look for people who are leading experts in the field—authorities who have published widely on the subject or who have had extensive experience working with the material. If possible, start with the top experts and then move to newcomers to the field but who bring hard work and new ideas to the discipline.

5. Carefully evaluate Internet sources. The four guidelines above—timeline, familiarity with previous research, bias, and reliability—apply to Internet sources as well. But because material on the Internet differs from print material in some significant ways, you need to apply these guidelines somewhat differently.

The most effective way to learn to evaluate sources is to gather enough information on the subject to become an expert yourself. From an informed perspective you can judge more easily the validity of others' opinions.

4 Credibility and Reliability of Electronic Sources

Most books and periodicals undergo a strenuous process of editorial and/or peer review before publication and distribution, a process that sometimes results in delays of a year or more before the information is available. The Internet offers the advantage of speed, making information available literally minutes after web pages are posted. But there are trade-offs.

The Internet is a largely unregulated living testimony to humankind's freedom of speech. Anyone with electronic access can post web pages. The challenge for both the technical writer and researcher, then, is to be certain that posted information is accurate and reliable. How can we do this?

First, acknowledge that, unlike print publications, most information on the Internet has not been critically reviewed for content or scrutinized for grammatical errors. Be skeptical. Assume the posted information is bogus and search for clues to establish its credibility and reliability. Guidelines at a Glance offer some questions you might ask yourself when reviewing electronic documents.

There is much reputable information available electronically. For instance, many professional electronic journals now available on-line have been subjected to a streamlined review process similar to that of print publications. But you must check diligently the accuracy and reliability of your sources. Whenever possible, apply a basic tenet of qualitative analysis when conducting research on the Internet: *triangulate data*—that is, confirm information you find from at least three unrelated sources, some of which are not electronic (e.g., books, articles, interviews, etc.)

TABLE	Guidelines at a Glance	Questions for reviewing electronic resources

- Is an author listed? If so, are credentials provided? Documents that give no author names may be less reliable or credible than those that do.

- Who allowed the posting? That is, who runs the server that posts the information. Is it a reputable government, military, or educational organization that might enforce strict guidelines for posting? Or is it a commercial provider that may be try to sell something? Just because a site is sponsored by a commercial server, however, does not mean it is not worth accessing. For instance, in pursuing the search for bucky balls, we accessed this commercial web page: <http://www.http.com/ARTLEX/Hobbies/Science/Index.htm>, which included a link to a bucky ball home page sponsored by the Department of Physics and Astronomy of SUNY at Stony Brook, <http://buckminsterphysics.sunysb.edu/>.

- If the server is reputable, does it restrict, screen, or endorse postings? An institution may endorse some web pages, but allow students or employees to post unscreened personal pages. (For example, a large research institution was recently embarrass when it learned that web surfers were accessing its home page from the home page of a woman who had posted nude photos of herself at various institutional landmarks.)

- Does the information appear accurate? Even if the server and author are reputable, some files on the Internet are not locked and can be altered by users. Be wary of false information.

- Does the information appear to be biased? Is a specific product or cause being promoted? If so, will this bias hinder your research?

- Is there a date indicating when the information was posted or updated? Much in formation on the Internet is already outdated. A current date may be an indication that the site is reviewed or updated regularly. However, a disadvantage to using Internet resources is that web sites can be withdrawn from public access without notice, with the result that another researcher (or teacher) is not able to confirm your source.

CHAPTER

5 Major Sources of Technical Information

Most technical information today is available in hard copy and online—you'll need to decide which is the most convenient way for you to find the material you need. To get started on your research, you may want to browse through the sources listed below to see which ones fit the type of research you're doing.

HARDCOPY SOURCES

The following are the major hardcopy sources of technical information found in most large libraries:

Guides to Technical Literature

Reference works that list individual articles, they can include bibliographies, indexes, abstracts, journals, and so forth. Most of these guides are listed in three reference books: Ching-Chih Chen's *Scientific and Technical Information Sources* (MIT Press); Eugene Paul Sheehy's *Guide to Reference Books* (American Library Association); and Malinowski and Richardson's *Science and Engineering Literature: A Guide to Reference Sources* (Libraries Unlimited).

Other hardcopy guides include the following:

American Statistics Index. Comprehensive guide to statistics published by federal agencies.

Chemical Abstracts. List of abstracts from journals.

Computer and Control Abstracts. List of abstracts from the computer field with many subfields represented.

Engineering Index. List of abstracts from journals.

Government Reports Announcements and Index. Comprehensive index to technical reports and federally sponsored contract research.

Index to International Statistics. Index to statistical publications of such intergovernmental organizations as the United Nations, the World Bank, OECD, and the International Monetary Fund. Abstracts are provided.

Index Medicus. Selective index to medical journals. Interdisciplinary,

NASA Scientific and Technical Aerospace Reports. Index to reports.

For more general information you may also check the Reader's Guide to Periodical Literature, available in most libraries.

Professional Journals

These are listed in the library's main catalogue and when you find an article in one of the literature guides, such as the *Engineering Index,* you then go the main catalogue to determine whether the particular library has copies of the journal. If the library does subscribe to the journal such as the *IEEE Transactions* series, the catalogue should establish the journal's call number and location in that library.

Books, Monographs, Conference Proceedings, and Review Series

All are listed in the library's main catalogue under the names of authors, editors, titles, subjects, and corporate authors. To find conference proceedings, you may need to know the conference title and date. Review series (such as *Advances in Bioengineering* or *Progress in Biochemical Pharmacology*) are listed under the series title.

Reports

These can be listed in the library's main catalogue under author, corporate author, title, or subject, but they can usually be found as well in a separate *reports checklist* prepared by the library and arranged by report number or government number. Be sure to check the *Government Reports Announcements Index* for a list of the government's reports on technical literature.

Dissertations

These can be valuable source of the latest information. The main guide to all dissertations in science and applied science is the *Dissertation Abstracts International. Volume B: The Sciences and Engineering.*

Standards and Patents

This vast area of technical literature can be accessed electronically. Consult your reference librarian.

ONLINE SOURCES

Many technical resources can be accessed through online databases that you can use at the library or access via modem from your home or office computer. The following are some of the database resources available:

Applied Science and Technology Index: In-depth coverage of journal citations in the sciences, engineering, and technology.

Business Periodicals Index: Journal citations in accounting, advertising, marketing, banking, computer technology, economics, finance and investments, industrial relations, management, occupational health and safety, and so on.

Compendex Plus: A computer searchable version of *Engineering Index.*

Compact D/SEC: Company data from annual and periodic reports filed with the Securities and Exchange Commission for almost all public companies.

Educational Resources Information Center (ERIC): Covers education, counseling, human social development, and other social science subjects.

GPO Monthly Catalog: Bibliographic citations for government publications such as books, reports, studies, serials, and maps. Published by the Government Printing Office.

Humanities Index: Indexes English language periodicals covering archaeology and classical studies, folklore, history, language and literature, performing arts, philosophy, and related subjects.

Newspaper Abstracts: Citations for articles from major newspapers *(Boston Globe, Chicago Tribune, Christian Science Monitor, Los Angeles Times, New York Times, Washington Post)* with brief abstracts or annotated headlines.

Psyclit: Journal citations in psychology and psychiatry.

Social Sciences Index: Journal citations in anthropology, economics, law and criminal justice, planning and public administration, political science, social aspects of medicine, sociology, etc.

The key to effective research is to be current. In business and industry, technical information gets released in many forms, and the material that ends up in hardcopy print is often dated information. The most current information flows orally between colleagues in laboratories and offices and sometimes ends up in printed form that gets filed as proposals, progress reports, memos, and internal reports, most of which are proprietary. The information may also be released in formal documents such as patents, specifications, manuals, and corporate reports.

In pure research circles, information is recorded in the laboratory but then moves through conference proceedings and formal reports. It may emerge in print as a refereed journal article (an article that has been accepted for publication by a group of experts whose job it is to screen all the articles submitted and recommend only the best ones). After that stage, several more years may pass before the information is printed in reference sources and textbooks.

If your research depends on these later stages of information, your material will be dated. The extent to which you can access information in its earlier stages improves the quality and accuracy of your research. Using electronic media to conduct your search increases your reach, speed, and versatility in locating information, and gives you access to the material before it is in the traditional printed format.

6 Researching in Cyberspace

If you have access to the Internet, you can add to your research capabilities by searching electronically from your home or office computer. But searching (or "surfing") the Net is different from researching in the library. The Net can yield a wealth of information or just a few bits of related material. In her book, *A Student's Guide to the Internet,* Carol Clark Powell gives these invaluable tips for working with the Internet:

- The Internet is not like a library where information has been categorized and organized with the use of a widely accepted set of rules. It is more like a rummage sale, where items of a similar nature usually get grouped together. In some cases, rather than putting costume jewelry with costume jewelry and old magazines with old magazines, someone will put all of the pink items in one pile and all of the green items in another. Although it is organized, the method of organization may not be useful to you.

- Everything on the Internet has a tendency to be updated, improved, uprooted, relocated, shuffled, cut, and redealt, and otherwise be taken from where it was and replaced or moved somewhere else. And this happens with some frequency. If you find something that is important to your research, do not expect it to be in the same place tomorrow—or even an hour later. If you find what you need, keep a copy of it.

- The Internet is vast. It is made up of hundreds of types of computers, utilizing a kinds of operating systems and connection options. It contains daily contribution from millions of people around the world, all with their own ideas and intellectual and cultural views. No one is an expert on all of the facets of the Internet. Although many people are very skilled with the tools and have a good idea where to look for information on many topics, no one can keep up. The Net simply grows and changes too quickly. Fortunately, it is not necessary to understand everything about the Internet to make use of its resources. Perhaps the essential quality needed to navigate the Internet is the willingness to try new avenues of exploration.*

With these tips in mind, and with the understanding that the information posted on the Internet is not always reliable or objective, you can use this electronic resource to your best advantage.

THE TOOLS OF ACCESS

The primary tool that most people use to access the information on the Internet is the World Wide Web (WWW). Other browsing services that can be useful if you do not have direct access to the Web include Gopher and Veronica, File Transfer Protocol (FTP), Telnet, and Wide Area Information Server (WAIS). It should be noted, however, that increasingly the World Wide Web is replacing these services; moreover, they often lead to resources that few organizations update consistently. These access tools are described briefly below; the World Wide Web—what it is, how it works, and how to write for it—is discussed in depth in Chapters 6 and 13.

WORLD WIDE WEB

The World Wide Web is a powerful hypermedia system that allows you to follow ideas from one hypertext link to another. The Web is made up of electronic documents called pages that combine images, text, and even sound. A "home page" is the entry point to access a collection of pages. By clicking on a linked word, image, or icon, you can jump to other pages. The Web is accessed through client programs called browsers. Some have full graphic interface—for example, Netscape Navigator and Communicator and the Microsoft Internet Explorer—which means that they access the graphic and audio dimensions of home pages. Others offer only text interface. Another way to reach the Web pages is to type in specific addresses (if you're not browsing and know the direct address). Locations on the Web can be accessed by Uniform Resource Locators (URLs), addresses that are made up of long strings of letters used to describe any Internet resource. For example, the address for the World Wide Web's own home page, which has everything you might want to know about the Web, is http://www.info.cern.ch/.

SEARCH ENGINES

A tool that can make your research a much easier and faster process is the search engine. There are many different search engines available, each with different features, although the basic function of all of them is the same: Search engines receive information from and scan Internet sites for resources and transfer this material into the individual engine's database. You request information by using a keyword or keywords for general topics (such as "hazardous waste sites") or by specific names (such as "Love Canal"). The search engine then scans its own database for items that match the term and returns a list of individual information sources. Usually, the more specific your keyword, the more potentially useful the sources on the list.

Some of the most common search engines are Yahoo (http://www.yahoo.com), a popular search tool often used as a beginning step for users new to researching the Internet; Galaxy (http://galaxyeinet.nettsearch/html), similar to Yahoo except that it prioritizes the search results in the order it assumes you wish to see them; Lycos (http://lycos.cs.cmu.edu/), from Carnegie Mellon is among the most powerful search engines with over four million records; and

*Carol Clark Powell, *A Student's Guide to the Internet*, 2nd ed. (New York: Prentice Hall, 1998).

Excite (http://www.excite.com), which provides the fullest range of services of all search engines to date including an online newspaper with reviews of Internet resources, a news wire service from Reuters, and its own columns.

GOPHER AND VERONICA

Gopher is an information browsing service for the Internet; it gives you access to electronic books, journals, databases, library catalogues, phone books, job listings, and so on. Although Gopher does not have hypertext links as does the World Wide Web, it does present the Internet as an organized directory system and offers links through menu items. You can access Gopher from most networked computer systems by selecting it from the Internet Services Menu or, if you work with UNIX or VMS, you can simply type gopher and the command prompt. Once you have connected to this service, follow the men trees to find the information you're looking for at Gopher sites around the world. Veronica (Very Easy Rodent-Oriented Net-Wide Index to Computerized Archives is a keyword search engine for gopher. Veronica maintains an index of Gopher menu items and allows you to search Gopher directories and files by entering title words.

FILE TRANSFER PROTOCOL (FTP) AND ARCHIE

FTP allows you to explore directories of remote computers on the Internet and to download files. To access an FTP program via your computer network, type **FTP** before the address of files you want to retrieve. For example, to find instructions on how to access computerized library systems of many universities around the world, type the following address ftp.unt.edu (/libraries/libraries.txt). If you don't know the address of the files you need, you can use Archie, the keyword search engine for FTP.

TELNET

Telnet is an Internet protocol that allows you to log onto a remote computer and use it as if you were physically present. For instance, through Telnet you can have real-time access to libraries and databases throughout the world. Accessing Telnet depends on the type of network connection and computer you're using. First, access Telnet via your computer system and then type telnet before the address of the site you wish to contact. You can use Telnet with Gopher or FTP to gain even greater access to various sites. For example, type telnet(libgopher.yale.edu) for access to library catalogues or telnet(locis.loc.gov) for the Library of Congress catalogue. Remember that once you've contacted the site, you may need to have the appropriate password to actually use the material.

WIDE AREA INFORMATION SERVER (WAIS)

WAIS allows you to perform keyword searches or searches for specific words or documents all over the Internet in different types of sites. WAIS is particularly good for searching complex textual databases in the sciences and technology. The best way to access WAIS is to use WWW browser such as Netscape and enter the address for WAIS, Inc. http://www.wais.com. From there, follow the directions to conduct the search.

Another way to keep in touch with the current thinking on almost any topic is to participate in electronic discussion groups. You can subscribe to LISTSERV Internet discussion groups and/or to USENET news groups. In both groups, you can join electronic conversations, by means of electronic mail, with experts and novices on topics ranging from architecture to quantum physics. Keep in mind that the information found through this route is sheer opinion and is the same as information you learn about by participating in or overhearing an informal discussion. Use these groups to spark ideas or discover interesting directions for new research. Don't use them as proof for any of your ideas.

EVALUATING INTERNET RESOURCES

Although the Internet offers seemingly endless research possibilities and provides access to many sources you would otherwise be unable to obtain easily, if at all, it's very diversity and multiplicity can also present problems and can lead you into time-consuming and frustrating detours from your research. As with print sources, but perhaps even more so, you need to exercise care in selecting and evaluating Internet material. The following guidelines will give you a framework for carrying out your evaluation.

1. Evaluate for reliability. Most print sources receive some form of external validation before they are published: the material is reviewed by authorities in the subject area or by other knowledgeable readers, and usually it undergoes an editing process in which the author(s) and editor(s) pay careful attention not only to the style in which the material is written but also to its clarity, accuracy, and significance. On the other hand, the Internet is not a library, as Carol Clark Powell notes earlier, and neither is it a vast electronic publishing house with editors and editorial and review procedures. Many Internet sources are self-published by their authors and have received no form of external validation. Many others are published by organization that have very specifically defined agendas that may have little or nothing to do with providing reliable information. In the process of conditioning your research, you may come across government documents, course materials from academic departments in universities, the opinions of self-proclaimed "experts," self-promotional material from groups or individuals directly or indirectly related to your topic, highly specialized or localized discussions of the topic, or even incomprehensible babblings by an obsessed individual. One way to evaluate all these potential sources is to check the abbreviation in the address that tells you where the source originates: edu means an educational institution; gov means a government body; com means a commercial organization; and org means a nonprofit organization. The letters immediately preceding these abbreviations give you the specific organization or institution. You can work backward from this information to discover whether the origin of the source is an appropriate authority. For example, an address including mit.edu (Massachusetts Institute of Technology) would likely be a good source for information on engineering.

2. Evaluate for probable usefulness. If you are locating sources though database searches, those searches will turn up sources that have either been published in print or, often, are online versions of a print medium (but note that some scholarly journals are published *online only*). These database sources still need to be subjected to the guidelines for evaluating print resources, but it's likely that some of them are potentially useful. On the other hand, if you are using a search engine to conduct keyword searches of the Internet, you can be overwhelmed by the number of matches your search turns up and thus the number of Internet addresses you might potentially link to. Although the matches are ostensibly listed in order of how precisely they conform to the specifications of your keywords, that does not mean that the first twenty (or thirty or one hundred) matches that are listed are necessarily taking you to useful sources. Read through the search results for a finite number of matches; twenty are probably sufficient to indicate whether you should pursue any of the links, refine your keyword search, or seek other sources. Carefully note the name and address of each listing(see "Evaluate for Reliability," above) and read the summary provided by the search engine. When you find a likely site, try linking to other similar sites, either by going to links provided in the site's homepage or by, referring to the "more like this" links on the sites address and summary on the search list.

Links in Internet sites are often good indications of the degree to which the material at the site is part of a network of knowledge and information. Unlinked sites may reveal a form of intellectual isolation.

3. Evaluate for timeliness. Remember that many Internet sites appear and disappear randomly, so a source you located last week may or may not be there when you log on this week. Be sure to check when a website was last updated: A site that hasn't been updated for a long time or that is infrequently updated may yield only stale or useless information.

4. Evaluate for authorship. Many online documents indicate the name and affiliation of their authors. You can thus note the credentials of an author, or, if you have not encountered references to the author's work in your previous research, you can research reference books, bibliographies, or other online sources to try to find some account of the author's work and expertise. The same criteria you would use to evaluate the authors of print sources should be used to evaluate online authors. Sometimes, however, an online source will not list a specific author: the authorship is essentially, then, "by" the organization or group that maintains the website. For example, if you conduct an online search for sources on aircraft carrier design, many of the addresses in your search results will lead you to Navy or other military websites, where the information may not bear the name of an individual author. Certainly the Navy knows a good deal about aircraft carrier design, but, as is the case in evaluating and using any sources—print or electronic from the government or from organizations, or academic institutions—you still need to achieve a balance of information and perspective from a number of sources, rather than rely too heavily on one or two.

Sources from electronic data bases often include online versions of scholarly journals, as well as many types of print resources. The articles in these journals generally have been externally reviewed for trustworthiness

FRAMING AN ARGUMENT: AVOIDING CUT AND PASTE

When you have finished researching your topic, the amount of information you are confronted with may seem overwhelming at first, and you might wonder how to put it all together to make a cogent argument. Your desk may be covered with note cards, photocopies, and computer printouts, and you may have several downloaded files stored in your computer's memory—all waiting to be incorporated into a final document. An easy mistake to make is to arrange all of the research you've collected into an organized package and assume that your job is done. If your assignment has been to do nothing more than research a topic and present the results of your search, then your task really is almost finished. All that remains is to type the material and organize it into logical categories (which you have already determined in the planning stages). But if you have been asked to research a topic and present a supported argument about it, then you still have some work to do.

As pointed out earlier in this chapter, the key to effective research is to begin with effective *research questions.* These questions should give direction to your research and allow you to collect information that helps you solve a problem. If you have followed these guidelines, you haven't simply gathered material aimlessly; you've conducted focused research that aims to provide a solution to the original problem. Instead of writing a document that merely describes your research findings without comment, you can suggest specific answers to the questions posed at the beginning and offer readers well-supported reasons for your opinions.

MOVING FROM RESEARCHING TO WRITING

Once your research is done, you might find it helpful to approach the writing task in this way:

1. Define the goal or problem. Begin by once again clearly defining the problem or issue you've been asked to research. (You've done this task in the planning stages but are revisiting it here.)

2. Review the results of research. Consider the material your research has generated. Is it complete? Are there new directions the research has suggested? If there are, you need to rethink the argument and perhaps continue the research phase until you're satisfied that you have fully explored the new areas. If the research is complete, you're ready to write.

3. Subdivide your presentation. Break down the problem into its various components and create subsections that contain specific conclusions supported by research.

4. State your focus. Organize each section by starting with the conclusions you've drawn followed by the research that led you there. This strategy saves you from making each section only a descriptive list of your research with neither focus nor point. For example, a subsection for a discussion on the role of the physical therapist in elderly patient care may begin this way:

> Mental status is a large influence in rehabilitation because it can change daily. With disorders such as cerebellar vascular insufficiencies, Alzheimer's, and other dementias, research has shown that the patient may be an active participant in treatment one day and totally passive the next. Four studies are seminal in illustrating this point.

5. Clarify the progression of your argument. Make sure you present the research you've done as logical steps toward the conclusions, instead of random, disconnected descriptions of the material. That means you need to use transitions that remind readers how the various sections connect to the larger argument, and you need to give frequent reader cues indicating how your discussion is moving in a clear direction.

6. Give each quotation a context. Introduce each quotation from the research with a statement about why the quotation is an important, supportive piece for your conclusion, and be sure to follow the quotation with a further explanation of how it fits into the larger argument you're making. Don't just drop quotations into your text and assume they're self-explanatory. For example, in the excerpt below, note how the writer integrates the quotation into her text, explaining how it supports her argument:

> Along with other factors mentioned previously, the psychological health of the patient can play a key role in recovery from hip fractures. Depression can be a major cause of the high mortality rate among elderly patients who have sustained such fractures (McKay 1992). A possible reason for this post-fracture depression is the high priority our country places on independence. In a recent issue of the *American Journal of Physical Medicine,* Janet Haas supports this theory:
>
> > Americans value independence and self sufficiency They place a high premium on physical mobility as a reflection of dignity and identity as well. Emphasis on mobility, appearance and vocational training promotes these values (Haas 1995, p. 57).

Haas's comments shed some light on post-fracture depression. With hip fractures comes temporary loss of functional independence and the potential for permanent disability. Patients face the reality that they can no longer take care of themselves, and such loss of independence can be interpreted as failure in the eyes of society and, consequently, also in the eyes of the patients. An interpretation of failure can lead to a reduction in self-esteem and produce serious depression (Haas 1995). When depression affects the patient's will to live, even the immune system can break down, leaving patients susceptible to a variety of diseases such as pneumonias or acute infections. In essence, patients may deem themselves failures and neglect the body until death (Duphrene 1995).

7. *Check for clear synthesis.* Your finished document should synthesize the research into a focused argument that says something not said by any of the individual pieces of your research. Only in the combination of the pieces is your point made.

If you follow these guidelines, you will avoid the common error that novice writers make of stringing together quotations with no clear purpose or direction. The resulting document is little more than a list of quotes cut and pasted together with no focus of original thought. Certainly the writer can say that the research has been done and displayed, but the situation is similar to that of a pastry chef who collects all the ingredients and spreads them out on the table without ever combining them into a pie. Your document is not complete until you've synthesized the research into focused and well-supported conclusions.

AVOIDING PLAGIARISM

Another problem that researchers face when they have collected information from a variety of sources is the possibility of plagiarism—representing someone else's words or ideas as your own. When you've read extensively what others have written about your subject, it can be tempting to simply lift their words and ideas and weave them into your document. A common form of such plagiarism occurs when writers either misunderstand how to properly document their sources, or they intentionally paste together direct quotations from several sources hoping that the mix of quoted material will disguise the fact that the information is stolen.

You have committed plagiarism when you

- copy a phrase, a sentence, or a longer passage from a source and do not give credit to the original author;
- summarize or paraphrase someone else's ideas without acknowledging the source;
- allow someone else to write significant portions of your document for you without admitting to the help; and
- forget to place quotations around another writer's words.

To avoid the problem of accidental plagiarism, get into the habit of clearly documenting the words and ideas you obtain from other sources during the research phase. Be sure to put quotation marks around any direct quotations as you write them down in your notes and identify the originator of any ideas you plan to use, even if you plan to summarize or paraphrase those ideas in your own words.

It's also important to be aware of when acknowledging your source is not necessary. You do not need to document the following:

- Your own independent ideas and words
- "Common knowledge" (information known and readily available to most people, such as information in encyclopedias or other reference guides that do not contain original thought or individual arguments)

- "Common sense" observations (something most people know—for example, that radioactive material is dangerous)

If you are unsure about whether to credit the original source, play it safe and provide documentation.

One rule of thumb should serve as a guideline: Plagiarism in any form is theft. If you understand that rule and you understand that plagiarism is not only illegal but vastly unfair to the original author, then the dilemma of whether to take someone else's work and use it without crediting the author is an ethical choice you make from your own moral standards (see Chapter 16: Considering Ethical Issues). If you get caught, your ethical credibility is undermined and you may face legal charges. If you plagiarize and don't get caught, you have missed the opportunity to contribute your own ideas to the body of knowledge about your subject, and you have essentially agreed to live as a fraud.

QUICK REVIEW

For technical communicators, there are usually two kinds of research: researching to solve a problem and researching to understand a product. Once you have determined the type of research you are doing, the approach to the research process is as follows:

- Plan the research by defining the problem clearly, brainstorming, clustering ideas into categories, organizing the categories, and developing research questions
- Determine the best sources of information (hardcopy material, interviews, questionnaires and surveys, product reviews, or other documents).
- Understand how to research using libraries, the Internet, and interviews with technical experts.
- Avoid "cut and paste" research that offers nothing original. Your finished document should synthesize the research into a focused argument that says something not said by any of the individual pieces of your research.
- Avoid plagiarism (the act of stealing someone else's words and ideas and presenting them as your own).

PART II

Preparing Technical and Laboratory Reports and Writing E-mail

7 Using Electronic Mail

erhaps the most widely used application on the Internet is electronic mail, which provides a connection to discussion forums on Listserv and Usenet and carries routine, day-to-day communication as well. E-mail transmits an electronic document via networked computer terminals to recipients in the same building or across the globe. Specific codes direct the message to any electronic mailbox designated by the sender—or to all the mailboxes on a mailing list. Alerted by an audio signal, the recipient opens the on-screen mailbox, reads the message, and then either responds, files the message, prints it out, forwards it, or deletes it. E-mail messages can be exchanged instantaneously or at the convenience of the communicating parties. Figure 7.1 shows a typical E-mail message.

E-MAIL BENEFITS

Compared to phone, fax, or conventional mail (or even face-to-face conversation, in some cases), E-mail offers benefits:

- *E-mail is fast, convenient, efficient, and relatively unintrusive.* Unlike conventional mail, which can take days to travel, E-mail travels instantly. Although a fax network can transmit printed copy rapidly, E-mail eliminates paper shuffling, dialing, and a host of other steps. Moreover, E-mail makes for efficiency by eliminating "telephone tag." It is less intrusive than the telephone, leaving the choice of when to read and respond to a message entirely up to the recipient.

- *E-mail is democratic.* With few exceptions, E-mail messages appear as plain print on a screen, with no special typestyles, fancy letterheads, page design, or paper texture—enabling readers to focus on the message instead of the medium.[1]

E-mail facilitates communication and collaboration

1. Recent multimedia developments allow charts, graphs, 3-D images, sound, voice, animation, or video to be added to certain E-mail messages, but many of these features would be considered inappropriate—if not frivolous—in routine correspondence.

```
┌─────────────────────────────────────────────────────────────────────┐
│ ┌─────────────────────────────────────────────────────────────────┐ │
│ │ Ihabicht@umassd.edu, powens@umassd.edu, rdumont@umassd.edu, td,8n/96 1 │ │
│ └─────────────────────────────────────────────────────────────────┘ │
│        To:  Ihabicht@umaxod.adu, povens@umased.eft, rdu-            │
│             mont@umased.edu, tdace@umasad.eda,                      │
│             ltravers@umasod.edu                                     │
│      From:  ethompson@umased.edu (Ed Thompson)                     │
│   Subject:  meeting of "program review" committee                  │
│                                                                     │
│ Hello All,                                                          │
│          You remember the Program Review Committee, don't           │
│ you?                                                                │
│          As far as I know, the department still needs to            │
│ produce two twelve-page review documents (one for under-            │
│ grad. and one for Professional Writing) and a five-page             │
│ document of 1, 2 and 3 year plan goals for the department.          │
│ The due date, unless it's pushed back, is October 2. I              │
│ guess we'd better got started soon.                                 │
│          Could everyone make an initial planning meeting            │
│ before we got into the hectic rush at the start of the              │
│ semester? I'd like to most on Wednesday, August 30th, at            │
│ 10:00 a.m. Let me know if you can't make it. By the way,            │
│ the long-awaited data books, finally arrived in late June.          │
│ I managed to get enough copies for everyone on the commit-          │
│ tee and will leave then with John to be picked up at your           │
│ convenience.                                                        │
│                                                                     │
│          Best,                                                      │
│          Ed                                                         │
│                                                                     │
└─────────────────────────────────────────────────────────────────────┘
```

FIGURE 7.1 A Typical E-mail Message

E-mail also allows for transmission of messages by anyone at any level in an organization to anyone at any other level. For instance, the maid clerk conceivably could E-mail the company president directly, whereas a conventional memo or phone call would be routed through the chain of management or screened by administrative assistants (Goodman 33–35). In addition, people who are ordinarily shy in face-to-face encounters may be more willing to express their views in an E- mail conversation.

- *E-mail can foster creative thinking.* E-mail dialogues involve a give-and-take, much like a conversation. Writers feel encouraged to express their thoughts spontaneously, thinking as they write, without worrying about page design, paragraph structure, perfect phrasing or the like. The focus is on conveying one's meaning to the recipients who in turn will respond with thoughts of their own. This relatively free exchange of views can lead to all sorts of new insights or ideas (Bruhn 43).

- *E-mail is excellent for collaborative work and research.* Collaborative teams keep in touch via E-mail, and researchers contact people who have the answers they need. Especially useful for collaborative work is the E-mail function that enables documents or electronic files of any length to be attached and sent for downloading by the receiver.

E-MAIL PRIVACY ISSUES

Although E-mail connects increasing millions of users, no specific laws protect any computer conversation from eavesdropping or snooping.[2] Gossip, personal messages, or complaints about the boss or a colleague—all might be read by unintended receivers. Employers often claim legal right to monitor *any* of their company's information, and some of these claims can be legitimate:

Monitoring of E-mail by an employer is legal

> In some instances it may be proper for an employee to monitor E-mail, if it has evidence of safety violations, illegal activity, racial discrimination, or sexual improprieties, for instance. Companies may also need access to business information, whether it is kept in an employee's drawer, file cabinet, or computer E-mail. (Bjerklie, "E-Mail" 15)

E-mail privacy can be compromised in other ways as well. Some notable examples:

E-mail offers no privacy

- Everyone on a group mailing list—intended reader or not—automatically receives a copy of the message.
- Even when deleted from the system, messages often live on for years, saved in a backup file.
- Anyone who gains access to your network and your private password can read your document, alter it, use parts of it out of context, pretend to be its author, forward it to whomever, plagiarize your ideas, or even author a document or conduct illegal activity in your name. (One partial safeguard is encryption software, which scrambles the message, enabling only those who possess the special code to unscramble it.)

E-MAIL QUALITY ISSUES

Free exchange via computer screen adds to the *quantity* of information exchanged, but not always to its *quality:*

A useful definition of "information"

> Claude Shannon, father of communication theory . . . once said, "Information is news that makes a difference. If it doesn't make a difference, it isn't information." A radio traffic report about a car crash up ahead is information if you can still change your route. But if you are already stuck in traffic, the message is . . . useless. (Rothschild 25)

Following are specific ways in which information quality can be compromised by E-mail communication:

E-mail does not always promote quality in communication

- The ease of sending and exchanging messages can generate overload and junk mail—a party announcement sent to 300 employees on a group mailing list, or the indiscriminate mailing of a political statement to dozens of newsgroups ("spamming").
- Some electronic messages may be poorly edited and long-winded.
- Off-the-cuff messages or responses might offend certain recipients. Email users often seem less restrained about making rude remarks ("flaming") than they would be in a face-to-face encounter.

---

2. Whereas the phone company and other private carriers are governed by FCC laws protecting privacy, no such legal protection has been developed for communication on the Internet (Peyser and Rhodes 82).

- Recipients might misinterpret the tone. "Emoticons" or "Smileys," punctuation cues that signify pleasure :-), displeasure :-(, sarcasm ;-), anger >:-< and other emotional states, offer some assistance but are not always an adequate or appropriate substitute for the subtle cues in spoken conversation. Also, common E-mail abbreviations (FYI, BTW, HAND—which mean "for your information," "by the way," and "have a nice day") might strike some readers as too informal.

E-MAIL GUIDELINES

Recipients who consider an E-mail message poorly written, irrelevant, offensive, or inappropriate will only end up resenting the sender. These guidelines offer suggestions for effective E-mail use.[3]

- Use E-mail to reach a lot of people quickly with a relatively brief, informal message.
- Don't use E-mail to send confidential information: employee evaluations, criticism of people, proprietary information, or anything that warrants privacy.
- Don't use E-mail to send formal correspondence to clients or customers, unless they request or approve this method beforehand.
- Don't use the company E-mail network for personal correspondence or for anything that is not work related.
- Check your distribution list before each mailing, to be sure the message reaches all intended primary and secondary readers but no unintended ones.
- Assume your E-mail correspondence is permanent and could be read by anyone anytime. Ask yourself whether you've written anything you couldn't say to another person face-to-face. Avoid spamming and flaming.
- Before you forward an incoming message to other recipients, be sure to obtain permission from the sender. (See page 176 for E-mail copyright issues.)
- Limit your message to a single topic, and keep the whole thing focused and concise. (Yours may be just one of many messages confronting the recipient.) Don't ramble.
- Use a clear subject line to identify your topic ("Subject: Request for Beta test data for Project #16"). This helps recipients decide whether to read the message immediately and makes it easier to file and retrieve for later reference.
- Refer dearly to the message to which you are responding ("Here are the Project 16 Beta test data you requested on Oct. 10").
- Try to keep the sentences and paragraphs short, for easy reading.
- Don't write in FULL CAPS—unless you want to SCREAM at the recipient!
- Close with a signature section that names your company or department, phone and fax number, and any other information the recipient might consider relevant.

3. Adapted from Bruhn 43; Goodman 33-35,167; Kawasaki 286, Nantz and Drexel 45–51; Peyser and Rhodes 82.

8 Short Reports

OBJECTIVES

At one time or another, you'll be asked to write a report. This report will satisfy one or all of the following needs:

- Supply a record of work accomplished
- Record and clarify complex information for future reference
- Present information to a large number of people
- Record problems encountered
- Document schedules, timetables, and milestones
- Recommend future action
- Document current status
- Record procedures

The most common types of reports include the following:

1. *Accident/incident reports:* What happened, how did it happen, when did it happen, why did it happen, who was involved?
2. *Feasibility reports:* Can we do it, should we do it?
3. *Inventory reports:* What's in storage, what's been sold, what needs to be ordered?
4. *Staff utilization reports:* Is labor sufficient and efficiently used?
5. *Progress/activity reports* (weekly, monthly, quarterly, annually): What's our status?
6. *Travel reports:* Where did I go, what did I learn, who did I meet, etc.?
7. *Lab reports:* How did we do it?
8. Performance appraisal reports: How's an employee doing on the job?
9. *Study reports:* What's wrong?
10. *Justification reports:* Here's why we need the material (or will pursue this action) on this date.

Although there are many different types of reports and individual companies have unique demands and requirements, certain traits are basic to all report writing.

CRITERIA FOR WRITING SHORT REPORTS

All reports share certain generic similarities in format, development, and style.

FORMAT

Every short report should contain four basic units: heading, introduction, discussion, and conclusion/recommendations.

Heading

The heading includes the date on which the report is written, the name(s) of the people to whom the report is written, the name(s) of the people from whom the report is sent, and the subject of the report (the subject line should contain a *topic* and a *focus*).

Sample Report Heading

```
DATE:      August 13, 1991

TO:        Shelley Stine

FROM:      Judy Simmons   J.S.

SUBJECT:   Report on Trip to Southwest Regional Confer-
           ence on English
           (Fort Worth, TX)
```

Introduction

The introduction supplies an overview of the report. It can include three optional subdivisions:

- Purpose—a topic sentence(s) explaining why you are submitting the report (rationale, justification, objectives) and exactly what the report's subject matter is.
- Personnel—names of others involved in the reporting activity.
- Dates—what period of time the report covers.

```
Introduction

Objectives: I attended the National Electronic Packaging
Conference in Anaheim, CA, to review innovations in vapor
phase soldering.

Dates: September 26-30, 1991

Personnel: Susan Lisk and Larry Rochelle
```

Some businesspeople omit the introductory comments in writing reports and begin with the discussion. They believe that introductions are unnecessary because the readers know why the reports are written and who is involved.

These assumptions are false for several reasons. First, it is false to assume that readers will know why you're writing the report, when the activities occurred, and who was involved. Perhaps if you are writing only to your immediate supervisor, there's no reason for introductory overviews. However, even in this situation you might have an unanticipated reader because:

- Immediate supervisors change-they are promoted, fired, retired, or go to work for another company.
- Immediate supervisors aren't always available-they're sick for the day, on vacation, or off site for some reason.

Second, avoiding introductory overviews assumes that your readers will remember the report's subject matter. This is false because reports are written not just for the present, when the topic is current, but for the future, when the topic is past history. Reports go on file—and return at a later date. At that later date,

- You won't remember the particulars of the reported subject matter.
- Your colleagues, many of whom weren't present when the report was originally written, won't be familiar with the subject.
- You might have outside, lay readers who need additional detail to understand the report.

An introduction—which seemingly states the obvious—is needed to satisfy multiple readers, readers other than those initially familiar with the subject matter, and future readers who are unaware of the original report.

Discussion

The discussion of the report summarizes your activities and the problems you encountered. This is the largest section of the report and involves development, organization, and style (more on these later).

Conclusion/Recommendations

The conclusion allows you to sum up, to relate what you've learned, or to state what decisions you have made regarding the activities reported. The recommendations allow you to suggest future action, to state what you believe you and/or your company should do next.

```
Conclusion/Recommendations:

The conference was beneficial. It not only taught me how
the computer can save us time and money, but also I
received hands-on training. Because the computer can
assist our billing and inventory control, let's buy and
install three terminals in bookkeeping before our next
quarter.
```

DEVELOPMENT

Now that you know what subdivisions are traditional in reports, your next question is, "What do I say in each section? How do I develop my ideas?"

First, answer the reporters' questions.

1. *Who* did you meet or contact, who was your liaison, who was involved in the accident, who comprised your technical team, etc.?
2. *When* did the activities documented occur (dates of travel, milestones, incidents, etc.)?
3. *Why* are you writing the report and/or why were you involved in the activity (rationale, justification, objectives)? Or, for a lab report, for example, why did the electrode, compound, equipment, or material act as it did?

4. *Where* did the activity take place?

5. *What* were the steps in the procedure, what conclusions have you reached, or what are your recommendations?

Second, when providing the foregoing information, *quantify!* Don't hedge or be vague or imprecise. Specify to the best of your abilities with photographic detail.

The following justification is an example of vague, imprecise writing.

Installation of the machinery is needed to replace a piece of equipment deemed unsatisfactory by an Equipment Engineering review.

Which machine are we purchasing? Which piece of equipment will it replace? Why is the equipment unsatisfactory (too old, too expensive, too slow)? When does it need to be replaced? Where does it need to be installed? Why is the installation important?

A department supervisor will not be happy with the preceding report. Instead, supervisors need information quantified, as follows:

The <u>exposure table</u> needs to be installed by <u>9/91</u> so that we can <u>manufacture printed wiring products with fine line paths and spacing (down to .0005 inches).</u> The table will replace the <u>outdated printer</u> in <u>Dept. 76.</u> Failure to install the table <u>will slow the production schedule by 27%.</u>

Note that the underlined words and phrases provide detail by quantifying.

STYLE

Style includes conciseness, simplicity, and highlighting techniques. As already discussed, you achieve conciseness by eliminating wordy phrases. Say *consider* rather than *take into consideration;* say *now* rather than *at this present time.* You achieve simplicity by avoiding old-fashioned words: *utilize* becomes *use, initiate* becomes *begin, supersedes* becomes *replaces.*

The value of highlighting has already been shown in this chapter. The parts of reports reviewed earlier use headings. (Introduction, Discussion, Conclusion/Recommendation). Graphics can also be used to help communicate content, as evident in the following example. A recent demographic study of Kansas City predicted growth patterns for Johnson County (a large county south of Kansas City):

Johnson County is expected to add 157,605 persons to its 1980 population of 270,269 by the year 2010. That population jump would be accompanied by a near doubling of the 96,925 households the county had in 1980. The addition of 131,026 jobs also is forecast for Johnson County by 2010, more than doubling its employment opportunity.

This report is difficult to access readily. We are overloaded with too much data. Luckily, the report provided a table (Table 8.1) for easier access to the data.

Through highlighting techniques (tables, white space, headings), the demographic forecast is made accessible at a glance.

TYPES OF SHORT REPORTS: CRITERIA

All short reports include a heading (date, to, from, subject), an introduction, a discussion, and conclusion/recommendations. However, different types of short reports customize these generic components to meet specific needs. Let's look at

TABLE **8.1**	Johnson County Predicted Growth by 2010		
	POPULATION	**HOUSEHOLDS**	**EMPLOYMENT**
1980	270,269	96,925	127,836
2010	427,874	192,123	258,862
% change	+58.3%	+98.2%	+102%

the criteria for five common types of short reports: trip reports, progress reports, lab reports, feasibility reports, and incident reports.

TRIP REPORTS

When you leave your work site to go to a conference, analyze problems in another work environment, give presentations, or make sales calls, you must report on these work-related travels. Your supervisors not only require that you document your expenses and time while off site, but they also want to be kept up to date on your work activities. Following is an overview of what you'll include in an effective trip report.

1. **Heading**
   Date
   To
   From
   Subject (Topic + focus)

2. **Introduction (overview, background)**

   Purpose: In the purpose section, document the date(s) and destination of your travel. Then comment on your objectives or rationale. What motivated the trip, what did you plan to achieve, what were your goals, why were you involved in job-related travel?

   You might also want to include these following optional subheadings:

   Personnel: With whom did you travel?

   Authorization: Who recommended or suggested that you leave your work site for job-related travel?

3. **Discussion (body, findings, agenda)**

   Using subheadings, document your activities. This can include a review of your observations, contacts, seminars attended, or difficulties encountered.

4. **Conclusion/recommendations**

   Conclusion: What did you accomplish—what did you learn, who did you meet, what sales did you make, what of benefit to yourself, colleagues, and/or your company occurred?

   Recommendations: What do you suggest next? Should the company continue on the present course (status quo) or should changes be made in personnel or in the approach to a particular situation? Would you suggest that other colleagues attend this conference in the future, or was the job-related travel not effective? In your opinion, what action should the company take?

Figure 8.1 presents an example of a trip report.

PROGRESS REPORTS

Your supervisors want to know what you're doing at work. They want to know what progress you're making on a project, whether you're on schedule, what difficulties you might have encountered, and/or what your plans are for the next reporting period. Because of this, supervisors ask you to write progress (or activity or status) reports—daily, weekly, monthly, quarterly, or annually. The following are components of an effective progress report.

1. Heading

Date

To

From

Subject: Include the topic about which you are reporting and the reporting interval (date).

```
Date:       May 31, 1992

To:         Joanna Faulkner

From:       Lupe Salinas  LS

Subject:    January Progress Report on Sales Calls
```

2. Introduction (overview, background)

Objectives: These can include the following:

- Why are you working on this project (what's the rationale)?
- What problems motivated the project?
- What do you hope to achieve?
- Who initiated the activity?

Personnel: With whom are you working on this project (i.e., work team, liaison, contacts)?

Previous activity: If this is the second, third, fourth, etc. report in a series, remind your readers what work has already been accomplished. Bring them up-to-date with background data or a reference to previous reports.

3. Discussion (findings, body, agenda)

Work accomplished: Using subheadings, itemize your work accomplished either through a chronological list or a discussion organized by importance.

Problems encountered: Inform your readers of any difficulties encountered (late shipments, delays, poor weather, labor shortages) not only to justify your possibly being behind schedule but also to show the readers where you'll need help to complete the project.

4. Conclusion/recommendations

Conclusion: Sum up what you've achieved during this reporting period and provide a prophetic conclusion—tell your readers what work you plan to complete next and what your anticipated date of completion is. Doing so will help your supervisors provide you with needed labor, allocate appropriate funds, and/or reschedule your milestones.

Recommendations: Suggest changes to be made which will allow you to meet your deadlines and/or request assistance.

Figure 8.2 presents an example of a progress report.

```
Date:      February 26, 1991
To:        Pat Berry
From:      Juliet Harris    Juliet
Subject:   Trip Report—Renton West Seminar on Electronic
Packaging
```

INTRODUCTION

```
On Tuesday, February 23, 1991, 1 attended the Renton West
National Electronic Packaging Seminar, held in Ruidoso,
NM. My goal was to acquire hands-on training and to learn
new techniques for electronic packaging.
```

OBSERVATIONS

```
The following were the most informative seminars attended:
```

Production Automation—Dr. Wang Hue

- Reviewed and provided hands-on training for foam-
  encapsulation automated techniques.

Vapor Phase Soldering—Garth McSwain

- Provided information on processing double-sided chip
  components.

Electronics for Extreme Temperatures—Xanadu Rand

- Presented scientific data on packaging under temperature
  extremes.

CONCLUSION/RECOMMENDATIONS

```
1. Dr. Hue's program was the most informative and useful.
   I suggest that we bring Dr. Hue to our site for further
   consultation.
2. Vapor phase soldering is too costly. We could not pay
   back our investment within this quarter.
3. Our new-hires would benefit from Xanadu Rand's scientific
   overview. Supervisors, however, will find the data
   remedial.
```

FIGURE 8.1 Trip Report

LAB REPORTS

Professionals in electronics, engineering, medical fields, chemistry technology, the computer industry, and other technologies often rank the ability to communicate as highly as they do technical skills. Conclusions derived from a technical procedure are worthless if they reside in a vacuum. The knowledge you acquire from a lab experiment *must* be communicated to your colleagues and supervisors so they can benefit from your discoveries. That's the purpose of a lab report—to document your findings. You write a lab report after you've performed a laboratory test to share with your readers:

- Why the test was performed
- How the test was performed

- What the test results were
- What follow-up action (if any) is required

The following are components of a successful lab report.

1. **Heading**

   Date

   To

   From

   Subject (topic + focus)

2. **Introduction (overview, background, purpose)**

   Why is this report being written? To answer this question, provide any or all of the following:

   - The rationale (What problem motivated this report?)
   - The objectives (What does this report hope to prove?)
   - Authorization (Under whose authority is this report being written?)

Date: April 2, 1991
To: Buddy Ramos
From: Pat Smith   *PS*
Subject: First Quarterly Report-Project 80 Construction

Background:

Department 93 is investigating our capability to support FY1991 build plans for Project 80. This activity is in response to a request from HEW.

Work Accomplished:

In this first quarter, we've studied the following:

1. Shipments: TDDS shipments this quarter have provided us a 30-day lead on scheduled part provisions.

2. Testing: Screening on high-voltage monitors has been completed with a pass/fail ratio of 76.4% pass to 23.6% fail. This meets our 75% goal.

3. Production: Rolanta production is costly ($84 per mil) and two weeks behind schedule.

Problems Encountered:

Rolanta production was delayed due to contamination in the assembly and fill areas, which led to a shutdown. An examination is still pending.

Conclusion:

We anticipate successful build for FY1991.

Recommendations:

We could use your leverage with the Rolanta problem. Please call Chuck Lyons and persuade him that timely production is mandatory to meet our schedules.

FIGURE 8.2 **Progress Report**

3. **Discussion (body, methodology)**

How was the test performed? To answer this question, provide the following

- Apparatus (What equipment, approach, or theory have you used to perform your test?)
- Procedure (What steps—chronologically organized—did you follow in performing the test?)

4. **Conclusion/ recommendations**

Conclusion: The conclusion of a lab report presents your findings. Now that you've performed the laboratory experiment, what have you learned or discovered or uncovered? How do you interpret your findings? What are the implications?

Recommendations: What follow-up action (if any) should be taken?

You might want to use graphics to supplement your lab report. Schematics and wiring diagrams are important in a lab report to clarify your activities, as shown in Figure 8.3.

FEASIBILITY REPORTS

Occasionally, your company plans a project but is uncertain whether the project is feasible. For example, your company might be considering the purchase of equipment but is concerned that the machinery will be too expensive, the wrong size for your facilities, or incapable of performing the desired tasks. Perhaps your company wants to expand and is considering new locations. The decision makers, however, are uncertain which locations would be best for the expansion. Maybe your company wants to introduce a new product to the marketplace, but your CEO wants to be sure that a customer base exists before funds are allocated.

One way a company determines the viability of a project is to perform a feasibility study and then write a feasibility report documenting the findings. The following are components of an effective feasibility report.

1. **Heading**

Date
To
From
Subject (topic + focus)

```
Subject: Feasibility Report on XYZ Project
                   (focus)         (topic)
```

2. **Introduction (overview, background)**

Objectives: Under this subheading, you can answer any of the following questions:

- What is the purpose of this feasibility report? Until you answer this question, your reader doesn't know. As mentioned earlier in this chapter, it's false to assume prior knowledge on the part of your audience. One of your responsibilities is to provide background data. To answer the question regarding the report's purpose, you should provide a clear and concise statement of intent.
- What problems motivated this study? To clarify for your readers the purposes behind the study, *briefly* explain either
  - What problems cause doubt about the feasibility of the project (i.e., is there a market, is there a piece of equipment available which would meet the company's needs, is land available for expansion?)

Date: July 18, 1991
To: Dr. Jones
From: Sam Ascendio, Lab Technician   *Sam*
Subject: Lab Report on the Accuracy of Decibel Voltage Gain
           (A) Measurements

Purpose

Technical Services has noted inaccuracies in recent measurements. In response to their request, this report will
present results of tested A (gain in decibels) of our ABC
voltage divider circuit. Measured A will be compared to
calculated A. This will determine the accuracy of the measuring device.

Apparatus

- Audio generator
- Decade resistance box
- 1/2 W resistor: four 470 ohm, two 1 kilohm, 100 kilohm.
- AC millivoltmeter

Procedure

1. Figure A shows a voltage divider. For each value of R
   (resistances) in Table 1, voltage gain was calculated
   (table attached).
2. An audio generator was adjusted to give a reading of 0
   dB for input voltage.
3. Output voltage was measured on the dB scale. This reading is the measured A and is recorded in Table 1.
4. Step 3 was repeated for each value of R listed in
   Table 1.
5. Figure B shows three cascaded voltage dividers. A1 =
   v2/v1, A2 = v3/v2, and A3 = v4/v3. These voltage gains
   added together give the total voltage, as recorded in
   Table 2.
6. The circuit in Figure B was connected.
7. Input voltage was set at 0 dB on the 1-V range of the
   AC millivoltmeter.
8. Values V2, V3, and V4 were read and recorded in Table 3.

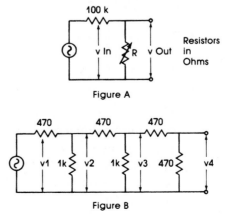

Figure A

Figure B

FIGURE 8.3 **Lab Report**

```
Table 1

R                    Calculated A          Measured A
240 kilohm             − 3.02 dB           − 3.21 dB
100 kilohm             − 6.02 dB           − 6.13 dB
46 kilohm             −10.00 dB            − 9.71 dB
11 kilohm             −20.00 dB            −20.00 dB
1 kilohm              −40.00 dB            −40.00 dB

Table 2                              Table 3

A1                     .500        V1               0 dB
A2                     .500        V2          − 5.97 dB
A3                     .500        V3          −11.97 dB
A                      .125        V4          −18.06 dB

Conclusions/Recommendations

Accuracy between measured and calculated A in Table 1 was
between .01 and .1 dB. This is acceptable. Accuracy
between measured A in Table 3 and calculated voltage gain
in Table 2 was between .01 and .1 dB—also very accurate.
These tests show minimal error. No further action should
be taken.
```

FIGURE 8.3 Lab Report (*Continued*)

— what problems led to the proposed project (i.e., current equipment is too costly or time consuming, current facilities are too limited for expansion, current net income is limited by an insufficient market)

• Who initiated the feasibility study.? List the name(s) of the managers or supervisors who requested this report.

Personnel: Document the names of your project team members, your liaison between your company and other companies involved, and/or your contacts at these other companies.

3. Discussion (body, findings)

Under this subheading, provide accessible and objective documentation regarding the following:

• What procedures did you use to analyze the feasibility of your project? Did you use technical research, site evaluations, vendor presentations, computer analyses, interviews, or outside consultants?

• What did your study reveal? State the facts about your project's cost, specifications, labor requirements, capabilities, availability, etc.

4. Conclusion/recommendations

Conclusion: In this section, you go beyond the mere facts as evident in the discussion section. You state the significance of the findings. You draw a conclusion from what you've found in your study. What does it mean?

Recommendations: Once you've drawn your conclusions, the next step is to recommend a course of action. What do you suggest that your company do next? Which piece of equipment should be purchased, where should the company locate its expansion, or is there a sufficient market for the product?

Figure 8.4 presents an example of a feasibility report.

INCIDENT REPORTS

If a problem occurs within your work environment which requires investigation and suggested solutions, you might be asked to prepare an incident report (also called a trouble report or accident report).

Engineering environments requiring maintenance reports rarely provide employees with easy-to-fill-in forms. To write an incident report when you have not been given a printed form, include the following components.

1. **Heading**
   Date
   To
   From
   Subject (topic + focus)

```
Subject: Report on Chilled Water Leaks in D/823
         (focus)          (topic)
```

2. **Introduction**
   Purpose: In this section, document when, where, and why you were called to perform maintenance. What motivated your visit to the scene of the problem?
3. **Discussion (body, findings, agenda, work accomplished)**
   Using subheadings or itemization, quantify what you saw (the problems motivating the activity) and what you did to solve the problem.
4. **Conclusion**
   Explain what caused the problem.
5. **Recommendations**
   Relate what could be done in the future to avoid similar problems.

Figure 8.5 presents an example of an incident report.

PROCESS

Now that you know the criteria for short reports in general and for specific types of short reports (trip reports, progress reports, lab reports, feasibility reports, and incident reports), the next step is to construct these reports. How do you begin? As always, *prewrite, write,* and *rewrite.*

LABORATORY REPORT

There are two kinds of laboratory reports: those written in industry to document laboratory research or tests on materials or equipment, and those written in academic institutions to record laboratory tests performed by students. The former are generally known as test reports or laboratory reports; those written by students are simply called lab reports.

```
Date:     October 20, 1991
To:       Jack Newton
From:     Hal Langston   Hal
Subject:  Report on the Feasibility of Purchasing an XYZ
          Mini-Computer
```

Introduction

Purpose: The purpose of this report is to determine whether three new XYZ 5200 mini-computers can replace our current time-sharing terminals.

Problems motivating the project: Our current time-sharing terminals are costing $2000 a month more than we budgeted due to high initial start-up and annual maintenance costs.

Principal: Kristen Jamberdino

Personnel: Hal Langston, Judi Simmons, and Rick Gambles (project team). Darwin Lawyer (contact at XYZ Corp.).

Discussion

Using research provided by our vendor contact and studies from our accounting department, we concluded the following (all figures in dollars):

	1st Year		2nd Year	
	Time-sharing	Mini-computer	Time-sharing	Mini-computer
Initial cost	—	30,000	—	—
Time-share Chg.	54,000	—	54,000	—
Info. storage	12,000	6,000	12,000	500
Cabinet	—	500	—	—
Maint. contract	—	9,000	—	9,000
Total	66,000	45,500	66,000	9,500

Conclusion

Three mini-computers save as much as 31% in the first year of installation and 85% in the second year.

```
1st Year  66,000 - 45, 500
   x 100 = 31.06%
   66,000
2nd Year  66,000 - 9,500
   x 100 = 85.61%
   66,000
```

In addition, mini-computers shorten response time, another cost savings.

Recommendations

We should purchase the three XYZ 5200 mini-computers to replace our current time-sharing system.

FIGURE 8.4 Feasibility Report

```
Date:    October 16, 1991
To:      Tom Warner
From:    Carlos Sandia  CS
Subject: Report on Chilled Water Leaks in D/823
```

Introduction

```
On October 15, 1991, a flood was reported in D/823. I was
called in to repair this leak. Following is a report on my
findings and maintenance activities.
```

Agenda

```
8:15 P.M.   The flood was reported.

8:25 P.M.   I arrived at D/823 and discovered that the
            water level in the expansion tank at HVAC unit
            #253R-01 was 4 feet above normal and the tank
            valve was open.

8:30 P.M.   A second flood was reported at HVAC unit #937-01
            in the same department.

8:35 P.M.   I shut off the open tank valves.

9:00 P.M.   The expansion tank levels and pressure returned
            to normal.

9:30 P.M.   All leaks were secure.
```

Conclusion

```
The chilled water leaks were caused by a malfunctioning
level switch on the chilled water expansion tanks in the
west boiler rooms. This caused the water level in the
tanks to rise, which increased the system pressure to
150 lb.
```

Recommendations

```
The level control switch will be repaired. An alarm system
will be installed by November 1, 1991.
```

FIGURE 8.5 Incident Report

Industrial laboratory reports can describe a wide range of topics, from tests of a piece of metal to determine its tensile strength, through analysis of a sample of soil (a "drill core") to identify its composition, to checks of a microwave oven to assess whether it emits radiation. Academic lab reports can also describe many topics, but their purpose is different since they describe tests which usually are intended to help students learn something or prove a theory rather than produce a result for a client.

Laboratory reports generally conform to a standard pattern, although emphasis differs depending on the purpose of the report and how its results will be used. Readers of industrial laboratory or test reports are usually more interested in results ("Is the enclosed sample of steel safe to use for construction of microwave towers which will be exposed to temperatures as low as $-400°C$ in a North Dakota winter?" a client may ask) than in how a test was carried out. Readers of academic lab reports are usually professors and instructors who are more likely to be interested in thoroughly documented details, from which they can assess the student report writer's understanding of the subject and what the test proved.

A laboratory report comprises several readily identifiable compartments, each usually preceded by a heading. These compartments are described briefly here:

Part	Section Title	Contents
Summary	**Summary**	A very brief statement of the purpose of the tests, the main findings, and what can be interpreted from them. (In short laboratory reports, the summary can be combined with the next compartment.)
Background	**Objective**	A more detailed description of why the tests were performed, on whose authority they were conducted, and what they were expected to achieve or prove.
Facts		*There are four parts here:*
	Equipment	A description of the test setup, plus a list of equipment and materials used. A drawing of the test hook-up may be inserted here. (If a series of tests is being performed, with a different equip-ment setup for each test, then a separate equip-ment description, materials list, and illustration should be inserted immediately before each test description.)
	Test Method	A detailed, step-by-step explanation of the tests. In industrial laboratory reports the depth of explanation depends on the reader's needs: if a reader is nontechnical and likely to be interested only in results, then the test description can be condensed. For lab reports written at a college or university, however, students are expected to provide a thorough description here.
	Test Results	Usually a brief statement of the test results or the findings evolving from the tests.
	Analysis (or Interpretation)	A detailed discussion of the results or findings, their implications, and what can be interpreted from them. (The analysis section is particularly important in academic lab reports.)
Outcome	**Conclusions**	A brief summing-up which shows how the test results, findings, and analysis meet the objective(s) established at the start of the report.
Backup	**Attachments**	These are pages of supporting data such as test measurements derived during the tests, or documentation such as specifications, procedures, instructions, and drawings, which would interrupt reading continuity if placed in the report narrative (in the test method section).

The compartments described here are those most likely to be used for either an industrial laboratory report or a college/university lab report. In practice, however, emphasis and labeling of the compartments probably win differ slightly, depending on the requirements of the organization employing the report writer or, in an academic setting, the professor or instructor who will evaluate the report.

9 Using Graphics

OBJECTIVES

Although your writing may have no grammatical or mechanical errors and you may present valuable information, you won't communicate effectively if your information is inaccessible. Consider the following paragraph:

In January 2000, the actual rainfall was 1.50", but the average for that month was 2.00". In February 2000, the actual rainfall was 1.50", but the annual average had been 2.50". In March 2000, the actual rainfall was 1.00", but the yearly average was 2.50". In April 2000, the actual rainfall was 1.00", but annual averages were 2.50". The May 2000 actual rainfall was 0.50", whereas the annual average had been 1.50". No rainfall was recorded in June 2000. Annually, the average had been 0.50". In July 2000, only 0.25" rain fell. Usually, July had 0.50" rain. In August 2000, again no rain fell, whereas the annual August rainfall measured 0.25". In September and October 2000, the actual rainfall (0.50") matched the annual average. Similarly, the November actual rainfall matched the annual average of 1.50". Finally, in December 2000, 2.00" rain fell, compared to the annual average of 1.50".

If you read the preceding paragraph in its entirety, you are an unusually dedicated reader. Such wall-to-wall words mixed with statistics do not create easily readable writing.

The goal of effective technical writing is to communicate information easily. The example paragraph fails to meet this goal. No reader can digest the data easily or see clearly the comparative changes from one month's precipitation to the next.

To present large blocks of data and/or reveal comparisons, you can supplement, if not replace, your text with graphics. In technical writing, visual aids accomplish several goals. Graphics (whether hand drawn, photographed, or computer generated) will help you achieve conciseness, clarity, and cosmetic appeal.

CONCISENESS

Visual aids allow you to provide large amounts of information in a small space. Words used to convey data (such as in the example paragraph) double, triple, or even quadruple the space needed to report information. By using graphics, you can also delete many dead words and phrases.

CLARITY

Visual aids can clarify complex information. Graphics help readers see the following:

- *Trends* (increasing or decreasing sales figures, for instance). These are most evident in line graphs.
- *Comparisons between like components* (as in actual monthly versus average rainfalls). These can be seen in grouped bar charts.
- *Percentages*. Pie charts help readers discern these.
- *Facts and figures*. A table states statistics more clearly than a wordy paragraph.

COSMETIC APPEAL

Visual aids help you break up the monotony of wall-to-wall words. If you only give unbroken text, your reader will tire, lose interest, and overlook key concerns. Graphics help you sustain your reader's interest. Let's face it; readers like to look at pictures.

The two types of graphics important for technical writing are tables and figures. This chapter helps you correctly use both.

COLOR

All graphics look best in color, don't they? Not necessarily. Without a doubt, a graphic depicted in vivid colors will attract your reader's attention. However, the colors might not aid communication. For example, colored graphics could have these drawbacks (Reynolds and Marchetta 5–7):

1. The colors might be distracting (glaring orange, red, and yellow combinations on a bar chart would do more harm than good).
2. Colors that look good today might go out of style in time (like those avocado green and autumn gold appliances in your family's kitchen).
3. Colored graphics increase production costs.
4. Colored graphics consume more disk space and computer memory than black-and-white graphics.
5. The colors you use might not look the same to all readers.

Let's expand on this last point. Just because you see the colors one way does not mean your readers will see them the same. We're not talking about people with vision problems. Instead, we're talking about what happens to your color graphics when someone reproduces them as black-and-white copies. We're also talking about computer monitor variations.

The color on a computer monitor depends on its resolution (the number of pixels displayed) and the monitor's RGB values (how much red, green, and blue light is displayed). Because all monitors do not display these same values, what you see on your monitor will not necessarily be the same as what your reader sees. To solve this problem, test your graphics on several monitors. Also, limit your choices to primary colors instead of the infinite array of other color possibilities. More important, use patterns to distinguish your information so that the color becomes secondary to the design.

THREE-DIMENSIONAL GRAPHICS

Many people are attracted to three-dimensional graphics. After all, they have obvious appeal. Three-dimensional graphics are more interesting and vivid than flat, one-dimensional graphics. However, 3-D graphics have drawbacks. A 3-D graphic is visually appealing, but it does not convey information quantifiably. A word of caution: use 3-D graphics sparingly. Better yet, use the 3-D graphic to create an impression; then, include a table to quantify your data.

CRITERIA FOR EFFECTIVE GRAPHICS

Figure 9.1 is an example of a cosmetically appealing, clear, and concise graphic. At a glance, the reader can pinpoint the comparative prices per barrel of crude oil between 1996 and 2000. Thus, the line graph is clear and concise. In addition, the writer has included an interesting artistic touch. The oil gushing out of the tower shades just the parts of the graph that emphasize the dollar amounts. Envision this graph without the shading. Only the line would exist. The shading provides the right touch of artistry to enhance the information communicated.

The graph shown in Figure 9.1, although successful, does not include all the traits common to effective visual aids. Whether hand drawn or computer generated, successful tables and figures

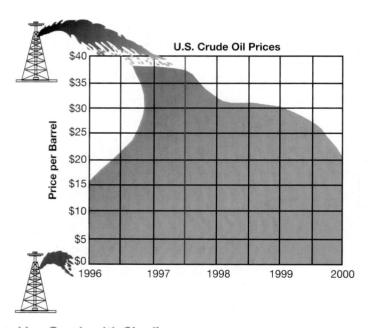

FIGURE 9.1 Line Graph with Shading

1. Are integrated with the text (i.e., the graphic complements the text; the text explains the graphic).
2. Are appropriately located (preferably immediately following the text referring to the graphic and not a page or pages later).
3. Add to the material explained in the text (without being redundant).
4. Communicate important information that could not be conveyed easily in a paragraph or longer text.

5. Do not contain details that detract from rather than enhance the information.
6. Are an effective size (not too small or too large).
7. Are neatly printed to be readable.
8. Are correctly labeled (with legends, headings, and titles).
9. Follow the style of other figures or tables in the text.
10. Are well conceived and carefully executed.

TYPES OF GRAPHICS

TABLES

Let's tabulate the information about rainfall in 2000 presented earlier. Because effective technical writing integrates text and graphic, you'll want to provide an introductory sentence prefacing the table, as follows:

> Table 9.1 reveals the actual amount of rainfall for each month in 2000 versus the average documented rainfall for those same months.

This table has advantages for both the writer and the reader. First, the headings eliminate needless repetition of words, thereby making the text more readable. Second, the audience can see easily the comparison between the actual amount of rainfall and the monthly averages. Thus, the table highlights the content's significant differences. Third, the table allows for easy future reference. Tables could be created for each year. Then the reader could compare quickly the changes in annual precipitation. Finally, if this information is included in a report, the writer will reference the table in the Table of Contents' List of Illustrations. This creates ease of access for the reader.

Criteria for Effective Tables

To construct tables correctly, do the following:

1. Number tables in order of presentation (i.e., Table 1, Table 2, Table 3, etc.).
2. Title every table. In your writing, refer to the table by its number, not its title. Simply say, "Table 1 shows...," "As seen in Table 1," or "The information in Table 1 reveals...."
3. Present the table as soon as possible after you've mentioned it in your text. Preferably, place the table on the same page as the appropriate text, not on a subsequent, unrelated page or in an appendix.

TABLE 9.1	2000 Monthly Rainfall versus Average Rainfall (All Figures in Inches)	
MONTH	**2000 RAINFALL**	**AVERAGE RAINFALL**
January	1.50	2.00
February	1.50	2.50
March	1.00	2.50
April	1.00	2.50
May	0.50	1.50
June	0.00	0.50
July	0.25	0.50
August	0.00	0.25
September	0.50	0.50
October	0.50	0.50
November	1.50	1.50
December	2.00	1.50

4. Don't present the table until you've mentioned it.

5. Use an introductory sentence or two to lead into the table.

6. After you've presented the table, explain its significance. You might write, "Thus, the average rainfall in both March and April exceeded the actual rainfall by 1.50 inches, reminding us of how dry the spring has been."

7. Write headings for each column. Choose terms that summarize the information in the columns. For example, you could write "% of Error," "Length in Ft.," or "Amount in $."

8. Since the size of columns is determined by the width of the data and/or headings, you may want to abbreviate terms (as shown in item 7). If you use abbreviations, however, be sure your audience understands your terminology.

9. Center tables between right and left margins. Don't crowd them on the page.

10. Separate columns with ample white space, vertical lines, or dashes.

11. Show that you've omitted information by printing two or three periods or a hyphen or dash in an empty column.

12. Be consistent when using numbers. Use either decimals or numerators and denominators for fractions. You could write 3 1/4 and 3 3/4 or 3.25 and 3.75. If you use decimal points for some numbers but other numbers are whole, include zeroes. For example, write 9.00 for 9.

TABLE 9.2	Student Headcount Enrollment by Age Group and Student Status, Fall 2000				
AGE GROUP	NEW STUDENTS	CONTINUING STUDENTS	READMITTED	OTHER	TOTAL
15–17	453	33	2	2	490
18–20	1,404	1,125	132	—	2,661
21–23	339	819	269	—	1,427
24–26	263	596	213	—	1,072
27–29	250	436	134	—	820
30–39	524	1,168	372	—	2,064
40–49	271	510	186	—	967
50–59	76	121	54	—	251
60+	19	48	16	—	83
Unknown	109	92	27	2	230
Total	3,708	4,948	1,405	4	10,065

13. If you do not conclude a table on one page, on the second page write *Continued* in parentheses after the number of the table and the table's title.

Table 9.2 is an excellent example of a correctly prepared table.

FIGURES

Another way to enhance your technical writing is to use figures. Whereas tables eliminate needless repetition of words, figures highlight and supplement important points in your writing. Like tables, figures help you communicate with your reader. Types of figures include the following:

- Bar charts
  - —Grouped bar charts
  - —3-D (tower) bar charts
  - —Pictographs
  - —Gantt charts
- Pie charts
- Line charts
  - —Broken line charts
  - —Curved line charts
- Flowcharts
- Organizational charts
- Schematics
- Line drawings
  - —Exploded views
  - —Cutaway views
  - —Super comic book look
- Photographs
- Icons
- Internet graphics

All of these types of figures (except photographs) can be computer generated using an assortment of computer programs. The program you use depends on your preference and hardware.

Criteria for Effective Figures

To construct figures correctly, do the following:

1. Number figures in order of presentation (i.e., Figure 1, Figure 2, Figure 3, etc.).
2. Title each figure. When you refer to the figure, use its number rather than its title: for example, "Figure 1 shows the relation between the average price for houses and the actual sales prices."
3. Preface each figure with an introductory sentence.
4. Don't use a figure until you've mentioned it in the text.
5. Present the figure as soon as possible after mentioning it instead of several paragraphs or pages later.
6. After you've presented the figure, explain its significance. Don't let the figure speak for itself. Remind the reader of the important facts you want to highlight.
7. Label the figure's components. For example, if you're using a bar or line chart, label the x- and y-axes clearly. If you're using line drawings, pie charts, or photographs, use clear *call-outs* (names or numbers that indicate particular parts) to label each component.

8. When necessary, provide a legend or key at the bottom of the figure to explain information. For example, a key in a bar or line chart will explain what each differently colored line or bar means. In line drawings and photographs, you can use numbered call-outs in place of names. If you do so, you'll need a legend at the bottom of the figure explaining what each number means.

9. If you abbreviate any labels, define these in a footnote. Place an asterisk (*) or a superscript number ([1], [2], [3], . .) after the term and then at the bottom of the figure where you explain your terminology.

10. If you've drawn information from another source, note this at the bottom of the figure.

11. Frame the figure. Center it between the left and right margins and/or window it in a box.

12. Size figures appropriately. Don't make them too small or too large.

13. Try the super comic book look (figures drawn in cartoon-like characters to highlight parts of the graphic and to interest readers).

Bar Charts

Bar charts show either vertical bars (as in Figure 9.2) or horizontal bars (as in Figure 9.3). These bars are scaled to reveal quantities and comparative values. You can shade, color, and/or crosshatch the bars to emphasize the contrasts. If you do so, include a key explaining what each symbolizes, as in Figure 9.4. *Pictographs* (as in Figure 9.5) use picture symbols instead of bars to show quantities. To create effective pictographs, do the following:

1. The picture should be representative of the topic discussed.

2. Each symbol equals a unit of measurement. The size of the units depends on your value selection as noted in the key or on the *x*- and *y*-axes.

3. Use more symbols of the same size to indicate a higher quantity; do not use larger symbols.

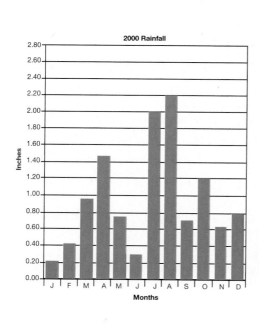

FIGURE 9.2 Vertical Bar Chart

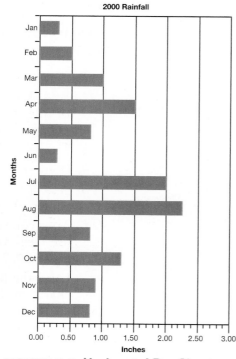

FIGURE 9.3 Horizontal Bar Chart

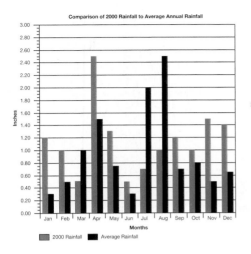

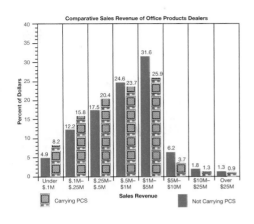

FIGURE 9.4 Grouped Bar Chart   **FIGURE 9.5** Pictograph

Gantt Charts

Gantt charts, or *schedule charts* (as in Figure 9.6), use bars to show chronological activities. For example, your goal might be to show a client phases of a project. This could include planned start dates, planned reporting milestones, planned completion dates, actual progress made toward completing the project, and work remaining. Gantt charts are an excellent way to represent these activities visually. They are often included in proposals to project schedules or in reports to show work completed. To create successful Gantt charts, do the following:

1. Label your *x*- and *y*-axes. For example, if the *y*-axis represents the various activities scheduled, then the *x*-axis represents time (either days, weeks, months, or years).

2. Provide grid lines (either horizontal or vertical) to help your readers pinpoint the time accurately.

3. Label your bars with exact dates for start or completion.

4. Quantify the percentages of work accomplished and work remaining.

5. Provide a legend or key to differentiate between planned activities and actual progress.

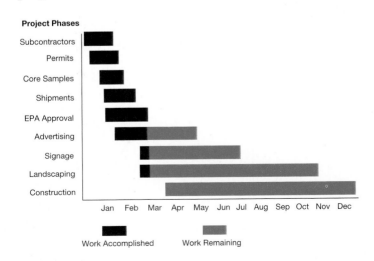

FIGURE 9.6 Gantt Chart

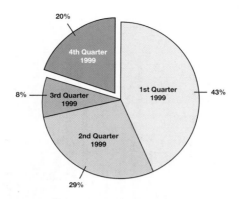

FIGURE 9.7 Pie Chart with Wedge Removed for Emphasis

Pie Charts

Use pie charts (as in Figure 9.7) to illustrate portions of a whole. The pie chart represents information as pie-shaped parts of a circle. The entire circle equals 100 percent, or 360 degrees. The pie pieces (the wedges) show the various divisions of the whole.

To create effective pie charts, do the following:

1. Be sure that the complete circle equals 100 percent, or 360 degrees.
2. Begin spacing wedges at the 12 o'clock position.
3. Use shading, color, and/or crosshatching to emphasize wedge distributions.
4. Use horizontal writing to label wedges.
5. If you don't have enough room for a label within each wedge, provide a key defining what each shade, color, and/or crosshatching symbolizes.
6. Provide percentages within wedges when possible.
7. Do not use too many wedges—this would crowd the chart and confuse readers.
8. Make sure that different sizes of wedges are fairly large and dramatic.

Line Charts

Line charts reveal relationships between sets of figures. To make a line chart, plot sets of numbers and connect the sets with lines. These lines create a picture showing the upward and downward movement of quantities. Line charts of more than one line (see Figure 9.8) are useful in showing comparisons between two sets of values. However, avoid creating line charts with too many lines, which will confuse your readers.

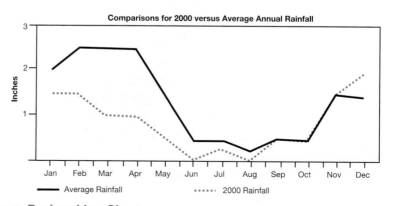

FIGURE 9.8 Broken Line Chart

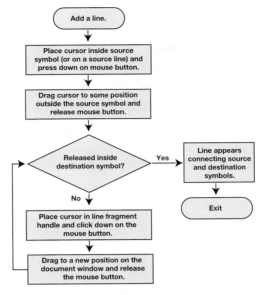

FIGURE 9.9 Flowchart

Flowcharts

You can show chronological sequence of activities using a flowchart. Flowcharts are especially useful for writing technical instructions (see Chapter 12). When using a flowchart, remember that ovals represent starts and stops, rectangles represent steps, and diamonds equal decisions (see Figure 9.9).

Organizational Charts

These charts (as in Figure 9.10) show the chain of command in an organization. You can use boxes around the information or use white space to distinguish among levels in the chart. An organizational chart helps your readers see where individuals work within a business and their relation to other workers.

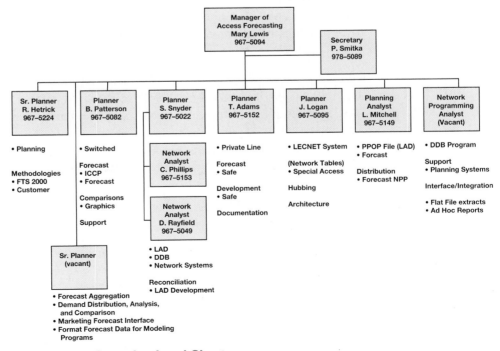

FIGURE 9.10 Organizational Chart

Schematics

Schematics are useful for presenting abstract information in technical fields such as electronics and engineering. A schematic diagrams the relationships among the parts of something such as an electrical circuit. The diagram uses symbols and abbreviations familiar to highly technical readers.

The schematic in Figure 9.11 shows various electronic parts (resistors, diodes, condensors) in a radio.

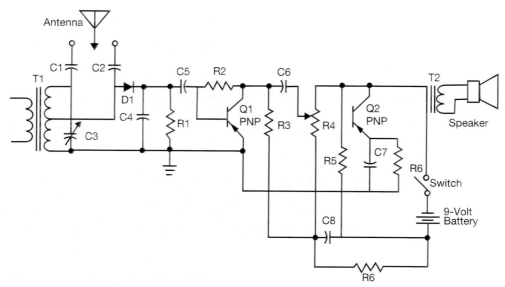

FIGURE 9.11 Schematic of a Radio

Line Drawings

Use line drawings to show the important parts of a mechanism or to enhance your text cosmetically. To create line drawings, do the following:

1. Maintain correct proportions in relation to each part of the object drawn.
2. If a sequence of drawings illustrates steps in a process, place the drawings in left-to-right or top-to-bottom order.
3. Using call-outs to name parts, label the components of the object drawn (see Figure 9.12).
4. If there are numerous components, use a letter or number to refer to each part. Then reference this letter or number in a key (see Figure 9.13).

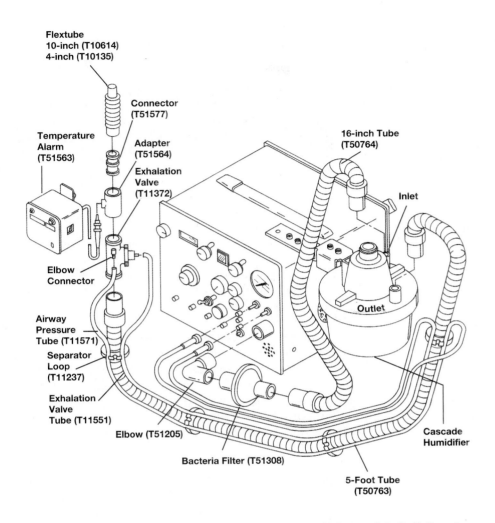

FIGURE 9.12 Line Drawing of Ventilator (Exploded View with Call-Outs)

Exhalation Valve Parts List		
Item	Part Number	Description
1	000723	Nut
2	003248	Cap
3	T50924	Diaphragm
4	Reference	Valve Body
5	Reference	Elbow Connector
–	T11372	Exhalation Valve

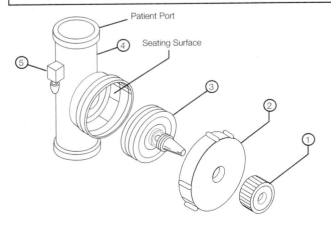

FIGURE 9.13 Line Drawing of Exhalation Valve (Exploded View with Key)

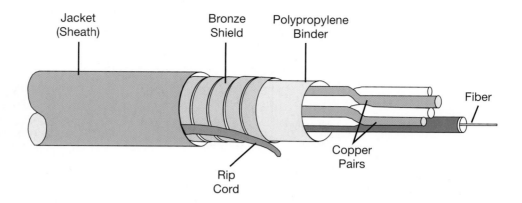

Jacket
(Sheath)

Bronze
Shield

Polypropylene
Binder

Fiber

Copper
Pairs

Rip
Cord

FIGURE 9.14 Line Drawing of Cable (Cutaway View)

5. Use exploded views (Figures 9.12 and 9.13) or cutaways (Figure 9.14) to highlight a particular part of the drawing.

Photographs

A photograph can illustrate your text effectively. Like a line drawing, a photograph can show the components of a mechanism. If you use a photo for this purpose, you'll need to label (name), number, or letter parts and provide a key. Photographs are excellent visual aids because they emphasize all parts equally. Their primary advantage is that they show something as it truly is.

Photographs have one disadvantage, however. They are difficult to reproduce. Whereas line drawings photocopy well, photographs do not.

Icons

Approximately 20 percent of America's population is functionally illiterate. In today's global economy, consumers speak diverse languages. Given these two facts, how can technical writers communicate to people who can't read and to people who speak different languages? Icons offer one solution. Icons (as in Figure 9.15) are visual representations of a capability, a danger, a direction, an acceptable behavior, or an unacceptable behavior.

For example, the computer industry uses icons (e.g., and open manila folders to represent a computer file). In manuals, a jagged lightning stroke iconically represents the danger of electrocution. On streets, an arrow represents the direction we should travel; on computers, the arrow shows us which direction to scroll. Universally depicted stick figures of men and women greet us on restroom doors to show us which rooms we can enter and which rooms we must avoid.

When used correctly, icons can save space, communicate rapidly, and help readers with language problems understand the writer's intent.

FIGURE 9.15 An Icon

To create effective icons, follow these suggestions:

1. *Keep it simple.* You should try to communicate a single idea. Icons are not appropriate for long discourse.

2. *Create a realistic image.* This could be accomplished by representing the idea as a photograph, drawing, caricature, outline, or silhouette.

3. *Make the image recognizable.* A top view of a telephone or computer terminal is confusing. A side view of a playing card is completely unrecognizable. Select the view of the object which best communicates your intent.

4. *Avoid cultural and gender stereotyping.* For example, if you're drawing a hand, you should avoid showing any skin color, and you should stylize the hand so it is neither clearly male nor female.

5. *Strive for universality.* Stick figures of men and women are recognizable worldwide. In contrast, letters—such as *P* for parking—will mean very little in China, Africa, or Europe. Even colors can cause trouble. In North America, red represents danger, but red is a joyous color in China. Yellow calls for caution in North America, but this color equals happiness and prosperity in the Arab culture (Horton, "Universal" 682–93).

INTERNET GRAPHICS

More and more, you will be writing online as the Internet, intranets, and extranets become prominent in technical writing. (We discuss the Internet in detail in Chapter 10.) You can create graphics for your Web site in three ways:

1. *Download existing online graphics for use in your site.* The Internet contains thousands of Web sites, which provide online clip art. These graphics include photographs, line drawings, cartoons, icons, animated images, arrows, buttons, horizontal lines, balls, letters, bullets, and hazard signs. In fact, you can download any image from any Web site. Many of these images are freeware, which you can download without cost and without infringing on copyright laws.

To download these images, just place your cursor arrow on the graphic you want, then right click on the mouse. A pop-up menu will appear. Scroll down to "Save image as." Once you have done this, a new menu will appear. You can save your image in the file of your choice, either on the hard drive or on your disk. The images from the Internet will already be *gif* (graphics interchange format) or *jpg* (joint photographic experts group) files. Thus, you will not have to convert them for use in your Web site.

2. *Modify and customize existing online graphics for use in your site.* If you plan to use an existing online graphic as your company's logo, for example, you will need to modify and/or customize the graphic. You'll want to do this for at least two reasons: to avoid infringing on copyright laws and to make the graphic uniquely yours.

To modify and customize graphics, you can download them in two ways. First, you can print the screen by pressing the "Print Screen" key (usually found on the upper right of your keyboard). This captures the entire screen image in a clipboard. Then you can open a graphics program and paste the captured image. Second, you can save the image in a file (as discussed above) and then open the graphic in a graphics package. Most graphics programs will allow you to customize a graphic. Popular programs include *Paint, Paint Shop Pro, PhotoShop, Corel Draw, Adobe Illustrator, Freehand,* and *Lview Pro.* In these graphics programs, you can manipulate the images: change colors, add text, reverse the images, crop, resize, redimension, rotate, retouch, delete or erase parts of the images, overlay multiple images, join multiple images, and so forth. After you make substantial changes, the new image becomes your property.

You could also take any existing graphic from hard-copy text (magazines, journals, books, newsletters, brochures, manuals, reports, etc.), scan the image, crop and retouch it, save it, then reopen this saved file in one of the above graphics programs for further manipulation.

Some graphics programs, such as *Paint*, save an image only as a *bmp*, a bitmap image. Once the image has been altered, you'll need to convert your bitmap file to a *gif* or *jpg* for use in your Web site. Doing so is important since the Internet will not read bmp images.

3. Create new graphics for use in your site. A final option is to create your own graphic. If you are artistic, draw your graphic in a program (see the list above), save the image as a *gif* or *jpg*, and then load the image into your Web site. This option might be more challenging and time-consuming. However, creating your own graphic gives you more control over the finished product, provides a graphic precisely suited to your company's needs, and helps to avoid infringement of copyright laws.

CHAPTER HIGHLIGHTS

1. Using graphics can allow you to create a more concise document.

2. Graphics often help you present information more clearly.

3. Graphics add variety to your text, breaking up wall-to-wall words.

4. Color and 3-D graphics can be effective. However, these two design elements also can cause problems. Your color choices might not be reproduced exactly as you planned, and a 3-D graphic could be misleading rather than informative.

5. Tables are effective for presenting numbers, dates, and columns of figures.

6. You can often communicate more easily with your audience when you use figures, such as bar charts, pie charts, line charts, flowcharts, and organizational charts, to highlight and supplement important parts of your text.

Virtual reality of a proposed building's interior stairwell.
Courtesy of the Office of Facility Planning, Johnson County Community College.

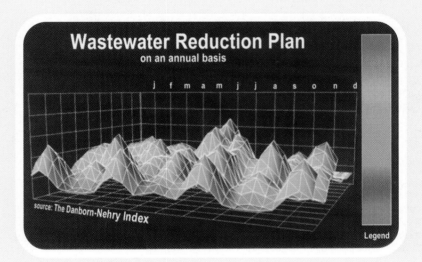

Topographical depiction of wastewater reduction.
Courtesy of Brandon Henry.

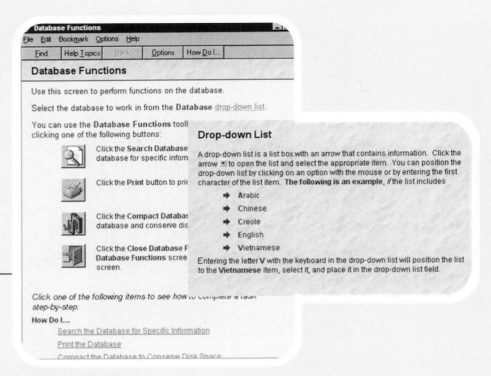

Online help screen.
Courtesy of Earl Eddings, Technical Communication Specialist.

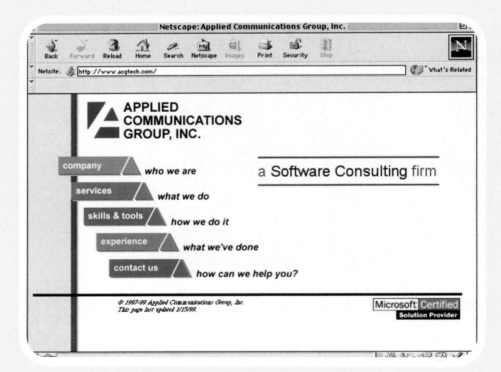

Screen capture of a web site home page.
Courtesy of Applied Communications Group, Inc.

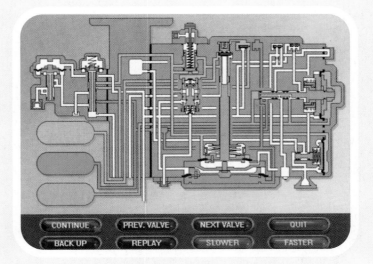

Cutaway view of a railcar braking system.
Courtesy of Burlington Northern Santa Fe Railroad.

Photograph of mechanical piping.
Courtesy of George Butler Associates, Inc.

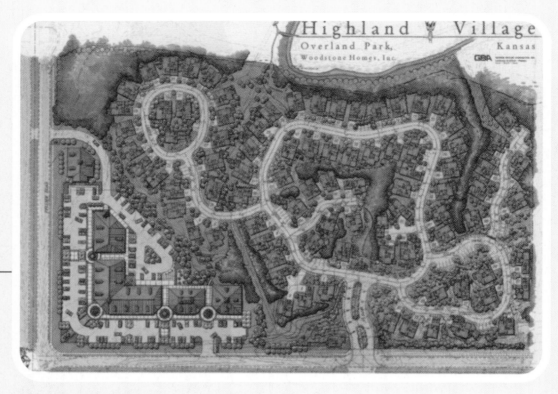

Architectural hand-drawn rendering.
Courtesy of George Butler Associates, Inc.

Comparison of planets with cutaway view.
Courtesy of Brandon Henry.

PART III

Critical Thinking and Problem Solving

10 Thinking Critically: Interacting with Information

You are a thinker. Your mind has extraordinary power that shows in everything you do, from small chores (comparing prices on gasoline or planning your day) to complex situations (writing a new computer program or sorting out a long-standing family argument). Your mind is able process, store, and create with the facts and ideas it encounters. Critical thinking means making the best use of these and other capacities of your mind.

Understanding how your mind works is the first step toward critical thinking. When you have that understanding, you can perform, the essential critical-thinking task: asking important questions about ideas and information. This chapter will show you the mind's basic actions, reintroducing you to the way your mind works. You will explore what it means to be an open-minded critical thinker, able to ask and understand questions that promote your success in college, career, and life.

In this chapter, you will explore answers to the following questions:

- What is critical thinking?
- Who is a critical thinker?
- How does your mind work?
- What are inductive and deductive reasoning?
- What are the types of knowledge?

WHAT IS CRITICAL THINKING?

If you read ten books on critical thinking, you probably will encounter ten different descriptions of what "critical thinking" means. You might see it referred to as "higher order" thinking or "metacognition" (thinking about thinking). You may wonder whether it is the same as any other kind of thinking, or whether it involves criticism. You might react negatively to it, figuring that the word "critical" implies something difficult and negative.

Actually, critical thinking is "critical" mainly in the sense of one particular dictionary definition of *critical*: "indispensable" and "important." It is not a secret, tricky process that you have never

"We do not live to think, but, on the contrary, we think in order that we may succeed in surviving."

JOSÉ ORTEGA Y GASSET

encountered or used before. It is not something that only "smart" people can accomplish. You think critically every day, even though you may not realize it.

A DEFINITION OF CRITICAL THINKING

The following is one way to define critical thinking, taking its many varied aspects into consideration.

> Critical thinking is thinking that *goes beyond the basic recall of information* but depends on the information recalled. It focuses on the *important, or critical*, aspects of the information. Critical thinking means *asking questions*. Critical thinking means that you *take in information, question it, and then use it* to create new ideas, solve problems, make decisions, construct arguments, make plans, and refine your view of the world.

The focus on critical thinking has increased as the world has become more saturated with information. You may hear the current time referred to as the "information age," in which each of us is bombarded every day with more information than anyone can possibly absorb. Information flows into your consciousness through television, newspapers, magazines, radio, the Internet, and the people with whom you interact. The more that comes your way, the more effort is necessary to sort through what's useful and not useful, what you believe and don't believe, and what is important in regard to what you already know and what is not. The more you improve your critical thinking, the better you will be able to swim in the ever-increasing sea of information—a skill that will help you both at school and on the job.

One way to clarify a concept is to take a look at its opposite. For example, if you were examining a gum condition in a dental hygiene class, you could begin to figure out what it is by ruling out other conditions with which you are familiar— "It's not gingivitis because. . . ." Similarly, you can clarify what critical thinking is by looking at what happens when people *don't* think critically. Not thinking critically means not examining critical or important aspects through questioning. A person who does not think critically tends to accept or reject information or ideas without examining them. Table 10.1 compares how a critical thinker and a non-critical thinker might respond to particular situations. Think about responses you or others have had to various situations in your life. Consider when you have seen critical thinking take place and when you haven't, and what resulted from each way of responding. This will help you begin to see what kind of an effect critical thinking can have on the way you live.

THE PATH OF CRITICAL THINKING

Look at Figure 10.1 to see a visual representation of critical thinking. Asking questions—the central part of the process—is the key to what makes you a critical thinker. Without asking questions about the information you take in, you would have no way to evaluate or change the information. Your output would mirror your input-unexamined and unchanged. For example, when you type a letter into a computer and print it, the computer does not think critically about your material. It spits out your letter exactly as you entered it (save, perhaps, a few changes made by the spell checker). The computer does not question you. Your mind, however, is far more discriminating than a computer. You have the power to question, use, and transform the information you encounter in your life.

Taking In Information

Most of the material in this book focuses on the second and third stages of critical thinking—questioning and using information. The first step of the process, however, is just as crucial. Your senses, especially your ability to hear and observe, allow you to take in information.

TABLE **10.1**		Not Thinking Critically vs. Thinking Critically	
YOUR ROLE	**SITUATION**	**NONQUESTIONING (UNCRITICAL) RESPONSE**	**QUESTIONING (CRITICAL) RESPONSE**
Student	Instructor is lecturing on the causes of the Vietnam war.	You assume everything your instructor says is true.	You consider what the instructor says, write questions about issues you want to clarify, discuss them with the instructor or classmates.
Spouse/partner	Your partner thinks he/she does not have enough quality time with you.	You think he/she is wrong and defend yourself.	You ask your partner why he/she thinks this is happening, and together you see how you can improve the situation.
Employee	Your supervisor is angry with you about something that happened.	You avoid your supervisor or deny responsibility for the incident.	You determine what caused your supervisor to place the blame on you; you talk with your supervisor about what happened and why.
Neighbor	People who differ from you move in next door.	You ignore or avoid them; you think their way of living is weird.	You introduce yourself and offer help if they need it; eventually, you respectfully explore your differences.
Consumer	You want to buy a car.	You decide on a brand new car and don't think through how you can pay for it.	You evaluate the effects of buying a new car versus a used car; you decide what kind of payment you can handle each month.

The information you receive is your raw material, the clay you will examine with questioning and mold into something new. The most crucial part of taking in information is to do so accurately and without judgment, because you want the material you work with to be as close to its natural state as possible. Just as the purest, finest ingredients allow a chef to make a rich cake, or the highest quality parts allow a car manufacturer to create a smooth-running automobile, the clearest, most complete information gives you the best material with which to work as you think. It gives you the most assurance that your final thought output will be as accurate and comprehensive as it can be.

For example, you are drowsy one day in your Biology course and you copy a biological principle inaccurately from the board. During a test later in the semester, you write a response to an essay question on this principle and feel confident about your work You then are shocked when you get your test back and receive no points for what you thought was a well-constructed, thoughtful essay. You had no idea that your original idea—the "material" with which you worked—was inaccurate.

In another scenario, say that you overhear part of a negative comment your spouse makes on the phone, and it sounds like it is about you. You spend the rest of the day fuming about it and decide not to come home for dinner after class and not to call to check in. By the time you return late in the evening, your spouse is worried sick about you. Only after an argument do you figure out that the comment had nothing to do with you at all. You made decisions based on inaccurate intake of information.

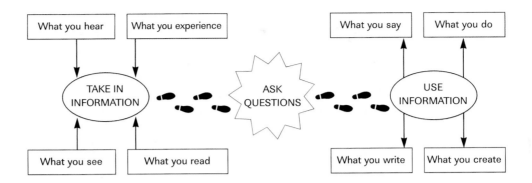

FIGURE 10.1 Critical Thinking Path

The first step of being a critical thinker is to take in information completely, accurately, and without judgment. Then, when you have collected your raw material, you can begin to sift through it—to discover its important or critical aspects through questioning.

Questioning Information

A critical thinker asks many kinds of questions about a given piece of information, such as: *Where did it come from? What idea or principle explains it? In what ways is it true or false? How do I feel about it, and why? How is this information similar to or different from what I already know? Is it useful or not? What caused it, and what effects does it have?* The center of the process is the questioning of what you have taken in.

Asking questions of information takes effort. Crucial to the questioning process are

- Knowing what questions to ask
- Knowing when to ask
- Wanting to ask

You need all three of these to make the most of your ability to question. It may not matter how much opportunity you have to question the information that comes your way if you don't know what to ask, have no idea when to ask it, and don't have any real desire to ask. What to ask and when to ask are covered in detail throughout this book. Whether to ask—the desire to ask—is up to you. Your desire to ask is often stimulated by the information and your perception of its usefulness.

Remember these types of questions. Most of this chapter will focus on how to use them to analyze the information that comes your way. When you explore the seven mind actions later in the chapter, refer to these questions to see how they illustrate the different actions your mind performs.

Using Information

After taking in and examining information, critical thinkers try to transform it into something they can use. They use information to help them solve a problem, make a decision, learn or create something new, or anticipate what will happen in the future. In this stage of the process, you can so synthesize what you know and have learned and present something completely new, such as an idea, a product, a process, or an approach.

This part of the critical thinking path is where you benefit from die hard work of asking questions, building new ideas through the creative action of your mind. This is where inventions happen, new processes are born, theories are created, and

From one or more examples (facts or events), you develop a general idea or ideas.
One way of coming up with an idea is to group similar facts or events together,

which may show you a pattern that allows you to make a general classification or statement about the group. Classifying a fact or event in this way helps you build knowledge (for example, you know what the Boston Tea Party was, but grouping it with other events of the Revolutionary War may help you learn the role it played in the larger concept of how the colonies became the United States). This mind action moves from the specific to the general, from the concrete to the abstract.

EXAMPLES:

- You have had trouble finding a baby-sitter. A classmate even brought her child to class once. Your brother drops his daughter at day care and doesn't like not seeing her all day. From these examples, you derive the idea that your school needs an on-campus day-care program.

- You examine what dogs, whales, and tigers have in common (classification as mammals).

- You see a movie and you decide it is mostly about pride.

Also called: generalization, classification, conceptualization, induction, inductive reasoning, abstracting, hypothesis

Your example: Name activities you enjoy; from them, come up with an idea of a class you would like to take.

The icon: The arrow and "Ex" pointing to a light bulb on their right indicate how an example or examples (the known) lead to the idea (the light bulb, representing the unknown or new idea).

Idea to Example

In a reverse of the previous action, you take an idea or ideas and think of examples (events or facts) that support or prove that idea. This mind action moves from the general to the specific, from the abstract to the concrete. Idea-to-example thinking builds knowledge as well. For example, for a presentation on the Revolutionary War, you want to illustrate the idea of rebellion against authority. You investigate facts and events and see that the Boston Tea Party is an example that supports your initial idea. Examining the important details within your examples will help you justify your idea.

EXAMPLES:

- For a paper, you start with this thesis statement: "Men still have an advantage over women in many areas of the modem workplace." To support that idea, you gather examples: Men make more money on average than women in the same jobs, there are more men in upper management positions than there are women, and so on.

- You tell your child that she should do her homework before going outside to play. You explain that she will be able to focus on concepts better right after she learns them in school that day, and that she will enjoy her evening more if she doesn't have the homework hanging over her head.

- You talk to your instructor about changing your major, giving examples that support your idea, such as the fact that you have worked in the field to which you want to change and you have fulfilled some of the requirements for that major already.

Also called: categorization, substantiation, proof, deduction, deductive reasoning, constructing support

Your example: Name an admirable person. Give examples that show how that person is admirable.

_____

_____

The icon: In a reverse of the previous icon, this one starts with the light bulb and has an arrow pointing to "Ex." This indicates that you start with the idea (the lit bulb, representing the known), and then move to the examples that support it (the unknown).

Evaluation

Here you judge whether something is useful or not useful, important or unimportant, good or bad, or right or wrong by identifying and weighing its positive and negative effects (pros and cons). Be sure to consider the context because the same choices might receive very different evaluations in different situations. For example, a cold drink might be good on the beach in August but not so good in the snowdrifts in January; and pizza for dinner might be great for one person but not fun for someone who can't digest cheese. With the facts you have gathered, you determine the value of something for you in terms of both predicted effects and your own needs. Cause-and-effect analysis always accompanies evaluation (in evaluating, you ask, "What is, or would be, the effect if . . .")

EXAMPLES:

- Your mother is not well and needs help with her day-to-day activities. You consider the possible effects of bringing in a home health aide, moving her to an assisted living facility, and setting up a schedule for her children to help her at her home. Together you decide that a home health aide would be the best choice right now.

- Someone offers you a chance to cheat on a test. You evaluate the potential effects if you are caught. You also evaluate the long-term effects of not actually learning the material and of doing something ethically wrong. You decide that it isn't right or worthwhile to cheat.

  Also called: value, judgment, rating

HOW MIND ACTIONS BUILD THINKING PROCESSES

The seven mind actions are the fundamental building blocks that are the foundation of the more complex thinking processes. Using these actions consciously as tools allows you to solve problems, make decisions, create, and learn. You rarely will use one at a time in a step-by-step process as they are presented here. You usually will combine them, overlap them, and repeat them, using different actions for different situations. For example:

- When a test question asks you to define or explain prejudice, you might give examples, similar to one another, that show your idea of prejudice (combining *similarity* with *example to idea*).

- When deciding whether to take a course that has been recommended to you, you find similarities and differences in what your friends have said about the course, think about what positive effects taking the course may have on your major or career choice, and evaluate whether it is a good decision (combining *similarity, difference, cause-and-effect, and evaluation*).

- When asked to defend your position on gun control, you discuss the causes of recent violent -events, the effects of the latest Congressional votes, and statistics about gun purchases and manufacturing (combining *idea to example, cause-and-effect,* and *similarity*).

"He knows enough who knows how to learn."

HENRY ADAMS

When you combine mind actions in working toward a specific goal, you are performing a thinking process. In the next few chapters you will find explorations of important critical-thinking processes: solving problems, making decisions, constructing and evaluating arguments, thinking logically, recognizing perspectives, and planning strategically. Each thinking process gives you the chance to put your thinking into action, directing your energy toward broadening your knowledge and achieving your goals. Figure 10.2 shows all of the mind actions and thinking processes together and reminds you that the mind actions form the core of the thinking processes.

WHAT ARE INDUCTIVE AND DEDUCTIVE THINKING?

Among the thinking terms you are most likely to encounter in your school and work are *inductive* thinking and *deductive* thinking. You already know how to think inductively and deductively; you have done so many times before. In fact, although you may not realize it, you have studied them in the previous section of this chapter. They merit their own separate section here because they are such widely used thinking patterns, and because the terms *induction* and *deduction* are commonly used in other material about thinking.

INDUCTIVE THINKING

Inductive thinking is the process of working from the known example(s) toward the previously unknown idea. You learned the basics of inductive thinking when you studied the example-to-idea mind action in this chapter. Inductive thinking helps you use existing information to come up with something new. The process, broken down into more detail, looks like this:

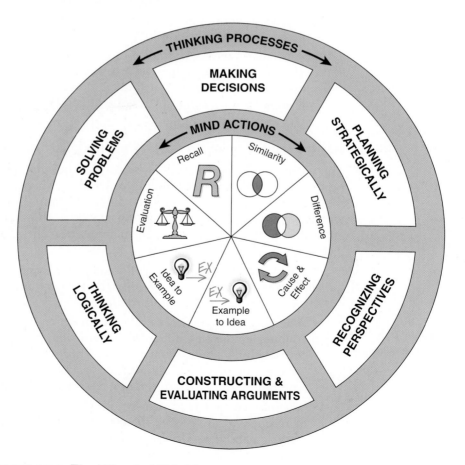

FIGURE 10.2 **The Wheel of Thinking**

1. Focus on available information
2. Look for a pattern or connection among the pieces of information
3. Formulate a classification of the pattern, or a general statement about its meaning
4. Test the classification or general statement to see how it applies to other information[3]

Following is a step-by-step example of inductive thinking, also called *induction*.

1. You have experienced stiffness in your right shoulder lately. You determine that it has happened primarily on Saturday and Sunday evenings. It seems to go away during the week.
2. You examine what you could be doing on the weekend that you don't do during the week to irritate your right shoulder muscle. You note that you spend your weekdays in class taking notes and at work as a courier, and you spend your weekends primarily at the computer doing assignments and papers.
3. You compare how you use your muscles for class and work with how you use your muscles for computer work. You decide that the only difference in how you use your shoulder on the weekend involves using your computer mouse with your right hand.
4. To test your idea, you spend a free weekday evening working on the computer, take a weekend night off, and observe how your shoulder reacts.

One more example:

1. You are late arriving to school almost every day.
2. You examine your mode of travel and see that you hit traffic at very nearly the same spot on any day that you leave the house after 8:00 A.M.
3. You decide that hitting that particular traffic at that particular time is preventing you from arriving on time to class.
4. To test your idea, you try two tactics. One morning you take a different route to school, and another morning you take your regular route but leave a half-hour earlier.

DEDUCTIVE THINKING

Deductive thinking is the process of supporting a given idea based on examples and evidence. You learned the basics of deductive thinking when you studied the idea-to-example mind action in this chapter. When you think deductively, you rely on what you know to support a hypothesis or generalization. In greater detail, the process looks like this:

1. Identify a situation or case.
2. Name a principle or generalization that you believe applies to the situation.
3. See if the situations or cases fit the principle.
4. If they fit, see what you can conclude about the specific situation or what you can predict about similar situations in the future.[4]

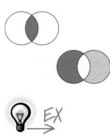

Following is a step-by-step example of deductive thinking, also called *deduction*.

1. The situation: You have trouble focusing during two of your classes (a Monday-Wednesday literature seminar and a Tuesday–Thursday sociology class in lecture format).

2. The generalization: Hunger distracts a student's attention in class.

3. You note that both classes meet in late morning and run over into the time when you normally would eat lunch. Because these classes are so different otherwise, you can't find any other reason why you would be distracted in both.

4. You conclude that your hunger is preventing you from concentrating in class and that you need to take classes at different times or eat something in the late morning before the classes meet.

One more example:

1. The generalization: People who like each other work well together.

2. The situations: Different ways in which you and your coworkers have been teamed up on projects.

3. You note that the teams that are accomplishing more have at least one pair of close friends in them and that they have fewer pairs of people who are known to dislike each other.

4. You conclude that in the future the teams should be chosen with an eye toward who enjoys working with one another.

Often, you probably perform inductive and deductive reasoning without thinking about it. The more you think about it, though, the more adept you will become at using it consciously and deliberately to make sense of the information you encounter. Being aware of the two main types of knowledge will help you take in the information with which you reason.

WHAT ARE THE TYPES OF KNOWLEDGE?

Your ability to acquire knowledge can benefit from an awareness of exactly what you are acquiring. Knowledge can be divided into two basic categories, *declarative* and *procedural*.[5]

Declarative knowledge refers to what the learner *knows* or *understands*—concepts or principles, facts, descriptions, episodes or events, cause-and-effect relationships, and sequences. Declarative knowledge focuses on the *who*, *what*, *where*, *when*, and *why*. Examples of declarative knowledge include the following:

- Types of numbers
- French verbs
- Leaders of Asian countries
- The events of the industrial revolution
- The causes of AIDS

"Minds are like para-
chutes. They only
function when they
are open."

SIR JAMES DEWAR

Procedural knowledge refers to what the learner is able to *do*—follow processes, perform actions, and demonstrate skills. Procedural knowledge focuses on the *how*. Examples of procedural knowledge include the following:

- How to balance a spreadsheet
- How to use mind actions to think critically
- How to make an omelet
- How to organize a date book

11 Problem-Solving Strategies

In technical fields you will encounter hypothetical problems to solve that require you to apply formulas and work with procedures. More important, you will face simulated problems in labs and actual problems throughout your career. A systematic approach is helpful to improve your problemsolving abilities.

It is easy to panic when a piece of equipment fails or a procedure does not produce the expected results. Using a systematic approach to solving problems will help you in these situations. A problem is basically a conflict between "what is" (the present state) and "what should be" or "what is desired" (the goal state). For example, a medical office assistant has a problem when she cannot calm a frightened child in order to take her blood pressure. The following steps can help you attack problem situations.

1. State the problem clearly. What are the facts? Recall the details of the situation. Don't include what caused the challenge or what the effects are or will be. Just state the basic details in one or two sentences. For example, an employee might be struggling with the new manager's style of micro-management where the employer is completely involved in all aspects of the employee's job.

2. Analyze the problem. What is happening that needs to change? What are the negative effects and how is it impacting what you value? Now is the time to analyze the causes and effects of the problem. Continuing with the example of the employee, the employee might see that this style is threatening because it makes the employee feel that he or she is not trusted. Trust is very important to the employee.

3. Brainstorm possible solutions. Examining a variety of solutions to a problem allows you to discover where you might need to alter your activities or mind set. If the employee could give the manager a daily update of his or her job performance, maybe the manager would relax a bit in his or her approach. Perhaps the employee could transfer to a different department.

4. Determine the criteria for your solution. What measure are you going to use to determine whether your solution is effective? You might decide that the solution has to coincide with your personal value system. You may need the solution to be cost-effective for your organization. Or, you may decide that it's important that your proposed solution doesn't end up causing a larger problem.

5. Explore each solution. Why might your ideas work? Why not? Could one of your solutions work? Evaluate the negative and positive effects of your ideas by applying the standards that you previously defined. Look as far as you can into your future to see if the solution will lead you where you want to go. How well does the solution fit the criteria? The employee might miss a great opportunity for advancement by leaving his department. Maybe if the employee keeps the employer informed, the working relationship will become trusting and supportive.

6. Choose and execute the solution you decide is best. Decide how you will put your solution to work. Then, apply it to your life. The employee may decide that the best solution is to speak directly with the manager about the issue. He writes down the ideas he wishes to discuss, the results he's seeking, and calls to arrange a meeting time.

Fill in the following chart using the suggestions from the list above.

Problem Solving Exercise	
STEP	**YOUR RESPONSE**
1. State the problem clearly.	State a problem you haven't resolved.
2. Analyze the problem.	How does this problem impact you?
3. Brainstorm possible solutions.	Explore as many options as you can.
4. Determine the criteria for your solution.	Name the measures you will use for determining the effectiveness of your solutions.
5. Explore each solution.	Discuss the negative and positive effects of your ideas. Match them to the criteria.
6. Choose and execute the solution you decide is best.	How will you apply the solution?

PART IV

Critical Reading

12 Critical Thinking, Reading, and Writing: Introduction

Critical thinking is essential for getting along in the world—in school, in work, and in life. It is a questioning attitude that is fundamental to both reading and writing. In reading, it allows us to understand better what a writer is saying; in writing, it helps clarify our thinking. This introduction helps you be more adept at all three: critical thinking, reading, and writing.

I1   FOSTERING A CRITICAL PERSPECTIVE

Critical thinking is a healthy skepticism that doesn't accept everything at face value, that looks at puzzling situations as problems to be solved, and that analyzes the unknown in order to understand it better. This skepticism is not necessarily negative and disapproving; more often it is discriminating, comparing something new with what you already know, examining the parts of the new thing or idea to see if they fit with what you know, so that you can make a judgment or an interpretation of the facts.

You already take a critical perspective in much of your life. If you wanted to buy a used car, for instance, you would select a dealer with care, carefully examine a number of cars for the sound of the engine, the condition of the tires, the presence of rust spots, and so on. If you have never bought a car before, you might consult with someone more knowledgeable than you. You wouldn't allow an overeager car dealer to sell you something you didn't want.

The same should be true with reading and listening in school: don't be sold an idea without examining it closely and asking questions. What does a speaker or writer mean? How does the idea compare with what I know is true? What else can I read or who else can I consult about this idea? You can think critically about almost anything.

I2   THINKING AND READING CRITICALLY

Critical thinking is most fruitful, most accurate, in subjects you know well. If you have read extensively about capital punishment, for example, you are better able to judge arguments on one side or the other. But you can increase your knowledge with a questioning attitude, by reading critically and applying the following techniques.

92

TECHNIQUES OF CRITICAL READING

- **Writing:** making notes on your reading throughout the process (12-a)
- **Previewing:** getting background; skimming (12-b)
- **Reading:** interacting with and absorbing the text (12-c)
- **Summarizing:** distilling and understanding content (12-d)
- **Forming your critical response** (12-e)

    Analyzing: separating into parts

    Interpreting: inferring meaning and assumptions

    Synthesizing: reassembling parts; making connections

    Evaluating: judging quality and value

You won't apply all these techniques every time you read, but the more you want to increase your understanding of the text (any piece of writing), the greater the number of techniques you will apply. For studying, you may apply them all; for reading a popular magazine, maybe only one or two. Apply as many techniques as necessary to achieve your purpose in reading.

a WRITING WHILE READING

You are probably familiar with note taking for the purpose of using the material later in writing. Another reason for writing while you read is to help you get more out of the text. By making notes, you relate what you read to what you already understand about the subject. You are not only reading more closely but making connections as well.

Some people choose to write directly in the margins of their books, thus making their notes readily accessible for later review. Others make their notes in separate notebooks or reading journals, linking them to the reading with page numbers. Still others, who like the convenience of notes in their textbooks but don't want to write on the pages, use self-stick removable sheets for attaching notes at appropriate places.

b PREVIEWING THE MATERIAL

You can greatly improve your comprehension of what you read by doing a little preliminary work. Skim the chapter or article, trying to grasp the context from headings, beginnings of paragraphs, illustrations, and so on. And as you skim, formulate questions that you can seek answers for as you read.

QUESTIONS FOR PREVIEWING A TEXT

- **Length:** Is the material brief enough to read in one sitting, or do you need more time?
- **Content clues:** What do the title, summary or abstract, headings, illustrations, and other features tell you? What questions do they raise in your mind?
- **Facts of publication:** Does the publisher or publication specialize in a particular kind of material—scholarly articles, say, or popular books? Does the date of publication suggest currency or datedness?
- **Author:** What does the biographical information tell you about the author's publications, interests, biases, and reputation in the field?
- **Yourself:** Do you anticipate particular difficulties with the content? What biases of your own may influence your response to the text—for instance, anxiety, curiosity, boredom, or a similar or opposed outlook?

c READING

To understand the material you read, plan to read at least twice. The first time through, read fast, not taking notes, just trying to get an overall idea of what the writer is saying—enjoying it, relating it to your own experiences, maybe marking places you don't understand. The second time through, read slowly—making notes, asking questions, looking up unfamiliar words, flipping pages backward and forward as you attempt to relate parts.

d SUMMARIZING

To think critically about what you read, you need to understand precisely what the writer has said. Writing a summary helps. Using your own words, condense the main points of a text, in this way seeing its strengths and weaknesses. Summaries of short pieces of writing are generally no more than one-fifth the length of the original, whereas summaries of longer pieces, such as books or chapters, are shorter proportionately.

In the box below is a procedure for summarizing. Note that your completed summary begins with an overall summary sentence and then adds supporting summary sentences. These sentences are not the original sentences; rather, they are sentences that you formulate to encapsulate the meaning of the original. You may find yourself using some of the same words the original writer used, but do not write your summary by selecting sentences or phrases from the original.

WRITING A SUMMARY

- Look up words or concepts you don't know so that you understand the author's sentences and how they relate to each other.
- Work through the text paragraph by paragraph to identify its sections—single paragraphs or groups of paragraphs focused on a single topic. To understand how paragraphs relate to each other, try outlining the text.
- Write a one- or two-sentence summary of each section you identify. Focus on the main point of the section, omitting examples, facts, and other supporting evidence.
- Write a sentence or two stating the author's central idea.
- Write a full paragraph (or more, if needed) that begins with the central idea and supports it with the sentences that summarize sections of the work. The paragraph should concisely and accurately state the thrust of the entire work.
- Use your own words. By writing, you re-create the meaning of the work in a way that makes sense for you.

An important aspect of summaries is that their sources are cited. A common way of telling your reader what you are summarizing is to mention the author and title in the first sentence. Other publishing information can be included in parentheses at the end of the summary or in a note at the end of your piece of writing.

The paragraph below is followed by a summary of it. Compare the summary with the original and note the objectivity, completeness, conciseness, and source citation.

Years ago when I was a high school student experiencing racial desegregation, there was a fierce current of resistance and militancy. It swept over and through our bodies as we, black students, stood, pressed against the red brick walls, watching the national guard with their guns, waiting for a moment when we would enter, when we would break through racism, waiting for the moment of change, of victory. And now even within myself, I find that spirit of militancy growing faint; all too often it is assaulted by feelings of despair and powerlessness. I find I must work to nourish it, to keep it strong. Feelings of despair and powerlessness are intensified by all the narratives of black self-hate that indicate that those militant 1960s did not have sustained radical impact-that the transformation of black self-consciousness did not become an ongoing revolutionary practice in black life. This causes such frustration and despair, for it means that we must return to this basic agenda, that we must renew efforts at transformation, that we must go over old ground. Perhaps what is worst about turning over old ground is the fear that the seeds, though planted again, will never survive, will never grow strong. Right now it is anger and rage at the continued racial genocide that rekindles within me that spirit of militancy. (bell hooks)

Summary

bell hooks, recalling the racial militancy of her 1960s school days, senses a need for a renewed revolutionary outlook that will replace the feelings of hopelessness that seem to characterize black consciousness today ("Overcoming White Supremacy," *Zeta* Jan. 1987: 26–27).

e FORMING YOUR CRITICAL RESPONSE

Critical reading requires you to go beyond understanding what authors of texts say to determining what they don't say, where additional meaning may lie. A poem about a road not taken, for example, is really about the poet's choices in life, and an essay about vegetarianism may really be an expression of the writer's opinion about animals' right to life. Critical thinking and reading consist of four overlapping operations: analyzing, interpreting, synthesizing, and (often) evaluating.

Analyzing

You *analyze* a text by examining its parts or elements. Who is the writer, for example, and for what readers is the publication intended? How is the text organized? What is its thesis, and how does the writer support that thesis? Is the vocabulary technical, professional, or common? What tone or attitude does the writer assume? What does the writer not say? If the writer uses long sentences and obscure words, is it to deliberately conceal meaning; if so, why?

Interpreting

As you analyze, you also *interpret*, trying to understand why the writer made particular choices. What assumptions does the writer make about the audience? What values and beliefs are promoted through the writer's position on the subject? Keeping in mind the evidence of the text, what conclusions can you draw? Be careful about allowing your own biases to cause a misinterpretation of the text or undue attention to a minor point.

Guidelines for Analysis, Interpretation, and Synthesis

- What is the purpose of your reading?
- What questions do you have about the work? What elements can you ignore?
- How do you interpret the meaning and significance of the elements, both individually and in relation to the whole text? What are your assumptions about the text? What do you conclude about the author's assumptions?
- What patterns can you see in (or synthesize from) the elements? How do the elements relate? How does this whole text relate to other texts?
- What do you conclude about the text? What does this conclusion add to the text?

Synthesizing

With interpretation, you put the pieces back together—you *synthesize*. For example, try to relate the writer's choice of words and selection of facts and examples to the purpose for writing. Relate the tone of the piece to who the writer and audience are. Make connections between the organization and the kinds of examples and other support the writer draws on. By making connections among pails, you arrive at a new understanding of the text. A synthesis is evidence of your mind working on another text.

Evaluating

At times your critical reading may require an *evaluation*—a judgment about the quality, value, or significance of the work. Evaluating what you read may be something you don't commonly do. Who are you, you might say, to judge the quality of work by an expert? That's a good question. The answer is that when you are asked to evaluate something you read, you need to judge it according to your own experience and knowledge. Here are some guidelines that may help.

GENERAL GUIDELINES FOR EVALUATION

- What are your reactions to the text? What in the text are you responding to?
- Is the work unified, with all the parts pertaining to a central idea? Is it coherent, with the parts relating clearly to each other?
- Is the work sound in its general idea? In its details and other evidence?
- Has the author achieved his or her purpose? Is the purpose worthwhile?
- Does the author seem authoritative? Trustworthy? Sincere?
- What is the overall quality of the work? What is its value or significance in the larger scheme of things?
- Do you agree or disagree with the work? Can you support, refute, or extend it?

13 WRITING CRITICALLY

Critical writing results from critical reading and critical thinking. It is more than a summary or a report of another text. It is your analysis, interpretation, synthesis, and perhaps evaluation of that text. You will find it useful in arguments, reviews of literature, and research writing.

To write a critical analysis of another piece of writing, use the notes and ideas you acquired as a result of your critical reading. First state your main point about the text and name the author and title. For support, use your analysis, interpretation, synthesis, and evaluation, concentrating not only on *what* is said but also on *how* it is said. Summarize as necessary for clarification, but don't limit yourself to summary.

CHAPTER 13

Reading in Technical and Applied Fields

LEARNING OBJECTIVES

- To learn what to expect In technical courses
- To develop specialized reading techniques for technical material
- To learn processes and develop problem-solving strategies
- To adapt your study strategies for technical material

Our society has become a technological one, in which there is a heavy reliance on automation and computerization. Consequently, more technical knowledge and expertise are required, and more students are pursuing degrees in a variety of technical fields, including computer information systems, mechanical and electrical technology, and computer-assisted drafting. Applied fields such as EKG and X-ray technology, air conditioning and refrigeration, food service, and horticulture are other examples.

TECHNICAL FIELDS: WHAT TO EXPECT

If you are earning a degree in a technical field, you will take two basic types of courses: the technical courses in your major and required courses in related disciplines. For example, if you are earning an associate's degree in data processing, you will take numerous technical courses such as COBOL programming, data processing logic, and systems design, but you are also required to complete courses such as English composition, business communication, and physical education. Here is what to expect in each of these categories.

TECHNICAL COURSES

The goal of many technical courses is to teach specific procedures and techniques that you will use on the job. These courses often present the theory and principles that govern the procedure, as well.

Grading and evaluation in technical courses is often performance-based. In addition to traditional exams and quizzes, instructors use "hands-on" exercises to evaluate your performance. In an air conditioning and refrigeration course, for example, you may be given a broken air conditioner to repair.

Work in technical fields often involves situations that require problem solving: a computer program has a "bug," a lab test produces inconsistent results, a number of landscape plantings die. Consequently, many instructors are careful to include problem-solving tasks in class, during labs, or on exams.

Many fields involve the use of instruments and equipment; others require measurement and recording of data. In either case, procedures must be followed exactly; measurements must be precise.

Although classroom lectures usually are a part of technical courses, most include some practical forms of instruction as well. Their purpose is to provide you with hands-on experience working with the procedures you are learning. As you work through lab assignments, remember that your skills are on display to your instructor just as they will be later to an employer. Develop systematic, organized routines to handle frequently used procedures and processes. Concentrate on following directions carefully. Always check and double-check your work. Labs also give you an opportunity to find out whether you actually like the career you have chosen. If you dislike labs or find them extremely difficult or too routine, you should question the appropriateness of your career choice.

NONTECHNICAL REQUIRED COURSES

Make a genuine effort to benefit from courses outside your technical field. If you are taking a required introductory psychology course, you may be the only student in the class majoring in nursing; in this case, your psychology professor can do little or nothing to relate the course to your field. As your instructor discusses various topics, you should consider how they can be applied in your field. As you learn about defense mechanisms, for example, you might consider how they might be exhibited by patients. As you study nontechnical courses, be sure to make connections and applications to your field.

READING TECHNICAL MATERIAL

Textbooks in technical fields are highly factual and packed with information. Compared to other textbooks, technical writing may seem "crowded" and difficult to read. In many technical courses, your instructor requires you to read manuals as well as textbooks. These are even more dense and, on occasion, poorly written. Use the following suggestions to help you read and learn from technical writing.

READ SLOWLY

Because technical writing is factual and contains numerous illustrations, diagrams, and sample problems, adjust your reading rate accordingly. Plan on spending twice as long reading a technical textbook as you spend on reading other, nontechnical texts.

REREAD WHEN NECESSARY

Do not expect to understand everything the first time you read the assignment. It is helpful to read an assignment once rather quickly to get an overview of the processes and procedures it presents. Then reread it to learn the exact steps or details.

HAVE A SPECIFIC PURPOSE

Reading technical material requires that you have a carefully defined purpose. Unless you know why you are reading and what you are looking for, it is easy to become lost or to lose your concentration. Previewing is particularly helpful in establishing purposes for reading.

PAY ATTENTION TO ILLUSTRATIONS AND DRAWINGS

Most technical books contain illustrations, diagrams, and drawings, as well as more common graphical aids such as tables, graphs, and charts. (Refer to Chapter 8 for suggestions on reading graphics). Although graphics can make the text appear more complicated than it really is, they actually are a form of visual explanation designed to make the text easier to understand. Read the following excerpt from a basic electronics textbook.

> Magnetic lines of force always take the shortest and easiest route. Sometimes the easiest is not the shortest, as shown in Figure 13-1. This figure shows what occurs when we place a piece of magnetic material into the field, in this case a small piece of soft iron. The field distorts to include the iron bar in its route. T' distortion of the magnetic field is due to the ability of the iron to conduct the magnetic field more easily than the surrounding air. The result of this field distortion by the iron bar is that the field intensity in the bar is greatly increased due to the greater concentration of lines of force. We can draw the following conclusion: *Magnetic fields always tend to arrange themselves in such a way that the maximum number of lines of force per unit area is established.* Another interesting thing occurs in that the piece of iron becomes a (temporary) magnet (with north and south poles) as long as it remains in the original magnetic field. Thus, the piece of iron has become a temporary magnet by induction.

FIGURE 13-1 **Deflection of lines of force by soft iron.**

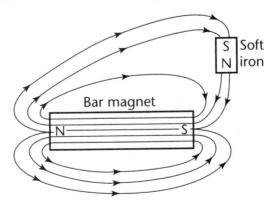

—Hand and Williams, *Basic Electronics: Components, Devices and Circuits,* pp. 15–16

Now, study the diagram and reread the passage above. Does the diagram make the passage easier to understand?

Here are a few suggestions on how to read illustrations and diagrams.

1. Note the type of illustrations or diagrams included in the assignment when you preview the chapter.

2. Look over each illustration and determine its purpose. The title or caption usually indicates what it is intended to show.

3. Examine the illustration first. Alternate between the text and the illustrations; illustrations are intended to be used with the paragraphs that refer to them. You may have to stop reading several times to refer to an illustration. For instance, when magnetic material is mentioned in the preceding example, stop reading and find where it is placed in the diagram. You may also have to reread parts of the explanation several times.

4. Look at each part of the illustration and note how the parts are connected. Note any abbreviations, symbols, arrows, or labels. In the example given, abbreviations are used for the north and south poles.

5. Test your understanding of illustrations by drawing and labeling an illustration of your own without looking at the one in the text. Then compare your drawing with the text. Note whether anything is left out. If so, continue drawing and checking until your drawing is complete and correct. Include these drawings in your notebook and use them for review and study.

USE VISUALIZATION

Visualization is a process of creating mental pictures or images. As you read, try to visualize the process or procedure that is being described. Make your image as specific and detailed as possible. Visualization will make reading these descriptions easier, as well as improving your ability to recall details. Here are a few examples of how students use visualization.

A nursing student learned the eight family life cycles by visualizing her sister's family at each stage.

A student taking a computer course was studying the two basic methods of organizing data on a magnetic disk: the sector method and the cylinder method. She visualized the sector method as slices of pie and the cylinder method as a stack of dinner plates.

Now read the following description of an optical disk system from a computer science textbook, and try to visualize as you read.

Optical Disks

Optical technology used with laser disk systems is providing a very high capacity storage medium with the **optical disk,** also called a **videodisk.** Videodisks will open new applications, since they can be used to store data, text, audio, and video images.

Optical disk systems look like magnetic disk systems. Each has a rotating platter and a head mechanism to record information. However, optical systems differ because they use light energy rather than magnetic fields to store data. A high-powered laser beam records data by one of two methods. With the **ablative method,** a hole is burned in the disk surface. With the **bubble method,** the disk surface is heated until a bubble forms.

The laser beam, in a lower power mode, reads the data by sensing the presence or absence of holes or bumps. The light beam will be reflected at different angles from a flat or disfigured surface. A series of mirrors is used to reflect the light beam to a photodiode, which transforms the light energy into an electric signal. The photodiode process works like the automatic doors at your local supermarket. As you walk toward the door, you deflect a light beam, which signals the doors to open.

—Athey, Zmud, and Day,
Computers and End-User Software, pp. 122–123

Did you visualize the disks with tiny holes or bumps?

MARK AND UNDERLINE

You may find your textbooks to be valuable reference sources for lab or onsite experiences; you may also use them when you are employed in your field. Take special care, then, to mark and underline your textbooks for future reference. Marking will also make previewing for exams easier. Develop a marking system that utilizes particular symbols or colors of ink to indicate procedures, important formulas, troubleshooting charts, and so forth.

LEARN TO READ TECHNICAL MANUALS

Many technical courses require students to operate equipment or become familiar with computer software. Study and frequent reference to a specific manual is an additional requirement in some technical courses. Unfortunately, many technical manuals are poorly written and organized, so you need to approach them differently than textbooks. Use the following suggestions when reading technical manuals:

1. Preview the manual to establish how it is arranged and exactly' what it contains. Does it have an index, a trouble-shooting section, a section with specific operating instructions? Study the table of contents carefully and mark sections that will be particularly useful.

2. Do not read the manual from cover to cover. First, locate and review those sections you identified as particularly useful. Concentrate on the parts that describe the overall operation of the machine: its purpose, capabilities, and functions.

3. Next, learn the codes, symbols, commands, or terminology used in the manual. Check to see if the manual provides a list of special terms. Many computer software manuals, for instance, contain a list of symbols, commands, or procedures used throughout the manual. If such a list is not included, begin making your own list on a separate sheet or on the inside cover of the manual.

4. Begin working with the manual and the equipment simultaneously, applying each step as you read it.

5. If the manual does not contain a useful index, make your own by jotting down page numbers of sections you know you'll need to refer to frequently.

6. If the manual is overly complicated or difficult to read, simplify it by writing your own step-by-step directions in the margin or on a separate sheet.

THOUGHT PATTERNS IN TECHNICAL FIELDS

The two thought patterns most commonly used in technical fields are process and problem/solution. Each is used in textbooks and manuals as well as in practical, hands-on situations.

Venipuncture

1. Wash hands, explain procedure to patient; assess patient status
2. Assemble equipment
3. Locate puncture site
4. Apply tourniquet, cleanse site
5. Place thumb distal to puncture site
6. Insert needle 30° angle, aspirate desired amount
7. Remove tourniquet, place dry compress on needle tip & withdraw
8. Remove needle from syringe and place specimen in container; label

FIGURE 13-2 Sample Summary Sheet

READING PROCESS DESCRIPTIONS

Testing procedures, directions, installations, repairs, instructions, and diagnostic checking procedures all follow the process pattern. To read materials written in this pattern, you must not only learn the steps but also learn them in the correct order. To study process material, use the following tips:

1. Prepare study sheets that summarize each process. For example, a nursing student learning the steps in venipuncture (taking a blood sample) wrote the summary sheet shown in Figure 13-2.

2. Test your recall by writing out the steps from memory. Recheck periodically by mentally reviewing each step.

3. For difficult or lengthy procedures, write each step on a separate index card. Shuffle the pack and practice putting the cards in the correct order.

4. Be certain you understand the logic behind the process. Figure out why each step is done in the specified order.

STUDY TECHNIQUES FOR TECHNICAL COURSES

Use the following suggestions to adapt your study skills to technical courses.

PRONOUNCE AND USE TECHNICAL VOCABULARY

Understanding the technical vocabulary in your discipline is essential. For technical and applied fields, it is especially important to learn to pronounce technical terms and to use them in your speech. To establish yourself as a professional in the field and to communicate effectively with other professionals, it is essential to speak the language. Use the suggestions in Chapter 4 for learning specialized terminology.

DRAW DIAGRAMS AND PICTURES

Although your textbook may include numerous drawings and illustrations, there is not enough space to include drawings for every process. An effective learning strategy is to draw diagrams and pictures whenever possible. These should be fast sketches; be concerned with describing parts or processes and do not worry about artwork or scale drawings. For example, a student studying air conditioning and refrigeration repair drew a quick sketch of a unit he was to repair in his lab before he began to disassemble it, and he referred to sketches he had drawn in his notebook as he diagnosed the problem.

RESERVE BLOCKS OF TIME EACH DAY FOR STUDY

Daily study and review are important in technical courses. Many technical courses require large blocks of time (two to three hours) to complete projects, problems, or drawings. Technical students find that taking less time is inefficient because if they leave a project unfinished, they have to spend time rethinking it and reviewing what they have already done when they return to the project.

FOCUS ON CONCEPTS AND PRINCIPLES

Because technical subjects are so detailed, many students focus on these details rather than on the concepts and principles to which they relate. Keep a sheet in the front of your notebook on which you record information to which you need to refer frequently. Include constants, conversions, formulas, metric equivalents, and commonly used abbreviations. Refer to this sheet so you won't interrupt your train of thought. Then you can focus on ideas rather than specific details.

MAKE USE OF THE GLOSSARY AND INDEX

Because of the large number of technical terms, formulas, and notations you will encounter, often it is necessary to refer to definitions and explanations. Place a paper clip at the beginning of the glossary and a second at the index so you can find them easily.

STUDY TIPS TEST-TAKING TIPS FOR TECHNICAL COURSES

Exams in technical courses may consist of objective questions or of problems to solve. (For suggestions on taking these types of exams, see the test-taking tips in Chapters 9 and 13.) Other times, exams may take the form of a practicum. A practicum is a simulation, or a rehearsal, or some problem or task you may face on the job. For example, a nursing student may be asked to perform a procedure while a supervisor observes. Or an EKG technologist may be evaluated in administering an EKG to a patient. Use these suggestions to prepare for practicum exams:

1. Identify possible tasks you may be asked to perform.

2. Learn the steps each task involves. Write summary notes and test your recall by writing or mentally rehearsing them without reference to your notes. Visualize yourself performing the task.

3. If possible, practice performing the task, mentally reviewing the steps you learned as you proceed.

4. Study with another student; test and evaluate each other.

SUMMARY

Students in technical or applied fields of study face two types of courses: technical courses and related courses in other disciplines. Specified techniques for reading in the sciences are

- reading slowly (and rereading)
- setting a specific reading purpose
- studying illustrations and drawings
- using visualization
- marking and underlining
- reading technical manuals

Two thought patterns predominate in technical and applied fields:

- process
- problem/solution

Suggestions are offered for adapting studying strategies to technical and applied courses.

CHAPTER 14 Reading Graphic Aids

"A picture is worth a thousand words."

ANONYMOUS

CHAPTER PREVIEW

Graphic aids are part of our everyday existence. In Chapter 14, you become familiar with the different types of graphic aids along with specific tips on how to read and interpret them. After reading Chapter 14, you should be able to answer the following questions:

WHAT ARE GRAPHIC AIDS?

Think for a moment about how graphic aids affect our lives. We see thousands of pictures, charts, graphs, and tables each day. Graphic aids are very useful for presenting information in visual form. Sometimes you may skip over graphics while reading because these aids take time to study and seem to slow down your learning. You may also have difficulty understanding or interpreting graphic aids clearly.

In reality, graphic aids are often more important than the information around them. They can present information more concisely and clearly than several paragraphs of writing. Graphic aids also show relationships between ideas that words alone could not effectively express. A familiar example would be a photo of a 300-pound man before he started a weight loss program and his dramatic photo one year later after he has lost 140 pounds.

A graphic aid is a visual representation in the text that displays complex information in a simplified format. Through graphics, authors summarize and condense material so that readers can more easily comprehend and remember important facts and details related to a concept or idea in the text.

WHAT ARE THE PURPOSES FOR GRAPHIC AIDS?

The main purposes for graphic aids are to:

1. Summarize information. With graphic aids, the author condenses several pages of text into one concise presentation. For example, most people learn sign language much faster from pictures illustrating the various hand positions than from words describing those positions.

105

2. Emphasize information. Through graphic aids, the author reinforces the key ideas and concepts that are essential for understanding the text material. For example, in an article about the decrease in the number of employees with health insurance, a graph illustrating the decline over a ten-year period would stress the author's points dramatically.

3. Persuade readers. Authors sometimes use graphic aids to convince readers that their idea or point of view is correct. For example, a political cartoon may make fun of the president appointing friends from his home state to the best government jobs in Washington, or a graph may show political candidates' positions in the polls.

HOW SHOULD I READ GRAPHIC AIDS?

1. Read the title to gain a general understanding of the topic. What is being described?

2. Determine what kind of graphic aid you are reading. If it is a bar graph, pay attention to the bar lengths and what they represent. if it is a table, notice the headings at the top, sides, or bottom of -this graphic aid.

3. Read any footnotes or wording that introduces the graphic aid. These items will tell you about who, when, where, how, and why this information was gathered.

4. Read the headings or labels for each part of the graphic aid. Do these headings help you understand the information presented? Look up in the surrounding text or in a dictionary the meaning of any words in the headings or labels that you do not know.

5. Make note of relationships, patterns, and trends in the graphic aid. Are there increases or decreases, gains or declines? What are the highest and lowest quantities?

6. Identify the source from which the information was taken. Is the source accurate, reliable, and unbiased?

7. In your own words, state the key points you learned from reading the graphic aid. What conclusions can you draw from the information? What questions are still not answered? Where would you go to seek further information on the topic?

WHAT ARE THE TYPES OF GRAPHIC AIDS?

Now let's look at explanations of the various types of graphic aids you may encounter.

GRAPHS

Graphs show relationships between two or more sets of information. These relationships usually involve numbers, such as quantities of things, percentages, and/or dates involving a certain time period.

Line Graphs

A line graph is a graph that usually shows a trend or changes over a period of time. A line graph compares two or more things measured on a vertical and horizontal line called an axis. The horizontal axis measures time, and the vertical axis measures amount.

Look at the graph (Figure 14.1) showing changing stock prices. You might be asked, "What was the closing price of one share of D.E.R stock in 1992?" To read the chart, locate 1992 on the horizontal line at the bottom of the graph. Then follow the arrow up and to the left and read how much it cost. The arrow points to a little more than halfway between $50 and $75; $65 would be a good estimate.

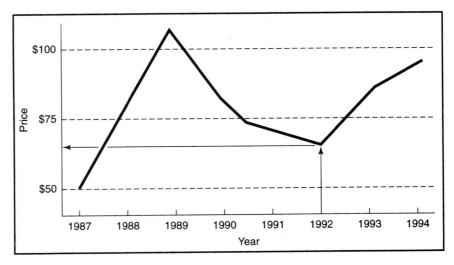

FIGURE 14.1 Closing Price for One Share of D.E.R. Stock Each Year

Study the line graph (Figure 14.1) further. Then answer the questions.

1. Which year was the price above $100 per share?
2. About how much was the closing price of stock in 1994?
3. Which year was the stock worth the least?
4. If the owner had sold fifty shares of stock in 1988, how much would she have been paid?

Multiple-Line Graphs

Multiple-line graphs are another way to present a large quantity of data in a small space. it would often take several paragraphs to explain in words the same information explained in just one graph. On the graph (Figure 14.2) there are three lines. Read the key to understand the meaning of each line. The key gives information that helps you interpret the graph.

Locate the key at the bottom of Figure 14.2. Notice that in 1970, total fish products were about twelve billion pounds. In 1994, imports were about six billion pounds, and the domestic catch was about eleven billion pounds.

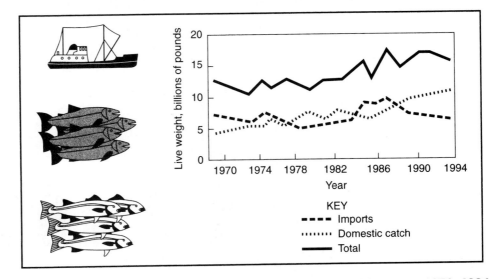

FIGURE 14.2 U.S. Fishery Products: Domestic Catch and Imports, 1970–1994
Source: U.S. Bureau of the Census.

Continue to study the graph. Then answer these questions:

1. Which year did the United States import about three billion more pounds of fish than were caught domestically?
2. Which year were the total pounds of fish products for the United States the lowest?
3. How many billion pounds of fish did the United States import in 1986?
4. In 1984, what was the total weight of fishery products?
5. How many billion pounds of fish were imported in 1988?
6. For what nine-year period did the United States import more pounds of fish products than were caught domestically?
7. How many pounds of fish products were used in the United States in 1994?

Bar Graphs

A bar graph is a series of horizontal or vertical bars; the length of each bar represents a quantity. The purpose of a bar graph is comparison. By looking at the different bar lengths, you can easily compare the data presented. Look at Figure 14.3. In 1995, almost 160 students at North Metro College studied Spanish. Approximately 70 students took French, 20 studied Russian and Italian, and 60 took German.

Now read and interpret the bar graph further and answer these questions:

1. What was the most popular language?
2. About how many students chose French as their foreign language?
3. Were there more German students or French students?
4. If the Russian and German students gathered in one room, how many students would be present?
5. Altogether, about how many students chose to take a foreign language course?

Bar graphs can be either vertical or horizontal. Sometimes each bar is divided into two or more parts. This section presents a variety of bar graphs. Be sure to read all titles, keys, and labels to completely understand all the data that are presented.

Note in Figure 14.4 that the state lottery income was about $1 billion in 1980. In 1985, the lottery income was almost $4 billion.

Now answer the questions about this bar graph:

1. In which five-year period was there the largest increase in revenue?

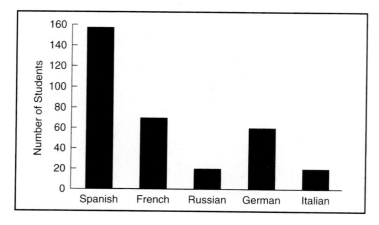

FIGURE 14.3 Students Taking Foreign Language Courses in 1995 at North Metro College

2. In which year was there a decrease in income from the year before?

3. Approximately how much growth in income was there between 1980 and 1993?

4. What was the approximate state lottery income in 1985?

Study Figure 14.5. This bar graph shows that the population of India was approximately 900 million in 1994. The population of Brazil was approximately 180 million and Japan's population was about 150 million.

1. Which country had over one billion people?

2. How many countries had less than 200 million people?

3. How many more people did India have than Japan?

4. If you added together the 1994 populations of the United States, Brazil, Russia, and Japan, would the total come closer to the population of India or the population of China?

Bar graphs can also be drawn with the bars going horizontally, as in the example of the swimming speeds of water creatures in Figure 14.6. In this case, the bars present the data more logically since animals are more apt to swim horizontally than vertically. In the graph, notice that humans swim five miles per hour. Dolphins can swim twenty miles per hour faster than humans, but humans can swim one mile per hour faster than goldfish.

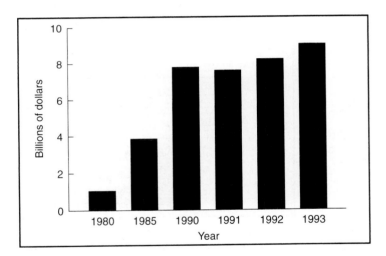

FIGURE 14.4 State Lottery Income, 1980–1993

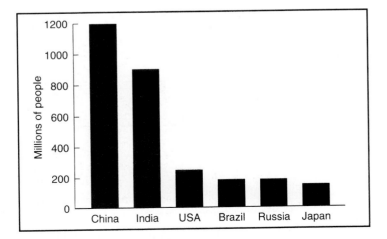

FIGURE 14.5 Population in 1994

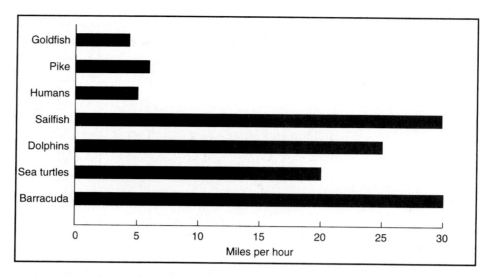

FIGURE 14.6 Speeds of Animals in the Water

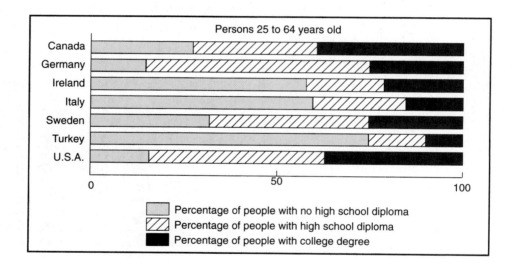

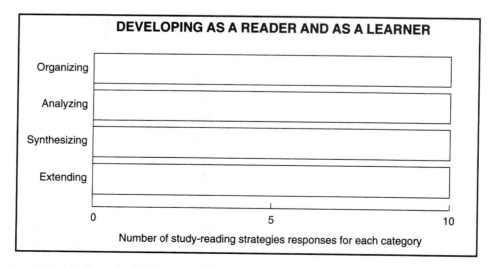

FIGURE 14.7 Level of Education in 1994
Source: Statistical Abstract of the United States, 1994.

Now answer these questions:

1. How fast is the sea turtle in the water?
2. Could a sea turtle catch a sailfish?
3. How much faster can the barracuda swim than the human being?
4. Which fish is the slowest swimmer?
5. How many times faster is the sea turtle than the human?

Multiple-bar graphs can also be presented horizontally. Each bar on a multiple-bar graph might also be divided into parts to show more information, as in the graph in Figure 14.7. Note the key at the bottom of the graph and how important the patterns are; they show that in Canada, about 20 percent had no high school diploma in 1994, approximately 30 percent had a high school diploma, and about 35 percent had a college degree.

Continue to study the graph carefully, and then answer the questions. In all of the questions, *population* refers to the twenty-five- to sixty-four-year-old population of each country as represented in the graph.

1. In how many countries did over 50 percent of the population not have a high school diploma?
2. Which country had the highest percentage of people with a college degree?
3. Which of the seven countries had the most people with at least a high school diploma?
4. Which country had the least educated population?
5. Which country had more people with a college degree than with only a high school diploma?
6. Which two countries had the same percentage of people with a college degree?

ACTIVITY: BAR GRAPH AND SELF-REFLECTION

Directions: Complete the accompanying bar graph. Shade the number of notches in which you checked "Yes" in your Organizing, Analyzing, Synthesizing, and Extending inventories. After you have completed your bar graph, assess where you are now as a learner. Think about what further organizing, analyzing, synthesizing, and extending reading and study strategies you will need to strengthen your learning. You may also be asked to complete the inventories and bar graph at the end of this book to measure your progress in reading and study strategies.

PICTOGRAPHS

A pictograph, or infographic, presents ideas or data by using pictures instead of bars or lines. The pictures are related thematically to the concepts being represented. In Figure 14.8, for instance, it takes 36.2 hours to make a car in Europe. Making a car takes 25.1 hours in the United States and 16.8 hours in Japan.

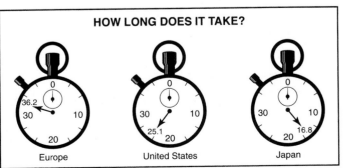

FIGURE 14.8 Average Number of Hours Needed to Make a Car

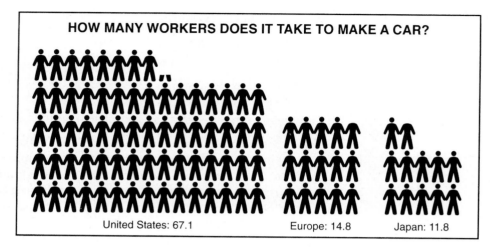

FIGURE 14.9 Average Number of Workers per Auto Plant

In Figure 14.9, notice that 67 Americans are needed to manufacture a car. In Europe, it takes approximately 15 people to make a car. About 12 people are needed in Japan. After studying two pictographs, answer the following questions:

1. Which country makes cars the fastest?
2. How many more people does it take to make a car in the United States than in Japan?
3. What will Americans have to do if they want to make cheaper cars?
4. Is the average number of hours needed to make a car in the United States closer to Europe's average or Japan's average?
5. How many hours does it take to make a car in the United States?

CIRCLE GRAPHS

A circle graph, or pie chart, is a circle divided into sections. These sections resemble the pieces of a pie. Each wedge represents a portion or percentage of the whole pie. When you add all the wedges together, they always equal 100 percent. Look at Figure 6.10 on page 184. Use a calculator and add up the percentages in all the wedges: 31 percent, 36 percent, 18 percent, 8 percent, and 7 percent, Your total is 100 percent.

Now note the word *source* at the bottom of Figure 14.10. A source is a book, a document, or some other published record that tells you where the information came from. Notice that the source for this graph is the Internal Revenue Service, Department of the Treasury, Washington, D.C. Sources are important because people must receive accurate information about a topic.

Continue to study the circle graph. Then respond to the questions:

1. What percentage of the government income for 1993 was borrowed money?
2. What was the total percentage of U.S. income from all the different sources of revenue?
3. How many billions of dollars did the government borrow in 1993?
4. Where did most of the government income come from?
5. Why do you think corporate income taxes were so low?

Where the income comes from:

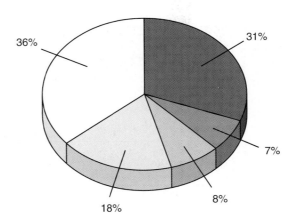

■	Social Security, Medicare, and unemployment and other retirement taxes
■	Corporate income taxes
▨	Excise, customs, estate, gift, and miscellaneous taxs
□	Borrowing to cover deficit
□	Personal income taxes

FIGURE 14.10 Fiscal Year 1993
Total U.S. income equals $1,154 billion.

Source: Internal Revenue Service, Department of the Treasury, *Fiscal Year 1994,* front cover. Washington, D.C.: Government Printing Office, 1993.

TABLES

A table is an orderly presentation of data arranged in columns and rows in a rectangular format. A table lists facts and figures in columns for fast and easy reference. Scientists, psychologists, educators, and journalists often use tables for condensing and comparing statistics. Notice the source in Figure 14.11. The Department of Agriculture is the source of this table.

In this table, note that the average crop production in bushels per acre in 1900 (50) had increased almost five times by 1990 (148). However, the number of farms in 1900 (5,737,000) had decreased by 1990 (2,140,000) by 3,597,000.

Study the table further and then answer the questions.

1. The average farm in 1990 was _____ times the size of a farm in 1930.

2. According to the table, the total number of acres of farmland was greatest in _____.

3. Annual farm income increased the most from _____ to _____.

4. The biggest decrease in the number of people farming was from _____ to _____.

5. Crop production increased the most from _____ to _____.

CHANGES IN FARMING SINCE 1900

	1900	1930	1960	1990
Farm population	29,875,000	30,529,000	15,699,000	4,591,000
Total land in farms (in acres)	839,000,000	990,112,000	1,176,000,000	987,420,000
Number of farms	5,737,000	6,295,000	3,963,000	2,140,000
Average size of farms (in acres)	146	157	297	461
Average crop production in bushels per acre	50	53	89	148
Average annual gross income per farm	$1,306	$1,527	$9,737	$91,179

FIGURE 14.11

Source: U.S. Department of Agriculture.

PHOTOGRAPHS

A photograph is a visual representation that sometimes accompanies a written description. It may evoke strong feelings, help us understand an idea, or arouse our curiosity about a topic. A short explanation called a caption sometimes appears with the photograph.

The old saying "A picture is worth a thousand words" certainly applies to the accompanying photographs. Look at them carefully and answer these questions:

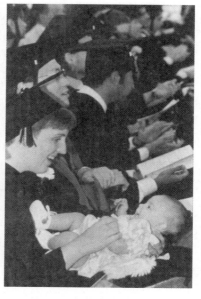

1. What effect does this photograph have on you? Why?
2. In what kind of a book would this photo be appropriate?

1. What is your reaction to this photograph? Why?
2. What important family values are suggested by this photograph?
3. Compare your responses to the two photographs. Are they similar or different? Why?

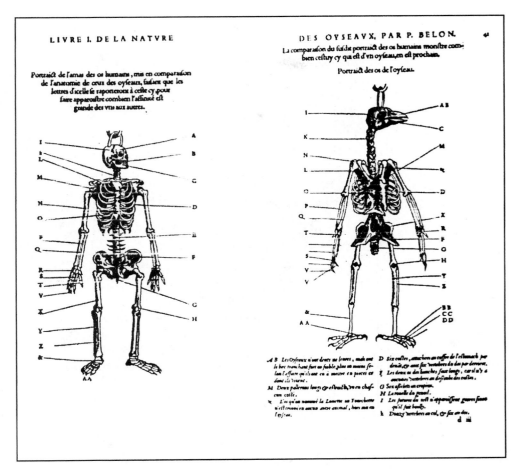

FIGURE 14.12 Pierre Belon displayed these drawings of the skeletons of a human being and a bird on opposite pages of a book to show homologous (similar) structures. As the originator of comparative anatomy, Belon demonstrated the remarkable skeletal similarities among the various vertebrates, from humans to fish—similarities which had been entirely unsuspected before.

Source: From Pierre Belon, *L'Histoire de la nature des oyseaux,* 1555. Reprinted in Leonard C. Bruno, *The Tradition of Science: Landmarks of Western Science in the Collections of the Library of Congress.* Washington, D.C.: Library of Congress, 1987.

DIAGRAMS

A diagram is a drawing, plan, or outline that helps you understand relationships between the parts of an object. Diagrams are frequently used in the sciences such as drawings of the human heart or layers in the earth's atmosphere.

Read the caption in Figure 14.12. Then answer the questions that follow.

List four homologous (similar) structures that Belon identified in his drawings.

	MAN	BIRD
Example:	mouth	beak
	1.	1.
	2.	2.
	3.	3.
	4.	4.

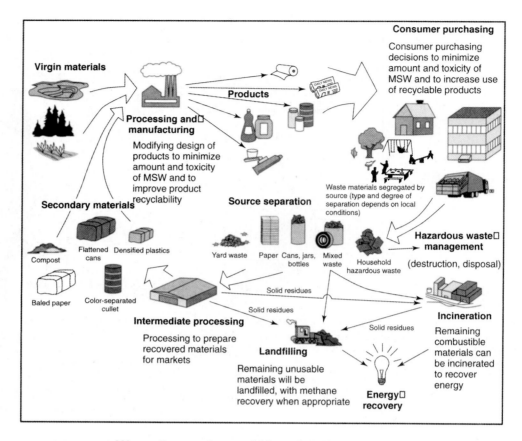

FIGURE 14.13 Waste Prevention and Materials Management in the Context of Materials Use

Other types of diagrams may explain a process or a sequence of events. They require more careful reading of each step of the sequence. The diagram in Figure 14.13 shows the process for manufacturing and recycling materials we use everyday.

Study the diagram (Figure 14.13) and then answer the questions.

1. Where does mixed waste go in the recycling process?
2. What kinds of products are made from virgin and recycled materials?
3. Give examples of how energy is created from waste materials.
4. Briefly explain how waste materials are recycled.

FIGURE 14.14

CARTOONS

A cartoon is a drawing that tells a story or presents a message in a serious or humorous manner. Cartoons are a popular form of communication and can be found in many kinds of texts and articles. Cartoonists sometimes exaggerate situations to make their point.

Look at the cartoon in Figure 14.14, and answer these questions:

1. In your own words, state the message of this cartoon.
2. Why do you think the car's appearance is so exaggerated?
3. Summarize the cartoon in your own words.

MAPS

A map is a pictorial representation of a geographic area. Maps often appear in history, geography, anthropology, and political science texts. They can also be found in astronomy, geology, ecology, and biology books.

Maps provide information about terrain, locations, and directions in an area or region. A legend, which is often in a box in the corner of a map, aids the reader in determining distance or direction.

Study the map in Figure 14.15, which was used during the 1991 Persian Gulf War. Then answer the questions.

1. What was the purpose of this map?
2. On what river is Baghdad located?
3. About how many kilometers are there between Baghdad and Kuwait?
4. Is Riyadh southeast or southwest of Jubayl?
5. Which city is farthest from the Persian Gulf—Baghdad or Medina?

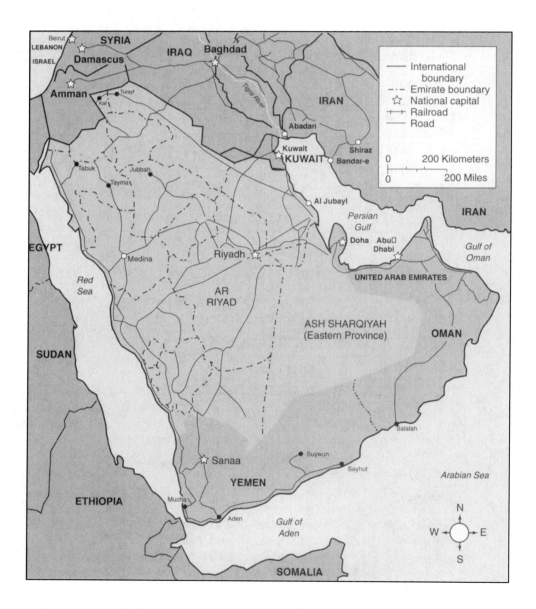

FIGURE 14.15 Countries of the Persian Gulf Region

Source: Based on *U.S. Marines in the Persian Gulf 1990–1991, with 2D Marine Division in Desert Shield and Desert Storm.* Washington, D.C.: History and Museum Division, Headquarters, U.S. Marine Corps, 1993.

SUMMARY

Graphics include tables, charts, graphs, diagrams, photographs, and maps. Here is a general step-by-step approach to reading graphics. As you read, apply each step to the graph shown in Figure 10-2 (Step 7 does not apply to this example).

1. Read the title or caption. The title will identify the subject and may suggest the relationship being described.

2. Discover how the graphic is organized. Read the column headings or labels on the horizontal and vertical axes.

3. Identify the variables. Decide what comparisons are being made or what relationship is being described.

4. Analyze the purpose. Based on what you have seen, predict what the graphic is intended to show. Is its purpose to show change over time, describe a process, compare costs, or present statistics?

5. Determine scale, values, or units of measurement. The scale is the ratio that a graphic has to the thing it represents. For example, a map may be scaled so that one inch on the map represents one mile.

6. Study the data to identify trends or patterns. Note changes, unusual statistics, or unexplained variations.

7. Read the graphic along with corresponding text. Refer to the paragraphs that discuss the graphic. These paragraphs may explain certain features of the graphic and identify trends or patterns.

8. Make a brief summary note. In the margin, jot a brief note summarizing the trend or pattern the graphic emphasizes. Writing will crystallize the idea in your mind and your note will be useful for reviewing.

15 Reading Mathematics

LEARNING OBJECTIVES

- To understand the sequential nature of mathematics
- To develop a systematic approach for reading mathematics textbooks
- To solve word problems in mathematics
- To learn common thought patterns used in mathematics
- To develop study techniques for mathematics

Many students think of mathematics as a separate discipline, and they are surprised to learn that they must take math classes as part of their degree requirements. Mathematics is essential to many academic disciplines, including nursing, business, technologies, and computer science. Although mathematics uses its own language, it is concerned with ideas, concepts, relationships, and problems—just like other disciplines.

Mathematics demands logical and critical thinking and the ability to deal with abstractions, relationships, and theoretical ideas. This subject is extremely rewarding because you see yourself learning and making progress; you can solve a problem today that you couldn't solve yesterday.

MATHEMATICS: A SEQUENTIAL THINKING PROCESS

Learning mathematics is a sequential process. You solve problems by using specific procedures; you verify theorems. In mathematics, much of what you learn is based on skills you have learned earlier; it is cumulative. In algebra, you have to understand radicals before you can solve quadratic equations, whose solutions involve radicals. In business math, to compute a compound interest schedule, you must understand simple interest.

Because mathematics is sequential and cumulative, be sure you begin with a course at the proper level. For example, a calculus course often is required for accounting majors. However, if you have not studied trigonometry, you lack the necessary background for calculus and should take a trigonometry course first. Check the course description in your college catalog for prerequisites, or consult your academic advisor for information on appropriate course placement. If it has been several years since your last math course, your college learning center may offer a placement test to assess your present level of skill.

Mathematics is a process of solving problems—a process of reasoning about situations and understanding the relationships between variables. Too many students learn steps, memorize procedures, and follow rules to solve problems without understanding why they are performing the operations. In mathematics, understanding the meaning of the various operations is essential. Make your learning practical and useful by understanding *how* and *why* the operations work.

READING MATHEMATICS TEXTBOOKS

To learn from mathematics texts, you must allow plenty of time and work at peak concentration. Mathematics texts are concise and to the point; nearly everything is important. Use the following suggestions to develop a systematic approach for reading in mathematics.

PREVIEW BEFORE READING

Previewing before reading is as important in mathematics as it is in every other discipline. For mathematics texts, however, your preview should include a brief review of your previous chapter assignment. Because learning new skills hinges on remembering what you have learned before, a brief review of previously learned material is valuable.

UNDERSTAND MATHEMATICAL LANGUAGE

One of the first steps to success with mathematics is learning to understand its language. Mathematics uses a symbolic language in which notations, symbols, numbers, and formulas are used to express ideas and relationships. Working with mathematics requires that you be able to convert mathematical language to everyday language. To understand a formula—$I = prt$, for example—you must translate mathematical language into everyday language: "Interest equals principal times rate times time." To solve a word problem expressed in everyday language, first you must convert it into mathematical language. You might think of equations as mathematical sentences. Just as an everyday sentence expresses a complete idea, an equation describes a mathematical relationship. Here are a few examples:

SENTENCE	EQUATION
The speed of train A is four times the speed of train B.	$A = 4B$
When I am as old as my mother (m), I shall be five times as old as my son (s) is now.	$5s = m$
In a sewing box there are three times as many pins as needles, and one-third as many needles as buttons; the total number of pins, needles, and buttons is 1872.	$3n + n + 1/3n = 1872$

When you learn a foreign language, it is not sufficient only to learn the new vocabulary; you must also learn the rules of word order, grammar, punctuation, and so forth. To read and understand mathematics, then, you must know not only the signs and symbols (the vocabulary) but also the basic rules for expressing relationships in mathematical form. Figure 15.1 shows five important types of symbols and mathematical language and gives examples of each, taken from introductory algebra.

Type	Function	Example
Punctuation marks	Make clear what parts of a statement are or are not separable; distinguish groups within groups	(...), [...], enclose members of a group
Models (graphs, charts, drawings)	Present a pictorial representation of a relationship or situation	
Numbers	Indicate size, order	3, 7, 9, 11
Symbols	Represent a number that is unknown or may vary	x, y
Signs for relations	Indicate relationships	a = b (equal) c > d (greater than)
Signs for operations	Give instructions	15 ÷ 3 division a − b subtraction

FIGURE 15.1 Aspects of Mathematical Language

Figure 15.2 shows a sample page from an intermediate algebra textbook. It has been marked to indicate the types of mathematical language used.

Much information is packed into small units of mathematical language. Some students find it helpful to translate formulas into words as they read, as is shown in the following equation.

$c^2 = a^2 = b^2$　　　means　　　the square of the hypotenuse of a right triangle is equal to the sum of the squares of the two remaining sides

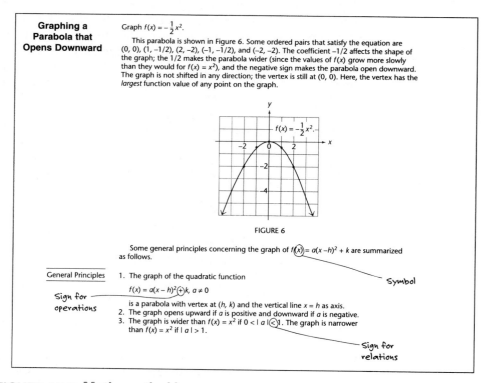

FIGURE 15.2 Mathematical Language

Source: Lial, Hornsby, and Miller, *Intermediate Algebra,* p. 518.

Front	Back
Median	A measure of central tendency; median is middle-most number: half are above it, half are below it Example: Given scores of 6, 8, 10, 14, 19, 21, 58, the median is *14*
Prime number	A whole number, larger than one, that cannot be divided evenly by any whole number except one and itself Example: 2, 3, 5, 7, 11, 13 . . .

FIGURE 15.3 Sample Index Cards

In mathematics you must know the exact meaning of both words and symbols. Use an index card system to help you learn the language of mathematics. Record the term or symbol, its meaning, an example, and a diagram, if possible. Two sample index cards are shown in Figure 15.3.

FOLLOW THE CHAPTER ORGANIZATION

Chapters in most mathematics textbooks contain four essential elements: the presentation or explanation, sample problems, graphs and diagrams, and exercises.

Reading Chapter Explanations. Each new operation, process, or term is explained in the text. As you read this explanation, focus on *how* and *why* the process works. Discover the reasoning behind the process. If the author refers to a sample problem in the explanation, it is necessary to move back and forth between the explanation and the sample problem. Read a sentence or two and then refer to the sample problem to see how the information is applied. Your purpose is to see how the sample problem illustrates the process being described. The steps to follow in computing the total installment cost and finance charge on a credit card are given in Figure 15.4. Then the authors give a sample problem and show how it is solved using the steps they have listed.

As you read, refer to previous chapters if an operation is unclear or if unfamiliar terms are used. In mathematics, you should expect to look back frequently because much of the material is sequential.

Reading Sample Problems. Sample problems demonstrate how an operation or process works. It may be tempting to skip over sample problems because they require time to work through or because they lack an accompanying verbal explanation. However, careful study of the sample problems is an essential part of learning in mathematics. Follow these steps in reading sample problems:

1. Before you read the solution, think of how you would solve the problem. More than one approach may be possible.

2. Read the solution and compare your answer with the textbooks.

3. Be sure you understand each step; you should know exactly what calculations were performed and why they were done.

4. When you have finished reading the sample problem, explain the steps in your own words. This will help you remember the method later. The best way to verbalize is to write the process down; this forces you to be clear and precise. Figure 15.5 presents two sample problems and shows how a student verbalized each process.

Since the enactment of the **Federal Truth-in-Lending Act** (Regulation Z) in 1969, lenders must report their **finance charge** (the charge for credit) and their **annual percentage rate.** The truth-in-lending law does not regulate interest rates or credit charges but merely requires a standardized and truthful report of what they are. The individual states set the allowable interest rates and charges. } back-ground

Find the annual percentage rate by first finding the **total installment cost** (or the **deferred payment price**) and the finance charge on the loan. Do this with the following steps. } overview of process

Finding total installment cost and finance charge

Step 1. Find the total installment cost.

Total installment cost = Down payment

+ (Amount of each payment x Number of payments)

Step 2. Find the finance charge.

Finance charge = Total installment cost – Cash price

Step 3. Finally, find the amount financed.

Amount financed = Cash price – Down payment

} step-by-step procedure

Example Diane Phillips bought a motorcycle for $980. She paid $200 down and then made 24 payments of $39.60 each. Find the (a) total installment cost, (b) finance charge, and (c) amount financed. } Sample problem

Solution (a) Find the total installment cost by multiplying the amount of each payment by the number of payments, and adding the down payment.

Total installment cost = $200 + ($39.60 x 24)
= $200 + $950.40
= $1150.40

(b) The finance charge is the difference between the total installment cost and the cash price.

Finance charge = $1150.40 – $980
= $170.40

Phillips pays an additional $170.40 for the motorcycle because it is bought on credit.

(c) The amount financed is
$980 – $200 = $780

} Solution

SOURCE: Miller and Salzman, *Business Mathematics,* 4th ed., p. 373

FIGURE 15.4 Problem Solving Step by Step

Source: Miller and Salzman, *Business Mathematics,* p. 373

Problem 1: Find the principal of a loan that gives an interest of $30 at 10% per year for 91 days.

Solution	Verbalization
1. $P = \frac{I}{RT}$	1. The formula for computing principal is interest divided by the product of the rate multiplied by the time.
2. $P = \frac{30}{.10 \times \frac{91}{365}}$	2. The interest is $30. The rate, 10% per year, is converted to a decimal, .10, and the time is expressed as a fraction of a year.
3. $P = \frac{30}{.025}$	3. The denominator is simplified.
4. $P = \$1200$	4. The principal is $1200.

Problem 2: $\sqrt{3x + 1} - \sqrt{x + 9} = 2$

Solution	Verbalization
1. $\sqrt{3x + 1} = 2 + \sqrt{x + 9}$	Isolate one radical on one side of the equation.
2. $(3x + 1)^2 = (2 + \sqrt{x + 9})^2$	Square both sides.
3. $3x + 1 = 4 + 4\sqrt{x + 9} + x + 9$	Use formula $(a + b)^2 = a^2 + 2ab + b^2$.
4. $2x - 12 = 4\sqrt{x + 9}$	Isolate the radical.
5. $x - 6 = 2\sqrt{x + 9}$	Factor out 2 on the left side. Divide the equation by 2.
6. $(x - 6)^2 (2\sqrt{x + 9})^2$ $x^2 - 12x + 36 = 4(x + 9)$	Square both sides.
7. $x^2 - 12x + 36 = 4x + 36$ $x^2 - 16x = 0$ $x(x - 16) = 0$ $x = 16$	Solve for x.

Ignore the solution $x = 0$ because it will not check in the original equation. Extraneous roots can occur when you square both sides of an equation.

$$= \frac{6 \pm \sqrt{4}}{2} = \frac{6 \pm 2}{2}$$

$$x = \frac{6 + 2}{2} = \frac{8}{2} = 4$$

or

$$x = \frac{6 - 2}{2} = \frac{4}{2} = 2$$

—Johnson and Steffensen, *Intermediate Algebra,* p. 178

FIGURE 15.5 Verbalizing a Process

Source: Johnson and Steffensen, *Elementary Algebra,* p. 178

5. Test your understanding by covering up the text's solution and solving the problem yourself. Finally, look over the solution, verifying its reasonableness and reviewing the process once again.

Reading and Drawing Graphs, Tables, and Diagrams.   Graphs, diagrams, and drawings are often included in textbook chapters. These are intended to help you understand processes and concepts by providing a visual representation. Treat these drawings as essential parts of the chapter. Here are a few suggestions on how to use them:

1. Study each drawing closely, frequently referring to the text that accompanies it. Test your understanding of the drawing by reconstructing and labeling it without reference to the text drawing; then compare drawings.

2. Use the drawings in the text as models on which to base your own drawings. As you solve end-of-chapter problems, create drawings similar to those included in the text. These may be useful as you decide how to solve the problem.

3. Draw your own diagrams to clarify or explore relationships. For example, an algebra student drew the following diagram of the trinomial equation:

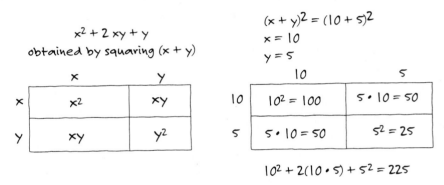

Another student drew the following diagram of the formula for finding the lengths of the sides of a right triangle, $c^2 = a^2 + b^2$.

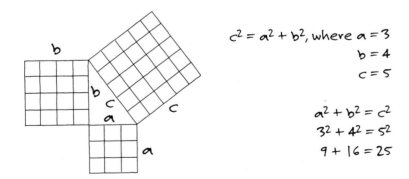

4. Use tables to organize and categorize large amounts of complicated data. For example, a student made the following table to solve a frequency distribution problem in statistics, in which she was asked to identify the frequency of female Democrats and male Republicans.

	DEMOCRAT	REPUBLICAN	TOTALS
Male	62	38	100
Female	31	19	50
Totals	93	57	150

Solving Word Problems.   Most textbook chapters conclude with problem-solving exercises; quizzes and exams also consist primarily of problems. Often the problems are expressed in words and are not conveniently set up for you in formulas or equations. Solving word problems is a seven-step process. Read the problem once to get an overview of the situation; then use the following steps:

1. Identify what is asked for. What are you supposed to find?

2. Locate the information that is provided to solve the problem. (Some math problems may include irrelevant information; if so, underline or circle pertinent data.)

3. Draw a diagram, if possible. Label the diagram.

4. Estimate your answer. If possible, make a reasonable guess about what the answer should be.

5. Decide on a procedure to solve the problem. Recall formulas you have learned that are related to the problem, and look for clue words that indicate a particular process. For example, the phrase "how fast" means *rate;* you may be able to use the formula $r = d/t$. If you do not know how to solve a problem, look for similarities between it and sample problems you have studied.

6. Solve the problem. Begin by setting up an equation.

7. Verify your answer. Compare your answer with your estimate. If there is a large discrepancy, this is a signal that you have made an error. Be sure to check your arithmetic.

Figure 15.6 gives an example of how this problem-solving process works.

USE WRITING TO LEARN

As in most disciplines, reading is not sufficient for learning. Underlining and marking are the strategies students generally use to enhance learning; however, these techniques do not work well in mathematics. Mathematics texts are very concise; everything is important. Writing, instead of underlining, is a useful reading and learning strategy in mathematics. Writing in your own words will force you to convert mathematical language to everyday language. It will also demonstrate what you understand and what you do not. The following list suggests a few ways to use writing to increase your understanding and learning in mathematics.

1. *Definitions.* Read the textbook definition; then close the book and write your own. Compare it to the textbook definition, noticing and collecting discrepancies. Rewrite your definition until it is correct and complete.

2. *Class notes.* Rewrite them, including more detail and explanation. Focus on process; include reasons and explanations. Include information from the corresponding textbook section.

3. *Questions.* Write lists of questions based on chapter assignments, homework assignments, and your class notes. Seek answers from classmates, your instructor, a review book, or the learning lab or tutorial services.

4. *Problems.* Once you think you understand a particular problem or process, write down what you understand. To test your recall, write several questions based on your written explanation. Put aside both your explanation and your questions for several days. Then take out the question sheet and, without reference to your explanation, try to answer your questions. Compare your answers with your original explanation.

5. *Tests.* When preparing for an exam, construct and answer sample questions and problems. It is also effective to exchange self-constructed problems with a classmate and solve them.

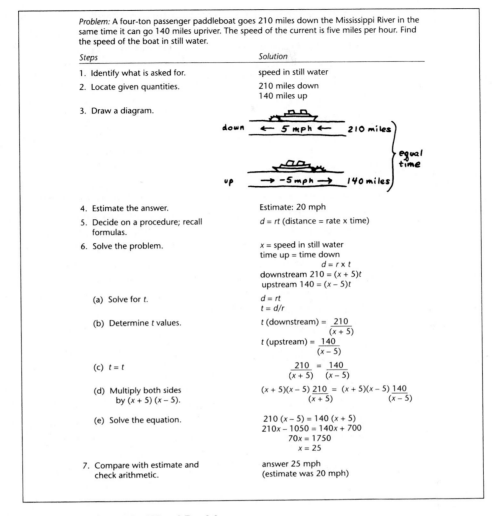

FIGURE 15.6 Sample Word Problem

6. *Diagrams.* Draw diagrams of sample problems as you read, and diagram actual problems before you attempt a solution. Describing the situation in visual terms often makes it more understandable.

7. *Review.* Review your course weekly. Write a description of what you have learned in the past week. You might compare your description with those of your classmates. Keep your weekly descriptions; they will be useful as you review for tests and final exams.

EXPECT GRADUAL UNDERSTANDING

Mathematics is a reasoning process: a process of understanding relationships and seeing similarities and differences. Consequently, mathematics is not an "either you understand it or you don't" discipline. Your understanding will develop by degrees; it grows as you work or "play with" problems. Here's how you can work toward developing your understanding as you read.

1. *Plan on reading, and then reading and solving, and finally re-solving, problems.* As you reread and re-solve problems, you often will come to a new understanding of the process involved. (This is similar to seeing a film or reading a novel a second time; you will notice and discover things you didn't see the first time.)

2. *Experiment with the chapter's content.* The style of most textbooks is quite concise, but this does not mean that there is no room for creativity or experimentation. Try various solutions to problems. As you experiment, you will come to a new understanding of the problems and solutions.

3. *Take risks.* Attempt a solution to a problem even if you do not fully understand it. As you work, you may discover more about how to arrive at the correct answer.

4. *Be active.* Active reading is essential in mathematics. Get involved with the ideas, ask questions, and search for applications.

THOUGHT PATTERNS IN MATHEMATICS

Many students think that memorizing formulas, entire problems, and complete theories is the key to success in mathematics. This does not usually work, because one predominant thought pattern in mathematics is *process*. Your goal in many situations is to see *how* a problem is solved or *how* a theory applies. As you make notes on chapters and as you rewrite class notes, explain, in your own words, how and why things work. Include reasons, explain relationships, and state why a particular operation was selected or why one problem-solving strategy was chosen over another.

Problem solving is also a primary pattern in most mathematics courses. Class activities, homework, and exams all require problem solving. Problem solving in mathematics involves creativity and even playfulness. It is not, as many students think, a matter of merely plugging numbers into a preselected formula and completing the necessary computations. Instead problem solving is the process of assessing a situation (problem) and assembling and applying what you have learned that fits the problem. As you attempt to solve problems, don't immediately reach for a formula. Instead, analyze, think, and experiment while you work. Try several approaches and decide which one works best.

A third common thought pattern evident in many mathematics courses is comparison and contrast. Understanding and solving problems often requires you to see the similarities and differences among problem types and to study variations of sample problems. As you read and study, then, make notes about similarities and differences as they occur to you. Write, in your own words, how one problem differs from, or is a variation of, a type of problem you have previously learned about.

STUDYING MATHEMATICS

Be certain to attend all classes. Because mathematics is sequential, if you miss one specific skill, that gap in your understanding may cause you trouble all term. Expect regular homework assignments and complete them on time even if your instructor does not require you to turn them in. Practice is an essential element in all mathematics courses. Never let yourself get behind or skip assignments. Try to study mathematics at least three times a week-more often, if possible. Use the following suggestions to learn mathematics more effectively.

PREPARING FOR CLASS

1. Emphasize accuracy and precision. In mathematics-, knowing how to solve the problem is not enough; you must produce the right answer. A small error in arithmetic can produce a wrong answer even when you know how to solve the problem. Use a calculator if your instructor allows it.

2. Read the chapter carefully before working on exercises. Don't worry if you don't understand everything right away. Then, as you work or solve problems, refer to the chapter frequently.

3. Before you begin a new chapter or assignment, always review the previous one. If you take a break while working on an assignment, do a brief one-minute review when you resume study.

4. Read the portion of your textbook that covers the next day's lecture before attending class. The lecture will be more meaningful if you have some idea of what it is about beforehand.

5. In your class notes, focus on recording key ideas, not every detail. If possible, record sample problems that your instructor solves in class. After class, review and organize your notes, rewriting them if necessary. Add your own observations and ideas from your textbook as well.

6. Find a study group and work together to solve problems immediately after class. Get the phone numbers of a few people in class whom you can call if you've missed an assignment or are stuck on a problem.

7. Keep your homework in a special notebook. Star the problems you have trouble with. Bring this notebook with you when you ask your instructor for help so that you can go immediately to those problems. When you review for a test, study the starred problems.

BUILDING YOUR CONFIDENCE

1. Approach mathematics confidently. Both men and women can suffer what has come to be known as "math anxiety." Math anxiety often reflects a negative self- concept: "I'm not good at math." Some students think, incorrectly, that one either has or does not have a mathematical mind. This is a myth. Some people may find the subject easier than others, but any average student is capable of learning mathematics.

2. If you feel uncomfortable about taking your first math course, consider taking a basic refresher course in which you are likely to be successful. You may not earn college credit, but you will build your confidence and prove to yourself that you can handle math. Other students find working with computerized review programs helpful when catching up on fundamentals. The machine is nonthreatening, offers no time pressures, and allows you to review a lesson as many times as you want. Many campuses offer workshops on overcoming math anxiety. To find out what help is available, check with your instructor, the learning lab, or the counseling center. A particularly useful book is *Overcoming Math Anxiety* by Sheila Tobias.

STUDY TIPS: PREPARING FOR EXAMS IN MATHEMATICS

Exams in mathematics are usually problems to solve. Use the following tips to prepare for exams in mathematics.

1. When studying for exams, pay attention to what your instructor has emphasized. Predict what will be on the exams, and make a study list that includes all the important topics.

2. When studying for an exam, review as many sample problems as possible. Don't just read the problems; practice solving them. Try to anticipate the variations that may appear. For example, a variation of the distance problem shown in a figure may give you the rate but ask you to compute the distance.

3. Identify problems that are most characteristic of the techniques presented in the chapter you are studying. Record these on a study sheet and summarize in your own words how you worked them. Compare your study sheet with that of a friend.

4. As you solve homework problems and review returned exams and quizzes, search for a pattern of errors. Is there one type of problem you frequently have trouble with? Do you make mistakes when setting up the equation, in factoring, or in computation? If you identify such a pattern, pay special attention to correcting these errors.

5. If you are having trouble with your course, get help immediately; once you get behind, it is very difficult to catch up. Consult with your instructor during his or her office hours. Check with the learning lab for tutoring or computer-assisted review programs.

6. If you find you are weak in a particular fundamental such as ratios or factoring, correct the problem as soon as possible. If you do not, it will interfere with your performance.

7. Obtain additional study aids. Schaum's *College Outline Series* offers excellent study guides. Check with your instructor for additional references.

8. When a test is returned, rework the problems on which you lost points to find out exactly what you did wrong.

SUMMARY

This chapter discusses reading strategies for mathematics. Mathematics is largely sequential and cumulative; each skill builds on—and hinges on—previously learned skills.

Techniques for reading mathematics are

- learning mathematical language
- following chapter organization (including explanations, sample problems, diagrams, and graphs)
- dealing with word problems
- using writing to learn

Thought patterns that are among the most common in mathematics are

- process
- problem/solution
- comparison and contrast

Suggestions are offered for adapting study techniques to mathematics.

PART V

Mathematics

Prealgebra Review

Mathematics is an important tool for everyday life. Knowing basic mathematical skills can simplify many tasks. For example, we use fractions to represent parts of a whole, such as "half an hour" or "third of a cup." Understanding decimals helps us work efficiently in our money system. Percent is a concept used virtually every day in ordinary and business life.

This optional review chapter covers basic topics and skills from prealgebra. Knowledge of these topics is needed for success in algebra.

Donald and Doris Fisher opened the first The Gap store (named after the "generation gap") in 1969 near the campus of what is now the San Francisco State University. This single store, which sold mostly Levi's blue jeans, has evolved into an international clothing giant. The Gap, Inc., now sells a variety of clothing through its store chains that include babyGap, Banana Republic, The Gap, GapKids, and Old Navy Clothing. In 1997, The Gap, Inc., had approximately $5.3 billion in revenue.

Problem Solving Notes

R.1   FACTORS AND THE LEAST COMMON MULTIPLE

A   FACTORING NUMBERS

> To factor means to write as a product.

In arithmetic we factor numbers, and in algebra we factor expressions containing variables. Throughout this text, you will encounter the word *factor* often. Always remember that factoring means writing as a product

Since $2 \cdot 3 = 6$, we say that 2 and 3 are **factors** of 6. Also, $2 \cdot 3$ is a **factorization** of 6.

Example 1   List the factors of 6.

   Solution: First we write the different factorizations of 6.

$$6 = 1 \cdot 6, \quad 6 = 2 \cdot 3$$

The factors of 6 are 1, 2, 3, and 6. ▬▬

Example 2   List the factors of 20.

   Solution: $20 = 1 \cdot 20, \quad 20 = 2 \cdot 10, \quad 20 = 4 \cdot 5, \quad 20 = 2 \cdot 2 \cdot 5$

The factors of 20 are 1, 2, 4, 5, 10, and 20. ▬▬

In this section, we will concentrate on **natural numbers** only. The natural numbers (also called counting numbers) are

Natural Numbers: 1, 2, 3, 4, 5, 6, 7, and so on

Every natural number except 1 is either a prime number or a composite number.

> **PRIME AND COMPOSITE NUMBERS**
>
> A **prime number** is a natural number greater than 1 whose only factors are 1 and itself. The first few prime numbers are 2, 3, 5, 7, 11, 13, 17, 19, 23, 29, . . .
> A **composite number** is a natural number greater than 1 that is not prime.

Example 3   Identify each number as prime or composite:

          3, 20, 7, 4

   Solution: 3 is a prime number. Its factors are 1 and 3 only.
20 is a composite number. Its factors are 1, 2, 4, 5, 10, and 20.
7 is a prime number. Its factors are 1 and 7 only.
4 is a composite number. Its factors are 1, 2, and 4.

B   WRITING PRIME FACTORIZATIONS

When a number is written as a product of primes, this product is called the **prime factorization** of the number. For example, the prime factorization of 12 is $2 \cdot 2 \cdot 3$ since

$$12 = 2 \cdot 2 \cdot 3$$

and all the factors are prime.

Objectives

A Write the factors of a number.
B Write the prime factorization of a number.
C Find the LCM of a list of numbers.

SSM   CD-ROM   Video R.1

Practice Problem 1

List the factors of 4.

Practice Problem 2

List the factors of 18.

Practice Problem 3

Identify each number as prime or composite:

     5, 18, 11, 6

Answers

1. 1, 2, 4,   **2.** 1, 2, 3, 6, 9, 18,   **3.** 5, 11 prime; 6, 18 composite

Practice Problem 4

Write the prime factorization of 28.

Practice Problem 5

Write the prime factorization of 60.

✓ CONCEPT CHECK

Suppose that you choose $80 = 4 \cdot 20$ as your first step in Example 5 and another student chooses $80 = 5 \cdot 16$. Will both end up with the same prime factorization as in Example 5? Explain.

Answers

4. $28 = 2 \cdot 2 \cdot 7$,   **5.** $60 = 2 \cdot 2 \cdot 3 \cdot 5$

✓ Concept Check: yes; answers may vary

Example 4   Write the prime factorization of 45.

Solution: We can begin by writing 45 as the product of two numbers, say 9 and 5.

$$45 = 9 \cdot 5$$

The number 5 is prime, but 9 is not. So we write 9 as $3 \cdot 3$.

$$45 = 9 \cdot 5$$
$$= 3 \cdot 3 \cdot 5$$

Each factor is now a prime number, so the prime factorization of 45 is $3 \cdot 3 \cdot 5$.

> **HELPFUL HINT**
>
> Recall that order is not important when multiplying numbers. For example,
> $$3 \cdot 3 \cdot 5 = 3 \cdot 5 \cdot 3 = 5 \cdot 3 \cdot 3 = 45$$
> For this reason, any of the products shown can be called *the* prime factorization of 45.

Example 5   Write the prime factorization of 80.

Solution: We first write 80 as a product of two numbers. We continue this process until all factors are prime.

$$80 = 8 \cdot 10$$
$$4 \cdot 2 \cdot 2 \cdot 5$$
$$= 2 \cdot 2 \cdot 2 \cdot 2 \cdot 5$$

All factors are now prime, so the prime factorization of 80 is

$$2 \cdot 2 \cdot 2 \cdot 2 \cdot 5$$

TRY THE CONCEPT CHECK IN THE MARGIN.

> **HELPFUL HINT**
>
> There are a few quick **divisibility tests** to determine if a number is divisible by the primes 2, 3, or 5.
> A whole number is divisible by
>
> ▲ **2** if the ones digit is 0, 2, 4, 6, or 8.
>
>   132 is divisible by 2
>
> ▲ **3** if the sum of the digits is divisible by 3.
>
>   144 is divisible by 3 since $1 + 4 + 4 = 9$ is divisible by 3.
>
> ▲ **5** if the ones digit is 0 or 5.
>
>   1115 is divisible by 5

When finding the prime factorization of larger numbers, you may want to use the procedure shown in Example 6.

Example 6   Write the prime factorization of 252.

Solution: Since the ones digit of 252 is 2, we know that 252 is divisible by 2.

$$\begin{array}{r} 126 \\ 2\overline{)252} \end{array}$$

126 is divisible by 2 also.

$$\begin{array}{r} 63 \\ 2\overline{)126} \\ 2\overline{)252} \end{array}$$

63 is not divisible by 2 but is divisible by 3. We divide 63 by 3 and continue in this same manner until the quotient is a prime number.

$$\begin{array}{r} 7 \\ 3\overline{)21} \\ 3\overline{)63} \\ 2\overline{)126} \\ 2\overline{)252} \end{array}$$

The prime factorization of 252 is $2 \cdot 2 \cdot 3 \cdot 3 \cdot 7$.

C   FINDING THE LEAST COMMON MULTIPLE

A **multiple** of a number is the product of that number and any natural number. For example, the multiples of 3 are

$$\underbrace{3 \cdot 1}_{3,} \quad \underbrace{3 \cdot 2}_{6,} \quad \underbrace{3 \cdot 3}_{9,} \quad \underbrace{3 \cdot 4}_{12,} \quad \underbrace{3 \cdot 5}_{15,} \quad \underbrace{3 \cdot 6}_{18,} \quad \underbrace{3 \cdot 7}_{21,} \text{ and so on}$$

The multiples of 2 are

$$\underbrace{2 \cdot 1}_{2,} \quad \underbrace{2 \cdot 2}_{4,} \quad \underbrace{2 \cdot 3}_{6,} \quad \underbrace{2 \cdot 4}_{8,} \quad \underbrace{2 \cdot 5}_{10,} \quad \underbrace{2 \cdot 6}_{12,} \quad \underbrace{2 \cdot 7}_{14,} \text{ and so on}$$

Notice that 2 and 3 have multiples that are common to both.

Multiples of 2:   2, 4, 6 , 8, 10, 12 , 14, 16, 18 , and so on

Multiples of 3:   3, 6 , 9, 12 , 15, 18 , 21, and so on

The least or smallest common multiple of 2 and 3 is 6. The number 6 is called the **least common multiple** or **LCM** of 2 and 3. It is the smallest number that is a multiple of both 2 and 3.

Finding the LCM by the method above can sometimes be time-consuming. Let's look at another method that uses prime factorization.

To find the LCM of 4 and 10, for example, we write the prime factorization of each.

$$4 = 2 \cdot 2$$
$$10 = 2 \cdot 5$$

If the LCM is to be a multiple of 4, it must contain the factors $2 \cdot 2$. If the LCM is to be a multiple of 10, it must contain the factors $2 \cdot 5$. Since we decide whether the LCM is a multiple of 4 and 10 separately, the LCM does not need to contain three factors of 2. The LCM only needs to contain

Practice Problem 6

Write the prime factorization of 297.

Answer
6. $3 \cdot 3 \cdot 3 \cdot 11$

a factor the greatest number of times that the factor appears in any **one** prime factorization.

The LCM is a
multiple of 4.

$$\text{LCM} = \overbrace{2 \cdot \underbrace{2 \cdot 5}} = 20$$

The LCM is a
multiple of 10.

The number 2 is a factor twice since that is the greatest number of times that 2 is a factor in either of the prime factorizations.

TO FIND THE LCM OF A LIST OF NUMBERS

Step 1. Write the prime factorization of each number.

Step 2. Write the product containing each different prime factor (from Step 1) the greatest number of times that it appears in any one factorization. This product is the LCM.

Practice Problem 7

Find the LCM of 14 and 35.

Example 7

Find the LCM of 18 and 24.

Solution: First we write the prime factorization of each number.

$$18 = 2 \cdot 3 \cdot 3$$
$$24 = 2 \cdot 2 \cdot 2 \cdot 3$$

Now we write each factor the greatest number of times that it appears in any **one** prime factorization.

The greatest number of times that 2 appears is **3** times.
The greatest number of times that 3 appears is **2** times.

$$\text{LCM} = \underbrace{2 \cdot 2 \cdot 2}_{\substack{2 \text{ is a factor} \\ 3 \text{ times.}}} \cdot \underbrace{3 \cdot 3}_{\substack{3 \text{ is a factor} \\ 2 \text{ times.}}} = 72$$

Practice Problem 8

Find the LCM of 5 and 9.

Example 8

Find the LCM of 11 and 6.

Solution: 11 is a prime number, so we simply rewrite it. Then we write the prime factorization of 6.

$$11 = 11$$
$$6 = 2 \cdot 3$$
$$\text{LCM} = 2 \cdot 3 \cdot 11 = 66.$$

Practice Problem 9

Find the LCM of 4, 15, and 10.

Example 9

Find the LCM of 5, 6, and 12.

Solution:
$$5 = 5$$
$$6 = 2 \cdot 3$$
$$12 = 2 \cdot 2 \cdot 3$$
$$\text{LCM} = 2 \cdot 2 \cdot 3 \cdot 5 = 60.$$

Answers

7. 70,   **8.** 45,   **9.** 60

R.2 FRACTIONS

A quotient of two numbers such as $\frac{2}{9}$ is called a **fraction**. The parts of a fraction are:

Fraction bar → $\dfrac{2}{9}$ ← Numerator
          ← Denominator

$\frac{2}{9}$ of the circle
is shaded.

A fraction may be used to refer to part of a whole. For example, $\frac{2}{9}$ of the circle in the figure is shaded. The denominator 9 tells us how many equal parts the whole circle is divided into and the numerator 2 tells us how many equal parts are shaded.

In this section, we will use **whole numbers**. The whole numbers consist of 0 and the natural numbers.

Whole Numbers: 0, 1, 2, 3, 4, 5, and so on

A WRITING EQUIVALENT FRACTIONS

More than one fraction can be used to name the same part of a whole. Such fractions are called **equivalent fractions**.

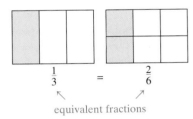

$\frac{1}{3}$ = $\frac{2}{6}$

equivalent fractions

> **EQUIVALENT FRACTIONS**
>
> Fractions that represent the same portion of a whole are called **equivalent fractions**.

To write equivalent fractions, we use the **fundamental principle of fractions**. This principle guarantees that, if we multiply both the numerator and the denominator by the same nonzero number, the result is an equivalent fraction. For example, if we multiply the numerator and denominator of $\frac{1}{3}$ by the same number, 2, the result is the equivalent fraction $\frac{2}{6}$.

$$\frac{1 \cdot 2}{3 \cdot 2} = \frac{2}{6}$$

> **FUNDAMENTAL PRINCIPLE OF FRACTIONS**
>
> If a, b, and c are numbers, then
>
> $$\frac{a}{b} = \frac{a \cdot c}{b \cdot c} \quad \text{or} \quad \frac{a \cdot c}{b \cdot c} = \frac{a}{b}$$
>
> as long as b and c are not 0.

Objectives

A Write equivalent fractions.
B Write fractions in simplest form.
C Multiply and divide fractions.
D Add and subtract fractions.

SSM   CD-ROM   Video
                 R.2

Practice Problem 1

Write $\dfrac{1}{4}$ as an equivalent fraction with a denominator of 20.

Practice Problem 2

Simplify: $\dfrac{20}{35}$

✓ CONCEPT CHECK

Explain the error in the following steps.

a. $\dfrac{15}{55} = \dfrac{1\,\cancel{5}}{5\,\cancel{5}} = \dfrac{1}{5}$

b. $\dfrac{6}{7} = \dfrac{5+1}{5+2} = \dfrac{1}{2}$

Practice Problems 3–4

Simplify each fraction.

3. $\dfrac{7}{20}$

4. $\dfrac{12}{40}$

Answers

1. $\dfrac{5}{20}$, 2. $\dfrac{4}{7}$, 3. $\dfrac{7}{20}$, 4. $\dfrac{3}{10}$

✓ **Concept Check:** answers may vary

Example 1

Write $\dfrac{2}{5}$ as an equivalent fraction with a denominator of 15.

Solution: Since $5 \cdot 3 = 15$, we use the fundamental principle of fractions and multiply the numerator and denominator of $\dfrac{2}{5}$ by 3.

$$\dfrac{2}{5} = \dfrac{2 \cdot 3}{5 \cdot 3} = \dfrac{6}{15}$$

Then $\dfrac{2}{5}$ is equivalent to $\dfrac{6}{15}$. They both represent the same part of a whole. ▬▬

B   SIMPLIFYING FRACTIONS

A fraction is said to be **simplified** or in **lowest terms** when the numerator and the denominator have no factors in common other than 1. For example, the fraction $\dfrac{5}{11}$ is in lowest terms since 5 and 11 have no common factors other than 1.

One way to simplify fractions is to write both the numerator and the denominator as a product of primes and then apply the fundamental principle of fractions.

Example 2

Simplify: $\dfrac{42}{49}$

Solution: We write the numerator and the denominator as products of primes. Then we apply the fundamental principle of fractions to the common factor 7.

$$\dfrac{42}{49} = \dfrac{2 \cdot 3 \cdot 7}{7 \cdot 7} = \dfrac{2 \cdot 3}{7} = \dfrac{6}{7}$$

TRY THE CONCEPT CHECK IN THE MARGIN.

Examples

Simplify each fraction.

3. $\dfrac{11}{27} = \dfrac{11}{3 \cdot 3 \cdot 3}$ There are no common factors other than 1, so $\dfrac{11}{27}$ is already simplified.

4. $\dfrac{88}{20} = \dfrac{2 \cdot 2 \cdot 2 \cdot 11}{2 \cdot 2 \cdot 5} = \dfrac{22}{5}$ ▬▬

The improper fraction $\dfrac{22}{5}$ from Example 4 may be written as the mixed number $4\dfrac{2}{5}$, but in this text, we will not do so.

Some fractions may be simplified by recalling that the fraction bar means division.

$$\dfrac{6}{6} = 6 \div 6 = 1 \quad \text{and} \quad \dfrac{3}{1} = 3 \div 1 = 3$$

Examples   Simplify by dividing the numerator by the denominator.

5. $\frac{3}{3} = 3 \div 3 = 1$

6. $\frac{4}{2} = 4 \div 2 = 2$

7. $\frac{7}{7} = 7 \div 7 = 1$

8. $\frac{8}{1} = 8 \div 1 = 8$

In general, if the numerator and the denominator are the same, the fraction is equivalent to 1. Also, if the denominator of a fraction is 1, the fraction is equivalent to the numerator.

If a is any number other than 0, then $\frac{a}{a} = 1$.

Also, if a is any number, $\frac{a}{1} = a$.

C MULTIPLYING AND DIVIDING FRACTIONS

To multiply two fractions, we multiply numerator times numerator to obtain the numerator of the product. Then we multiply denominator times denominator to obtain the denominator of the product.

MULTIPLYING FRACTIONS

$\frac{a}{b} \cdot \frac{c}{d} = \frac{a \cdot c}{b \cdot d}$,   if $b \neq 0$ and $d \neq 0$

Example 9   Multiply: $\frac{2}{15} \cdot \frac{5}{13}$. Simplify the product if possible.

Solution: $\frac{2}{15} \cdot \frac{5}{13} = \frac{2 \cdot 5}{15 \cdot 13}$ Multiply numerators.
Multiply denominators.

To simplify the product, we divide the numerator and the denominator by any common factors.

$= \frac{2 \cdot 5}{3 \cdot 5 \cdot 13}$

$= \frac{2}{39}$

Practice Problems 5–8

Simplify by dividing the numerator by the denominator.

5. $\frac{4}{4}$ **6.** $\frac{9}{3}$

7. $\frac{10}{10}$ **8.** $\frac{5}{1}$

Practice Problem 9

Multiply: $\frac{3}{7} \cdot \frac{3}{5}$. Simplify the product if possible.

Answers

5. 1,   **6.** 3,   **7.** 1,   **8.** 5,   **9.** $\frac{9}{35}$

Before we divide fractions, we first define **reciprocals**. Two numbers are reciprocals of each other if their product is 1.

The reciprocal of $\frac{2}{3}$ is $\frac{3}{2}$ because $\frac{2}{3} \cdot \frac{3}{2} = \frac{6}{6} = 1$.

The reciprocal of 5 is $\frac{1}{5}$ because $5 \cdot \frac{1}{5} = \frac{5}{1} \cdot \frac{1}{5} = \frac{5}{5} = 1$.

To divide fractions, we multiply the first fraction by the reciprocal of the second fraction. For example,

$$\frac{1}{2} \div \frac{5}{7} = \frac{1}{2} \cdot \frac{7}{5} = \frac{1 \cdot 7}{2 \cdot 5} = \frac{7}{10}$$

HELPFUL HINT	To divide, multiply by the reciprocal.

DIVIDING FRACTIONS

$$\frac{a}{b} \div \frac{c}{d} = \frac{a}{b} \cdot \frac{d}{c}, \qquad \text{if } b \neq 0, d \neq 0, \text{ and } c \neq 0$$

Practice Problems 10–12

Divide and simplify.

10. $\frac{2}{9} \div \frac{3}{4}$

11. $\frac{8}{11} \div 24$

12. $\frac{5}{4} \div \frac{5}{8}$

Examples   Divide and simplify.

10. $\frac{4}{5} \div \frac{5}{16} = \frac{4}{5} \cdot \frac{16}{5} = \frac{4 \cdot 16}{5 \cdot 5} = \frac{64}{25}$

11. $\frac{7}{10} \div 14 = \frac{7}{10} \div \frac{14}{1} = \frac{7}{10} \cdot \frac{1}{14} = \frac{7 \cdot 1}{2 \cdot 5 \cdot 2 \cdot 7} = \frac{1}{20}$

12. $\frac{3}{8} \div \frac{3}{10} = \frac{3}{8} \cdot \frac{10}{3} = \frac{3 \cdot 2 \cdot 5}{2 \cdot 2 \cdot 2 \cdot 3} = \frac{5}{4}$

D   ADDING AND SUBTRACTING FRACTIONS

To add or subtract fractions with the same denominator, we combine numerators and place the sum or difference over the common denominator.

ADDING AND SUBTRACTING FRACTIONS WITH THE SAME DENOMINATOR

$$\frac{a}{b} + \frac{c}{b} = \frac{a + c}{b}, \qquad \text{if } b \neq 0$$

$$\frac{a}{b} - \frac{c}{b} = \frac{a - c}{b}, \qquad \text{if } b \neq 0$$

Answers

10. $\frac{8}{27}$,   **11.** $\frac{1}{33}$,   **12.** 2

Examples

Add or subtract as indicated. Then simplify if possible.

13. $\dfrac{2}{7} + \dfrac{4}{7} = \dfrac{2+4}{7} = \dfrac{6}{7}$

14. $\dfrac{3}{10} + \dfrac{2}{10} = \dfrac{3+2}{10} = \dfrac{5}{10} = \dfrac{5}{2 \cdot 5} = \dfrac{1}{2}$

15. $\dfrac{9}{7} - \dfrac{2}{7} = \dfrac{9-2}{7} = \dfrac{7}{7} = 1$

16. $\dfrac{5}{3} - \dfrac{1}{3} = \dfrac{5-1}{3} = \dfrac{4}{3}$

To add or subtract with different denominators, we first write the fractions as **equivalent fractions** with the same denominator. We will use the smallest or least common denominator, the LCD. The LCD is the same as the least common multiple we reviewed in Section R.1.

Example 17 Add: $\dfrac{2}{5} + \dfrac{1}{4}$

Solution: We first must find the least common denominator before the fractions can be added. The least common multiple for the denominators 5 and 4 is 20. This is the LCD we will use.

We write both fractions as equivalent fractions with denominators of 20. Since

$$\dfrac{2}{5} = \dfrac{2 \cdot 4}{5 \cdot 4} = \dfrac{8}{20} \quad \text{and} \quad \dfrac{1}{4} = \dfrac{1 \cdot 5}{4 \cdot 5} = \dfrac{5}{20}$$

then

$$\dfrac{2}{5} + \dfrac{1}{4} = \dfrac{8}{20} + \dfrac{5}{20} = \dfrac{13}{20}$$

Example 18 Subtract and simplify: $\dfrac{19}{6} - \dfrac{23}{12}$

Solution: The LCD is 12. We write both fractions as equivalent fractions with denominators of 12.

$$\dfrac{19}{6} - \dfrac{23}{12} = \dfrac{19 \cdot 2}{6 \cdot 2} - \dfrac{23}{12}$$

$$= \dfrac{38}{12} - \dfrac{23}{12}$$

$$= \dfrac{15}{12} = \dfrac{3 \cdot 5}{2 \cdot 2 \cdot 3} = \dfrac{5}{4}$$

Practice Problems 13–16

Add or subtract as indicated. Then simplify if possible.

13. $\dfrac{2}{11} + \dfrac{5}{11}$

14. $\dfrac{1}{8} + \dfrac{3}{8}$

15. $\dfrac{13}{10} - \dfrac{3}{10}$

16. $\dfrac{7}{6} - \dfrac{2}{6}$

Practice Problem 17

Add: $\dfrac{3}{8} + \dfrac{1}{20}$

Practice Problem 18

Subtract and simplify: $\dfrac{8}{15} - \dfrac{1}{3}$

Answers

13. $\dfrac{7}{11}$, **14.** $\dfrac{1}{2}$, **15.** 1, **16.** $\dfrac{5}{6}$, **17.** $\dfrac{17}{40}$,

18. $\dfrac{1}{5}$

Focus On Study Skills

CRITICAL THINKING

What Is Critical Thinking?

Although exact definitions often vary, thinking critically usually refers to evaluating, analyzing, and interpreting information in order to make a decision, draw a conclusion, reach a goal, make a prediction, or form an opinion. Critical thinking often involves problem solving, communication, and reasoning skills. Critical thinking is more than a technique that helps students pass their courses. Critical thinking skills are life skills. Developing these skills can help you solve problems in your workplace and in everyday life. For instance, well-developed critical thinking skills would be useful in the following situation:

> Suppose you work as a medical lab technician. Your lab supervisor has decided that some lab equipment should be replaced. She asks you to collect information on several different models from equipment manufacturers. Your assignment is to study the data and then make a recommendation on which model the lab should buy.

How Can Critical Thinking Be Developed?

Just as physical exercise can help to develop and strengthen certain muscles of the body, mental exercise can help to develop critical thinking skills. Mathematics is ideal for helping to develop critical thinking skills because it requires using logic and reasoning, recognizing patterns, making conjectures and educated guesses, and drawing conclusions. You will find many opportunities to build your critical thinking skills throughout this course:

- ▲ In real-life application problems (see Exercise 24 in Section 2.5)
- ▲ In conceptual and writing exercises marked with the ✎ icon (see Exercise 43 in Section 1.2)
- ▲ In the Combining Concepts subsection of exercise sets (see Exercise 57 in Section 4.4)
- ▲ In the Chapter Activities (see the Chapter 8 Activity)
- ▲ In the Critical Thinking and Group Activities found in Focus On features like this one throughout the book (see page 46).

EXERCISE SET R.2

A *Write each fraction as an equivalent fraction with the given denominator. See Example 1.*

1. $\frac{4}{5}$ with a denominator of 20

2. $\frac{4}{5}$ with a denominator of 25

B *Simplify each fraction. See Examples 2 through 8.*

3. $\frac{3}{6}$

4. $\frac{15}{20}$

5. $\frac{3}{7}$

6. $\frac{5}{9}$

7. $\frac{20}{20}$

8. $\frac{35}{7}$

9. $\frac{42}{6}$

10. $\frac{18}{30}$

11. $\frac{42}{45}$

12. $\frac{16}{20}$

13. $\frac{8}{40}$

14. $\frac{64}{24}$

15. $\frac{120}{244}$

16. $\frac{360}{700}$

C *Multiply or divide as indicated. See Examples 9 through 12.*

17. $\frac{7}{8} \cdot \frac{3}{21}$

18. $\frac{7}{10} \cdot \frac{5}{21}$

19. $\frac{3}{35} \cdot \frac{10}{63}$

20. $\frac{9}{20} \div 12$

21. $\frac{25}{36} \div 10$

1. _____

2. _____

3. _____

4. _____

5. _____

6. _____

7. _____

8. _____

9. _____

10. _____

11. _____

12. _____

13. _____

14. _____

15. _____

16. _____

17. _____

18. _____

19. _____

20. _____

21. _____

Name _____

D *Add or subtract as indicated. See Examples 13 through 18.*

22. $\dfrac{13}{132} + \dfrac{35}{132}$

23. $\dfrac{17}{21} - \dfrac{10}{21}$

24. $\dfrac{18}{35} - \dfrac{11}{35}$

25. $\dfrac{2}{3} + \dfrac{3}{7}$

26. $\dfrac{3}{4} + \dfrac{1}{6}$

27. $\dfrac{10}{3} - \dfrac{5}{21}$

28. $\dfrac{11}{7} - \dfrac{3}{35}$

29. $\dfrac{10}{21} + \dfrac{5}{21}$

30. $\dfrac{11}{35} + \dfrac{3}{35}$

31. $\dfrac{5}{22} - \dfrac{5}{33}$

32. $\dfrac{7}{10} - \dfrac{8}{15}$

33. $\dfrac{12}{5} - 1$

34. $2 - \dfrac{3}{8}$

35. $\dfrac{2}{3} - \dfrac{5}{9} + \dfrac{5}{6}$

36. $\dfrac{8}{11} - \dfrac{1}{4} + \dfrac{1}{2}$

R.3   DECIMALS AND PERCENTS

A   WRITING DECIMALS AS FRACTIONS

Like fractional notation, decimal notation is used to denote a part of a whole. Below is a place value chart that shows the value of each place.

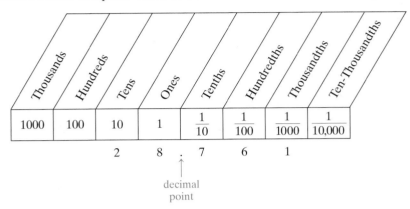

TRY THE CONCEPT CHECK IN THE MARGIN.

The next chart shows decimals written as fractions.

Decimal Form	Fractional Form
0.1 ——tenths——	$\dfrac{1}{10}$
0.07 ——hundredths——	$\dfrac{7}{100}$
2.31 ——hundredths——	$\dfrac{231}{100}$
0.9862 ——ten-thousandths——	$\dfrac{9862}{10,000}$

Examples   Write each decimal as a fraction. Do not simplify.

1. $0.37 = \dfrac{37}{100}$
2 decimal places   2 zeros

2. $1.3 = \dfrac{13}{10}$
1 decimal place   1 zero

3. $2.649 = \dfrac{2649}{1000}$
3 decimal places   3 zeros

Objectives

A Write decimals as fractions.

B Add, subtract, multiply, and divide decimals.

C Round decimals to a given decimal place.

D Write fractions as decimals.

E Write percents as decimals and decimals as percents.

SSM   CD-ROM   Video R.3

✓ CONCEPT CHECK

Fill in the blank: In the number 52.634, the 3 is in the _____ place.

a. Tens

b. Ones

c. Tenths

d. Hundredths

e. Thousandths

Practice Problems 1–3

Write each decimal as a fraction. Do not simplify.

1. 0.27

2. 5.1

3. 7.685

Answers

✓Concept Check: d

1. $\dfrac{27}{100}$, **2.** $\dfrac{51}{10}$, **3.** $\dfrac{7685}{1000}$

B ADDING, SUBTRACTING, MULTIPLYING, AND DIVIDING DECIMALS

To **add** or **subtract** decimals, we write the numbers vertically with decimal points lined up. We then add the like place values from right to left. We place the decimal point in the answer directly below the decimal points in the problem.

Practice Problem 4

Add.

a. $7.19 + 19.782 + 1.006$

b. $12 + 0.79 + 0.03$

Example 4   Add.

a. $5.87 + 23.279 + 0.003$ b. $7 + 0.23 + 0.6$

Solution: a.
$$\begin{array}{r} 5.87 \\ 23.279 \\ + \ 0.003 \\ \hline 29.152 \end{array}$$

b.
$$\begin{array}{r} 7. \\ 0.23 \\ + \ 0.6 \\ \hline 7.83 \end{array}$$

Practice Problem 5

Subtract.

a. $84.23 - 26.982$

b. $90 - 0.19$

Example 5   Subtract.

a. $32.15 - 11.237$ b. $70 - 0.48$

Solution: a.
$$\begin{array}{r} {\scriptstyle 1 \ \ 11 \ 4 \ 10} \\ 3\,2\,.\,1\,5\,0 \\ -\,1\,1\,.\,2\,3\,7 \\ \hline 2\,0\,.\,9\,1\,3 \end{array}$$

b.
$$\begin{array}{r} {\scriptstyle 6 \ 9 \ \ 9 \ 10} \\ 7\,0\,.\,0\,0 \\ -\ \ \ 0\,.\,4\,8 \\ \hline 6\,9\,.\,5\,2 \end{array}$$

Now let's study the following product of decimals. Notice the pattern in the decimal points.

$$0.\underset{\underset{\substack{2 \text{ decimal} \\ \text{places}}}{\uparrow}}{03} \times 0.\underset{\underset{\substack{1 \text{ decimal} \\ \text{place}}}{\uparrow}}{6} = \frac{3}{100} \times \frac{6}{10} = \frac{18}{1000} \quad \text{or} \quad 0.\underset{\underset{\substack{3 \text{ decimal} \\ \text{places}}}{\uparrow}}{018}$$

In general, to **multiply** decimals we multiply the numbers as if they were whole numbers. The decimal point in the product is placed so that the number of decimal places in the product is the same as the *sum* of the number of decimal places in the factors.

Practice Problem 6

Multiply.

a. 0.31×4.6

b. 1.26×0.03

Example 6   Multiply.

a. 0.072×3.5 b. 0.17×0.02

Solution: a.
$$\begin{array}{r} 0.072 \quad {\scriptstyle 3 \text{ decimal places}}\\ \times \quad 3.5 \quad {\scriptstyle 1 \text{ decimal place}}\\ \hline 360 \\ 216 \quad\;\; \\ \hline 0.2520 \quad {\scriptstyle 4 \text{ decimal places}} \end{array}$$

b.
$$\begin{array}{r} 0.17 \quad {\scriptstyle 2 \text{ decimal places}}\\ \times \ 0.02 \quad {\scriptstyle 2 \text{ decimal places}}\\ \hline 0.0034 \quad {\scriptstyle 4 \text{ decimal places}} \end{array}$$

To divide a decimal by a whole number using long division, we place the decimal point in the quotient directly above the decimal point in the dividend. For example,

$$\begin{array}{r} 2.47 \\ 3\overline{\smash{)}7.41} \\ \underline{-6} \\ 1\,4 \\ \underline{-1\,2} \\ 2\,1 \\ \underline{-2\,1} \\ 0 \end{array}$$

To check, see that

$2.47 \times 3 = 7.41$

Answers

4. a. 27.978, **b.** 12.82, **5. a.** 57.248, **b.** 89.81, **6. a.** 1.426, **b.** 0.0378

In general, to **divide** decimals we move the decimal point in the divisor to the right until the divisor is a whole number. Then we move the decimal point in the dividend the same number of places that the decimal point in the divisor was moved. The decimal point in the quotient lies directly above the decimal point in the dividend.

Example 7   Divide.

  a. $9.46 \div 0.04$ **b.** $31.5 \div 0.007$

Solution: **a.**

$$
\begin{array}{r}
236.5 \\
04.\overline{)946.0} \\
-8 \\
\hline
14 \\
-12 \\
\hline
26 \\
-24 \\
\hline
20 \\
-20 \\
\hline
0
\end{array}
$$

b.

$$
\begin{array}{r}
4500. \\
0007.\overline{)31500.} \\
-28 \\
\hline
35 \\
-35 \\
\hline
0
\end{array}
$$

Practice Problem 7

Divide.

a. $21.75 \div 0.5$

b. $15.6 \div 0.006$

C   ROUNDING DECIMALS

We **round** the decimal part of a decimal number in nearly the same way as we round the whole numbers. The only difference is that we drop digits to the right of the rounding place, instead of replacing these digits by 0s. For example,

  24.954 rounded to the nearest hundredth is 24.95
      ↑

TO ROUND DECIMALS TO A PLACE VALUE TO THE RIGHT OF THE DECIMAL POINT

Step 1. Locate the digit to the right of the given place value.

Step 2.   ▲ If this digit is 5 or greater, add 1 to the digit in the given place value and drop all digits to its right.

   ▲ If this digit is less than 5, drop all digits to the right of the given place.

Example 8   Round 7.8265 to the nearest hundredth.

Solution: ┌ hundredths place
     7.8265
       ↑____ Step 1. Locate the digit to the right of the hundredths place.

      Step 2. This digit is 5 or greater, so we add 1 to the hundredths place digit and drop all digits to its right.

Thus, 7.8265 rounded to the nearest hundredth is 7.83.

Practice Problem 8

Round 12.9187 to the nearest hundredth.

Example 9   Round 19.329 to the nearest tenth.

Solution: ┌ tenths place
     19.329
       ↑____ Step 1. Locate the digit to the right of the tenths place.

      Step 2. This digit is less than 5, so we drop this digit and all digits to its right.

Thus, 19.329 rounded to the nearest tenth is 19.3.

Practice Problem 9

Round 245.348 to the nearest tenth.

Answers

7. a. 43.5, **b.** 2600, **8.** 12.92, **9.** 245.3

D WRITING FRACTIONS AS DECIMALS

To write fractions as decimals, interpret the fraction bar as division and find the quotient.

> **WRITING FRACTIONS AS DECIMALS**
>
> To write fractions as decimals, divide the numerator by the denominator.

Practice Problem 10

Write $\frac{2}{5}$ as a decimal.

Example 10 Write $\frac{1}{4}$ as a decimal.

Solution:

$$
\begin{array}{r}
0.25 \\
4\,\overline{)1.00} \\
-8 \\
\hline
20 \\
-20 \\
\hline
0
\end{array}
$$

$$\frac{1}{4} = 0.25$$

Practice Problem 11

Write $\frac{5}{6}$ as a decimal.

Example 11 Write $\frac{2}{3}$ as a decimal.

Solution:

$$
\begin{array}{r}
0.666 \\
3\,\overline{)2.000} \\
-1\,8 \\
\hline
20 \\
-18 \\
\hline
20 \\
-18 \\
\hline
2
\end{array}
$$

This pattern will continue so that $\frac{2}{3} = 0.6666\ldots$

A bar can be placed over the digit 6 to indicate that it repeats.

$$\frac{2}{3} = 0.666\ldots = 0.\overline{6}$$

We can also write a decimal approximation for $\frac{2}{3}$. For example, $\frac{2}{3}$ rounded to the nearest hundredth is 0.67. This can be written as $\frac{2}{3} \approx 0.67$. The $\approx$ sign means "is approximately equal to."

TRY THE CONCEPT CHECK IN THE MARGIN.

✓ **CONCEPT CHECK**

The notation $0.5\overline{2}$ is the same as

a. $\frac{52}{100}$

b. $\frac{52\ldots}{100}$

c. $0.52222222\ldots$

Practice Problem 12

Write $\frac{1}{9}$ as a decimal. Round to the nearest thousandth.

Example 12 Write $\frac{22}{7}$ as a decimal. Round to the nearest hundredth.

(The fraction $\frac{22}{7}$ is an approximation for π.)

Answers

10. 0.4, **11.** 0.8$\overline{3}$, **12.** 0.111

✓Concept Check: c

Solution:

$$3.142 \approx 3.14$$
$$7 \overline{)22.000}$$
$$\underline{-21}$$
$$10$$
$$\underline{-7}$$
$$30$$
$$\underline{-28}$$
$$20$$
$$\underline{-14}$$
$$6$$

If rounding to the nearest hundredth, carry the division process out to one more decimal place, the thousandths place.

The fraction $\frac{22}{7}$ in decimal form is approximately 3.14.

E WRITING PERCENTS AS DECIMALS AND DECIMALS AS PERCENTS

The word **percent** comes from the Latin phrase *per centum*, which means **"per 100."** Thus, 53% means 53 per 100, or

$$53\% = \frac{53}{100}$$

When solving problems containing percents, it is often necessary to write a percent as a decimal. To see how this is done, study the chart below.

Percent	Fraction	Decimal
7%	$\frac{7}{100}$	0.07
63%	$\frac{63}{100}$	0.63
109%	$\frac{109}{100}$	1.09

To convert directly from a percent to a decimal, notice that

$$7\% = 0.07$$

TO WRITE A PERCENT AS A DECIMAL

Drop the percent symbol and move the decimal point two places to the left.

Example 13   Write each percent as a decimal.

a. 25%   **b.** 2.6%   **c.** 195%

Solution: We drop the % and move the decimal point two places to the left. Recall that the decimal point of a whole number is to the right of the ones place digit.

a. $25\% = 25.\% = 0.25$

b. $2.6\% = 02.6\% = 0.026$

c. $195\% = 195.\% = 1.95$

Practice Problem 13

Write each percent as a decimal.

a. 20%

b. 1.2%

c. 465%

Answers

13. a. 0.20, **b.** 0.12, **c.** 4.65

To write a decimal as a percent, we simply reverse the preceding steps. That is, we move the decimal point two places to the right and attach the percent symbol, %.

TO WRITE A DECIMAL AS A PERCENT

Move the decimal point two places to the right and attach the percent symbol, %.

Practice Problem 14

Write each decimal as a percent.

a. 0.42

b. 0.003

c. 2.36

d. 0.7

Example 14   Write each decimal as a percent.

   a. 0.85   **b.** 1.25   **c.** 0.012   **d.** 0.6

Solution: We move the decimal point two places to the right and attach the percent symbol, %.

   a. $0.85 = 0.85 = 85\%$

   b. $1.25 = 1.25 = 125\%$

   c. $0.012 = 0.012 = 1.2\%$

   d. $0.6 = 0.60 = 60\%$

Answers

14. a. 42%,   **b.** 0.3%,   **c.** 236%,   **d.** 70%

EXERCISE SET R.3

A *Write each decimal as a fraction. Do not simplify. See Examples 1 through 3.*

1. 7.23   **2.** 0.114   **3.** 0.239   **4.** 123.1

B *Add or subtract as indicated. See Examples 4 and 5.*

5. $2.31 + 6.4$ **6.** $32.4 + 1.58 + 0.0934$ **7.** $8.8 - 2.3$ **8.** $7.6 - 2.1$

9. $18 - 2.78$ **10.** $28 - 3.31$

11. $\begin{array}{r} 45.02 \\ 3.006 \\ +\ 8.405 \\ \hline \end{array}$

12. $\begin{array}{r} 65.0028 \\ 5.0903 \\ +\ 6.9 \\ \hline \end{array}$ **13.** $\begin{array}{r} 654.9 \\ -\ 56.67 \\ \hline \end{array}$ **14.** $\begin{array}{r} 863.2 \\ -\ 39.45 \\ \hline \end{array}$

Multiply or divide as indicated. See Examples 6 and 7.

15. $\begin{array}{r} 0.2 \\ \times\ 0.6 \\ \hline \end{array}$ **16.** $\begin{array}{r} 0.7 \\ \times\ 0.9 \\ \hline \end{array}$ **17.** $\begin{array}{r} 6.75 \\ \times\ 10 \\ \hline \end{array}$ **18.** $\begin{array}{r} 8.91 \\ \times\ 100 \\ \hline \end{array}$

19. $\begin{array}{r} 5.62 \\ \times\ 7.7 \\ \hline \end{array}$ **20.** $\begin{array}{r} 8.03 \\ \times\ 5.5 \\ \hline \end{array}$ **21.** $\begin{array}{r} 16.003 \\ \times\ 5.31 \\ \hline \end{array}$

22. $\begin{array}{r} 31.006 \\ \times\ 3.71 \\ \hline \end{array}$ **23.** $5\overline{)0.47}$ **24.** $2\overline{)11.7}$

25. $0.6\overline{)42}$ **26.** $0.9\overline{)36}$ **27.** $0.82\overline{)4.756}$

28. $0.92\overline{)3.312}$ **29.** $0.063\overline{)52.92}$ **30.** $0.054\overline{)51.84}$

1. _____
2. _____
3. _____
4. _____
5. _____
6. _____
7. _____
8. _____
9. _____
10. _____
11. _____
12. _____
13. _____
14. _____
15. _____
16. _____
17. _____
18. _____
19. _____
20. _____
21. _____
22. _____
23. _____
24. _____
25. _____
26. _____
27. _____
28. _____
29. _____
30. _____

31. _____

32. _____

33. _____

34. _____

35. _____

36. _____

37. _____

38. _____

39. _____

40. _____

41. _____

42. _____

43. _____

44. _____

45. _____

46. _____

47. _____

48. _____

49. _____

50. _____

51. _____

52. _____

53. _____

154

C *Round each decimal to the given place value. See Examples 8 and 9.*

31. 0.57, nearest tenth

32. 0.58, nearest tenth

33. 0.234, nearest hundredth

34. 0.452, nearest hundredth

35. 0.5942, nearest thousandth

36. 63.4523, nearest thousandth

37. 98,207.23, nearest tenth

38. 68,936.543, nearest tenth

39. 12.347, nearest tenth

40. 42.9878, nearest thousandth

D *Write each fraction as a decimal. If the decimal is a repeating decimal, write using the bar notation and then round to the nearest hundredth. See Examples 10 through 12.*

41. $\frac{1}{3}$

42. $\frac{7}{16}$

43. $\frac{5}{8}$

44. $\frac{6}{11}$

45. $\frac{1}{6}$

E *Write each percent as a decimal. See Example 13.*

46. 36%

47. 3.1%

48. 2.2%

49. 135%

50. 417%

51. 81.49%

52. In a recent telephone survey, approximately 61% of the respondents said that they are better off financially than their parents were at the same age. Write this percent as a decimal. (*Source: Reader's Digest*, December 1996)

53. The average one-year survival rate for a heart transplant recipient is 82.3%. The average one-year survival rate for a liver transplant patient is 81.6%. Write each percent as a decimal. (*Source*: Bureau of Health Resources Development)

Write each decimal as a percent. See Example 14.

54. 0.876   **55.** 0.521   **56.** 0.5   **57.** 0.1

 COMBINING CONCEPTS

58. The estimated life expectancy at birth for female Canadians was 82.65 years in 1996. The estimated life expectancy at birth for male Canadians was only 75.67 years in 1996. How much longer is a female Canadian born in 1996 expected to live than a male Canadian born in the same year? (*Source*: The Central Intelligence Agency, *1996 World Factbook*)

59. The chart shows the average number of pounds of meats consumed by each United States citizen in 1995. (*Source*: National Agricultural Statistics Service)

Meat	Pounds
Chicken	71.3
Turkey	17.9
Beef	67.3
Pork	52.5
Veal	1.0
Lamb/Mutton	1.2

a. How much more beef than pork did the average U.S. citizen consume in 1995?

b. How much poultry (chicken and turkey) did the average U.S. citizen consume in 1995?

c. What was the total amount of meat consumed by the average U.S. citizen in 1995?

60. An estimated $\frac{16}{25}$ of Americans own at least one credit card. What percent of Americans own credit cards? (*Sources*: *Bank Advertising News* and The Gallup Organization, 1996)

54. _____

55. _____

56. _____

57. _____

58. _____

59. a. _____

b. _____

c. _____

60. _____

Focus On History

FACTORING MACHINE

Small numbers can be broken down into their prime factors relatively easily. However, factoring larger numbers can be difficult and time-consuming. The first known successful attempt to automate the process of factoring whole numbers is credited to a French infantry officer and mathematics enthusiast, Eugène Olivier Carissan. In 1919, he designed and built a machine that uses gears and a hand crank to factor numbers.

Carissan's factoring machine had been all but forgotten after his death in 1925. In 1989, a Canadian researcher came across a description of the machine in an article printed in an obscure French journal in 1920. This led to a five-year search for traces of the machine. Eventually, the factoring machine was found in a French astronomical observatory which had received the invention from Carissan's family after his death.

Mathematical historians agree that the factoring machine was a remarkable achievement in its precomputer era. Up to 40 numbers per second could be processed by the machine while its operator turned the crank at two revolutions per minute. Carissan was able to use his machine to prove that the number 708,158,977 was prime in under 10 minutes. He could also find the prime factorizations of up to 13-digit numbers with the machine.

Chapter R Highlights

Definitions and Concepts	Examples

Section R.1   Factors and the Least Common Multiple

To **factor** means to write as a product.

When a number is written as a product of primes, this product is called the **prime factorization** of a number.

The factors of 12 are

$$1, 2, 3, 4, 6, 12$$

Write the prime factorization of 60.

$$60 = 6 \cdot 10$$
$$2 \cdot 3 \cdot 2 \cdot 5$$

The prime factorization of 60 is $2 \cdot 2 \cdot 3 \cdot 5$.

The least common multiple (LCM) of a list of numbers is the smallest number that is a multiple of all the numbers in the list.

To Find the LCM of a List of Numbers

Step 1.   Write the prime factorization of each number.
Step 2.   Write the product containing each different prime factor (from Step 1) the greatest number of times that it appears in any one factorization. This product is the LCM.

Find the LCM of 12 and 40.

$$12 = 2 \cdot 2 \cdot 3$$
$$40 = 2 \cdot 2 \cdot 2 \cdot 5$$
$$\text{LCM} = 2 \cdot 2 \cdot 2 \cdot 3 \cdot 5 = 120$$

Section R.2   Fractions

Fractions that represent the same quantity are called **equivalent fractions**.

$$\frac{1}{5} = \frac{1 \cdot 4}{5 \cdot 4} = \frac{4}{20}$$

$\frac{1}{5}$ and $\frac{4}{20}$ are equivalent fractions.

Fundamental Principle of Fractions

If a, b, and c are numbers, then

$$\frac{a}{b} = \frac{a \cdot c}{b \cdot c} \quad \text{or} \quad \frac{a \cdot c}{b \cdot c} = \frac{a}{b}$$

as long as b and c are not 0.

A fraction is **simplified** when the numerator and the denominator have no factors in common other than 1.

To simplify a fraction, factor the numerator and the denominator; then apply the fundamental principle of fractions.

$\frac{13}{17}$ is simplified.

Simplify.

$$\frac{6}{14} = \frac{2 \cdot 3}{2 \cdot 7} = \frac{3}{7}$$

SECTION R.2 (CONTINUED)

Two fractions are **reciprocals** if their product is 1. The reciprocal of $\frac{a}{b}$ is $\frac{b}{a}$, as long as a and b are not 0.

The reciprocal of $\frac{6}{25}$ is $\frac{25}{6}$.

To multiply fractions, numerator times numerator is the numerator of the product and denominator times denominator is the denominator of the product.

$$\frac{2}{5} \cdot \frac{3}{7} = \frac{6}{35}$$

To divide fractions, multiply the first fraction by the reciprocal of the second fraction.

$$\frac{5}{9} \div \frac{2}{7} = \frac{5}{9} \cdot \frac{7}{2} = \frac{35}{18}$$

To add fractions with the same denominator, add the numerators and place the sum over the common denominator.

$$\frac{5}{11} + \frac{3}{11} = \frac{8}{11}$$

To subtract fractions with the same denominator, subtract the numerators and place the difference over the common denominator.

$$\frac{13}{15} - \frac{3}{15} = \frac{10}{15} = \frac{2}{3}$$

To add or subtract fractions with different denominators, first write each fraction as an equivalent fraction with the LCD as denominator.

$$\frac{2}{9} + \frac{3}{6} = \frac{2 \cdot 2}{9 \cdot 2} + \frac{3 \cdot 3}{6 \cdot 3} = \frac{4 + 9}{18} = \frac{13}{18}$$

SECTION R.3 DECIMALS AND PERCENTS

To write decimals as fractions, use place values.

$$0.12 = \frac{12}{100}$$

TO ADD OR SUBTRACT DECIMALS

Step 1. Write the decimals so that the decimal points line up vertically.

Step 2. Add or subtract as for whole numbers.

Step 3. Place the decimal point in the sum or difference so that it lines up vertically with the decimal points in the problem.

Subtract: $2.8 - 1.04$ Add: $25 + 0.02$

$$\begin{array}{r} {}^{7\ 10} \\ 2.8\cancel{0} \\ -1.04 \\ \hline 1.76 \end{array} \qquad \begin{array}{r} 25. \\ +\ 0.02 \\ \hline 25.02 \end{array}$$

TO MULTIPLY DECIMALS

Step 1. Multiply the decimals as though they are whole numbers.

Step 2. The decimal point in the product is placed so that the number of decimal places in the product is equal to the **sum** of the number of decimal places in the factors.

Multiply: 1.48×5.9

$$\begin{array}{r} 1.4\,8 \quad \leftarrow 2 \text{ decimal places} \\ \times \quad 5.9 \quad \leftarrow 1 \text{ decimal place} \\ \hline 1\,3\,3\,2 \\ 7\,4\,0 \\ \hline 8.7\,3\,2 \quad \leftarrow 3 \text{ decimal places} \end{array}$$

SECTION R.3   (CONTINUED)

TO DIVIDE DECIMALS

Step 1.   Move the decimal point in the divisor to the right until the divisor is a whole number.

Step 2.   Move the decimal point in the dividend to the right the **same number of places** as the decimal point was moved in Step 1.

Step 3.   Divide. The decimal point in the quotient is directly over the moved decimal point in the dividend.

To write fractions as decimals, divide the numerator by the denominator.

Divide: $1.118 \div 2.6$

$$
\begin{array}{r}
0.43 \\
2.6\overline{)1.118} \\
-104 \\
\hline
78 \\
-78 \\
\hline
0
\end{array}
$$

Write $\dfrac{3}{8}$ as a decimal.

$$
\begin{array}{r}
0.375 \\
8\overline{)3.000} \\
-2\,4 \\
\hline
60 \\
-56 \\
\hline
40 \\
-40 \\
\hline
0
\end{array}
$$

To write a percent as a decimal, drop the % symbol and move the decimal point two places to the left.

To write a decimal as a percent, move the decimal point two places to the right and attach the % symbol.

$25\% = 25.\% = 0.25$

$0.7 = 0.70 = 70\%$

Real Numbers and Introduction to Algebra

CHAPTER 1

The power of mathematics is its flexibility. We apply numbers to almost every aspect of our lives. The power of algebra is its generality. In algebra, we use letters to represent numbers.

In this chapter, we begin with a review of the basic symbols—the language—of arithmetic. We then introduce the use of a variable in place of a number. From there, we translate phrases to algebraic expressions and sentences to equations. This is the beginning of problem solving, which we formally study in Chapter 2.

The stars have been a source of interest to different cultures for centuries. Polaris, the North Star, guided ancient sailors. The Egyptians honored Sirius, the brightest star in the sky, in temples. Around 150 B.C., a Greek astronomer, Hipparchus, devised a system of classifying the brightness of stars. He called the brightest stars "first magnitude" and the faintest stars "sixth magnitude." Hipparchus's system is the basis of the apparent magnitude scale used by modern astronomers. This modern scale has been modified to include negative numbers.

Problem Solving Notes

1.1   SYMBOLS AND SETS OF NUMBERS

We begin with a review of the set of natural numbers and the set of whole numbers and how we use symbols to compare these numbers. A **set** is a collection of objects, each of which is called a **member** or **element** of the set. A pair of brace symbols { } encloses the list of elements and is translated as "the set of" or "the set containing."

Objectives

A Define the meaning of the symbols $=$, $\neq$, $<$, $>$, $\leq$, and $\geq$.

B Translate sentences into mathematical statements.

C Identify integers, rational numbers, irrational numbers, and real numbers.

D Find the absolute value of a real number.

SSM   CD-ROM   Video 1.1

NATURAL NUMBERS

$$\{1, 2, 3, 4, 5, 6, \ldots\}$$

WHOLE NUMBERS

$$\{0, 1, 2, 3, 4, 5, 6, \ldots\}$$

⌐ HELPFUL HINT

The three dots (an ellipsis) at the end of the list of elements of a set means that the list continues in the same manner indefinitely.

These numbers can be pictured on a **number line**. To draw a number line, first draw a line. Choose a point on the line and label it 0. To the right of 0, label any other point 1. Being careful to use the same distance as from 0 to 1, mark off equally spaced distances. Label these points 2, 3, 4, 5, and so on. Since the whole numbers continue indefinitely, it is not possible to show every whole number on the number line. The arrow at the right end of the line indicates that the pattern continues indefinitely.

A EQUALITY AND INEQUALITY SYMBOLS

Picturing natural numbers and whole numbers on a number line helps us to see the order of the numbers. Symbols can be used to describe in writing the order of two quantities. We will use equality symbols and inequality symbols to compare quantities.

Below is a review of these symbols. The letters a and b are used to represent quantities. Letters such as a and b that are used to represent numbers or quantities are called **variables**.

		MEANING
Equality symbol:	$a = b$	a is equal to b.
Inequality symbols:	$a \neq b$	a is not equal to b.
	$a < b$	a is less than b.
	$a > b$	a is greater than b.
	$a \leq b$	a is less than or equal to b.
	$a \geq b$	a is greater than or equal to b.

These symbols may be used to form **mathematical statements** such as

$$2 = 2 \quad \text{and} \quad 2 \neq 6$$

On the number line, we see that a number **to the right of** another number is **larger**. Similarly, a number **to the left of** another number is smaller. For example, 3 is to the left of 5 on the number line, which means that 3 is less than 5, or $3 < 5$. Similarly, 2 is to the right of 0 on the number line, which means 2 is greater than 0, or $2 > 0$. Since 0 is to the left of 2, we can also say that 0 is less than 2, or $0 < 2$.

 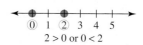

3 < 5   2 > 0 or 0 < 2

> **HELPFUL HINT**
>
> Notice that $2 > 0$ has exactly the same meaning as $0 < 2$. Switching the order of the numbers and reversing the "direction of the inequality symbol" does not change the meaning of the statement.
>
> $5 > 3$ has the same meaning as $3 < 5$.
>
> Also notice that when the statement is true, the inequality arrow points to the smaller number.

Examples   Determine whether each statement is true or false.

1. $2 < 3$ True.   Since 2 is to the left of 3 on the number line

2. $72 > 27$ True.   Since 72 is to the right of 27 on the number line

3. $8 \geq 8$ True.   Since 8 = 8 is true

4. $8 \leq 8$ True.   Since 8 = 8 is true

5. $23 \leq 0$ False.   Since neither 23 < 0 nor 23 = 0 is true

6. $0 \leq 23$ True.   Since 0 < 23 is true

> **HELPFUL HINT**
>
> If either $3 < 3$ or $3 = 3$ is true, then $3 \leq 3$ is true.

B TRANSLATING SENTENCES INTO MATHEMATICAL STATEMENTS

Now, let's use the symbols discussed above to translate sentences into mathematical statements.

Example 7   Translate each sentence into a mathematical statement.

a. Nine is less than or equal to eleven.
b. Eight is greater than one.
c. Three is not equal to four.

Solution:

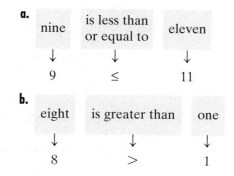

a.

nine	is less than or equal to	eleven
↓	↓	↓
9	≤	11

b.

eight	is greater than	one
↓	↓	↓
8	>	1

c.

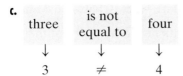

three	is not equal to	four
↓	↓	↓
3	≠	4

C IDENTIFYING COMMON SETS OF NUMBERS

Whole numbers are not sufficient to describe many situations in the real world. For example, quantities smaller than zero must sometimes be represented, such as temperatures, less than 0 degrees.

We can place numbers less than zero on the number line as follows: Numbers less than 0 are to the left of 0 and are labeled $-1, -2, -3$, and so on. The numbers we have labeled on the number line below are called the set of **integers**.

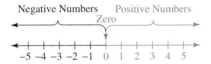

Integers to the left of 0 are called **negative integers**; integers to the right of 0 are called **positive integers**. The integer 0 is neither positive nor negative.

INTEGERS

$$\{\ldots, -3, -2, -1, 0, 1, 2, 3, \ldots\}$$

Example 8 Use an integer to express the number in the following. "Pole of Inaccessibility, Antarctica, is the coldest location in the world, with an average annual temperature of 72 degrees below zero." (*Source: The Guinness Book of Records*)

Solution: The integer -72 represents 72 degrees below zero.

A problem with integers in real-life settings arises when quantities are smaller than some integer but greater than the next smallest integer. On the number line, these quantities may be visualized by points between integers. Some of these quantities between integers can be represented as a quotient of integers. For example,

The point on the number line halfway between 0 and 1 can be represented by $\frac{1}{2}$, a quotient of integers.

Practice Problem 8

Use an integer to express the number in the following. "The lowest altitude in North America is found in Death Valley, California. Its altitude is 282 feet below sea level." (*Source: The World Almanac, 1997*)

Answer

8. -282

Practice Problem 9

Graph the numbers on the number
line.

$$-2.5, \quad -\frac{2}{3}, \quad \frac{1}{5}, \quad \frac{5}{4}, \quad 2.25$$

```
◄──┼──┼──┼──┼──┼──┼──┼──┼──┼──┼──►
  -5 -4 -3 -2 -1  0  1  2  3  4  5
```

1 unit

$\sqrt{2}$ units

Answer

9.
```
     -2.5  -2/3  1/5  5/4  2.25
◄──┼──┼──┼──┼──┼──┼──┼──┼──┼──┼──►
  -5 -4 -3 -2 -1  0  1  2  3  4  5
```

The point on the number line halfway between 0 and −1 can be represented by $-\frac{1}{2}$. Other quotients of integers and their graphs are shown in the margin.

These numbers, each of which can be represented as a quotient of integers, are examples of rational numbers. It's not possible to list the set of rational numbers using the notation that we have been using. For this reason, we will use a different notation.

┌─ **RATIONAL NUMBERS** ──────────────────────────┐

$$\left\{ \frac{a}{b} \,\middle|\, a \text{ and } b \text{ are integers and } b \neq 0 \right\}$$

└──┘

We read this set as "the set of numbers $\frac{a}{b}$ such that a and b are integers and **b is not 0.**"

Notice that every integer is also a rational number since each integer can be written as a quotient of integers. For example, the integer 5 is also a rational number since $5 = \frac{5}{1}$. In this rational number, $\frac{5}{1}$, recall that the top number, 5, is called the numerator and the bottom number, 1, is called the denominator.

Let's practice graphing numbers on a number line.

Example 9   Graph the numbers on the number line.

$$-\frac{4}{3}, \quad \frac{1}{4}, \quad \frac{3}{2}, \quad 2\frac{1}{8}, \quad 3.5$$

Solution: To help graph the improper fractions in the list, we first write them as mixed numbers.

```
  -4/3 or -1 1/3   1/4   3/2 or 1 1/2  2 1/8    3.5
◄──┼────●────┼──────●────┼───●───┼──●──┼──●──┼──►
  -2       -1       0    1       2     3     4
```

Every rational number has a point on the number line that corresponds to it. But not every point on the number line corresponds to a rational number. Those points that do not correspond to rational numbers correspond instead to **irrational numbers**.

┌─ **IRRATIONAL NUMBERS** ────────────────────────────┐

{nonrational numbers that correspond to points on the number line}

└──┘

An irrational number that you have probably seen is π. Also, $\sqrt{2}$, the length of the diagonal of the square shown, is an irrational number.

Both rational and irrational numbers can be written as decimal numbers. The decimal equivalent of a rational number will either terminate or repeat in a pattern. For example, upon dividing we find that

$$\frac{3}{4} = 0.75 \qquad \text{(decimal number terminates or ends)}$$

$$\frac{2}{3} = 0.66666\ldots \qquad \text{(decimal number repeats in a pattern)}$$

The decimal representation of an irrational number will neither terminate nor repeat. (For further review of decimals, see Section R.3.)

The set of numbers, each of which corresponds to a point on the number line, is called the set of **real numbers**. One and only one point on the number line corresponds to each real number.

REAL NUMBERS

{numbers that correspond to points on the number line}

Several different sets of numbers have been discussed in this section. The following diagram shows the relationships among these sets of real numbers. Notice that, together, the rational numbers and the irrational numbers make up the real numbers.

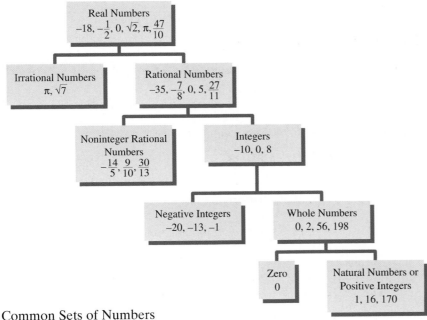

Common Sets of Numbers

Example 10   Given the set $\left\{-2, 0, \frac{1}{4}, 112, -3, 11, \sqrt{2}\right\}$, list the numbers in this set that belong to the set of:

    a. Natural numbers   **b.** Whole numbers
    c. Integers   **d.** Rational numbers
    e. Irrational numbers   **f.** Real numbers

Solution: **a.** The natural numbers are 11 and 112.
    b. The whole numbers are 0, 11, and 112.
    c. The integers are $-3, -2, 0, 11,$ and 112.
    d. Recall that integers are rational numbers also. The rational numbers are $-3, -2, 0, \frac{1}{4}, 11,$ and 112.
    e. The irrational number is $\sqrt{2}$.
    f. The real numbers are all numbers in the given set.

Practice Problem 10

Given the set $\left\{-100, -\frac{2}{5}, 0, \pi, 6, 913\right\}$, list the numbers in this set that belong to the set of:

a. Natural numbers

b. Whole numbers

c. Integers

d. Rational numbers

e. Irrational numbers

f. Real numbers

Answers

10. a. $6, 913,$ **b.** $0, 6, 913,$ **c.** $-100, 0, 6, 913,$
d. $-100, -\frac{2}{5}, 0, 6, 913,$ **e.** $\pi,$ **f.** all numbers in the given set

D FINDING THE ABSOLUTE VALUE OF A NUMBER

The number line not only gives us a picture of the real numbers, it also helps us visualize the distance between numbers. The distance between a real number a and 0 is given a special name called the **absolute value** of a. "The absolute value of a" is written in symbols as $|a|$.

> ### ABSOLUTE VALUE
>
> The **absolute value** of a real number a, denoted by $|a|$, is the distance between a and 0 on a number line.

For example, $|3| = 3$ and $|-3| = 3$ since both 3 and -3 are a distance of 3 units from 0 on the number line.

> **HELPFUL HINT**
>
> Since $|a|$ is a distance, $|a|$ is always either positive or 0, never negative. That is, **for any real number a, $|a| \geq 0$.**

Practice Problem 11

Find the absolute value of each number.

a. $|7|$,   b. $|-8|$,   c. $\left|-\frac{2}{3}\right|$

Example 11   Find the absolute value of each number.

a. $|4|$ b. $|-5|$ c. $|0|$

Solution: a. $|4| = 4$ since 4 is 4 units from 0 on the number line.
b. $|-5| = 5$ since -5 is 5 units from 0 on the number line.
c. $|0| = 0$ since 0 is 0 units from 0 on the number line.

Practice Problem 12

Insert $<$, $>$, or $=$ in the appropriate space to make each statement true.

a. $|-4|$ 4,   b. -3 $|0|$,
c. $|-2.7|$ $|-2|$,   d. $|6|$ $|16|$,
e. $|-6|$ $|-16|$

Example 12   Insert $<$, $>$, or $=$ in the appropriate space to make each statement true.

a. $|0|$ 2   b. $|-5|$ 5   c. $|-3|$ $|-2|$
d. $|5|$ $|6|$ e. $|-7|$ $|6|$

Solution: a. $|0| < 2$ since $|0| = 0$ and $0 < 2$.
b. $|-5| = 5$.
c. $|-3| > |-2|$ since $3 > 2$.
d. $|5| < |6|$ since $5 < 6$.
e. $|-7| > |6|$ since $7 > 6$.

1.2 INTRODUCTION TO VARIABLE EXPRESSIONS AND EQUATIONS

A EXPONENTS AND THE ORDER OF OPERATIONS

Frequently in algebra, products occur that contain repeated multiplication of the same factor. For example, the volume of a cube whose sides each measure 2 centimeters is $(2 \cdot 2 \cdot 2)$ cubic centimeters. We may use **exponential notation** to write such products in a more compact form. For example.

$$2 \cdot 2 \cdot 2 \quad \text{may be written as} \quad 2^3$$

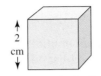

2
cm

Volume is $(2 \cdot 2 \cdot 2)$
cubic centimeters.

The 2 in 2^3 is called the **base;** it is the repeated factor. The 3 in 2^3 is called the **exponent** and is the number of times the base is used as a factor. The expression 2^3 is called an **exponential expression**.

$$\overset{\text{exponent}}{2^3} = 2 \cdot 2 \cdot 2 = 8$$

base────↑   2 is a factor 3 times.

Example 1   Evaluate (find the value of) each expression.

   a. 3^2 [read as "3 squared" or as "3 to the second power"]
   b. 5^3 [read as "5 cubed" or as "5 to the third power"]
   c. 2^4 [read as "2 to the fourth power"]
   d. 7^1 **e.** $\left(\dfrac{3}{7}\right)^2$

Solution: **a.** $3^2 = 3 \cdot 3 = 9$
   b. $5^3 = 5 \cdot 5 \cdot 5 = 125$
   c. $2^4 = 2 \cdot 2 \cdot 2 \cdot 2 = 16$
   d. $7^1 = 7$
   e. $\left(\dfrac{3}{7}\right)^2 = \left(\dfrac{3}{7}\right)\left(\dfrac{3}{7}\right) = \dfrac{3 \cdot 3}{7 \cdot 7} = \dfrac{9}{49}$

HELPFUL HINT

$2^3 \neq 2 \cdot 3$ since 2^3 indicates repeated **multiplication** of the same factor.
   $2^3 = 2 \cdot 2 \cdot 2 = 8$, whereas $2 \cdot 3 = 6$

Using symbols for mathematical operations is a great convenience. The more operation symbols presented in an expression, the more careful we must be when performing the indicated operation. For example, in the expression $2 + 3 \cdot 7$, do we add first or multiply first? To eliminate confusion, **grouping symbols** are used. Examples of grouping symbols are parentheses ( ), brackets [ ], braces { }, and the fraction bar. If we wish $2 + 3 \cdot 7$ to be simplified by adding first, we enclose $2 + 3$ in parentheses.

$$(2 + 3) \cdot 7 = 5 \cdot 7 = 35$$

Objectives

A Define and use exponents and the order of operations.

B Evaluate algebraic expressions, given replacement values for variables.

C Determine whether a number is a solution of a given equation.

D Translate phrases into expressions and sentences into equations.

SSM   CD-ROM   Video
1.2

Practice Problem 1

Evaluate each expression.

a. 4^2

b. 2^2

c. 3^4

d. 9^1

e. $\left(\dfrac{2}{5}\right)^2$

If we wish to multiply first, $3 \cdot 7$ may be enclosed in parentheses.

$$2 + (3 \cdot 7) = 2 + 21 = 23$$

To eliminate confusion when no grouping symbols are present, we use the following agreed-upon order of operations.

ORDER OF OPERATIONS

Simplify expressions using the following order. If grouping symbols such as parentheses are present, simplify expressions within those first, starting with the innermost set. If fraction bars are present, simplify the numerator and the denominator separately.

1. Evaluate exponential expressions.
2. Perform multiplications or divisions in order from left to right.
3. Perform additions or subtractions in order from left to right.

Using this order of operations, we now simplify $2 + 3 \cdot 7$. There are no grouping symbols and no exponents, so we multiply and then add.

$$2 + 3 \cdot 7 = 2 + 21 \qquad \text{Multiply.}$$
$$= 23 \qquad \text{Add.}$$

Practice Problems 2–4

Simplify each expression.

2. $3 + 2 \cdot 4^2$

3. $\dfrac{9}{5} \cdot \dfrac{1}{3} - \dfrac{1}{3}$

4. $8[2(6 + 3) - 9]$

Examples   Simplify each expression.

2. $6 \div 3 + 5^2 = 6 \div 3 + 25 \qquad$ Evaluate 5^2.
$ = 2 + 25 \qquad$ Divide.
$ = 27 \qquad$ Add.

3. $\dfrac{3}{2} \cdot \dfrac{1}{2} - \dfrac{1}{2} = \dfrac{3}{4} - \dfrac{1}{2} \qquad$ Multiply.

$\phantom{\dfrac{3}{2} \cdot \dfrac{1}{2} - \dfrac{1}{2}} = \dfrac{3}{4} - \dfrac{2}{4} \qquad$ The least common denominator is 4.

$\phantom{\dfrac{3}{2} \cdot \dfrac{1}{2} - \dfrac{1}{2}} = \dfrac{1}{4} \qquad$ Subtract.

4. $3[4(5 + 2) - 10] = 3[4(7) - 10] \qquad$ Simplify the expression in parentheses. They are the innermost grouping symbols.
$ = 3[28 - 10] \qquad$ Multiply 4 and 7.
$ = 3[18] \qquad$ Subtract inside the brackets.
$ = 54 \qquad$ Multiply.

In the next example, the fraction bar serves as a grouping symbol and separates the numerator and denominator. Simplify each separately.

Practice Problem 5

Simplify.

$$\dfrac{1 + |7 - 4| + 3^2}{8 - 5}$$

Example 5   Simplify: $\dfrac{3 + |4 - 3| + 2^2}{6 - 3}$

Solution:

$$\dfrac{3 + |4 - 3| + 2^2}{6 - 3} = \dfrac{3 + |1| + 2^2}{6 - 3} \qquad \text{Simplify the expression inside the absolute value bars.}$$

$$= \dfrac{3 + 1 + 2^2}{3} \qquad \text{Find the absolute value and simplify the denominator.}$$

$$= \dfrac{3 + 1 + 4}{3} \qquad \text{Evaluate the exponential expression.}$$

$$= \dfrac{8}{3} \qquad \text{Simplify the numerator.}$$

Answers

2. 35, **3.** $\dfrac{4}{15}$, **4.** 72, **5.** $\dfrac{13}{3}$

Be careful when evaluating an exponential expression.

$$3 \cdot 4^2 = 3 \cdot 16 = 48 \qquad (3 \cdot 4)^2 = (12)^2 = 144$$

↑   ↑

Base is 4.   Base is $3 \cdot 4$.

B EVALUATING ALGEBRAIC EXPRESSIONS

An **algebraic expression** is a collection of numbers, variables, operation symbols, and grouping symbols. For example,

$$2x, \qquad -3, \qquad 2x - 10, \qquad 5(p^2 + 1), \qquad \text{and} \qquad \frac{3y^2 - 6y + 1}{5}$$

are algebraic expressions. The expression $2x$ means $2 \cdot x$. Also, $5(p^2 + 1)$ means $5 \cdot (p^2 + 1)$ and $3y^2$ means $3 \cdot y^2$. If we give a specific value to a variable, we can **evaluate an algebraic expression**. To evaluate an algebraic expression means to find its numerical value once we know the value of the variables.

Algebraic expressions often occur during problem solving. For example, the expression

$$16t^2$$

gives the distance in feet (neglecting air resistance) that an object will fall in t seconds. (See Exercise 63 in this section.)

Example 6   Evaluate each expression when $x = 3$ and $y = 2$.

   a. $2x - y$ **b.** $\dfrac{3x}{2y}$ **c.** $\dfrac{x}{y} + \dfrac{y}{2}$ **d.** $x^2 - y^2$

Solution: **a.** Replace x with 3 and y with 2.

$$2x - y = 2(3) - 2 \quad \text{Let } x = 3 \text{ and } y = 2.$$
$$= 6 - 2 \qquad \text{Multiply.}$$
$$= 4 \qquad \text{Subtract.}$$

b. $\dfrac{3x}{2y} = \dfrac{3 \cdot 3}{2 \cdot 2} = \dfrac{9}{4} \quad \text{Let } x = 3 \text{ and } y = 2.$

c. Replace x with 3 and y with 2. Then simplify.

$$\frac{x}{y} + \frac{y}{2} = \frac{3}{2} + \frac{2}{2} = \frac{5}{2}$$

d. Replace x with 3 and y with 2.

$$x^2 - y^2 = 3^2 - 2^2 = 9 - 4 = 5$$

Practice Problem 6

Evaluate each expression when $x = 1$ and $y = 4$.

a. $2y - x$

b. $\dfrac{8x}{3y}$

c. $\dfrac{x}{y} + \dfrac{5}{y}$

d. $y^2 - x^2$

Answers

6. a. 7, **b.** $\dfrac{2}{3}$, **c.** $\dfrac{3}{2}$, **d.** 15

C SOLUTIONS OF EQUATIONS

Many times a problem-solving situation is modeled by an equation. An **equation** is a mathematical statement that two expressions have equal value. The equal symbol "=" is used to equate the two expressions. For example, $3 + 2 = 5$, $7x = 35$, $\frac{2(x-1)}{3} = 0$, and $I = PRT$ are all equations.

HELPFUL HINT

An equation contains the equal symbol "=". An algebraic expression does not.

TRY THE CONCEPT CHECK IN THE MARGIN.

✓ **CONCEPT CHECK**

Which of the following are equations? Which are expressions?

a. $5x = 8$
b. $5x - 8$
c. $12y + 3x$
d. $12y = 3x$

When an equation contains a variable, deciding which values of the variable make an equation a true statement is called **solving** an equation for the variable. A **solution** of an equation is a value for the variable that makes the equation true. For example, 3 is a solution of the equation $x + 4 = 7$, because if x is replaced with 3 the statement is true.

$$x + 4 = 7$$
$$\downarrow$$
$$3 + 4 \stackrel{?}{=} 7 \qquad \text{Replace } x \text{ with 3.}$$
$$7 = 7 \qquad \text{True.}$$

Similarly, 1 is not a solution of the equation $x + 4 = 7$, because $1 + 4 = 7$ is **not** a true statement.

Practice Problem 7

Decide whether 3 is a solution of $5x - 10 = x + 2$.

Example 7   Decide whether 2 is a solution of $3x + 10 = 8x$.

Solution: Replace x with 2 and see if a true statement results.

$$3x + 10 = 8x \qquad \text{Original equation}$$
$$3(2) + 10 \stackrel{?}{=} 8(2) \qquad \text{Replace } x \text{ with 2.}$$
$$6 + 10 \stackrel{?}{=} 16 \qquad \text{Simplify each side.}$$
$$16 = 16 \qquad \text{True.}$$

Since we arrived at a true statement after replacing x with 2 and simplifying both sides of the equation, 2 is a solution of the equation. ▄

D TRANSLATING WORDS TO SYMBOLS

Now that we know how to represent an unknown number by a variable, let's practice translating phrases into algebraic expressions and sentences into equations. Oftentimes solving problems involves the ability to translate word phrases and sentences into symbols. Below is a list of key words and phrases to help us translate.

Addition $(+)$	Subtraction $(-)$	Multiplication $(\cdot)$	Division $(\div)$	Equality $(=)$
Sum	Difference of	Product	Quotient	Equals
Plus	Minus	Times	Divide	Gives
Added to	Subtracted from	Multiply	Into	Is/was/ should be
More than	Less than	Twice	Ratio	Yields
Increased by	Decreased by	Of	Divided by	Amounts to
Total	Less			Represents
				Is the same as

Example 8   Write an algebraic expression that represents each phrase. Let the variable x represent the unknown number.

    a. The sum of a number and 3
    b. The product of 3 and a number
    c. Twice a number
    d. 10 decreased by a number
    e. 5 times a number increased by 7

Solution: **a.** $x + 3$ since "sum" means to add
    b. $3 \cdot x$ and $3x$ are both ways to denote the product of 3 and x
    c. $2 \cdot x$ or $2x$
    d. $10 - x$ because "decreased by" means to subtract
    e. $5x + 7$
        5 times
        a number

HELPFUL HINT

Make sure you understand the difference when translating phrases containing "decreased by," "subtracted from" and "less than."

Phrase	Translation	
A number decreased by 10	$x - 10$	
A number subtracted from 10	$10 - x$	Notice the order.
10 less than a number	$x - 10$	

Now let's practice translating sentences into equations.

Example 9   Write each sentence as an equation. Let x represent the unknown number.

    a. The quotient of 15 and a number is 4.
    b. Three subtracted from 12 is a number.
    c. Four times a number added to 17 is 21.

Solution: **a.** In words: the quotient of 15 and a number   is   4

            ↓   ↓ ↓

        Translate: $\dfrac{15}{x}$ $=$ 4

    b. In words: three subtracted **from** 12 is a number

                   ↓   ↓   ↓

        Translate: $12 - 3$ $=$ x

    Care must be taken when the operation is subtraction. The expression $3 - 12$ would be incorrect. Notice that $3 - 12 \neq 12 - 3$.

    c. In words: four times a number added to 17 is 21

                ↓   ↓   ↓ ↓ ↓

        Translate: $4x$ $+$ $17 = 21$

Practice Problem 8

Write an algebraic expression that represents each phrase. Let the variable x represent the unknown number.

a. The product of a number and 5

b. A number added to 7

c. Three times a number

d. A number subtracted from 8

e. Twice a number plus 1

Practice Problem 9

Write each sentence as an equation. Let x represent the unknown number.

a. The product of a number and 6 is 24.

b. The difference of 10 and a number is 18.

c. Twice a number decreased by 1 is 99.

Answers
8. a. $5x$, **b.** $7 + x$, **c.** $3x$, **d.** $8 - x$, **e.** $2x + 1$, **9. a.** $6x = 24$, **b.** $10 - x = 18$, **c.** $2x - 1 = 99$

CALCULATOR EXPLORATIONS
EXPONENTS

To evaluate exponential expressions on a calculator, find the key marked $\boxed{y^x}$ or $\boxed{\wedge}$. To evaluate, for example, 3^5, press the following keys: $\boxed{3}\ \boxed{y^x}\ \boxed{5}\ \boxed{=}$ or $\boxed{3}\ \boxed{\wedge}\ \boxed{5}\ \boxed{=}$.

⇕ or
$\boxed{\text{ENTER}}$

The display should read $\boxed{\qquad 243}$.

ORDER OF OPERATIONS

Some calculators follow the order of operations, and others do not. To see whether or not your calculator has the order of operations built in, use your calculator to find $2 + 3 \cdot 4$. To do this, press the following sequence of keys:

$\boxed{2}\ \boxed{+}\ \boxed{3}\ \boxed{\times}\ \boxed{4}\ \boxed{=}$.

⇕ or
$\boxed{\text{ENTER}}$

The correct answer is 14 because the order of operations is to multiply before we add. If the calculator displays $\boxed{\qquad 14}$, then it has the order of operations built in.

Even if the order of operations is built in, parentheses must sometimes be inserted. For example, to simplify $\dfrac{5}{12 - 7}$, press the keys

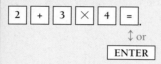

⇕ or
$\boxed{\text{ENTER}}$

The display should read $\boxed{\qquad 1}$.

Use a calculator to evaluate each expression.

1. 5^3

2. 7^4

3. 9^5

4. 8^6

5. $2(20 - 5)$

6. $3(14 - 7) + 21$

7. $24(862 - 455) + 89$

8. $99 + (401 + 962)$

9. $\dfrac{4623 + 129}{36 - 34}$

10. $\dfrac{956 - 452}{89 - 86}$

EXERCISE SET 1.2

B *Evaluate each expression when x = 1, y = 3, and z = 5. See Example 6.*

1. $3y$

2. $4x$

3. $\dfrac{z}{5x}$

4. $\dfrac{y}{2z}$

5. $3x - 2$

6. $6y - 8$

7. $|2x + 3y|$

8. $|5z - 2y|$

9. $xy + z$

10. $yz - x$

11. $5y^2$

12. $2z^2$

1. _____

2. _____

3. _____

4. _____

5. _____

6. _____

7. _____

8. _____

9. _____

10. _____

11. _____

12. _____

Problem Solving Notes

1.3 ADDING REAL NUMBERS

Real numbers can be added, subtracted, multiplied, divided, and raised to powers, just as whole numbers can.

A ADDING REAL NUMBERS

To begin, we will use the number line to help picture the addition of real numbers.

Example 1   Add: $3 + 2$

Solution: We start at 0 on a number line, and draw an arrow representing 3. This arrow is three units long and points to the right since 3 is positive. From the tip of this arrow, we draw another arrow representing 2. The number below the tip of this arrow is the sum, 5.

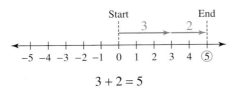

$$3 + 2 = 5$$

Example 2   Add: $-1 + (-2)$

Solution: We start at 0 on a number line, and draw an arrow representing -1. This arrow is one unit long and points to the left since -1 is negative. From the tip of this arrow, we draw another arrow representing -2. The number below the tip of this arrow is the sum, -3.

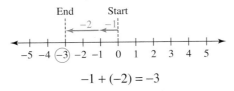

$$-1 + (-2) = -3$$

Thinking of integers as money earned or lost might help make addition more meaningful. Earnings can be thought of as positive numbers. If $1 is earned and later another $3 is earned, the total amount earned is $4. In other words, $1 + 3 = 4$.

On the other hand, losses can be thought of as negative numbers. If $1 is lost and later another $3 is lost, a total of $4 is lost. In other words, $(-1) + (-3) = -4$.

In Examples 1 and 2, we added numbers with the same sign. Adding numbers whose signs are not the same can be pictured on a number line also.

Example 3   Add: $-4 + 6$

Solution:

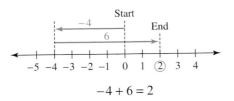

$$-4 + 6 = 2$$

Objectives

A Add real numbers.
B Solve problems that involve addition of real numbers.
C Find the opposite of a number.

SSM   CD-ROM   Video
1.3

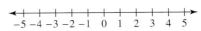
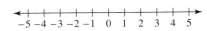

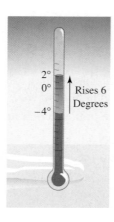

Practice Problem 4

Add using a number line: $5 + (-4)$

$$\xleftarrow{\hspace{0.3cm}}\overset{+\;+\;+\;+\;+\;+\;+\;+\;+\;+\;+}{\underset{-5\;-4\;-3\;-2\;-1\;\;\;0\;\;\;1\;\;\;2\;\;\;3\;\;\;4\;\;\;5}{}}\xrightarrow{\hspace{0.3cm}}$$

Practice Problem 5

Add without using a number line:
$(-8) + (-5)$

Practice Problem 6

Add without using a number line:
$(-14) + 6$

Practice Problems 7–12

Add without using a number line.

7. $(-17) + (-10)$

8. $(-4) + 12$

9. $1.5 + (-3.2)$

10. $-\dfrac{6}{11} + \left(-\dfrac{3}{11}\right)$

11. $12.8 + (-3.6)$

12. $-\dfrac{4}{5} + \dfrac{2}{3}$

Practice Problem 13

Find each sum.

a. $16 + (-9) + (-9)$

b. $[3 + (-13)] + [-4 + (-7)]$

Answers

4. 1, **5.** -13, **6.** -8, **7.** -27, **8.** 8,
9. -1.7, **10.** $-\dfrac{9}{11}$, **11.** 9.2, **12.** $-\dfrac{2}{15}$,
13. a. -2, **b.** -21

Using temperature as an example, if the thermometer registers 4 degrees below 0 degrees and then rises 6 degrees, the new temperature is 2 degrees above 0 degrees. Thus, it is reasonable that $-4 + 6 = 2$.

Example 4   Add: $4 + (-6)$

Solution:

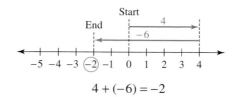

$$4 + (-6) = -2$$

Using a number line each time we add two numbers can be time consuming. Instead, we can notice patterns in the previous examples and write rules for adding real numbers.

> **ADDING REAL NUMBERS**
>
> To add two real numbers
>
> 1. with the *same sign*, add their absolute values. Use their common sign as the sign of the answer.
> 2. with *different signs*, subtract their absolute values. Give the answer the same sign as the number with the larger absolute value.

Example 5   Add without using a number line: $(-7) + (-6)$

Solution: Here, we are adding two numbers with the same sign.

$$(-7) + (-6) = -13$$
↑ sum of absolute values
same sign

Example 6   Add without using a number line: $(-10) + 4$

Solution: Here, we are adding two numbers with different signs.

$$(-10) + 4 = -6$$
↑ difference of absolute values
sign of number with larger absolute value, (-10)

Examples   Add without using a number line.

7. $(-8) + (-11) = -19$

8. $(-2) + 10 = 8$

9. $0.2 + (-0.5) = -0.3$

10. $-\dfrac{7}{10} + \left(-\dfrac{1}{10}\right) = -\dfrac{8}{10} = -\dfrac{4}{5}$

11. $11.4 + (-4.7) = 6.7$

12. $-\dfrac{3}{8} + \dfrac{2}{5} = -\dfrac{15}{40} + \dfrac{16}{40} = \dfrac{1}{40}$

Example 13   Find each sum.

a. $3 + (-7) + (-8)$

b. $[7 + (-10)] + [-2 + (-4)]$

Solution: **a.** Perform the additions from left to right.

$$3 + (-7) + (-8) = -4 + (-8)$$

Adding numbers with different signs

$$= -12$$

Adding numbers with like signs

HELPFUL HINT:
Don't forget that brackets are grouping symbols. We simplify within them first.

b. Simplify inside the brackets first.

$$[7 + (-10)] + [-2 + (-4)] = [-3] + [-6]$$
$$= -9 \quad \text{Add.} \quad \blacksquare$$

B SOLVING PROBLEMS THAT INVOLVE ADDITION

Positive and negative numbers are used in everyday life. Stock market returns show gains and losses as positive and negative numbers. Temperatures in cold climates often dip into the negative range, commonly referred to as "below zero" temperatures. Bank statements report deposits and withdrawals as positive and negative numbers.

Example 14   Calculating Gain or Loss

During a three-day period, a share of Lamplighter's International stock recorded the following gains and losses:

Monday	Tuesday	Wednesday
a gain of $2	a loss of $1	a loss of $3

Find the overall gain or loss for the stock for the three days.

Solution: Gains can be represented by positive numbers. Losses can be represented by negative numbers. The overall gain or loss is the sum of the gains and losses.

In words: gain plus loss plus loss
          ↓   ↓   ↓   ↓   ↓
Translate: 2   +   (−1) +   (−3) = −2

The overall loss is $2. ■

C FINDING OPPOSITES

To help us subtract real numbers in the next section, we first review what we mean by opposites. To help us, the graphs of 4 and −4 are shown on the number line below.

Notice that the graph of 4 and −4 lie on opposite sides of 0, and each is 4 units away from 0. Such numbers are known as **opposites** or **additive inverses** of each other.

Practice Problem 14

During a four-day period, a share of Walco stock recorded the following gains and losses:

Tuesday	Wednesday
a loss of $2	a loss of $1
Thursday	Friday
a gain of $3	a gain of $3

Find the overall gain or loss for the stock for the four days.

Answer

14. a gain of $3

OPPOSITE OR ADDITIVE INVERSE

Two numbers that are the same distance from 0 but lie on opposite sides of 0 are called **opposites** or **additive inverses** of each other.

Practice Problems 15–18

Find the opposite of each number.

15. -35

16. 12

17. $-\dfrac{3}{11}$

18. 1.9

Examples   Find the opposite of each number.

15. 10 The opposite of 10 is -10.

16. -3 The opposite of -3 is 3.

17. $\dfrac{1}{2}$ The opposite of $\dfrac{1}{2}$ is $-\dfrac{1}{2}$.

18. -4.5 The opposite of -4.5 is 4.5.   ▬▬▬

We use the symbol " $-$ " to represent the phrase "the opposite of" or "the additive inverse of." In general, if a is a number, we write the opposite or additive inverse of a as $-a$. We know that the opposite of -3 is 3. Notice that this translates as

$$\underset{\downarrow}{\text{the opposite of}} \quad \underset{\downarrow}{-3} \quad \underset{\downarrow}{\text{is}} \quad \underset{\downarrow}{3}$$

$$- \qquad (-3) \quad = \quad 3$$

This is true in general.

> If a is a number, then $-(-a) = a$.

Practice Problem 19

Simplify each expression.

a. $-(-22)$

b. $-\left(-\dfrac{2}{7}\right)$

c. $-(-x)$

d. $-|-14|$

Example 19   Simplify each expression.

a. $-(-10)$ b. $-\left(-\dfrac{1}{2}\right)$

c. $-(-2x)$ d. $-|-6|$

Solution: a. $-(-10) = 10$ b. $-\left(-\dfrac{1}{2}\right) = \dfrac{1}{2}$

c. $-(-2x) = 2x$

d. Since $|-6| = 6$, then $-|-6| = -6$.   ▬▬▬

Let's discover another characteristic about opposites. Notice that the sum of a number and its opposite is 0.

$$10 + (-10) = 0$$
$$-3 + 3 = 0$$
$$\dfrac{1}{2} + \left(-\dfrac{1}{2}\right) = 0$$

In general, we can write the following:

> The sum of a number a and its opposite $-a$ is 0.
> $$a + (-a) = 0$$

Notice that this means that the opposite of 0 is then 0 since $0 + 0 = 0$.

EXERCISE SET 1.3

A *Add. See Examples 1 through 13.*

1. $6 + 3$

2. $9 + (-12)$

3. $-6 + (-8)$

4. $-6 + (-14)$

5. $8 + (-7)$

6. $6 + (-4)$

7. $-14 + 2$

8. $-10 + 5$

9. $-2 + (-3)$

10. $-7 + (-4)$

11. $-9 + (-3)$

12. $7 + (-5)$

13. $-7 + 3$

14. $-5 + 9$

15. $10 + (-3)$

16. $8 + (-6)$

17. $5 + (-7)$

18. $3 + (-6)$

19. $-16 + 16$

20. $23 + (-23)$

ANSWERS

1. _____

2. _____

3. _____

4. _____

5. _____

6. _____

7. _____

8. _____

9. _____

10. _____

11. _____

12. _____

13. _____

14. _____

15. _____

16. _____

17. _____

18. _____

19. _____

20. _____

Problem Solving Notes

1.4 SUBTRACTING REAL NUMBERS

A SUBTRACTING REAL NUMBERS

Now that addition of real numbers has been discussed, we can explore subtraction. We know that $9 - 7 = 2$. Notice that $9 + (-7) = 2$, also. This means that

$$9 - 7 = 9 + (-7)$$

Notice that the *difference* of 9 and 7 is the same as the *sum* of 9 and the opposite of 7. This is how we can subtract real numbers.

> **SUBTRACTING REAL NUMBERS**
>
> If a and b are real numbers, then $a - b = a + (-b)$.

In other words, to find the difference of two numbers, we add the opposite of the number being subtracted.

Example 1 Subtract.

    a. $-13 - 4$ **b.** $5 - (-6)$ **c.** $3 - 6$ **d.** $-1 - (-7)$

Solution:

    a. $-13 - 4 = -13 + (-4)$ Add -13 to the opposite of 4,
                           which is -4.
                 $= -17$

    b. $5 - (-6) = 5 + (6)$ Add 5 to the opposite of -6,
                          which is 6.
               $= 11$

    c. $3 - 6 = 3 + (-6)$ Add 3 to the opposite of 6, which is -6.
            $= -3$

    d. $-1 - (-7) = -1 + (7) = 6$

HELPFUL HINT

Study the patterns indicated.

No change ⌐   Change to addition.
               Change to opposite.

$$5 - 11 = 5 + (-11) = -6$$
$$-3 - 4 = -3 + (-4) = -7$$
$$7 - (-1) = 7 + (1) = 8$$

Examples Subtract.

  2. $5.3 - (-4.6) = 5.3 + (4.6) = 9.9$

  3. $-\dfrac{3}{10} - \dfrac{5}{10} = -\dfrac{3}{10} + \left(-\dfrac{5}{10}\right) = -\dfrac{8}{10} = -\dfrac{4}{5}$

  4. $-\dfrac{2}{3} - \left(-\dfrac{4}{5}\right) = -\dfrac{2}{3} + \left(\dfrac{4}{5}\right) = -\dfrac{10}{15} + \dfrac{12}{15} = \dfrac{2}{15}$

Objectives

A Subtract real numbers.

B Evaluate algebraic expressions using real numbers.

C Determine whether a number is a solution of a given equation.

D Solve problems that involve subtraction of real numbers.

E Find complementary and supplementary angles.

SSM   CD-ROM   Video
                  1.4

Practice Problem 1

Subtract.

a. $-20 - 6$

b. $3 - (-5)$

c. $7 - 17$

d. $-4 - (-9)$

Practice Problems 2–4

Subtract.

2. $9.6 - (-5.7)$ 3. $-\dfrac{4}{9} - \dfrac{2}{9}$

4. $-\dfrac{1}{4} - \left(-\dfrac{2}{5}\right)$

Answers

1. a. -26, **b.** 8, **c.** -10, **d.** 5,

2. 15.3, **3.** $-\dfrac{2}{3}$, **4.** $\dfrac{3}{20}$

Practice Problem 5

Subtract 7 from -11.

Practice Problem 6

Simplify each expression.

a. $-20 - 5 + 12 - (-3)$

b. $5.2 - (-4.4) + (-8.8)$

Practice Problem 7

Simplify each expression.

a. $-9 + [(-4 - 1) - 10]$

b. $5^2 - 20 + [-11 - (-3)]$

Example 5   Subtract 8 from -4.

Solution: Be careful when interpreting this: The order of numbers in subtraction is important. 8 is to be subtracted **from** -4.

$$-4 - 8 = -4 + (-8) = -12$$ ▬▬

If an expression contains additions and subtractions, just write the subtractions as equivalent additions. Then simplify from left to right.

Example 6   Simplify each expression.

   a. $-14 - 8 + 10 - (-6)$

   b. $1.6 - (-10.3) + (-5.6)$

Solution: **a.** $-14 - 8 + 10 - (-6) =$

$$-14 + (-8) + 10 + 6 = -6$$

   b. $1.6 - (-10.3) + (-5.6) =$

$$1.6 + 10.3 + (-5.6) = 6.3$$ ▬▬

When an expression contains parentheses and brackets, remember the order of operations. Start with the innermost set of parentheses or brackets and work your way outward.

Example 7   Simplify each expression.

   a. $-3 + [(-2 - 5) - 2]$

   b. $2^3 - 10 + [-6 - (-5)]$

Solution: **a.** Start with the innermost set of parentheses. Rewrite $-2 - 5$ as an addition.

$$-3 + [(-2 - 5) - 2] = -3 + [(-2 + (-5)) - 2]$$
$$= -3 + [(-7) - 2] \quad \text{Add: } -2 + (-5).$$
$$= -3 + [-7 + (-2)] \quad \text{Write } -7 - 2 \text{ as an addition.}$$
$$= -3 + [-9] \quad \text{Add.}$$
$$= -12 \quad \text{Add.}$$

   b. Start simplifying the expression inside the brackets by writing $-6 - (-5)$ as an addition.

$$2^3 - 10 + [-6 - (-5)] = 2^3 - 10 + [-6 + 5]$$
$$= 2^3 - 10 + [-1] \quad \text{Add.}$$
$$= 8 - 10 + (-1) \quad \text{Evaluate } 2^3.$$
$$= 8 + (-10) + (-1) \quad \text{Write } 8 - 10 \text{ as an addition.}$$
$$= -2 + (-1) \quad \text{Add.}$$
$$= -3 \quad \text{Add.}$$ ▬▬

B   EVALUATING ALGEBRAIC EXPRESSIONS

It is important to be able to evaluate expressions for given replacement values. This helps, for example, when checking solutions of equations.

Answers

5. -18,   **6. a.** -10,   **b.** 0.8,
7. a. -24,   **b.** -3

Example 8   Find the value of each expression when $x = 2$ and $y = -5$.

      a. $\dfrac{x - y}{12 + x}$ **b.** $x^2 - y$

Solution: **a.** Replace x with 2 and y with -5. Be sure to put parentheses around -5 to separate signs. Then simplify the resulting expression.

$$\frac{x - y}{12 + x} = \frac{2 - (-5)}{12 + 2} = \frac{2 + 5}{14} = \frac{7}{14} = \frac{1}{2}$$

      b. Replace the x with 2 and y with -5 and simplify.

$$x^2 - y = 2^2 - (-5) = 4 - (-5) = 4 + 5 = 9$$

C | SOLUTIONS OF EQUATIONS

Recall from Section 1.2 that a solution of an equation is a value for the variable that makes the equation true.

Example 9   Determine whether -4 is a solution of $x - 5 = -9$.

Solution: Replace x with -4 and see if a true statement results.

$$\begin{aligned} x - 5 &= -9 && \text{Original equation} \\ -4 - 5 &\stackrel{?}{=} -9 && \text{Replace } x \text{ with } -4. \\ -4 + (-5) &\stackrel{?}{=} -9 \\ -9 &\stackrel{?}{=} -9 && \text{True.} \end{aligned}$$

Thus -4 is a solution of $x - 5 = -9$.

D | SOLVING PROBLEMS THAT INVOLVE SUBTRACTION

Another use of real numbers is in recording altitudes above and below sea level, as shown in the next example.

Example 10   Finding the Difference in Elevations

The lowest point in North America is in Death Valley, at an elevation of 282 feet below sea level. Nearby, Mount Whitney reaches 14,494 feet, the highest point in the United States outside Alaska. How much of a variation in elevation is there between these two extremes? (*Source*: U.S. Geological Survey)

Solution: To find the variation in elevation between the two heights, find the difference of the high point and the low point.

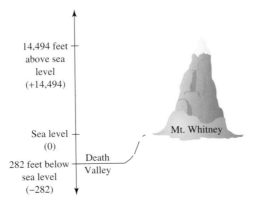

14,494 feet above sea level (+14,494)

Sea level (0)

282 feet below sea level (−282)

Death Valley

Mt. Whitney

Practice Problem 8

Find the value of each expression when $x = 1$ and $y = -4$.

    a. $\dfrac{x - y}{14 + x}$

    b. $x^2 - y$

Practice Problem 9

Determine whether -2 is a solution of $-1 + x = 1$.

Practice Problem 10

At 6.00 P.M., the temperature at the Winter Olympics was $14°$; by morning the temperature dropped to $-23°$. Find the overall change in temperature.

Answers

8. a. $\dfrac{1}{3}$, **b.** 5, **9.** -2 is not a solution,

10. $-37°$

In words: | high point | minus | low point |
|---|---|---|
| ↓ | ↓ | ↓ |

Translate: $14{,}494 \quad - \quad (-282) = 14{,}494 + 282$

$$= 14{,}776 \text{ feet}$$

Thus, the variation in elevation is 14,776 feet. ▬▬

E   FINDING COMPLEMENTARY AND SUPPLEMENTARY ANGLES

A knowledge of geometric concepts is needed by many professionals, such as doctors, carpenters, electronic technicians, gardeners, machinists, and pilots, just to name a few. With this in mind, we review the geometric concepts of **complementary** and **supplementary angles**.

COMPLEMENTARY AND SUPPLEMENTARY ANGLES

Two angles are **complementary** if their sum is 90°.

$$x + y = 90°$$

Two angles are **supplementary** if their sum is 180°.

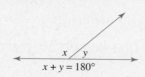

$$x + y = 180°$$

Practice Problem 11

Find each unknown complementary or supplementary angle.

a.

b.

Example 11   Find each unknown complementary or supplementary angle.

a.

b.

Solution: **a.** These angles are complementary, so their sum is 90°. This means that x is $90° - 38°$.

$$x = 90° - 38° = 52°$$

b. These angles are supplementary, so their sum is 180°. This means that y is $180° - 62°$.

$$y = 180° - 62° = 118°$$

▬▬▬

Answer

11. a. 102°,   **b.** 9°

EXERCISE SET 1.4

A *Subtract. See Examples 1 through 4.*

1. $-6 - 4$

2. $-12 - 8$

3. $4 - 9$

4. $8 - 11$

5. $16 - (-3)$

6. $12 - (-5)$

7. $\dfrac{1}{2} - \dfrac{1}{3}$

8. $\dfrac{3}{4} - \dfrac{7}{8}$

9. $-16 - (-18)$

10. $-20 - (-48)$

11. $-6 - 5$

12. $-8 - 4$

13. $7 - (-4)$

14. $3 - (-6)$

15. $-6 - (-11)$

16. $-4 - (-16)$

17. $16 - (-21)$

18. $15 - (-33)$

19. $9.7 - 16.1$

20. $8.3 - 11.2$

1. _____

2. _____

3. _____

4. _____

5. _____

6. _____

7. _____

8. _____

9. _____

10. _____

11. _____

12. _____

13. _____

14. _____

15. _____

16. _____

17. _____

18. _____

19. _____

20. _____

Focus On Study Skills

STUDY TIPS

Have you wondered what you can do to be successful in your algebra course? If so, that may well be your first step to success in algebra! Here are some tips on how to use this text and how to study mathematics in general.

Using this Text

1. Each example in the section has a parallel Practice Problem. As you read a section, try each Practice Problem after you've finished the corresponding example. This "learn-by-doing" approach will help you grasp ideas before you move on to other concepts.

2. The main section of exercises in an exercise set are referenced by an objective, such as **A** or **B** and also an example(s). Use this referencing in case you have trouble completing an assignment from the exercise set.

3. If you need extra help in a particular section, check at the beginning of the section to see what videotapes and software are available.

4. Integrated Reviews in each chapter offer you a chance to practice—in one place—the many concepts that you have learned separately over several sections.

5. There are many opportunities at the end of each chapter to help you understand the concepts of the chapter.

   Chapter Highlights contain chapter summaries with examples.
   Chapter Reviews contain review problems organized by section.
   Chapter Tests are sample tests to help you prepare for an exam.
   Cumulative Reviews are reviews consisting of material from the beginning of the book to the end of the particular chapter.

General Tips

1. *Choose to attend all class periods*. If possible, sit near the front of the classroom. This way, you will see and hear the presentation better. It may also be easier for you to participate in classroom activities.

2. *Do your homework*. You've probably heard the phrase "practice makes perfect" in relation to music and sports. It also applies to mathematics. You will find that the more time you spend solving mathematics problems, the easier the process becomes. Be sure to block out enough time to complete your assignments.

3. *Check your work*. Review the steps you made while working a problem. Learn to check your answers in the original problems. You can also compare your answers to the answers to selected exercises listed in the back of the book. If you have made a mistake, figure out what went wrong. Then correct your mistake.

4. *Learn from your mistakes*. Everyone, even your instructors, makes mistakes. You can use your mistakes to become a better math student. The key is finding and understanding your mistakes. Was your mistake a careless mistake? If so, you can try to work more slowly and make a conscious effort to carefully check your work. Did you make a mistake because you don't understand a concept? If so, take the time to review the concept or ask questions to better understand the concept.

5. *Know how to get help if you need it*. It's OK to ask for help. In fact, it's a good idea to ask for help whenever there is something that you don't understand. Make sure you know when your instructor has office hours and how to find his or her office. Find out if math tutoring services are available on your campus. Check out the hours, location, and requirements of the tutoring service. You might also want to find another student in your class that you can call to discuss your assignment.

6. *Organize your class materials*, including homework assignments, graded quizzes and tests, and notes from your class or lab. All of these items will make valuable references throughout your course and as you study for upcoming tests and your final exam. Make sure you can locate any of these materials when you need them.

1.5 MULTIPLYING REAL NUMBERS

A MULTIPLYING REAL NUMBERS

Multiplication of real numbers is similar to multiplication of whole numbers. We just need to determine when the answer is positive, when it is negative, and when it is zero. To discover sign patterns for multiplication, recall that multiplication is repeated addition. For example, 3(2) means that 2 is added to itself three times, or

$$3(2) = 2 + 2 + 2 = 6$$

Also,

$$3(-2) = (-2) + (-2) + (-2) = -6$$

Since $3(-2) = -6$, this suggests that the product of a positive number and a negative number is a negative number.

What about the product of two negative numbers? To find out, consider the following pattern.

Factor decreases by 1 each time.

$$-3 \cdot 2 = -6$$
$$-3 \cdot 1 = -3 \quad \text{Product increases by 3 each time}$$
$$-3 \cdot 0 = 0$$
$$-3 \cdot -1 = 3$$
$$-3 \cdot -2 = 6$$

This suggests that the product of two negative numbers is a positive number. Our results are given below.

MULTIPLYING REAL NUMBERS

1. The product of two numbers with the same sign is a positive number.
2. The product of two numbers with different signs is a negative number.

Examples   Multiply.

1. $-6(4) = -24$

2. $2(-10) = -20$

3. $-5(-10) = 50$

4. $-\dfrac{2}{3} \cdot \dfrac{4}{7} = -\dfrac{2 \cdot 4}{3 \cdot 7} = -\dfrac{8}{21}$

5. $5(-1.7) = -8.5$

6. $-18(-3) = 54$

We already know that the product of 0 and any whole number is 0. This is true of all real numbers.

PRODUCTS INVOLVING ZERO

If b is a real number, then $b \cdot 0 = 0$. Also $0 \cdot b = 0$.

Objectives

A Multiply real numbers.

B Evaluate algebraic expressions using real numbers.

C Determine whether a number is a solution of a given equation.

SSM   CD-ROM   Video
1.5

Practice Problems 1–6

Multiply.

1. $-8(3)$ 2. $5(-30)$

3. $-4(-12)$ 4. $-\dfrac{5}{6} \cdot \dfrac{1}{4}$

5. $6(-2.3)$ 6. $-15(-2)$

Answers

1. -24, **2.** -150, **3.** 48, **4.** $-\dfrac{5}{24}$,

5. -13.8, **6.** 30

Practice Problem 7

Use the order of operations and simplify each expression.

a. $5(0)(-3)$ b. $(-1)(-6)(-7)$

c. $(-2)(4)(-8)$ d. $(-2)^2$

e. $-3(-9) - 4(-4)$

HELPFUL HINT

You may have noticed from the example that if we multiply:

▲ an *even* number of negative numbers, the product is *positive*.

▲ an *odd* number of negative numbers, the product is *negative*.

Practice Problem 8

Evaluate each expression when $x = -1$ and $y = -5$.

a. $3x - y$ b. $x^2 - y^3$

Practice Problem 9

Determine whether -10 is a solution of $3x + 4 = -26$.

Answers

7. a. 0, **b.** -42, **c.** 64, **d.** 4, **e.** 43,

8. a. 2 **b.** 126, **9.** -10 is a solution.

Example 7   Use the order of operations and simplify each expression.

a. $7(0)(-6)$ **b.** $(-2)(-3)(-4)$ **c.** $(-1)(5)(-9)$

d. $(-2)^3$ **e.** $(-4)(-11) - 5(-2)$

Solution: **a.** By the order of operations, we multiply from left to right. Notice that because one of the factors is 0, the product is 0.

$$7(0)(-6) = 0(-6) = 0$$

b. Multiply two factors at a time, from left to right.

$$(-2)(-3)(-4) = (6)(-4) \quad \text{Multiply } (-2)(-3).$$
$$= -24$$

c. Multiply from left to right.

$$(-1)(5)(-9) = (-5)(-9) \quad \text{Multiply } (-1)(5).$$
$$= 45$$

d. The exponent 3 means 3 factors of the base -2.

$$(-2)^3 = (-2)(-2)(-2)$$
$$= -8 \qquad \text{Multiply.}$$

e. Follow the order of operations.

$$(-4)(-11) - 5(-2) = 44 - (-10) \quad \text{Find the products.}$$
$$= 44 + 10 \qquad \text{Add 44 to the opposite of } -10.$$
$$= 54 \qquad \text{Add.}$$

B   EVALUATING ALGEBRAIC EXPRESSIONS

Now that we know how to multiply positive and negative numbers, we continue to practice evaluating algebraic expressions.

Example 8   Evaluate each expression when $x = -2$ and $y = -4$.

a. $5x - y$ **b.** $x^3 - y^2$

Solution: **a.** Replace x with -2 and y with -4 and simplify.

$$5x - y = 5(-2) - (-4) = -10 - (-4) = -10 + 4 = -6$$

b. Replace x with -2 and y with -4.

$$x^3 - y^2 = (-2)^3 - (-4)^2 \quad \text{Substitute the given values for the variables.}$$
$$= -8 - (16) \qquad \text{Evaluate } (-2)^3 \text{ and } (-4)^2.$$
$$= -8 + (-16) \qquad \text{Write as a sum.}$$
$$= -24 \qquad \text{Add.}$$

C   SOLUTIONS OF EQUATIONS

To prepare for solving equations, we continue to check possible solutions for an equation.

Example 9   Determine whether -3 is a solution of $-7x + 2 = 23$.

Solution: Replace x with -3 and see if a true statement results.

$$-7x + 2 = 23 \quad \text{Original equation}$$
$$-7(-3) + 2 \stackrel{?}{=} 23 \quad \text{Replace } x \text{ with } -3.$$
$$21 + 2 \stackrel{?}{=} 23 \quad \text{Multiply.}$$
$$23 = 23 \quad \text{True.}$$

Thus, -3 is a solution of $-7x + 2 = 23$.

EXERCISE SET 1.5

A *Multiply. See Examples 1 through 6.*

1. $-6(4)$

2. $-8(5)$

3. $2(-1)$

4. $7(-4)$

5. $-5(-10)$

6. $-6(-11)$

7. $-3 \cdot 4$

8. $-2 \cdot 8$

9. $-6(-7)$

10. $-6(-9)$

11. $2(-9)$

12. $3(-5)$

13. $-\dfrac{1}{2}\left(-\dfrac{3}{5}\right)$

14. $-\dfrac{1}{8}\left(-\dfrac{1}{3}\right)$

15. $-\dfrac{3}{4}\left(-\dfrac{8}{9}\right)$

16. $-\dfrac{5}{6}\left(-\dfrac{3}{10}\right)$

17. $5(-1.4)$

18. $6(-2.5)$

19. $-0.2(-0.7)$

20. $-0.5(-0.3)$

ANSWERS

1. _____

2. _____

3. _____

4. _____

5. _____

6. _____

7. _____

8. _____

9. _____

10. _____

11. _____

12. _____

13. _____

14. _____

15. _____

16. _____

17. _____

18. _____

19. _____

20. _____

Numbers have a long history. The numbers we are accustomed to using probably originated in India in the 3rd century and were later adapted by Arabic cultures. Many other ancient civilizations developed their own unique number systems.

Roman Numerals

I	V	X	L	C	D	M
1	5	10	50	100	500	1000

If numerals decrease in value from left to right, the values are added. If a smaller numeral appears to the left of a larger numeral, the smaller value is subtracted. For example:

$$XVII = 10 + 5 + 1 + 1 = 17 \quad \text{but} \quad XLIV = 50 - 10 + 5 - 1 = 44$$

Chinese Numerals

一	二	三	四	五	六	七	八	九	十	百	千	萬	億
1	2	3	4	5	6	7	8	9	10	100	1000	10,000	100,000

Numerals are written vertically. If a digit representing 2–9 appears before a digit representing 10, 100, 1000, 10,000, or 100,000, multiplication is indicated.

$$\left. \begin{array}{c} 七 \\ 千 \end{array} \right\} (7 \times 1000)$$

$$\left. \begin{array}{c} 三 \\ 百 \end{array} \right\} (3 \times 100)$$

$$\left. \begin{array}{c} 八 \\ 十 \end{array} \right\} (8 \times 10)$$

$$\left. \begin{array}{c} 五 \end{array} \right\} 5 \qquad \text{is } 7000 + 300 + 80 + 5 = 7385$$

Egyptian Hieroglyphic Numerals

I	∩	9	⚱	↾	𓆤
1	10	100	1000	10,000	100,000

The Egyptian system is also multiplicative. For example, 3 ∩ hieroglyphs represents 3×10.

$$IIII \, ∩∩ \, 99999 \, ⚱⚱ \, ↾ = 4 \times 1 + 2 \times 10 + 5 \times 100 + 2 \times 1000 + 1 \times 10,000$$
$$= 12,524$$

▲ Write several numbers using each of the Roman, Chinese, and Egyptian hieroglyphic systems. Trade your numbers with another student in your group to translate into our numerals. Check one another's work.

▲ Research the number system of another ancient culture (such as Babylonian, Mayan Indian, or Ionic Greek). Write the numbers 712, 4690, 5113, and 208 using that system. Demonstrate the system to the rest of your group.

1.6 Dividing Real Numbers

A Finding Reciprocals

Addition and subtraction are related. Every difference of two numbers $a - b$ can be written as the sum $a + (-b)$. Multiplication and division are related also. For example, the quotient $6 \div 3$ can be written as the product $6 \cdot \frac{1}{3}$. Recall that the pair of numbers 3 and $\frac{1}{3}$ has a special relationship. Their product is 1 and they are called **reciprocals** or **multiplicative inverses** of each other.

Reciprocal or Multiplicative Inverse

Two numbers whose product is 1 are called **reciprocals** or **multiplicative inverses** of each other.

Examples Find the reciprocal of each number.

1. 22 The reciprocal of 22 is $\frac{1}{22}$ since $22 \cdot \frac{1}{22} = 1$.

2. $\frac{3}{16}$ The reciprocal of $\frac{3}{16}$ is $\frac{16}{3}$ since $\frac{3}{16} \cdot \frac{16}{3} = 1$.

3. -10 The reciprocal of -10 is $-\frac{1}{10}$ since $-10 \cdot -\frac{1}{10} = 1$.

4. $-\frac{9}{13}$ The reciprocal of $-\frac{9}{13}$ is $-\frac{13}{9}$ since $-\frac{9}{13} \cdot -\frac{13}{9} = 1$.

5. 1.7 The reciprocal of 1.7 is $\frac{1}{1.7}$ since $1.7 \cdot \frac{1}{1.7} = 1$.

Does the number 0 have a reciprocal? If it does, it is a number n such that $0 \cdot n = 1$. Notice that this can never be true since $0 \cdot n = 0$. This means that 0 has no reciprocal.

Quotients Involving Zero

The number 0 does not have a reciprocal.

B Dividing Real Numbers

We may now write a quotient as an equivalent product.

Quotient of Two Real Numbers

If a and b are real numbers and b is not 0, then

$$a \div b = \frac{a}{b} = a \cdot \frac{1}{b}$$

In other words, the quotient of two real numbers is the product of the first number and the multiplicative inverse or reciprocal of the second number.

Example 6 Use the definition of the quotient of two numbers to find each quotient.

a. $-18 \div 3$ **b.** $\frac{-14}{-2}$ **c.** $\frac{20}{-4}$

Objectives

A Find the reciprocal of a real number.

B Divide real numbers.

C Evaluate algebraic expressions using real numbers.

D Determine whether a number is a solution of a given equation.

SSM   CD-ROM   Video 1.6

Practice Problems 1–5

Find the reciprocal of each number.

1. 13   2. $\frac{7}{15}$ 3. -5

4. $-\frac{8}{11}$ 5. 7.9

Practice Problem 6

Use the definition of the quotient of two numbers to find each quotient.

a. $-12 \div 4$ b. $\frac{-20}{-10}$

c. $\frac{36}{-4}$

Answers

1. $\frac{1}{13}$, 2. $\frac{15}{7}$, 3. $-\frac{1}{5}$, 4. $-\frac{11}{8}$, 5. $\frac{1}{7.9}$,

6. a. -3, b. 2, c. -9

Solution: **a.** $-18 \div 3 = -18 \cdot \frac{1}{3} = -6$

b. $\frac{-14}{-2} = -14 \cdot -\frac{1}{2} = 7$

c. $\frac{20}{-4} = 20 \cdot -\frac{1}{4} = -5$　━━━

Since the quotient $a \div b$ can be written as the product $a \cdot \frac{1}{b}$, it follows that sign patterns for dividing two real numbers are the same as sign patterns for multiplying two real numbers.

DIVIDING REAL NUMBERS

1. The quotient of two numbers with the same sign is a positive number.
2. The quotient of two numbers with different signs is a negative number.

Practice Problems 7–10

Divide.

7. $\frac{-25}{5}$　　8. $\frac{-48}{-6}$

9. $\frac{50}{-2}$　　10. $\frac{-72}{0.2}$

Examples　Divide.

7. $\frac{-30}{-10} = 3$ 　 Same sign, so the quotient is positive.

8. $\frac{-100}{5} = -20$

9. $\frac{20}{-2} = -10$ 　 Unlike signs, so the quotient is negative.

10. $\frac{42}{-0.6} = -70$

$0.6 \overline{)42.0}^{\;70.}$ 　━━━

In the examples above, we divided mentally or by long division. When we divide by a fraction, it is usually easier to multiply by its reciprocal.

Examples　Divide.

11. $\frac{2}{3} \div \left(-\frac{5}{4}\right) = \frac{2}{3} \cdot \left(-\frac{4}{5}\right) = -\frac{8}{15}$

12. $-\frac{1}{6} \div \left(-\frac{2}{3}\right) = -\frac{1}{6} \cdot \left(-\frac{3}{2}\right) = \frac{3}{12} = \frac{1}{4}$ 　━━━

Practice Problems 11–12

Divide.

11. $-\frac{5}{9} \div \frac{2}{3}$

12. $-\frac{2}{7} \div \left(-\frac{1}{5}\right)$

Our definition of the quotient of two real numbers does not allow for division by 0 because 0 does not have a reciprocal. How then do we interpret $\frac{3}{0}$? We say that an expression such as this one is undefined. Can we divide 0 by a number other than 0? Yes; for example,

$$\frac{0}{3} = 0 \cdot \frac{1}{3} = 0$$

DIVISION INVOLVING ZERO

Division by 0 is undefined. For example, $\frac{-5}{0}$ is undefined.

0 divided by a nonzero number is 0. For example, $\frac{0}{-5} = 0$.

Answers

7. -5, **8.** 8, **9.** -25, **10.** -360, **11.** $-\frac{5}{6}$,

12. $\frac{10}{7}$

Examples   Perform each indicated operation.

13. $\frac{1}{0}$ is undefined.   **14.** $\frac{0}{-3} = 0$

15. $\frac{0(-8)}{2} = \frac{0}{2} = 0$

Notice that $\frac{12}{-2} = -6$, $-\frac{12}{2} = -6$, and $\frac{-12}{2} = -6$. This means that

$$\frac{12}{-2} = -\frac{12}{2} = \frac{-12}{2}.$$

In other words, a single negative sign in a fraction can be written in the denominator, in the numerator, or in front of the fraction without changing the value of the fraction.

> In general, if a and b are real numbers, $b \neq 0$, $\frac{a}{-b} = \frac{-a}{b} = -\frac{a}{b}$.

Examples combining basic arithmetic operations along with the principles of the order of operations help us to review these concepts.

Example 16   Simplify each expression.

a. $\frac{(-12)(-3) + 4}{-7 - (-2)}$

b. $\frac{2(-3)^2 - 20}{-5 + 4}$

Solution: **a.** First, simplify the numerator and denominator separately, then divide.

$$\frac{(-12)(-3) + 4}{-7 - (-2)} = \frac{36 + 4}{-7 + 2}$$
$$= \frac{40}{-5}$$
$$= -8 \quad \text{Divide.}$$

b. Simplify the numerator and denominator separately, then divide.

$$\frac{2(-3)^2 - 20}{-5 + 4} = \frac{2 \cdot 9 - 20}{-5 + 4} = \frac{18 - 20}{-5 + 4} = \frac{-2}{-1} = 2$$

C EVALUATING ALGEBRAIC EXPRESSIONS

Using what we have learned about dividing real numbers, we continue to practice evaluating algebraic expressions.

Example 17   Evaluate $\frac{3x}{2y}$ when $x = -2$ and $y = -4$.

Solution: Replace x with -2 and y with -4 and simplify.

$$\frac{3x}{2y} = \frac{3(-2)}{2(-4)} = \frac{-6}{-8} = \frac{3}{4}$$

Practice Problems 13–15

Perform each indicated operation.

13. $\frac{-7}{0}$

14. $\frac{0}{-2}$

15. $\frac{0(-5)}{3}$

Practice Problem 16

Simplify each expression.

a. $\frac{-7(-4) + 2}{-10 - (-5)}$

b. $\frac{5(-2)^3 + 52}{-4 + 1}$

Practice Problem 17

Evaluate $\frac{x + y}{3x}$ when $x = -1$ and $y = -5$.

Answers

13. undefined, **14.** 0, **15.** 0,
16. a. -6, **b.** -4, **17.** 2

Practice Problem 18

Determine whether -8 is a solution of $\frac{x}{4} - 3 = x + 3$.

D   SOLUTIONS OF EQUATIONS

We use our skills in dividing real numbers to check possible solutions for an equation.

Example 18   Determine whether -10 is a solution of $\frac{-20}{x} + 5 = x$.

Solution:

$$\frac{-20}{x} + 5 = x \qquad \text{Original equation}$$

$$\frac{-20}{-10} + 5 \overset{?}{=} -10 \qquad \text{Replace } x \text{ with } -10.$$

$$2 + 5 \overset{?}{=} -10 \qquad \text{Divide.}$$

$$7 = -10 \qquad \textbf{False.}$$

Since we have a false statement, -10 is *not* a solution of the equation.

🖩 CALCULATOR EXPLORATIONS

ENTERING NEGATIVE NUMBERS ON A SCIENTIFIC CALCULATOR

To enter a negative number on a scientific calculator, find a key marked $\boxed{+/-}$. (On some calculators, this key is marked $\boxed{\text{CHS}}$ for "change sign.") To enter -8, for example, press the keys $\boxed{8}$ $\boxed{+/-}$. The display will read $\boxed{-8}$.

ENTERING NEGATIVE NUMBERS ON A GRAPHING CALCULATOR

To enter a negative number on a graphing calculator, find a key marked $\boxed{(-)}$. Do not confuse this key with the key $\boxed{-}$, which is used for subtraction. To enter -8, for example, press the keys $\boxed{(-)}$ $\boxed{8}$. The display will read $\boxed{-8}$.

OPERATIONS WITH REAL NUMBERS

To evaluate $-2(7-9) - 20$ on a calculator, press the keys

$\boxed{2}$ $\boxed{+/-}$ $\boxed{\times}$ $\boxed{(}$ $\boxed{7}$ $\boxed{-}$ $\boxed{9}$ $\boxed{)}$ $\boxed{-}$ $\boxed{2}$ $\boxed{0}$ $\boxed{=}$, or

$\boxed{(-)}$ $\boxed{2}$ $\boxed{(}$ $\boxed{7}$ $\boxed{-}$ $\boxed{9}$ $\boxed{)}$ $\boxed{-}$ $\boxed{2}$ $\boxed{0}$ $\boxed{\text{ENTER}}$.

The display will read $\boxed{-16}$ or

$$\begin{array}{l} -2(7-9) - 20 \\ \qquad\qquad\quad -16 \end{array}$$

Use a calculator to simplify each expression.

1. $-38(26 - 27)$

2. $-59(-8) + 1726$

3. $134 + 25(68 - 91)$

4. $45(32) - 8(218)$

5. $\dfrac{-50(294)}{175 - 265}$

6. $\dfrac{-444 - 444.8}{-181 - 324}$

7. $9^5 - 4550$

8. $5^8 - 6259$

9. $(-125)^2$ (Be careful.)

10. -125^2 (Be careful.)

Answer

18. -8 is a solution.

EXERCISE SET 1.6

B *Divide. See Examples 6 through 15.*

1. $\dfrac{18}{-2}$

2. $\dfrac{20}{-10}$

3. $\dfrac{-16}{-4}$

4. $\dfrac{-18}{-6}$

5. $\dfrac{-48}{12}$

6. $\dfrac{-60}{5}$

7. $\dfrac{0}{-4}$

8. $\dfrac{0}{-9}$

9. $-\dfrac{15}{3}$

10. $-\dfrac{24}{8}$

11. $\dfrac{5}{0}$

12. $\dfrac{3}{0}$

13. $\dfrac{-12}{-4}$

14. $\dfrac{-45}{-9}$

15. $\dfrac{30}{-2}$

16. $\dfrac{14}{-2}$

17. $\dfrac{6}{7} \div \left(-\dfrac{1}{3}\right)$

18. $\dfrac{4}{5} \div \left(-\dfrac{1}{2}\right)$

19. $-\dfrac{5}{9} \div \left(-\dfrac{3}{4}\right)$

20. $-\dfrac{1}{10} \div \left(-\dfrac{8}{11}\right)$

21. $-\dfrac{4}{9} \div \dfrac{4}{9}$

1. _____
2. _____
3. _____
4. _____
5. _____
6. _____
7. _____
8. _____
9. _____
10. _____
11. _____
12. _____
13. _____
14. _____
15. _____
16. _____
17. _____
18. _____
19. _____
20. _____
21. _____

Problem Solving Notes

1.7 PROPERTIES OF REAL NUMBERS

A USING THE COMMUTATIVE AND ASSOCIATIVE PROPERTIES

In this section we give names to properties of real numbers with which we are already familiar. Throughout this section, the variables a, b, and c represent real numbers.

We know that order does not matter when adding numbers. For example, we know that $7 + 5$ is the same as $5 + 7$. This property is given a special name—the **commutative property of addition**. We also know that order does not matter when multiplying numbers. For example, we know that $-5(6) = 6(-5)$. This property means that multiplication is commutative also and is called the **commutative property of multiplication**.

COMMUTATIVE PROPERTIES

Addition: $a + b = b + a$

Multiplication: $a \cdot b = b \cdot a$

These properties state that the *order* in which any two real numbers are added or multiplied does not change their sum or product. For example, if we let $a = 3$ and $b = 5$, then the commutative properties guarantee that

$$3 + 5 = 5 + 3 \quad \text{and} \quad 3 \cdot 5 = 5 \cdot 3$$

HELPFUL HINT

Is subtraction also commutative? Try an example. Is $3 - 2 = 2 - 3$? **No!** The left side of this statement equals 1; the right side equals -1. There is no commutative property of subtraction. Similarly, there is no commutative property for division. For example, $10 \div 2$ does not equal $2 \div 10$.

Example 1 Use a commutative property to complete each statement.

    a. $x + 5 =$ _____   **b.** $3 \cdot x =$ _____

Solution: **a.** $x + 5 = 5 + x$ By the commutative property of addition
          b. $3 \cdot x = x \cdot 3$ By the commutative property of multiplication

TRY THE CONCEPT CHECK IN THE MARGIN.

Let's now discuss grouping numbers. We know that when we add three numbers, the way in which they are grouped or associated does not change their sum. For example, we know that $2 + (3 + 4) = 2 + 7 = 9$. This result is the same if we group the numbers differently. In other words, $(2 + 3) + 4 = 5 + 4 = 9$, also. Thus, $2 + (3 + 4) = (2 + 3) + 4$. This property is called the **associative property of addition**.

We also know that changing the grouping of numbers when multiplying does not change their product. For example, $2 \cdot (3 \cdot 4) = (2 \cdot 3) \cdot 4$ (check it). This is the **associative property of multiplication**.

Objectives

A Use the commutative and associative properties.

B Use the distributive property.

C Use the identity and inverse properties.

  SSM   CD-ROM   Video
                    1.7

✓ CONCEPT CHECK

Which of the following pairs of actions are commutative?

a. "taking a test" and "studying for the test"

b. "putting on your left shoe" and "putting on your right shoe"

c. "putting on your shoes" and "putting on your socks"

d. "reading the sports section" and "reading the comics section"

Practice Problem 1

Use a commutative property to complete each statement.

a. $7 \cdot y =$ _____
b. $4 + x =$ _____

Answers

1. a. $y \cdot 7$, **b.** $x + 4$
✓ **Concept Check:**   b, d

Practice Problem 2

Use an associative property to complete each statement.

a. $5 \cdot (-3 \cdot 6) = $ _____

b. $(-2 + 7) + 3 = $ _____

These properties state that the way in which three numbers are *grouped* does not change their sum or their product.

Example 2   Use an associative property to complete each statement.

a. $5 + (4 + 6) = $ _____

b. $(-1 \cdot 2) \cdot 5 = $ _____

Solution: **a.** $5 + (4 + 6) = (5 + 4) + 6$ By the associative property of addition

b. $(-1 \cdot 2) \cdot 5 = -1 \cdot (2 \cdot 5)$ By the associative property of multiplication

> **HELPFUL HINT**
>
> Remember the difference between the commutative properties and the associative properties. The commutative properties have to do with the *order* of numbers and the associative properties have to do with the *grouping* of numbers.

Examples   Determine whether each statement is true by an associative property or a commutative property.

3. $(7 + 10) + 4 = (10 + 7) + 4$ Since the order of two numbers was changed and their grouping was not, this is true by the commutative property of addition.

4. $2 \cdot (3 \cdot 1) = (2 \cdot 3) \cdot 1$ Since the grouping of the numbers was changed and their order was not, this is true by the associative property of multiplication.

Practice Problems 3–4

Determine whether each statement is true by an associative property or a commutative property.

3. $5 \cdot (4 \cdot 7) = 5 \cdot (7 \cdot 4)$

4. $-2 + (4 + 9) = (-2 + 4) + 9$

Let's now illustrate how these properties can help us simplify expressions.

Practice Problems 5–6

Simplify each expression.

5. $(-3 + x) + 17$

6. $4(5x)$

Examples   Simplify each expression.

5. $10 + (x + 12) = 10 + (12 + x)$ By the commutative property of addition

$= (10 + 12) + x$ By the associative property of addition

$= 22 + x$ Add.

6. $-3(7x) = (-3 \cdot 7)x$ By the associative property of multiplication

$= -21x$ Multiply.

B   USING THE DISTRIBUTIVE PROPERTY

The **distributive property of multiplication over addition** is used repeatedly throughout algebra. It is useful because it allows us to write a product as a sum or a sum as a product.

We know that $7(2 + 4) = 7(6) = 42$. Compare that with $7(2) + 7(4) = 14 + 28 = 42$. Since both original expressions equal 42, they must equal each other, or

$$7(2 + 4) = 7(2) + 7(4)$$

This is an example of the distributive property. The product on the left side of the equal sign is equal to the sum on the right side. We can think of the 7 as being distributed to each number inside the parentheses.

Answers

2. a. $(5 \cdot -3) \cdot 6$, **b.** $-2 + (7 + 3)$, **3.** commutative, **4.** associative, **5.** $14 + x$, **6.** $20x$

Since multiplication is commutative, this property can also be written as

$$(\overset{\frown}{b + c})a = ba + ca$$

The distributive property can also be extended to more than two numbers inside the parentheses. For example,

$$3(x + y + z) = 3(x) + 3(y) + 3(z)$$
$$= 3x + 3y + 3z$$

Since we define subtraction in terms of addition, the distributive property is also true for subtraction. For example,

$$2(\overset{\frown}{x - y}) = 2(x) - 2(y)$$
$$= 2x - 2y$$

Examples
Use the distributive property to write each expression without parentheses. Then simplify the result.

7. $2(x + y) = 2(x) + 2(y)$
$= 2x + 2y$

8. $-5(-3 + 2z) = -5(-3) + (-5)(2z)$
$= 15 - 10z$

9. $5(x + 3y - z) = 5(x) + 5(3y) - 5(z)$
$= 5x + 15y - 5z$

10. $-1(2 - y) = (-1)(2) - (-1)(y)$
$= -2 + y$

11. $-(3 + x - w) = -1(3 + x - w)$
$= (-1)(3) + (-1)(x) - (-1)(w)$
$= -3 - x + w$

> **HELPFUL HINT**
>
> Notice in Example 11 that $-(3 + x - w)$ is first rewritten as $-1(3 + x - w)$.

12. $4(3x + 7) + 10 = 4(3x) + 4(7) + 10$ Apply the distributive property.
$= 12x + 28 + 10$ Multiply.
$= 12x + 38$ Add.   ▬▬▬

The distributive property can also be used to write a sum as a product.

Examples
Use the distributive property to write each sum as a product.

13. $8 \cdot 2 + 8 \cdot x = 8(2 + x)$
14. $7s + 7t = 7(s + t)$ ▬▬▬

C USING THE IDENTITY AND INVERSE PROPERTIES

Next, we look at the **identity properties**.

The number 0 is called the identity for addition because when 0 is added to any real number, the result is the same real number. In other words, the *identity* of the real number is not changed.

The number 1 is called the identity for multiplication because when a real number is multiplied by 1, the result is the same real number. In other words, the *identity* of the real number is not changed.

> ### IDENTITIES FOR ADDITION AND MULTIPLICATION
>
> 0 is the identity element for addition.
> $$a + 0 = a \quad \text{and} \quad 0 + a = a$$
> 1 is the identity element for multiplication.
> $$a \cdot 1 = a \quad \text{and} \quad 1 \cdot a = a$$

Practice Problems 7–12

Use the distributive property to write each expression without parentheses. Then simplify the result.

7. $5(x + y)$

8. $-3(2 + 7x)$

9. $4(x + 6y - 2z)$

10. $-1(3 - a)$

11. $-(8 + a - b)$

12. $9(2x + 4) + 9$

Practice Problems 13–14

Use the distributive property to write each sum as a product.

13. $9 \cdot 3 + 9 \cdot y$

14. $4x + 4y$

Answers

7. $5x + 5y$, **8.** $-6 - 21x$,
9. $4x + 24y - 8z$, **10.** $-3 + a$,
11. $-8 - a + b$, **12.** $18x + 45$,
13. $9(3 + y)$, **14.** $4(x + y)$

Notice that 0 is the *only* number that can be added to any real number with the result that the sum is the same real number. Also, 1 is the *only* number that can be multiplied by any real number with the result that the product is the same real number.

Additive inverses or **opposites** were introduced in Section 1.3. Two numbers are called additive inverses or opposites if their sum is 0. The additive inverse or opposite of 6 is -6 because $6 + (-6) = 0$. The additive inverse or opposite of -5 is 5 because $-5 + 5 = 0$.

Reciprocals or **multiplicative inverses** were introduced in Section R.2. Two nonzero numbers are called reciprocals or multiplicative inverses if their product is 1. The reciprocal or multiplicative inverse of $\frac{2}{3}$ is $\frac{3}{2}$ because $\frac{2}{3} \cdot \frac{3}{2} = 1$. Likewise, the reciprocal of -5 is $-\frac{1}{5}$ because $-5\left(-\frac{1}{5}\right) = 1$.

ADDITIVE OR MULTIPLICATIVE INVERSES

The numbers a and $-a$ are additive inverses or opposites of each other because their sum is 0; that is,
$$a + (-a) = 0$$
The numbers b and $\frac{1}{b}$ (for $b \neq 0$) are reciprocals or multiplicative inverses of each other because their product is 1; that is,
$$b \cdot \frac{1}{b} = 1$$

✓ CONCEPT CHECK

Which of the following is the

a. opposite of $-\frac{3}{10}$, and which is the

b. reciprocal of $-\frac{3}{10}$?

1, $\quad -\frac{10}{3}, \quad \frac{3}{10}, \quad 0, \quad \frac{10}{3}, \quad -\frac{3}{10}$

TRY THE CONCEPT CHECK IN THE MARGIN.

Practice Problems 15–21

Name the property illustrated by each true statement.

15. $5 + (-5) = 0$

16. $12 + y = y + 12$

17. $-4 \cdot (6 \cdot x) = (-4 \cdot 6) \cdot x$

18. $6 + (z + 2) = 6 + (2 + z)$

19. $3\left(\frac{1}{3}\right) = 1$

20. $(x + 0) + 23 = x + 23$

21. $(7 \cdot y) \cdot 10 = y \cdot (7 \cdot 10)$

Examples   Name the property illustrated by each true statement.

15. $3 \cdot y = y \cdot 3$ Commutative property of multiplication (order changed)

16. $(x + 7) + 9 = x + (7 + 9)$ Associative property of addition (grouping changed)

17. $(b + 0) + 3 = b + 3$ Identity element for addition

18. $2 \cdot (z \cdot 5) = 2 \cdot (5 \cdot z)$ Commutative property of multiplication (order changed)

19. $-2 \cdot \left(-\frac{1}{2}\right) = 1$ Multiplicative inverse property

20. $-2 + 2 = 0$ Additive inverse property

21. $-6 \cdot (y \cdot 2) = (-6 \cdot 2) \cdot y$ Commutative and associative properties of multiplication (order and grouping changed)

Answers

✓ Concept Check: a. $\frac{3}{10}$,   b. $-\frac{10}{3}$

15. additive inverse property, **16.** commutative property of addition, **17.** associative property of multiplication, **18.** commutative property of addition, **19.** multiplicative inverse property, **20.** identity element for addition, **21.** commutative and associative properties of multiplication

EXERCISE SET 1.7

B *Use the distributive property to write each expression without parentheses. Then simplify the result. See Examples 7 through 12.*

1. $4(x + y)$

2. $7(a + b)$

3. $9(x - 6)$

4. $11(y - 4)$

5. $2(3x + 5)$

6. $5(7 + 8y)$

7. $7(4x - 3)$

8. $3(8x - 1)$

9. $3(6 + x)$

10. $2(x + 5)$

11. $-2(y - z)$

12. $-3(z - y)$

13. $-7(3y + 5)$

14. $-5(2r + 11)$

15. $5(x + 4m + 2)$

16. $8(3y + z - 6)$

17. $-4(1 - 2m + n)$

18. $-4(4 + 2p + 5)$

19. $-(5x + 2)$

20. $-(9r + 5)$

21. $-(r - 3 - 7p)$

1. _____

2. _____

3. _____

4. _____

5. _____

6. _____

7. _____

8. _____

9. _____

10. _____

11. _____

12. _____

13. _____

14. _____

15. _____

16. _____

17. _____

18. _____

19. _____

20. _____

21. _____

Problem Solving Notes

1.8   READING GRAPHS

In today's world, where the exchange of information must be fast and entertaining, graphs are becoming increasingly popular. They provide a quick way of making comparisons, drawing conclusions, and approximating quantities.

A   READING BAR GRAPHS

A **bar graph** consists of a series of bars arranged vertically or horizontally. The bar graph in Example 1 shows a comparison of the rates charged by selected electricity companies. The names of the companies are listed horizontally and a bar is shown for each company. Corresponding to the height of the bar for each company is a number along a vertical axis. These vertical numbers are cents charged for each kilowatt-hour of electricity used.

Example 1   The following bar graph shows the cents charged per kilowatt-hour for selected electricity companies.

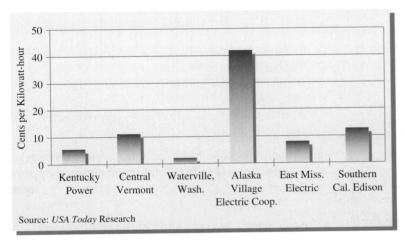

Source: *USA Today* Research

a. Which company charges the highest rate?
b. Which company charges the lowest rate?
c. Approximate the electricity rate charged by the first four companies listed.
d. Approximate the difference in the rates charged by the companies in parts (a) and (b).

Solution: **a.** The tallest bar corresponds to the company that charges the highest rate. Alaska Village Electric Cooperative charges the highest rate.

b. The shortest bar corresponds to the company that charges the lowest rate. Waterville, Washington charges the lowest rate.

c. To approximate the rate charged by Kentucky Power, we go to the top of the bar that corresponds to this company. From the top of the bar, we move horizontally to the left until the vertical axis is reached.

Objectives

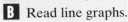

A Read bar graphs.
B Read line graphs.

SSM   CD-ROM   Video
1.8

Practice Problem 1

Use the bar graph from Example 1 to answer the following.

a. Approximate the rate charged by East Mississippi Electric.

b. Approximate the rate charged by Southern California Edison.

c. Find the difference in rates charged by Southern California Edison and East Mississippi Electric.

Answers

1.a. 8¢ per kilowatt-hour,   **b.** 12¢ per kilowatt-hour,   **c.** 4¢

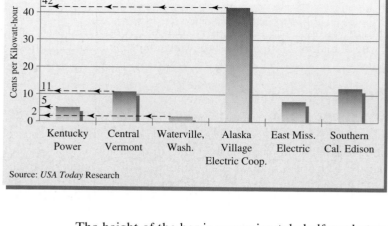

Source: *USA Today* Research

The height of the bar is approximately halfway between the 0 and 10 marks. We therefore conclude that

Kentucky Power charges approximately 5¢ per kilowatt-hour.
Central Vermont charges approximately 11¢ per kilowatt-hour.
Waterville, Washington charges approximately 2¢ per kilowatt-hour.
Alaska Village Electric charges approximately 42¢ per kilowatt-hour.

d. The difference in rates for Alaska Village Electric Cooperative and Waterville, Washington is approximately 42¢ − 2¢ or 40¢.

Practice Problem 2

Use the graph from Example 2 to answer the following.

a. How much money did the film *Snow White and the Seven Dwarfs* generate?

b. How much more money did the film *The Jungle Book* make than the film *Bambi*?

Example 2   The following bar graph shows Disney's top eight animated films and the amount of money they generated at theaters.

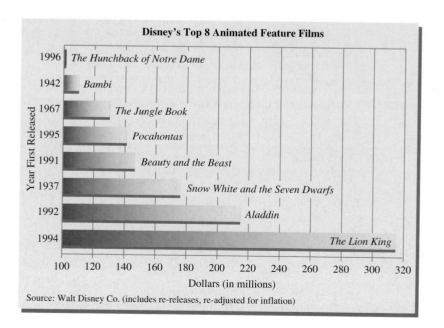

Disney's Top 8 Animated Feature Films

Source: Walt Disney Co. (includes re-releases, re-adjusted for inflation)

Answers

2. a. 175 million,   **b.** 20 million

a. Find the film shown that generated the most income for Disney and approximate the income.

b. How much more money did the film *Aladdin* make than the film *Beauty and the Beast*?

Solution: **a.** Since these bars are arranged horizontally, we look for the longest bar, which is the bar representing the film *The Lion King*. To approximate the income from this film, we move from the right edge of this bar vertically downward to the dollars axis. This film generated approximately 315 million dollars, or $315,000,000, the most income for Disney.

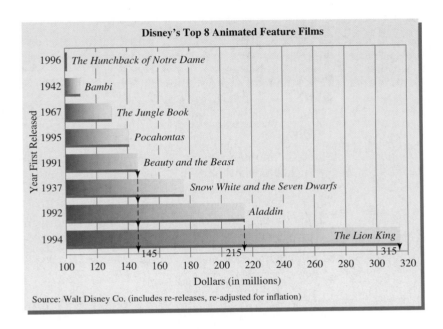

Disney's Top 8 Animated Feature Films

Source: Walt Disney Co. (includes re-releases, re-adjusted for inflation)

b. *Aladdin* generated approximately 215 million dollars. *Beauty and the Beast* generated approximately 145 million dollars. To find how much more money *Aladdin* generated than *Beauty and the Beast*, we subtract $215 - 145 = 70$ million dollars, or $70,000,000.

B READING LINE GRAPHS

A **line graph** consists of a series of points connected by a line. The graph in Example 3 is a line graph.

Example 3   The line graph below shows the relationship between the distance driven in a 14-foot U-Haul truck in one day and the total cost of renting this truck for that day. Notice that the horizontal axis is labeled Distance and the vertical axis is labeled Total Cost.

Practice Problem 3

Use the graph from Example 3 to answer the following.

a. Find the total cost of renting the truck if 50 miles are driven.

b. Find the total number of miles driven if the total cost of renting is $100.

Answers

3. a. $50, **b.** 180 miles

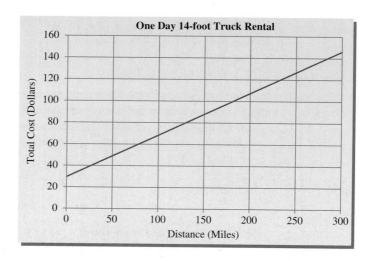

a. Find the total cost of renting the truck if 100 miles are driven.

b. Find the number of miles driven if the total cost of renting is $140.

Solution: **a.** Find the number 100 on the horizontal scale and move vertically upward until the line is reached. From this point on the line, we move horizontally to the left until the vertical scale is reached. We find that the total cost of renting the truck if 100 miles are driven is approximately $70.

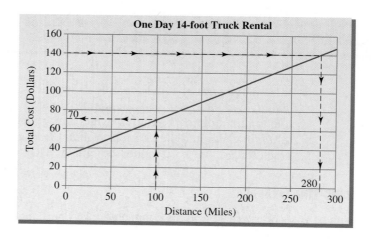

b. We find the number 140 on the vertical scale and move horizontally to the right until the line is reached. From this point on the line, we move vertically downward until the horizontal scale is reached. We find that the truck is driven approximately 280 miles.   ▬▬▬

From the previous example, we can see that graphing provides a quick way to approximate quantities. In Chapter 6 we show how we can use equations to find exact answers to the questions posed in Example 3. The next graph is another example of a line graph. It is also sometimes called a **broken line graph**.

Example 4   The line graph shows the relationship between time spent smoking a cigarette and pulse rate. Time is recorded along the horizontal axis in minutes, with 0 minutes being the moment a smoker lights a cigarette. Pulse is recorded along the vertical axis in heartbeats per minute.

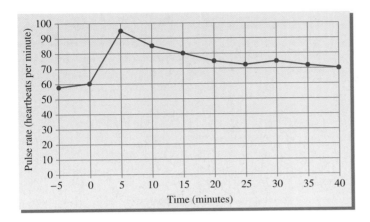

a. What is the pulse rate 15 minutes after lighting a cigarette?
b. When is the pulse rate the lowest?
c. When does the pulse rate show the greatest change?

Solution: a. We locate the number 15 along the time axis and move vertically upward until the line is reached. From this point on the line, we move horizontally to the left until the pulse rate axis is reached. Reading the number of beats per minute, we find that the pulse rate is 80 beats per minute 15 minutes after lighting a cigarette.

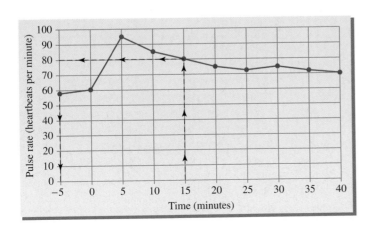

Practice Problem 4

Use the graph from Example 4 to answer the following.

a. What is the pulse rate 40 minutes after lighting a cigarette?

b. What is the pulse rate when the cigarette is being lit?

c. When is the pulse rate the highest?

Answers
4. a. 70,   **b.** 60,   **c.** 5 min. after lighting

b. We find the lowest point of the line graph, which represents the lowest pulse rate. From this point, we move vertically downward to the time axis. We find that the pulse rate is the lowest at −5 minutes, which means 5 minutes *before* lighting a cigarette.

c. The pulse rate shows the greatest change during the 5 minutes between 0 and 5. Notice that the line graph is *steepest* between 0 and 5 minutes.   ▬▬▬▬

EXERCISE SET 1.8

A *The following bar graph shows the number of teenagers expected to use the Internet for the years shown. Use this graph to answer Exercises 1–4. See Example 1.*

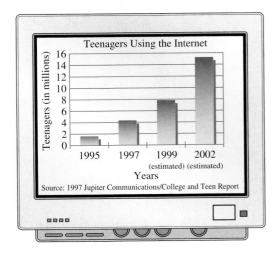

1. Approximate the number of teenagers expected to use the Internet in 1999.

2. Approximate the number of teenagers who use the Internet in 1995.

3. What year shows the greatest *increase* in number of teenagers using the Internet?

4. How many more teenagers are expected to use the Internet in 2002 than in 1999?

B *Many fires are deliberately set. An increasing number of those arrested for arson are juveniles (age 17 and under). The following line graph shows the percent of deliberately set fires started by juveniles. Use this graph to answer Exercises 5–12. See Example 3 and 4.*

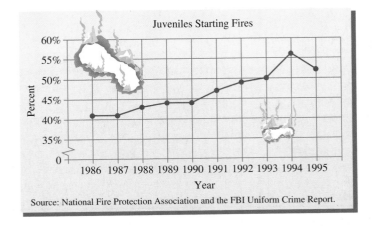

5. What year shows the highest percent of arson fires started by juveniles?

6. What year since 1990 shows a decrease in the percent of fires started by juveniles?

7. Name two consecutive years where the percent remained the same.

8. What year(s) shows the lowest percent of arson fires started by juveniles?

9. Estimate the percent of arson fires started by juveniles in 1995.

10. Estimate the percent of arson fires started by juveniles in 1992.

11. What year shows the greatest increase in the percent of fires started by juveniles?

12. What trend do you notice from this graph?

1. _____

2. _____

3. _____

4. _____

5. _____

6. _____

7. _____

8. _____

9. _____

10. _____

11. _____

12. _____

CHAPTER 1 HIGHLIGHTS

DEFINITIONS AND CONCEPTS	EXAMPLES

SECTION 1.1   SYMBOLS AND SETS OF NUMBERS

A **set** is a collection of objects, called **elements**, enclosed in braces.

$\{a, c, e\}$

Natural numbers: $\{1, 2, 3, 4, \ldots\}$

Whole numbers: $\{0, 1, 2, 3, 4, \ldots\}$

Integers: $\{\ldots, -3, -2, -1, 0, 1, 2, 3, \ldots\}$

Rational numbers: {real numbers that can be expressed as a quotient of integers}

Irrational numbers: {real numbers that cannot be expressed as a quotient of integers}

Real numbers: {all numbers that correspond to a point on the number line}

Given the set $\{-3.4, \sqrt{3}, 0, \frac{2}{3}, 5, -4\}$ list the numbers that belong to the set of

Natural numbers   5

Whole numbers   0, 5

Integers $-4, 0, 5$

Rational numbers $-3.4, 0, \frac{2}{3}, 5, -4$

Irrational numbers $\sqrt{3}$

Real numbers $-3.4, \sqrt{3}, 0, \frac{2}{3}, 5, -4$

A line used to picture numbers is called a **number line**.

The **absolute value** of a real number a denoted by $|a|$ is the distance between a and 0 on the number line.

$|5| = 5 \quad |0| = 0 \quad |-2| = 2$

SYMBOLS:
= is equal to
≠ is not equal to
> is greater than
< is less than
≤ is less than or equal to
≥ is greater than or equal to

$-7 = -7$
$3 \neq -3$
$4 > 1$
$1 < 4$
$6 \leq 6$
$18 \geq -\dfrac{1}{3}$

ORDER PROPERTY FOR REAL NUMBERS

For any two real numbers a and b, a is less than b if a is to the left of b on the number line.

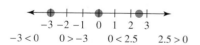

$-3 < 0 \qquad 0 > -3 \qquad 0 < 2.5 \qquad 2.5 > 0$

SECTION 1.2   INTRODUCTION TO VARIABLE EXPRESSIONS AND EQUATIONS

The expression a^n is an **exponential expression**. The number a is called the **base**; it is the repeated factor. The number n is called the **exponent**; it is the number of times that the base is a factor.

$4^3 = 4 \cdot 4 \cdot 4 = 64$
$7^2 = 7 \cdot 7 = 49$

Section 1.2	**(Continued)**

Order of Operations

Simplify expressions in the following order. If grouping symbols are present, simplify expressions within those first, starting with the innermost set. Also, simplify the numerator and the denominator of a fraction separately.

1. Simplify exponential expressions.
2. Multiply or divide in order from left to right.
3. Add or subtract in order from left to right.

A symbol used to represent a number is called a **variable**.

An **algebraic expression** is a collection of numbers, variables, operation symbols, and grouping symbols.

To **evaluate an algebraic expression** containing a variable, substitute a given number for the variable and simplify.

A mathematical statement that two expressions are equal is called an **equation**.

A **solution** of an equation is a value for the variable that makes the equation a true statement.

$$\frac{8^2 + 5(7-3)}{3 \cdot 7} = \frac{8^2 + 5(4)}{21}$$
$$= \frac{64 + 5(4)}{21}$$
$$= \frac{64 + 20}{21}$$
$$= \frac{84}{21}$$
$$= 4$$

Examples of variables are
$$q, \quad x, \quad z$$

Examples of algebraic expressions are

$$5x, \quad 2(y-6), \quad \frac{q^2 - 3q + 1}{6}$$

Evaluate $x^2 - y^2$ when $x = 5$ and $y = 3$.
$$x^2 - y^2 = (5)^2 - 3^2$$
$$= 25 - 9$$
$$= 16$$

Equations:
$$3x - 9 = 20$$
$$A = \pi r^2$$

Determine whether 4 is a solution of $5x + 7 = 27$.

$$5x + 7 = 27$$
$$5(4) + 7 \stackrel{?}{=} 27$$
$$20 + 7 \stackrel{?}{=} 27$$
$$27 = 27 \quad \text{True.}$$

4 is a solution.

Section 1.3	**Adding Real Numbers**

To Add Two Numbers with the Same Sign

1. Add their absolute values.
2. Use their common sign as the sign of the sum.

To Add Two Numbers with Different Signs

1. Subtract their absolute values.
2. Use the sign of the number whose absolute value is larger as the sign of the sum.

Add.

$$10 + 7 = 17$$
$$-3 + (-8) = -11$$

$$-25 + 5 = -20$$
$$14 + (-9) = 5$$

SECTION 1.3 (CONTINUED)	
Two numbers that are the same distance from 0 but lie on opposite sides of 0 are called **opposites** or **additive inverses**. The opposite of a number a is denoted by $-a$.	The opposite of -7 is 7. The opposite of 123 is -123.

SECTION 1.4 SUBTRACTING REAL NUMBERS	
To subtract two numbers a and b, add the first number a to the opposite of the second number b. $a - b = a + (-b)$	Subtract. $3 - (-44) = 3 + 44 = 47$ $-5 - 22 = -5 + (-22) = -27$ $-30 - (-30) = -30 + 30 = 0$

SECTION 1.5 MULTIPLYING REAL NUMBERS	
MULTIPLYING REAL NUMBERS The product of two numbers with the same sign is a positive number. The product of two numbers with different signs is a negative number. PRODUCTS INVOLVING ZERO The product of 0 and any number is 0. $b \cdot 0 = 0$ and $0 \cdot b = 0$	Multiply. $7 \cdot 8 = 56 \qquad -7 \cdot (-8) = 56$ $-2 \cdot 4 = -8 \qquad 2 \cdot (-4) = -8$ $-4 \cdot 0 = 0 \qquad 0 \cdot \left(-\dfrac{3}{4}\right) = 0$

SECTION 1.6 DIVIDING REAL NUMBERS	
QUOTIENT OF TWO REAL NUMBERS $\dfrac{a}{b} = a \cdot \dfrac{1}{b}$ DIVIDING REAL NUMBERS The quotient of two numbers with the same sign is a positive number. The quotient of two numbers with different signs is a negative number.	Divide. $\dfrac{42}{2} = 42 \cdot \dfrac{1}{2} = 21$ $\dfrac{90}{10} = 9 \qquad \dfrac{-90}{-10} = 9$ $\dfrac{42}{-6} = -7 \qquad \dfrac{-42}{6} = -7$

SECTION 1.6   (CONTINUED)

QUOTIENTS INVOLVING ZERO

The quotient of a nonzero number and 0 is undefined.

$\dfrac{b}{0}$ is undefined.

$\dfrac{-85}{0}$ is undefined.

The quotient of 0 and any nonzero number is 0.

$\dfrac{0}{b} = 0$

$\dfrac{0}{18} = 0 \qquad \dfrac{0}{-47} = 0$

SECTION 1.7   PROPERTIES OF REAL NUMBERS

COMMUTATIVE PROPERTIES

Addition: $a + b = b + a$
Multiplication: $a \cdot b = b \cdot a$

$3 + (-7) = -7 + 3$
$-8 \cdot 5 = 5 \cdot (-8)$

ASSOCIATIVE PROPERTIES

Addition: $(a + b) + c = a + (b + c)$
Multiplication: $(a \cdot b) \cdot c = a \cdot (b \cdot c)$

$(5 + 10) + 20 = 5 + (10 + 20)$
$(-3 \cdot 2) \cdot 11 = -3 \cdot (2 \cdot 11)$

Two numbers whose product is 1 are called **multiplicative inverses** or **reciprocals**. The reciprocal of a nonzero number a is $\frac{1}{a}$ because $a \cdot \frac{1}{a} = 1$.

The reciprocal of 3 is $\dfrac{1}{3}$.

The reciprocal of $-\dfrac{2}{5}$ is $-\dfrac{5}{2}$.

DISTRIBUTIVE PROPERTY

$a(b + c) = a \cdot b + a \cdot c$

$5(6 + 10) = 5 \cdot 6 + 5 \cdot 10$
$-2(3 + x) = -2 \cdot 3 + (-2)(x)$

IDENTITIES

$a + 0 = a \qquad 0 + a = a$
$a \cdot 1 = a \qquad 1 \cdot a = a$

$5 + 0 = 5 \qquad 0 + (-2) = -2$
$-14 \cdot 1 = -14 \qquad 1 \cdot 27 = 27$

INVERSES

Additive or opposite: $a + (-a) = 0$

Multiplicative or reciprocal: $b \cdot \dfrac{1}{b} = 1$

$7 + (-7) = 0$

$3 \cdot \dfrac{1}{3} = 1$

SECTION 1.8   READING GRAPHS

To find the value on the vertical axis representing a location on a graph, move horizontally from the location on the graph until the vertical axis is reached. To find the value on the horizontal axis representing a location on a graph, move vertically from the location on the graph until the horizontal axis is reached.

This broken line graph shows the average public classroom teachers' salaries for the school year ending in the years shown.

Estimate the average public teacher's salary for the school year ending in 1989.   The average salary is approximately $29,500.

Find the earliest year that the average salary rose above $32,000.   The year was 1991.

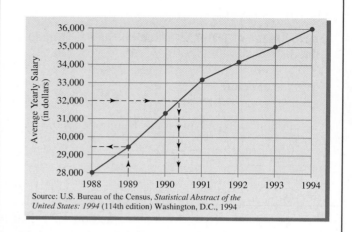

Source: U.S. Bureau of the Census, *Statistical Abstract of the United States: 1994* (114th edition) Washington, D.C., 1994

Equations, Inequalities, and Problem Solving

In this chapter, we solve equations and inequalities. Once we know how to solve equations and inequalities, we may solve problems. Of course, problem solving is an integral topic in algebra and its discussion is continued throughout this text.

A glacier is a giant mass of rocks and ice that flows downhill like a river. Alaska alone has an estimated 100,000 glaciers. They form high in the mountains where snow does not melt. After years of accumulated snowfall, the weight of the compacted ice causes it to slide downhill, and a glacier is formed. In Example 1 on page 93, we will find the time it takes for ice at the beginning of a glacier to reach its face, or ending wall.

Glaciers account for approximately 77 percent of the world's fresh water. They also contain records of past environments between their layers of snow. Scientists study glaciers to learn about past global climates and then use this information to predict future environmental changes.

(*Source: Blue Ice in Motion—The Story of Alaska's Glaciers*, by Sally D. Wiley, published in 1990 by The Alaska Natural History Association and Judith Ann Rose)

Problem Solving Notes

2.1   SIMPLIFYING EXPRESSIONS

As we explore in this section, we will see that an expression such as $3x + 2x$ is not as simple as possible. This is because—even without replacing x by a value—we can perform the indicated addition.

A   IDENTIFYING TERMS, LIKE TERMS, AND UNLIKE TERMS

Before we practice simplifying expressions, some new language is presented. A **term** is a number or the product of a number and variables raised to powers.

Terms

$$-y, \quad 2x^3, \quad -5, \quad 3xz^2, \quad \frac{2}{y}, \quad 0.8z$$

The **numerical coefficient** of a term is the numerical factor. The numerical coefficient of $3x$ is 3. Recall that $3x$ means $3 \cdot x$.

Term	Numerical Coefficient
$3x$	3
$\frac{y^3}{5}$	$\frac{1}{5}$ since $\frac{y^3}{5}$ means $\frac{1}{5} \cdot y^3$
$-0.7ab^3c^5$	-0.7
z	1
$-y$	-1
-5	-5

> **HELPFUL HINT**
>
> The term $-y$ means $-1y$ and thus has a numerical coefficient of -1. The term z means $1z$ and thus has a numerical coefficient of 1.

Example 1   Identify the numerical coefficient in each term.

    a. $-3y$ **b.** $22z^4$ **c.** y **d.** $-x$ **e.** $\frac{x}{7}$

Solution: **a.** The numerical coefficient of $-3y$ is -3.
        b. The numerical coefficient of $22z^4$ is 22.
        c. The numerical coefficient of y is 1, since y is $1y$.
        d. The numerical coefficient of $-x$ is -1, since $-x$ is $-1x$.
        e. The numerical coefficient of $\frac{x}{7}$ is $\frac{1}{7}$, since $\frac{x}{7}$ is $\frac{1}{7} \cdot x$.

Terms with the same variables raised to exactly the same powers are called **like terms**. Terms that aren't like terms are called **unlike terms**.

Like Terms	Unlike Terms	
$3x, 2x$	$5x, 5x^2$	Why? Same variable x, but different powers x and x^2
$-6x^2y, 2x^2y, 4x^2y$	$7y, 3z, 8x^2$	Why? Different variables
$2ab^2c^3, ac^3b^2$	$6abc^3, 6ab^2$	Why? Different variables and different powers

Objectives

A Identify terms, like terms, and unlike terms.

B Combine like terms.

C Simplify expressions containing parentheses.

D Write word phrases as algebraic expressions.

SSM   CD-ROM   Video 2.1

Practice Problem 1

Identify the numerical coefficient in each term.

a. $-4x$ b. $15y^3$ c. x d. $-y$ e. $\frac{z}{4}$

Answers

1. a. -4,   **b.** 15,   **c.** 1,   **d.** -1,   **e.** $\frac{1}{4}$

HELPFUL HINT

In like terms, each variable and its exponent must match exactly, but these factors don't need to be in the same order.

$2x^2y$ and $3yx^2$ are like terms.

Practice Problem 2

Determine whether the terms are like or unlike.

a. $7x, -6x$ b. $3x^2y^2, -x^2y^2, 4x^2y^2$

c. $-5ab, 3ba$

Example 2   Determine whether the terms are like or unlike.

     a. $2x, 3x^2$ b. $4x^2y, x^2y, -2x^2y$ c. $-2yz, -3zy$
     d. $-x^4, x^4$

Solution: a. Unlike terms, since the exponents on x are not the same.
        b. Like terms, since each variable and its exponent match.
        c. Like terms, since $zy = yz$ by the commutative property.
        d. Like terms.

B   COMBINING LIKE TERMS

An algebraic expression containing the sum or difference of like terms can be simplified by applying the distributive property. For example, by the distributive property, we rewrite the sum of the like terms $3x + 2x$ as

$$3x + 2x = (3 + 2)x = 5x$$

Also,

$$-y^2 + 5y^2 = (-1 + 5)y^2 = 4y^2$$

Simplifying the sum or difference of like terms is called **combining like terms**.

Practice Problem 3

Simplify each expression by combining like terms.

a. $9y - 4y$ b. $11x^2 + x^2$

c. $5y - 3x + 6x$

Example 3   Simplify each expression by combining like terms.

     a. $7x - 3x$ b. $10y^2 + y^2$ c. $8x^2 + 2x - 3x$

Solution: a. $7x - 3x = (7 - 3)x = 4x$
        b. $10y^2 + y^2 = (10 + 1)y^2 = 11y^2$
        c. $8x^2 + 2x - 3x = 8x^2 + (2 - 3)x = 8x^2 - x$

Practice Problems 4–7

Simplify each expression by combining like terms.

4. $7y + 2y + 6 + 10$

5. $-2x + 4 + x - 11$

6. $3z - 3z^2$

7. $8.9y + 4.2y - 3$

Examples   Simplify each expression by combining like terms.

4. $2x + 3x + 5 + 2 = (2 + 3)x + (5 + 2)$
                  $= 5x + 7$

5. $-5a - 3 + a + 2 = -5a + 1a + (-3 + 2)$
                    $= (-5 + 1)a + (-3 + 2)$
                     $= -4a - 1$

6. $4y - 3y^2$ These two terms cannot be combined because they are unlike terms.

7. $2.3x + 5x - 6 = (2.3 + 5)x - 6$
                  $= 7.3x - 6$

The examples above suggest the following.

Answers
2. a. like,   **b.** like,   **c.** like,
3. a. $5y$,   **b.** $12x^2$,   **c.** $5y + 3x$,
4. $9y + 16$,   **5.** $-x - 7$,   **6.** $3z - 3z^2$,
7. $13.1y - 3$

Combining Like Terms

To **combine like terms**, combine the numerical coefficients and multiply the result by the common variable factors.

C Simplifying Expressions Containing Parentheses

In simplifying expressions we make frequent use of the distributive property to remove parentheses.

Examples   Find each product by using the distributive property to remove parentheses.

8. $5(x + 2) = 5(x) + 5(2)$ Apply the distributive property.
$= 5x + 10$ Multiply.

9. $-2(y + 0.3z - 1) = -2(y) + (-2)(0.3z)$ Apply the distributive property.
$+ (-2)(-1)$
$= -2y - 0.6z + 2$ Multiply.

10. $-(x + y - 2z + 6) = -1(x + y - 2z + 6)$ Distribute -1 over each term.
$= -1(x) - 1(y) - 1(-2z)$
$- 1(6)$
$= -x - y + 2z - 6$

Find each product by using the distributive property to remove parentheses.

8. $3(y + 6)$

9. $-4(x + 0.2y - 3)$

10. $-(3x + 2y + z - 1)$

HELPFUL HINT

If a "$-$" sign precedes parentheses, the sign of each term inside the parentheses is changed when the distributive property is applied to remove parentheses.

Examples:
$-(2x + 1) = -2x - 1$
$-(x - 2y) = -x + 2y$
$-(-5x + y - z) = 5x - y + z$
$-(-3x - 4y - 1) = 3x + 4y + 1$

When simplifying an expression containing parentheses, we often use the distributive property first to remove parentheses and then again to combine any like terms.

Examples   Simplify each expression.

11. $3(2x - 5) + 1 = 6x - 15 + 1$ Apply the distributive property.
$= 6x - 14$ Combine like terms.

12. $8 - (7x + 2) + 3x = 8 - 7x - 2 + 3x$ Apply the distributive property.
$= -7x + 3x + 8 - 2$
$= -4x + 6$ Combine like terms.

Simplify each expression.

11. $4(x - 6) + 20$

12. $5 - (3x + 9)$

13. $-3(7x + 1) - (4x - 2)$

Answers
8. $3y + 18$, **9.** $-4x - 0.8y + 12$,
10. $-3x - 2y - z + 1$, **11.** $4x - 4$,
12. $-3x - 4$, **13.** $-25x - 1$

13. $-2(4x + 7) - (3x - 1) = -8x - 14 - 3x + 1$ Apply the distributive property.

$$= -11x - 13$$ Combine like terms.

Practice Problem 14

Subtract $9x - 10$ from $4x - 3$.

Example 14   Subtract $4x - 2$ from $2x - 3$.

Solution: We first note that "subtract $4x - 2$ **from** $2x - 3$" translates to $(2x - 3) - (4x - 2)$. Next, we simplify the algebraic expression.

$$(2x - 3) - (4x - 2) = 2x - 3 - 4x + 2$$ Apply the distributive property.

$$= -2x - 1$$ Combine like terms.

Practice Problems 15–17

Write each phrase as an algebraic expression and simplify if possible. Let x represent the unknown number.

15. Three times a number, *subtracted from* 10

16. The sum of a number and 2, divided by 5

17. Three times a number, added to the sum of twice a number and 6

D   WRITING ALGEBRAIC EXPRESSIONS

To prepare for problem solving, we next practice writing word phrases as algebraic expressions.

Examples

Write each phrase as an algebraic expression and simplify if possible. Let x represent the unknown number.

15. Twice a number, plus 6

$$2x \qquad + \quad 6$$

This expression cannot be simplified.

16. The difference of a number and 4, divided by 7

$$(x - 4) \qquad \div \quad 7$$

This expression cannot be simplified.

17. Five plus the sum of a number and 1

$$5 \quad + \qquad (x + 1)$$

Next, we simplify this expression.

$$5 + (x + 1) = 5 + x + 1$$
$$= 6 + x$$

Answers

14. $-5x + 7$,   **15.** $10 - 3x$,   **16.** $\dfrac{(x + 2)}{5}$,

17. $5x + 6$

EXERCISE SET 2.1

A *Identify the numerical coefficient of each term. See Example 1.*

1. $-7y$ **2.** x

Indicate whether each list of terms are like or unlike. See Example 2.

3. $5y, -y$ **4.** $-2x^2y, 6xy$

B *Simplify each expression by combining any like terms. See Examples 3 through 7.*

5. $7y + 8y$ **6.** $3x + 2x$ **7.** $8w - w + 6w$

8. $c - 7c + 2c$ **9.** $3b - 5 - 10b - 4$ **10.** $6g + 5 - 3g - 7$

11. $m - 4m + 2m - 6$ **12.** $a + 3a - 2 - 7a$ **13.** $5g - 3 - 5 - 5g$

14. $8p + 4 - 8p - 15$

C *Simplify each expression. Use the distributive property to remove any parentheses. See Examples 8 through 10.*

15. $5(y + 4)$ **16.** $7(r + 3)$

Remove parentheses and simplify each expression. See Examples 11 through 13.

17. $7(d - 3) + 10$

Write each phrase as an algebraic expression. Simplify if possible. See Example 14.

18. Add $6x + 7$ to $4x - 10$ **19.** Add $3y - 5$ to $y + 16$

20. Subtract $7x + 1$ from $3x - 8$

ANSWERS

1. _____
2. _____
3. _____
4. _____
5. _____
6. _____
7. _____
8. _____
9. _____
10. _____
11. _____
12. _____
13. _____
14. _____
15. _____
16. _____
17. _____
18. _____
19. _____
20. _____

Problem Solving Notes

2.2   THE ADDITION PROPERTY OF EQUALITY

A USING THE ADDITION PROPERTY

Recall from Section 1.2 that an equation is a statement that two expressions have the same value. Also, a value of the variable that makes an equation a true statement is called a solution or root of the equation. The process of finding the solution of an equation is called **solving** the equation for the variable. In this section, we concentrate on solving **linear equations** in one variable.

LINEAR EQUATION IN ONE VARIABLE

A linear equation in one variable can be written in the form
$$Ax + B = C$$
where A, B, and C are real numbers and $A \neq 0$.

Evaluating a linear equation for a given value of the variable, as we did in Section 1.2, can tell us whether that value is a solution. But we can't rely on evaluating an equation as our method of solving it—with what value would we start?

Instead, to solve a linear equation in x, we write a series of simpler equations, all *equivalent* to the original equation, so that the final equation has the form

$$x = \text{number} \qquad \text{or} \qquad \text{number} = x$$

Equivalent equations are equations that have the same solution. This means that the "number" above is the solution to the original equation.

The first property of equality that helps us write simpler equivalent equations is the **addition property of equality**.

ADDITION PROPERTY OF EQUALITY

If a, b, and c are real numbers, then
$$a = b \qquad \text{and} \qquad a + c = b + c$$
are equivalent equations.

This property guarantees that adding the same number to both sides of an equation does not change the solution of the equation. Since subtraction is defined in terms of addition, we may also **subtract the same number from both sides** without changing the solution.

A good way to picture a true equation is as a balanced scale. Since it is balanced, each side of the scale weighs the same amount.

$x - 2$ 5

Objectives

A Use the addition property of equality to solve linear equations.

B Simplify an equation and then use the addition property of equality.

C Write word phrases as algebraic expressions.

SSM   CD-ROM   Video
                                2.2

If the same weight is added to or subtracted from each side, the scale remains balanced.

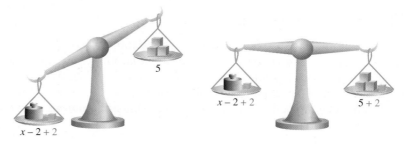

We use the addition property of equality to write equivalent equations until the variable is alone (by itself on one side of the equation) and the equation looks like "x = number" or "number = x."

TRY THE CONCEPT CHECK IN THE MARGIN.

✓ **CONCEPT CHECK**

Use the addition property to fill in the blank so that the middle equation simplifies to the last equation.

$$x - 5 = 3$$
$$x - 5 + \underline{} = 3 + \underline{}$$
$$x = 8$$

Practice Problem 1

Solve $x - 5 = 8$ for x.

Example 1   Solve $x - 7 = 10$ for x.

Solution: To solve for x, we first get x alone on one side of the equation. To do this, we add 7 to both sides of the equation.

$$x - 7 = 10$$
$$x - 7 + 7 = 10 + 7 \qquad \text{Add 7 to both sides.}$$
$$x = 17 \qquad \text{Simplify.}$$

The solution of the equation $x = 17$ is obviously 17. Since we are writing equivalent equations, the solution of the equation $x - 7 = 10$ is also 17.

Check: To check, replace x with 17 in the original equation.

$$x - 7 = 10 \qquad \text{Original equation.}$$
$$17 - 7 \stackrel{?}{=} 10 \qquad \text{Replace } x \text{ with 17.}$$
$$10 = 10 \qquad \text{True.}$$

Since the statement is true, 17 is the solution.   ▬▬▬

Practice Problem 2

Solve: $y + 1.7 = 0.3$

Example 2   Solve: $y + 0.6 = -1.0$

Solution: To solve for y, we subtract 0.6 from both sides of the equation.

$$y + 0.6 = -1.0$$
$$y + 0.6 - 0.6 = -1.0 - 0.6 \qquad \text{Subtract 0.6 from both sides.}$$
$$y = -1.6 \qquad \text{Combine like terms.}$$

Check: $$y + 0.6 = -1.0 \qquad \text{Original equation.}$$
$$-1.6 + 0.6 \stackrel{?}{=} -1.0 \qquad \text{Replace } y \text{ with } -1.6.$$
$$-1.0 = -1.0 \qquad \text{True.}$$

The solution is -1.6.   ▬▬▬

Example 3 Solve: $\frac{1}{2} = x - \frac{3}{4}$

Solution: To get x alone, we add $\frac{3}{4}$ to both sides.

$$\frac{1}{2} = x - \frac{3}{4}$$

$$\frac{1}{2} + \frac{3}{4} = x - \frac{3}{4} + \frac{3}{4} \quad \text{Add } \frac{3}{4} \text{ to both sides.}$$

$$\frac{1}{2} \cdot \frac{2}{2} + \frac{3}{4} = x \quad \text{The LCD is 4.}$$

$$\frac{2}{4} + \frac{3}{4} = x \quad \text{Add the fractions.}$$

$$\frac{5}{4} = x$$

Check: $\quad \frac{1}{2} = x - \frac{3}{4} \quad$ Original equation.

$$\frac{1}{2} \overset{?}{=} \frac{5}{4} - \frac{3}{4} \quad \text{Replace } x \text{ with } \frac{5}{4}.$$

$$\frac{1}{2} \overset{?}{=} \frac{2}{4} \quad \text{Subtract.}$$

$$\frac{1}{2} = \frac{1}{2} \quad \text{True.}$$

The solution is $\frac{5}{4}$.

Practice Problem 3

Solve: $\frac{7}{8} = y - \frac{1}{3}$

HELPFUL HINT

We may solve an equation so that the variable is alone on *either* side of the equation. For example, $\frac{5}{4} = x$ is equivalent to $x = \frac{5}{4}$.

Example 4 Solve: $5t - 5 = 6t$

Solution: To solve for t, we first want all terms containing t on one side of the equation. To do this, we subtract $5t$ from both sides of the equation.

$$5t - 5 = 6t$$

$$5t - 5 - 5t = 6t - 5t \quad \text{Subtract } 5t \text{ from both sides.}$$

$$-5 = t \quad \text{Combine like terms.}$$

Check: $\quad 5t - 5 = 6t \quad$ Original equation.

$$5(-5) - 5 \overset{?}{=} 6(-5) \quad \text{Replace } t \text{ with } -5.$$

$$-25 - 5 \overset{?}{=} -30$$

$$-30 = -30 \quad \text{True.}$$

The solution is -5.

Practice Problem 4

Solve: $3x + 10 = 4x$

B SIMPLIFYING EQUATIONS

Many times, it is best to simplify one or both sides of an equation before applying the addition property of equality.

Practice Problem 5

Solve: $10w + 3 - 4w + 4$
$$= -2w + 3 + 7w$$

Example 5

Solve: $2x + 3x - 5 + 7 = 10x + 3 - 6x - 4$

Solution: First we simplify both sides of the equation.

$2x + 3x - 5 + 7 = 10x + 3 - 6x - 4$
$5x + 2 = 4x - 1$ Combine like terms on each side of the equation.

Next, we want all terms with a variable on one side of the equation and all numbers on the other side.

$5x + 2 - 4x = 4x - 1 - 4x$ Subtract $4x$ from both sides.
$x + 2 = -1$ Combine like terms.
$x + 2 - 2 = -1 - 2$ Subtract 2 from both sides to get x alone.
$x = -3$ Combine like terms.

Check: $2x + 3x - 5 + 7 = 10x + 3 - 6x - 4$ Original equation.
$2(-3) + 3(-3) - 5 + 7 \stackrel{?}{=} 10(-3) + 3 - 6(-3) - 4$

Replace x with -3.

$-6 - 9 - 5 + 7 \stackrel{?}{=} -30 + 3 + 18 - 4$ Multiply.
$-13 = -13$ True.

The solution is -3.   ▬▬▬

If an equation contains parentheses, we use the distributive property to remove them, as before. Then we combine any like terms.

Practice Problem 6

Solve: $3(2w - 5) - (5w + 1) = -3$

Example 6

Solve: $6(2a - 1) - (11a + 6) = 7$

Solution:

$\overset{\frown}{6(2a} - 1) - \overset{\frown}{1(11a} + 6) = 7$

$6(2a) + 6(-1) - 1(11a) - 1(6) = 7$ Apply the distributive property.
$12a - 6 - 11a - 6 = 7$ Multiply.
$a - 12 = 7$ Combine like terms.
$a - 12 + 12 = 7 + 12$ Add 12 to both sides.
$a = 19$ Simplify.

Check: Check by replacing a with 19 in the original equation.   ▬▬▬

Practice Problem 7

Solve: $12 - y = 9$

Example 7

Solve: $3 - x = 7$

Solution: First we subtract 3 from both sides.

$3 - x = 7$
$3 - x - 3 = 7 - 3$ Subtract 3 from both sides.
$-x = 4$ Simplify.

Answers
5. $w = -4$,   **6.** $w = 13$,   **7.** $y = 3$

We have not yet solved for x since x is not alone. However, this equation does say that the opposite of x is 4. If the opposite of x is 4, then x is the opposite of 4, or $x = -4$.

If $-x = 4$,

then $x = -4$.

Check: $3 - x = 7$ Original equation.

$3 - (-4) \stackrel{?}{=} 7$ Replace x with -4.

$3 + 4 \stackrel{?}{=} 7$ Add.

$7 = 7$ True.

The solution is -4.

C WRITING ALGEBRAIC EXPRESSIONS

In this section, we continue to practice writing algebraic expressions.

Example 8   **a.** The sum of two numbers is 8. If one number is 3, find the other number.

b. The sum of two numbers is 8. If one number is x, write an expression representing the other number.

Solution: **a.** If the sum of two numbers is 8 and one number is 3, we find the other number by subtracting 3 from 8. The other number is $8 - 3$, or 5.

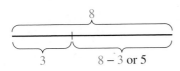

b. If the sum of two numbers is 8 and one number is x, we find the other number by subtracting x from 8. The other number is represented by $8 - x$.

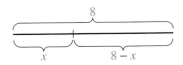

Focus On History

THE GOLDEN RECTANGLE IN ART

The golden rectangle is a rectangle whose length is approximately 1.6 times its width. The early Greeks thought that a rectangle with these dimensions was the most pleasing to the eye. Examples of the golden rectangle are found in many ancient, as well as modern, works of art. For example, the Parthenon in Athens, Greece, shows the golden rectangle in many aspects of its design. Modern-era artists, including Piet Mondrian (1872–1944) and Georges Seurat (1859–1891), also frequently used the proportions of a golden rectangle in their paintings.

Mondrian

Composition with Gray and Light Brown 1918, Oil on canvas, 80.2 x 49.9 cm (31 9/16 x 19 5/8 in); Museum of Fine Arts, Houston, Texas

To test whether a rectangle is a golden rectangle, divide the rectangle's length by its width. If the result is approximately 1.6, we can consider the rectangle to be a golden rectangle. For instance, consider Mondrian's *Composition with Gray and Light Brown*, which was painted on an 80.2 × 49.9 cm canvas. Because $\frac{80.2}{49.9} \approx 1.6$, the dimensions of the canvas form a golden rectangle. In what other ways are golden rectangles connected with this painting?

Examples of golden rectangles can be found in the designs of many everyday objects. Visual artists, from architects to product and package designers, use the golden rectangle shape in such things as the face of a building, the floor of a room, the front of a food package, the front cover of a book, and even the shape of a credit card.

GROUP ACTIVITY

Find an example of a golden rectangle in a building or an everyday object. Use a ruler to measure its dimensions and verify that the length is approximately 1.6 times the width.

EXERCISE SET 2.2

A *Solve each equation. Check each solution. See Examples 1 through 4.*

1. $x + 7 = 10$

2. $x + 14 = 25$

▣ **3.** $x - 2 = -4$

4. $y - 9 = 1$

5. $3 + x = -11$

6. $8 + z = -8$

7. $r - 8.6 = -8.1$

8. $t - 9.2 = -6.8$

9. $\dfrac{1}{3} + f = \dfrac{3}{4}$

10. $c + \dfrac{1}{6} = \dfrac{3}{8}$

Problem Solving Notes

2.3   THE MULTIPLICATION PROPERTY OF EQUALITY

A   USING THE MULTIPLICATION PROPERTY

As useful as the addition property of equality is, it cannot help us solve every type of linear equation in one variable. For example, adding or subtracting a value on both sides of the equation does not help solve

$$\frac{5}{2}x = 15$$

Instead, we apply another important property of equality, the **multiplication property of equality**.

MULTIPLICATION PROPERTY OF EQUALITY

If a, b, and c are real numbers and $c \neq 0$, then

$$a = b \quad \text{and} \quad ac = bc$$

are equivalent equations.

This property guarantees that multiplying both sides of an equation by the same nonzero number does not change the solution of the equation. Since division is defined in terms of multiplication, we may also **divide both sides of the equation by the same nonzero number** without changing the solution.

Example 1   Solve: $\frac{5}{2}x = 15$

Solution: To get x alone, we multiply both sides of the equation by the reciprocal of $\frac{5}{2}$, which is $\frac{2}{5}$.

$$\frac{5}{2}x = 15$$

$$\frac{2}{5} \cdot \left(\frac{5}{2}x \right) = \frac{2}{5} \cdot 15 \quad \text{Multiply both sides by } \frac{2}{5}.$$

$$\left(\frac{2}{5} \cdot \frac{5}{2} \right)x = \frac{2}{5} \cdot 15 \quad \text{Apply the associative property.}$$

$$1x = 6 \quad \text{Simplify.}$$

or

$$x = 6$$

Check: Replace x with 6 in the original equation.

$$\frac{5}{2}x = 15 \quad \text{Original equation.}$$

$$\frac{5}{2}(6) \stackrel{?}{=} 15 \quad \text{Replace } x \text{ with 6.}$$

$$15 = 15 \quad \text{True.}$$

The solution is 6.

Objectives

A Use the multiplication property of equality to solve linear equations.

B Use both the addition and multiplication properties of equality to solve linear equations.

C Write word phrases as algebraic expressions.

SSM   CD-ROM   Video
2.3

Practice Problem 1

Solve: $\frac{3}{7}x = 9$

Answer

1. $x = 21$

In the equation $\frac{5}{2}x = 15$, $\frac{5}{2}$ is the coefficient of x. When the coefficient of x is a *fraction*, we will get x alone by multiplying by the reciprocal. When the coefficient of x is an integer or a decimal, it is usually more convenient to divide both sides by the coefficient. (Dividing by a number is, of course, the same as multiplying by the reciprocal of the number.)

Practice Problem 2

Solve: $7x = 42$

Example 2

Solve: $5x = 30$

Solution: To get x alone, we divide both sides of the equation by 5, the coefficient of x.

$$5x = 30$$

$$\frac{5x}{5} = \frac{30}{5} \quad \text{Divide both sides by 5.}$$

$$1 \cdot x = 6 \quad \text{Simplify.}$$

$$x = 6$$

Check: $5x = 30$ Original equation.

$5 \cdot 6 \stackrel{?}{=} 30$ Replace x with 6.

$30 = 30$ True.

The solution is 6.   ▬▬▬

Practice Problem 3

Solve: $-4x = 52$

Example 3

Solve: $-3x = 33$

Solution: Recall that $-3x$ means $-3 \cdot x$. To get x alone, we divide both sides by the coefficient of x, that is, -3.

$$-3x = 33$$

$$\frac{-3x}{-3} = \frac{33}{-3} \quad \text{Divide both sides by } -3.$$

$$1x = -11 \quad \text{Simplify.}$$

$$x = -11$$

Check: $-3x = 33$ Original equation.

$-3(-11) \stackrel{?}{=} 33$ Replace x with -11.

$33 = 33$ True.

The solution is -11.   ▬▬▬

Practice Problem 4

Solve: $\frac{y}{5} = 13$

Example 4

Solve: $\frac{y}{7} = 20$

Solution: Recall that $\frac{y}{7} = \frac{1}{7}y$. To get y alone, we multiply both sides of the equation by 7, the reciprocal of $\frac{1}{7}$.

$$\frac{y}{7} = 20$$

$$\frac{1}{7}y = 20$$

$$7 \cdot \frac{1}{7}y = 7 \cdot 20 \quad \text{Multiply both sides by 7.}$$

$$1y = 140 \quad \text{Simplify.}$$

$$y = 140$$

Answers

2. $x = 6$,   **3.** $x = -13$,   **4.** $y = 65$

Check: $\dfrac{y}{7} = 20$ Original equation.

$\dfrac{140}{7} \overset{?}{=} 20$ Replace y with 140.

$20 = 20$ True.

The solution is 140.

Example 5   Solve: $3.1x = 4.96$

Solution: $3.1x = 4.96$

$\dfrac{3.1x}{3.1} = \dfrac{4.96}{3.1}$ Divide both sides by 3.1.

$1x = 1.6$ Simplify.

$x = 1.6$

Check: Check by replacing x with 1.6 in the original equation. The solution is 1.6.

Example 6   Solve: $-\dfrac{2}{3}x = -\dfrac{5}{2}$

Solution: To get x alone, we multiply both sides of the equation by $-\dfrac{3}{2}$, the reciprocal of the coefficient of x.

$-\dfrac{2}{3}x = -\dfrac{5}{2}$

$-\dfrac{3}{2} \cdot -\dfrac{2}{3}x = -\dfrac{3}{2} \cdot -\dfrac{5}{2}$ Multiply both sides by $-\dfrac{3}{2}$, the reciprocal of $-\dfrac{2}{3}$.

$x = \dfrac{15}{4}$ Simplify.

Check: Check by replacing x with $\dfrac{15}{4}$ in the original equation. The solution is $\dfrac{15}{4}$.

B USING BOTH THE ADDITION AND MULTIPLICATION PROPERTIES

We are now ready to combine the skills learned in the last section with the skills learned from this section to solve equations by applying more than one property.

Example 7   Solve: $-z - 4 = 6$

Solution: First, to get $-z$, the term containing the variable alone, we add 4 to both sides of the equation.

$-z - 4 + 4 = 6 + 4$ Add 4 to both sides.

$-z = 10$ Simplify.

Next, recall that $-z$ means $-1 \cdot z$. Thus to get z alone, we either multiply or divide both sides of the equation by -1. In this example, we divide.

$-z = 10$

$\dfrac{-z}{-1} = \dfrac{10}{-1}$ Divide both sides by the coefficient -1.

$1z = -10$ Simplify.

$z = -10$

Practice Problem 8

Solve: $-7x + 2x + 3 - 20 = -2$

Check:

$$-z - 4 = 6 \quad \text{Original equation.}$$
$$-(-10) - 4 \stackrel{?}{=} 6 \quad \text{Replace z with } -10.$$
$$10 - 4 \stackrel{?}{=} 6$$
$$6 = 6 \quad \text{True.}$$

The solution is -10.

Example 8   Solve: $a + a - 10 + 7 = -13$

Solution: First, we simplify both sides of the equation by combining like terms.

$$a + a - 10 + 7 = -13$$
$$2a - 3 = -13 \qquad \text{Combine like terms.}$$
$$2a - 3 + 3 = -13 + 3 \qquad \text{Add 3 to both sides.}$$
$$2a = -10 \qquad \text{Simplify.}$$
$$\frac{2a}{2} = \frac{-10}{2} \qquad \text{Divide both sides by 2.}$$
$$a = -5 \qquad \text{Simplify.}$$

Check: To check, replace a with -5 in the original equation. The solution is -5.

C   WRITING ALGEBRAIC EXPRESSIONS

We continue to sharpen our problem-solving skills by writing algebraic expressions.

Example 9   Writing an Expression for Consecutive Integers

If x is the first of three consecutive integers, express the sum of the three integers in terms of x. Simplify if possible.

Practice Problem 9

If x is the first of two consecutive integers, express the sum of the first and the second integer in terms of x. Simplify if possible.

Solution: An example of three consecutive integers is

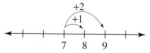

The second consecutive integer is always 1 more than the first, and the third consecutive integer is 2 more than the first. If x is the first of three consecutive integers, the three consecutive integers are

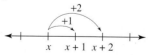

Their sum is

In words: | first integer | + | second integer | + | third integer |

Translate: x $+$ $(x + 1)$ $+$ $(x + 2)$

which simplifies to $3x + 3$.

Study these examples of consecutive even and odd integers.

Even integers:

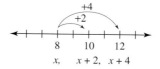

Odd integers:

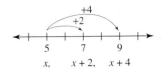

$x,\quad x+2,\quad x+4$

⌐HELPFUL HINT

If x is an odd integer, then $x + 2$ is the next odd integer. This 2 simply means that odd integers are always 2 units from each other.

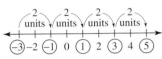

Focus On The Real World

SURVEYS

Recall that the golden rectangle is a rectangle whose length is approximately 1.6 times its width. It is thought that for about 75% of adults, a rectangle in the shape of the golden rectangle is most pleasing to the eye.

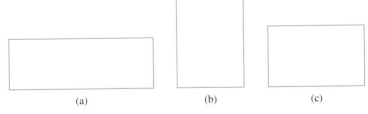

(a)   (b)   (c)

GROUP ACTIVITIES

1. Measure the dimensions of each of the three rectangles shown above and decide which one best approximates the shape of the golden rectangle.
2. Using the three rectangles shown above, conduct a survey asking students which rectangle they prefer. (To avoid bias, take care not to reveal which rectangle is the golden rectangle.) Tally your results and find the percent of survey respondents who preferred each rectangle. Do your results agree with the percent suggested above?

Problem Solving Notes

Name _____ **Section** _____ **Date** _____

MENTAL MATH

Solve each equation mentally. See Examples 2 and 3.

1. $3a = 27$ **2.** $9c = 54$ **3.** $5b = 10$

4. $7t = 14$ **5.** $6x = -30$

EXERCISE SET 2.3

A Solve each equation. Check each solution. See Examples 1 through 6.

1. $-5x = 20$ **2.** $7x = 49$ **3.** $3x = 0$

4. $2x = 0$ **5.** $-x = -12$ **6.** $-y = 8$

7. $\frac{2}{3}x = -8$ **8.** $\frac{3}{4}n = -15$ **9.** $\frac{1}{6}d = \frac{1}{2}$

10. $\frac{1}{8}v = \frac{1}{4}$

B Solve each equation. Check each solution. See Examples 7 and 8.

11. $2x - 4 = 16$ **12.** $3x - 1 = 26$ **13.** $-x + 2 = 22$

14. $-x + 4 = -24$ **15.** $6a + 3 = 3$

Problem Solving Notes

2.4   Further Solving Linear Equations

A   Solving Linear Equations

We now combine our knowledge from the previous sections into a general strategy for solving linear equations. One new piece in this strategy is a suggestion to "clear an equation of fractions" as a first step. Doing so makes the equation more manageable, since working with integers is more convenient than working with fractions.

Objectives

A Apply the general strategy for solving a linear equation.

B Solve equations containing fractions or decimals.

C Recognize identities and equations with no solution.

SSM   CD-ROM   Video 2.4

To Solve Linear Equations in One Variable

Step 1. Multiply on both sides to clear the equation of fractions if they occur.

Step 2. Use the distributive property to remove parentheses if they occur.

Step 3. Simplify each side of the equation by combining like terms.

Step 4. Get all variable terms on one side and all numbers on the other side by using the addition property of equality.

Step 5. Get the variable alone by using the multiplication property of equality.

Step 6. Check the solution by substituting it into the original equation.

Example 1   Solve: $4(2x - 3) + 7 = 3x + 5$

Solution: There are no fractions, so we begin with Step 2.

$$4(2x - 3) + 7 = 3x + 5$$

Step 2. $8x - 12 + 7 = 3x + 5$ Apply the distributive property.

Step 3. $\quad 8x - 5 = 3x + 5$ Combine like terms.

Step 4. Get all variable terms on the same side of the equation by subtracting $3x$ from both sides; then adding 5 to both sides.

$$8x - 5 - 3x = 3x + 5 - 3x \quad \text{Subtract } 3x \text{ from both sides.}$$
$$5x - 5 = 5 \quad \text{Simplify.}$$
$$5x - 5 + 5 = 5 + 5 \quad \text{Add 5 to both sides.}$$
$$5x = 10 \quad \text{Simplify.}$$

Step 5. Use the multiplication property of equality to get x alone.

$$\frac{5x}{5} = \frac{10}{5} \quad \text{Divide both sides by 5.}$$
$$x = 2 \quad \text{Simplify.}$$

Step 6. Check.

$$4(2x - 3) + 7 = 3x + 5 \quad \text{Original equation}$$
$$4[2(2) - 3] + 7 \stackrel{?}{=} 3(2) + 5 \quad \text{Replace } x \text{ with 2.}$$
$$4(4 - 3) + 7 \stackrel{?}{=} 6 + 5$$
$$4(1) + 7 \stackrel{?}{=} 11$$
$$4 + 7 \stackrel{?}{=} 11$$
$$11 = 11 \quad \text{True.}$$

The solution is 2.

Practice Problem 1

Solve: $5(3x - 1) + 2 = 12x + 6$

Answer

1. $x = 3$

> **HELPFUL HINT**
>
> When checking solutions, use the original written equation.

Practice Problem 2

Solve: $9(5 - x) = -3x$

Example 2   Solve: $8(2 - t) = -5t$

Solution: First, we apply the distributive property.

$$\overset{\frown}{8(2 - t)} = -5t$$

Step 2.	$16 - 8t = -5t$	Use the distributive property.
Step 4.	$16 - 8t + 8t = -5t + 8t$	Add $8t$ to both sides.
	$16 = 3t$	Combine like terms.
Step 5.	$\dfrac{16}{3} = \dfrac{3t}{3}$	Divide both sides by 3.
	$\dfrac{16}{3} = t$	Simplify.

Step 6. Check.

$8(2 - t) = -5t$	Original equation
$8\left(2 - \dfrac{16}{3}\right) \overset{?}{=} -5\left(\dfrac{16}{3}\right)$	Replace t with $\dfrac{16}{3}$.
$8\left(\dfrac{6}{3} - \dfrac{16}{3}\right) \overset{?}{=} -\dfrac{80}{3}$	The LCD is 3.
$8\left(-\dfrac{10}{3}\right) \overset{?}{=} -\dfrac{80}{3}$	Subtract fractions.
$-\dfrac{80}{3} \overset{?}{=} -\dfrac{80}{3}$	True.

The solution is $\dfrac{16}{3}$.

B   SOLVING EQUATIONS CONTAINING FRACTIONS OR DECIMALS

If an equation contains fractions, we can clear the equation of fractions by multiplying both sides by the LCD of all denominators. By doing this, we avoid working with time-consuming fractions.

Practice Problem 3

Solve: $\dfrac{5}{2}x - 1 = \dfrac{3}{2}x - 4$

Example 3   Solve: $\dfrac{x}{2} - 1 = \dfrac{2}{3}x - 3$

Solution: We begin by clearing fractions. To do this, we multiply both sides of the equation by the LCD of 2 and 3, which is 6.

$$\frac{x}{2} - 1 = \frac{2}{3}x - 3$$

Step 1.	$6\left(\dfrac{x}{2} - 1\right) = 6\left(\dfrac{2}{3}x - 3\right)$	Multiply both sides by the LCD, 6.

Answers

2. $x = \dfrac{15}{2}$,   **3.** $x = -3$

<table>
<tr><td>

HELPFUL HINT
Don't forget to multiply *each* term by the LCD.

</td><td>

Step 2. $6\left(\dfrac{x}{2}\right) - 6(1) = 6\left(\dfrac{2}{3}x\right) - 6(3)$ Apply the distributive property.

$$3x - 6 = 4x - 18 \qquad \text{Simplify.}$$

</td></tr>
</table>

There are no longer grouping symbols and no like terms on either side of the equation, so we continue with Step 4.

$$3x - 6 = 4x - 18$$

Step 4. $3x - 6 - 3x = 4x - 18 - 3x$ Subtract $3x$ from both sides.

$-6 = x - 18$ Simplify.

$-6 + 18 = x - 18 + 18$ Add 18 to both sides.

$12 = x$ Simplify.

Step 5. The variable is now alone, so there is no need to apply the multiplication property of equality.

Step 6. Check.

$$\dfrac{x}{2} - 1 = \dfrac{2}{3}x - 3 \qquad \text{Original equation}$$

$$\dfrac{12}{2} - 1 \overset{?}{=} \dfrac{2}{3} \cdot 12 - 3 \qquad \text{Replace } x \text{ with 12.}$$

$$6 - 1 \overset{?}{=} 8 - 3 \qquad \text{Simplify.}$$

$$5 = 5 \qquad \text{True.}$$

The solution is 12.

Example 4   Solve: $\dfrac{2(a + 3)}{3} = 6a + 2$

Solution: We clear the equation of fractions first.

$$\dfrac{2(a + 3)}{3} = 6a + 2$$

Step 1. $3 \cdot \dfrac{2(a + 3)}{3} = 3(6a + 2)$ Clear the fraction by multiplying both sides by the LCD, 3.

Step 2. Next, we use the distributive property and remove parentheses.

$$2a + 6 = 18a + 6 \qquad \text{Apply the distributive property.}$$

Step 4. $2a + 6 - 6 = 18a + 6 - 6$ Subtract 6 from both sides.

$2a = 18a$

$2a - 18a = 18a - 18a$ Subtract $18a$ from both sides.

$-16a = 0$

Step 5. $\dfrac{-16a}{-16} = \dfrac{0}{-16}$ Divide both sides by -16.

$a = 0$ Write the fraction in simplest form.

Step 6. To check, replace a with 0 in the original equation. The solution is 0.

Practice Problem 4

Solve: $\dfrac{3(x - 2)}{5} = 3x + 6$

Answer
4. $x = -3$

When solving a problem about money, you may need to solve an equation containing decimals. If you choose, you may multiply to clear the equation of decimals.

Example 5 Solve: $0.25x + 0.10(x - 3) = 0.05(22)$

Solution: First we clear this equation of decimals by multiplying both sides of the equation by 100. Recall that multiplying a decimal number by 100 has the effect of moving the decimal point 2 places to the right.

$$0.25x + 0.10(x - 3) = 0.05(22)$$

Step 1. $0.25x + 0.10(x - 3) = 0.05(22)$ Multiply both sides by 100.

$$25x + 10(x - 3) = 5(22)$$

Step 2. $\quad 25x + 10x - 30 = 110$ Apply the distributive property.

Step 3. $\quad\quad\quad\quad 35x - 30 = 110$ Combine like terms.

Step 4. $\quad 35x - 30 + 30 = 110 + 30$ Add 30 to both sides.

$$35x = 140$$ Combine like terms.

Step 5. $\quad\quad\quad \dfrac{35x}{35} = \dfrac{140}{35}$ Divide both sides by 35.

$$x = 4$$

Step 6. To check, replace x with 4 in the original equation. The solution is 4.

C RECOGNIZING IDENTITIES AND EQUATIONS WITH NO SOLUTION

So far, each equation that we have solved has had a single solution. However, not every equation in one variable has a single solution. Some equations have no solution, while others have an infinite number of solutions. For example,

$$x + 5 = x + 7$$

has no solution since no matter which **real number** we replace x with, the equation is false.

real number $+ 5 =$ same real number $+ 7$ FALSE

On the other hand,

$$x + 6 = x + 6$$

has infinitely many solutions since x can be replaced by any real number and the equation is always true.

real number $+ 6 =$ same real number $+ 6$ TRUE

The equation $x + 6 = x + 6$ is called an **identity**. The next few examples illustrate special equations like these.

Example 6 Solve: $-2(x - 5) + 10 = -3(x + 2) + x$

Solution:

$$-2(x - 5) + 10 = -3(x + 2) + x$$
$$-2x + 10 + 10 = -3x - 6 + x \quad \text{Apply the distributive property on both sides.}$$
$$-2x + 20 = -2x - 6 \quad \text{Combine like terms.}$$
$$-2x + 20 + 2x = -2x - 6 + 2x \quad \text{Add } 2x \text{ to both sides.}$$
$$20 = -6 \quad \text{Combine like terms.}$$

The final equation contains no variable terms, and there is no value for x that makes $20 = -6$ a true equation. We conclude that there is **no solution** to this equation. ━━━━

Example 7 Solve: $3(x - 4) = 3x - 12$

Solution:

$$3(x - 4) = 3x - 12$$
$$3x - 12 = 3x - 12 \quad \text{Apply the distributive property.}$$

The left side of the equation is now identical to the right side. Every real number may be substituted for x and a true statement will result. We arrive at the same conclusion if we continue.

$$3x - 12 = 3x - 12$$
$$3x - 12 + 12 = 3x - 12 + 12 \quad \text{Add 12 to both sides.}$$
$$3x = 3x \quad \text{Combine like terms.}$$
$$3x - 3x = 3x - 3x \quad \text{Subtract } 3x \text{ from both sides.}$$
$$0 = 0$$

Again, one side of the equation is identical to the other side. Thus, $3(x - 4) = 3x - 12$ is an **identity** and **every real number** is a solution.

TRY THE CONCEPT CHECK IN THE MARGIN.

Practice Problem 6

Solve: $5(2 - x) + 8x = 3(x - 6)$

Practice Problem 7

Solve: $-6(2x + 1) - 14 = -10(x + 2) - 2x$

✓ CONCEPT CHECK

Suppose you have simplified several equations and obtain the following results. What can you conclude about the solutions to the original equation?
a. $7 = 7$ b. $x = 0$ c. $7 = -4$

Answers

6. no solution,   **7.** Every real number is a solution.

✓ Concept Check:

a. Every real number is a solution.,   **b.** The solution is 0.,   **c.** There is no solution.

CALCULATOR EXPLORATIONS
CHECKING EQUATIONS

We can use a calculator to check possible solutions of equations. To do this, replace the variable by the possible solution and evaluate both sides of the equation separately.

Equation: $3x - 4 = 2(x + 6)$ Solution: $x = 16$

$$3x - 4 = 2(x + 6) \quad \text{Original equation}$$

$$3(16) - 4 \stackrel{?}{=} 2(16 + 6) \quad \text{Replace } x \text{ with 16.}$$

Now evaluate each side with your calculator.

Evaluate left side: $\boxed{3}\;\boxed{\times}\;\boxed{16}\;\boxed{-}\;\boxed{4}\;\boxed{=}$ Display: $\boxed{\qquad 44}$

or $\boxed{\text{ENTER}}$

Evaluate right side: $\boxed{2}\;\boxed{(}\;\boxed{16}\;\boxed{+}\;\boxed{6}\;\boxed{)}\;\boxed{=}$ Display: $\boxed{\qquad 44}$

or $\boxed{\text{ENTER}}$

Since the left side equals the right side, the equation checks.

Use a calculator to check the possible solutions to each equation.

1. $2x = 48 + 6x; \quad x = -12$

2. $-3x - 7 = 3x - 1; \quad x = -1$

3. $5x - 2.6 = 2(x + 0.8); \quad x = 4.4$

4. $-1.6x - 3.9 = -6.9x - 25.6; \quad x = 5$

5. $\dfrac{564x}{4} = 200x - 11(649); \quad x = 121$

6. $20(x - 39) = 5x - 432; \quad x = 23.2$

2.5 AN INTRODUCTION TO PROBLEM SOLVING

In the preceding sections, we practiced translating phrases into expressions and sentences into equations as well as solving linear equations. We are now ready to put our skills to practical use. To begin, we present a general strategy for problem solving.

Objective

A Translate a problem to an equation, then use the equation to solve the problem.

SSM   CD-ROM   Video
2.5

GENERAL STRATEGY FOR PROBLEM SOLVING

1. UNDERSTAND the problem. During this step, become comfortable with the problem. Some ways of doing this are:

   Read and reread the problem.
   Choose a variable to represent the unknown.
   Construct a drawing.
   Propose a solution and check. Pay careful attention to how you check your proposed solution. This will help when writing an equation to model the problem.

2. TRANSLATE the problem into an equation.
3. SOLVE the equation.
4. INTERPRET the results: *Check* the proposed solution in the stated problem and *state* your conclusion.

A TRANSLATING AND SOLVING PROBLEMS

Much of problem solving involves a direct translation from a sentence to an equation.

Example 1   Finding an Unknown Number

Twice the sum of a number and 4 is the same as four times the number decreased by 12. Find the number.

Solution: 1. UNDERSTAND. Read and reread the problem. If we let

$$x = \text{the unknown number, then}$$

"the sum of a number and 4" translates to "$x + 4$" and
"four times the number" translates to "$4x$"

2. TRANSLATE.

twice	sum of a number and 4	is the same as	four times the number	decreased by	12
↓	↓	↓	↓	↓	↓
2	$(x + 4)$	=	$4x$	−	12

Practice Problem 1

Three times the difference of a number and 5 is the same as twice the number decreased by 3. Find the number.

3. SOLVE.

$$2(x + 4) = 4x - 12$$
$$2x + 8 = 4x - 12 \qquad \text{Apply the distributive property.}$$
$$2x + 8 - 4x = 4x - 12 - 4x \qquad \text{Subtract } 4x \text{ from both sides.}$$
$$-2x + 8 = -12$$
$$-2x + 8 - 8 = -12 - 8 \qquad \text{Subtract 8 from both sides.}$$
$$-2x = -20$$
$$\frac{-2x}{-2} = \frac{-20}{-2} \qquad \text{Divide both sides by } -2.$$
$$x = 10$$

4. INTERPRET.

Check: Check this solution in the problem as it was originally stated. To do so, replace "number" with 10. Twice the sum of "10" and 4 is 28, which is the same as 4 times "10" decreased by 12.

State: The number is 10.   ▬▬▬▬

Practice Problem 2

An 18-foot wire is to be cut so that the longer piece is 5 times longer than the shorter piece. Find the length of each piece.

Example 2   Finding the Length of a Board

A 10-foot board is to be cut into two pieces so that the longer piece is 4 times the shorter. Find the length of each piece.

Solution: **1.** UNDERSTAND the problem. To do so, read and re-read the problem. You may also want to propose a solution. For example, if 3 feet represents the length of the shorter piece, then $4(3) = 12$ feet is the length of the longer piece, since it is 4 times the length of the shorter piece. This guess gives a total board length of 3 feet + 12 feet = 15 feet, too long. However, the purpose of proposing a solution is not to guess correctly, but to help better understand the problem and how to model it.

In general, if we let

x = length of shorter piece, then

$4x$ = length of longer piece

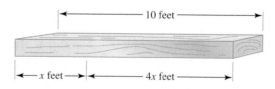

2. TRANSLATE the problem. First, we write the equation in words.

length of shorter piece	added to	length of longer piece	equals	total length of board
↓	↓	↓	↓	↓
x	$+$	$4x$	$=$	10

3. SOLVE.

$$x + 4x = 10$$
$$5x = 10 \qquad \text{Combine like terms.}$$
$$\frac{5x}{5} = \frac{10}{5} \qquad \text{Divide both sides by 5.}$$
$$x = 2$$

4. INTERPRET.

Check: Check the solution in the stated problem. If the shorter piece of board is 2 feet, the longer piece is $4 \cdot (2 \text{ feet}) = 8$ feet and the sum of the two pieces is 2 feet $+$ 8 feet $= 10$ feet.

State: The shorter piece of board is 2 feet and the longer piece of board is 8 feet.

> **HELPFUL HINT:**
>
> Make sure that units are included in your answer, if appropriate.

Example 3   Finding the Number of Republican and Democratic Senators

In a recent year, Congress had 8 more Republican senators than Democratic. If the total number of senators is 100, how many senators of each party were there?

Solution:

1. UNDERSTAND the problem. Read and reread the problem. Let's suppose that there are 40 Democratic senators. Since there are 8 more Republicans than Democrats, there must be $40 + 8 = 48$ Republicans. The total number of Democrats and Republicans is then $40 + 48 = 88$. This is incorrect since the total should be 100, but we now have a better understanding of the problem.

In general, if we let

$x =$ number of Democrats, then

$x + 8 =$ number of Republicans

2. TRANSLATE the problem. First, we write the equation in words.

number of Democrats	added to	number of Republicans	equals	100
↓	↓	↓	↓	↓
x	$+$	$(x + 8)$	$=$	100

Practice Problem 3

In a recent year, the total number of Democrats and Republicans in the U.S. House of Representatives was 433. There were 39 more Republicans than Democrats. Find the number of representatives from each party.

Answer

3. Democrats $= 197$ representatives; Republicans $= 236$ representatives

3. SOLVE.

$$x + (x + 8) = 100$$

$$2x + 8 = 100 \quad \text{Combine like terms.}$$

$$2x + 8 - 8 = 100 - 8 \quad \text{Subtract 8 from both sides.}$$

$$2x = 92$$

$$\frac{2x}{2} = \frac{92}{2} \quad \text{Divide both sides by 2.}$$

$$x = 46$$

4. INTERPRET.

Check: If there are 46 Democratic senators, then there are $46 + 8 = 54$ Republican senators. The total number of senators is then $46 + 54 = 100$. The results check.

State: There were 46 Democratic and 54 Republican senators.

Practice Problem 4

Enterprise Car Rental charges a daily rate of $34 plus $0.20 per mile. Suppose that you rent a car for a day and your bill (before taxes) is $104. How many miles did you drive?

Example 4   **Calculating Cellular Phone Usage**

A local cellular phone company charges Elaine Chapoton $50 per month and $0.36 per minute of phone use in her usage category. If Elaine was charged $99.68 for a month's cellular phone use, determine the number of whole minutes of phone use.

Solution: **1.** UNDERSTAND. Read and reread the problem. Let's propose that Elaine uses the phone for 70 minutes. Pay careful attention as to how we calculate her bill. For 70 minutes of use, Elaine's phone bill will be $50 plus $0.36 per minute of use. This is $50 + 0.36(70) = $75.20, less than $99.68. We now understand the problem and know that the number of minutes is greater than 70.

If we let

$$x = \text{number of minutes, then}$$

$$0.36x = \text{charge per minute of phone use}$$

2. TRANSLATE.

$50	added to	minute charge	is equal to	$99.68
↓	↓	↓	↓	↓
50	+	0.36x	=	99.68

3. SOLVE.

$$50 + 0.36x = 99.68$$

$$50 + 0.36x - 50 = 99.68 - 50 \quad \text{Subtract 50 from both sides.}$$

$$0.36x = 49.68 \quad \text{Simplify.}$$

$$\frac{0.36x}{0.36} = \frac{49.68}{0.36} \quad \text{Divide both sides by 0.36.}$$

$$x = 138 \quad \text{Simplify.}$$

4. INTERPRET.

Check: If Elaine spends 138 minutes on her cellular phone, her bill is $50 + $0.36(138) = $99.68.

State: Elaine spent 138 minutes on her cellular phone this month.

Example 5   Finding Angle Measures

If the two walls of the Vietnam Veterans Memorial in Washington D.C. were connected, an isosceles triangle would be formed. The measure of the third angle is 97.5° more than the measure of either of the other two equal angles. Find the measure of the third angle. (*Source*: National Park Service)

Solution: **1.** UNDERSTAND. Read and reread the problem. We then draw a diagram (recall that an isosceles triangle has two angles with the same measure) and let

x = degree measure of one angle

x = degree measure of the second equal angle

$x + 97.5$ = degree measure of the third angle

$$(x + 97.5)°$$
$$x° \qquad\qquad x°$$

Practice Problem 5

The measure of the second angle of a triangle is twice the measure of the smallest angle. The measure of the third angle of the triangle is three times the measure of the smallest angle. Find the measures of the angles.

Answer

5. smallest = 30°; second = 60°; third = 90°

2. TRANSLATE. Recall that the sum of the measures of the angles of a triangle equals 180.

measure of first angle	measure of second angle	measure of third angle	equals	180
↓	↓	↓	↓	↓
x +	x +	$(x + 97.5)$	=	180

3. SOLVE.

$$x + x + (x + 97.5) = 180$$
$$3x + 97.5 = 180 \qquad \text{Combine like terms.}$$
$$3x + 97.5 - 97.5 = 180 - 97.5 \qquad \text{Subtract 97.5 from both sides.}$$
$$3x = 82.5$$
$$\frac{3x}{3} = \frac{82.5}{3} \qquad \text{Divide both sides by 3.}$$
$$x = 27.5$$

4. INTERPRET.

Check: If $x = 27.5$, then the measure of the third angle is $x + 97.5 = 125$. The sum of the angles is then $27.5 + 27.5 + 125 = 180$, the correct sum.

State: The third angle measures 125°.*

(*This is rounded to the nearest whole degree. The two walls actually meet at an angle of 125 degrees 12 minutes.)

2.6   FORMULAS AND PROBLEM SOLVING

A   USING FORMULAS TO SOLVE PROBLEMS

A **formula** describes a known relationship among quantities. Many formulas are given as equations. For example, the formula

$$d = r \cdot t$$

stands for the relationship

$$\text{distance} = \text{rate} \cdot \text{time}$$

Let's look at one way that we can use this formula.

If we know we traveled a distance of 100 miles at a rate of 40 miles per hour, we can replace the variables d and r in the formula $d = rt$ and find our time, t.

$$d = rt \qquad \text{Formula.}$$
$$100 = 40t \qquad \text{Replace } d \text{ with 100 and } r \text{ with 40.}$$

To solve for t, we divide both sides of the equation by 40.

$$\frac{100}{40} = \frac{40t}{40} \qquad \text{Divide both sides by 40.}$$

$$\frac{5}{2} = t \qquad \text{Simplify.}$$

The time traveled is $\frac{5}{2}$ hours, or $2\frac{1}{2}$ hours.

In this section, we solve problems that can be modeled by known formulas. We use the same problem-solving strategy that was introduced in the previous section.

Example 1   Finding Time Given Rate and Distance

A glacier is a giant mass of rocks and ice that flows downhill like a river. Portage Glacier in Alaska is about 6 miles, or 31,680 *feet*, long and moves 400 *feet* per year. Icebergs are created when the front end of the glacier flows into Portage Lake. How long does it take for ice at the head (beginning) of the glacier to reach the lake?

Objectives

A Use formulas to solve problems.
B Solve a formula or equation for one of its variables.

SSM   CD-ROM   Video
2.6

Practice Problem 1

A family is planning their vacation to Disney World. They will drive from a small town outside New Orleans, Louisiana, to Orlando, Florida, a distance of 700 miles. They plan to average a rate of 55 miles per hour. How long will this trip take?

Answer

1. approximately $12\frac{8}{11}$ hours

Solution:

1. UNDERSTAND. Read and reread the problem. The appropriate formula needed to solve this problem is the distance formula, $d = rt$. To become familiar with this formula, let's find the distance that ice traveling at a rate of 400 feet per year travels in 100 years. To do so, we let time t be 100 years and rate r be the given 400 feet per year, and substitute these values into the formula $d = rt$. We then have that distance $d = 400(100) = 40{,}000$ feet. Since we are interested in finding how long it takes ice to travel 31,680 feet, we now know that it is less than 100 years.

Since we are using the formula $d = rt$, we let

t = the time in years for ice to reach the lake

r = rate or speed of ice

d = distance from beginning of glacier to lake

2. TRANSLATE. To translate to an equation, we use the formula $d = rt$ and let distance $d = 31{,}680$ feet and rate $r = 400$ feet per year.

$$d = r \cdot t$$
$$31{,}680 = 400 \cdot t \quad \text{Let } d = 31{,}680 \text{ and } r = 400.$$

3. SOLVE. Solve the equation for t. To solve for t, divide both sides by 400.

$$\frac{31{,}680}{400} = \frac{400 \cdot t}{400} \quad \text{Divide both sides by 400.}$$
$$79.2 = t \quad \text{Simplify.}$$

4. INTERPRET.

Check: To check, substitute 79.2 for t and 400 for r in the distance formula and check to see that the distance is 31,680 feet.

State: It takes 79.2 years for the ice at the head of Portage Glacier to reach the lake.

HELPFUL HINT

Don't forget to include units, if appropriate.

Practice Problem 2

A wood deck is being built behind a house. The width of the deck must be 18 feet because of the shape of the house. If there is 450 square feet of decking material, find the length of the deck.

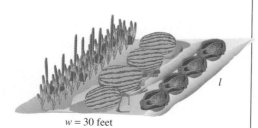

Example 2   Calculating the Length of a Garden

Charles Pecot can afford enough fencing to enclose a rectangular garden with a perimeter of 140 feet. If the width of his garden is to be 30 feet, find the length.

Solution:

1. UNDERSTAND. Read and reread the problem. The formula needed to solve this problem is the formula for the perimeter of a rectangle, $P = 2l + 2w$. Before continuing, let's become familar with this formula.

l = the length of the rectangular garden

w = the width of the rectangular garden

P = perimeter of the garden

Answer

2. 25 feet

2. TRANSLATE. To translate to an equation, we use the formula $P = 2l + 2w$ and let perimeter $P = 140$ feet and width $w = 30$ feet.

$$P = 2l + 2w \qquad \text{Let } P = 140 \text{ and } w = 30.$$
$$140 = 2l + 2(30)$$

3. SOLVE.

$$140 = 2l + 2(30)$$
$$140 = 2l + 60 \qquad \text{Multiply } 2(30).$$
$$140 - 60 = 2l + 60 - 60 \qquad \text{Subtract 60 from both sides.}$$
$$80 = 2l \qquad \text{Combine like terms.}$$
$$40 = l \qquad \text{Divide both sides by 2.}$$

4. INTERPRET.

Check: Substitute 40 for l and 30 for w in the perimeter formula and check to see that the perimeter is 140 feet.

State: The length of the rectangular garden is 40 feet. ▬▬▬

B SOLVING A FORMULA FOR A VARIABLE

We say that the formula

$$d = rt$$

is solved for d because d is alone on one side of the equation and the other side contains no d's. Suppose that we have a large number of problems to solve where we are given distance d and rate r and asked to find time t. In this case, it may be easier to first solve the formula $d = rt$ for t. To solve for t, we divide both sides of the equation by r.

$$d = rt$$
$$\frac{d}{r} = \frac{rt}{r} \qquad \text{Divide both sides by } r.$$
$$\frac{d}{r} = t \qquad \text{Simplify.}$$

To solve a formula or an equation for a specified variable, we use the same steps as for solving a linear equation. These steps are listed next.

TO SOLVE EQUATIONS FOR A SPECIFIED VARIABLE

Step 1. Multiply on both sides to clear the equation of fractions if they occur.

Step 2. Use the distributive property to remove parentheses if they occur.

Step 3. Simplify each side of the equation by combining like terms.

Step 4. Get all terms containing the specified variable on one side and all other terms on the other side by using the addition property of equality.

Step 5. Get the specified variable alone by using the multiplication property of equality.

Practice Problem 3

Solve $C = 2\pi r$ for r. (This formula is used to find the circumference C of a circle given its radius r.)

Example 3

Solve $V = lwh$ for l.

Solution: This formula is used to find the volume of a box. To solve for l, we divide both sides by wh.

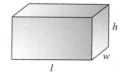

$$V = lwh$$

$$\frac{V}{wh} = \frac{lwh}{wh} \quad \text{Divide both sides by } wh.$$

$$\frac{V}{wh} = l \quad \text{Simplify.}$$

Since we have l alone on one side of the equation, we have solved for l in terms of V, w, and h. Remember that it does not matter on which side of the equation we get the variable alone.

Practice Problem 4

Solve $P = 2l + 2w$ for w.

Example 4

Solve $y = mx + b$ for x.

Solution: First we get mx alone by subtracting b from both sides.

$$y = mx + b$$

$$y - b = mx + b - b \quad \text{Subtract } b \text{ from both sides.}$$

$$y - b = mx \quad \text{Combine like terms.}$$

Next we solve for x by dividing both sides by m.

$$\frac{y - b}{m} = \frac{mx}{m}$$

$$\frac{y - b}{m} = x \quad \text{Simplify.}$$

Practice Problem 5

Solve $A = \dfrac{a + b}{2}$ for b.

Example 5

Solve $A = \dfrac{bh}{2}$ for h.

Solution: First let's clear the equation of fractions by multiplying both sides by 2.

$$A = \frac{bh}{2}$$

$$2 \cdot A = 2\left(\frac{bh}{2}\right) \quad \text{Multiply both sides by 2 to clear fractions.}$$

$$2A = bh$$

$$\frac{2A}{b} = \frac{bh}{b} \quad \text{Divide both sides by } b \text{ to get } h \text{ alone.}$$

$$\frac{2A}{b} = h \quad \text{Simplify.}$$

Answers

3. $r = \dfrac{C}{2\pi}$,   **4.** $w = \dfrac{P - 2l}{2}$,

5. $b = 2A - a$

EXERCISE SET 2.6

A *Substitute the given values into each given formula and solve for the unknown variable. See Examples 1 and 2.*

1. $A = bh$; $\quad A = 45, b = 15$
(Area of a parallelogram)

2. $d = rt$; $\quad d = 195, t = 3$
(Distance formula)

3. $S = 4lw + 2wh$; $\quad S = 102, l = 7$, $w = 3$
(Surface area of a special rectangular box)

4. $V = lwh$; $\quad l = 14, w = 8, h = 3$
(Volume of a rectangular box)

5. $A = \frac{1}{2}(B + b)h$; $\quad A = 180, B = 11$, $b = 7$
(Area of a trapezoid)

6. $A = \frac{1}{2}(B + h)h$; $\quad A = 60, B = 7$, $b = 3$
(Area of a trapezoid)

7. $P = a + b + c$; $\quad P = 30, a = 8$, $b = 10$
(Perimeter of a triangle)

8. $V = \frac{1}{3}Ah$; $\quad V = 45, h = 5$
(Volume of a pyramid)

9. $C = 2\pi r$; $\quad C = 15.7$ (use the approximation 3.14 for π)
(Circumference of a circle)

10. $A = \pi r^2$; $\quad r = 4.5$ (use the approximation 3.14 for π)
(Area of a circle)

11. $I = PRT$; $\quad I = 3750, P = 25,000$, $R = 0.05$
(Simple interest formula)

12. $I = PRT$; $\quad I = 1,056,000, R = 0.055$, $T = 6$
(Simple interest formula)

13. $V = \frac{1}{3}\pi r^2 h$; $\quad V = 565.2, r = 6$ (use the approximation 3.14 for π)
(Volume of a cone)

14. $V = \frac{4}{3}\pi r^3$; $\quad r = 3$ (use the approximation 3.14 for π)
(Volume of a sphere)

B *Solve each formula for the specified variable. See Examples 3 through 5.*

15. $f = 5gh$ for h

16. $C = 2\pi r$ for r

17. $V = LWH$ for W

18. $T = mnr$ for n

19. $3x + y = 7$ for y

20. $-x + y = 13$ for y

ANSWERS

1. _____
2. _____
3. _____
4. _____
5. _____
6. _____
7. _____
8. _____
9. _____
10. _____
11. _____
12. _____
13. _____
14. _____
15. _____
16. _____
17. _____
18. _____
19. _____
20. _____

257

Problem Solving Notes

2.7 PERCENT, RATIO, AND PROPORTION

A SOLVING PERCENT EQUATIONS

Much of today's statistics is given in terms of percent: a basketball player's free throw percent, current interest rates, stock market trends, and nutrition labeling, just to name a few. In this section, we first explore percent, percent equations, and applications involving percents. See Section R.3 if a further review of percents is needed.

Example 1   The number 63 is what percent of 72?

Solution: **1.** UNDERSTAND. Read and reread the problem. Next, let's suppose that the percent is 80%. To check, we find 80% of 72.

80% of 72 = 0.80(72) = 57.6

Close, but not 63. At this point, though, we have a better understanding of the problem, we know the correct answer is close to and greater than 80%, and we know how to check our proposed solution later.

Let $x =$ the unknown percent.

2. TRANSLATE. Recall that "is" means "equals" and "of" signifies multiplying. Let's translate the sentence directly.

the number 63	is	what percent	of	72
↓	↓	↓	↓	↓
63	=	x	·	72

3. SOLVE.

$$63 = 72x$$
$$0.875 = x \qquad \text{Divide both sides by 72.}$$
$$87.5\% = x \qquad \text{Write as a percent.}$$

4. INTERPRET.

Check: Verify that 87.5% of 72 is 63.

State: The number 63 is 87.5% of 72.

Example 2   The number 120 is 15% of what number?

Solution: **1.** UNDERSTAND. Read and reread the problem.

Let $x =$ the unknown number.

2. TRANSLATE.

the number 120	is	15%	of	what number
↓	↓	↓	↓	↓
120	=	15%	·	x

3. SOLVE.

$$120 = 0.15x \qquad \text{Write 15\% as 0.15.}$$
$$800 = x \qquad \text{Divide both sides by 0.15.}$$

4. INTERPRET.

Objectives

A Solve percent equations.
B Solve problems involving percents.
C Write ratios as fractions.
D Solve proportions.
E Solve problems modeled by proportions.

SSM   CD-ROM   Video 2.7

Practice Problem 1

The number 22 is what percent of 40?

Practice Problem 2

The number 150 is 40% of what number?

Answers

1. 55%, **2.** 375

Check: Check the proposed solution by finding 15% of 800 and verifying that the result is 120.

State: Thus, 120 is 15% of 800.

B SOLVING PROBLEMS INVOLVING PERCENT

As mentioned earlier, percents are often used in statistics. Recall that the graph below is called a circle graph or a pie chart. The circle or pie represents a whole, or 100%. Each circle is divided into sectors (shaped like pieces of a pie) that represent various parts of the whole 100%.

Practice Problem 3

Use the circle graph and part (c) of Example 3 to answer each question.

a. What percent of homeowners spend $250–$4999 on yearly home maintenance?

b. What percent of homeowners spend $250 or more on yearly home maintenance?

c. How many of the homeowners in the town of Fairview might we expect to spend from $250–$999 per year on home maintenance.

Example 3 The circle graph below shows how much money homeowners in the United States spend annually on maintaining their homes. Use this graph to answer the questions below.

Yearly Home Maintenance in the U.S.

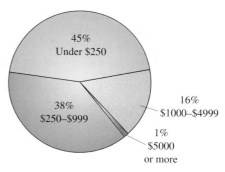

a. What percent of homeowners spend under $250 on yearly home maintenance?

b. What percent of homeowners spend less than $1000 per year on home maintenance?

c. How many of the 22,000 homeowners in a town called Fairview might we expect to spend under $250 a year on home maintenance?

Solution:

a. From the circle graph, we see that 45% of homeowners spend under $250 per year on home maintenance.

b. From the circle graph, we know that 45% of homeowners spend under $250 per year and 38% of homeowners spend $250–$999 per year, so that the sum 45% + 38% or 83% of homeowners spend less than $1000 per year.

c. Since 45% of homeowners spend under $250 per year on maintenance, we find 45% of 22,000.

$$45\% \text{ of } 22{,}000 = 0.45(22{,}000)$$
$$= 9900$$

We might then expect that 9900 homeowners in Fairview spend under $250 per year on home maintenance.

C WRITING RATIOS AS FRACTIONS

A **ratio** is the quotient of two numbers or two quantities. For example, a percent can be thought of as a ratio, since it is the quotient of a number and 100.

Answers
3. a. 54%, **b.** 55%, **c.** 8360

$$53\% = \frac{53}{100} \quad \text{or} \quad \text{the ratio of 53 to 100}$$

RATIO

The ratio of a number a to a number b is their quotient. Ways of writing ratios are

$$a \text{ to } b, \quad a : b, \quad \frac{a}{b}$$

Example 4   Write a ratio for each phrase. Use fractional notation.

a. The ratio of 2 parts salt to 5 parts water
b. The ratio of 18 inches to 2 feet

Solution: **a.** The ratio of 2 parts salt to 5 parts water is $\frac{2}{5}$.
b. First we convert to the same unit of measurement. For example,

$$2 \text{ feet} = 2 \cdot 12 \text{ inches} = 24 \text{ inches}$$

The ratio of 18 inches to 2 feet is then

$$\frac{18}{24}, \text{ or } \frac{3}{4} \text{ in lowest terms.}$$

D   SOLVING PROPORTIONS

Ratios can be used to form proportions. A **proportion** is a mathematical statement that two ratios are equal.

For example, the equation

$$\frac{1}{2} = \frac{4}{8}$$

is a proportion that says that the ratios $\frac{1}{2}$ and $\frac{4}{8}$ are equal.

Notice that a proportion contains four numbers. If any three numbers are known, we can solve and find the fourth number. One way to do so is to use cross products. To understand cross products, let's start with the proportion

$$\frac{a}{b} = \frac{c}{d}$$

and multiply both sides by the LCD, bd.

$$\frac{a}{b} = \frac{c}{d}$$

$$bd\left(\frac{a}{b}\right) = bd\left(\frac{c}{d}\right) \quad \text{Multiply both sides by the LCD, } bd.$$

$$\underbrace{ad}_{\text{cross product}} = \underbrace{bc}_{\text{cross product}} \quad \text{Simplify.}$$

Practice Problem 4

Write a ratio for each phrase. Use fractional notation.

a. The ratio of 3 parts oil to 7 parts gasoline

b. The ratio of 40 minutes to 3 hours

Answers

4. a. $\frac{3}{7}$,   **b.** $\frac{2}{9}$

Notice why ad and bc are called cross products.

$$\frac{a}{b} \,\,\overset{bc}{\underset{ad}{=}}\,\, \frac{c}{d}$$

> **CROSS PRODUCTS**
>
> If $\dfrac{a}{b} = \dfrac{c}{d}$, then $ad = bc$.

Practice Problem 5

Solve for x: $\dfrac{3}{8} = \dfrac{63}{x}$

Example 5   Solve for x: $\dfrac{45}{x} = \dfrac{5}{7}$

Solution: To solve, we set cross products equal.

$$\frac{45}{x} \,\,=\,\, \frac{5}{7}$$

$45 \cdot 7 = x \cdot 5$	Set cross products equal.
$315 = 5x$	Multiply.
$\dfrac{315}{5} = \dfrac{5x}{5}$	Divide both sides by 5.
$63 = x$	Simplify.

Check: To check, substitute 63 for x in the original proportion. The solution is 63.

Practice Problem 6

Solve for x: $\dfrac{2x + 1}{7} = \dfrac{x - 3}{5}$

Example 6   Solve for x: $\dfrac{x - 5}{3} = \dfrac{x + 2}{5}$

Solution:

$$\frac{x - 5}{3} \,\,=\,\, \frac{x + 2}{5}$$

$5(x - 5) = 3(x + 2)$	Set cross products equal.
$5x - 25 = 3x + 6$	Multiply.
$5x = 3x + 31$	Add 25 to both sides.
$2x = 31$	Subtract $3x$ from both sides.
$\dfrac{2x}{2} = \dfrac{31}{2}$	Divide both sides by 2.
$x = \dfrac{31}{2}$	

Check: Verify that $\dfrac{31}{2}$ is the solution.

Answers

5. $x = 168$,   **6.** $x = -\dfrac{26}{3}$

TRY THE CONCEPT CHECK IN THE MARGIN.

E SOLVING PROBLEMS MODELED BY PROPORTIONS

Proportions can be used to model and solve many real-life problems. When using proportions in this way, it is important to judge whether the solution is reasonable. Doing so helps us to decide if the proportion has been formed correctly, We use the same problem-solving strategy that was introduced in Section 2.5.

Example 7   Calculating Cost with a Proportion

Three boxes of 3.5-inch high-density diskettes cost $37.47. How much should 5 boxes cost?

Solution: **1.** UNDERSTAND. Read and reread the problem. We know that the cost of 5 boxes is more than the cost of 3 boxes, or $37.47, and less than the cost of 6 boxes, which is double the cost of 3 boxes, or 2($37.47) = $74.94. Let's suppose that 5 boxes cost $60.00. To check, we see if 3 boxes is to 5 boxes as the *price* of 3 boxes is to the *price* of 5 boxes. In other words, we see if

$$\frac{3 \text{ boxes}}{5 \text{ boxes}} = \frac{\text{price of 3 boxes}}{\text{price of 5 boxes}}$$

or

$$\frac{2}{3} = \frac{8}{12}$$

$$3(60.00) = 5(37.47) \quad \text{Set cross products equal.}$$

or

$$180.00 = 187.35 \quad \text{Not a true statement.}$$

Thus, $60 is not correct but we now have a better understanding of the problem.

Let x = price of 5 boxes of diskettes.

2. TRANSLATE.

$$\frac{3 \text{ boxes}}{5 \text{ boxes}} = \frac{\text{price of 3 boxes}}{\text{price of 5 boxes}}$$

$$\frac{3}{5} = \frac{37.47}{x}$$

3. SOLVE.

$$\frac{3}{5} = \frac{37.47}{x}$$

$$3x = 5(37.47) \quad \text{Set cross products equal.}$$

$$3x = 187.35$$

$$x = 62.45 \quad \text{Divide both sides by 3.}$$

4. INTERPRET.

✓ **CONCEPT CHECK**

For which of the following equations can we immediately use cross products to solve for x?

a. $\dfrac{2 - x}{5} = \dfrac{1 + x}{3}$

b. $\dfrac{2}{5} - x = \dfrac{1 + x}{3}$

Practice Problem 7

To estimate the number of people in Jackson, population 50,000, who have no health insurance, 250 people were polled. Of those polled, 39 had no insurance. How many people in the city might we expect to be uninsured?

Answers

✓ Concept Check: a

7. 7800 people

Check: Verify that 3 boxes is to 5 boxes as $37.47 is to $62.45. Also, notice that our solution is a reasonable one as discussed in Step 1.

State: Five boxes of high-density diskettes cost $62.45. ▪━━

HELPFUL HINT

The proportion $\dfrac{5 \text{ boxes}}{3 \text{ boxes}} = \dfrac{\text{price of 5 boxes}}{\text{price of 3 boxes}}$ could also have been used to solve the problem above. Notice that the cross products are the same.

When shopping for an item offered in many different sizes, it is important to be able to determine the best buy, or the best price per unit. To find the **unit price** of an item, divide the total price of the item by the total number of units.

$$\text{unit price} = \frac{\text{total price}}{\text{number of units}}$$

For example, if a 16-ounce can of green beans is priced at $0.88, its unit price is

$$\text{unit price} = \frac{\$0.88}{16} = \$0.055$$

Practice Problem 8

Which is the better buy for the same brand of toothpaste?

8 ounces for $2.59

10 ounces for $3.11

Example 8 Finding the Better Buy

A supermarket offers a 14-ounce box of cereal for $3.79 and an 18-ounce box of the same brand of cereal for $4.99. Which is the better buy?

Solution: To find the better buy, we compare unit prices. The following unit prices were rounded to three decimal places.

Size	Price	Unit Price
14 ounce	$3.79	$\dfrac{\$3.79}{14} \approx \$0.271$
18 ounce	$4.99	$\dfrac{\$4.99}{18} \approx \$0.277$

The 14-ounce box of cereal has the lower unit price so it is the better buy. ━━━

Answer

8. 10 ounces

EXERCISE SET 2.7

A *Find each number described. See Examples 1 and 2.*

1. What number is 16% of 70?

2. What number is 88% of 1000?

3. The number 28.6 is what percent of 52?

4. The number 87.2 is what percent of 436?

5. The number 45 is 25% of what number?

6. The number 126 is 35% of what number?

7. Find 23% of 20.

8. Find 140% of 86.

9. The number 40 is 80% of what number?

10. The number 56.25 is 45% of what number?

11. The number 144 is what percent of 480?

12. The number 42 is what percent of 35?

D *Solve each proportion. See Examples 5 and 6.*

13. $\frac{2}{3} = \frac{x}{6}$

14. $\frac{x}{2} = \frac{16}{6}$

15. $\frac{x}{10} = \frac{5}{9}$

16. $\frac{9}{4x} = \frac{6}{2}$

17. $\frac{4x}{6} = \frac{7}{2}$

18. $\frac{a}{5} = \frac{3}{2}$

19. $\frac{x-3}{x} = \frac{4}{7}$

20. $\frac{y}{y-16} = \frac{5}{3}$

1. _____

2. _____

3. _____

4. _____

5. _____

6. _____

7. _____

8. _____

9. _____

10. _____

11. _____

12. _____

13. _____

14. _____

15. _____

16. _____

17. _____

18. _____

19. _____

20. _____

Problem Solving Notes

2.8 LINEAR INEQUALITIES AND PROBLEM SOLVING

Relationships among measurable quantities are not always described by equations. For example, suppose that a salesperson earns a base of $600 per month plus a commission of 20% of sales. Find the minimum amount of sales needed to receive a total income of *at least* $1500 per month. Here, the phrase "at least" implies that an income of $1500 *or more* is acceptable. In symbols, we can write

$$\text{income} \geq 1500$$

This is an example of an inequality, which we will solve in Example 11.

A **linear inequality** is similar to a linear equation except that the equality symbol is replaced with an inequality symbol, such as $<$, $>$, $\leq$, or $\geq$.

LINEAR INEQUALITY IN ONE VARIABLE

A linear inequality in one variable is an inequality that can be written in the form

$$ax + b < c$$

where a, b, and c are real numbers and $a \neq 0$. For example,

$$3x + 5 \geq 4 \qquad 2y < 0 \qquad 4n \geq n - 3$$

$$3(x - 4) < 5x \qquad \frac{x}{3} \leq 5$$

In this section, when we make definitions, state properties, or list steps about an inequality containing the symbol $<$, we mean that the definition, property, or steps apply to an inequality containing the symbols $>$, $\leq$, and $\geq$, also.

A USING INTERVAL NOTATION

A **solution** of an inequality is a value of the variable that makes the inequality a true statement. The **solution set** of an inequality is the set of all solutions. For example, the solution set of the inequality $x > 2$ contains all numbers greater than 2. In **set notation,** this solution set is written as $\{x \mid x > 2\}$. Its graph is an interval on the number line since an infinite number of values satisfy the variable. If we use open/closed-circle notation, the graph of $\{x \mid x > 2\}$ looks like the following:

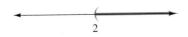

$$\{x \mid x > 2\}$$

In this text, a different graphing notation will be used to help us understand **interval notation**. Instead of an open circle, we use a parenthesis; instead of a closed circle, we use a bracket. With this new notation, the graph of $\{x \mid x > 2\}$ now looks like

2

and can be represented in interval notation as $(2, \infty)$. The symbol ∞ is read "infinity" and indicates that the interval includes *all* numbers greater

---

Objective

A Use interval notation.

B Solve linear inequalities using the addition property of inequality.

C Solve linear inequalities using the multiplication property of inequality.

D Solve linear inequalities using both properties of inequality.

E Solve problems that can be modeled by linear inequalities.

SSM   CD-ROM   Video
2.8

than 2. The left parenthesis indicates that 2 *is not* included in the interval. Using a left bracket, $[$, would indicate that 2 *is* included in the interval. The following table shows three equivalent ways to describe an interval: in set notation, as a graph, and in interval notation.

Set Notation	Graph	Interval Notation
$\{x\|x < a\}$	⟵————)————⟶ a	$(-\infty, a)$
$\{x\|x > a\}$	⟵————(————⟶ a	(a, ∞)
$\{x\|x \le a\}$	⟵————]————⟶ a	$(-\infty, a]$
$\{x\|x \ge a\}$	⟵————[————⟶ a	$[a, \infty)$
$\{x\|a < x < b\}$	⟵——(—)——⟶ a b	(a, b)
$\{x\|a \le x \le b\}$	⟵——[—]——⟶ a b	$[a, b]$
$\{x\|a < x \le b\}$	⟵——(—]——⟶ a b	$(a, b]$
$\{x\|a \le x < b\}$	⟵——[—)——⟶ a b	$[a, b)$

Practice Problems 1–3

Graph each set on a number line and then write it in interval notation.

1. $\{x\|x > -3\}$

   ⟵┼┼┼┼┼┼┼┼┼┼┼⟶
   −5 −4 −3 −2 −1 0 1 2 3 4 5

2. $\{x\|x \le 0\}$

   ⟵┼┼┼┼┼┼┼┼┼┼┼⟶
   −5 −4 −3 −2 −1 0 1 2 3 4 5

3. $\{x\|-0.5 \le x < 2\}$

   ⟵┼┼┼┼┼┼┼┼┼┼┼⟶
   −5 −4 −3 −2 −1 0 1 2 3 4 5

✓ CONCEPT CHECK

Explain what is wrong with writing the interval $(5, \infty]$.

Answers

1. $(-3, \infty)$,

   ⟵┼┼┼(┼┼┼┼┼┼┼⟶
   −5 −4 −3 −2 −1 0 1 2 3 4 5

2. $(-\infty, 0]$,

   ⟵┼┼┼┼┼┼┼┼┼┼┼⟶
   −5 −4 −3 −2 −1 0 1 2 3 4 5

3. $[-0.5, 2)$

   −0.5
   ⟵┼┼┼┼┼┼┼┼┼┼┼⟶
   −5 −4 −3 −2 −1 0 1 2 3 4 5

✓ **Concept Check:** Should be $(5, \infty)$ since a parenthesis is always used to enclose ∞.

> ## HELPFUL HINT
>
> Notice that a parenthesis is always used to enclose ∞ and $-\infty$.

Examples

Graph each set on a number line and then write it in interval notation.

1. $\{x\|x \ge 2\}$

   ⟵┼┼┼┼┼[———⟶
   −2 −1 0 1 2 3 4 $[2, \infty)$

2. $\{x\|x < -1\}$

   ⟵———)┼┼┼┼⟶
   −3 −2 −1 0 1 2 3 $(-\infty, -1)$

3. $\{x\|0.5 < x \le 3\}$

   0.5
   ⟵┼(——]┼⟶
   −1 0 1 2 3 $(0.5, 3]$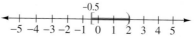

TRY THE CONCEPT CHECK IN THE MARGIN.

B USING THE ADDITION PROPERTY OF INEQUALITY

Interval notation can be used to write solutions of linear inequalities. To solve a linear inequality, we use a process similar to the one used to solve a linear equation. We use properties of inequalities to write equivalent inequalities until the variable is alone on one side of the inequality.

> ### ADDITION PROPERTY OF INEQUALITY
>
> If a, b, and c are real numbers, then
>
> $$a < b \quad \text{and} \quad a + c < b + c$$
>
> are equivalent inequalities.

In other words, we may add the same real number to both sides of an inequality, and the resulting inequality will have the same solution set. This property also allows us to subtract the same real number from both sides.

Example 4   Solve: $x - 2 < 5$. Graph the solution set.

Solution:

$$x - 2 < 5$$
$$x - 2 + 2 < 5 + 2 \quad \text{Add 2 to both sides.}$$
$$x < 7 \quad\quad \text{Simplify.}$$

The solution set is $\{x \mid x < 7\}$, which in interval notation is $(-\infty, 7)$. The graph of the solution set is

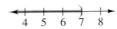

HELPFUL HINT

In Example 4, the solution set is $\{x \mid x < 7\}$. This means that *all* numbers less than 7 are solutions. For example, 6.9, 0, $-\pi$, 1, and -56.7 are solutions, just to name a few. To see this, replace x in $x - 2 < 5$ with each of these numbers and see that the result is a true inequality.

Example 5   Solve: $4x - 2 < 5x$. Graph the solution set.

Solution: To get x alone on one side of the inequality, we subtract $4x$ from both sides.

$$4x - 2 < 5x$$
$$4x - 2 - 4x < 5x - 4x \quad \text{Subtract } 4x \text{ from both sides.}$$
$$-2 < x \quad \text{or} \quad x > -2 \quad \text{Simplify.}$$

HELPFUL HINT

Don't forget that $-2 < x$ means the same as $x > -2$.

The solution set is $\{x \mid x > -2\}$, which in interval notation is $(-2, \infty)$. The graph is

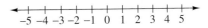

Practice Problem 4

Solve: $x + 3 < 1$. Graph the solution set.

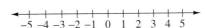

Practice Problem 5

Solve: $3x - 4 < 4x$. Graph the solution set.

Answers

4. $\{x \mid x < -2\}, (-\infty, -2),$

5. $\{x \mid x > -4\}, (-4, \infty),$

Practice Problem 6

Solve: $5x - 1 \geq 4x + 4$. Graph the solution set.

Example 6   Solve: $3x + 4 \geq 2x - 6$. Graph the solution set.

Solution:

$$3x + 4 \geq 2x - 6$$

$$3x + 4 - 2x \geq 2x - 6 - 2x \qquad \text{Subtract } 2x \text{ from both sides.}$$

$$x + 4 \geq -6 \qquad \text{Combine like terms.}$$

$$x + 4 - 4 \geq -6 - 4 \qquad \text{Subtract 4 from both sides.}$$

$$x \geq -10 \qquad \text{Simplify.}$$

The solution set is $\{x \mid x \geq -10\}$, which in interval notation is $[-10, \infty)$. The graph of the solution set is

C   USING THE MULTIPLICATION PROPERTY OF INEQUALITY

Next, we introduce and use the multiplication property of inequality to solve linear inequalities. To understand this property, let's start with the true statement $-3 < 7$ and multiply both sides by 2.

$$-3 < 7$$

$$-3(2) < 7(2) \qquad \text{Multiply both sides by 2.}$$

$$-6 < 14 \qquad \text{True.}$$

The statement remains true.

Notice what happens if both sides of $-3 < 7$ are multiplied by -2.

$$-3 < 7$$

$$-3(-2) < 7(-2)$$

$$6 < -14 \qquad \text{False.}$$

The inequality $6 < -14$ is a false statement. However, *if the direction of the inequality sign is reversed*, the result is

$$6 > -14 \qquad \text{True.}$$

These examples suggest the following property.

MULTIPLICATION PROPERTY OF INEQUALITY

If a, b, and c are real numbers and c is **positive**, then $a < b$ and $ac < bc$ are equivalent inequalities.

If a, b, and c are real numbers and c is **negative**, then $a < b$ and $ac > bc$ are equivalent inequalities.

In other words, we may multiply both sides of an inequality by the same positive real number, and the result is an equivalent inequality. We may also multiply both sides of an inequality by the same *negative number* and *reverse the direction of the inequality symbol*, and the result is an equivalent inequality. The multiplication property holds for division also, since division is defined in terms of multiplication.

HELPFUL HINT

Whenever both sides of an inequality are multiplied or divided by a negative number, the direction of the inequality symbol *must be* reversed to form an equivalent inequality.

Answer

6. $\{x \mid x \geq 5\}$, $[5, \infty)$

Example 7   Solve: $\frac{1}{4}x \le \frac{3}{2}$. Graph the solution set.

Solution:

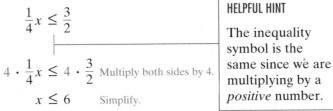

$$\frac{1}{4}x \le \frac{3}{2}$$

$$4 \cdot \frac{1}{4}x \le 4 \cdot \frac{3}{2} \quad \text{Multiply both sides by 4.}$$

$$x \le 6 \qquad \text{Simplify.}$$

> **HELPFUL HINT**
>
> The inequality symbol is the same since we are multiplying by a *positive* number.

The solution set is $\{x | x \le 6\}$, which in interval notation is $(-\infty, 6]$. The graph of this solution set is

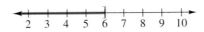

Example 8   Solve: $-2.3x < 6.9$. Graph the solution set.

Solution: $-2.3x < 6.9$

> **HELPFUL HINT**
>
> The inequality symbol is *reversed* since we divided by a *negative* number.

$$\frac{-2.3x}{-2.3} > \frac{6.9}{-2.3} \quad \begin{array}{l}\text{Divide both sides by } -2.3 \text{ and} \\ \text{reverse the inequality symbol.}\end{array}$$

$$x > -3 \qquad \text{Simplify.}$$

The solution set is $\{x | x > -3\}$, which is $(-3, \infty)$ in interval notation. The graph of the solution set is

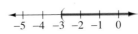

TRY THE CONCEPT CHECK IN THE MARGIN.

To solve linear inequalities in general, we follow steps similar to those for solving linear equations.

SOLVING A LINEAR INEQUALITY IN ONE VARIABLE

Step 1.   Clear the equation of fractions by multiplying both sides of the inequality by the least common denominator (LCD) of all fractions in the inequality.

Step 2.   Use the distributive property to remove grouping symbols such as parentheses.

Step 3.   Combine like terms on each side of the inequality.

Step 4.   Use the addition property of inequality to write the inequality as an equivalent inequality with variable terms on one side and numbers on the other side.

Step 5.   Use the multiplication property of inequality to get the variable alone on one side of the inequality.

D   USING BOTH PROPERTIES OF INEQUALITY

Many problems require us to use both properties of inequality.

Example 9   Solve: $5 - x \le 4x - 15$. Write the solution set in interval notation.

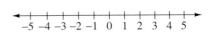
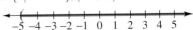

Solution:

$$5 - x \leq 4x - 15$$

$$5 - x + x \leq 4x - 15 + x \quad \text{Add } x \text{ to both sides.}$$

$$5 \leq 5x - 15 \quad \text{Combine like terms.}$$

$$5 + 15 \leq 5x - 15 + 15 \quad \text{Add 15 to both sides.}$$

$$20 \leq 5x \quad \text{Combine like terms.}$$

$$\frac{20}{5} \leq \frac{5x}{5} \quad \text{Divide both sides by 5.}$$

$$4 \leq x \quad \text{or} \quad x \geq 4 \quad \text{Simplify.}$$

The solution set is $[4, \infty)$.

Practice Problem 10

Solve: $\frac{3}{4}(x + 2) \geq x - 6$. Write the solution set in interval notation.

Example 10 Solve: $\frac{2}{5}(x - 6) \geq x - 1$. Write the solution set in interval notation.

Solution:

$$\frac{2}{5}(x - 6) \geq x - 1$$

$$5\left[\frac{2}{5}(x - 6)\right] \geq 5(x - 1) \quad \text{Multiply both sides by 5 to eliminate fractions.}$$

$$2x - 12 \geq 5x - 5 \quad \text{Use the distributive property.}$$

$$-3x - 12 \geq -5 \quad \text{Subtract } 5x \text{ from both sides.}$$

$$-3x \geq 7 \quad \text{Add 12 to both sides.}$$

$$\frac{-3x}{-3} \leq \frac{7}{-3} \quad \text{Divide both sides by } -3 \text{ and reverse the inequality symbol.}$$

$$x \leq -\frac{7}{3} \quad \text{Simplify.}$$

The solution set is $\left(-\infty, -\frac{7}{3}\right]$.

E LINEAR INEQUALITIES AND PROBLEM SOLVING

Problems containing words such as "at least," "at most," "between," "no more than," and "no less than" usually indicate that an inequality is to be solved instead of an equation. In solving applications involving linear inequalities, we use the same four-step strategy as when we solved applications involving linear equations.

Practice Problem 11

A salesperson earns $1000 a month plus a commission of 15% of sales. Find the minimum amount of sales needed to receive a total income of at least $4000 per month.

Example 11 Calculating Income with Commission

A salesperson earns $600 per month plus a commission of 20% of sales. Find the minimum amount of sales needed to receive a total income of at least $1500 per month.

Solution:

1. UNDERSTAND. Read and reread the problem. Let

   x = amount of sales

2. TRANSLATE. As stated in the beginning of this section, we want the income to be greater than or equal to $1500. To write an inequality, notice that the salesperson's income consists of $600 plus a commission (20% of sales).

Answers

10. $(-\infty, 30]$, **11.** $20,000

In words:

	600	+	commission (20% of sales)	$\geq$	1500
	$\downarrow$		$\downarrow$		$\downarrow$
Translate:	600	+	$0.20x$	$\geq$	1500

3. SOLVE the inequality for x.

$$600 + 0.20x \geq 1500$$
$$600 + 0.20x - 600 \geq 1500 - 600$$
$$0.20x \geq 900$$
$$x \geq 4500$$

4. INTERPRET.

Check: The income for sales of $4500 is

$$600 + 0.20(4500), \text{ or } 1500$$

Thus, if sales are greater than or equal to $4500, income is greater than or equal to $1500.

State: The minimum amount of sales needed for the salesperson to earn at least $1500 per month is $4500.   ▬▬

Example 12   Finding the Annual Consumption

In the United States, the annual consumption of cigarettes is declining. The consumption c in billions of cigarettes per year since the year 1985 can be approximated by the formula

$$c = -14.25t + 598.69$$

where t is the number of years after 1985. Use this formula to predict the years that the consumption of cigarettes will be less than 200 billion per year.

Solution: **1.** UNDERSTAND. Read and reread the problem. To become familiar with the given formula, let's find the cigarette consumption after 20 years, which would be the year $1985 + 20$, or 2005. To do so, we substitute 20 for t in the given formula.

$$c = -14.25(20) + 598.69 = 313.69$$

Thus, in 2005, we predict cigarette consumption to be about 313.69 billion.

Variables have already been assigned in the given formula. For review, they are

c = the annual consumption of cigarettes in the United States in billions of cigarettes

t = the number of years after 1985

2. TRANSLATE. We are looking for the years that the consumption of cigarettes c is less than 200. Since we are finding years t, we substitute the expression in the formula given for c, or

$$-14.25t + 598.69 < 200$$

Practice Problem 12

Use the formula given in Example 12 to predict when the consumption of cigarettes will be less than 100 billion per year.

Answer

12. after the year 2020

3. SOLVE the inequality.

$$-14.25t + 598.69 < 200$$

Subtract 598.69 from both sides.

$$-14.25t < -398.69$$

Divide both sides by -14.25 and round the result.

$$t > 27.98$$

4. INTERPRET.

Check: We substitute a number greater than 27.98 and see that c is less than 200.

State: The annual consumption of cigarettes will be less than 200 billion for the years more than 27.98 years after 1985, or in approximately $28 + 1985 = 2013$. ■

Focus On Study Skills

STUDYING FOR AND TAKING A MATH EXAM

Remember that one of the best ways to start preparing for an exam is to keep current with your assignments as they are made. Make an effort to clear up any confusion on topics as you cover them. Begin reviewing for your exam a few days in advance. If you find a topic during your review that you still don't understand, you'll have plenty of time to ask your instructor, another student in your class, or a math tutor for help. Don't wait until the last minute to "cram" for the test.

▲ Reread your notes and carefully review the Chapter Highlights at the end of each chapter to be covered.
▲ Try solving a few exercises from each section.
▲ Pay special attention to any new terminology or definitions in the chapter. Be sure you can state the meanings of definitions in your own words.
▲ Find a quiet place to take the Chapter Test found at the end of the chapter to be covered. This gives you a chance to practice taking the real exam, so try the Chapter Test without referring to your notes or looking up anything in your book. Give yourself the same amount of time to take the Chapter Test as you will have to take the exam for which you are preparing. If your exam covers more than one chapter, you should try taking the Chapter Tests for each chapter covered. You may also find working through the Cumulative Reviews helpful when preparing for a multi-chapter test.
▲ When you have finished taking the Chapter Test, check your answers against those in the back of the book. Redo any of the problems you missed. Then spend extra time solving similar problems.
▲ If you tend to get anxious while taking an exam, try to visualize yourself taking the exam in advance. Picture yourself being calm, clearheaded, and successful. Picture yourself remembering concepts and definitions with no trouble. When you are well prepared for an exam, a lot of nervousness can be avoided through positive thinking.
▲ Get lots of rest the night before the exam. It's hard to show how well you know the material if your brain is foggy from lack of sleep.

When it's time to take your exam, remember these hints:

▲ Make sure you have all the tools you will need to take the exam, including an extra pencil and eraser, paper (if needed), and calculator (if allowed).
▲ Try to relax. Taking a few deep breaths, inhaling and then exhaling slowly before you begin, might help.
▲ Are there any special definitions or solution steps that you'll need to remember during the exam? As soon as you get your exam, write these down at the top, bottom, or on the back of your paper.
▲ Scan the entire test to get an idea of what questions are being asked.
▲ Start with the questions that are easiest for you. This will help build your confidence. Then return to the harder ones.
▲ Read all directions carefully. Make sure that your final result answers the question being asked.
▲ Show all of your work. Try to work neatly.
▲ Don't spend too much time on a single problem. If you get stuck, try moving on to other problems so you can increase your chances of finishing the test. If you have time, you can return to the problem giving you trouble.
▲ Before turning in your exam, check your work carefully if time allows. Be on the lookout for careless mistakes.

MENTAL MATH

Solve each inequality mentally.

1. $x - 2 < 4$ **2.** $x - 1 > 6$ **3.** $x + 5 \geq 15$ **4.** $x + 1 \leq 8$

5. $3x > 12$ **6.** $5x < 20$ **7.** $\dfrac{x}{2} \leq 1$ **8.** $\dfrac{x}{4} \geq 2$

EXERCISE SET 2.8

D *Solve. Write the solution set using interval notation. See Examples 9 and 10.*

1. $-2x + 7 \geq 9$ **2.** $8 - 5x \leq 23$ **3.** $15 + 2x \geq 4x - 7$

4. $20 + x < 6x$ **5.** $3(x - 5) < 2(2x - 1)$

6. $5(x + 4) \leq 4(2x + 3)$ **7.** $\dfrac{1}{2} + \dfrac{2}{3} \geq \dfrac{x}{6}$

8. $\dfrac{3}{4} - \dfrac{2}{3} > \dfrac{x}{6}$ **9.** $-5x + 4 \leq -4(x - 1)$

10. $-6x + 2 < -3(x + 4)$ **11.** $\dfrac{1}{4}(x - 7) \geq x + 2$

12. $\dfrac{3}{5}(x + 1) \leq x + 1$

MENTAL MATH ANSWERS

1. _____
2. _____
3. _____
4. _____
5. _____
6. _____
7. _____
8. _____

ANSWERS

1. _____
2. _____
3. _____
4. _____
5. _____
6. _____
7. _____
8. _____
9. _____
10. _____
11. _____
12. _____

CHAPTER 2 HIGHLIGHTS

DEFINITIONS AND CONCEPTS	EXAMPLES

SECTION 2.1 SIMPLIFYING EXPRESSIONS

The **numerical coefficient** of a **term** is its numerical factor.

TERM	NUMERICAL COEFFICIENT
$-7y$	-7
x	1
$\frac{1}{5}a^2b$	$\frac{1}{5}$

Terms with the same variables raised to exactly the same powers are **like terms**.

LIKE TERMS	UNLIKE TERMS
$12x, -x$	$3y, 3y^2$
$-2xy, 5yx$	$7a^2b, -2ab^2$

To combine like terms, add the numerical coefficients and multiply the result by the common variable factor.

To remove parentheses, apply the distributive property.

$$9y + 3y = 12y$$
$$-4z^2 + 5z^2 - 6z^2 = -5z^2$$

$$-4(x + 7) + 10(3x - 1)$$
$$= -4x - 28 + 30x - 10$$
$$= 26x - 38$$

SECTION 2.2 THE ADDITION PROPERTY OF EQUALITY

A **linear equation in one variable** can be written in the form $Ax + B = C$ where A, B, and C are real numbers and $A \neq 0$.

Equivalent equations are equations that have the same solution.

$$-3x + 7 = 2$$
$$3(x - 1) = -8(x + 5) + 4$$

$x - 7 = 10$ and $x = 17$ are equivalent equations.

ADDITION PROPERTY OF EQUALITY

Adding the same number to or subtracting the same number from both sides of an equation does not change its solution.

$$y + 9 = 3$$
$$y + 9 - 9 = 3 - 9$$
$$y = -6$$

SECTION 2.3 THE MULTIPLICATION PROPERTY OF EQUALITY

MULTIPLICATION PROPERTY OF EQUALITY

Multiplying both sides or dividing both sides of an equation by the same nonzero number does not change its solution.

$$\frac{2}{3}a = 18$$
$$\frac{3}{2}\left(\frac{2}{3}a\right) = \frac{3}{2}(18)$$
$$a = 27$$

SECTION 2.4 FURTHER SOLVING LINEAR EQUATIONS

TO SOLVE LINEAR EQUATIONS

1. Clear the equation of fractions.

Solve: $\dfrac{5(-2x + 9)}{6} + 3 = \dfrac{1}{2}$

1. $6 \cdot \dfrac{5(-2x + 9)}{6} + 6 \cdot 3 = 6 \cdot \dfrac{1}{2}$

2. Remove any grouping symbols such as parentheses.

3. Simplify each side by combining like terms.

2. $5(-2x + 9) + 18 = 3$ Apply the distributive property.
$$-10x + 45 + 18 = 3$$

3. $-10x + 63 = 3$ Combine like terms.

SECTION 2.4 (CONTINUED)	
4. Get all variable terms on one side and all numbers on the other side by using the addition property of equality. **5.** Get the variable alone by using the multiplication property of equality. **6.** Check the solution by substituting it into the original equation.	**4.** $-10x + 63 - 63 = 3 - 63$ Subtract 63. $-10x = -60$ **5.** $\dfrac{-10x}{-10} = \dfrac{-60}{-10}$ Divide by -10. $x = 6$

SECTION 2.5 AN INTRODUCTION TO PROBLEM SOLVING	
PROBLEM-SOLVING STEPS	The height of the Hudson volcano in Chili is twice the height of the Kiska volcano in the Aleutian Islands. If the sum of their heights is 12,870 feet, find the height of each.
1. UNDERSTAND the problem.	**1.** Read and reread the problem. Guess a solution and check your guess. Let x be the height of the Kiska volcano. Then $2x$ is the height of the Hudson volcano. x $2x$ Kiska Hudson
2. TRANSLATE the problem.	**2.** | height of Kiska	added to	height of Hudson	is	12,870
↓	↓	↓	↓	↓
x	$+$	$2x$	$=$	12,870	
3. SOLVE the equation.	**3.** $x + 2x = 12{,}870$ $3x = 12{,}870$ $x = 4290$
4. INTERPRET the results.	**4.** *Check*: If x is 4290 then $2x$ is $2(4290)$ or 8580. Their sum is $4290 + 8580$ or $12{,}870$, the required amount. *State*: The Kiska volcano is 4290 feet high and the Hudson volcano is 8580 feet high.

SECTION 2.6 FORMULAS AND PROBLEM SOLVING	
An equation that describes a known relationship among quantities is called a **formula**. **To solve a formula for a specified variable**, use the same steps as for solving a linear equation. Treat the specified variable as the only variable of the equation.	$A = lw$ (area of a rectangle) $I = PRT$ (simple interest) Solve $P = 2l + 2w$ for l. $P = 2l + 2w$ $P - 2w = 2l + 2w - 2w$ Subtract $2w$. $P - 2w = 2l$ $\dfrac{P - 2w}{2} = \dfrac{2l}{2}$ Divide by 2. $\dfrac{P - 2w}{2} = l$ Simplify.

SECTION 2.7 PERCENT, RATIO, AND PROPORTION

Use the same problem-solving steps to solve a problem containing percents.

1. UNDERSTAND.

2. TRANSLATE.

3. SOLVE.

4. INTERPRET.

A **ratio** is the quotient of two numbers or two quantities.

The ratio of *a* to *b* can also be written as

$$\frac{a}{b} \quad \text{or} \quad a:b$$

A **proportion** is a mathematical statement that two ratios are equal.

In the proportion $\frac{a}{b} = \frac{c}{d}$, the products ad and bc are called **cross products**.

If $\frac{a}{b} = \frac{c}{d}$, then $ad = bc$.

32% of what number is 36.8?

1. Read and reread. Propose a solution and check. Let x = the unknown number.

2.

32%	of	what number	is	36.8
↓	↓	↓	↓	↓
32%	·	x	=	36.8

3. Solve
$$32\% \cdot x = 36.8$$
$$0.32x = 36.8$$
$$\frac{0.32x}{0.32} = \frac{36.8}{0.32} \quad \text{Divide by 0.32.}$$
$$x = 115 \quad \text{Simplify.}$$

4. 32% of 115 is 36.8.

Write the ratio of 5 hours to 1 day using fractional notation.

$$\frac{5 \text{ hours}}{1 \text{ day}} = \frac{5 \text{ hours}}{24 \text{ hours}} = \frac{5}{24}$$

$$\frac{2}{3} = \frac{8}{12} \qquad \frac{x}{7} = \frac{15}{35}$$

$$\frac{2}{3} \diagdown \frac{8}{12} \longrightarrow \quad 3 \cdot 8 \text{ or } 24$$
$$\qquad \longrightarrow \quad 2 \cdot 12 \text{ or } 24$$

Solve: $\frac{3}{4} = \frac{x}{x - 1}$

$$\frac{3}{4} \diagup \frac{x}{x - 1}$$
$$3(x - 1) = 4x \quad \text{Set cross products equal.}$$
$$3x - 3 = 4x$$
$$-3 = x$$

SECTION 2.8 LINEAR INEQUALITIES AND PROBLEM SOLVING

Properties of inequalities are similar to properties of equations. Don't forget that if you multiply or divide both sides of an inequality by the same *negative* number, you must reverse the direction of the inequality symbol.

TO SOLVE LINEAR INEQUALITIES

1. Clear the inequality of fractions.
2. Remove grouping symbols.
3. Simplify each side by combining like terms.
4. Write all variable terms on one side and all numbers on the other side using the addition property of inequality.
5. Get the variable alone by using the multiplication property of inequality.

$$-2x \le 4$$
$$\frac{-2x}{-2} \ge \frac{4}{-2} \quad \text{Divide by } -2, \text{ reverse the inequality symbol.}$$
$$x \ge -2$$

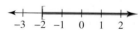

Solve: $3(x + 2) \le -2 + 8$

1. $3(x + 2) \le -2 + 8$ No fractions to clear.
2. $3x + 6 \le -2 + 8$ Apply the distributive property.
3. $3x + 6 \le 6$ Combine like terms.
4. $3x + 6 - 6 \le 6 - 6$ Subtract 6.
$$3x \le 0$$
5. $\frac{3x}{3} \le \frac{0}{3}$ Divide by 3.
$$x \le 0$$

Graphing Equations and Inequalities

CHAPTER 3

In Chapter 2 we learned to solve and graph the solutions of linear equations and inequalities in one variable on number lines. Now we define and present techniques for solving and graphing linear equations and inequalities in two variables on grids.

OPEC, the Organization of Petroleum Exporting Countries, was established in 1960. OPEC's goal is to control the price of crude oil worldwide by controlling oil production. For example, if OPEC countries agree to limit the amount of oil they produce, an oil shortage is created and oil prices rise. In 1997, the members of OPEC were Algeria, Indonesia, Iran, Iraq, Kuwait, Libya, Nigeria, Qatar, Saudi Arabia, United Arab Emirates, and Venezuela. OPEC's headquarters is in Vienna, Austria.

Problem Solving Notes

3.1 THE RECTANGULAR COORDINATE SYSTEM

In Section 1.8, we learned how to read graphs. Example 4 in Section 1.8 presented the broken line graph below showing the relationship between time spent smoking a cigarette and pulse rate. The horizontal line or axis shows time in minutes and the vertical line or axes shows the pulse rate in heartbeats per minute. Notice in this graph that there are two numbers associated with each point of the graph. For example, we discussed earlier that 15 minutes after "lighting up," the pulse rate is 80 beats per minute. If we agree to write the time first and the pulse rate second, we can say there is a point on the graph corresponding to the **ordered pair** of numbers (15, 80). A few more ordered pairs are shown alongside their corresponding points.

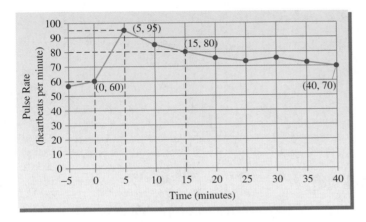

A PLOTTING ORDERED PAIRS OF NUMBERS

In general, we use the idea of ordered pairs to describe the location of a point in a plane (such as a piece of paper). We start with a horizontal and a vertical axis. Each axis is a number line, and for the sake of consistency we construct our axes to intersect at the 0 coordinate of both. This point of intersection is called the **origin**. Notice that these two number lines or axes divide the plane into four regions called **quadrants**. The quadrants are usually numbered with Roman numerals as shown. The axes are not considered to be in any quadrant.

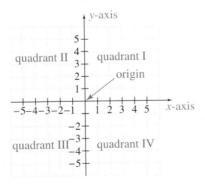

It is helpful to label axes, so we label the horizontal axis the ***x*-axis** and the vertical axis the ***y*-axis**. We call the system described above the **rectangular coordinate system**, or the **coordinate plane**. Just as with other graphs shown, we can then describe the locations of points by ordered pairs of numbers. We list the horizontal ***x*-axis** measurement first and the vertical ***y*-axis** measurement second.

To plot or graph the point corresponding to the ordered pair

$$(a, b)$$

We start at the origin. We then move a units left or right (right if a is positive, left if a is negative). From there, we move b units up or down (up if b is positive, down if b is negative). For example, to plot the point corresponding to the ordered pair $(3, 2)$, we start at the origin, move 3 units right, and from there move 2 units up. (See the figure below.) The x-value, 3, is also called the **x-coordinate** and the y-value, 2, is also called the **y-coordinate**. From now on, we will call the point with coordinates $(3, 2)$ simply the point $(3, 2)$. The point $(-2, 5)$ is graphed below also.

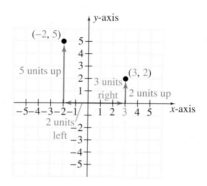

✓ CONCEPT CHECK

Is the graph of the point $(-5, 1)$ in the same location as the graph of the point $(1, -5)$? Explain.

Practice Problem 1

On a single coordinate system, plot each ordered pair. State in which quadrant, if any, each point lies.

a. $(4, 2)$ b. $(-1, -3)$

c. $(2, -2)$ d. $(-5, 1)$

e. $(0, 3)$ f. $(3, 0)$

g. $(0, -4)$ h. $\left(-2\frac{1}{2}, 0\right)$

HELPFUL HINT

Don't forget that **each ordered pair corresponds to exactly one point in the plane and that each point in the plane corresponds to exactly one ordered pair.**

TRY THE **CONCEPT CHECK** IN THE MARGIN.

Example 1

On a single coordinate system, plot each ordered pair. State in which quadrant, if any, each point lies.

a. $(5, 3)$ b. $(-2, -4)$ c. $(1, -2)$ d. $(-5, 3)$

e. $(0, 0)$ f. $(0, 2)$ g. $(-5, 0)$ h. $\left(0, -5\frac{1}{2}\right)$

Solution:

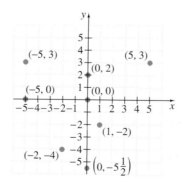

Point $(5, 3)$ lies in quadrant I.
Point $(-5, 3)$ lies in quadrant II.
Point $(-2, -4)$ lies in quadrant III.
Point $(1, -2)$ lies in quadrant IV.
Points $(0, 0)$, $(0, 2)$, $(-5, 0)$,
and $\left(0, -5\frac{1}{2}\right)$ lie on axes, so
they are not in any quadrant. ▬▬▬

Answers

1.

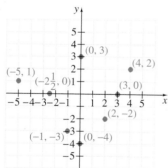

Point $(4, 2)$ lies in quadrant I.
Point $(-5, 1)$ lies in quadrant II.
Point $(-1, -3)$ lies in quadrant III.
Point $(2, -2)$ lies in quadrant IV.

Points $(0, 3)$, $(3, 0)$, $(0, -4)$, and $\left(-2\frac{1}{2}, 0\right)$

lie on axes, so they are not in any quadrant.

✓ Concept Check:

The graph of point $(-5, 1)$ lies in quadrant II and the graph of point $(1, -5)$ lies in quadrant IV. They are *not* in the same location.

TRY THE CONCEPT CHECK IN THE MARGIN.

From Example 1, notice that the *y*-coordinate of any point on the *x*-axis is 0. For example, the point $(-5, 0)$ lies on the *x*-axis. Also, the *x*-coordinate of any point on the *y*-axis is 0. For example, the point $(0, 2)$ lies on the *y*-axis.

B CREATING SCATTER DIAGRAMS

Data that can be represented as an ordered pair is called **paired data**. Many types of data collected from the real world are paired data. For instance, the annual measurement of a child's height can be written as an ordered pair of the form (year, height in inches) and is paired data. The graph of paired data as points in the rectangular coordinate system is called a **scatter diagram**. Scatter diagrams can be used to look for patterns and trends in paired data.

Example 2   The table gives the annual revenues for Wal-Mart Stores for the years shown. (*Source*: Wal-Mart Stores, Inc.)

Year	Wal-Mart Revenue (in billions of dollars)
1993	56
1994	68
1995	83
1996	95
1997	106

a. Write this paired data as a set of ordered pairs of the form (year, revenue in billions of dollars).

b. Create a scatter diagram of the paired data.

c. What trend in the paired data does the scatter diagram show?

Solution: **a.** The ordered pairs are $(1993, 56)$, $(1994, 68)$, $(1995, 83)$, $(1996, 95)$, and $(1997, 106)$.

b. We begin by plotting the ordered pairs. Because the *x*-coordinate in each ordered pair is a year, we label the *x*-axis "Year" and mark the horizontal axis with the years given. Then we label the *y*-axis or vertical axis "Wal-Mart Revenue (in billions of dollars)." It is convenient to mark the vertical axis in multiples of 20, starting with 0.

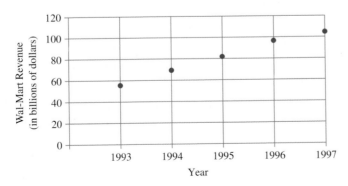

c. The scatter diagram shows that Wal-Mart revenue steadily increased over the years 1993–1997.

✓ CONCEPT CHECK

For each description of a point in the rectangular coordinate system, write an ordered pair that represents it.

a. Point A is located three units to the left of the *y*-axis and five units above the *x*-axis.

b. Point B is located six units below the origin.

Practice Problem 2

The table gives the number of cable TV subscribers (in millions) for the years shown. (*Source*: *Television and Cable Factbook*, Warren Publishing, Inc., Washington, D.C.)

Year	Cable TV Subscribers (in millions)
1986	38
1988	44
1990	50
1992	53
1994	57
1996	62

a. Write this paired data as a set of ordered pairs of the form (year, number of cable TV subscribers in millions).

b. Create a scatter diagram of the paired data.

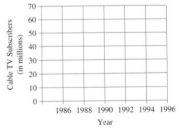

c. What trend in the paired data does the scatter diagram show?

Answers

2. a. $(1986, 38)$, $(1988, 44)$, $(1990, 50)$, $(1992, 53)$, $(1994, 57)$, $(1996, 62)$,

b.

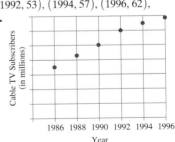

c. The number of cable TV subscribers has steadily increased.

✓ Concept Check: **a.** $(-3, 5)$,   **b.** $(0, -6)$

C COMPLETING ORDERED PAIRS SOLUTIONS

Let's see how we can use ordered pairs to record solutions of equations containing two variables. An equation in one variable such as $x + 1 = 5$ has one solution, which is 4: the number 4 is the value of the variable x that makes the equation true.

An equation in two variables, such as $2x + y = 8$, has solutions consisting of two values, one for x and one for y. For example, $x = 3$ and $y = 2$ is a solution of $2x + y = 8$ because, if x is replaced with 3 and y with 2, we get a true statement.

$$2x + y = 8$$
$$2(3) + 2 \stackrel{?}{=} 8$$
$$8 = 8 \quad \text{True.}$$

The solution $x = 3$ and $y = 2$ can be written as $(3, 2)$, an ordered pair of numbers.

In general, an ordered pair is a **solution** of an equation in two variables if replacing the variables by the values of the ordered pair results in a *true statement*. For example, another ordered pair solution of $2x + y = 8$ is $(5, -2)$. Replacing x with 5 and y with -2 results in a true statement.

$$2x + y = 8$$
$$2(5) + (-2) \stackrel{?}{=} 8 \quad \text{Replace } x \text{ with 5 and } y \text{ with } -2.$$
$$10 - 2 \stackrel{?}{=} 8$$
$$8 = 8 \quad \text{True.}$$

Example 3   Complete each ordered pair so that it is a solution to the equation $3x + y = 12$.

a. $(0, \ )$ **b.** $(\ , 6)$ **c.** $(-1, \ )$

Solution: **a.** In the ordered pair $(0, \ )$, the x-value is 0. We let $x = 0$ in the equation and solve for y.

$$3x + y = 12$$
$$3(0) + y = 12 \quad \text{Replace } x \text{ with 0.}$$
$$0 + y = 12$$
$$y = 12$$

The completed ordered pair is $(0, 12)$.

b. In the ordered pair $(\ , 6)$, the y-value is 6. We let $y = 6$ in the equation and solve for x.

$$3x + y = 12$$
$$3x + 6 = 12 \quad \text{Replace } y \text{ with 6.}$$
$$3x = 6 \quad \text{Subtract 6 from both sides.}$$
$$x = 2 \quad \text{Divide both sides by 3.}$$

The ordered pair is $(2, 6)$.

c. In the ordered pair $(-1, \ )$, the x-value is -1. We let $x = -1$ in the equation and solve for y.

$$3x + y = 12$$
$$3(-1) + y = 12 \quad \text{Replace } x \text{ with } -1.$$
$$-3 + y = 12$$
$$y = 15 \quad \text{Add 3 to both sides.}$$

The ordered pair is $(-1, 15)$.

Practice Problem 3

Complete each ordered pair so that it is a solution to the equation $x + 2y = 8$.

a. $(0, \ )$

b. $(\ , 3)$

c. $(-4, \ )$

Answers

3. a. $(0, 4)$,   **b.** $(2, 3)$,   **c.** $(-4, 6)$

Solutions of equations in two variables can also be recorded in a **table of paired values**, as shown in the next example.

Example 4 Complete the table for the equation $y = 3x$.

	x	y
a.	-1	
b.		0
c.		-9

Solution:

a. We replace x with -1 in the equation and solve for y.

$$y = 3x$$
$$y = 3(-1) \quad \text{Let } x = -1.$$
$$y = -3$$

The ordered pair is $(-1, -3)$

b. We replace y with 0 in the equation and solve for x.

$$y = 3x$$
$$0 = 3x \quad \text{Let } y = 0.$$
$$0 = x \quad \text{Divide both sides by 3.}$$

The ordered pair is $(0, 0)$.

c. We replace y with -9 in the equation and solve for x.

$$y = 3x$$
$$-9 = 3x \quad \text{Let } y = -9.$$
$$-3 = x \quad \text{Divide both sides by 3.}$$

The ordered pair is $(-3, -9)$. The completed table is shown to the right.

x	y
-1	-3
0	0
-3	-9

Example 5 Complete the table for the equation $y = 3$.

x	y
-2	
0	
-5	

Solution: The equation $y = 3$ is the same as $0x + y = 3$. No matter what value we replace x by, y always equals 3. The completed table is shown to the right.

x	y
-2	3
0	3
-5	3

By now, you have noticed that equations in two variables often have more than one solution. We discuss this more in the next section.

A table showing ordered pair solutions may be written vertically, or horizontally as shown in the next example.

Practice Problem 4

Complete the table for the equation $y = -2x$.

	x	y
a.	-3	
b.		0
c.		10

Practice Problem 5

Complete the table for the equation $x = 5$.

x	y
	-2
	0
	4

Answers

4.

	x	y
a.	-3	6
b.	0	0
c.	-5	10

5.

x	y
5	-2
5	0
5	4

Practice Problem 6

A company purchased a fax machine for $400. The business manager of the company predicts that the fax machine will be used for 7 years and the value in dollars y of the machine in x years is $y = -50x + 400$. Complete the table.

x	1	2	3	4	5	6	7
y							

Example 6   A small business purchased a computer for $2000. The business predicts that the computer will be used for 5 years and the value in dollars y of the computer in x years is $y = -300x + 2000$. Complete the table.

x	0	1	2	3	4	5
y						

Solution: To find the value of y when x is 0, we replace x with 0 in the equation. We use this same procedure to find y when x is 1 and when x is 2.

When $x = 0$,	**When $x = 1$,**	**When $x = 2$,**
$y = -300x + 2000$	$y = -300x + 2000$	$y = -300x + 2000$
$y = -300 \cdot 0 + 2000$	$y = -300 \cdot 1 + 2000$	$y = -300 \cdot 2 + 2000$
$y = 0 + 2000$	$y = -300 + 2000$	$y = -600 + 2000$
$y = 2000$	$y = 1700$	$y = 1400$

We have the ordered pairs $(0, 2000)$, $(1, 1700)$, and $(2, 1400)$. This means that in 0 years the value of the computer is $2000, in 1 year the value of the computer is $1700, and in 2 years the value is $1400. To complete the table of values, we continue the procedure for $x = 3$, $x = 4$, and $x = 5$.

When $x = 3$,	**When $x = 4$,**	**When $x = 5$,**
$y = -300x + 2000$	$y = -300x + 2000$	$y = -300x + 2000$
$y = -300 \cdot 3 + 2000$	$y = -300 \cdot 4 + 2000$	$y = -300 \cdot 5 + 2000$
$y = -900 + 2000$	$y = -1200 + 2000$	$y = -1500 + 2000$
$y = 1100$	$y = 800$	$y = 500$

The completed table is

x	0	1	2	3	4	5
y	2000	1700	1400	1100	800	500

The ordered pair solutions recorded in the completed table for Example 6 are another set of paired data. They are graphed next. Notice that this scatter diagram gives a visual picture of the decrease in value of the computer.

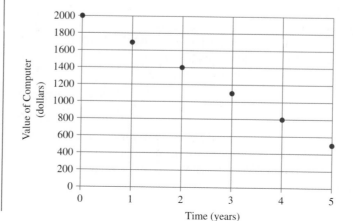

x	y
0	2000
1	1700
2	1400
3	1100
4	800
5	500

Answer

6.

x	1	2	3	4	5	6	7
y	350	300	250	200	150	100	50

Name _____ Section _____ Date _____

MENTAL MATH

Give two ordered pair solutions for each linear equation.

1. $x + y = 10$

2. $x + y = 6$

EXERCISE SET 3.1

A Plot each ordered pair. State in which quadrant, if any, each point lies. See Example 1.

1. $(1, 5)$ $(-5, -2)$ $(-3, 0)$
$(0, -1)$ $(2, -4)$ $\left(-1, 4\frac{1}{2}\right)$

2. $(2, 4)$ $(0, 2)$ $(-2, 1)$
$(-3, -3)$ $\left(3\frac{3}{4}, 0\right)$ $(5, -4)$

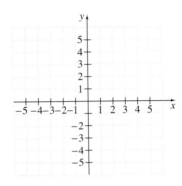

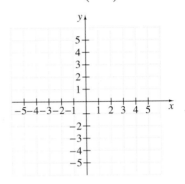

3. When is the graph of the ordered pair (a, b) the same as the graph of the ordered pair (b, a)?

4. In your own words, describe how to plot an ordered pair.

Find the x- and y-coordinates of each labeled point. See Example 1.

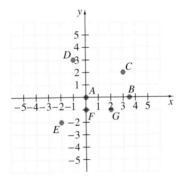

5. A

6. B

7. C

8. D

9. E

10. F

11. G

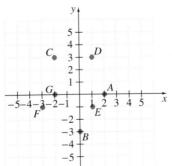

12. A

13. B

14. C

15. D

16. E

17. F

18. G

1. _____

2. _____

3. _____

4. _____

5. _____

6. _____

7. _____

8. _____

9. _____

10. _____

11. _____

12. _____

13. _____

14. _____

15. _____

16. _____

17. _____

18. _____

Problem Solving Notes

3.2 GRAPHING LINEAR EQUATIONS

In the previous section, we found that equations in two variables may have more than one solution. For example, both $(2, 2)$ and $(0, 4)$ are solutions of the equation $x + y = 4$. In fact, this equation has an infinite number of solutions. Other solutions include $(-2, 6)$, $(4, 0)$, and $(6, -2)$. Notice the pattern that appears in the graph of these solutions.

Objectives

A Graph a linear equation by finding and plotting ordered pair solutions.

SSM   CD-ROM   Video
3.2

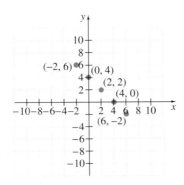

These solutions all appear to lie on the same line, as seen in the second graph. It can be shown that every ordered pair solution of the equation corresponds to a point on this line, and every point on this line corresponds to an ordered pair solution. Thus, we say that this line is the **graph of the equation** $x + y = 4$.

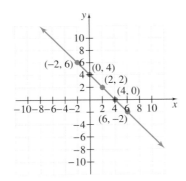

The equation $x + y = 4$ is called a **linear equation in two variables** and **the graph of every linear equation in two variables is a straight line**.

LINEAR EQUATION IN TWO VARIABLES

A linear equation in two variables is an equation that can be written in the form

$$Ax + By = C$$

where A, B, and C are real numbers and A and B are not both 0. The graph of a linear equation in two variables is a straight line.

The form $Ax + By = C$ is called **standard form.** Following are examples of linear equation in two variables.

$$2x + y = 8 \qquad -2x = 7y \qquad y = \frac{1}{3}x + 2 \qquad y = 7$$

A GRAPHING LINEAR EQUATIONS

From geometry, we know that a straight line is determined by just two points. Thus, to graph a linear equation in two variables we need to find just two of its infinitely many solutions. Once we do so, we plot the solution points and draw the line connecting the points. Usually, we find a third solution as well, as a check.

Practice Problem 1

Graph the linear equation $x + 3y = 6$.

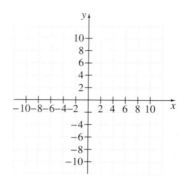

Example 1   Graph the linear equation $2x + y = 5$.

Solution: To graph this equation, we find three ordered pair solutions of $2x + y = 5$. To do this, we choose a value for one variable, x or y, and solve for the other variable. For example, if we let $x = 1$, then $2x + y = 5$ becomes

$$2x + y = 5$$
$$2(\mathbf{1}) + y = 5 \quad \text{Replace } x \text{ with 1.}$$
$$2 + y = 5 \quad \text{Multiply.}$$
$$y = \mathbf{3} \quad \text{Subtract 2 from both sides.}$$

Since $y = 3$ when $x = 1$, the ordered pair $(1, 3)$ is a solution of $2x + y = 5$. Next, we let $x = 0$.

$$2x + y = 5$$
$$2(\mathbf{0}) + y = 5 \quad \text{Replace } x \text{ with 0.}$$
$$0 + y = 5$$
$$y = \mathbf{5}$$

The ordered pair $(0, 5)$ is a second solution.

The two solutions found so far allow us to draw the straight line that is the graph of all solutions of $2x + y = 5$. However, we will find a third ordered pair as a check. Let $y = -1$.

$$2x + y = 5$$
$$2x + (\mathbf{-1}) = 5 \quad \text{Replace } y \text{ with } -1.$$
$$2x - 1 = 5$$
$$2x = 6 \quad \text{Add 1 to both sides.}$$
$$x = \mathbf{3} \quad \text{Divide both sides by 2.}$$

The third solution is $(3, -1)$. These three ordered pair solutions are listed in the table and plotted on the coordinate plane. The graph of $2x + y = 5$ is the line through the three points.

x	y
1	3
0	5
3	-1

Answer

1.

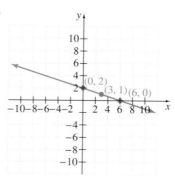

Helpful Hint

All three points should fall on the same straight line. If not, check your ordered pair solutions for a mistake.

Example 2   Graph the linear equation $-5x + 3y = 15$.

Solution: We find three ordered pair solutions of $-5x + 3y = 15$.

Let x = 0.

$-5x + 3y = 15$
$-5 \cdot 0 + 3y = 15$
$0 + 3y = 15$
$3y = 15$
$y = 5$

Let y = 0.

$-5x + 3y = 15$
$-5x + 3 \cdot 0 = 15$
$-5x + 0 = 15$
$-5x = 15$
$x = -3$

Let x = -2.

$-5x + 3y = 15$
$-5 \cdot -2 + 3y = 15$
$10 + 3y = 15$
$3y = 5$
$y = \dfrac{5}{3}$

The ordered pairs are $(0, 5)$, $(-3, 0)$, and $\left(-2, \dfrac{5}{3}\right)$. The graph of $-5x + 3y = 15$ is the line through the three points.

x	y
0	5
−3	0
−2	$\frac{5}{3} = 1\frac{2}{3}$

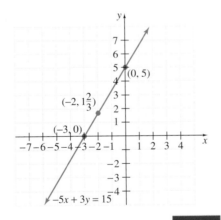

Example 3   Graph the linear equation $y = 3x$.

Solution: We find three ordered pair solutions. Since this equation is solved for y, we'll choose three x values.

If $x = 2$, $y = 3 \cdot 2 = 6$.
If $x = 0$, $y = 3 \cdot 0 = 0$.
If $x = -1$, $y = 3 \cdot -1 = -3$.

Practice Problem 2

Graph the linear equation $-2x + 4y = 8$.

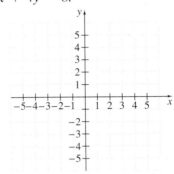

Practice Problem 3

Graph the linear equation $y = 2x$.

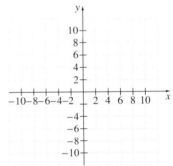

Answers

2.

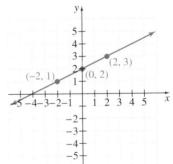

3.

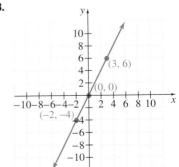

Next, we plot the ordered pair solutions and draw a line through the plotted points. The line is the graph of $y = 3x$. Every point on the graph represents an ordered pair solution of the equation and every ordered pair solution is a point on this line.

x	y
2	6
0	0
−1	−3

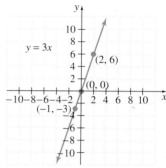

Practice Problem 4

Graph the linear equation $y = -\dfrac{1}{2}x$.

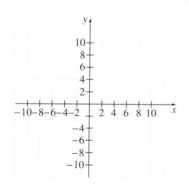

Example 4

Graph the linear equation $y = -\dfrac{1}{3}x$.

Solution   We find three ordered pair solutions, plot the solutions, and draw a line through the plotted solutions. To avoid fractions, we'll choose x values that are multiples of 3 to substitute into the equation.

If $x = 6$, then $y = -\dfrac{1}{3} \cdot 6 = -2$.

If $x = 0$, then $y = -\dfrac{1}{3} \cdot 0 = 0$.

If $x = -3$, then $y = -\dfrac{1}{3} \cdot -3 = 1$.

x	y
6	−2
0	0
−3	1

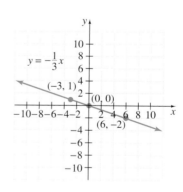

Let's compare the graphs in Examples 3 and 4. The graph of $y = 3x$ tilts upward (as we follow the line from left to right) and the graph of $y = -\dfrac{1}{3}x$ tilts downward (as we follow the line from left to right). Also notice that both lines go through the origin or that $(0, 0)$ is an ordered pair solution of both equations. We will learn more about the tilt, or slope, of a line in Section 3.4.

Answer

4.

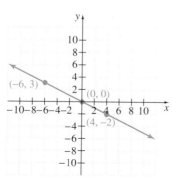

Example 5
Graph the linear equation $y = 3x + 6$ and compare this graph with the graph of $y = 3x$ in Example 3.

Solution
We find three ordered pair solutions, plot the solutions, and draw a line through the plotted solutions. We choose x values and substitute into the equation $y = 3x + 6$.

If $x = -3$, then $y = 3(-3) + 6 = -3$.
If $x = 0$, then $y = 3(0) + 6 = 6$.
If $x = 1$, then $y = 3(1) + 6 = 9$.

x	y
-3	-3
0	6
1	9

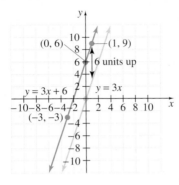

The most startling similarity is that both graphs appear to have the same upward tilt as we move from left to right. Also, the graph of $y = 3x$ crosses the y-axis at the origin, while the graph of $y = 3x + 6$ crosses the y-axis at 6. It appears that the graph of $y = 3x + 6$ is the same as the graph of $y = 3x$ except that the graph of $y = 3x + 6$ is moved 6 units upward.

Example 6
Graph the linear equation $y = -2$.

Solution:
Recall from Section 3.1 that the equation $y = -2$ is the same as $0x + y = -2$. No matter what value we replace x with, y is -2.

x	y
0	-2
3	-2
-2	-2

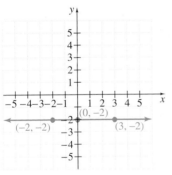

Notice that the graph of $y = -2$ is a horizontal line.

Graph the linear equation $y = 2x + 3$ and compare this graph with the graph of $y = 2x$ in Practice Problem 3.

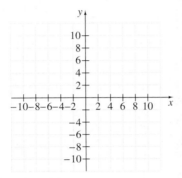

Practice Problem 6

Graph the linear equation $x = 3$.

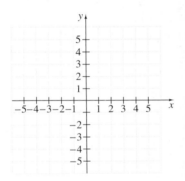

Answers
5.

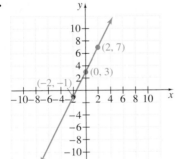

Same as the graph of $y = 2x$ except that the graph of $y = 2x + 3$ is moved 3 units upward.
6.

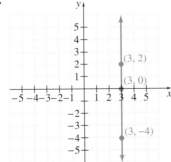

GRAPHING CALCULATOR EXPLORATIONS

In this section, we begin an optional study of graphing calculators and graphing software packages for computers. These graphers use the same point plotting technique that was introduced in this section. The advantage of this graphing technology is, of course, that graphing calculators and computers can find and plot ordered pair solutions much faster than we can. Note, however, that the features described in these boxes may not be available on all graphing calculators.

The rectangular screen where a portion of the rectangular coordinate system is displayed is called a **window**. We call it a **standard window** for graphing when both the x- and y-axes show coordinates between -10 and 10. This information is often displayed in the window menu on a graphing calculator as

Xmin $= -10$
Xmax $= 10$
Xscl $= 1$ The scale on the x-axis is one unit per tick mark.
Ymin $= -10$
Ymax $= 10$
Yscl $= 1$ The scale on the y-axis is one unit per tick mark.

To use a graphing calculator to graph the equation $y = 2x + 3$, press the $\boxed{\text{Y=}}$ key and enter the keystrokes $\boxed{2}\,\boxed{x}\,\boxed{+}\,\boxed{3}$. The top row should now read $Y_1 = 2x + 3$. Next press the $\boxed{\text{GRAPH}}$ key, and the display should look like this:

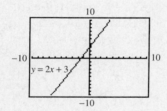

Use a standard window and graph the following linear equations. (Unless otherwise stated, use a standard window when graphing.)

1. $y = -3x + 7$ **2.** $y = -x + 5$ **3.** $y = 2.5x - 7.9$

4. $y = -1.3x + 5.2$ **5.** $y = -\dfrac{3}{10}x + \dfrac{32}{5}$ **6.** $y = \dfrac{2}{9}x - \dfrac{22}{3}$

EXERCISE SET 3.2

A *For each equation, find three ordered pair solutions by completing the table. Then use the ordered pairs to graph the equation. See Examples 1 through 6.*

1. $x - y = 6$

x	y
	0
4	
	−1

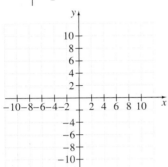

2. $x - y = 4$

x	y
0	
	2
−1	

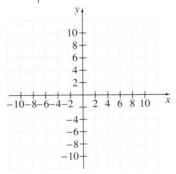

3. $y = -4x$

x	y
1	
0	
−1	

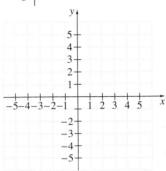

4. $y = -5x$

x	y
1	
0	
−1	

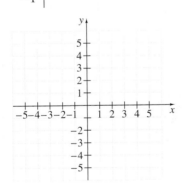

5. $y = \dfrac{1}{3}x$

x	y
0	
6	
−3	

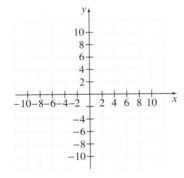

6. $y = \dfrac{1}{2}x$

x	y
0	
−4	
2	

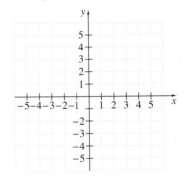

7. $y = -4x + 3$

x	y
0	
1	
2	

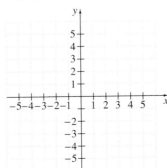

8. $y = -5x + 2$

x	y
0	
1	
2	

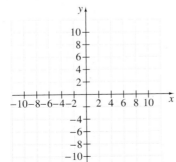

Name _____

Graph each linear equation. See Examples 1 through 6.

9. $x + y = 1$

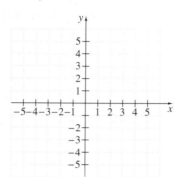

10. $x + y = 7$

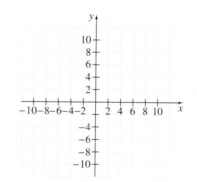

11. $x - y = -2$

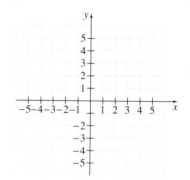

12. $-x + y = 6$

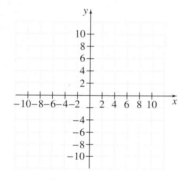

13. $x - 2y = 6$

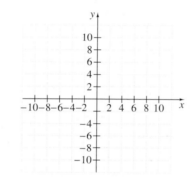

14. $-x + 5y = 5$

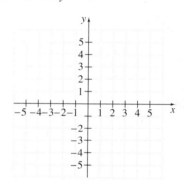

15. $y = 6x + 3$

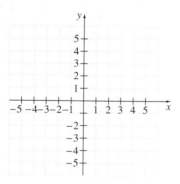

16. $y = -2x + 7$

17. $x = -4$

18. $y = 5$

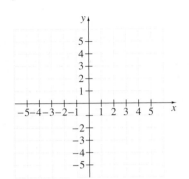

19. $y = 3$

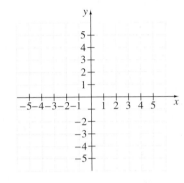

20. $x = -1$

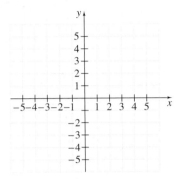

21. $y = x$

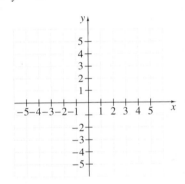

22. $y = -x$

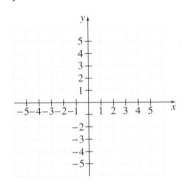

23. $y = 5x$

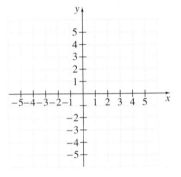

24. $y = 4x$

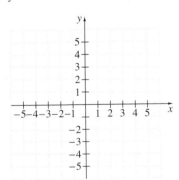

25. $x + 3y = 9$

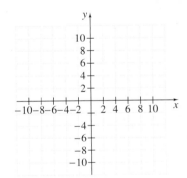

Problem Solving Notes

3.3 INTERCEPTS

A IDENTIFYING INTERCEPTS

The graph of $y = 4x - 8$ is shown below. Notice that this graph crosses the y-axis at the point $(0, -8)$. This point is called the **y-intercept point** and -8 is called the **y-intercept**. Likewise the graph crosses the x-axis at $(2, 0)$. This point is called the **x-intercept point** and 2 is the **x-intercept**.

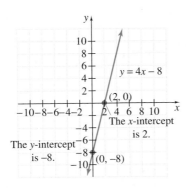

The intercept points are $(2, 0)$ and $(0, -8)$.

HELPFUL HINT

If a graph crosses the x-axis at $(-3, 0)$ and the y-axis at $(0, 7)$, then

Intercept points

$$
\underbrace{(-3, 0) \qquad (0, 7)}
$$
$\qquad \uparrow \qquad\qquad \uparrow$
$\quad$ x-intercept $\quad$ y-intercept

Notice that if y is 0, the corresponding x-value is the x-intercept. Likewise, if x is 0, the corresponding y-value is the y-intercept.

Examples   Identify the x- and y-intercepts and the intercept points.

1.

Solution: x-intercept: -3
y-intercept: 2
intercept points: $(-3, 0)$ and $(0, 2)$

Objectives

A Identify intercepts of a graph.

B Graph a linear equation by finding and plotting intercept points.

C Identify and graph vertical and horizontal lines.

SSM   CD-ROM   Video
3.3

Practice Problems 1–3

Identify the x- and y-intercepts and the intercept points.

1.
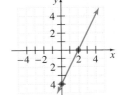

2.
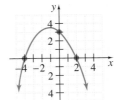

3.
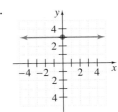

Answers

1. x-intercept: 2; y-intercept: -4; intercept points: $(2, 0)$ and $(0, -4)$, **2.** x-intercepts: $-4, 2$; y-intercept: 3; intercept points: $(-4, 0)$, $(2, 0)$, and $(0, 3)$, **3.** x-intercept: none; y-intercept: 3; intercept point: $(0, 3)$

2.

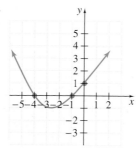

Solution: x-intercepts: $-4, -1$
y-intercept: 1
intercept points: $(-4, 0), (-1, 0), (0, 1)$

3.

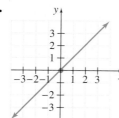

Solution: x-intercept: 0
y-intercept: 0
intercept point: $(0, 0)$

B FINDING AND PLOTTING INTERCEPT POINTS

Given an equation of a line, intercept points are usually easy to find since one coordinate is 0.

One way to find the y-intercept of a line, given its equation, is to let $x = 0$, since a point on the y-axis has an x-coordinate of 0. To find the x-intercept of a line, let $y = 0$, since a point on the x-axis has a y-coordinate of 0.

FINDING x- AND y-INTERCEPTS

To find the x-intercept, let $y = 0$ and solve for x.
To find the y-intercept, let $x = 0$ and solve for y.

Example 4

Graph $x - 3y = 6$ by finding and plotting its intercept points.

Solution: We let $y = 0$ to find the x-intercept and $x = 0$ to find the y-intercept.

Let $y = 0$.	Let $x = 0$.
$x - 3y = 6$	$x - 3y = 6$
$x - 3(0) = 6$	$0 - 3y = 6$
$x - 0 = 6$	$-3y = 6$
$x = 6$	$y = -2$

The x-intercept is 6 and the y-intercept is -2. We find a third ordered pair solution to check our work. If we let $y = -1$, then $x = 3$. We plot the points $(6, 0)$, $(0, -2)$,

---

Practice Problem 4

Graph $2x - y = 4$ by finding and plotting its intercept points.

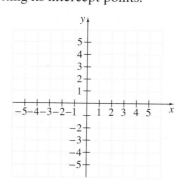

Answer

4.

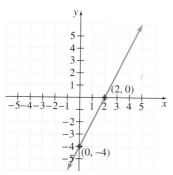

and $(3, -1)$. The graph of $x - 3y = 6$ is the line drawn through these points as shown.

x	y
6	0
0	−2
3	−1

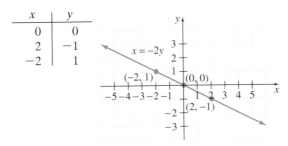

Example 5

Graph $x = -2y$ by finding and plotting its intercept points.

Solution:

We let $y = 0$ to find the x-intercept and $x = 0$ to find the y-intercept.

Let $y = 0$.   Let $x = 0$.

$x = -2y$ $x = -2y$

$x = -2(0)$ $0 = -2y$

$x = 0$ $0 = y$

Both the x-intercept and y-intercept are 0. In other words, when $x = 0$, then $y = 0$, which gives the ordered pair $(0, 0)$. Also, when $y = 0$, then $x = 0$, which gives the same ordered pair $(0, 0)$. This happens when the graph passes through the origin. Since two points are needed to determine a line, we must find at least one more ordered pair that satisfies $x = -2y$. We let $y = -1$ to find a second ordered pair solution and let $y = 1$ as a checkpoint.

Let $y = -1$.   Let $y = 1$.

$x = -2(-1)$ $x = -2(1)$

$x = 2$ $x = -2$

The ordered pairs are $(0, 0)$, $(2, -1)$, and $(-2, 1)$. We plot these points to graph $x = -2y$.

x	y
0	0
2	−1
−2	1

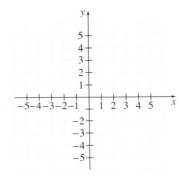

C GRAPHING VERTICAL AND HORIZONTAL LINES

The equation $x = 2$, for example, is a linear equation in two variables because it can be written in the form $x + 0y = 2$. The graph of this equation is a vertical line, as shown in the next example.

Practice Problem 5

Graph $y = 3x$ by finding and plotting its intercept points.

Answer

5.

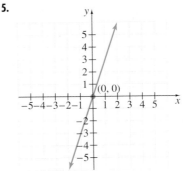

Practice Problem 6

Graph: $x = -3$

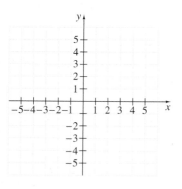

Practice Problem 7

Graph: $y = 4$

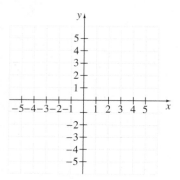

Answers

6.

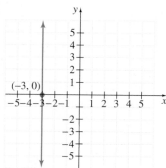

7.

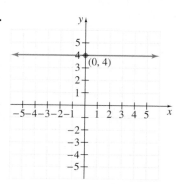

⊡ **Example 6**   Graph: $x = 2$.

Solution: The equation $x = 2$ can be written as $x + 0y = 2$. For any y-value chosen, notice that x is 2. No other value for x satisfies $x + 0y = 2$. Any ordered pair whose x-coordinate is 2 is a solution of $x + 0y = 2$. We will use the ordered pair solutions $(2, 3)$, $(2, 0)$, and $(2, -3)$ to graph $x = 2$.

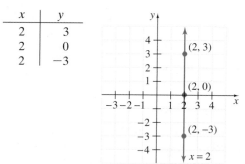

x	y
2	3
2	0
2	-3

The graph is a vertical line with x-intercept 2. Note that this graph has no y-intercept because x is never 0.

In general, we have the following.

> **VERTICAL LINES**
>
> The graph of $x = c$, where c is a real number, is a vertical line with x-intercept c.

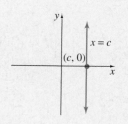

⊡ **Example 7**   Graph: $y = -3$

Solution: The equation $y = -3$ can be written as $0x + y = -3$. For any x-value chosen, y is -3. If we chose 4, 1, and -2 as x-values, the ordered pair solutions are $(4, -3)$, $(1, -3)$, and $(-2, -3)$. We use these ordered pairs to graph $y = -3$. The graph is a horizontal line with y-intercept -3 and no x-intercept.

x	y
4	-3
1	-3
-2	-3

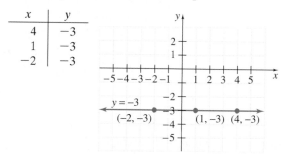

In general, we have the following.

> **HORIZONTAL LINES**
>
> The graph of $y = c$, where c is a real number, is a horizontal line with y-intercept c.
>
>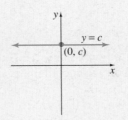

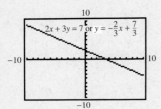 # GRAPHING CALCULATOR EXPLORATIONS

You may have noticed that to use the $\boxed{Y=}$ key on a grapher to graph an equation, the equation must be solved for y. For example, to graph $2x + 3y = 7$, we solve this equation for y.

$$2x + 3y = 7$$

$$3y = -2x + 7 \qquad \text{Subtract } 2x \text{ from both sides.}$$

$$\frac{3y}{3} = -\frac{2x}{3} + \frac{7}{3} \qquad \text{Divide both sides by 3.}$$

$$y = -\frac{2}{3}x + \frac{7}{3} \qquad \text{Simplify.}$$

To graph $2x + 3y = 7$ or $y = -\frac{2}{3}x + \frac{7}{3}$, press the $\boxed{Y=}$ key and enter

$$Y_1 = -\frac{2}{3}x + \frac{7}{3}$$

Graph each linear equation.

1. $x = 3.78y$

2. $-2.61y = x$

3. $-2.2x + 6.8y = 15.5$

4. $5.9x - 0.8y = -10.4$

Focus On The Real World

READING A MAP

How do you find a location on a map? Most maps we use today have a grid that is based on the rectangular coordinate system we use in algebra. After finding the coordinates of cities and other landmarks from the map index, the grid can help us find places on the map. To eliminate confusion, many maps use letters to label the grid along one edge and numbers along the other. However, the coordinates are still pairs of numbers and letters. For instance, the coordinates for Toledo on the map are A-2.

CRITICAL THINKING

1. Find the coordinates of the following cities: Hamilton, Columbus, Youngstown, and Cincinnati.
2. What cities correspond to the following coordinates: F-3, A-3, B-4, and D-2?
3. How are the map's coordinate system and the rectangular coordinate system we use in algebra the same? How are they different? What are the advantages of each?

EXERCISE SET 3.3

B *Graph each linear equation by finding and plotting its intercept points. See Examples 4 and 5.*

1. $x - y = 3$

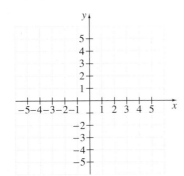

2. $x - y = -4$

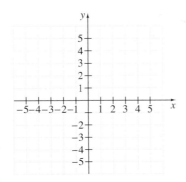

3. $x = 5y$

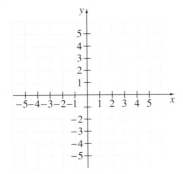

4. $2x = y$

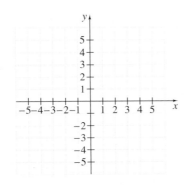

5. $-x + 2y = 6$

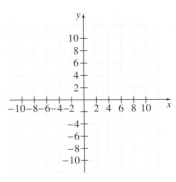

6. $x - 2y = -8$

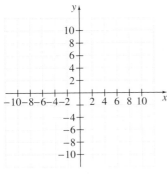

7. $2x - 4y = 8$

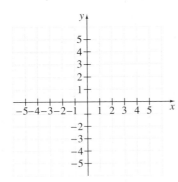

8. $2x + 3y = 6$

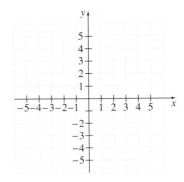

9. $x = 2y$

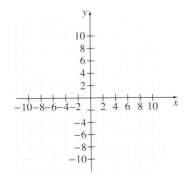

10. $y = -2x$

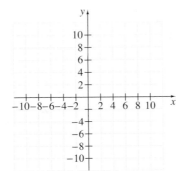

11. $y = 3x + 6$

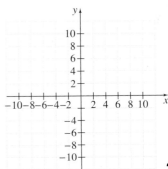

12 $y = 2x + 10$

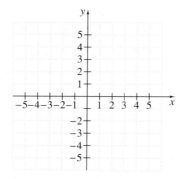

13. $x = y$

14. $x = -y$

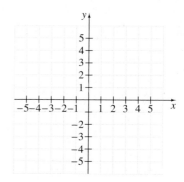

C *Graph each linear equation. See Examples 6 and 7.*

15. $x = -1$

16. $y = 5$

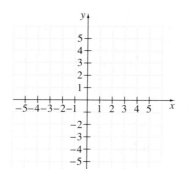

17. $y = 0$

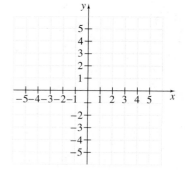

18. $x = 0$

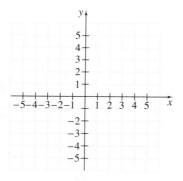

Focus On The Real World

ROAD GRADES

Have you ever driven on a hilly highway and seen a sign like the one below? The 7% on the sign refers to the grade of the road. The grade of a road is the same as its slope given as a percent. A 7 percent grade means that for every rise of 7 units there is a run of 100 units. The type of units doesn't matter as long as they are the same. For instance, we could say that a 7 percent grade represents a rise of 7 feet for every run of 100 feet or a rise of 7 meters for every run of 100 meters, and so on.

$$7\% = 0.07 = \frac{7}{100} \qquad\qquad 7\% \text{ grade} = \frac{\text{rise of } 7}{\text{run of } 100}$$

Most highways are designed to have grades of 6 percent or less. If a portion of a highway has a grade that is steeper than 6 percent, a sign is usually posted giving the grade and the number of miles for the grade. Truck drivers need to know when the road is particularly steep. They may need to take precautions such as using a different gear, reducing their speed, or testing their brakes.

Here is a sampling of road grades:
- ▲ A portion of the John Scott Highway in Steubenville, Ohio, has a 10% grade. (*Source:* Ohio Department of Transportation)
- ▲ Joaquin Road in Portola Valley, California, has an average grade of 15%. (*Source:* Western Wheelers Bicycle Club)
- ▲ The steepest grade in Seattle, Washington, is 26% on East Roy Street between 25th Avenue and 26th Avenue. (*Source: Seattle Post-Intelligencer*, Nov. 21, 1994)
- ▲ The steepest street in Pittsburgh, Pennsylvania, is Canton Avenue with a 37% grade. (*Source:* Pittsburgh Department of Public Works)
- ▲ The steepest street in the world is Baldwin Street in Dunedin, New Zealand. Its maximum grade is 79%. (*Source: Guinness Book of Records*, 1996)

COOPERATIVE LEARNING ACTIVITY

Try to find a road sign with a percent grade warning or the name and grade of a steep road in your area. Describe its slope and make a scale drawing to represent its grade.

Problem Solving Notes

3.4   SLOPE

A   FINDING THE SLOPE OF A LINE GIVEN TWO POINTS

Thus far, much of this chapter has been devoted to graphing lines. You have probably noticed by now that a key feature of a line is its slant or steepness. In mathematics, the slant or steepness of a line is formally known as its **slope**. We measure the slope of a line by the ratio of vertical change (rise) to the corresponding horizontal change (run) as we move along the line.

On the line below, for example, suppose that we begin at the point $(1, 2)$ and move to the point $(4, 6)$. The vertical change is the change in y-coordinates: $6 - 2$ or 4 units. The corresponding horizontal change is the change in x-coordinates: $4 - 1 = 3$ units. The ratio of these changes is

$$\text{slope} = \frac{\text{change in } y \text{ (vertical change)}}{\text{change in } x \text{ (horizontal change)}} = \frac{4}{3}$$

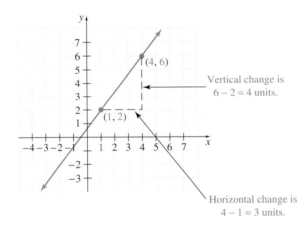

Vertical change is
$6 - 2 = 4$ units.

Horizontal change is
$4 - 1 = 3$ units.

The slope of this line, then, is $\frac{4}{3}$. This means that for every 4 units of change in y-coordinates, there is a corresponding change of 3 units in x-coordinates.

HELPFUL HINT

It makes no difference which two points of a line are chosen to find its slope. The slope of a line is the same everywhere on the line.

SLOPE OF A LINE

The slope m of the line containing the points (x_1, y_1) and (x_2, y_2) is given by

$$m = \frac{\text{rise}}{\text{run}} = \frac{\text{change in } y}{\text{change in } x} = \frac{y_2 - y_1}{x_2 - x_1}, \qquad \text{as long as } x_2 \neq x_1$$

Objectives

A Find the slope of a line given two points of the line.

B Find the slope of a line given its equation.

C Find the slopes of horizontal and vertical lines.

D Compare the slopes of parallel and perpendicular lines.

SSM   CD-ROM   Video
3.4

Practice Problem 1

Find the slope of the line through $(-2, 3)$ and $(4, -1)$. Graph the line.

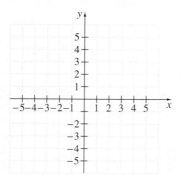

✓ **CONCEPT CHECK**

The points $(-2, -5)$, $(0, -2)$, $(4, 4)$, and $(10, 13)$ all lie on the same line. Work with a partner and verify that the slope is the same no matter which points are used to find slope.

Answers

1. $-\dfrac{2}{3}$

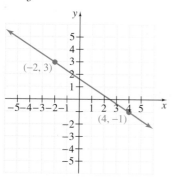

✓ Concept Check: $m = \dfrac{3}{2}$

Example 1   Find the slope of the line through $(-1, 5)$ and $(2, -3)$. Graph the line.

Solution: Let (x_1, y_1) be $(-1, 5)$ and (x_2, y_2) be $(2, -3)$. Then, by the definition of slope,

$$m = \frac{y_2 - y_1}{x_2 - x_1}$$

$$= \frac{-3 - 5}{2 - (-1)}$$

$$= \frac{-8}{3} = -\frac{8}{3}$$

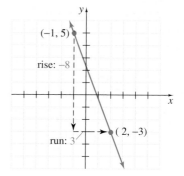

The slope of the line is $-\dfrac{8}{3}$.

In Example 1, we could just as well have identified (x_1, y_1) with $(2, -3)$ and (x_2, y_2) with $(-1, 5)$. It makes no difference which point is called (x_1, y_1) or (x_2, y_2).

TRY THE CONCEPT CHECK IN THE MARGIN.

HELPFUL HINT

When finding the slope of a line through two given points, it makes no difference which given point is called (x_1, y_1) and which is called (x_2, y_2). However, once an x-coordinate is called x_1, make sure its corresponding y-coordinate is called y_1.

Example 2

Find the slope of the line through $(-1, -2)$ and $(2, 4)$. Graph the line.

Solution: Let (x_1, y_1) be $(2, 4)$ and (x_2, y_2) be $(-1, -2)$.

$$m = \frac{y_2 - y_1}{x_2 - x_1}$$

$$= \frac{-2 - 4}{-1 - 2}$$

$$= \frac{-6}{-3} = 2$$

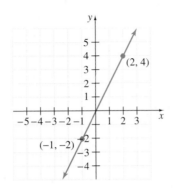

The slope is 2.

TRY THE CONCEPT CHECK IN THE MARGIN.

Notice that the slope of the line in Example 1 is negative, and the slope of the line in Example 2 is positive. Let your eye follow the line with negative slope from left to right and notice that the line "goes down." If you follow the line with positive slope from left to right, you will notice that the line "goes up." This is true in general.

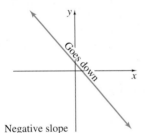

Negative slope

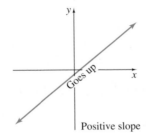

Positive slope

B FINDING THE SLOPE OF A LINE GIVEN ITS EQUATION

As we have seen, the slope of a line is defined by two points on the line. Thus, if we know the equation of a line, we can find its slope by finding two of its points. For example, let's find the slope of the line

$$y = 3x - 2$$

To find two points, we can choose two values for x and substitute to find corresponding y-values. If $x = 0$, for example, $y = 3 \cdot 0 - 2$ or $y = -2$. If $x = 1$, $y = 3 \cdot 1 - 2$ or $y = 1$. This gives the ordered pairs $(0, -2)$ and $(1, 1)$. Using the definition for slope, we have

$$m = \frac{1 - (-2)}{1 - 0} = \frac{3}{1} = 3 \quad \text{The slope is 3.}$$

Notice that the slope, 3, is the same as the coefficient of x in the equation $y = 3x - 2$. This is true in general.

> If a linear equation is solved for y, the coefficient of x is its slope. In other words, the slope of the line given by $y = mx + b$ is m, the coefficient of x.

Practice Problem 2

Find the slope of the line through $(-2, 1)$ and $(3, 5)$. Graph the line.

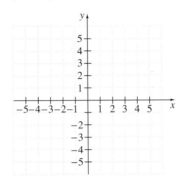

✓ CONCEPT CHECK

What is wrong with the following slope calculation?

$(3, 5)$ and $(-2, 6)$

$$m = \frac{5 - 6}{-2 - 3} = \frac{-1}{-5} = \frac{1}{5}$$

Answers

2. $\frac{4}{5}$

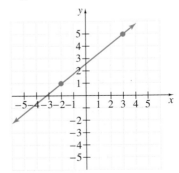

✓ Concept Check:

$$m = \frac{5 - 6}{3 - (-2)} = \frac{-1}{5} = -\frac{1}{5}$$

Practice Problem 3

Find the slope of the line
$5x + 4y = 10$.

Example 3

Find the slope of the line $-2x + 3y = 11$.

Solution: When we solve for y, the coefficient of x is the slope.

$$-2x + 3y = 11$$
$$3y = 2x + 11 \qquad \text{Add } 2x \text{ to both sides.}$$
$$y = \frac{2}{3}x + \frac{11}{3} \qquad \text{Divide both sides by 3.}$$

The slope is $\frac{2}{3}$. ↰

INTERPRETING SLOPE

The slope of a line can be thought of as a rate of change. For example, the slope of the line given by the equation $y = 2x$ is 2. Thinking of a slope of $2 \left(\text{or } \frac{2}{1}\right)$ as a rate of change means that for every change of 1 unit in the value of x, the value of y will change similarly by 2 units. A slope of -3 means that for every increase of 1 unit in the value of x, the value of y decreases by 3 units.

Interpreting slope as a rate of change is also meaningful in real-life applications. For example, the cost y (in cents) of an in-state long-distance telephone call in Massachusetts is given by the linear equation $y = 11x + 27$, where x is the length of the call in minutes (*Source*: Based on data from Bell Atlantic). By looking at the equation, we can see that the slope of this line is $11 \left(\text{or } \frac{11}{1}\right)$. This slope means that the cost of such a telephone call increases 11 cents for each additional 1 minute spent on the telephone. Another way to say this is that the cost of the telephone call is increasing at a rate of 11 cents per minute.

C   FINDING SLOPES OF HORIZONTAL AND VERTICAL LINES

Practice Problem 4

Find the slope of $y = 3$.

Example 4

Find the slope of the line $y = -1$.

Solution: Recall that $y = -1$ is a horizontal line with y-intercept -1. To find the slope, we find two ordered pair solutions of $y = -1$, knowing that solutions of $y = -1$ must have a y-value of -1. We will use $(2, -1)$ and $(-3, -1)$. We let (x_1, y_1) be $(2, -1)$ and (x_2, y_2) be $(-3, -1)$.

$$m = \frac{y_2 - y_1}{x_2 - x_1} = \frac{-1 - (-1)}{-3 - 2} = \frac{0}{-5} = 0$$

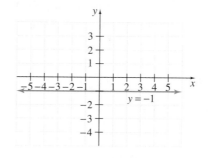

The slope of the line $y = -1$ is 0. Since the y-values will have a difference of 0 for all horizontal lines, we can say that all **horizontal lines have a slope of 0**.

Example 5   Find the slope of the line $x = 5$.

Solution: Recall that the graph of $x = 5$ is a vertical line with x-intercept 5. To find the slope, we find two ordered pair solutions of $x = 5$. Ordered pair solutions of $x = 5$ must have an x-value of 5. We will use $(5, 0)$ and $(5, 4)$. We let $(x_1, y_1) = (5, 0)$ and $(x_2, y_2) = (5, 4)$.

$$m = \frac{y_2 - y_1}{x_2 - x_1} = \frac{4 - 0}{5 - 5} = \frac{4}{0}$$

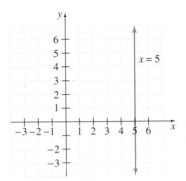

Since $\frac{4}{0}$ is undefined, we say the slope of the vertical line $x = 5$ is undefined. Since the x-values will have a difference of 0 for all vertical lines, we can say that all **vertical lines have undefined slope**. ▬▬▬

HELPFUL HINT

Slope of 0 and undefined slope are not the same. Vertical lines have undefined slope or no slope, while horizontal lines have a slope of 0.

D SLOPES OF PARALLEL AND PERPENDICULAR LINES

Two lines in the same plane are **parallel** if they do not intersect. Slopes of lines can help us determine whether lines are parallel. Since parallel lines have the same steepness, it follows that they have the same slope.

For example, the graphs of

$$y = -2x + 4$$

and

$$y = -2x - 3$$

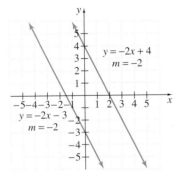

are shown. These lines have the same slope, -2. They also have different y-intercepts, so the lines are parallel. (If the y-intercepts are the same also, the lines are the same.)

Practice Problem 5

Find the slope of the line $x = -2$.

Answer

5. undefined slope

PARALLEL LINES

Nonvertical parallel lines have the same slope and different *y*-intercepts.

Two lines are **perpendicular** if they lie in the same plane and meet at a 90° (right) angle. How do the slopes of perpendicular lines compare? The product of the slopes of two perpendicular lines is −1.

For example, the graphs of

$$y = 4x + 1$$

and

$$y = -\frac{1}{4}x - 3$$

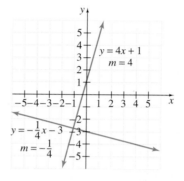

are shown. The slopes of the lines are 4 and $-\frac{1}{4}$. Their product is $4\left(-\frac{1}{4}\right) = -1$, so the lines are perpendicular.

PERPENDICULAR LINES

If the product of the slopes of two lines is −1, then the lines are perpendicular. (Two nonvertical lines are perpendicular if the slope of one is the negative reciprocal of the slope of the other.)

HELPFUL HINT

Here are examples of numbers that are negative reciprocals.

Number	Negative Reciprocal
$\frac{1}{3}$	$-\frac{3}{1}$ or -3
-5 or $-\frac{5}{1}$	$\frac{1}{5}$

HELPFUL HINT

Here are a few reminders about vertical and horizontal lines.

- ▲ Two distinct vertical lines are parallel.
- ▲ Two distinct horizontal lines are parallel.
- ▲ A horizontal line and a vertical line are always perpendicular.

Example 6 Determine whether each pair of lines is parallel, perpendicular, or neither.

a. $y = -\dfrac{1}{5}x + 1$ **b.** $x + y = 3$ **c.** $3x + y = 5$

  $2x + 10y = 3$ $-x + y = 4$ $2x + 3y = 6$

Solution: **a.** The slope of the line $y = -\dfrac{1}{5}x + 1$ is $-\dfrac{1}{5}$. We find the slope of the second line by solving it for y.

$$2x + 10y = 3$$
$$10y = -2x + 3 \qquad \text{Subtract } 2x \text{ from both sides.}$$
$$y = \frac{-2}{10}x + \frac{3}{10} \qquad \text{Divide both sides by 10.}$$
$$y = -\frac{1}{5}x + \frac{3}{10} \qquad \text{Simplify.}$$

The slope of this line is $-\dfrac{1}{5}$ also. Since the lines have the same slope and different y-intercepts, they are parallel, as shown in the figure.

b. To find each slope, we solve each equation for y.

$x + y = 3$ $-x + y = 4$
$y = -x + 3$ $y = x + 4$
$\uparrow$ $\uparrow$
The slope is -1.   The slope is 1.

The slopes are not the same, so the lines are not parallel. Next we check the product of the slopes: $(-1)(1) = -1$. Since the product is -1, the lines are perpendicular, as shown in the figure.

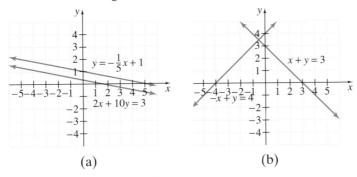

(a)   (b)

c. We solve each equation for y to find each slope. The slopes are -3 and $-\dfrac{2}{3}$. The slopes are not the same and their product is not -1. Thus, the lines are neither parallel nor perpendicular.

TRY THE CONCEPT CHECK IN THE MARGIN.

Practice Problem 6

Determine whether each pair of lines is parallel, perpendicular, or neither.

a. $x + y = 5$
    $2x + y = 5$

b. $5y = 2x - 3$
    $5x + 2y = 1$

c. $y = 2x + 1$
    $4x - 2y = 8$

✓ **CONCEPT CHECK**

Write the equations of three parallel lines.

Answers

6. a. neither, **b.** perpendicular, **c.** parallel

✓ **Concept Check**

For example, $y = 2x + 3$, $y = 2x - 1$, $y = 2x$

GRAPHING CALCULATOR EXPLORATIONS

It is possible to use a grapher and sketch the graph of more than one equation on the same set of axes. This feature can be used to see parallel lines with the same slope. For example, graph the equations $y = \frac{2}{5}x$, $y = \frac{2}{5}x + 7$, and $y = \frac{2}{5}x - 4$ on the same set of axes. To do so, press the $\boxed{\text{Y=}}$ key and enter the equations on the first three lines.

$$Y_1 = \left(\frac{2}{5}\right)x$$

$$Y_2 = \left(\frac{2}{5}\right)x + 7$$

$$Y_3 = \left(\frac{2}{5}\right)x - 4$$

The displayed equations should look like:

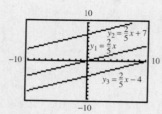

These lines are parallel as expected since they all have a slope of $\frac{2}{5}$. The graph of $y = \frac{2}{5}x + 7$ is the graph of $y = \frac{2}{5}x$ moved 7 units upward with a y-intercept of 7. Also, the graph of $y = \frac{2}{5}x - 4$ is the graph of $y = \frac{2}{5}x$ moved 4 units downward with a y-intercept of -4.

Graph the parallel lines on the same set of axes. Describe the similarities and differences in their graphs.

1. $y = 3.8x$, $y = 3.8x - 3$, $y = 3.8x + 9$

2. $y = -4.9x$, $y = -4.9x + 1$, $y = -4.9x + 8$

3. $y = \frac{1}{4}x$, $y = \frac{1}{4}x + 5$, $y = \frac{1}{4}x - 8$

4. $y = -\frac{3}{4}x$, $y = -\frac{3}{4}x - 5$, $y = -\frac{3}{4}x + 6$

Name _____ **Section** _____ **Date** _____

MENTAL MATH

Decide whether a line with the given slope is upward-sloping, downward-sloping, horizontal, or vertical.

1. $m = \dfrac{7}{6}$ **2.** $m = -3$ **3.** $m = 0$ **4.** m is undefined

EXERCISE SET 3.4

A *Use the points shown on each graph to find the slope of each line. See Examples 1 and 2.*

1.
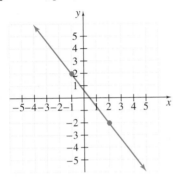

2.
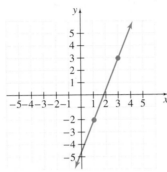

3.
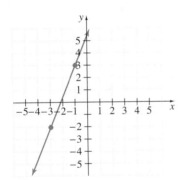

4.
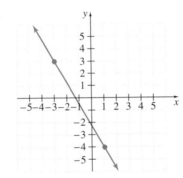

Find the slope of the line that passes through the given points. See Examples 1 and 2.

5. $(0, 0)$ and $(7, 8)$ **6.** $(-1, 5)$ and $(0, 0)$ ▣ **7.** $(-1, 5)$ and $(6, -2)$

8. $(-1, 9)$ and $(-3, 4)$ **9.** $(1, 4)$ and $(5, 3)$ **10.** $(3, 1)$ and $(2, 6)$

11. $(-2, 8)$ and $(1, 6)$

B *Find the slope of each line. See Example 3.*

12. $y = 5x - 2$ **13.** $y = -2x + 6$ **14.** $2x + y = 7$

15. $-5x + y = 10$ ▣ **16.** $2x - 3y = 10$

ANSWERS

1. _____

2. _____

3. _____

4. _____

5. _____

6. _____

7. _____

8. _____

9. _____

10. _____

11. _____

12. _____

13. _____

14. _____

15. _____

16. _____

317

Problem Solving Notes

3.5 GRAPHING LINEAR INEQUALITIES IN TWO VARIABLES

Recall that a linear equation in two variables is an equation that can be written in the form $Ax + By = C$ where A, B, and C are real numbers and A and B are not both 0. A **linear inequality in two variables** is an inequality that can be written in one of the forms

$$Ax + By < C \qquad Ax + By \leq C$$
$$Ax + By > C \qquad Ax + By \geq C$$

where A, B, and C are real numbers and A and B are not both 0.

A DETERMINING SOLUTIONS OF LINEAR INEQUALITIES IN TWO VARIABLES

Just as for linear equations in x and y, an ordered pair is a **solution** of an inequality in x and y if replacing the variables with the coordinates of the ordered pair results in a true statement.

Example 1 Determine whether each ordered pair is a solution of the equation $2x - y < 6$.

a. $(5, -1)$ **b.** $(2, 7)$

Solution: **a.** We replace x with 5 and y with -1 and see if a true statement results.

$$2x - y < 6$$
$$2(5) - (-1) < 6 \qquad \text{Replace } x \text{ with 5 and } y \text{ with } -1.$$
$$10 + 1 < 6$$
$$11 < 6 \qquad \text{False.}$$

The ordered pair $(5, -1)$ is not a solution since $11 < 6$ is a false statement.

b. We replace x with 2 and y with 7 and see if a true statement results.

$$2x - y < 6$$
$$2(2) - 7 < 6 \qquad \text{Replace } x \text{ with 2 and } y \text{ with 7.}$$
$$4 - 7 < 6$$
$$-3 < 6 \qquad \text{True.}$$

The ordered pair $(2, 7)$ is a solution since $-3 < 6$ is a true statement.

B GRAPHING LINEAR INEQUALITIES IN TWO VARIABLES

The linear equation $x - y = 1$ is graphed next. Recall that all points on the line correspond to ordered pairs that satisfy the equation $x - y = 1$. It can be shown that all the points above the line $x - y = 1$ have coordinates that satisfy the inequality $x - y < 1$. Similarly, all points below the line have coordinates that satisfy the inequality $x - y > 1$.

Practice Problem 1

Determine whether each ordered pair is a solution of $x - 4y > 8$.

a. $(-3, 2)$ b. $(9, 0)$

Answers

1. a. no,   **b.** yes

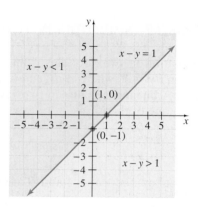

The region above the line and the region below the line are called **half-planes**. Every line divides the plane (similar to a sheet of paper extending indefinitely in all directions) into two half-planes; the line is called the **boundary**.

Recall that the inequality $x - y \leq 1$ means

$$x - y = 1 \qquad \text{or} \qquad x - y < 1$$

Thus, the graph of $x - y \leq 1$ is the half-plane $x - y < 1$ along with the boundary line $x - y = 1$.

TO GRAPH A LINEAR INEQUALITY IN TWO VARIABLES

Step 1. Graph the boundary line found by replacing the inequality sign with an equal sign. If the inequality sign is $>$ or $<$, graph a dashed boundary line (indicating that the points on the line are not solutions of the inequality). If the inequality sign is $\geq$ or $\leq$, graph a solid boundary line (indicating that the points on the line are solutions of the inequality).

Step 2. Choose a point, not on the boundary line, as a test point. Substitute the coordinates of this test point into the original inequality.

Step 3. If a true statement is obtained in Step 2, shade the half-plane that contains the test point. If a false statement is obtained, shade the half-plane that does not contain the test point.

Practice Problem 2

Graph: $x - y > 3$

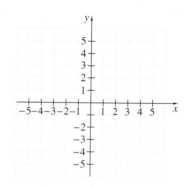

Answer

2.

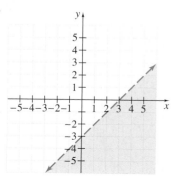

Example 2   Graph: $x + y < 7$

Solution: First we graph the boundary line by graphing the equation $x + y = 7$. We graph this boundary as a dashed line because the inequality sign is $<$, and thus the points on the line are not solutions of the inequality $x + y < 7$.

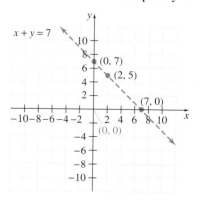

Next we choose a test point, being careful not to choose a point on the boundary line. We choose $(0, 0)$, and substitute the coordinates of $(0, 0)$ into $x + y < 7$.

$x + y < 7$ Original inequality

$0 + 0 < 7$ Replace x with 0 and y with 0.

$0 < 7$ True.

Since the result is a true statement, $(0, 0)$ is a solutuion of $x + y < 7$, and every point in the same half-plane as $(0, 0)$ is also a solution. To indicate this, we shade the entire half-plane containing $(0, 0)$, as shown.

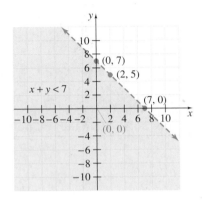

Example 3   Graph: $2x - y \geq 3$

Solution: We graph the boundary line by graphing $2x - y = 3$. We draw this line as a solid line because the inequality sign is $\geq$, and thus the points on the line are solutions of $2x - y \geq 3$. Once again, $(0, 0)$ is a convenient test point since it is not on the boundary line.

We substitute 0 for x and 0 for y into the original inequality.

$2x - y \geq 3$

$2(0) - 0 \geq 3$ Let $x = 0$ and $y = 0$.

$0 \geq 3$ False.

Since the statement is false, no point in the half-plane containing $(0, 0)$ is a solution. Therefore, we shade the half-plane that does not contain $(0, 0)$. Every point in the shaded half-plane and every point on the boundary line is a solution of $2x - y \geq 3$.

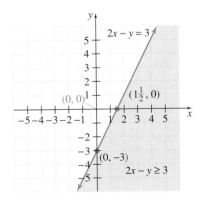

Practice Problem 3

Graph: $x - 4y \leq 4$

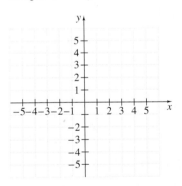

Answer

3.

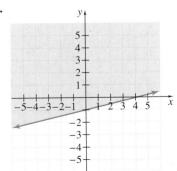

Practice Problem 4

Graph: $y < 3x$

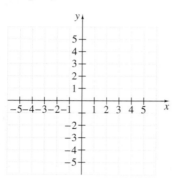

Practice Problem 5

Graph: $3x + 2y \geq 12$

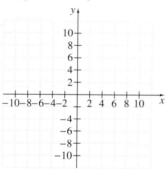

Answers

4.

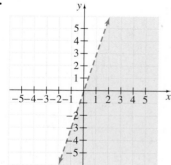

5.

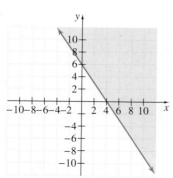

HELPFUL HINT

When graphing an inequality, make sure the test point is substituted into the **original inequality**. For Example 3, we substituted the test point $(0, 0)$ into the **original inequality** $2x - y \geq 3$, *not* $2x - y = 3$.

Example 4   Graph: $x > 2y$

Solution: We find the boundary line by graphing $x = 2y$. The boundary line is a dashed line since the inequality symbol is $>$. We cannot use $(0, 0)$ as a test point because it is a point on the boundary line. We choose instead $(0, 2)$.

$x > 2y$

$0 > 2(2)$ Let $x = 0$ and $y = 2$.

$0 > 4$ False.

Since the statement is false, we shade the half-plane that does not contain the test point $(0, 2)$, as shown.

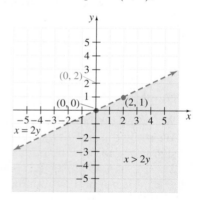

Example 5   Graph: $5x + 4y \leq 20$

Solution: We graph the solid boundary line $5x + 4y = 20$ and choose $(0, 0)$ as the test point.

$5x + 4y \leq 20$

$5(0) + 4(0) \leq 20$ Let $x = 0$ and $y = 0$.

$0 \leq 20$ True.

We shade the half-plane that contains $(0, 0)$, as shown.

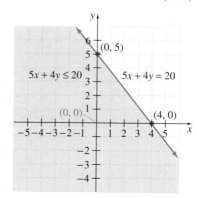

Example 6   Graph: $y > 3$

Solution: We graph the dashed boundary line $y = 3$ and choose $(0, 0)$ as the test point. (Recall that the graph of $y = 3$ is a horizontal line with y-intercept 3.)

$y > 3$

$0 > 3$ Let $y = 0$.

$0 > 3$ False.

We shade the half-plane that does not contain $(0, 0)$, as shown.

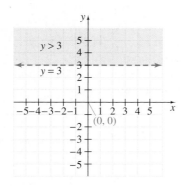

Graph: $x < 2$

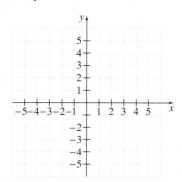

Answer

6.

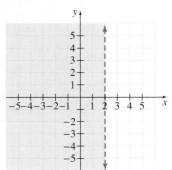

Focus On The Real World

MISLEADING GRAPHS

Graphs are very common in magazines and in newspapers such as *USA Today*. Graphs can be a convenient way to get an idea across because, as the old saying goes, "a picture is worth a thousand words." However, some graphs can be deceptive, which may or may not be intentional. It is important to know some of the ways that graphs can be misleading.

Beware of graphs like the one at the right. Notice that the graph shows a company's profit for various months. It appears that profit is growing quite rapidly. However, this impressive picture tells us little without knowing what units of profit are being graphed. Does the graph show profit in dollars or millions of dollars? An unethical company with profit increases of only a few pennies could use a graph like this one to make the profit increase seem much more substantial than it really is. A truthful graph describes the size of the units used along the vertical axis.

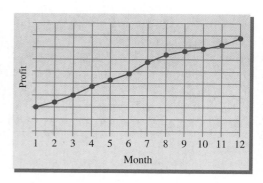

Another type of graph to watch for is one that misrepresents relationships. For example, the bar graph at the right shows the number of men and women employees in the accounting and shipping departments of a certain company. In the accounting department, the bar representing the number of women is shown twice as tall as the bar representing the number of men. However, the number of women (13) is not twice the number of men (10). This set of bars misrepresents the relationship between the number of men and women. Do you

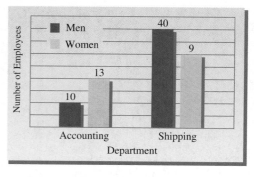

see how the relationship between the number of men and women in the shipping department is distorted by the heights of the bars used? A truthful graph will use bar heights that are proportional to the numbers they represent.

The impression a graph can give also depends on its vertical scale. The two graphs below represent exactly the same data. The only difference between the two graphs is the vertical scale—one shows enrollments from 246 to 260 students and the other shows enrollments between 0 and 300 students. If you were trying to convince readers that algebra enrollment at UPH had changed drastically over the period 1996–2000, which graph would you use? Which graph do you think gives the more honest representation?

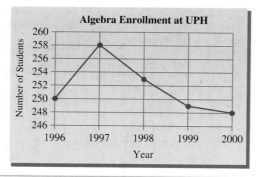

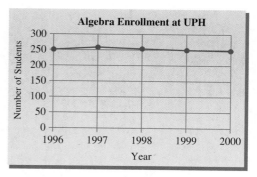

CHAPTER 3 HIGHLIGHTS

DEFINITIONS AND CONCEPTS	**EXAMPLES**

SECTION 3.1   THE RECTANGULAR COORDINATE SYSTEM

The **rectangular coordinate system** consists of a plane and a vertical and a horizontal number line intersecting at their 0 coordinate. The vertical number line is called the **y-axis** and the horizontal number line is called the **x-axis**. The point of intersection of the axes is called the **origin**.

To **plot** or **graph** an ordered pair means to find its corresponding point on a rectangular coordinate system.

To plot or graph an ordered pair such as $(3, -2)$, start at the origin. Move 3 units to the right and from there, 2 units down.

To plot or graph $(-3, 4)$; start at the origin. Move 3 units to the left and from there, 4 units up.

An ordered pair is a **solution** of an equation in two variables if replacing the variables with the coordinates of the ordered pair results in a true statement.

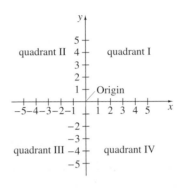

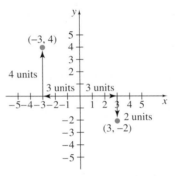

If one coordinate of an ordered pair solution is known, the other value can be determined by substitution.

Complete the ordered pair $(0, \ )$ for the equation $x - 6y = 12$.

$$x - 6y = 12$$
$$0 - 6y = 12 \quad \text{Let } x = 0.$$
$$\frac{-6y}{-6} = \frac{12}{-6} \quad \text{Divide by } -6.$$
$$y = -2$$

The ordered pair solution is $(0, -2)$.

SECTION 3.2   GRAPHING LINEAR EQUATIONS

A **linear equation in two variables** is an equation that can be written in the form $Ax + By = C$, where A and B are not both 0. The form $Ax + By = C$ is called **standard form**.

$$3x + 2y = -6 \qquad x = -5$$
$$y = 3 \qquad y = -x + 10$$

$x + y = 10$ is in standard form.

SECTION 3.2   (CONTINUED)

To graph a linear equation in two variables, find three ordered pair solutions. Plot the solution points and draw the line connecting the points.

Graph: $x - 2y = 5$

x	y
5	0
1	-2
-1	-3

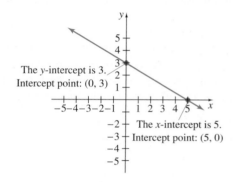

(5, 0)

$(-1, -3)$ $(1, -2)$

SECTION 3.3   INTERCEPTS

An **intercept point** of a graph is a point where the graph intersects an axis. If a graph intersects the x-axis at a, then a is the **x-intercept** and the corresponding intercept point is $(a, 0)$. If a graph intersects the y-axis at b, then b is the **y-intercept** and the corresponding intercept point is $(0, b)$.

The y-intercept is 3.
Intercept point: $(0, 3)$

The x-intercept is 5.
Intercept point: $(5, 0)$

To find the x-intercept, let $y = 0$ and solve for x.
To find the y-intercept, let $x = 0$ and solve for y.

Find the intercepts for $2x - 5y = -10$.

If $y = 0$, then
$$2x - 5 \cdot 0 = -10$$
$$2x = -10$$
$$\frac{2x}{2} = \frac{-10}{2}$$
$$x = -5$$
The x-intercept is -5.
Intercept point: $(-5, 0)$.

If $x = 0$, then
$$2 \cdot 0 - 5y = -10$$
$$-5y = -10$$
$$\frac{-5y}{-5} = \frac{-10}{-5}$$
$$y = 2$$
The y-intercept is 2.
Intercept point: $(0, 2)$.

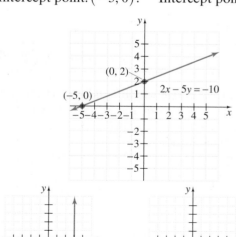

$(0, 2)$ $(-5, 0)$ $2x - 5y = -10$

The graph of $x = c$ is a vertical line with x-intercept c.

The graph of $y = c$ is a horizontal line with y-intercept c.

$x = 3$ $y = -1$

SECTION 3.4 SLOPE

The **slope m** of the line through points (x_1, y_1) and (x_2, y_2) is given by

$$m = \frac{y_2 - y_1}{x_2 - x_1} \qquad \text{as long as } x_2 \neq x_1$$

A horizontal line has slope 0.
The slope of a vertical line is undefined.
Nonvertical parallel lines have the same slope.
Two nonvertical lines are perpendicular if the slope of one is the negative reciprocal of the slope of the other.

The slope of the line through points $(-1, 6)$ and $(-5, 8)$ is

$$m = \frac{y_2 - y_1}{x_2 - x_1} = \frac{8 - 6}{-5 - (-1)} = \frac{2}{-4} = -\frac{1}{2}$$

The slope of the line $y = -5$ is 0.
The line $x = 3$ has undefined slope.

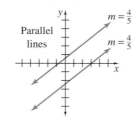

Parallel lines

$m = \frac{4}{5}$
$m = \frac{4}{5}$

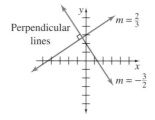

Perpendicular lines

$m = \frac{2}{3}$
$m = -\frac{3}{2}$

SECTION 3.5 GRAPHING LINEAR INEQUALITIES IN TWO VARIABLES

A **linear inequality in two variables** is an inequality that can be written in one of these forms:

$Ax + By < C \qquad Ax + By \leq C$
$Ax + By > C \qquad Ax + By \geq C$

TO GRAPH A LINEAR INEQUALITY

1. Graph the boundary line by graphing the related equation. Draw the line solid if the inequality symbol is $\leq$ or $\geq$. Draw the line dashed if the inequality symbol is $<$ or $>$.
2. Choose a test point not on the line. Substitute its coordinates into the original inequality.
3. If the resulting inequality is true, shade the half-plane that contains the test point. If the inequality is not true, shade the half-plane that does not contain the test point.

$2x - 5y < 6 \qquad x \geq -5$
$\qquad y > -8x \qquad y \leq 2$

Graph: $2x - y \leq 4$
1. Graph $2x - y = 4$. Draw a solid line because the inequality symbol is $\leq$.
2. Check the test point $(0, 0)$ in the original inequality, $2x - y \leq 4$.

$\quad 2 \cdot 0 - 0 \leq 4 \qquad$ Let $x = 0$ and $y = 0$.

$\qquad 0 \leq 4 \qquad$ True.

3. The inequality is true, so shade the half-plane containing $(0, 0)$ as shown.

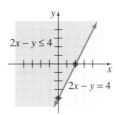

$2x - y \leq 4$
$2x - y = 4$

Exponents and Polynomials

CHAPTER 4

Recall from Chapter 1 that an exponent is a shorthand notation for repeated factors. This chapter explores additional concepts about exponents and exponential expressions. An especially useful type of exponential expression is a polynomial. Polynomials model many real-world phenomena. Our goal in this chapter is to become proficient with operations on polynomials.

One way exponents are used is to express very large or very small numbers in scientific notation. The U.S. Bureau of the Census conducts a census every 10 years to count U.S. citizens. However, government agencies and private businesses need more up-to-date population figures between censuses. From its research, the Census Bureau devised a way to estimate the U.S. population based on a few simple observations. There is a birth in the U.S. every 8 seconds and there is a death in the U.S. every 14 seconds. The estimate also takes into account the arrival of immigrants (one approximately every 42 seconds) and U.S. citizens returning from abroad (one approximately every 4108 seconds). These facts along with the most recent census figures can be used to estimate the current population. Visit the Bureau of the Census Web site, at http://www.census.gov/cgi-bin/popclock.

329

Problem Solving Notes

4.1   EXPONENTS

A   EVALUATING EXPONENTIAL EXPRESSIONS

In this section, we continue our work with integer exponents. As we reviewed in Section 1.2, for example,

$$2 \cdot 2 \cdot 2 \cdot 2 \cdot 2 = 2^5$$

The exponent 5 tells us how many times that 2 is a factor. The expression 2^5 is called an **exponential expression**. It is also called the fifth **power** of 2, or we can say that 2 is **raised** to the fifth power.

$$5^6 = \underbrace{5 \cdot 5 \cdot 5 \cdot 5 \cdot 5 \cdot 5}_{\text{6 factors; each factor is 5}} \quad \text{and} \quad (-3)^4 = \underbrace{(-3) \cdot (-3) \cdot (-3) \cdot (-3)}_{\text{4 factors; each factor is } -3}$$

The base of an exponential expression is the repeated factor. The exponent is the number of times that the base is used as a factor.

$$\overset{\text{exponent or power}}{a^n} = \underbrace{a \cdot a \cdot a \dots a}_{n \text{ factors of } a}$$
$$\underset{\text{base}}{\uparrow}$$

Examples   Evaluate (find the value of) each expression.

  1. $2^3 = 2 \cdot 2 \cdot 2 = 8$
  2. $3^1 = 3$. To raise 3 to the first power means to use 3 as a factor only once. When no exponent is shown, the exponent is assumed to be 1.
  3. $(-4)^2 = (-4)(-4) = 16$
  4. $-4^2 = -(4 \cdot 4) = -16$
  5. $\left(\dfrac{1}{2}\right)^4 = \dfrac{1}{2} \cdot \dfrac{1}{2} \cdot \dfrac{1}{2} \cdot \dfrac{1}{2} = \dfrac{1}{16}$
  6. $4 \cdot 3^2 = 4 \cdot 9 = 36$

Notice how similar -4^2 is to $(-4)^2$ in the examples above. The difference between the two is the parentheses. In $(-4)^2$, the parentheses tell us that the base, or the repeated factor, is -4. In -4^2, only 4 is the base.

┌─ **HELPFUL HINT**

Be careful when identifying the base of an exponential expression. Pay close attention to the use of parentheses.

$(-3)^2$	-3^2	$2 \cdot 3^2$
The base is -3.	The base is 3.	The base is 3.
$(-3)^2 = (-3)(-3) = 9$	$-3^2 = -(3 \cdot 3) = -9$	$2 \cdot 3^2 = 2 \cdot 3 \cdot 3 = 18$

An exponent has the same meaning whether the base is a number or a variable. If x is a real number and n is a positive integer, then x^n is the product of n factors, each of which is x.

$$x^n = \underbrace{x \cdot x \cdot x \cdot x \cdot x \dots x}_{n \text{ factors; each factor is } x}$$

Objectives

A Evaluate exponential expressions.
B Use the product rule for exponents.
C Use the power rule for exponents.
D Use the power rules for products and quotients.
E Use the quotient rule for exponents, and define a number raised to the 0 power.
F Decide which rule(s) to use to simplify an expression.

SSM   CD-ROM   Video 4.1

Practice Problems 1–6

Evaluate (find the value of) each expression.

1. 3^4 2. 7^1

3. $(-2)^3$ 4. -2^3

5. $\left(\dfrac{2}{3}\right)^2$ 6. $5 \cdot 6^2$

Answers

1. 81, **2.** 7, **3.** -8, **4.** -8, **5.** $\dfrac{4}{9}$, **6.** 180

Practice Problem 7

Evaluate each expression for the given value of x.

a. $3x^2$ when x is 4

b. $\dfrac{x^4}{-8}$ when x is -2

Example 7

Evaluate each expression for the given value of x.

a. $2x^3$ when x is 5   **b.** $\dfrac{9}{x^2}$ when x is -3

Solution:

a. When x is 5, $2x^3 = 2 \cdot 5^3$

$$= 2 \cdot (5 \cdot 5 \cdot 5)$$
$$= 2 \cdot 125$$
$$= 250$$

b. When x is -3, $\dfrac{9}{x^2} = \dfrac{9}{-3^2}$

$$= \dfrac{9}{(-3)(-3)}$$
$$= \dfrac{9}{9} = 1$$

B USING THE PRODUCT RULE

Exponential expressions can be multiplied, divided, added, subtracted, and themselves raised to powers. By our definition of an exponent,

$$5^4 \cdot 5^3 = \underbrace{(5 \cdot 5 \cdot 5 \cdot 5)}_{\text{4 factors of 5}} \cdot \underbrace{(5 \cdot 5 \cdot 5)}_{\text{3 factors of 5}}$$

$$= \underset{\text{7 factors of 5}}{5 \cdot 5 \cdot 5 \cdot 5 \cdot 5 \cdot 5 \cdot 5}$$

$$= 5^7$$

Also,

$$x^2 \cdot x^3 = (x \cdot x) \cdot (x \cdot x \cdot x)$$
$$= x \cdot x \cdot x \cdot x \cdot x$$
$$= x^5$$

In both cases, notice that the result is exactly the same if the exponents are added.

$$5^4 \cdot 5^3 = 5^{4+3} = 5^7 \quad \text{and} \quad x^2 \cdot x^3 = x^{2+3} = x^5$$

This suggests the following rule.

PRODUCT RULE FOR EXPONENTS

If m and n are positive integers and a is a real number, then
$$a^m \cdot a^n = a^{m+n}$$
For example:
$$3^5 \cdot 3^7 = 3^{5+7} = 3^{12}$$

In other words, to multiply two exponential expressions with the **same base**, we keep the base and add the exponents. We call this *simplifying* the exponential expression.

Practice Problems 8–12

Use the product rule to simplify each expression.

8. $7^3 \cdot 7^2$ 9. $x^4 \cdot x^9$

10. $r^5 \cdot r$ 11. $s^6 \cdot s^2 \cdot s^3$

12. $(-3)^9 \cdot (-3)$

Examples

Use the product rule to simplify each expression.

8. $4^2 \cdot 4^5 = 4^{2+5} = 4^7$

9. $x^2 \cdot x^5 = x^{2+5} = x^7$

10. $y^3 \cdot y = y^3 \cdot y^1$
$$= y^{3+1}$$
$$= y^4$$

HELPFUL HINT

Don't forget that if no exponent is written, it is assumed to be 1.

Answers

7. a. 48, **b.** -2, **8.** 7^5, **9.** x^{13}, **10.** r^6,
11. s^{11}, **12.** $(-3)^{10}$

11. $y^3 \cdot y^2 \cdot y^7 = y^{3+2+7} = y^{12}$

12. $(-5)^7 \cdot (-5)^8 = (-5)^{7+8} = (-5)^{15}$

Example 13   Use the product rule to simplify $(2x^2)(-3x^5)$.

Solution: Recall that $2x^2$ means $2 \cdot x^2$ and $-3x^5$ means $-3 \cdot x^5$.

$$(2x^2)(-3x^5) = 2 \cdot x^2 \cdot -3 \cdot x^5 \quad \text{Remove parentheses.}$$
$$= 2 \cdot -3 \cdot x^2 \cdot x^5 \quad \text{Group factors with common bases.}$$
$$= -6x^7 \quad \text{Simplify.}$$

Practice Problem 13

Use the product rule to simplify $(6x^3)(-2x^9)$.

HELPFUL HINT

These examples will remind you of the difference between adding and multiplying terms.

Addition
$$5x^3 + 3x^3 = (5 + 3)x^3 = 8x^3$$
$$7x + 4x^2 = 7x + 4x^2$$

Multiplication
$$(5x^3)(3x^3) = 5 \cdot 3 \cdot x^3 \cdot x^3 = 15x^{3+3} = 15x^6$$
$$(7x)(4x^2) = 7 \cdot 4 \cdot x \cdot x^2 = 28x^{1+2} = 28x^3$$

C USING THE POWER RULE

Exponential expressions can themselves be raised to powers. Let's try to discover a rule that simplifies an expression like $(x^2)^3$. By the definition of a^n,

$$(x^2)^3 = (x^2)(x^2)(x^2) \quad (x^2)^3 \text{ means 3 factors of } (x^2).$$

which can be simplified by the product rule for exponents.

$$(x^2)^3 = (x^2)(x^2)(x^2) = x^{2+2+2} = x^6$$

Notice that the result is exactly the same if we multiply the exponents.

$$(x^2)^3 = x^{2 \cdot 3} = x^6$$

The following rule states this result.

POWER RULE FOR EXPONENTS

If m and n are positive integers and a is a real number, then
$$(a^m)^n = a^{mn}$$
For example:
$$(7^2)^5 = 7^{2 \cdot 5} = 7^{10}$$

In other words, to raise an exponential expression to a power, we keep the base and multiply the exponents.

Examples   Use the power rule to simplify each expression.

14. $(5^3)^6 = 5^{3 \cdot 6} = 5^{18}$

15. $(y^8)^2 = y^{8 \cdot 2} = y^{16}$

Practice Problems 14–15

Use the power rule to simplify each expression.

14. $(9^4)^{10}$ **15.** $(z^6)^3$

Answers
13. $-12x^{12}$,   **14.** 9^{40},   **15.** z^{18}

Take a moment to make sure that you understand when to apply the product rule and when to apply the power rule.

Product Rule → Add Exponents

$$x^5 \cdot x^7 = x^{5+7} = x^{12}$$
$$y^6 \cdot y^2 = y^{6+2} = y^8$$

Power Rule → Multiply Exponents

$$(x^5)^7 = x^{5 \cdot 7} = x^{35}$$
$$(y^6)^2 = y^{6 \cdot 2} = y^{12}$$

D   USING THE POWER RULES FOR PRODUCTS AND QUOTIENTS

When the base of an exponential expression is a product, the definition of a^n still applies. To simplify $(xy)^3$, for example,

$$(xy)^3 = (xy)(xy)(xy) \quad \text{\small $(xy)^3$ means 3 factors of (xy).}$$
$$= x \cdot x \cdot x \cdot y \cdot y \cdot y \quad \text{\small Group factors with common bases.}$$
$$= x^3 y^3 \quad \text{\small Simplify.}$$

Notice that to simplify the expression $(xy)^3$, we raise each factor within the parentheses to a power of 3.

$$(xy)^3 = x^3 y^3$$

In general, we have the following rule.

> **POWER OF A PRODUCT RULE**
>
> If n is a positive integer and a and b are real numbers, then
> $$(ab)^n = a^n b^n$$
> For example:
> $$(3x)^5 = 3^5 x^5$$

In other words, to raise a product to a power, we raise each factor to the power.

Practice Problems 16–18

Simplify each expression.

16. $(xy)^7$ 17. $(3y)^4$

18. $(-2p^4 q^2 r)^3$

Examples   Simplify each expression.

16. $(st)^4 = s^4 \cdot t^4 = s^4 t^4$ Use the power of a product rule.

17. $(2a)^3 = 2^3 \cdot a^3 = 8a^3$ Use the power of a product rule.

18. $(-5x^2 y^3 z)^2 = (-5)^2 \cdot (x^2)^2 \cdot (y^3)^2 \cdot (z^1)^2$ Use the power of a product rule.
$$= 25x^4 y^6 z^2 \quad \text{Use the power rule for exponents.}$$

Let's see what happens when we raise a quotient to a power. To simplify $\left(\dfrac{x}{y}\right)^3$, for example,

$$\left(\frac{x}{y}\right)^3 = \left(\frac{x}{y}\right)\left(\frac{x}{y}\right)\left(\frac{x}{y}\right) \quad \text{\small $\left(\frac{x}{y}\right)^3$ means 3 factors of $\left(\frac{x}{y}\right)$.}$$
$$= \frac{x \cdot x \cdot x}{y \cdot y \cdot y} \quad \text{\small Multiply fractions.}$$
$$= \frac{x^3}{y^3} \quad \text{\small Simplify.}$$

Notice that to simplify the expression, $\left(\dfrac{x}{y}\right)^3$, we raise both the numerator and the denominator to a power of 3.

$$\left(\frac{x}{y}\right)^3 = \frac{x^3}{y^3}$$

In general, we have the rule shown in the margin.

POWER OF A QUOTIENT RULE

If n is a positive integer and a and c are real numbers, then
$$\left(\frac{a}{c}\right)^n = \frac{a^n}{c^n}, \quad c \neq 0$$
For example:
$$\left(\frac{y}{7}\right)^3 = \frac{y^3}{7^3}$$

In other words, to raise a quotient to a power, we raise both the numerator and the denominator to the power.

Answers

16. $x^7 y^7$,   **17.** $81y^4$,   **18.** $-8p^{12} q^6 r^3$

Examples   Simplify each expression.

19. $\left(\dfrac{m}{n}\right)^7 = \dfrac{m^7}{n^7}, \quad n \neq 0$ Use the power of a quotient rule.

20. $\left(\dfrac{2x^4}{3y^5}\right)^4 = \dfrac{2^4 \cdot (x^4)^4}{3^4 \cdot (y^5)^4}$ Use the power of a quotient rule.

$\qquad = \dfrac{16x^{16}}{81y^{20}}, \quad y \neq 0$ Use the power rule for exponents.

Practice Problems 19–20

Simplify each expression.

19. $\left(\dfrac{r}{s}\right)^6$ 20. $\left(\dfrac{5x^6}{9y^3}\right)^2$

E   USING THE QUOTIENT RULE AND DEFINING THE ZERO EXPONENT

Another pattern for simplifying exponential expressions involves quotients.

$$\dfrac{x^5}{x^3} = \dfrac{x \cdot x \cdot x \cdot x \cdot x}{x \cdot x \cdot x}$$

$$= \dfrac{x \cdot x \cdot x \cdot x \cdot x}{x \cdot x \cdot x}$$

$$= x \cdot x$$

$$= x^2$$

Notice that the result is exactly the same if we subtract exponents of the common bases.

$$\dfrac{x^5}{x^3} = x^{5-3} = x^2$$

The following rule states this result in a general way.

QUOTIENT RULE FOR EXPONENTS

If m and n are positive integers and a is a real number, then

$$\dfrac{a^m}{a^n} = a^{m-n}, \quad a \neq 0$$

For example:

$$\dfrac{x^6}{x^2} = x^{6-2} = x^4, \quad x \neq 0$$

In other words, to divide one exponential expression by another with a common base, we keep the base and subtract the exponents.

Examples   Simplify each quotient.

21. $\dfrac{x^5}{x^2} = x^{5-2} = x^3$ Use the quotient rule.

22. $\dfrac{4^7}{4^3} = 4^{7-3} = 4^4 = 256$ Use the quotient rule.

23. $\dfrac{(-3)^5}{(-3)^2} = (-3)^3 = -27$ Use the quotient rule.

24. $\dfrac{2x^5y^2}{xy} = 2 \cdot \dfrac{x^5}{x^1} \cdot \dfrac{y^2}{y^1}$

$\qquad = 2 \cdot (x^{5-1}) \cdot (y^{2-1})$ Use the quotient rule.

$\qquad = 2x^4y^1 \quad \text{or} \quad 2x^4y$

Practice Problems 21–24

Simplify each quotient.

21. $\dfrac{y^7}{y^3}$ 22. $\dfrac{5^9}{5^6}$

23. $\dfrac{(-2)^{14}}{(-2)^{10}}$ 24. $\dfrac{7a^4b^{11}}{ab}$

Answers

19. $\dfrac{r^6}{s^6}, \quad s \neq 0,$ **20.** $\dfrac{25x^{12}}{81y^6}, \quad y \neq 0,$ **21.** $y^4,$
22. $125,$ **23.** $16,$ **24.** $7a^3b^{10}$

Let's now give meaning to an expression such as x^0. To do so, we will simplify $\dfrac{x^3}{x^3}$ in two ways and compare the results.

$$\dfrac{x^3}{x^3} = x^{3-3} = x^0 \qquad \text{Apply the quotient rule.}$$

$$\dfrac{x^3}{x^3} = \dfrac{x \cdot x \cdot x}{x \cdot x \cdot x} = 1 \qquad \text{Apply the fundamental principle for fractions.}$$

Since $\dfrac{x^3}{x^3} = x^0$ and $\dfrac{x^3}{x^3} = 1$, we define that $x^0 = 1$ as long as x is not 0.

ZERO EXPONENT

$a^0 = 1$, as long as a is not 0.

For example: $5^0 = 1$.

In other words, a base raised to the 0 power is 1, as long as the base is not 0.

Examples   Simplify each expression.

25. $3^0 = 1$

26. $(5x^3y^2)^0 = 1$

27. $(-4)^0 = 1$

28. $-4^0 = -1 \cdot 4^0 = -1 \cdot 1 = -1$

TRY THE CONCEPT CHECK IN THE MARGIN.

F DECIDING WHICH RULE TO USE

Let's practice deciding which rule to use to simplify.

Example 29   Simplify each expression.

a. $x^7 \cdot x^4$

b. $\left(\dfrac{1}{2}\right)^4$

c. $(9y^5)^2$

Solution: **a.** Here, we have a product, so we use the product rule to simplify.

$$x^7 \cdot x^4 = x^{7+4} = x^{11}$$

b. This is a quotient raised to a power, so we use the power of a quotient rule.

$$\left(\dfrac{1}{2}\right)^4 = \dfrac{1^4}{2^4} = \dfrac{1}{16}$$

c. This is a product raised to a power, so we use the power of a product rule.

$$(9y^5)^2 = 9^2(y^5)^2 = 81y^{10}$$

Practice Problems 25–28

Simplify each expression.

25. 8^0 26. $(2r^2s)^0$

27. $(-5)^0$ 28. -5^0

✓ CONCEPT CHECK

To simplify each expression, tell whether you would *add* the exponents, *subtract* the exponents, *multiply* the exponents, or *divide* the exponents, or *none of these*.

a. $\left(x^{63}\right)^{21}$ b. $\dfrac{y^{15}}{y^3}$

c. $z^{16} + z^8$ d. $w^{45} \cdot w^9$

Practice Problem 29

Simplify each expression.

a. $\dfrac{x^7}{x^4}$ b. $(3y^4)^4$ c. $\left(\dfrac{x}{4}\right)^3$

Answers

25. 1, **26.** 1, **27.** 1, **28.** -1, **29. a.** x^3,

b. $81y^{16}$, **c.** $\dfrac{x^3}{64}$

✓Concept Check

a. multiply, **b.** subtract, **c.** none of these, **d.** add

Name _____ Section _____ Date _____

MENTAL MATH

State the bases and the exponents for each expression.

1. 3^2 **2.** 5^4 **3.** $(-3)^6$ **4.** -3^7

5. -4^2

EXERCISE SET 4.1

A Evaluate each expression. See Examples 1 through 6.

1. 7^2 **2.** -3^2 **3.** $(-5)^1$ **4.** $(-3)^2$

5. -2^4

Evaluate each expression with the given replacement values. See Example 7.

6. x^2 when $x = -2$ **7.** x^3 when $x = -2$ **8.** $5x^3$ when $x = 3$

9. $4x^2$ when $x = -1$ **10.** $2xy^2$ when $x = 3$ and $y = 5$

B Use the product rule to simplify each expression. Write the results using exponents. See Examples 8 through 13.

11. $x^2 \cdot x^5$ **12.** $y^2 \cdot y$

C Use the power rule and the power of a product or quotient rule to simplify each expression. See Examples 14 through 20.

13. $(x^9)^4$ **14.** $(y^7)^5$ **15.** $(pq)^7$

E Use the quotient rule and simplify each expression. See Examples 21 through 24.

16. $\dfrac{x^3}{x}$ **17.** $\dfrac{y^{10}}{y^9}$ **18.** $\dfrac{p^7 q^{20}}{pq^{15}}$ **19.** $\dfrac{x^8 y^6}{xy^5}$

20. $\dfrac{7x^2 y^6}{14x^2 y^3}$

ANSWERS

1. _____

2. _____

3. _____

4. _____

5. _____

6. _____

7. _____

8. _____

9. _____

10. _____

11. _____

12. _____

13. _____

14. _____

15. _____

16. _____

17. _____

18. _____

19. _____

20. _____

Focus On Study Skills

Remember that one of the best ways to start preparing for an exam is to keep current with your assignments as they are made. Make an effort to clear up any confusion on topics as you cover them.

Begin reviewing for your exam a few days in advance. This way, if you find a topic that you still don't understand, you'll have plenty of time to ask your instructor, another student in your class, or a math tutor for help. Don't wait until the last minute to "cram" for an exam.

▲ Reread your notes and carefully review the Chapter Highlights at the end of each chapter to be covered.
▲ Pay special attention to any new terminology or definitions in the chapter. Be sure you can state the meanings of definitions in your own words.
▲ If you tend to get anxious while taking an exam, try to visualize yourself taking the exam in advance. Picture yourself being calm, clearheaded, and successful. Picture yourself remembering formulas and definitions with no trouble. When you are well prepared for an exam, a lot of nervousness can be avoided through positive thinking.
▲ Get lots of rest the night before the exam. It's hard to show how well you know the material if your brain is foggy from lack of sleep.
▲ Give yourself enough time so that you will arrive early for the exam. This way, if you run into difficulties on the way, you should still arrive on time.

4.2   NEGATIVE EXPONENTS AND SCIENTIFIC NOTATION

A   SIMPLIFYING EXPRESSIONS CONTAINING NEGATIVE EXPONENTS

Our work with exponential expressions so far has been limited to exponents that are positive integers or 0. Here we expand to give meaning to an expression like x^{-3}.

Suppose that we wish to simplify the expression $\dfrac{x^2}{x^5}$. If we use the quotient rule for exponents, we subtract exponents:

$$\frac{x^2}{x^5} = x^{2-5} = x^{-3}, \quad x \neq 0$$

But what does x^{-3} mean? Let's simplify $\dfrac{x^2}{x^5}$ using the definition of a^n.

$$\frac{x^2}{x^5} = \frac{x \cdot x}{x \cdot x \cdot x \cdot x \cdot x}$$

$$= \frac{x \cdot x}{x \cdot x \cdot x \cdot x \cdot x} \qquad \text{Divide numerator and denominator by common factors by applying the fundamental principle for fractions.}$$

$$= \frac{1}{x^3}$$

If the quotient rule is to hold true for negative exponents, then x^{-3} must equal $\dfrac{1}{x^3}$.

From this example, we state the definition for negative exponents.

NEGATIVE EXPONENTS

If a is a real number other than 0 and n is an integer, then

$$a^{-n} = \frac{1}{a^n}$$

For example,

$$x^{-3} = \frac{1}{x^3}$$

In other words, another way to write a^{-n} is to take its reciprocal and change the sign of its exponent.

Examples   Simplify by writing each expression with positive exponents only.

    1. $3^{-2} = \dfrac{1}{3^2} = \dfrac{1}{9}$ Use the definition of negative exponent.

    2. $2x^{-3} = 2 \cdot \dfrac{1}{x^3} = \dfrac{2}{x^3}$ Use the definition of negative exponent.

> **┌ HELPFUL HINT**
>
> Don't forget that since there are no parentheses, only x is the base for the exponent -3.

    3. $2^{-1} + 4^{-1} = \dfrac{1}{2} + \dfrac{1}{4} = \dfrac{2}{4} + \dfrac{1}{4} = \dfrac{3}{4}$

    4. $(-2)^{-4} = \dfrac{1}{(-2)^4} = \dfrac{1}{(-2)(-2)(-2)(-2)} = \dfrac{1}{16}$

Objectives

A Simplify expressions containing negative exponents.

B Use the rules and definitions for exponents to simplify exponential expressions.

C Write numbers in scientific notation.

D Convert numbers in scientific notation to standard form.

   SSM   CD-ROM   Video
                        4.2

Practice Problems 1–4

Simplify by writing each expression with positive exponents only.

1. 5^{-3} 2. $7x^{-4}$

3. $5^{-1} + 3^{-1}$ 4. $(-3)^{-4}$

Answers

1. $\dfrac{1}{125}$,   2. $\dfrac{7}{x^4}$,   3. $\dfrac{8}{15}$,   4. $\dfrac{1}{81}$

HELPFUL HINT

A negative exponent *does not affect* the sign of its base.
Remember: Another way to write a^{-n} is to take its reciprocal and
change the sign of its exponent, $a^{-n} = \dfrac{1}{a^n}$. For example,

$$x^{-2} = \frac{1}{x^2}, \qquad\qquad 2^{-3} = \frac{1}{2^3} \quad \text{or} \quad \frac{1}{8}$$

$$\frac{1}{y^{-4}} = \frac{1}{\frac{1}{y^4}} = y^4, \qquad \frac{1}{5^{-2}} = 5^2 \quad \text{or} \quad 25$$

Practice Problems 5–8

Simplify each expression. Write each result using positive exponents only.

5. $\left(\dfrac{6}{7}\right)^{-2}$ **6.** $\dfrac{x}{x^{-4}}$

7. $\dfrac{y^{-9}}{z^{-5}}$ **8.** $\dfrac{y^{-4}}{y^6}$

Examples

Simplify each expression. Write each result using positive exponents only.

5. $\left(\dfrac{2}{3}\right)^{-3} = \dfrac{2^{-3}}{3^{-3}} = \dfrac{3^3}{2^3} = \dfrac{27}{8}$ Use the negative exponent rule.

6. $\dfrac{y}{y^{-2}} = \dfrac{y^1}{y^{-2}} = y^{1-(-2)} = y^3$ Use the quotient rule.

7. $\dfrac{p^{-4}}{q^{-9}} = \dfrac{q^9}{p^4}$ Use the negative exponent rule.

8. $\dfrac{x^{-5}}{x^7} = x^{-5-7} = x^{-12} = \dfrac{1}{x^{12}}$

B SIMPLIFYING EXPONENTIAL EXPRESSIONS

All the previously stated rules for exponents apply for negative exponents also. Here is a summary of the rules and definitions for exponents.

SUMMARY OF EXPONENT RULES

If m and n are integers and a, b, and c are real numbers, then:

Product rule for exponents:	$a^m \cdot a^n = a^{m+n}$
Power rule for exponents:	$(a^m)^n = a^{m \cdot n}$
Power of a product:	$(ab)^n = a^n b^n$
Power of a quotient:	$\left(\dfrac{a}{c}\right)^n = \dfrac{a^n}{c^n}, \quad c \neq 0$
Quotient rule for exponents:	$\dfrac{a^m}{a^n} = a^{m-n}, \quad a \neq 0$
Zero exponent:	$a^0 = 1, \quad a \neq 0$
Negative exponent:	$a^{-n} = \dfrac{1}{a^n}, \quad a \neq 0$

Examples

Simplify each expression. Write each result using positive exponents only.

9. $\dfrac{(x^3)^4 x}{x^7} = \dfrac{x^{12} \cdot x}{x^7} = \dfrac{x^{12+1}}{x^7} = \dfrac{x^{13}}{x^7} = x^{13-7} = x^6$ Use the power rule.

10. $\left(\dfrac{3a^2}{b}\right)^{-3} = \dfrac{3^{-3}(a^2)^{-3}}{b^{-3}}$ Raise each factor in the numerator and the denominator to the -3 power.

$= \dfrac{3^{-3}a^{-6}}{b^{-3}}$ Use the power rule.

$= \dfrac{b^3}{3^3a^6}$ Use the negative exponent rule.

$= \dfrac{b^3}{27a^6}$ Write 3^3 as 27.

11. $(y^{-3}z^6)^{-6} = (y^{-3})^{-6}(z^6)^{-6}$ Raise each factor to the -6 power.

$= y^{18}z^{-36} = \dfrac{y^{18}}{z^{36}}$

12. $\dfrac{(2x)^5}{x^3} = \dfrac{\overbrace{2^5 \cdot x^5}}{x^3} = 2^5 \cdot x^{5-3} = 32x^2$ Raise each factor in the numerator to the fifth power.

13. $\dfrac{x^{-7}}{(x^4)^3} = \dfrac{x^{-7}}{x^{12}} = x^{-7-12} = x^{-19} = \dfrac{1}{x^{19}}$

14. $(5y^3)^{-2} = 5^{-2}(y^3)^{-2}$ Raise each factor to the -2 power.

$= 5^{-2}y^{-6} = \dfrac{1}{5^2y^6} = \dfrac{1}{25y^6}$

▬▬▬▬

C WRITING NUMBERS IN SCIENTIFIC NOTATION

Both very large and very small numbers frequently occur in many fields of science. For example, the distance between the sun and the planet Pluto is approximately 5,906,000,000 kilometers, and the mass of a proton is approximately 0.00000000000000000000000165 gram. It can be tedious to write these numbers in this standard decimal notation, so **scientific notation** is used as a convenient shorthand for expressing very large and very small numbers.

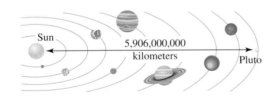

SCIENTIFIC NOTATION

A positive number is written in scientific notation if it is written as the product of a number a, where $1 \le a < 10$, and an integer power r of 10:

$a \times 10^r$

The following numbers are written in scientific notation. The $\times$ sign for multiplication is used as part of the notation.

2.03×10^2 7.362×10^7 5.906×10^9 (Distance between the sun and Pluto)

1×10^{-3} 8.1×10^{-5} 1.65×10^{-24} (Mass of a proton)

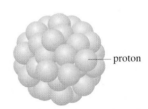

proton

Mass of proton is approximately
0.000 000 000 000 000 000 000 001 65 gram

The following steps are useful when writing numbers in scientific notation.

> **TO WRITE A NUMBER IN SCIENTIFIC NOTATION**
>
> **Step 1.** Move the decimal point in the original number so that the new number has a value between 1 and 10.
>
> **Step 2.** Count the number of decimal places the decimal point is moved in Step 1. If the decimal point is moved to the left, the count is positive. If the decimal point is moved to the right, the count is negative.
>
> **Step 3.** Multiply the new number in Step 1 by 10 raised to an exponent equal to the count found in Step 2.

Practice Problem 15

Write each number in scientific notation.

a. 420,000   b. 0.00017

c. 9,060,000,000   d. 0.000007

Example 15   Write each number in scientific notation.

a. 367,000,000   b. 0.000003
c. 20,520,000,000   d. 0.00085

Solution: **a. Step 1.** Move the decimal point until the number is between 1 and 10.

367,000,000   8 places

Step 2. The decimal point is moved to the left 8 places, so the count is positive 8.

Step 3. $367,000,000 = 3.67 \times 10^8$.

b. Step 1. Move the decimal point until the number is between 1 and 10.

0.000003   6 places

Step 2. The decimal point is moved 6 places to the right, so the count is -6.

Step 3. $0.000003 = 3.0 \times 10^{-6}$

c. $20,520,000,000 = 2.052 \times 10^{10}$

d. $0.00085 = 8.5 \times 10^{-4}$

D CONVERTING NUMBERS TO STANDARD FORM

A number written in scientific notation can be rewritten in standard form. For example, to write 8.63×10^3 in standard form, recall that $10^3 = 1000$.

$$8.63 \times 10^3 = 8.63(1000) = 8630$$

Notice that the exponent on the 10 is positive 3, and we moved the decimal point 3 places to the right.

To write 7.29×10^{-3} in standard form, recall that $10^{-3} = \frac{1}{10^3} = \frac{1}{1000}$.

$$7.29 \times 10^{-3} = 7.29\left(\frac{1}{1000}\right) = \frac{7.29}{1000} = 0.00729$$

The exponent on the 10 is negative 3, and we moved the decimal to the left 3 places.

In general, **to write a scientific notation number in standard form**, move the decimal point the same number of places as the exponent on 10. If the exponent is positive, move the decimal point to the right; if the exponent is negative, move the decimal point to the left.

TRY THE CONCEPT CHECK IN THE MARGIN.

Example 16   Write each number in standard notation, without exponents.

a. 1.02×10^5 b. 7.358×10^{-3}

c. 8.4×10^7 d. 3.007×10^{-5}

Solution: a. Move the decimal point 5 places to the right.

$$1.02 \times 10^5 = 102,000.$$

b. Move the decimal point 3 places to the left.

$$7.358 \times 10^{-3} = 0.007358$$

c. $8.4 \times 10^7 = 84,000,000.$ 7 places to the right

d. $3.007 \times 10^{-5} = 0.00003007$ 5 places to the left

Performing operations on numbers written in scientific notation makes use of the rules and definitions for exponents.

Example 17   Perform each indicated operation. Write each result in standard decimal notation.

a. $(8 \times 10^{-6})(7 \times 10^3)$

b. $\dfrac{12 \times 10^2}{6 \times 10^{-3}}$

Solution: a. $(8 \times 10^{-6})(7 \times 10^3) = 8 \cdot 7 \cdot 10^{-6} \cdot 10^3$

$$= 56 \times 10^{-3}$$

$$= 0.056$$

b. $\dfrac{12 \times 10^2}{6 \times 10^{-3}} = \dfrac{12}{6} \times 10^{2-(-3)} = 2 \times 10^5 = 200,000$

CALCULATOR EXPLORATIONS
SCIENTIFIC NOTATION

To enter a number written in scientific notation on a scientific calculator, locate the scientific notation key, which may be marked $\boxed{\text{EE}}$ or $\boxed{\text{EXP}}$. To enter 3.1×10^7, press $\boxed{3.1}$ $\boxed{\text{EE}}$ $\boxed{7}$. The display should read $\boxed{3.1 \quad 07}$.

Enter each number written in scientific notation on your calculator.

1. 5.31×10^3

2. -4.8×10^{14}

3. 6.6×10^{-9}

4. -9.9811×10^{-2}

Multiply each of the following on your calculator. Notice the form of the result.

5. $3,000,000 \times 5,000,000$

6. $230,000 \times 1,000$

Multiply each of the following on your calculator. Write the product in scientific notation.

7. $(3.26 \times 10^6)(2.5 \times 10^{13})$

8. $(8.76 \times 10^{-4})(1.237 \times 10^9)$

MENTAL MATH

State each expression using positive exponents only.

1. $5x^{-2}$ **2.** $3x^{-3}$ **3.** $\dfrac{1}{y^{-6}}$

4. $\dfrac{1}{x^{-3}}$ **5.** $\dfrac{4}{y^{-3}}$ **6.** $\dfrac{16}{y^{-7}}$

EXERCISE SET 4.2

A *Simplify each expression. Write each result using positive exponents only. See Examples 1 through 8.*

1. 4^{-3} **2.** 6^{-2} **3.** $7x^{-3}$ **4.** $(7x)^{-3}$

B *Simplify each expression. Write each result using positive exponents only. See Examples 9 through 14.*

5. $\dfrac{x^2 x^5}{x^3}$ **6.** $\dfrac{y^4 y^5}{y^6}$ **7.** $\dfrac{p^2 p}{p^{-1}}$ **8.** $\dfrac{y^3 y}{y^{-2}}$

C *Write each number in scientific notation. See Example 15.*

9. 78,000 **10.** 9,300,000,000 **11.** 0.00000167 **12.** 0.00000017

13. 0.00635 **14.** 0.00194

MENTAL MATH ANSWERS

1. _____

2. _____

3. _____

4. _____

5. _____

6. _____

ANSWERS

1. _____

2. _____

3. _____

4. _____

5. _____

6. _____

7. _____

8. _____

9. _____

10. _____

11. _____

12. _____

13. _____

14. _____

Problem Solving Notes

4.3 INTRODUCTION TO POLYNOMIALS

A DEFINING TERM AND COEFFICIENT

In this section, we introduce a special algebraic expression called a polynomial. Let's first review some definitions presented in Section 2.1.

Recall that a term is a number or the product of a number and variables raised to powers. The terms of the expression $4x^2 + 3x$ are $4x^2$ and $3x$. The terms of the expression $9x^4 - 7x - 1$ are $9x^4$, $-7x$, and -1.

Expression	Terms
$4x^2 + 3x$	$4x^2, 3x$
$9x^4 - 7x - 1$	$9x^4, -7x, -1$
$7y^3$	$7y^3$

The **numerical coefficient** of a term, or simply the **coefficient**, is the numerical factor of each term. If no numerical factor appears in the term, then the coefficient is understood to be 1. If the term is a number only, it is called a **constant** term or simply a constant.

Term	Coefficient
x^5	1
$3x^2$	3
$-4x$	-4
$-x^2y$	-1
3 (constant)	3

Example 1   Complete the table for the expression $7x^5 - 8x^4 + x^2 - 3x + 5$.

Term	Coefficient
x^2	
	-8
$-3x$	
	7
5	

Solution: The completed table is

Term	Coefficient
x^2	1
$-8x^4$	-8
$-3x$	-3
$7x^5$	7
5	5

Objectives

A Define term and coefficient of a term.

B Define polynomial, monomial, binomial, trinomial, and degree.

C Evaluate polynomials for given replacement values.

D Simplify a polynomial by combining like terms.

E Simplify a polynomial in several variables.

SSM   CD-ROM   Video 4.3

Practice Problem 1

Complete the table for the expression $-6x^6 + 4x^5 + 7x^3 - 9x^2 - 1$.

Term	Coefficient
$7x^3$	
	-9
$-6x^6$	
	4
-1	

Answers

1. term: $-9x^2$; $4x^5$,   coefficient: 7; -6; -1

B DEFINING POLYNOMIAL, MONOMIAL, BINOMIAL, TRINOMIAL, AND DEGREE

Now we are ready to define what we mean by a polynomial.

> **POLYNOMIAL**
>
> A **polynomial in x** is a finite sum of terms of the form ax^n, where a is a real number and n is a whole number.

For example,

$$x^5 - 3x^3 + 2x^2 - 5x + 1$$

is a polynomial in x. Notice that this polynomial is written in **descending powers** of x because the powers of x decrease from left to right. (Recall that the term 1 can be thought of as $1x^0$.)

On the other hand,

$$x^{-5} + 2x - 3$$

is **not** a polynomial because one of its terms contains a variable with an exponent, -5, that is not a whole number.

> **TYPES OF POLYNOMIALS**
>
> A **monomial** is a polynomial with exactly one term.
> A **binomial** is a polynomial with exactly two terms.
> A **trinomial** is a polynomial with exactly three terms.

The following are examples of monomials, binomials, and trinomials. Each of these examples is also a polynomial.

POLYNOMIALS			
Monomials	Binomials	Trinomials	None of These
ax^2	$x + y$	$x^2 + 4xy + y^2$	$5x^3 - 6x^2 + 3x - 6$
$-3z$	$3p + 2$	$x^5 + 7x^2 - x$	$-y^5 + y^4 - 3y^3 - y^2 + y$
4	$4x^2 - 7$	$-q^4 + q^3 - 2q$	$x^6 + x^4 - x^3 + 1$

Each term of a polynomial has a degree. The **degree** of a term in one variable is the exponent on the variable.

Practice Problem 2

Identify the degree of each term of the trinomial $-15x^3 + 2x^2 - 5$.

Example 2   Identify the degree of each term of the trinomial $12x^4 - 7x + 3$.

Solution: The term $12x^4$ has degree 4.
The term $-7x$ has degree 1 since $-7x$ is $-7x^1$.
The term 3 has degree 0 since 3 is $3x^0$.

Each polynomial also has a degree.

> **DEGREE OF A POLYNOMIAL**
>
> The **degree of a polynomial** is the greatest degree of any term of the polynomial.

Practice Problem 3

Find the degree of each polynomial and tell whether the polynomial is a monomial, binomial, trinomial, or none of these.

a. $-6x + 14$

b. $9x - 3x^6 + 5x^4 + 2$

c. $10x^2 - 6x - 6$

Example 3   Find the degree of each polynomial and tell whether the polynomial is a monomial, binomial, trinomial, or none of these.

a. $-2t^2 + 3t + 6$ b. $15x - 10$ c. $7x + 3x^3 + 2x^2 - 1$

Answers

2. 3; 2; 0,   **3. a.** binomial, 1,
b. none of these, 6,   **c.** trinomial, 2

Solution:

a. The degree of the trinomial $-2t^2 + 3t + 6$ is 2, the greatest degree of any of its terms.

b. The degree of the binomial $15x - 10$ or $15x^1 - 10$ is 1.

c. The degree of the polynomial $7x + 3x^3 + 2x^2 - 1$ is 3.   ▬

Ⓒ EVALUATING POLYNOMIALS

Polynomials have different values depending on replacement values for the variables. When we find the value of a polynomial for a given replacement value, we are evaluating the polynomial for that value.

Example 4   Evaluate each polynomial when $x = -2$.
  a. $-5x + 6$
  b. $3x^2 - 2x + 1$

Solution: **a.** $-5x + 6 = -5(-2) + 6$ Replace x with -2.
$$= 10 + 6$$
$$= 16$$

b. $3x^2 - 2x + 1 = 3(-2)^2 - 2(-2) + 1$ Replace x with
$$= 3(4) + 4 + 1 \qquad -2.$$
$$= 12 + 4 + 1$$
$$= 17$$ ▬

Many physical phenomena can be modeled by polynomials.

Example 5   Finding Free-Fall Time

The CN Tower in Toronto, Ontario, is 1821 feet tall and is the world's tallest self-supporting structure. An object is dropped from the top of this building. Neglecting air resistance, the height of the object at time t seconds is given by the polynomial $-16t^2 + 1821$. Find the height of the object when $t = 1$ second and when $t = 10$ seconds. (*Source*: World Almanac)

Solution: To find each height, we evaluate the polynomial when $t = 1$ and when $t = 10$.

$$-16t^2 + 1821 = -16(1)^2 + 1821 \qquad \text{Replace } t \text{ with 1.}$$
$$= -16(1) + 1821$$
$$= -16 + 1821$$
$$= 1805$$

The height of the object at 1 second is 1805 feet.

$$-16t^2 + 1821 = -16(10)^2 + 1821 \qquad \text{Replace } t \text{ with 10.}$$
$$= -16(100) + 1821$$
$$= -1600 + 1821$$
$$= 221$$

The height of the object at 10 seconds is 221 feet.

Practice Problem 4

Evaluate each polynomial when $x = -1$.

a. $-2x + 10$

b. $6x^2 + 11x - 20$

Practice Problem 5

From Example 5, find the height of the object when $t = 3$ seconds and when $t = 7$ seconds.

Answers
4. a. 12,   **b.** -25,   **5.** 1677 feet; 1037 feet

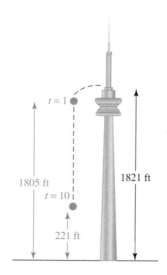

D SIMPLIFYING POLYNOMIALS BY COMBINING LIKE TERMS

Polynomials with like terms can be simplified by combining like terms. Recall that like terms are terms that contain exactly the same variables raised to exactly the same powers.

Like Terms	Unlike Terms
$5x^2, -7x^2$	$3x, 3y$
$y, 2y$	$-2x^2, -5x$
$\frac{1}{2}a^2b, -a^2b$	$6st^2, 4s^2t$

Only like terms can be combined. We combine like terms by applying the distributive property.

Examples   Simplify each polynomial by combining any like terms.

6. $-3x + 7x = (-3 + 7)x = 4x$

7. $11x^2 + 5 + 2x^2 - 7 = 11x^2 + 2x^2 + 5 - 7$
$$= 13x^2 - 2$$

8. $9x^3 + x^3 = 9x^3 + 1x^3$ Write x^3 as $1x^3$.
$$= 10x^3$$

9. $5x^2 + 6x - 9x - 3 = 5x^2 - 3x - 3$ Combine like terms $6x$ and $-9x$.

10. $\frac{2}{5}x^4 + \frac{2}{3}x^3 - x^2 + \frac{1}{10}x^4 - \frac{1}{6}x^3$

$$= \left(\frac{2}{5} + \frac{1}{10}\right)x^4 + \left(\frac{2}{3} - \frac{1}{6}\right)x^3 - x^2$$

$$= \left(\frac{4}{10} + \frac{1}{10}\right)x^4 + \left(\frac{4}{6} - \frac{1}{6}\right)x^3 - x^2$$

$$= \frac{5}{10}x^4 + \frac{3}{6}x^3 - x^2$$

$$= \frac{1}{2}x^4 + \frac{1}{2}x^3 - x^2$$

Practice Problems 6–10

Simplify each polynomial by combining any like terms.

6. $-6y + 8y$

7. $14y^2 + 3 - 10y^2 - 9$

8. $7x^3 + x^3$

9. $23x^2 - 6x - x - 15$

10. $\frac{2}{7}x^3 - \frac{1}{4}x + 2 - \frac{1}{2}x^3 + \frac{3}{8}x$

Example 11   Write a polynomial that describes the total area of the squares and rectangles shown on the next page. Then simplify the polynomial.

Answers

6. $2y$,   **7.** $4y^2 - 6$,   **8.** $8x^3$,

9. $23x^2 - 7x - 15$,   **10.** $-\frac{3}{14}x^3 + \frac{1}{8}x + 2$

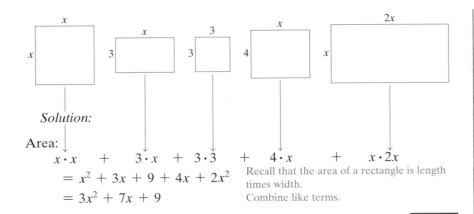

Solution:

Area:

$$x \cdot x \quad + \quad 3 \cdot x \quad + \quad 3 \cdot 3 \quad + \quad 4 \cdot x \quad + \quad x \cdot 2x$$
$$= x^2 + 3x + 9 + 4x + 2x^2 \qquad \text{Recall that the area of a rectangle is length}$$
$$\qquad\qquad\qquad\qquad\qquad\qquad \text{times width.}$$
$$= 3x^2 + 7x + 9 \qquad\qquad \text{Combine like terms.}$$

E SIMPLIFYING POLYNOMIALS CONTAINING SEVERAL VARIABLES

A polynomial may contain more than one variable, such as

$$5x + 3xy^2 - 6x^2y^2 + x^2y - 2y + 1$$

We call this expression a polynomial in several variables.

The **degree of a term** with more than one variable is the sum of the exponents on the variables. The **degree of the polynomial** in several variables is still the greatest degree of the terms of the polynomial.

Example 12   Identify the degrees of the terms and the degree of the polynomial $5x + 3xy^2 - 6x^2y^2 + x^2y - 2y + 1$.

Solution: To organize our work, we use a table.

Terms of Polynomial	Degree of Term	Degree of Polynomial
$5x$	1	
$3xy^2$	1 + 2 or 3	
$-6x^2y^2$	2 + 2 or 4	4 (highest degree)
x^2y	2 + 1 or 3	
$-2y$	1	
1	0	

To simplify a polynomial containing several variables, we combine any like terms.

Examples   Simplify each polynomial by combining any like terms.

13. $3xy - 5y^2 + 7xy - 9x^2 = (3 + 7)xy - 5y^2 - 9x^2$
$$= 10xy - 5y^2 - 9x^2$$

> **HELPFUL HINT**
>
> This term can be written as $10xy$ or $10yx$

14. $9a^2b - 6a^2 + 5b^2 + a^2b - 11a^2 + 2b^2$
$$= 10a^2b - 17a^2 + 7b^2$$

Practice Problem 11

Write a polynomial that describes the total area of the squares and rectangles shown below. Then simplify the polynomial.

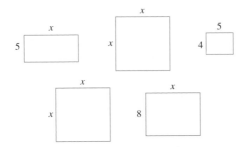

Practice Problem 12

Identify the degrees of the terms and the degree of the polynomial $-2x^3y^2 + 4 - 8xy + 3x^3y + 5xy^2$.

Practice Problems 13–14

Simplify each polynomial by combining any like terms.

13. $11ab - 6a^2 - ba + 8b^2$

14. $7x^2y^2 + 2y^2 - 4y^2x^2 + x^2 - y^2 + 5x^2$

Answers

11. $2x^2 + 13x + 20$,   **12.** $5, 0, 2, 4, 3; 5$
13. $10ab - 6a^2 + 8b^2$,   **14.** $3x^2y^2 + y^2 + 6x^2$

Focus On History

EXPONENTIAL NOTATION

The French mathematician and philosopher René Descartes (1596–1650) is generally credited with devising the system of exponents that we use in math today. His book *La Géométrie* was the first to show successive powers of an unknown quantity x as x, xx, x^3, x^4, x^5, and so on. No one knows why Descartes preferred to write xx instead of x^2. However, the use of xx for the square of the quantity x continued to be popular. Those who used the notation defended it by saying that xx takes up no more space when written than x^2 does.

Before Descartes popularized the use of exponents to indicate powers, other less convenient methods were used. Some mathematicians preferred to write out the Latin words *quadratus* and *cubus* whenever they wanted to indicate that a quantity was to be raised to the second power or the third power. Other mathematicians used the abbreviations of *quadratus* and *cubus*, Q and C, to indicate second and third powers of a quantity.

EXERCISE SET 4.3

A *Complete each table for each polynomial. See Example 1.*

1. $x^2 - 3x + 5$

Term	Coefficient
x^2	
	-3
5	

2. $2x^3 - x + 4$

Term	Coefficient
	2
$-x$	
4	

3. $-5x^4 + 3.2x^2 + x - 5$

Term	Coefficient
$-5x^4$	
$3.2x^2$	
x	
-5	

4. $9.7x^7 - 3x^5 + x^3 - \dfrac{1}{4}x^2$

Term	Coefficient
$9.7x^7$	
$-3x^5$	
x^3	
$-\dfrac{1}{4}x^2$	

C *Evaluate each polynomial when* **(a)** $x = 0$ *and* **(b)** $x = -1$. *See Examples 4 and 5.*

5. $x + 6$

6. $2x - 10$

7. $x^2 - 5x - 2$

8. $x^2 - 4$

9. $x^3 - 15$

10. $-2x^3 + 3x^2 - 6$

D *Simplify each expression by combining like terms. See Examples 6 through 10.*

11. $14x^2 + 9x^2$

12. $18x^3 - 4x^3$

13. $15x^2 - 3x^2 - y$

14. $12k^3 - 9k^3 + 11$

15. $8s - 5s + 4s$

E *Simplify each polynomial by combining any like terms. See Examples 13 and 14.*

16. $3ab - 4a + 6ab - 7a$

17. $-9xy + 7y - xy - 6y$

18. $4x^2 - 6xy + 3y^2 - xy$

19. $3a^2 - 9ab + 4b^2 - 7ab$

20. $5x^2y + 6xy^2 - 5yx^2 + 4 - 9y^2x$

ANSWERS

1. see table

2. see table

3. see table

4. see table

5. _____

6. _____

7. _____

8. _____

9. _____

10. _____

11. _____

12. _____

13. _____

14. _____

15. _____

16. _____

17. _____

18. _____

19. _____

20. _____

Problem Solving Notes

4.4 ADDING AND SUBTRACTING POLYNOMIALS

A ADDING POLYNOMIALS

To add polynomials, we use commutative and associative properties and then combine like terms. To see if you are ready to add polynomials,

TRY THE CONCEPT CHECK IN THE MARGIN.

> **TO ADD POLYNOMIALS**
>
> To add polynomials, combine all like terms.

Examples   Add.

1. $(4x^3 - 6x^2 + 2x + 7) + (5x^2 - 2x)$

$= 4x^3 - 6x^2 + 2x + 7 + 5x^2 - 2x$ Remove parentheses.

$= 4x^3 + (-6x^2 + 5x^2) + (2x - 2x) + 7$ Combine like terms.

$= 4x^3 - x^2 + 7$ Simplify.

2. $(-2x^2 + 5x - 1)$ and $(-2x^2 + x + 3)$ translates to

$(-2x^2 + 5x - 1) + (-2x^2 + x + 3)$

$= -2x^2 + 5x - 1 - 2x^2 + x + 3$ Remove parentheses.

$= (-2x^2 - 2x^2) + (5x + 1x) + (-1 + 3)$ Combine like terms.

$= -4x^2 + 6x + 2$ Simplify. ▬▬

Polynomials can be added vertically if we line up like terms underneath one another.

Example 3   Add $(7y^3 - 2y^2 + 7)$ and $(6y^2 + 1)$ using a vertical format.

Solution: Vertically line up like terms and add.

$$7y^3 - 2y^2 + 7$$
$$\underline{6y^2 + 1}$$
$$7y^3 + 4y^2 + 8$$ ▬▬

B SUBTRACTING POLYNOMIALS

To subtract one polynomial from another, recall the definition of subtraction. To subtract a number, we add its opposite: $a - b = a + (-b)$. To subtract a polynomial, we also add its opposite. Just as $-b$ is the opposite of b, $-(x^2 + 5)$ is the opposite of $(x^2 + 5)$.

Example 4   Subtract: $(5x - 3) - (2x - 11)$

Solution: From the definition of subtraction, we have

$(5x - 3) - (2x - 11) = (5x - 3) + [-(2x - 11)]$ Add the opposite.

$= (5x - 3) + (-2x + 11)$ Apply the distributive property.

$= 3x + 8$ Combine like terms. ▬▬

Objectives

A Add polynomials.

B Subtract polynomials.

C Add or subtract polynomials in one variable.

D Add or subtract polynomials in several variables.

SSM   CD-ROM   Video 4.4

✓ CONCEPT CHECK

When combining like terms in the expression $5x - 8x^2 - 8x$, which of the following is the proper result?

a. $-11x^2$ b. $-3x - 8x^2$

c. $-11x$ d. $-11x^4$

Practice Problems 1–2

Add.

1. $(3x^5 - 7x^3 + 2x - 1) + (3x^3 - 2x)$

2. $(5x^2 - 2x + 1)$ and $(-6x^2 + x - 1)$

Practice Problem 3

Add: $(9y^2 - 6y + 5)$ and $(4y + 3)$ using a vertical format.

> **TO SUBTRACT POLYNOMIALS**
>
> To subtract two polynomials, change the signs of the terms of the polynomial being subtracted and then add.

Practice Problem 4

Subtract: $(9x + 5) - (4x - 3)$

Answers

1. $3x^5 - 4x^3 - 1$, **2.** $-x^2 - x$,

3. $9y^2 - 2y + 8$, **4.** $5x + 8$

✓ Concept Check: b

Practice Problem 5

Subtract: $(4x^3 - 10x^2 + 1) - (-4x^3 + x^2 - 11)$

Example 5

Subtract: $(2x^3 + 8x^2 - 6x) - (2x^3 - x^2 + 1)$

Solution: First, we change the sign of each term of the second polynomial, then we add.

$$(2x^3 + 8x^2 - 6x) - (2x^3 - x^2 + 1) =$$
$$(2x^3 + 8x^2 - 6x) + (-2x^3 + x^2 - 1)$$
$$= 2x^3 - 2x^3 + 8x^2 + x^2 - 6x - 1$$
$$= 9x^2 - 6x - 1 \quad \text{Combine like terms.}$$

Just as polynomials can be added vertically, so can they be subtracted vertically.

Practice Problem 6

Subtract $(6y^2 - 3y + 2)$ from $(2y^2 - 2y + 7)$ using a vertical format.

Example 6

Subtract $(5y^2 + 2y - 6)$ from $(-3y^2 - 2y + 11)$ using a vertical format.

Solution: Arrange the polynomials in a vertical format, lining up like terms.

$$\begin{array}{r} -3y^2 - 2y + 11 \\ -(5y^2 + 2y - 6) \\ \hline \end{array} \qquad \begin{array}{r} -3y^2 - 2y + 11 \\ -5y^2 - 2y + 6 \\ \hline -8y^2 - 4y + 17 \end{array}$$

> **HELPFUL HINT**
>
> Don't forget to change the sign of each term in the polynomial being subtracted.

C ADDING AND SUBTRACTING POLYNOMIALS IN ONE VARIABLE

Let's practice adding and subtracting polynomials in one variable.

Practice Problem 7

Subtract $(3x + 1)$ from the sum of $(4x - 3)$ and $(12x - 5)$.

Example 7

Subtract $(5z - 7)$ from the sum of $(8z + 11)$ and $(9z - 2)$.

Solution: Notice that $(5z - 7)$ is to be subtracted **from** a sum. The translation is

$$[(8z + 11) + (9z - 2)] - (5z - 7)$$
$$= 8z + 11 + 9z - 2 - 5z + 7 \quad \text{Remove grouping symbols.}$$
$$= 8z + 9z - 5z + 11 - 2 + 7 \quad \text{Group like terms.}$$
$$= 12z + 16 \quad \text{Combine like terms.}$$

D ADDING AND SUBTRACTING POLYNOMIALS IN SEVERAL VARIABLES

Now that we know how to add or subtract polynomials in one variable, we can also add and subtract polynomials in several variables.

Practice Problems 8–9

Add or subtract as indicated.

8. $(2a^2 - ab + 6b^2) - (-3a^2 + ab - 7b^2)$

9. $(5x^2y^2 + 3 - 9x^2y + y^2) - (-x^2y^2 + 7 - 8xy^2 + 2y^2)$

Examples

Add or subtract as indicated.

8. $(3x^2 - 6xy + 5y^2) + (-2x^2 + 8xy - y^2)$
$= 3x^2 - 6xy + 5y^2 - 2x^2 + 8xy - y^2$
$= x^2 + 2xy + 4y^2 \quad \text{Combine like terms.}$

9. $(9a^2b^2 + 6ab - 3ab^2) - (5b^2a + 2ab - 3 - 9b^2)$ — Change the sign of
$= 9a^2b^2 + 6ab - 3ab^2 - 5b^2a - 2ab + 3 + 9b^2$ — each term of the polynomial being subtracted.

$= 9a^2b^2 + 4ab - 8ab^2 + 3 + 9b^2 \quad \text{Combine like terms.}$

Answers

5. $8x^3 - 11x^2 + 12$, **6.** $-4y^2 + y + 5$,
7. $13x - 9$, **8.** $5a^2 - 2ab + 13b^2$,
9. $6x^2y^2 - 4 - 9x^2y + 8xy^2 - y^2$

EXERCISE SET 4.4

A *Add. See Examples 1 through 3.*

1. $(3x + 7) + (9x + 5)$

2. $(3x^2 + 7) + (3x^2 + 9)$

3. $(-7x + 5) + (-3x^2 + 7x + 5)$

4. $(3x - 8) + (4x^2 - 3x + 3)$

5. $(-5x^2 + 3) + (2x^2 + 1)$

B *Subtract. See Examples 4 and 5.*

6. $(2x + 5) - (3x - 9)$

7. $(5x^2 + 4) - (-2y^2 + 4)$

8. $3x - (5x - 9)$

9. $4 - (-y - 4)$

10. $(2x^2 + 3x - 9) - (-4x + 7)$

C *Add or subtract as indicated. See Example 7.*

11. $(3x + 5) + (2x - 14)$

12. $(9x - 1) - (5x + 2)$

13. $(7y + 7) - (y - 6)$

14. $(14y + 12) + (-3y - 5)$

15. $(x^2 + 2x + 1) - (3x^2 - 6x + 2)$

D *Add or subtract as indicated. See Examples 8 and 9.*

16. $(9a + 6b - 5) + (-11a - 7b + 6)$

17. $(3x - 2 + 6y) + (7x - 2 - y)$

18. $(4x^2 + y^2 + 3) - (x^2 + y^2 - 2)$

19. $(7a^2 - 3b^2 + 10) - (-2a^2 + b^2 - 12)$

20. $(x^2 + 2xy - y^2) + (5x^2 - 4xy + 20y^2)$

ANSWERS
1.
2.
3.
4.
5.
6.
7.
8.
9.
10.
11.
12.
13.
14.
15.
16.
17.
18.
19.
20.

357

$$= 6x^2 - 11x - 10$$

Combine like terms.

This idea can be expanded so that we can multiply any two polynomials.

Practice Problem 3

Multiply: $(2y^2 + 3)(y - 4)$

Practice Problem 4

Multiply: $(2x + 9)^2$

Example 3   Multiply: $(y^2 + 6)(2y - 1)$

Solution: $\overset{\text{F}\qquad\text{O}\qquad\text{I}\qquad\text{L}}{(y^2 + 6)(2y - 1) = 2y^3 - 1y^2 + 12y - 6}$

Notice in this example that there are no like terms that can be combined, so the product is $2y^3 - y^2 + 12y - 6$.

B   SQUARING BINOMIALS

An expression such as $(3y + 1)^2$ is called the square of a binomial. Since $(3y + 1)^2 = (3y + 1)(3y + 1)$, we can use the FOIL method to find this product.

Example 4   Multiply: $(3y + 1)^2$

Solution: $(3y + 1)^2 = (3y + 1)(3y + 1)$

$\qquad\qquad\overset{\text{F}\qquad\qquad\text{O}\qquad\quad\text{I}\qquad\quad\text{L}}{= (3y)(3y) + (3y)(1) + 1(3y) + 1(1)}$

$\qquad\qquad = 9y^2 + 3y + 3y + 1$

$\qquad\qquad = 9y^2 + 6y + 1$

Notice the pattern that appears in Example 4.

$(3y + 1)^2 = 9y^2 + 6y + 1$

$9y^2$ is the first term of the binomial squared. $(3y)^2 = 9y^2$.

$6y$ is 2 times the product of both terms of the binomial. $(2)(3y)(1) = 6y$.

1 is the second term of the binomial squared. $(1)^2 = 1$.

This pattern leads to the following, which can be used when squaring a binomial. We call these **special products**.

> **SQUARING A BINOMIAL**
>
> A binomial squared is equal to the square of the first term plus or minus twice the product of both terms plus the square of the second term.
>
> $$(a + b)^2 = a^2 + 2ab + b^2$$
> $$(a - b)^2 = a^2 - 2ab + b^2$$

This product can be visualized geometrically.

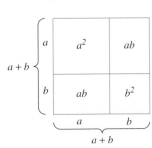

The area of the large square is side · side.

Area $= (a + b)(a + b) = (a + b)^2$

The area of the large square is also the sum of the areas of the smaller rectangles.

Area $= a^2 + ab + ab + b^2 = a^2 + 2ab + b^2$

Thus, $(a + b)^2 = a^2 + 2ab + b^2$.

Answers

3. $2y^3 - 8y^2 + 3y - 12$,   **4.** $4x^2 + 36x + 81$

Examples   Use a special product to square each binomial.

first term squared	plus or minus	twice the product of the terms	plus	second term squared

5. $(t + 2)^2 = t^2 + 2(t)(2) + 2^2 = t^2 + 4t + 4$
6. $(p - q)^2 = p^2 - 2(p)(q) + q^2 = p^2 - 2pq + q^2$
7. $(2x + 5)^2 = (2x)^2 + 2(2x)(5) + 5^2 = 4x^2 + 20x + 25$
8. $(x^2 - 7y)^2 = (x^2)^2 - 2(x^2)(7y) + (7y)^2 = x^4 - 14x^2y + 49y^2$

HELPFUL HINT

Notice that

$(a + b)^2 \neq a^2 + b^2$ The middle term $2ab$ is missing.

$(a + b)^2 = (a + b)(a + b) = a^2 + 2ab + b^2$

Likewise,

$(a - b)^2 \neq a^2 - b^2$

$(a - b)^2 = (a - b)(a - b) = a^2 - 2ab + b^2$

C MULTIPLYING THE SUM AND DIFFERENCE OF TWO TERMS

Another special product is the product of the sum and difference of the same two terms, such as $(x + y)(x - y)$. Finding this product by the FOIL method, we see a pattern emerge.

$$(x + y)(x - y) = x^2 - xy + xy - y^2$$

$$= x^2 - y^2$$

Notice that the middle two terms subtract out. This is because the **O**uter product is the opposite of the **I**nner product. Only the **difference of squares** remains.

> MULTIPLYING THE SUM AND DIFFERENCE OF TWO TERMS
>
> The product of the sum and difference of two terms is the square of the first term minus the square of the second term.
>
> $$(a + b)(a - b) = a^2 - b^2$$

Examples   Use a special product to multiply.

first term squared	minus	second term squared

9. $(x + 4)(x - 4) = x^2 - 4^2 = x^2 - 16$
10. $(6t + 7)(6t - 7) = (6t)^2 - 7^2 = 36t^2 - 49$

Practice Problems 5–8

Use a special product to square each binomial.

5. $(y + 3)^2$ 6. $(r - s)^2$

7. $(6x + 5)^2$ 8. $(x^2 - 3y)^2$

Practice Problems 9–13

Use a special product to multiply.

9. $(x + 7)(x - 7)$

10. $(4y + 5)(4y - 5)$

11. $\left(x - \dfrac{1}{3}\right)\left(x + \dfrac{1}{3}\right)$

12. $(3a - b)(3a + b)$

13. $(2x^2 - 6y)(2x^2 + 6y)$

Answers

5. $y^2 + 6y + 9$, 6. $r^2 - 2rs + s^2$,
7. $36x^2 + 60x + 25$, 8. $x^4 - 6x^2y + 9y^2$,
9. $x^2 - 49$, 10. $16y^2 - 25$, 11. $x^2 - \dfrac{1}{9}$,
12. $9a^2 - b^2$, 13. $4x^4 - 36y^2$

11. $\left(x - \dfrac{1}{4}\right)\left(x + \dfrac{1}{4}\right) = x^2 - \left(\dfrac{1}{4}\right)^2 = x^2 - \dfrac{1}{16}$

12. $(2p - q)(2p + q) = (2p)^2 - q^2 = 4p^2 - q^2$

13. $(3x^2 - 5y)(3x^2 + 5y) = (3x^2)^2 - (5y)^2 = 9x^4 - 25y^2$ ▬▬▬

TRY THE CONCEPT CHECK IN THE MARGIN.

D USING SPECIAL PRODUCTS

Let's now practice using our special products on a variety of multiplications. This practice will help us recognize when to apply what special product.

Examples   Use a special product to multiply.

14. $(x - 9)(x + 9)$
     $= x^2 - 9^2 = x^2 - 81$

> This is the sum and difference of the same two terms.

15. $(3y + 2)^2$
     $= (3y)^2 + 2(3y)(2) + 2^2$
     $= 9y^2 + 12y + 4$

> This is a binomial squared.

16. $(6a + 1)(a - 7)$

> No special product applies.

$$\overset{\text{F}}{}\quad\overset{\text{O}}{}\quad\overset{\text{I}}{}\quad\overset{\text{L}}{}$$
$$= 6a \cdot a + 6a(-7) + 1 \cdot a + 1(-7)$$

> Use the FOIL method.

$$= 6a^2 - 42a + a - 7$$
$$= 6a^2 - 41a - 7$$ ▬▬▬

✓ **CONCEPT CHECK**

Match each expression on the left to the equivalent expression or expressions in the list on the right.

$(a + b)^2$ a. $(a + b)(a + b)$
$(a + b)(a - b)$ b. $a^2 - b^2$

c. $a^2 + b^2$
d. $a^2 - 2ab + b^2$
e. $a^2 + 2ab + b^2$

Practice Problems 14–16

Use a special product to multiply.

14. $(7x - 1)^2$

15. $(5y + 3)(2y - 5)$

16. $(2a - 1)(2a + 1)$

Answers

14. $49x^2 - 14x + 1$, **15.** $10y^2 - 19y - 15$,
16. $4a^2 - 1$

✓ **Concept Check:** a or e, b

EXERCISE SET 4.6

A *Multiply using the FOIL method. See Examples 1 through 3.*

1. $(x + 3)(x + 4)$ **2.** $(x + 5)(x + 1)$ **3.** $(x - 5)(x + 10)$

4. $(y - 12)(y + 4)$ **5.** $(5x - 6)(x + 2)$ **6.** $(3y - 5)(2y - 7)$

7. $(y - 6)(4y - 1)$ **8.** $(2x - 9)(x - 11)$ **9.** $(2x + 5)(3x - 1)$

10. $(6x + 2)(x - 2)$

C *Multiply. See Examples 9 through 13.*

11. $(a - 7)(a + 7)$ **12.** $(b + 3)(b - 3)$ **13.** $(x + 6)(x - 6)$

14. $(x - 8)(x + 8)$ **15.** $(3x - 1)(3x + 1)$ **16.** $(4x - 5)(4x + 5)$

17. $(x^2 + 5)(x^2 - 5)$ **18.** $(a^2 + 6)(a^2 - 6)$ **19.** $(2y^2 - 1)(2y^2 + 1)$

20. $(3x^2 + 1)(3x^2 - 1)$

1. _____
2. _____
3. _____
4. _____
5. _____
6. _____
7. _____
8. _____
9. _____
10. _____
11. _____
12. _____
13. _____
14. _____
15. _____
16. _____
17. _____
18. _____
19. _____
20. _____

Focus On The Real World

SPACE EXPLORATION

From scientific observations on Earth, we know that Saturn, the second largest planet in our solar system, is a giant ball of gas surrounded by rings and orbited by 19 moons. We also know that Saturn has a diameter of 120,000 kilometers and a mass of 569,000,000,000,000,000,000,000,000 kilograms. But what is Saturn like below its outer layer of clouds? What are Saturn's rings made of? What is the surface of Saturn's largest moon, Titan, like? Could life ever be supported on Titan?

NASA is hoping to answer these questions and more about the sixth planet from the sun in our solar system with its $3,400,000,000 Cassini mission. The Cassini spacecraft is scheduled to arrive in orbit around Saturn in June 2004. The goal of this mission is to study Saturn, its rings, and its moons. The Cassini spacecraft will also launch the Huygens probe to study Titan.

The Cassini mission began on October 15, 1997, with the launch of a mighty Titan IV booster rocket. The entire launch vehicle including rocket fuel weighed more than 2,000,000 pounds before launch. The Titan IV flung Cassini into space at a speed of 14,400 kilometers per hour. To take advantage of something called *gravity assist*, Cassini is taking a roundabout path to Saturn past Venus (twice), Earth, and Jupiter. Altogether, Cassini will travel 3,540,000,000 kilometers before reaching Saturn, a planet that is only 1,430,000,000 kilometers from the sun.

Once Cassini reaches Saturn, it will begin collecting all kinds of data about the planet and its moons. Over the course of Cassini's mission, it will collect over 2,000,000,000,000 bits of scientific information, or about the same amount of data in 800 sets of the *Encyclopedia Britannica*. About once per day, Cassini will use its 4-meter antenna to transmit the latest data that it has collected back to Earth at a frequency of 8,400,000,000 cycles per second. For comparison, the FM band on a radio is centered around 100,000,000 cycles per second. It will take from 70 to 90 minutes for Cassini's transmissions to reach Earth and by then the signals are very weak. The power of the signal transmitted by the spacecraft is 20 watts but, even with the huge antennas used on Earth, only 0.0000000000000001 watt can be received. (*Source*: based on data from National Aeronautics and Space Administration)

CRITICAL THINKING

1. Make a list of the numbers (other than those in dates) used in the article. Rewrite each number in scientific notation.
2. What are the advantages of scientific notation?
3. What are the disadvantages of scientific notation?
4. In your opinion, how large or small should a number be to make using scientific notation worthwhile?

4.7 DIVIDING POLYNOMIALS

A DIVIDING BY A MONOMIAL

To divide a polynomial by a monomial, recall addition of fractions. Fractions that have a common denominator are added by adding the numerators:

$$\frac{a}{c} + \frac{b}{c} = \frac{a+b}{c}$$

If we read this equation from right to left and let a, b, and c be monomials, $c \neq 0$, we have the following:

> **TO DIVIDE A POLYNOMIAL BY A MONOMIAL**
>
> Divide each term of the polynomial by the monomial.
>
> $$\frac{a+b}{c} = \frac{a}{c} + \frac{b}{c}, \quad c \neq 0$$

Throughout this section, we assume that denominators are not 0.

Example 1   Divide: $6m^2 + 2m$ by $2m$

Solution: We begin by writing the quotient in fraction form. Then we divide each term of the polynomial $6m^2 + 2m$ by the monomial $2m$.

$$\frac{6m^2 + 2m}{2m} = \frac{6m^2}{2m} + \frac{2m}{2m}$$

$$= 3m + 1 \qquad \text{Simplify.}$$

Check: To check, we multiply.

$$2m(3m + 1) = 2m(3m) + 2m(1) = 6m^2 + 2m$$

The quotient $3m + 1$ checks.

TRY THE CONCEPT CHECK IN THE MARGIN.

Example 2   Divide: $\dfrac{9x^5 - 12x^2 + 3x}{3x^2}$

Solution:

$$\frac{9x^5 - 12x^2 + 3x}{3x^2} = \frac{9x^5}{3x^2} - \frac{12x^2}{3x^2} + \frac{3x}{3x^2} \qquad \text{Divide each term by } 3x^2.$$

$$= 3x^3 - 4 + \frac{1}{x} \qquad \text{Simplify.}$$

Notice that the quotient is not a polynomial because of the term $\dfrac{1}{x}$. This expression is called a rational expression— we will study rational expressions further in Chapter 5. Although the quotient of two polynomials is not always a polynomial, we may still check by multiplying.

Check: $3x^2\left(3x^3 - 4 + \dfrac{1}{x}\right) = 3x^2(3x^3) - 3x^2(4) + 3x^2\left(\dfrac{1}{x}\right)$

$$= 9x^5 - 12x^2 + 3x$$

Objectives

A Divide a polynomial by a monomial.

B Use long division to divide a polynomial by a polynomial other than a monomial.

C Use synthetic division.

SSM   CD-ROM   Video
                          4.7

Practice Problem 1

Divide: $25x^3 + 5x^2$ by $5x^2$

✓ CONCEPT CHECK

In which of the following is $\dfrac{x+5}{5}$ simplified correctly?

a. $\dfrac{x}{5} + 1$ b. x c. $x + 1$

Practice Problem 2

Divide: $\dfrac{30x^7 + 10x^2 - 5x}{5x^2}$

Answers

1. $5x + 1$,   **2.** $6x^5 + 2 - \dfrac{1}{x}$

✓ Concept Check: a

Practice Problem 3

Divide: $\dfrac{12x^3y^3 - 18xy + 6y}{3xy}$

Example 3 Divide: $\dfrac{8x^2y^2 - 16xy + 2x}{4xy}$

Solution:

$$\dfrac{8x^2y^2 - 16xy + 2x}{4xy} = \dfrac{8x^2y^2}{4xy} - \dfrac{16xy}{4xy} + \dfrac{2x}{4xy} \quad \text{Divide each term by } 4xy.$$

$$= 2xy - 4 + \dfrac{1}{2y} \quad \text{Simplify.}$$

Check:

$$4xy\left(2xy - 4 + \dfrac{1}{2y}\right) = 4xy(2xy) - 4xy(4) + 4xy\left(\dfrac{1}{2y}\right)$$

$$= 8x^2y^2 - 16xy + 2x$$

B DIVIDING BY A POLYNOMIAL OTHER THAN A MONOMIAL

To divide a polynomial by a polynomial other than a monomial, we use a process known as long division. Polynomial long division is similar to number long division, so we review long division by dividing 13 into 3660.

$$
\begin{array}{r}
281 \\
13\overline{)3660} \\
\underline{26}\downarrow \\
106 \\
\underline{104}\downarrow \\
20 \\
\underline{13} \\
7
\end{array}
$$

$2 \cdot 13 = 26$

Subtract and bring down the next digit in the dividend.

$8 \cdot 13 = 104$

Subtract and bring down the next digit in the dividend.

$1 \cdot 13 = 13$

Subtract. There are no more digits to bring down, so the remainder is 7.

The quotient is 281 R 7, which can be written as $281\dfrac{7}{13} \begin{smallmatrix}\leftarrow \text{remainder} \\ \leftarrow \text{divisor}\end{smallmatrix}$.

Recall that division can be checked by multiplication. To check a division problem such as this one, we see that

$$13 \cdot 281 + 7 = 3660$$

Now we demonstrate long division of polynomials.

Example 4 Divide $x^2 + 7x + 12$ by $x + 3$ using long division.

Solution:

Practice Problem 4

Divide: $x^2 + 12x + 35$ by $x + 5$

To subtract, change the signs of these terms and add.

$$
\begin{array}{r}
x \\
x + 3\overline{)x^2 + 7x + 12} \\
\underline{x^2 + 3x}\downarrow \\
4x + 12
\end{array}
$$

How many times does x divide x^2? $\dfrac{x^2}{x} = x$.

Multiply: $x(x + 3)$.

Subtract and bring down the next term.

Now we repeat this process.

To subtract, change the signs of these terms and add.

$$
\begin{array}{r}
x + 4 \\
x + 3\overline{)x^2 + 7x + 12} \\
\underline{x^2 + 3x} \\
4x + 12 \\
\underline{4x + 12} \\
0
\end{array}
$$

How many times does x divide $4x$? $\dfrac{4x}{x} = 4$.

Multiply: $4(x + 3)$.

Subtract. The remainder is 0.

The quotient is $x + 4$.

Check: We check by multiplying.

$$\underbrace{\text{divisor}}\cdot\underbrace{\text{quotient}}+\underbrace{\text{remainder}}=\underbrace{\text{dividend}}$$

or

$$(x+3)\cdot(x+4)+\quad 0\quad=x^2+7x+12$$

The quotient checks.   ▬▬▬

Example 5   Divide $6x^2 + 10x - 5$ by $3x - 1$ using long division.

Solution:

$$
\begin{array}{r}
2x + 4 \\
3x - 1\overline{)6x^2 + 10x - 5} \\
\underline{6x^2 - 2x} \\
12x - 5 \\
\underline{12x - 4} \\
-1
\end{array}
$$

$\frac{6x^2}{3x} = 2x$, so $2x$ is a term of the quotient.

$2x(3x - 1)$.
Subtract and bring down the next term.

$\frac{12x}{3x} = 4, 4(3x - 1)$

Subtract. The remainder is -1.

Thus $(6x^2 + 10x - 5)$ divided by $(3x - 1)$ is $(2x + 4)$ with a remainder of -1. This can be written as

$$\frac{6x^2 + 10x - 5}{3x - 1} = 2x + 4 + \frac{-1}{3x - 1} \quad\begin{array}{l}\leftarrow \text{remainder}\\ \leftarrow \text{divisor}\end{array}$$

Check: To check, we multiply $(3x - 1)(2x + 4)$. Then we add the remainder, -1, to this product.

$$(3x - 1)(2x + 4) + (-1) = (6x^2 + 12x - 2x - 4) - 1$$
$$= 6x^2 + 10x - 5$$

The quotient checks.   ▬▬▬

Notice that the division process is continued until the degree of the remainder polynomial is less than the degree of the divisor polynomial.

Example 6   Divide: $\dfrac{4x^2 + 7 + 8x^3}{2x + 3}$

Solution: Before we begin the division process, we rewrite $4x^2 + 7 + 8x^3$ as $8x^3 + 4x^2 + 0x + 7$. Notice that we have written the polynomial in descending order and have represented the missing x term by $0x$.

$$
\begin{array}{r}
4x^2 - 4x + 6 \\
2x + 3\overline{)8x^3 + 4x^2 + 0x + 7} \\
\underline{8x^3 + 12x^2} \\
-8x^2 + 0x \\
\underline{-8x^2 - 12x} \\
12x + 7 \\
\underline{12x + 18} \\
-11
\end{array}
$$

Remainder.

Thus, $\dfrac{4x^2 + 7 + 8x^3}{2x + 3} = 4x^2 - 4x + 6 + \dfrac{-11}{2x + 3}$.

Practice Problem 5

Divide: $6x^2 + 7x - 5$ by $2x - 1$

Practice Problem 6

Divide: $\dfrac{5 - x + 9x^3}{3x + 2}$

Answers

5. $3x + 5$,   **6.** $3x^2 - 2x + 1 + \dfrac{3}{3x + 2}$

C USING SYNTHETIC DIVISION

When a polynomial is to be divided by a binomial of the form $x - c$, a shortcut process called **synthetic division** may be used. On the left is an example of long division, and on the right is the same example showing the coefficients of the variables only.

$$
\begin{array}{r}
2x^2 + 5x + 2 \\
x - 3\,)\overline{2x^3 - x^2 - 13x + 1} \\
\underline{2x^3 - 6x^2} \\
5x^2 - 13x \\
\underline{5x^2 - 15x} \\
2x + 1 \\
\underline{2x - 6} \\
7
\end{array}
\qquad
\begin{array}{r}
2 \quad 5 \quad 2 \\
1 - 3\,)\overline{2 - 1 - 13 + 1} \\
\underline{2 - 6} \\
5 - 13 \\
\underline{5 - 15} \\
2 + 1 \\
\underline{2 - 6} \\
7
\end{array}
$$

Notice that as long as we keep coefficients of powers of x in the same column, we can perform division of polynomials by performing algebraic operations on the coefficients only. This shortest process of dividing with coefficients only in a special format is called synthetic division. To find $(2x^3 - x^2 - 13x + 1) \div (x - 3)$ by synthetic division, follow the next example.

Practice Problem 7

Use synthetic division to divide $3x^3 - 2x^2 + 5x + 4$ by $x - 2$.

Example 7   Use synthetic division to divide $2x^3 - x^2 - 13x + 1$ by $x - 3$.

Solution: To use synthetic division, the divisor must be in the form $x - c$. Since we are dividing by $x - 3$, c is 3. We write down 3 and the coefficients of the dividend.

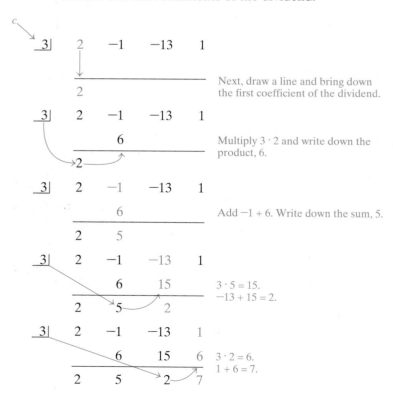

Answers

7. $3x^2 + 4x + 13 + \dfrac{30}{x - 2}$

The quotient is found in the bottom row. The numbers 2, 5, and 2 are the coefficients of the quotient polynomial, and the number 7 is the remainder. The degree of the quotient polynomial is one less than the degree of the dividend. In our example, the degree of the dividend is 3, so the degree of the quotient polynomial is 2. As we found when we performed the long division, the quotient is

$$2x^2 + 5x + 2, \quad \text{remainder } 7$$

or

$$2x^2 + 5x + 2 + \frac{7}{x - 3}$$

When using synthetic division, if there are missing powers of the variable, insert 0s as coefficients.

Example 8 Use synthetic division to divide $x^4 - 2x^3 - 11x^2 + 34$ by $x + 2$.

Solution: The divisior is $x + 2$, which in the form $x - c$ is $x - (-2)$. Thus, c is -2. There is no x-term in the dividend, so we insert a coefficient of 0. The dividend coefficients are $1, -2, -11, 0$, and 34.

$$
\begin{array}{r|rrrrr}
-2 & 1 & -2 & -11 & 0 & 34 \\
   &   & -2 & 8 & 6 & -12 \\
\hline
   & 1 & -4 & -3 & 6 & 22 \\
\end{array}
$$

The dividend is a fourth-degree polynomial, so the quotient polynomial is a third-degree polynomial. The quotient is $x^3 - 4x^2 - 3x + 6$ with a remainder of 22. Thus,

$$\frac{x^4 - 2x^3 - 11x^2 + 34}{x + 2} = x^3 - 4x^2 - 3x + 6 + \frac{22}{x + 2}$$

HELPFUL HINT

Before dividing by synthetic division, write the dividend in descending order of variable exponents. Any "missing powers" of the variable must be represented by 0 times the variable raised to the missing power.

TRY THE CONCEPT CHECK IN THE MARGIN.

Practice Problem 8

Use synthetic division to divide $x^4 + 3x^3 - 5x + 4$ by $x + 1$.

✓ CONCEPT CHECK

Which division problems are candidates for the synthetic division process?
a. $(3x^2 + 5) \div (x + 4)$

b. $(x^3 - x^2 + 2) \div (3x^3 - 2)$

c. $(y^4 + y - 3) \div (x^2 + 1)$

d. $x^5 \div (x - 5)$

Answers

8. $x^3 + 2x^2 - 2x - 3 + \dfrac{7}{x + 1}$

✓ Concept Check: a and d

Focus On Business and Career

BUSINESS TERMS

For most businesses, a financial goal is to "make money." But what does that mean from a mathematical point of view? To find out, we must first discuss some common business terms.

▲ **Revenue** is the amount of money a business takes in. A company's annual revenue is the amount of money it collects during its fiscal, or business, year. For most companies, the largest source of revenue is from the sales of their products or services. For instance, a computer manufacturer's annual revenue is the amount of money it collects during the year from selling computers to customers. Large companies may also have revenues from investment interest or leases. When revenue can be expressed as a function of another variable, it is often denoted as $R(x)$.

▲ **Expenses** are the costs of doing business. For instance, a large part of a computer manufacturer's expenses include the cost of the computer components it buys from wholesalers to use in the manufacturing or assembling process. Other expenses include salaries, mortgage payments, equipment, taxes, advertising, and so on. Some businesses refer to their expenses simply as cost. When cost can be expressed as a function of another variable, it is often denoted as $C(x)$.

▲ **Net income/loss** is the difference between a company's annual revenues and expenses. This difference may also be referred to as net earnings. Positive net earnings—that is, a positive difference—result in a net income or net profit. Posting a net income can be interpreted as "making money." Negative net earnings—that is, a negative difference—result in a net loss. Posting a net loss can be interpreted as "losing money." A profit function can be expressed as $P(x) = R(x) - C(x)$. In this case, a negative profit is interpreted as a net loss.

GROUP ACTIVITY

Locate several corporate annual reports. Using the data in the reports, verify that the net income or net earnings given in a report was calculated as the difference between revenue and expenses. If this was not the case, can you tell what caused the variation? If so, explain.

Name _____ **Section** _____ **Date** _____

MENTAL MATH

Simplify each expression.

1. $\dfrac{a^6}{a^4}$

2. $\dfrac{y^2}{y}$

3. $\dfrac{a^3}{a}$

4. $\dfrac{p^8}{p^3}$

5. $\dfrac{k^5}{k^2}$

6. $\dfrac{k^7}{k^5}$

EXERCISE SET 4.7

A *Perform each division. See Examples 1 through 3.*

1. $\dfrac{8x^3 - 4x^2 + 6x + 2}{2}$

2. $\dfrac{12x^4 + 3x^2}{x}$

3. $\dfrac{15x^2 - 9x^5}{x}$

4. $\dfrac{15p^3 + 18p^2}{3p}$

5. $\dfrac{14m^2 - 27m^3}{7m}$

6. $\dfrac{6x^5 + 3x^4}{3x^4}$

7. $\dfrac{a^2b^2 - ab^3}{ab}$

8. $\dfrac{m^3n^2 - mn^4}{mn}$

CHAPTER 4 HIGHLIGHTS

DEFINITIONS AND CONCEPTS	EXAMPLES

SECTION 4.1 EXPONENTS

a^n means the product of n factors, each of which is a.	$3^2 = 3 \cdot 3 = 9$ $(-5)^3 = (-5)(-5)(-5) = -125$ $\left(\dfrac{1}{2}\right)^4 = \dfrac{1}{2} \cdot \dfrac{1}{2} \cdot \dfrac{1}{2} \cdot \dfrac{1}{2} = \dfrac{1}{16}$
If m and n are integers and no denominators are 0, **Product Rule:** $a^m \cdot a^n = a^{m+n}$ **Power Rule:** $(a^m)^n = a^{mn}$ **Power of a Product Rule:** $(ab)^n = a^n b^n$ **Power of a Quotient Rule:** $\left(\dfrac{a}{b}\right)^n = \dfrac{a^n}{b^n}$ **Quotient Rule:** $\dfrac{a^m}{a^n} = a^{m-n}$ **Zero Exponent:** $a^0 = 1, a \neq 0$	$x^2 \cdot x^7 = x^{2+7} = x^9$ $(5^3)^8 = 5^{3 \cdot 8} = 5^{24}$ $(7y)^4 = 7^4 y^4$ $\left(\dfrac{x}{8}\right)^3 = \dfrac{x^3}{8^3}$ $\dfrac{x^9}{x^4} = x^{9-4} = x^5$ $5^0 = 1, x^0 = 1, x \neq 0$

SECTION 4.2 NEGATIVE EXPONENTS AND SCIENTIFIC NOTATION

If $a \neq 0$ and n is an integer, $a^{-n} = \dfrac{1}{a^n}$	$3^{-2} = \dfrac{1}{3^2} = \dfrac{1}{9}; \ 5x^{-2} = \dfrac{5}{x^2}$ Simplify: $\left(\dfrac{x^{-2}y}{x^5}\right)^{-2} = \dfrac{x^4 y^{-2}}{x^{-10}}$ $= x^{4-(-10)}y^{-2}$ $= \dfrac{x^{14}}{y^2}$
A positive number is written in scientific notation if it is written as the product of a number a, $1 \leq a < 10$, and an integer power r of 10. $a \times 10^r$	$12000 = 1.2 \times 10^4$ $0.00000568 = 5.68 \times 10^{-6}$

SECTION 4.3 INTRODUCTION TO POLYNOMIALS

A **term** is a number or the product of a number and variables raised to powers.	$-5x, 7a^2b, \dfrac{1}{4}y^4, 0.2$
The **numerical coefficient** or **coefficient** of a term is its numerical factor.	TERM COEFFICIENT $7x^2$ 7 y 1 $-a^2b$ -1
A **polynomial** is a finite sum of terms of the form ax^n where a is a real number and n is a whole number.	$5x^3 - 6x^2 + 3x - 6$ (Polynomial)
A **monomial** is a polynomial with exactly 1 term.	$\dfrac{5}{6}y^3$ (Monomial)
A **binomial** is a polynomial with exactly 2 terms.	$-0.2a^2b - 5b^2$ (Binomial)

SECTION 4.3 CONTINUED	
A **trinomial** is a polynomial with exactly 3 terms.	$3x^2 - 2x + 1$ (Trinomial)
The **degree of a polynomial** is the greatest degree of any term of the polynomial.	POLYNOMIAL DEGREE $5x^2 - 3x + 2$ 2 $7y + 8y^2z^3 - 12$ $2 + 3 = 5$

SECTION 4.4   ADDING AND SUBTRACTING POLYNOMIALS

To add polynomials, combine like terms.	Add. $(7x^2 - 3x + 2) + (-5x - 6)$ $= 7x^2 - 3x + 2 - 5x - 6$ $= 7x^2 - 8x - 4$
To subtract two polynomials, change the signs of the terms of the second polynomial, then add.	Subtract. $(17y^2 - 2y + 1) - (-3y^3 + 5y - 6)$ $= (17y^2 - 2y + 1) + (3y^3 - 5y + 6)$ $= 17y^2 - 2y + 1 + 3y^3 - 5y + 6$ $= 3y^3 + 17y^2 - 7y + 7$

SECTION 4.5   MULTIPLYING POLYNOMIALS

To multiply two polynomials, multiply each term of one polynomial by each term of the other polynomial, and then combine like terms.	Multiply. $(2x + 1)(5x^2 - 6x + 2)$ $= 2x(5x^2 - 6x + 2) + 1(5x^2 - 6x + 2)$ $= 10x^3 - 12x^2 + 4x + 5x^2 - 6x + 2$ $= 10x^3 - 7x^2 - 2x + 2$

SECTION 4.6   SPECIAL PRODUCTS

The **FOIL method** may be used when multiplying two binomials.	Multiply: $(5x - 3)(2x + 3)$ $(5x - 3)(2x + 3)$ $= (5x)(2x) + (5x)(3) + (-3)(2x) + (-3)(3)$ $= 10x^2 + 15x - 6x - 9$ $= 10x^2 + 9x - 9$
Squaring a Binomial $(a + b)^2 = a^2 + 2ab + b^2$ $(a - b)^2 = a^2 - 2ab + b^2$	Square each binomial. $(x + 5)^2 = x^2 + 2(x)(5) + 5^2$ $= x^2 + 10x + 25$ $(3x - 2y)^2 = (3x)^2 - 2(3x)(2y) + (2y)^2$ $= 9x^2 - 12xy + 4y^2$
Multiplying the Sum and Difference of Two Terms $(a + b)(a - b) = a^2 - b^2$	Multiply. $(6y + 5)(6y - 5) = (6y)^2 - 5^2$ $= 36y^2 - 25$

SECTION 4.7 DIVIDING POLYNOMIALS

To divide a polynomial by a monomial:

$$\frac{a + b}{c} = \frac{a}{c} + \frac{b}{c}, c \neq 0$$

Divide.

$$\frac{15x^5 - 10x^3 + 5x^2 - 2x}{5x^2}$$

$$= \frac{15x^5}{5x^2} - \frac{10x^3}{5x^2} + \frac{5x^2}{5x^2} - \frac{2x}{5x^2}$$

$$= 3x^3 - 2x + 1 - \frac{2}{5x}$$

To divide a polynomial by a polynomial other than a monomial, use long division.

$$5x - 1 + \frac{-4}{2x + 3} \text{ or}$$

$$2x + 3 \overline{)10x^2 + 13x - 7} \qquad 5x - 1 - \frac{4}{2x + 3}$$
$$\underline{10x^2 + 15x}$$
$$-2x - 7$$
$$\underline{-2x - 3}$$
$$-4$$

A shortcut method called **synthetic division** may be used to divide a polynomial by a binomial of the form $x - c$.

Use synthetic division to divide $2x^3 - x^2 - 8x - 1$ by $x - 2$.

$$\begin{array}{r|rrrr} 2 & 2 & -1 & -8 & -1 \\ & & 4 & 6 & -4 \\ \hline & 2 & 3 & -2 & -5 \end{array}$$

The quotient is $2x^2 + 3x - 2 - \dfrac{5}{x - 2}$.

Factoring Polynomials

CHAPTER 5

In Chapter 4, we learned how to multiply polynomials. Now we will deal with an operation that is the reverse process of multiplying—factoring. Factoring is an important algebraic skill because it allows us to write a sum as a product. As we will see in Sections 5.6 and 5.7, factoring can be used to solve equations other than linear equations. In Chapter 6, we will also use factoring to simplify and perform arithmetic operations on rational expressions.

The America's Cup sailing competition is held roughly every three years. This international yacht race dates from 1851. The sailboats were originally built from wood and later from aluminum. Today's International America's Cup Class sailboats are made of carbon fiber. This strong but lightweight material reduced the boats' weight by 50% and made them much faster.

Problem Solving Notes

5.1   THE GREATEST COMMON FACTOR

In the product $2 \cdot 3 = 6$, 2 and 3 are called **factors** of 6 and $2 \cdot 3$ is a **factored form** of 6. This is true of polynomials also. Since $(x + 2)(x + 3) = x^2 + 5x + 6$, then $(x + 2)$ and $(x + 3)$ are factors of $x^2 + 5x + 6$, and $(x + 2)(x + 3)$ is a factored form of the polynomial.

> The process of writing a polynomial as a product is called **factoring** the polynomial.

Study the examples below and look for a pattern.

TRY THE CONCEPT CHECK IN THE MARGIN.

Multiplying: $5(x^2 + 3) = 5x^2 + 15$ $2x(x - 7) = 2x^2 - 14x$

Factoring: $5x^2 + 15 = 5(x^2 + 3)$ $2x^2 - 14x = 2x(x - 7)$

Do you see that factoring is the reverse process of multiplying?

$$\overset{\text{factoring}}{x^2 + 5x + 6 = (x + 2)(x + 3)}$$
$$\underset{\text{multiplying}}{}$$

A   FINDING THE GREATEST COMMON FACTOR

The first step in factoring a polynomial is to see whether the terms of the polynomial have a common factor. If there is one, we can write the polynomial as a product by **factoring out** the common factor. We will usually factor out the *greatest* common factor (GCF).

The **greatest common factor (GCF) of a list of terms** is the product of the GCF of the numerical coefficients and the GCF of the variable factors.

$$20x^2y^2 = 2 \cdot 2 \cdot 5 \cdot x \cdot x \cdot y \cdot y$$
$$6xy^3 = 2 \cdot 3 \cdot x \cdot y \cdot y \cdot y$$
$$\text{GCF} = 2 \cdot x \cdot y \cdot y = 2xy^2$$

Notice that the GCF of a list of variables raised to powers is the variables raised to the *smallest* exponent in the list. For example,

the GCF of x^2, x^5, and x^3 is x^2,

because x^2 is the largest *common* factor.

Example 1   Find the greatest common factor of each list of terms.

    a. $6x^2$, $10x^3$, and $-8x$
    b. $-18y^2$, $-63y^3$, and $27y^4$
    c. a^3b^2, a^5b, and a^6b^2

Solution: **a.** $6x^2 = 2 \cdot 3 \cdot x^2$
         $10x^3 = 2 \cdot 5 \cdot x^3$
         $-8x = -1 \cdot 2 \cdot 2 \cdot 2 \cdot x^1$ → The GCF of x^2, x^3, and x is x.
         GCF $= 2 \cdot x^1$ or $2x$

    b. $-18y^2 = -1 \cdot 2 \cdot 3 \cdot 3 \cdot y^2$
         $-63y^3 = -1 \cdot 3 \cdot 3 \cdot 7 \cdot y^3$ → The GCF of y^2, y^3, and y^4 is y^2.
         $27y^4 = 3 \cdot 3 \cdot 3 \cdot y^4$
         GCF $= 3 \cdot 3 \cdot y^2$ or $9y^2$

Objectives

A Find the greatest common factor of a list of terms.

B Factor out the greatest common factor from the terms of a polynomial.

C Factor by grouping.

    SSM   CD-ROM   Video
                          5.1

✓ CONCEPT CHECK

Multiply: $2(x - 4)$
What do you think the result of factoring $2x - 8$ would be? Why?

Practice Problem 1

Find the greatest common factor of each list of terms.

a. $6x^2$, $9x^4$, and $-12x^5$

b. $-16y$, $-20y^6$, and $40y^4$

c. a^5b^4, ab^3, and a^3b^2

Answers

1. a. $3x^2$,   **b.** $4y$,   **c.** ab^2

✓ Concept Check

$2x - 8$; The result would be $2(x - 4)$ because factoring is the reverse process of multiplying.

c. The GCF of a^3, a^5, and a^6 is a^3.
The GCF of b^2, b, and b^2 is b. Thus,
the GCF of a^3b^2, a^5b, and a^6b^2 is a^3b.

B FACTORING OUT THE GREATEST COMMON FACTOR

To factor a polynomial such as $8x + 14$, we first see whether the terms have a greatest common factor other than 1. In this case, they do: The GCF of $8x$ and 14 is 2.

We factor out 2 from each term by writing each term as a product of 2 and the term's remaining factors.

$$8x + 14 = 2 \cdot 4x + 2 \cdot 7$$

Using the distributive property, we can write

$$8x + 14 = 2 \cdot 4x + 2 \cdot 7$$
$$= 2(4x + 7)$$

Thus, a factored form of $8x + 14$ is $2(4x + 7)$. We can check by multiplying: $2(4x + 7) = 2 \cdot 4x + 2 \cdot 7 = 8x + 14$.

TRY THE CONCEPT CHECK IN THE MARGIN.

✓ **CONCEPT CHECK**

Which of the following is/are factored form(s) of $6t + 18$?

a. 6

b. $6 \cdot t + 6 \cdot 3$

c. $6(t + 3)$

d. $3(t + 6)$

HELPFUL HINT

A factored form of $8x + 14$ is *not*
$$2 \cdot 4x + 2 \cdot 7$$
Although the *terms* have been factored (written as a product), the *polynomial* $8x + 14$ has not been factored. A factored form of $8x + 14$ is the *product* $2(4x + 7)$.

Practice Problem 2

Factor each polynomial by factoring out the greatest common factor (GCF).

a. $10y + 25$

b. $x^4 - x^9$

Example 2 Factor each polynomial by factoring out the greatest common factor (GCF).

  a. $6t + 18$ **b.** $y^5 - y^7$

Solution: **a.** The GCF of terms $6t$ and 18 is 6. Thus,

$$6t + 18 = 6 \cdot t + 6 \cdot 3$$
$$= 6(t + 3) \quad \text{Apply the distributive property.}$$

We can check our work by multiplying 6 and $(t + 3)$.

$6(t + 3) = 6 \cdot t + 6 \cdot 3 = 6t + 18$, the original polynomial.

  b. The GCF of y^5 and y^7 is y^5. Thus,

$$y^5 - y^7 = (y^5)1 - (y^5)y^2$$
$$= y^5(1 - y^2)$$

HELPFUL HINT   Don't forget the 1.

Answers

2. a. $5(2y + 5)$, **b.** $x^4(1 - x^5)$

✓ **Concept Check:** c

Example 3
Factor: $-9a^5 + 18a^2 - 3a$

Solution:

$$-9a^5 + 18a^2 - 3a = (3a)(-3a^4) + (3a)(6a) + (3a)(-1)$$
$$= 3a(-3a^4 + 6a - 1)$$

> **HELPFUL HINT**   Don't forget the -1.

In Example 3, we could have chosen to factor out $-3a$ instead of $3a$. If we factor out $-3a$, we have

$$-9a^5 + 18a^2 - 3a = (-3a)(3a^4) + (-3a)(-6a) + (-3a)(1)$$
$$= -3a(3a^4 - 6a + 1)$$

> **HELPFUL HINT**   Notice the changes in signs when factoring out $-3a$.

Examples
Factor.

4. $6a^4 - 12a = 6a(a^3 - 2)$

5. $\frac{3}{7}x^4 + \frac{1}{7}x^3 - \frac{5}{7}x^2 = \frac{1}{7}x^2(3x^2 + x - 5)$

6. $15p^2q^4 + 20p^3q^5 + 5p^3q^3 = 5p^2q^3(3q + 4pq^2 + p)$

Example 7
Factor: $5(x + 3) + y(x + 3)$

Solution: The binomial $(x + 3)$ is present in both terms and is the greatest common factor. We use the distributive property to factor out $(x + 3)$.

$$5(x + 3) + y(x + 3) = (x + 3)(5 + y)$$

C FACTORING BY GROUPING

Once the GCF is factored out, we can often continue to factor the polynomial, using a variety of techniques. We discuss here a technique called **factoring by grouping**. This technique can be used to factor some polynomials with four terms.

Example 8
Factor $xy + 2x + 3y + 6$ by grouping.

Solution: The GCF of the first two terms is x, and the GCF of the last two terms is 3.

$$xy + 2x + 3y + 6 = (xy + 2x) + (3y + 6)$$
$$= x(y + 2) + 3(y + 2)$$

> **HELPFUL HINT**
>
> Notice that this is *not* a factored form of the original polynomial. It is a sum, not a product.

Next we factor out the common binomial factor, $(y + 2)$.

$$x(y + 2) + 3(y + 2) = (y + 2)(x + 3)$$

Check: Multiply $(y + 2)$ by $(x + 3)$.

$$(y + 2)(x + 3) = xy + 2x + 3y + 6,$$

the original polynomial.

Thus, the factored form of $xy + 2x + 3y + 6$ is the product $(y + 2)(x + 3)$ ▬▬▬

Practice Problems 9–10

Factor by grouping.

9. $28x^3 - 7x^2 + 12x - 3$

10. $2xy + 5y^2 - 4x - 10y$

Examples   Factor by grouping.

9. $15x^3 - 10x^2 + 6x - 4$
   $= (15x^3 - 10x^2) + (6x - 4)$
   $= 5x^2(3x - 2) + 2(3x - 2)$ Factor each group.
   $= (3x - 2)(5x^2 + 2)$ Factor out the common factor, $(3x - 2)$.

10. $3x^2 + 4xy - 3x - 4y$
    $= (3x^2 + 4xy) + (-3x - 4y)$
    $= x(3x + 4y) - 1(3x + 4y)$ Factor each group. A -1 is factored from the second pair of terms so that there is a common factor, $(3x + 4y)$.
    $= (3x + 4y)(x - 1)$ Factor out the common factor, $(3x + 4y)$.

▬▬▬

Practice Problems 11–13

Factor by grouping.

11. $4x^3 + x - 20x^2 - 5$

12. $2x - 2 + x^3 - 3x^2$

13. $3xy - 4 + x - 12y$

Examples   Factor by grouping.

11. $3x^3 - 2x - 9x^2 + 6$ Factor each group. A -3 is factored from the second pair of terms so that there is a common factor, $(3x^2 - 2)$.
    $= x(3x^2 - 2) - 3(3x^2 - 2)$
    $= (3x^2 - 2)(x - 3)$ Factor out the common factor, $(3x^2 - 2)$.

12. $5x - 10 + x^3 - x^2 = 5(x - 2) + x^2(x - 1)$

There is no common binomial factor that can now be factored out. No matter how we rearrange the terms, no grouping will lead to a common factor. Thus, this polynomial is not factorable by grouping.

13. $3xy + 2 - 3x - 2y$

Notice that the first two terms have no common factor other than 1. However, if we rearrange these terms, a grouping emerges that does lead to a common factor.

$3xy + 2 - 3x - 2y$
$= (3xy - 3x) + (-2y + 2)$
$= 3x(y - 1) - 2(y - 1)$ Factor -2 from the second group.
$= (y - 1)(3x - 2)$ Factor out the common factor, $(y - 1)$.

▬▬▬

> **HELPFUL HINT**
>
> Throughout this chapter, we will be factoring polynomials. Even when the instructions do not so state, it is always a good idea to check your answers by multiplying.

Answers

9. $(4x - 1)(7x^2 + 3)$, **10.** $(2x + 5y)(y - 2)$,
11. $(4x^2 + 1)(x - 5)$, **12.** Can't be factored,
13. $(3y + 1)(x - 4)$

EXERCISE SET 5.1

B *Factor out the GCF from each polynomial. See Examples 2 through 7.*

1. $3a + 6$

2. $18a + 12$

3. $30x - 15$

4. $42x - 7$

5. $x^3 + 5x^2$

6. $y^5 - 6y^4$

7. $6y^4 - 2y$

8. $5x^2 + 10x^6$

9. $32xy - 18x^2$

10. $10xy - 15x^2$

11. $a^7b^6 - a^3b^2 + a^2b^5 - a^2b^2$

12. $x^9y^6 + x^3y^5 - x^4y^3 + x^3y^3$

13. $5x^3y - 15x^2y + 10xy$

14. $14x^3y + 7x^2y - 7xy$

15. $8x^5 + 16x^4 - 20x^3 + 12$

16. $9y^6 - 27y^4 + 18y^2 + 6$

17. $\dfrac{1}{3}x^4 + \dfrac{2}{3}x^3 - \dfrac{4}{3}x^5 + \dfrac{1}{3}x$

18. $\dfrac{2}{5}y^7 - \dfrac{4}{5}y^5 + \dfrac{3}{5}y^2 - \dfrac{2}{5}y$

19. $y(x + 2) + 3(x + 2)$

20. $z(y + 4) + 3(y + 4)$

1. _____
2. _____
3. _____
4. _____
5. _____
6. _____
7. _____
8. _____
9. _____
10. _____
11. _____
12. _____
13. _____
14. _____
15. _____
16. _____
17. _____
18. _____
19. _____
20. _____

Problem Solving Notes

5.2   FACTORING TRINOMIALS OF THE FORM $x^2 + bx + c$

A   FACTORING TRINOMIALS OF THE FORM $x^2 + bx + c$

In this section, we factor trinomials of the form $x^2 + bx + c$, such as

$$x^2 + 4x + 3, \quad x^2 - 8x + 15, \quad x^2 + 4x - 12, \quad r^2 - r - 42$$

Notice that for these trinomials, the coefficient of the squared variable is 1.

Recall that factoring means to write as a product and that factoring and multiplying are reverse processes. Using the FOIL method of multiplying binomials, we have that

$$
\overset{\text{F}\quad\text{O}\quad\text{I}\quad\text{L}}{(x + 3)(x + 1) = x^2 + 1x + 3x + 3}
$$
$$
= x^2 + 4x + 3
$$

Thus, a factored form of $x^2 + 4x + 3$ is $(x + 3)(x + 1)$.

Notice that the product of the first terms of the binomials is $x \cdot x = x^2$, the first term of the trinomial. Also, the product of the last two terms of the binomials is $3 \cdot 1 = 3$, the third term of the trinomial. The sum of these same terms is $3 + 1 = 4$, the coefficient of the middle, x, term of the trinomial.

The product of these numbers is 3.

$$x^2 + 4x + 3 = (x + 3)(x + 1)$$

The sum of these numbers is 4.

Many trinomials, such as the one above, factor into two binomials. To factor $x^2 + 7x + 10$, let's assume that it factors into two binomials and begin by writing two pairs of parentheses. The first term of the trinomial is x^2, so we use x and x as the first terms of the binomial factors.

$$x^2 + 7x + 10 = (x + \square)(x + \square)$$

To determine the last term of each binomial factor, we look for two integers whose product is 10 and whose sum is 7. The integers are 2 and 5. Thus,

$$x^2 + 7x + 10 = (x + 2)(x + 5)$$

To see if we have factored correctly, we multiply.

$$(x + 2)(x + 5) = x^2 + 5x + 2x + 10$$
$$= x^2 + 7x + 10 \qquad \text{Combine like terms.}$$

HELPFUL HINT

Since multiplication is commutative, the factored form of $x^2 + 7x + 10$ can be written as either $(x + 2)(x + 5)$ or $(x + 5)(x + 2)$.

Objectives

A Factor trinomials of the form $x^2 + bx + c$.

B Factor out the greatest common factor and then factor a trinomial of the form $x^2 + bx + c$.

SSM   CD-ROM   Video 5.2

TO FACTOR A TRINOMIAL OF THE FORM $x^2 + bx + c$

The product of these numbers is c.

$$x^2 + bx + c = (x + \square)(x + \square)$$

The sum of these numbers is b.

Practice Problem 1

Factor: $x^2 + 9x + 20$

Example 1   Factor: $x^2 + 7x + 12$

Solution: We begin by writing the first terms of the binomial factors.

$$(x + \square)(x + \square)$$

Next we look for two numbers whose product is 12 and whose sum is 7. Since our numbers must have a positive product and a positive sum, we look at pairs of positive factors of 12 only.

Factors of 12	Sum of Factors
1, 12	13
2, 6	8
3, 4	7

Correct sum, so the numbers are 3 and 4.

$$x^2 + 7x + 12 = (x + 3)(x + 4)$$

Check: Multiply $(x + 3)$ by $(x + 4)$.

Practice Problem 2

Factor each trinomial.

a. $x^2 - 13x + 22$

b. $x^2 - 27x + 50$

Example 2   Factor: $x^2 - 8x + 15$

Solution: Again, we begin by writing the first terms of the binomials.

$$(x + \square)(x + \square)$$

Now we look for two numbers whose product is 15 and whose sum is -8. Since our numbers must have a positive product and a negative sum, we look at pairs of negative factors of 15 only.

Factors of 15	Sum of Factors
$-1, -15$	-16
$-3, -5$	-8

Correct sum, so the numbers are -3 and -5.

$$x^2 - 8x + 15 = (x - 3)(x - 5)$$

Practice Problem 3

Factor: $x^2 + 5x - 36$

Answers

1. $(x + 4)(x + 5)$,   **2. a.** $(x - 2)(x - 11)$,
b. $(x - 2)(x - 25)$,   **3.** $(x + 9)(x - 4)$

Example 3   Factor: $x^2 + 4x - 12$

Solution: $x^2 + 4x - 12 = (x + \square)(x + \square)$

We look for two numbers whose product is -12 and whose sum is 4. Since our numbers must have a negative product, we look at pairs of factors with opposite signs.

Factors of -12	Sum of Factors
$-1, 12$	11
$1, -12$	-11
$-2, 6$	4
$2, -6$	-4
$-3, 4$	1
$3, -4$	-1

Correct sum, so the numbers are -2 and 6.

$$x^2 + 4x - 12 = (x - 2)(x + 6)$$

Example 4   Factor: $r^2 - r - 42$

Solution: Because the variable in this trinomial is r, the first term of each binomial factor is r.

$$r^2 - r - 42 = (r + \square)(r + \square)$$

Now we look for two numbers whose product is -42 and whose sum is -1, the numerical coefficient of r. The numbers are 6 and -7. Therefore,

$$r^2 - r - 42 = (r + 6)(r - 7)$$

Example 5   Factor: $a^2 + 2a + 10$

Solution: Look for two numbers whose product is 10 and whose sum is 2. Neither 1 and 10 nor 2 and 5 give the required sum, 2. We conclude that $a^2 + 2a + 10$ is not factorable with integers. A polynomial such as $a^2 + 2a + 10$ is called a **prime polynomial**.

Example 6   Factor: $x^2 + 5xy + 6y^2$

Solution: $x^2 + 5xy + 6y^2 = (x + \square)(x + \square)$

Recall that the middle term $5xy$ is the same as $5yx$. Notice that $5y$ is the "coefficient" of x. We then look for two terms whose product is $6y^2$ and whose sum is $5y$. The terms are $2y$ and $3y$ because $2y \cdot 3y = 6y^2$ and $2y + 3y = 5y$. Therefore,

$$x^2 + 5xy + 6y^2 = (x + 2y)(x + 3y)$$

Example 7   Factor: $x^4 + 5x^2 + 6$

Solution: As usual, we begin by writing the first terms of the binomials. Since the greatest power of x in this polynomial is x^4, we write

$$(x^2 + \square)(x^2 + \square) \quad \text{since } x^2 \cdot x^2 = x^4$$

Now we look for two factors of 6 whose sum is 5. The numbers are 2 and 3. Thus,

$$x^4 + 5x^2 + 6 = (x^2 + 2)(x^2 + 3)$$

Practice Problem 4

Factor each trinomial.

a. $q^2 - 3q - 40$

b. $y^2 + 2y - 48$

Practice Problem 5

Factor: $x^2 + 6x + 15$

Practice Problem 6

Factor each trinomial.

a. $x^2 + 6xy + 8y^2$

b. $a^2 - 13ab + 30b^2$

Practice Problem 7

Factor: $x^4 + 8x^2 + 12$

Answers

4. a. $(q - 8)(q + 5)$, **b.** $(y + 8)(y - 6)$, **5.** prime polynomial, **6. a.** $(x + 2y)(x + 4y)$, **b.** $(a - 3b)(a - 10b)$, **7.** $(x^2 + 6)(x^2 + 2)$

The following sign patterns may be useful when factoring trinomials.

HELPFUL HINT—SIGN PATTERNS

A positive constant in a trinomial tells us to look for two numbers with the same sign. The sign of the coefficient of the middle term tells us whether the signs are both positive or both negative.

$$\underset{\downarrow}{\overset{\text{both}}{\text{positive}}} \quad \underset{\downarrow}{\overset{\text{same}}{\text{sign}}}$$

$$x^2 + 10x + 16 = (x + 2)(x + 8)$$

$$\underset{\downarrow}{\overset{\text{both}}{\text{negative}}} \quad \underset{\downarrow}{\overset{\text{same}}{\text{sign}}}$$

$$x^2 - 10x + 16 = (x - 2)(x - 8)$$

A negative constant in a trinomial tells us to look for two numbers with opposite signs.

$$\underset{\downarrow}{\overset{\text{opposite}}{\text{signs}}} \qquad\qquad\qquad \underset{\downarrow}{\overset{\text{opposite}}{\text{signs}}}$$

$$x^2 + 6x - 16 = (x + 8)(x - 2) \qquad x^2 - 6x - 16 = (x - 8)(x + 2)$$

B   **FACTORING OUT THE GREATEST COMMON FACTOR**

Remember that the first step in factoring any polynomial is to factor out the greatest common factor (if there is one other than 1 or -1).

Practice Problem 8

Factor each trinomial.

a. $x^3 + 3x^2 - 4x$

b. $4x^2 - 24x + 36$

Example 8   Factor: $3m^2 - 24m - 60$

Solution: First we factor out the greatest common factor, 3, from each term.

$$3m^2 - 24m - 60 = 3(m^2 - 8m - 20)$$

Now we factor $m^2 - 8m - 20$ by looking for two factors of -20 whose sum is -8. The factors are -10 and 2. Therefore, the complete factored form is

$$3m^2 - 24m - 60 = 3(m + 2)(m - 10)$$

HELPFUL HINT

Remember to write the common factor 3 as part of the factored form.

Answers

8. a. $x(x + 4)(x - 1)$,   **b.** $4(x - 3)(x - 3)$

EXERCISE SET 5.2

A *Factor each trinomial completely. If a polynomial can't be factored, write "prime." See Examples 1 through 7.*

■ **1.** $x^2 + 7x + 6$ **2.** $x^2 + 6x + 8$ **3.** $x^2 - 10x + 9$

4. $x^2 - 6x + 9$ ■ **5.** $x^2 - 3x - 18$ **6.** $x^2 - x - 30$

7. $x^2 + 3x - 70$ **8.** $x^2 + 4x - 32$ **9.** $x^2 + 5x + 2$

10. $x^2 - 7x + 5$

B *Factor each trinomial completely. See Examples 1 through 8.*

11. $2z^2 + 20z + 32$ **12.** $3x^2 + 30x + 63$ **13.** $2x^3 - 18x^2 + 40x$

14. $x^3 - x^2 - 56x$ ■ **15.** $x^2 - 3xy - 4y^2$ **16.** $x^2 - 4xy - 77y^2$

17. $x^2 + 15x + 36$ **18.** $x^2 + 19x + 60$ **19.** $x^2 - x - 2$

20. $x^2 - 5x - 14$

1. _____
2. _____
3. _____
4. _____
5. _____
6. _____
7. _____
8. _____
9. _____
10. _____
11. _____
12. _____
13. _____
14. _____
15. _____
16. _____
17. _____
18. _____
19. _____
20. _____

Problem Solving Notes

5.3 FACTORING TRINOMIALS OF THE FORM $ax^2 + bx + c$

A FACTORING TRINOMIALS OF THE FORM $ax^2 + bx + c$

In this section, we factor trinomials of the form $ax^2 + bx + c$, such as

$$3x^2 + 11x + 6, \qquad 8x^2 - 22x + 5, \qquad 2x^2 + 13x - 7$$

Notice that the coefficient of the squared variable in these trinomials is a number other than 1. We will factor these trinomials using a trial-and-check method based on our work in the last section.

To begin, let's review the relationship between the numerical coefficients of the trinomial and the numerical coefficients of its factored form. For example, since $(2x + 1)(x + 6) = 2x^2 + 13x + 6$, the factored form of $2x^2 + 13x + 6$ is

$$2x^2 + 13x + 6 = (2x + 1)(x + 6)$$

Notice that $2x$ and x are factors of $2x^2$, the first term of the trinomial. Also, 6 and 1 are factors of 6, the last term of the trinomial, as shown:

$$\overset{\overset{\displaystyle 2x \cdot x}{\frown}}{2x^2 + 13x + 6 = (2x + 1)(x + 6)}$$
$$\underset{1 \cdot 6}{\smile}$$

Also notice that $13x$, the middle term, is the sum of the following products:

$$2x^2 + 13x + 6 = (2x + 1)(x + 6)$$

$$\begin{array}{r} 1x \\ + 12x \\ \hline 13x \end{array} \quad \text{Middle term}$$

Let's use this pattern to factor $5x^2 + 7x + 2$. First, we find factors of $5x^2$. Since all numerical coefficients in this trinomial are positive, we will use factors with positive numerical coefficients only. Thus, the factors of $5x^2$ are $5x$ and x. Let's try these factors as first terms of the binomials. Thus far, we have

$$5x^2 + 7x + 2 = (5x + \square)(x + \square)$$

Next, we need to find positive factors of 2. Positive factors of 2 are 1 and 2. Now we try possible combinations of these factors as second terms of the binomials until we obtain a middle term of $7x$.

$$(5x + 1)(x + 2) = 5x^2 + 11x + 2$$

$$\begin{array}{r} 1x \\ + 10x \\ \hline 11x \end{array} \longrightarrow \textbf{Incorrect} \text{ middle term}$$

Let's try switching factors 2 and 1.

$$(5x + 2)(x + 1) = 5x^2 + 7x + 2$$

$$\begin{array}{r} 2x \\ + 5x \\ \hline 7x \end{array} \longrightarrow \textbf{Correct} \text{ middle term}$$

Thus the factored form of $5x^2 + 7x + 2$ is $(5x + 2)(x + 1)$. To check, we multiply $(5x + 2)$ and $(x + 1)$. The product is $5x^2 + 7x + 2$.

Objectives

A Factor trinomials of the form $ax^2 + bx + c$, where $a \neq 1$.

B Factor out the GCF before factoring a trinomial of the form $ax^2 + bx + c$.

SSM   CD-ROM   Video 5.3

Practice Problem 1

Factor each trinomial.

a. $4x^2 + 12x + 5$

b. $5x^2 + 27x + 10$

✓ **CONCEPT CHECK**

Do the terms of $3x^2 + 29x + 18$ have a common factor? Without multiplying, decide which of the following factored forms could not be a factored form of $3x^2 + 29x + 18$.

a. $(3x + 18)(x + 1)$

b. $(3x + 2)(x + 9)$

c. $(3x + 6)(x + 3)$

d. $(3x + 9)(x + 2)$

Practice Problem 2

Factor each trinomial.

a. $6x^2 - 5x + 1$

b. $2x^2 - 11x + 12$

Answers

1. a. $(2x + 5)(2x + 1)$, **b.** $(5x + 2)(x + 5)$,
2. a. $(3x - 1)(2x - 1)$, **b.** $(2x - 3)(x - 4)$

✓ Concept Check: no; a, c, d

Example 1

Factor: $3x^2 + 11x + 6$

Solution: Since all numerical coefficients are positive, we use factors with positive numerical coefficients. We first find factors of $3x^2$.

Factors of $3x^2$: $3x^2 = 3x \cdot x$

If factorable, the trinomial will be of the form

$$3x^2 + 11x + 6 = (3x + \square)(x + \square)$$

Next we factor 6.

Factors of 6: $6 = 1 \cdot 6$, $6 = 2 \cdot 3$

Now we try combinations of factors of 6 until a middle term of $11x$ is obtained. Let's try 1 and 6 first.

$$(3x + 1)(x + 6) = 3x^2 + 19x + 6$$

$$\underbrace{}_{\substack{1x \\ + 18x \\ \hline 19x}} \longrightarrow \textbf{Incorrect} \text{ middle term}$$

Now let's next try 6 and 1.

$$(3x + 6)(x + 1)$$

Before multiplying, notice that the terms of the factor $3x + 6$ have a common factor of 3. The terms of the original trinomial $3x^2 + 11x + 6$ have no common factor other than 1, so the terms of the factored form of $3x^2 + 11x + 6$ can contain no common factor other than 1. This means that $(3x + 6)(x + 1)$ is not a factored form.

Next let's try 2 and 3 as last terms.

$$(3x + 2)(x + 3) = 3x^2 + 11x + 6$$

$$\underbrace{}_{\substack{2x \\ + 9x \\ \hline 11x}} \longrightarrow \textbf{Correct} \text{ middle term}$$

Thus the factored form of $3x^2 + 11x + 6$ is $(3x + 2)(x + 3)$.

HELPFUL HINT

If the terms of a trinomial have no common factor (other than 1), then the terms of neither of its binomial factors will contain a common factor (other than 1).

TRY THE CONCEPT CHECK IN THE MARGIN.

Example 2

Factor: $8x^2 - 22x + 5$

Solution: Factors of $8x^2$: $8x^2 = 8x \cdot x$, $8x^2 = 4x \cdot 2x$

We'll try $8x$ and x.

$$8x^2 - 22x + 5 = (8x + \square)(x + \square)$$

Since the middle term, $-22x$, has a negative numerical coefficient, we factor 5 into negative factors.

Factors of 5: $5 = -1 \cdot -5$

Let's try -1 and -5.

$$(8x - 1)(x - 5) = 8x^2 - 41x + 5$$

$$\underbrace{\begin{array}{c} -1x \\ + (-40x) \\ \hline -41x \end{array}}_{} \longrightarrow \textbf{Incorrect } \text{middle term}$$

Now let's try -5 and -1.

$$(8x - 5)(x - 1) = 8x^2 - 13x + 5$$

$$\underbrace{\begin{array}{c} -5x \\ + (-8x) \\ \hline -13x \end{array}}_{} \longrightarrow \textbf{Incorrect } \text{middle term}$$

Don't give up yet! We can still try other factors of $8x^2$. Let's try $4x$ and $2x$ with -1 and -5.

$$(4x - 1)(2x - 5) = 8x^2 - 22x + 5$$

$$\underbrace{\begin{array}{c} -2x \\ + (-20x) \\ \hline -22x \end{array}}_{} \longrightarrow \textbf{Correct } \text{middle term}$$

The factored form of $8x^2 - 22x + 5$ is $(4x - 1)(2x - 5)$.

━━━━

Example 3   Factor: $2x^2 + 13x - 7$

Solution: Factors of $2x^2$: $2x^2 = 2x \cdot x$

Factors of -7: $-7 = -1 \cdot 7$, $-7 = 1 \cdot -7$

We try possible combinations of these factors:

$$(2x + 1)(x - 7) = 2x^2 - 13x - 7 \quad \textbf{Incorrect } \text{middle term}$$
$$(2x - 1)(x + 7) = 2x^2 + 13x - 7 \quad \textbf{Correct } \text{middle term}$$

The factored form of $2x^2 + 13x - 7$ is $(2x - 1)(x + 7)$.

━━━━

Example 4   Factor: $10x^2 - 13xy - 3y^2$

Solution: Factors of $10x^2$: $10x^2 = 10x \cdot x$, $10x^2 = 2x \cdot 5x$

Factors of $-3y^2$: $-3y^2 = -3y \cdot y$,
$-3y^2 = 3y \cdot -y$

We try some combinations of these factors:

$$(10x - 3y)(x + y) = 10x^2 + 7xy - 3y^2$$
$$(x + 3y)(10x - y) = 10x^2 + 29xy - 3y^2$$
$$(5x + 3y)(2x - y) = 10x^2 + xy - 3y^2$$
$$(2x - 3y)(5x + y) = 10x^2 - 13xy - 3y^2$$

                Correct middle term

Practice Problem 3

Factor each trinomial.

a. $35x^2 + 4x - 4$

b. $4x^2 + 3x - 7$

Practice Problem 4

Factor each trinomial.

a. $14x^2 - 3xy - 2y^2$

b. $12a^2 - 16ab - 3b^2$

Answers

3. a. $(5x + 2)(7x - 2)$,   **b.** $(4x + 7)(x - 1)$,
4. a. $(7x + 2y)(2x - y)$,
b. $(6a + b)(2a - 3b)$

The factored form of $10x^2 - 13xy - 3y^2$ is $(2x - 3y)(5x + y)$.

B   FACTORING OUT THE GREATEST COMMON FACTOR

Don't forget that the first step in factoring any polynomial is to look for a common factor to factor out.

Practice Problem 5

Factor each trinomial.

a. $3x^3 + 17x^2 + 10x$

b. $6xy^2 + 33xy - 18x$

Example 5   Factor: $24x^4 + 40x^3 + 6x^2$

Solution: Notice that all three terms have a common factor of $2x^2$. Thus we factor out $2x^2$ first.

$$24x^4 + 40x^3 + 6x^2 = 2x^2(12x^2 + 20x + 3)$$

Next we factor $12x^2 + 20x + 3$.

Factors of $12x^2$: $12x^2 = 4x \cdot 3x$, $12x^2 = 12x \cdot x$, $12x^2 = 6x \cdot 2x$

Since all terms in the trinomial have positive numerical coefficients, we factor 3 using positive factors only.

Factors of 3: $3 = 1 \cdot 3$

We try some combinations of the factors.

$$2x^2(4x + 3)(3x + 1) = 2x^2(12x^2 + 13x + 3)$$
$$2x^2(12x + 1)(x + 3) = 2x^2(12x^2 + 37x + 3)$$
$$2x^2(2x + 3)(6x + 1) = 2x^2(12x^2 + 20x + 3)$$

Correct middle term

The factored form of $24x^4 + 40x^3 + 6x^2$ is $2x^2(2x + 3)(6x + 1)$.

> **HELPFUL HINT**
>
> Don't forget to include the common factor in the factored form.

Answers

5. a. $x(3x + 2)(x + 5)$,
b. $3x(2y - 1)(y + 6)$

EXERCISE SET 5.3

A *Complete each factored form.*

1. $5x^2 + 22x + 8 = (5x + 2)($ $\quad)$

2. $2y^2 + 15y + 25 = (2y + 5)($ $\quad)$

3. $50x^2 + 15x - 2 = (5x + 2)($ $\quad)$

4. $6y^2 + 11y - 10 = (2y + 5)($ $\quad)$

5. $20x^2 - 7x - 6 = (5x + 2)($ $\quad)$

Factor each trinomial completely. See Examples 1 through 4.

6. $2x^2 + 13x + 15$

7. $3x^2 + 8x + 4$

8. $8y^2 - 17y + 9$

9. $21x^2 - 41x + 10$

10. $2x^2 - 9x - 5$

11. $36r^2 - 5r - 24$

12. $20r^2 + 27r - 8$

13. $3x^2 + 20x - 63$

14. $10x^2 + 17x + 3$

15. $2x^2 + 7x + 5$

B *Factor each trinomial completely. See Examples 1 through 5.*

16. $12x^3 + 11x^2 + 2x$

17. $8a^3 + 14a^2 + 3a$

18. $21x^2 - 48x - 45$

19. $12x^2 - 14x - 10$

20. $12x^2 + 7x - 12$

1. _____
2. _____
3. _____
4. _____
5. _____
6. _____
7. _____
8. _____
9. _____
10. _____
11. _____
12. _____
13. _____
14. _____
15. _____
16. _____
17. _____
18. _____
19. _____
20. _____

Problem Solving Notes

5.4 Factoring Trinomials of the Form $ax^2 + bx + c$ by Grouping

Objectives

A Use the grouping method to factor trinomials of the form $ax^2 + bx + c$, $a \neq 1$.

SSM   CD-ROM   Video
                        5.4

A Using the Grouping Method

There is an alternative method that can be used to factor trinomials of the form $ax^2 + bx + c$, $a \neq 1$. This method is called the **grouping method** because it uses factoring by grouping as we learned in Section 5.1.

To see how this method works, let's multiply the following:

$$(2x + 1)(3x + 5) = 6x^2 + 10x + 3x + 5$$
$$10 \cdot 3 = 30$$
$$6 \cdot 5 = 30$$
$$= 6x^2 + 13x + 5$$

Notice that the product of the coefficients of the first and last terms is $6 \cdot 5 = 30$. This is the same as the product of the coefficients of the two middle terms, $10 \cdot 3 = 30$.

Let's use this pattern to write $2x^2 + 11x + 12$ as a four-term polynomial. We will then factor the polynomial by grouping.

$$2x^2 + 11x + 12$$
$$= 2x^2 + \square x + \square x + 12$$

Find two numbers whose product is $2 \cdot 12 = 24$ and whose sum is 11.

Since we want a positive product and a positive sum, we consider pairs of positive factors of 24 only.

Factors of 24	Sum of Factors
1, 24	25
2, 12	14
3, 8	11

Correct sum

The factors are 3 and 8. Now we use these factors to write the middle term $11x$ as $3x + 8x$ (or $8x + 3x$). We replace $11x$ with $3x + 8x$ in the original trinomial and then we can factor by grouping.

$$2x^2 + 11x + 12 = 2x^2 + 3x + 8x + 12$$
$$= (2x^2 + 3x) + (8x + 12) \quad \text{Group the terms.}$$
$$= x(2x + 3) + 4(2x + 3) \quad \text{Factor each group.}$$
$$= (2x + 3)(x + 4) \quad \text{Factor out } (2x + 3).$$

In general, we have the following procedure.

To Factor Trinomials of the Form $ax^2 + bx + c$ by Grouping

Step 1. Factor out a greatest common factor, if there is one other than 1 (or -1).

Step 2. Find two numbers whose product is $a \cdot c$ and whose sum is b.

Step 3. Write the middle term, bx using the factors found in Step 2.

Step 4. Factor by grouping.

Practice Problem 1

Factor each trinomial by grouping.

a. $3x^2 + 14x + 8$

b. $12x^2 + 19x + 5$

Example 1

Factor $8x^2 - 14x + 5$ by grouping.

Solution:

Step 1. The terms of this trinomial contain no greatest common factor other than 1 (or -1).

Step 2. This trinomial is of the form $ax^2 + bx + c$ with $a = 8$, $b = -14$, and $c = 5$. Find two numbers whose product is $a \cdot c$ or $8 \cdot 5 = 40$, and whose sum is b or -14. The numbers are -4 and -10.

Step 3. Write $-14x$ as $-4x - 10x$ so that

$$8x^2 - 14x + 5 = 8x^2 - 4x - 10x + 5$$

Step 4. Factor by grouping.

$$8x^2 - 4x - 10x + 5 = 4x(2x - 1) - 5(2x - 1)$$
$$= (2x - 1)(4x - 5)$$

Practice Problem 2

Factor each trinomial by grouping.

a. $6x^2y - 7xy - 5y$

b. $30x^2 - 26x + 4$

Example 2

Factor $6x^2 - 2x - 20$ by grouping.

Solution:

Step 1. First factor out the greatest common factor, 2.

$$6x^2 - 2x - 20 = 2(3x^2 - x - 10)$$

Step 2. Next notice that $a = 3$, $b = -1$, and $c = -10$ in this trinomial. Find two numbers whose product is $a \cdot c$ or $3(-10) = -30$ and whose sum is b, -1. The numbers are -6 and 5.

Step 3. $3x^2 - x - 10 = 3x^2 - 6x + 5x - 10$

Step 4.
$$= 3x(x - 2) + 5(x - 2)$$
$$= (x - 2)(3x + 5)$$

The factored form of $6x^2 - 2x - 20 = 2(x - 2)(3x + 5)$.

Don't forget to include the common factor of 2.

Answers

1. a. $(x + 4)(3x + 2)$, **b.** $(4x + 5)(3x + 1)$,
2. a. $y(2x + 1)(3x - 5)$,
b. $2(5x - 1)(3x - 2)$

EXERCISE SET 5.4

A *Factor each polynomial by grouping. Notice that Step 3 has already been done in these exercises. See Examples 1 and 2.*

1. $x^2 + 3x + 2x + 6$

2. $x^2 + 5x + 3x + 15$

3. $x^2 - 4x + 7x - 28$

4. $x^2 - 6x + 2x - 12$

5. $y^2 + 8y - 2y - 16$

6. $z^2 + 10z - 7z - 70$

7. $3x^2 + 4x + 12x + 16$

8. $2x^2 + 5x + 14x + 35$

9. $8x^2 - 5x - 24x + 15$

10. $4x^2 - 9x - 32x + 72$

1. _____

2. _____

3. _____

4. _____

5. _____

6. _____

7. _____

8. _____

9. _____

10. _____

Problem Solving Notes

5.5 FACTORING PERFECT SQUARE TRINOMIALS AND THE DIFFERENCE OF TWO SQUARES

A RECOGNIZING PERFECT SQUARE TRINOMIALS

A trinomial that is the square of a binomial is called a **perfect square trinomial**. For example,

$$(x + 3)^2 = (x + 3)(x + 3)$$
$$= x^2 + 6x + 9$$

Thus $x^2 + 6x + 9$ is a perfect square trinomial.

In Chapter 4, we discovered special product formulas for squaring binomials, recognizing that

$$(a + b)^2 = a^2 + 2ab + b^2 \quad \text{and} \quad (a - b)^2 = a^2 - 2ab + b^2$$

Because multiplication and factoring are reverse processes, we can now use these special products to help us factor perfect square trinomials. If we reverse these equations, we have the following.

FACTORING PERFECT SQUARE TRINOMIALS

$$a^2 + 2ab + b^2 = (a + b)^2$$
$$a^2 - 2ab + b^2 = (a - b)^2$$

To use these equations to help us factor, we must first be able to recognize a perfect square trinomial. A trinomial is a perfect square when

1. Two terms, a^2 and b^2, are squares, and
2. another term is $2 \cdot a \cdot b$ or $-2 \cdot a \cdot b$. That is, this term is twice the product of a and b, or its opposite.

Example 1   Decide whether $x^2 + 8x + 16$ is a perfect square trinomial.

Solution: **1.** Two terms, x^2 and 16, are squares $(16 = 4^2)$.

   2. Twice the product of x and 4 is the other term of the trinomial.

   $$2 \cdot x \cdot 4 = 8x$$

   Thus, $x^2 + 8x + 16$ is a perfect square trinomial.

Example 2   Decide whether $4x^2 + 10x + 9$ is a perfect square trinomial.

Solution: **1.** Two terms, $4x^2$ and 9, are squares.

   $$4x^2 = (2x)^2 \quad \text{and} \quad 9 = 3^2$$

   2. Twice the product of $2x$ and 3 is not the other term of the trinomial.

   $$2 \cdot 2x \cdot 3 = 12x, \textit{ not } 10x$$

   The trinomial is *not* a perfect square trinomial.

Objectives

A Recognize perfect square trinomials.

B Factor perfect square trinomials.

C Factor the difference of two squares.

SSM   CD-ROM   Video 5.5

Practice Problem 1

Decide whether each trinomial is a perfect square trinomial.

a. $x^2 + 12x + 36$

b. $x^2 + 20x + 100$

Practice Problem 2

Decide whether each trinomial is a perfect square trinomial.

a. $9x^2 + 20x + 25$

b. $4x^2 + 8x + 11$

Answers

1. a. yes, **b.** yes, **2. a.** no, **b.** no

Practice Problem 3

Decide whether each trinomial is a perfect square trinomial.

a. $25x^2 - 10x + 1$

b. $9x^2 - 42x + 49$

Example 3   Decide whether $9x^2 - 12x + 4$ is a perfect square trinomial.

Solution:

1. Two terms, $9x^2$ and 4, are squares.

$$9x^2 = (3x)^2 \quad \text{and} \quad 4 = 2^2$$

2. Twice the product of $3x$ and 2 is the opposite of the other term of the trinomial.

$2 \cdot 3x \cdot 2 = 12x$, the opposite of $-12x$

Thus, $9x^2 - 12x + 4$ is a perfect square trinomial.

B FACTORING PERFECT SQUARE TRINOMIALS

Now that we can recognize perfect square trinomials, we are ready to factor them.

Practice Problem 4

Factor: $x^2 + 16x + 64$

Example 4   Factor: $x^2 + 12x + 36$

Solution:

$$x^2 + 12x + 36 = x^2 + 2 \cdot x \cdot 6 + 6^2 \qquad \begin{array}{l} 36 = 6^2 \text{ and} \\ 12x = 2 \cdot x \cdot 6 \end{array}$$

$$a^2 + 2 \cdot a \cdot b + b^2$$

$$= (x + 6)^2$$

$$(a + b)^2$$

Practice Problem 5

Factor: $9r^2 + 24rs + 16s^2$

Example 5   Factor: $25x^2 + 20xy + 4y^2$

Solution:

$$25x^2 + 20xy + 4y^2 = (5x)^2 + 2 \cdot 5x \cdot 2y + (2y)^2$$
$$= (5x + 2y)^2$$

Practice Problem 6

Factor: $9n^2 - 6n + 1$

Example 6   Factor: $4m^2 - 4m + 1$

Solution:

$$4m^2 - 4m + 1 = (2m)^2 - 2 \cdot 2m \cdot 1 + 1^2$$

$$a^2 - 2 \cdot a \cdot b + b^2$$

$$= (2m - 1)^2$$

$$(a - b)^2$$

Practice Problem 7

Factor: $9x^2 + 15x + 4$

Example 7   Factor: $25x^2 + 50x + 9$

Solution:

Notice that this trinomial is not a perfect square trinomial.

$$25x^2 = (5x)^2, \ 9 = 3^2$$
but

$$2 \cdot 5x \cdot 3 = 30x$$

and $30x$ is not the middle term $50x$.

Although $25x^2 + 50x + 9$ is not a perfect square trinomial, it is factorable. Using techniques we learned in Section 4.3, we find that

$$25x^2 + 50x + 9 = (5x + 9)(5x + 1)$$

Answers

3. a. yes, **b.** yes, **4.** $(x + 8)^2$,
5. $(3r + 4s)^2$, **6.** $(3n - 1)^2$,
7. $(3x + 1)(3x + 4)$

Example 8   Factor: $162x^3 - 144x^2 + 32x$

Solution: Don't forget to first look for a common factor. There is a greatest common factor of $2x$ in this trinomial.

$$162x^3 - 144x^2 + 32x = 2x(81x^2 - 72x + 16)$$
$$= 2x[(9x)^2 - 2 \cdot 9x \cdot 4 + 4^2]$$
$$= 2x(9x - 4)^2$$

Practice Problem 8

Factor: $12x^3 - 84x^2 + 147x$

C FACTORING THE DIFFERENCE OF TWO SQUARES

In Chapter 3, we discovered another special product, the product of the sum and difference of two terms a and b:

$$(a + b)(a - b) = a^2 - b^2$$

Reversing this equation gives us another factoring pattern, which we use to factor the difference of two squares.

> **FACTORING THE DIFFERENCE OF TWO SQUARES**
>
> $$a^2 - b^2 = (a + b)(a - b)$$

Let's practice using this pattern.

Examples   Factor each binomial.

9. $x^2 - 4 = x^2 - 2^2 = (x + 2)(x - 2)$
$$a^2 - b^2 = (a + b)(a - b)$$

10. $y^2 - 25 = y^2 - 5^2 = (y + 5)(y - 5)$

11. $y^2 - \dfrac{4}{9} = y^2 - \left(\dfrac{2}{3}\right)^2 = \left(y + \dfrac{2}{3}\right)\left(y - \dfrac{2}{3}\right)$

12. $x^2 + 4$

Note that the binomial $x^2 + 4$ is the *sum* of two squares since we can write $x^2 + 4$ as $x^2 + 2^2$. We might try to factor using $(x + 2)(x + 2)$ or $(x - 2)(x - 2)$. But when we multiply to check, we find that neither factoring is correct.

$$(x + 2)(x + 2) = x^2 + 4x + 4$$
$$(x - 2)(x - 2) = x^2 - 4x + 4$$

In both cases, the product is a trinomial, not the required binomial. In fact, $x^2 + 4$ is a prime polynomial.

Practice Problems 9–12

Factor each binomial.

9. $x^2 - 9$ 10. $a^2 - 16$

11. $c^2 - \dfrac{9}{25}$ 12. $s^2 + 9$

Answers

8. $3x(2x - 7)^2$,   **9.** $(x - 3)(x + 3)$,

10. $(a - 4)(a + 4)$,   **11.** $\left(c - \dfrac{3}{5}\right)\left(c + \dfrac{3}{5}\right)$,

12. prime polynomial

Practice Problems 13–15

Factor each difference of two squares.

13. $9s^2 - 1$
14. $16x^2 - 49y^2$
15. $p^4 - 81$

Examples   Factor each difference of two squares.

13. $4x^2 - 1 = (2x)^2 - 1^2 = (2x + 1)(2x - 1)$

14. $25a^2 - 9b^2 = (5a)^2 - (3b)^2 = (5a + 3b)(5a - 3b)$

15. $y^4 - 16 = (y^2)^2 - 4^2$
$= (y^2 + 4)(y^2 - 4)$ Factor the difference of two squares.
$= (y^2 + 4)(y + 2)(y - 2)$ Factor the difference of two squares.

HELPFUL HINTS

1. Don't forget to first see whether there's a greatest common factor (other than 1 or −1) that can be factored out.
2. Factor completely. In other words, check to see whether any factors can be factored further (as in Example 15).

Practice Problems 16–17

Factor each difference of two squares.

16. $9x^3 - 25x$

17. $48x^4 - 3$

Examples   Factor each difference of two squares.

16. $4x^3 - 49x = x(4x^2 - 49)$ Factor out the common factor x.
$= x[(2x)^2 - 7^2]$
$= x(2x + 7)(2x - 7)$ Factor the difference of two squares.

17. $162x^4 - 2 = 2(81x^4 - 1)$ Factor out the common factor, 2.
$= 2(9x^2 + 1)(9x^2 - 1)$ Factor the difference of two squares.
$= 2(9x^2 + 1)(3x + 1)(3x - 1)$ Factor the difference of two squares.

Answers

13. $(3s - 1)(3s + 1)$,
14. $(4x - 7y)(4x + 7y)$,
15. $(p^2 + 9)(p + 3)(p - 3)$,
16. $x(3x - 5)(3x + 5)$,
17. $3(4x^2 + 1)(2x + 1)(2x - 1)$

CALCULATOR EXPLORATION
GRAPHING

A graphing calculator is a convenient tool for evaluating an expression at a given replacement value. For example, let's evaluate $x^2 - 6x$ when $x = 2$. To do so, store the value 2 in the variable x and then enter and evaluate the algebraic expression.

```
2 → X
                2
X² − 6X
             −8
```

The value of $x^2 - 6x$ when $x = 2$ is -8. You may want to use this method for evaluating expressions as you explore the following.

We can use a graphing calculator to explore factoring patterns numerically. Use your calculator to evaluate $x^2 - 2x + 1$, $x^2 - 2x - 1$, and $(x - 1)^2$ for each value of x given in the table. What do you observe?

	$x^2 - 2x + 1$	$x^2 - 2x - 1$	$(x - 1)^2$
$x = 5$			
$x = -3$			
$x = 2.7$			
$x = -12.1$			
$x = 0$			

Notice in each case that $x^2 - 2x - 1 \neq (x - 1)^2$. Because for each x in the table the value of $x^2 - 2x + 1$ and the value of $(x - 1)^2$ are the same, we might guess that $x^2 - 2x + 1 = (x - 1)^2$. We can verify our guess algebraically with multiplication:

$$(x - 1)(x - 1) = x^2 - x - x + 1 = x^2 - 2x + 1$$

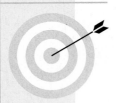

Focus On Mathematical Connections

GEOMETRY

Factoring polynomials can be visualized using areas of rectangles. To see this, let's first find the areas of the following squares and rectangles. (Recall that Area = Length · Width)

Area: $x \cdot x = x^2$ square units

Area: $1 \cdot x = x$ square units

Area: $1 \cdot 1 = 1$ square unit

To use these areas to visualize factoring the polynomial $x^2 + 3x + 2$, for example, use the shapes below to form a rectangle. The factored form is found by reading the length and the width of the rectangle as shown below.

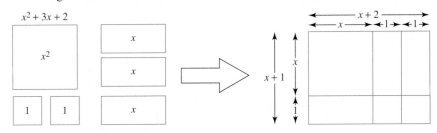

Thus, $x^2 + 3x + 2 = (x + 2)(x + 1)$.

Try using this method to visualize the factored form of each polynomial below.

GROUP ACTIVITY

Work in a group and use tiles to find the factored form of the polynomials below. (Tiles can be hand made from index cards.)

1. $x^2 + 6x + 5$

4. $x^2 + 4x + 3$

2. $x^2 + 5x + 6$

5. $x^2 + 6x + 9$

3. $x^2 + 5x + 4$

6. $x^2 + 4x + 4$

Name _____ **Section** _____ **Date** _____

MENTAL MATH ANSWERS

1. _____

2. _____

3. _____

4. _____

ANSWERS

1. _____

2. _____

3. _____

4. _____

5. _____

6. _____

7. _____

8. _____

9. _____

10. _____

11. _____

12. _____

13. _____

14. _____

15. _____

16. _____

17. _____

18. _____

19. _____

20. _____

21. _____

22. _____

23. _____

MENTAL MATH

State each number as a square.

1. 1 **2.** 25

State each term as a square.

3. $9x^2$ **4.** $16y^2$

EXERCISE SET 5.5

A *Determine whether each trinomial is a perfect square trinomial. See Examples 1 through 3.*

1. $x^2 + 16x + 64$ **2.** $x^2 + 22x + 121$ **3.** $y^2 + 5y + 25$

4. $y^2 + 4y + 16$

B *Factor each trinomial completely. See Examples 4 through 8.*

5. $x^2 + 22x + 121$ **6.** $x^2 + 18x + 81$ **7.** $x^2 - 16x + 64$

8. $x^2 - 12x + 36$ **9.** $16a^2 - 24a + 9$ **10.** $25x^2 + 20x + 4$

11. $x^4 + 4x^2 + 4$ **12.** $m^4 + 10m^2 + 25$

C *Factor each binomial completely. See Examples 9 through 17.*

13. $x^2 - 4$ **14.** $x^2 - 36$ **15.** $81 - p^2$ **16.** $100 - t^2$

17. $4r^2 - 1$ **18.** $9t^2 - 1$ **19.** $n^4 - 16$ **20.** $x^3y - 4xy^3$

21. $16 - a^2b^2$ **22.** $x^2 - \dfrac{1}{4}$ **23.** $y^2 - \dfrac{1}{16}$

Problem Solving Notes

5.6   SOLVING QUADRATIC EQUATIONS BY FACTORING

In this section, we introduce a new type of equation—the **quadratic equation**.

QUADRATIC EQUATION

A quadratic equation is one that can be written in the form

$$ax^2 + bx + c = 0$$

where a, b, and c are real numbers and $a \neq 0$.

Objectives

A Solve quadratic equations by factoring.

B Solve equations with degree greater than 2 by factoring.

SSM   CD-ROM   Video
5.6

Some examples of quadratic equations are

$$3x^2 + 5x + 6 = 0 \qquad x^2 = 9 \qquad y^2 + y = 1$$

The form $ax^2 + bx + c = 0$ is called the **standard form** of a quadratic equation. The quadratic equation $3x^2 + 5x + 6 = 0$ is the only equation above that is in standard form.

   Quadratic equations model many real-life situations. For example, let's suppose an object is dropped from the top of a 256-foot cliff and we want to know how long before the object strikes the ground. The answer to this question is found by solving the quadratic equation $-16t^2 + 256 = 0$. (See Example 1 in Section 5.7.)

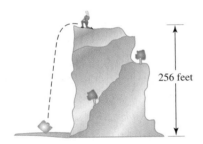

256 feet

A SOLVING QUADRATIC EQUATIONS BY FACTORING

Some quadratic equations can be solved by making use of factoring and the **zero factor property**.

ZERO FACTOR PROPERTY

If a and b are real numbers and if $ab = 0$, then $a = 0$, or $b = 0$.

In other words, if the product of two numbers is 0, then at least one of the numbers must be 0.

Example 1   Solve: $(x - 3)(x + 1) = 0$

   Solution: If this equation is to be a true statement, then either the factor $x - 3$ must be 0 or the factor $x + 1$ must be 0. In other words, either

$$x - 3 = 0 \qquad \text{or} \qquad x + 1 = 0$$

If we solve these two linear equations, we have

$$x = 3 \qquad \text{or} \qquad x = -1$$

Practice Problem 1

Solve: $(x - 7)(x + 2) = 0$

Answer

1. 7 and -2

Thus, 3 and -1 are both solutions of the equation $(x - 3)(x + 1) = 0$. To check, we replace x with 3 in the original equation. Then we replace x with -1 in the original equation.

Check:

$$(x - 3)(x + 1) = 0 \qquad\qquad (x - 3)(x + 1) = 0$$
$$(3 - 3)(3 + 1) = 0 \quad \text{Replace } x \text{ with 3.} \qquad (-1 - 3)(-1 + 1) = 0 \quad \text{Replace } x \text{ with } -1.$$
$$0(4) = 0 \quad \text{True.} \qquad\qquad (-4)(0) = 0 \quad \text{True.}$$

The solutions are 3 and -1.

HELPFUL HINT

The zero factor property says that *if a product is 0, then a factor is 0.*

If $a \cdot b = 0$, then $a = 0$ or $b = 0$.
If $x(x + 5) = 0$, then $x = 0$ or $x + 5 = 0$.
If $(x + 7)(2x - 3) = 0$, then $x + 7 = 0$ or $2x - 3 = 0$.

Use this property only when the product is 0. For example, if $a \cdot b = 8$, we do not know the value of a or b. The values may be $a = 2, b = 4$ or $a = 8, b = 1$, or any other two numbers whose product is 8.

Practice Problem 2

Solve: $(x - 10)(3x + 1) = 0$

Example 2 Solve: $(x - 5)(2x + 7) = 0$

Solution: The product is 0. By the zero factor property, this is true only when a factor is 0. To solve, we set each factor equal to 0 and solve the resulting linear equations.

$$(x - 5)(2x + 7) = 0$$
$$x - 5 = 0 \qquad \text{or} \qquad 2x + 7 = 0$$
$$x = 5 \qquad\qquad\qquad 2x = -7$$
$$x = -\frac{7}{2}$$

Check: Let $x = 5$.
$$(x - 5)(2x + 7) = 0$$
$$(5 - 5)(2 \cdot 5 + 7) \stackrel{?}{=} 0 \quad \text{Replace } x \text{ with 5.}$$
$$0 \cdot 17 \stackrel{?}{=} 0$$
$$0 = 0 \quad \text{True.}$$

Let $x = -\frac{7}{2}$.
$$(x - 5)(2x + 7) = 0$$
$$\left(-\frac{7}{2} - 5\right)\left(2\left(-\frac{7}{2}\right) + 7\right) \stackrel{?}{=} 0 \quad \text{Replace } x \text{ with } -\frac{7}{2}.$$
$$\left(-\frac{17}{2}\right)(-7 + 7) \stackrel{?}{=} 0$$
$$\left(-\frac{17}{2}\right) \cdot 0 \stackrel{?}{=} 0$$
$$0 = 0 \quad \text{True.}$$

The solutions are 5 and $-\frac{7}{2}$.

Example 3

Solve: $x(5x - 2) = 0$

Solution:

$$x(5x - 2) = 0$$

$x = 0$ or $5x - 2 = 0$ Use the zero factor property.

$$5x = 2$$

$$x = \frac{2}{5}$$

Check these solutions in the original equation. The solutions are 0 and $\frac{2}{5}$.

Practice Problem 3

Solve each equation.

a. $y(y + 3) = 0$

b. $x(4x - 3) = 0$

Example 4

Solve: $x^2 - 9x - 22 = 0$

Solution:

One side of the equation is 0. However, to use the zero factor property, one side of the equation must be 0 *and* the other side must be written as a product (must be factored). Thus, we must first factor this polynomial.

$$x^2 - 9x - 22 = 0$$
$$(x - 11)(x + 2) = 0 \quad \text{Factor.}$$

Now we can apply the zero factor property.

$x - 11 = 0$ or $x + 2 = 0$

$x = 11$ $x = -2$

Check:

Let $x = 11$.

$$x^2 - 9x - 22 = 0$$
$$11^2 - 9 \cdot 11 - 22 \stackrel{?}{=} 0$$
$$121 - 99 - 22 \stackrel{?}{=} 0$$
$$22 - 22 \stackrel{?}{=} 0$$
$$0 = 0 \quad \text{True.}$$

Let $x = -2$.

$$x^2 - 9x - 22 = 0$$
$$(-2)^2 - 9(-2) - 22 \stackrel{?}{=} 0$$
$$4 + 18 - 22 \stackrel{?}{=} 0$$
$$22 - 22 \stackrel{?}{=} 0$$
$$0 = 0 \quad \text{True.}$$

The solutions are 11 and -2.

Practice Problem 4

Solve: $x^2 - 3x - 18 = 0$

Example 5

Solve: $x^2 - 9x = -20$

Solution:

First we rewrite the equation in standard form so that one side is 0. Then we factor the polynomial.

$$x^2 - 9x = -20$$
$$x^2 - 9x + 20 = 0 \quad \text{Write in standard form by adding 20 to both sides.}$$
$$(x - 4)(x - 5) = 0 \quad \text{Factor.}$$

Next we use the zero factor property and set each factor equal to 0.

$x - 4 = 0$ or $x - 5 = 0$ Set each factor equal to 0.

$x = 4$ $x = 5$ Solve.

Check: Check these solutions in the original equation. The solutions are 4 and 5.

Practice Problem 5

Solve: $x^2 - 14x = -24$

Answers

3. a. 0 and -3,   **b.** 0 and $\frac{3}{4}$,   **4.** 6 and -3,

5. 12 and 2

The following steps may be used to solve a quadratic equation by factoring.

TO SOLVE QUADRATIC EQUATIONS BY FACTORING

Step 1. Write the equation in standard form so that one side of the equation is 0.

Step 2. Factor the quadratic equation completely.

Step 3. Set each factor containing a variable equal to 0.

Step 4. Solve the resulting equations.

Step 5. Check each solution in the original equation.

Since it is not always possible to factor a quadratic polynomial, not all quadratic equations can be solved by factoring. Other methods of solving quadratic equations are presented in Chapter 10.

Practice Problem 6

Solve each equation.

a. $x(x - 4) = 5$

b. $x(3x + 7) = 6$

Example 6   Solve: $x(2x - 7) = 4$

Solution: First we write the equation in standard form; then we factor.

$$x(2x - 7) = 4$$
$$2x^2 - 7x = 4 \qquad \text{Multiply.}$$
$$2x^2 - 7x - 4 = 0 \qquad \text{Write in standard form.}$$
$$(2x + 1)(x - 4) = 0 \qquad \text{Factor.}$$

$2x + 1 = 0 \quad$ or $\quad x - 4 = 0 \qquad$ Set each factor equal to zero.

$2x = -1 \qquad\qquad x = 4 \qquad$ Solve.

$$x = -\frac{1}{2}$$

Check the solutions in the original equation. The solutions are $-\frac{1}{2}$ and 4.

HELPFUL HINT

To solve the equation $x(2x - 7) = 4$, do **not** set each factor equal to 4. Remember that to apply the zero factor property, one side of the equation must be 0 and the other side of the equation must be in factored form.

B SOLVING EQUATIONS WITH DEGREE GREATER THAN TWO BY FACTORING

Some equations with degree greater than 2 can be solved by factoring and then using the zero factor property.

Practice Problem 7

Solve: $2x^3 - 18x = 0$

Example 7   Solve: $3x^3 - 12x = 0$

Solution: To factor the left side of the equation, we begin by factoring out the greatest common factor, $3x$.

Answers

6. a. 5 and -1, **b.** $\frac{2}{3}$ and -3, **7.** 0, 3, and -3

$$3x^3 - 12x = 0$$
$$3x(x^2 - 4) = 0 \qquad \text{Factor out the GCF, } 3x.$$
$$3x(x + 2)(x - 2) = 0 \qquad \begin{array}{l}\text{Factor } x^2 - 4, \text{ a differ-} \\ \text{ence of two squares.}\end{array}$$

$3x = 0 \qquad \text{or} \qquad x + 2 = 0 \qquad \text{or} \qquad x - 2 = 0 \qquad \begin{array}{l}\text{Set each factor equal} \\ \text{to 0.}\end{array}$

$x = 0 \qquad\qquad\qquad x = -2 \qquad\qquad\qquad x = 2 \qquad \text{Solve.}$

Thus, the equation $3x^3 - 12x = 0$ has three solutions: 0, −2, and 2.

Check: Replace x with each solution in the original equation.

Let $x = 0$.   Let $x = -2$.   Let $x = 2$.

$3(0)^3 - 12(0) \stackrel{?}{=} 0 \quad 3(-2)^3 - 12(-2) \stackrel{?}{=} 0 \quad 3(2)^3 - 12(2) \stackrel{?}{=} 0$

$\qquad\qquad 0 = 0 \qquad\qquad 3(-8) + 24 \stackrel{?}{=} 0 \qquad\qquad 3(8) - 24 \stackrel{?}{=} 0$

$\qquad\qquad \text{True.} \qquad\qquad\qquad\quad 0 = 0 \ \text{True.} \qquad\qquad\quad 0 = 0 \ \text{True.}$

The solutions are 0, −2, and 2. ▬▬▬

Example 8   Solve: $(5x - 1)(2x^2 + 15x + 18) = 0$

Solution:

$(5x - 1)(2x^2 + 15x + 18) = 0$
$(5x - 1)(2x + 3)(x + 6) = 0 \qquad \text{Factor the trinomial.}$

$5x - 1 = 0 \quad \text{or} \quad 2x + 3 = 0 \quad \text{or} \quad x + 6 = 0 \qquad \begin{array}{l}\text{Set each factor} \\ \text{equal to 0.}\end{array}$

$\qquad 5x = 1 \qquad\qquad 2x = -3 \qquad\qquad x = -6 \quad \text{Solve.}$

$\qquad x = \dfrac{1}{5} \qquad\qquad x = -\dfrac{3}{2}$

Check each solution in the original equation. The solutions are $\dfrac{1}{5}$, $-\dfrac{3}{2}$, and −6. ▬▬▬

Practice Problem 8

Solve: $(x + 3)(3x^2 - 20x - 7) = 0$

Answer

8. -3, $-\dfrac{1}{3}$, and 7

Focus On Mathematical Connections

NUMBER THEORY

By now, you have realized that being able to write a number as the product of prime numbers is very useful in the process of factoring polynomials. You probably also know at least a few numbers that are prime (such as 2, 3, and 5). But what about the other prime numbers? When we come across a number, how will we know if it is a prime number? Apparently, the ancient Greek mathematician Eratosthenes had a similar question because in the third century B.C. he devised a simple method for identifying primes. The method is called the Sieve of Eratosthenes because it "sifts out" the primes in a list of numbers. The Sieve of Eratosthenes is generally considered to be the most useful for identifying primes less than 1,000,000.

Here's how the sieve works: suppose you want to find the prime numbers in the first n natural numbers. Write the numbers, in order, from 2 to n. We know that 2 is prime, so circle it. Now, cross out each number greater than 2 that is a multiple of 2. Consider the next number in the list that is not crossed out. This number is 3, which we know is prime. Circle 3 and then cross out all multiples of 3 in the remainder of the list. Continue considering each uncircled number in the list. Once you have reached the largest prime number less than or equal to $\sqrt{n}$, and you have eliminated all its multiples in the remainder of the list, you can stop. Circle all the numbers left in the list that have not yet been circled. Now all of the circled numbers in the list are prime numbers. The list below demonstrates the Sieve of Eratosthenes on the numbers 2 through 30. Because $\sqrt{30} \approx 5.477$, we need only check and eliminate the multiples of primes up to and including 5, which is the largest prime less than or equal to the square root of 30.

We can see that the prime numbers less than 30 are 2, 3, 5, 7, 11, 13, 17, 19, 23, and 29.

GROUP ACTIVITY

Work with your group to identify the prime numbers less than 300 using the Sieve of Eratosthenes. What is the largest prime number that you will need to check in this process?

Exercise Set 5.6

A *Solve each equation. See Examples 1 through 3.*

1. $(x - 2)(x + 1) = 0$　　　**2.** $(x + 3)(x + 2) = 0$　　　**3.** $(x - 6)(x - 7) = 0$

4. $(x + 4)(x - 10) = 0$　　　**5.** $(x + 9)(x + 17) = 0$

B *Solve each equation. See Examples 4 through 8.*

6. $x^2 - 13x + 36 = 0$　　　**7.** $x^2 + 2x - 63 = 0$　　　**8.** $x^2 + 2x - 8 = 0$

9. $x^2 - 5x + 6 = 0$　　　**10.** $x^2 - 7x = 0$　　　**11.** $x^2 - 3x = 0$

12. $x^2 + 20x = 0$　　　**13.** $x^2 + 15x = 0$　　　**14.** $x^2 = 16$

15. $x^2 = 9$　　　**16.** $x^2 - 4x = 32$　　　**17.** $x^2 - 5x = 24$

18. $x(3x - 1) = 14$　　　**19.** $x(4x - 11) = 3$　　　**20.** $3x^2 + 19x - 72 = 0$

Answers

1. _____

2. _____

3. _____

4. _____

5. _____

6. _____

7. _____

8. _____

9. _____

10. _____

11. _____

12. _____

13. _____

14. _____

15. _____

16. _____

17. _____

18. _____

19. _____

20. _____

Problem Solving Notes

5.7 QUADRATIC EQUATIONS AND PROBLEM SOLVING

A SOLVING PROBLEMS MODELED BY QUADRATIC EQUATIONS

Some problems may be modeled by quadratic equations. To solve these problems, we use the same problem-solving steps that were introduced in Section 2.5. When solving these problems, keep in mind that a solution of an equation that models a problem may not be a solution to the problem. For example, a person's age or the length of a rectangle is always a positive number. Thus we discard solutions that do not make sense as solutions of the problem.

Example 1   **Finding Free-Fall Time**

For a TV commercial, a piece of luggage is dropped from a cliff 256 feet above the ground to show the durability of the luggage. Neglecting air resistance, the height h in feet of the luggage above the ground after t seconds is given by the quadratic equation

$$h = -16t^2 + 256$$

Find how long it takes for the luggage to hit the ground.

Solution: **1.** UNDERSTAND. Read and reread the problem. Then draw a picture of the problem.

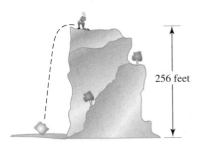

256 feet

The equation $h = -16t^2 + 256$ models the height of the falling luggage at time t. Familiarize yourself with this equation by finding the height of the luggage at $t = 1$ second and $t = 2$ seconds.

When $t = 1$ second, the height of the suitcase is
$h = -16(1)^2 + 256 = 240$ feet.
When $t = 2$ seconds, the height of the suitcase is
$h = -16(2)^2 + 256 = 192$ feet.

2. TRANSLATE. To find how long it takes the luggage to hit the ground, we want to know the value of t for which the height $h = 0$.

$$0 = -16t^2 + 256$$

3. SOLVE. We solve the quadratic equation by factoring.

$$0 = -16t^2 + 256$$
$$0 = -16(t^2 - 16)$$
$$0 = -16(t - 4)(t + 4)$$
$$t - 4 = 0 \qquad \text{or} \qquad t + 4 = 0$$
$$t = 4 \qquad\qquad\qquad t = -4$$

Objectives

A Solve problems that can be modeled by quadratic equations.

SSM   CD-ROM   Video
5.7

Practice Problem 1

An object is dropped from the roof of a 144-foot-tall building. Neglecting air resistance, the height h in feet of the object above ground after t seconds is given by the quadratic equation

$$h = -16t^2 + 144$$

Find how long it takes the object to hit the ground.

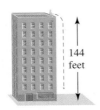

144 feet

Answer
1. 3 seconds

4. INTERPRET. Since the time t cannot be negative, the proposed solution is 4 seconds.

Check: Verify that the height of the luggage when t is 4 seconds is 0.

When $t = 4$ seconds, $h = -16(4)^2 + 256 = -256 + 256 = 0$ feet.

State: The solution checks and the luggage hits the ground 4 seconds after it is dropped.

Practice Problem 2

The square of a number minus twice the number is 63. Find the number.

Example 2 Finding a Number

The square of a number plus three times the number is 70. Find the number.

Solution: **1.** UNDERSTAND. Read and reread the problem. Suppose that the number is 5. The square of 5 is 5^2 or 25. Three times 5 is 15. Then $25 + 15 = 40$, not 70, so the number must be greater than 5. Remember, the purpose of proposing a number, such as 5, is to better understand the problem. Now that we do, we will let $x =$ the number.

2. TRANSLATE.

the square of a number	plus	three times the number	is	70
↓	↓	↓	↓	↓
x^2	$+$	$3x$	$=$	70

3. SOLVE.

$$x^2 + 3x = 70$$
$$x^2 + 3x - 70 = 0 \qquad \text{Subtract 70 from both sides.}$$
$$(x + 10)(x - 7) = 0 \qquad \text{Factor.}$$
$$x + 10 = 0 \quad \text{or} \quad x - 7 = 0 \qquad \text{Set each factor equal to 0.}$$
$$x = -10 \qquad\qquad x = 7 \qquad \text{Solve.}$$

4. INTERPRET.

Check: The square of -10 is $(-10)^2$, or 100. Three times -10 is $3(-10)$ or -30. Then $100 + (-30) = 70$, the correct sum, so -10 checks.

The square of 7 is 7^2 or 49. Three times 7 is $3(7)$, or 21. Then $49 + 21 = 70$, the correct sum, so 7 checks.

State: There are two numbers. They are -10 and 7.

Practice Problem 3

The length of a rectangle is 5 feet more than its width. The area of the rectangle is 176 square feet. Find the length and the width of the rectangle.

Example 3 Finding the Dimensions of a Sail

The height of a triangular sail is 2 meters less than twice the length of the base. If the sail has an area of 30 square meters, find the length of its base and the height.

Solution: **1.** UNDERSTAND. Read and reread the problem. Since we are finding the length of the base and the height, we let

$x =$ the length of the base

Answers

2. 9 and -7, **3.** length $= 16$ ft; width $= 11$ ft

and since the height is 2 meters less than twice the base,

$2x - 2 =$ the height

An illustration is shown to the left.

Height $= 2x - 2$

Base $= x$

2. TRANSLATE. We are given that the area of the triangle is 30 square meters, so we use the formula for area of a triangle.

$$\begin{array}{ccccccc} \text{area of} \\ \text{triangle} & = & \dfrac{1}{2} & \cdot & \text{base} & \cdot & \text{height} \\ \downarrow & & \downarrow & & \downarrow & & \downarrow \\ 30 & = & \dfrac{1}{2} & \cdot & x & \cdot & (2x - 2) \end{array}$$

3. SOLVE. Now we solve the quadratic equation.

$$30 = \frac{1}{2}x(2x - 2)$$

$30 = x^2 - x$ Multiply.

$x^2 - x - 30 = 0$ Write in standard form.

$(x - 6)(x + 5) = 0$ Factor.

$x - 6 = 0$ or $x + 5 = 0$ Set each factor equal to 0.

$x = 6$ $x = -5$

4. INTERPRET. Since x represents the length of the base, we discard the solution -5. The base of a triangle cannot be negative. The base is then 6 feet and the height is $2(6) - 2 = 10$ feet.

Check: To check this problem, we recall that $\dfrac{1}{2}$ base $\cdot$ height $=$ area, or

$$\frac{1}{2}(6)(10) = 30 \quad \text{The required area}$$

State: The base of the triangular sail is 6 meters and the height is 10 meters.   ▬▬▬

The next example makes use of the **Pythagorean theorem** and consecutive integers. Before we review this theorem, recall that a **right triangle** is a triangle that contains a 90° or right angle. The **hypotenuse** of a right triangle is the side opposite the right angle and is the longest side of the triangle. The **legs** of a right triangle are the other sides of the triangle.

PYTHAGOREAN THEOREM

In a right triangle, the sum of the squares of the lengths of the two legs is equal to the square of the length of the hypotenuse.

$$(\text{leg})^2 + (\text{leg})^2 = (\text{hypotenuse})^2 \quad \text{or} \quad a^2 + b^2 = c^2$$

Leg b Hypotenuse c

Leg a

Study the following diagrams for a review of consecutive integers.

Consecutive integers: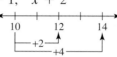

If x is the first integer: x, $x + 1$, $x + 2$

Consecutive even integers: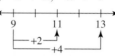

If x is the first even integer: x, $x + 2$, $x + 4$

Consecutive odd integers:

If x is the first odd integer: x, $x + 2$, $x + 4$

Practice Problem 4

Solve.

a. Find two consecutive odd integers whose product is 23 more than their sum.

b. The length of one leg of a right triangle is 7 meters less than the length of the other leg. The length of the hypotenuse is 13 meters. Find the lengths of the legs.

Example 4   Finding the Dimensions of a Triangle

Find the lengths of the sides of a right triangle if the lengths can be expressed as three consecutive even integers.

Solution:

1. UNDERSTAND. Read and reread the problem. Let's suppose that the length of one leg of the right triangle is 4 units. Then the other leg is the next even integer, or 6 units, and the hypotenuse of the triangle is the next even integer, or 8 units. Remember that the hypotenuse is the longest side. Let's see if a triangle with sides of these lengths forms a right triangle. To do this, we check to see whether the Pythagorean theorem holds true.

$$4^2 + 6^2 \stackrel{?}{=} 8^2$$

$$16 + 36 \stackrel{?}{=} 64$$

$$52 = 64 \quad \text{False.}$$

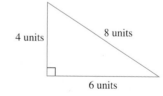

Our proposed numbers do not check, but we now have a better understanding of the problem.

We let x, $x + 2$, and $x + 4$ be three consecutive even integers. Since these integers represent lengths of the sides of a right triangle, we have

$$x = \text{one leg}$$
$$x + 2 = \text{other leg}$$
$$x + 4 = \text{hypotenuse (longest side)}$$

Answers

4. a. 5 and 7 or -5 and -3,   **b.** 5 meters, 12 meters

2. TRANSLATE. By the Pythagorean theorem, we have that

$$(\text{hypotenuse})^2 = (\text{leg})^2 + (\text{leg})^2$$
$$(x + 4)^2 = (x)^2 + (x + 2)^2$$

3. SOLVE. Now we solve the equation.

$$(x + 4)^2 = x^2 + (x + 2)^2$$

$x^2 + 8x + 16 = x^2 + x^2 + 4x + 4$ Multiply.

$x^2 + 8x + 16 = 2x^2 + 4x + 4$ Combine like terms.

$x^2 - 4x - 12 = 0$ Write in standard form.

$(x - 6)(x + 2) = 0$ Factor.

$x - 6 = 0$ or $x + 2 = 0$ Set each factor equal to 0.

$x = 6$ $x = -2$

4. INTERPRET. We discard $x = -2$ since length cannot be negative. If $x = 6$, then $x + 2 = 8$ and $x + 4 = 10$.

Check: Verify that $(\text{hypotenuse})^2 = (\text{leg})^2 + (\text{leg})^2$, or $10^2 = 6^2 + 8^2$, or $100 = 36 + 64$.

State: The sides of the right triangle have lengths 6 units, 8 units, and 10 units.

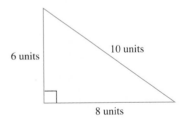

Focus On The Real World

Now that you have discovered a way to identify relatively small prime numbers with the Sieve of Eratosthenes, perhaps you are wondering whether there are very large primes. The answer is yes. Another ancient Greek mathematician, Euclid, proved that there are an infinite number of primes. Thus, there do exist huge numbers that are prime. Researchers call prime numbers with more than 1000 digits, *titanic primes*.

Knowing that very large prime numbers exist and finding them and proving that they are in fact prime are two very different things. David Slowinski, a research scientist at Silicon Graphics Inc.'s Cray Research unit and a discoverer or co-discoverer of seven titanic primes, compares the hunt for large primes with looking for a needle in a haystack. At the time that this book was written, the largest-known prime number, discovered November 13, 1996, had 420,921 digits which would fill up about 13.5 newspaper pages in standard-size type. The following newspaper article from the *San Jose Mercury News* describes how this prime number was found.

San Jose Mercury News
Published: Nov. 23, 1996

**"Move Over, Supercomputers"
Net-linked band of off-the shelf PC
users finds largest example of
Mersenne prime number**

By Dan Gillmor
Mercury News Computing Editor

In an undertaking previously reserved for powerful supercomputers, a worldwide band of personal computer users has found the largest known example of a special kind of prime number. Combining the Internet's global reach with hundreds of off-the-shelf PCs running Intel microprocessors, they divided an enormous problem into megabyte-sized tasks—and showed again how PCs are moving rapidly up computing's evolutionary scale.

Ten days ago in Paris, a PC operated by 29-year-old programmer Joel Armengaud found what is now the biggest-known Mersenne prime number. Mersenne primes are named after a 17th-century French monk, Father Marin Mersenne, who was fascinated by mathematics.

To test whether the nearly 421,000-digit number was such a prime, Armengaud was running a program written by Florida-based programmer George Woltman. Early this year, Woltman organized what he dubbed "The Great Internet Mersenne Prime Search," a hunt that has attracted more than 750 people from around the globe, some of whom have devoted more than one machine to the task.

A prime number is an integer greater than zero whose divisors are only itself and 1. (The number 2 is prime because it can only be divided evenly by 1 and 2, for example.) Mersenne primes take the form 2 to some power, minus 1—in other words, 2 multiplied by itself a certain number of times with 1 subtracted from the result.

The smallest Mersenne prime is 3, or 2 to the 2nd power ($2 \times 2 = 4$) minus 1. The next largest "regular" prime number is 5, and the next largest Mersenne prime is 7, or 2 to the 3rd power ($2 \times 2 \times 2 = 8$) minus 1.

The latest discovery—only the 35th known Mersenne prime—is 2 to the 1,398,269th power (2 multiplied by itself 1,398,269 times) minus 1. It's 420,921 digits long, or more than 150 pages of single-spaced text.

The previous largest-known Mersenne prime, discovered earlier this year, is 2 to the 1,257,787th power minus 1. It was found by David Slowinski and Paul Gage, computer scientists at Silicon Graphics Inc.'s Cray Research unit, using a Cray supercomputer.

While Armengaud, Woltman and others in the search are in it mainly for fun, finding more efficient ways to crunch huge numbers is more than an abstract exercise. Related techniques have helped people handle other important computing tasks—such as modeling complex weather patterns and exploring for oil. And cryptography, the art of scrambling messages to ensure privacy, relies on the difficulty of performing complex operations on extremely large numbers.

And the way the latest Mersenne prime was found represents another step in computing's evolution.

GROUP ACTIVITIES

1. The following World Wide Web site contains information on the current status of the largest-known prime numbers: http://www.utm.edu/research/primes/largest.html. Visit this site and report on the five largest-known prime numbers. Be sure to include information about the numbers themselves, who found them, when they were found, and how they were found (if possible). How many numbers larger than that found by Armengaud, Woltman, et. al. in November 1996 have been found?

2. Explore some of the links from the Web site listed in Activity 1 to find information on the history of the search for the largest-known primes (look for information on the largest known prime by year). Summarize the trends in the methods used to hunt for titanic primes. When is a prime number with over 1,000,000 digits expected to be discovered?

Name _____ **Section** _____ **Date** _____

EXERCISE SET 5.7

A *See Examples 1 through 4 for all exercises. Represent each given condition using a single variable, x.*

1. The length and width of a rectangle whose length is 4 centimeters more than its width

2. The length and width of a rectangle whose length is twice its width

3. Two consecutive odd integers

4. Two consecutive even integers

5. The base and height of a triangle whose height is one more than four times its base

6. The base and height of a trapezoid whose base is three less than five times its height

base

Use the information given to find the dimensions of each figure.

7.

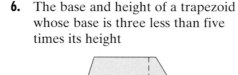
x

8. $x + 3$ ... $x - 2$

The *area* of the square is 121 square units. Find the length of its sides.

The *area* of the rectangle is 84 square inches. Find its length and width.

9.

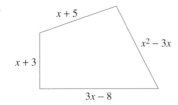
$x + 5$
$x^2 - 3x$
$x + 3$
$3x - 8$

10.

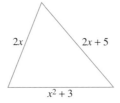
$2x$ $2x + 5$
$x^2 + 3$

The *perimeter* of the quadrilateral is 120 centimeters. Find the lengths of the sides.

The *perimeter* of the triangle is 85 feet. Find the lengths of its sides.

CHAPTER 5 HIGHLIGHTS

DEFINITIONS AND CONCEPTS	EXAMPLES

SECTION 5.1   THE GREATEST COMMON FACTOR

Factoring is the process of writing an expression as a product.

The GCF of a list of common variables raised to powers is the variable raised to the smallest exponent in the list.

The GCF of a list of terms is the product of all common factors.

Factor: $6 = 2 \cdot 3$

Factor: $x^2 + 5x + 6 = (x + 2)(x + 3)$

The GCF of z^5, z^3, and z^{10} is z^3.

Find the GCF of $8x^2y$, $10x^3y^2$, and $50x^2y^3$.

$$8x^2y = 2 \cdot 2 \cdot 2 \cdot x^2 \cdot y$$
$$10x^3y^2 = 2 \cdot 5 \cdot x^3 \cdot y^2$$
$$50x^2y^3 = 2 \cdot 5 \cdot 5 \cdot x^2 \cdot y^3$$
$$\text{GCF} = 2 \cdot x^2 \cdot y \quad \text{or} \quad 2x^2y$$

TO FACTOR BY GROUPING

Step 1. Group the terms into two groups of two terms.

Step 2. Factor out the GCF from each group.

Step 3. If there is a common binomial factor, factor it out.

Step 4. If not, rearrange the terms and try Steps 1–3 again.

Factor: $10ax + 15a - 6xy - 9y$

Step 1. $(10ax + 15a) + (-6xy - 9y)$

Step 2. $5a(2x + 3) - 3y(2x + 3)$

Step 3. $(2x + 3)(5a - 3y)$

SECTION 5.2   FACTORING TRINOMIALS OF THE FORM $x^2 + bx + c$

The sum of these numbers is b.

$$x^2 + bx + c = (x + \square)(x + \square)$$

The product of these numbers is c.

Factor: $x^2 + 7x + 12$

$$3 + 4 = 7 \qquad 3 \cdot 4 = 12$$

$$x^2 + 7x + 12 = (x + 3)(x + 4)$$

SECTION 5.3   FACTORING TRINOMIALS OF THE FORM $ax^2 + bx + c$

To factor $ax^2 + bx + c$, try various combinations of factors of ax^2 and c until a middle term of bx is obtained when checking.

Factor: $3x^2 + 14x - 5$

Factors of $3x^2$: $3x$, x

Factors of -5: $-1, 5$ and $1, -5$.

$$(3x - 1)(x + 5)$$

$$\begin{array}{r} -1x \\ + 15x \\ \hline 14x \end{array} \quad \textbf{Correct} \text{ middle term}$$

SECTION 5.4 FACTORING TRINOMIALS OF THE FORM $ax^2 + bx + c$ BY GROUPING

TO FACTOR $ax^2 + bx + c$ BY GROUPING

Step 1. Find two numbers whose product is $a \cdot c$ and whose sum is b.

Step 2. Rewrite bx, using the factors found in step 1.

Step 3. Factor by grouping.

Factor: $3x^2 + 14x - 5$

Step 1. Find two numbers whose product is $3 \cdot (-5)$ or -15 and whose sum is 14. They are 15 and -1.

Step 2. $3x^2 + 14x - 5$
$= 3x^2 + 15x - 1x - 5$

Step 3. $= 3x(x + 5) - 1(x + 5)$
$= (x + 5)(3x - 1)$

SECTION 5.5 FACTORING PERFECT SQUARE TRINOMIALS AND THE DIFFERENCE OF TWO SQUARES

A **perfect square trinomial** is a trinomial that is the square of some binomial.

PERFECT SQUARE TRINOMIAL = SQUARE OF BINOMIAL

$$x^2 + 4x + 4 = (x + 2)^2$$
$$25x^2 - 10x + 1 = (5x - 1)^2$$

Factoring Perfect Square Trinomials

$$a^2 + 2ab + b^2 = (a + b)^2$$
$$a^2 - 2ab + b^2 = (a - b)^2$$

Factor:

$x^2 + 6x + 9 =$
$x^2 + 2(x \cdot 3) + 3^2 = (x + 3)^2$
$4x^2 - 12x + 9 =$
$(2x)^2 - 2(2x \cdot 3) + 3^2 = (2x - 3)^2$

Difference of Two Squares

$$a^2 - b^2 = (a + b)(a - b)$$

Factor:

$$x^2 - 9 = x^2 - 3^2 = (x + 3)(x - 3)$$

SECTION 5.6 SOLVING QUADRATIC EQUATIONS BY FACTORING

A **quadratic equation** is an equation that can be written in the form $ax^2 + bx + c = 0$ with a not 0. The form $ax^2 + bx + c = 0$ is called the **standard form** of a quadratic equation.

Zero Factor Property
If a and b are real numbers and if $ab = 0$, then $a = 0$ or $b = 0$.

Quadratic Equation	Standard Form
$x^2 = 16$	$x^2 - 16 = 0$
$y = -2y^2 + 5$	$2y^2 + y - 5 = 0$

If $(x + 3)(x - 1) = 0$, then $x + 3 = 0$ or $x - 1 = 0$.

TO SOLVE QUADRATIC EQUATIONS BY FACTORING

Step 1. Write the equation in standard form so that one side of the equation is 0.

Step 2. Factor completely.

Step 3. Set each factor containing a variable equal to 0.

Step 4. Solve the resulting equations.

Step 5. Check in the original equation.

Solve: $3x^2 = 13x - 4$

Step 1. $3x^2 - 13x + 4 = 0$

Step 2. $(3x - 1)(x - 4) = 0$

Step 3. $3x - 1 = 0$ or $x - 4 = 0$

Step 4. $3x = 1$ $\qquad x = 4$
$x = \dfrac{1}{3}$

Step 5. Check both $\frac{1}{3}$ and 4 in the original equation.

SECTION 5.7   QUADRATIC EQUATIONS AND PROBLEM SOLVING

PROBLEM-SOLVING STEPS

A garden is in the shape of a rectangle whose length is two feet more than its width. If the area of the garden is 35 square feet, find its dimensions.

1. UNDERSTAND the problem.

1. Read and reread the problem. Guess a solution and check your guess. Draw a diagram.

Let x be the width of the rectangular garden. Then $x + 2$ is the length.

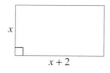

2. TRANSLATE.

2.

length	·	width	=	area
↓		↓		↓
$(x + 2)$	·	x	=	35

3. SOLVE.

3.
$$(x + 2)x = 35$$
$$x^2 + 2x - 35 = 0$$
$$(x - 5)(x + 7) = 0$$
$$x - 5 = 0 \quad \text{or} \quad x + 7 = 0$$
$$x = 5 \qquad\qquad x = -7$$

4. INTERPRET.

4. Discard the solution of -7 since x represents width.

Check: If x is 5 feet then $x + 2 = 5 + 2 = 7$ feet. The area of a rectangle whose width is 5 feet and whose length is 7 feet is (5 feet)(7 feet) or 35 square feet.

State: The garden is 5 feet by 7 feet.

Rational Expressions

In this chapter, we expand our knowledge of algebraic expressions to include algebraic fractions, called **rational expressions**. We explore the operations of addition, subtraction, multiplication, and division using principles similar to the principles for number fractions.

Rational expressions can be found in formulas for many real-world situations. Rainfall intensity describes the depth of rain that falls during a certain time period. According to the National Geographic Society, the rainiest location in the world is Mount Waialeale, Hawaii. Mount Waialeale receives an average of 460 inches of rain per year. Most of the United States only receives an average of 10 to 60 inches of rain per year.

Rainfall also varies in intensity from storm to storm. Most rainfall can be classified on a scale ranging from 0.10 inches per hour (light rain) to 0.30 inches per hour (heavy rain). However, the intensities on this scale are occasionally exceeded by particularly strong storms.

Problem Solving Notes

6.1   SIMPLIFYING RATIONAL EXPRESSIONS

A   EVALUATING RATIONAL EXPRESSIONS

A rational number is a number that can be written as a quotient of integers. A **rational expression** is also a quotient; it is a quotient of polynomials. Examples are

$$\frac{3y^3}{8}, \quad \frac{-4p}{p^3 + 2p + 1}, \quad \frac{5x^2 - 3x + 2}{3x + 7}$$

RATIONAL EXPRESSION

A rational expression is an expression that can be written in the form

$$\frac{P}{Q}$$

where P and Q are polynomials and $Q \neq 0$.

Rational expressions have different numerical values depending on what value replaces the variable.

Example 1   Find the numerical value of $\frac{x + 4}{2x - 3}$ for each replacement value.

　　　　a. $x = 5$　　　　　　**b.** $x = -2$

Solution:　　**a.** We replace each x in the expression with 5 and then simplify.

$$\frac{x + 4}{2x - 3} = \frac{5 + 4}{2(5) - 3} = \frac{9}{10 - 3} = \frac{9}{7}$$

　　　　b. We replace each x in the expression with -2 and then simplify.

$$\frac{x + 4}{2x - 3} = \frac{-2 + 4}{2(-2) - 3} = \frac{2}{-7} \quad \text{or} \quad -\frac{2}{7}$$

In the example above, we wrote $\frac{2}{-7}$ as $-\frac{2}{7}$. For a negative fraction such as $\frac{2}{-7}$, recall from Section 1.6 that

$$\frac{2}{-7} = \frac{-2}{7} = -\frac{2}{7}$$

In general, for any fraction

$$\frac{-a}{b} = \frac{a}{-b} = -\frac{a}{b}, \quad b \neq 0$$

This is also true for rational expressions. For example,

$$\frac{-(x + 2)}{\underbrace{x}} = \frac{x + 2}{-x} = -\frac{x + 2}{x}$$

↑
Notice the parentheses.

B   IDENTIFYING WHEN A RATIONAL EXPRESSION IS UNDEFINED

In the preceding box, notice that we wrote $b \neq 0$ for the denominator b. The denominator of a rational expression must not equal 0 since division by

Practice Problem 1

Find the value of $\frac{x - 3}{5x + 1}$ for each replacement value.

a. $x = 4$

b. $x = -3$

Answers

1. a. $\frac{1}{21}$,　**b.** $\frac{-6}{-14} = \frac{3}{7}$

0 is not defined. This means we must be careful when replacing the variable in a rational expression by a number. For example, suppose we replace x with 5 in the rational expression $\frac{2 + x}{x - 5}$. The expression becomes

$$\frac{2 + x}{x - 5} = \frac{2 + 5}{5 - 5} = \frac{7}{0}$$

But division by 0 is undefined. Therefore, in this expression we can allow x to be any real number *except* 5. **A rational expression is undefined for values that make the denominator 0.**

Practice Problem 2

Are there any values for x for which each rational expression is undefined?

a. $\frac{x}{x + 2}$ b. $\frac{x - 3}{x^2 + 5x + 4}$

c. $\frac{x^2 - 3x + 2}{5}$

Example 2 Are there any values for x for which each expression is undefined?

a. $\frac{x}{x - 3}$ b. $\frac{x^2 + 2}{x^2 - 3x + 2}$

c. $\frac{x^3 - 6x^2 - 10x}{3}$

Solution: To find values for which a rational expression is undefined, we find values that make the denominator 0.

a. The denominator of $\frac{x}{x - 3}$ is 0 when $x - 3 = 0$ or when $x = 3$. Thus, when $x = 3$, the expression $\frac{x}{x - 3}$ is undefined.

b. We set the denominator equal to 0.

$$x^2 - 3x + 2 = 0$$
$$(x - 2)(x - 1) = 0 \qquad \text{Factor.}$$
$$x - 2 = 0 \quad \text{or} \quad x - 1 = 0 \qquad \text{Set each factor equal to 0.}$$
$$x = 2 \qquad\qquad x = 1 \qquad \text{Solve.}$$

Thus, when $x = 2$ or $x = 1$, the denominator $x^2 - 3x + 2$ is 0. So the rational expression $\frac{x^2 + 2}{x^2 - 3x + 2}$ is undefined when $x = 2$ or when $x = 1$.

c. The denominator of $\frac{x^3 - 6x^2 - 10x}{3}$ is never 0, so there are no values of x for which this expression is undefined.

C SIMPLIFYING RATIONAL EXPRESSIONS

A fraction is said to be written in lowest terms or simplest form when the numerator and denominator have no common factors other than 1 (or -1). For example, the fraction $\frac{7}{10}$ is in lowest terms since the numerator and denominator have no common factors other than 1 (or -1).

The process of writing a rational expression in lowest terms or simplest form is called **simplifying** a rational expression. The following **fundamental principle of rational expressions** is used to simplify a rational expression.

> FUNDAMENTAL PRINCIPLE OF RATIONAL EXPRESSIONS
>
> If $\frac{P}{Q}$ is a rational expression and R is a nonzero polynomial, then
> $$\frac{PR}{QR} = \frac{P}{Q}$$

Answers

2. a. $x = -2$,   **b.** $x = -4, x = -1$,   **c.** no

Simplifying a rational expression is similar to simplifying a fraction.

Simplify: $\dfrac{15}{20}$

$\dfrac{15}{20} = \dfrac{3 \cdot 5}{2 \cdot 2 \cdot 5}$ Factor the numerator and the denominator.

$\qquad = \dfrac{3 \cdot 5}{2 \cdot 2 \cdot 5}$ Look for common factors.

$\qquad = \dfrac{3}{2 \cdot 2} = \dfrac{3}{4}$ Apply the fundamental principle.

Simplify: $\dfrac{x^2 - 9}{x^2 + x - 6}$

$\dfrac{x^2 - 9}{x^2 + x - 6} = \dfrac{(x - 3)(x + 3)}{(x - 2)(x + 3)}$ Factor the numerator and the denominator.

$\qquad = \dfrac{(x - 3)(x + 3)}{(x - 2)(x + 3)}$ Look for common factors.

$\qquad = \dfrac{x - 3}{x - 2}$ Apply the fundamental principle.

Thus, the rational expression $\dfrac{x^2 - 9}{x^2 + x - 6}$ has the same value as the rational expression $\dfrac{x - 3}{x - 2}$ for all values of x except 2 and -3. (Remember that when x is 2, the denominator of both rational expressions is 0 and when x is -3, the original rational expression has a denominator of 0.)

As we simplify rational expressions, we will assume that the simplified rational expression is equal to the original rational expression for all real numbers except those for which either denominator is 0. The following steps may be used to simplify rational expressions.

To Simplify a Rational Expression

Step 1. Completely factor the numerator and denominator.

Step 2. Apply the fundamental principle of rational expressions to divide out common factors.

Example 3   Simplify: $\dfrac{5x - 5}{x^3 - x^2}$

Solution: To begin, we factor the numerator and denominator if possible. Then we apply the fundamental principle.

$$\dfrac{5x - 5}{x^3 - x^2} = \dfrac{5(x - 1)}{x^2(x - 1)} = \dfrac{5}{x^2}$$

Example 4   Simplify: $\dfrac{x^2 + 8x + 7}{x^2 - 4x - 5}$

Solution: We factor the numerator and denominator and then apply the fundamental principle.

$$\dfrac{x^2 + 8x + 7}{x^2 - 4x - 5} = \dfrac{(x + 7)(x + 1)}{(x - 5)(x + 1)} = \dfrac{x + 7}{x - 5}$$

Practice Problem 3

Simplify: $\dfrac{x^4 + x^3}{5x + 5}$

Practice Problem 4

Simplify: $\dfrac{x^2 + 11x + 18}{x^2 + x - 2}$

Answers

3. $\dfrac{x^3}{5}$,   **4.** $\dfrac{x + 9}{x - 1}$

Practice Problem 5

Simplify: $\dfrac{x^2 + 10x + 25}{x^2 + 5x}$

Example 5   Simplify: $\dfrac{x^2 + 4x + 4}{x^2 + 2x}$

Solution: We factor the numerator and denominator and then apply the fundamental principle.

$$\frac{x^2 + 4x + 4}{x^2 + 2x} = \frac{(x + 2)(x + 2)}{x(x + 2)} = \frac{x + 2}{x}$$

HELPFUL HINT

When simplifying a rational expression, the fundamental principle apples to **common *factors***, **not common *terms***.

$$\frac{x \cdot (x + 2)}{x \cdot x} = \frac{x + 2}{x} \qquad\qquad \frac{x + 2}{x}$$

Common factors. These can be divided out.   Common terms. Fundamental principle does not apply. This is in simplest form.

TRY THE CONCEPT CHECK IN THE MARGIN.

Practice Problem 6

Simplify: $\dfrac{x + 5}{x^2 - 25}$

Example 6   Simplify: $\dfrac{x + 9}{x^2 - 81}$

Solution: We factor and then apply the fundamental principle.

$$\frac{x + 9}{x^2 - 81} = \frac{x + 9}{(x + 9)(x - 9)} = \frac{1}{x - 9}$$

Practice Problem 7

Simplify each rational expression.

a. $\dfrac{x + 4}{4 + x}$ b. $\dfrac{x - 4}{4 - x}$

Example 7   Simplify each rational expression.

     a. $\dfrac{x + y}{y + x}$ **b.** $\dfrac{x - y}{y - x}$

Solution: **a.** The expression $\dfrac{x + y}{y + x}$ can be simplified by using the commutative property of addition to rewrite the denominator $y + x$ as $x + y$.

$$\frac{x + y}{y + x} = \frac{x + y}{x + y} = 1$$

b. The expression $\dfrac{x - y}{y - x}$ can be simplified by recognizing that $y - x$ and $x - y$ are opposites. In other words, $y - x = -1(x - y)$. We proceed as follows:

$$\frac{x - y}{y - x} = \frac{1 \cdot (x - y)}{(-1)(x - y)} = \frac{1}{-1} = -1$$

✓ CONCEPT CHECK

Recall that the fundamental principle applies to common factors only. Which of the following are *not* true? Explain why.

a. $\dfrac{3 - 1}{3 + 5} = -\dfrac{1}{5}$

b. $\dfrac{2x + 10}{2} = x + 5$

c. $\dfrac{37}{72} = \dfrac{3}{2}$

d. $\dfrac{2x + 3}{2} = x + 3$

Answers

5. $\dfrac{x + 5}{x}$,   **6.** $\dfrac{1}{x - 5}$,   **7. a.** 1,   **b.** -1

✓ Concept Check: a, c, d

EXERCISE SET 6.1

C *Simplify each expression. See Examples 3 through 7.*

1. $\dfrac{2}{8x + 16}$

2. $\dfrac{3}{9x + 6}$

3. $\dfrac{x - 2}{x^2 - 4}$

4. $\dfrac{x + 5}{x^2 - 25}$

5. $\dfrac{2x - 10}{3x - 30}$

6. $\dfrac{3x - 12}{4x - 16}$

7. $\dfrac{x + 7}{7 + x}$

8. $\dfrac{y + 9}{9 + y}$

9. $\dfrac{x - 7}{7 - x}$

10. $\dfrac{y - 9}{9 - y}$

11. $\dfrac{-5a - 5b}{a + b}$

12. $\dfrac{7x + 35}{x^2 + 5x}$

13. $\dfrac{x + 5}{x^2 - 4x - 45}$

14. $\dfrac{x - 3}{x^2 - 6x + 9}$

15. $\dfrac{5x^2 + 11x + 2}{x + 2}$

16. $\dfrac{12x^2 + 4x - 1}{2x + 1}$

17. $\dfrac{x + 7}{x^2 + 5x - 14}$

18. $\dfrac{x - 10}{x^2 - 17x + 70}$

19. $\dfrac{2x^2 + 3x - 2}{2x - 1}$

20. $\dfrac{4x^2 + 24x}{x + 6}$

1. _____

2. _____

3. _____

4. _____

5. _____

6. _____

7. _____

7. _____

9. _____

10. _____

11. _____

12. _____

13. _____

14. _____

15. _____

16. _____

17. _____

18. _____

19. _____

20. _____

Problem Solving Notes

6.2 MULTIPLYING AND DIVIDING RATIONAL EXPRESSIONS

A MULTIPLYING RATIONAL EXPRESSIONS

Just as simplifying rational expressions is similar to simplifying number fractions, multiplying and dividing rational expressions is similar to multiplying and dividing number fractions.

Multiply: $\dfrac{3}{5} \cdot \dfrac{10}{11}$ Multiply: $\dfrac{x - 3}{x + 5} \cdot \dfrac{2x + 10}{x^2 - 9}$

Multiply numerators and then multiply denominators.

$\dfrac{3}{5} \cdot \dfrac{10}{11} = \dfrac{3 \cdot 10}{5 \cdot 11}$ $\dfrac{x - 3}{x + 5} \cdot \dfrac{2x + 10}{x^2 - 9} = \dfrac{(x - 3) \cdot (2x + 10)}{(x + 5) \cdot (x^2 - 9)}$

Simplify by factoring numerators and denominators.

$= \dfrac{3 \cdot 2 \cdot 5}{5 \cdot 11}$ $= \dfrac{(x - 3) \cdot 2(x + 5)}{(x + 5)(x + 3)(x - 3)}$

Apply the fundamental principle.

$= \dfrac{3 \cdot 2}{11} \quad \text{or} \quad \dfrac{6}{11}$ $= \dfrac{2}{x + 3}$

MULTIPLYING RATIONAL EXPRESSIONS

If $\dfrac{P}{Q}$ and $\dfrac{R}{S}$ are rational expressions, then

$$\frac{P}{Q} \cdot \frac{R}{S} = \frac{PR}{QS}$$

To multiply rational expressions, multiply the numerators and then multiply the denominators.

Example 1   Multiply.

a. $\dfrac{25x}{2} \cdot \dfrac{1}{y^3}$ b. $\dfrac{-7x^2}{5y} \cdot \dfrac{3y^5}{14x^2}$

Solution: To multiply rational expressions, we first multiply the numerators and then multiply the denominators of both expressions. Then we write the product in lowest terms.

a. $\dfrac{25x}{2} \cdot \dfrac{1}{y^3} = \dfrac{25x \cdot 1}{2 \cdot y^3} = \dfrac{25x}{2y^3}$

The expression $\dfrac{25x}{2y^3}$ is in lowest terms.

b. $\dfrac{-7x^2}{5y} \cdot \dfrac{3y^5}{14x^2} = \dfrac{-7x^2 \cdot 3y^5}{5y \cdot 14x^2}$ Multiply.

The expression $\dfrac{-7x^2 \cdot 3y^5}{5y \cdot 14x^2}$ is not in lowest terms, so we factor the numerator and the denominator and apply the fundamental principle.

Objectives

A Multiply rational expressions.

B Divide rational expressions.

C Multiply and divide rational expressions.

D Converting between units of measure.

SSM   CD-ROM   Video
6.2

Practice Problem 1

Multiply.

a. $\dfrac{16y}{3} \cdot \dfrac{1}{x^2}$ b. $\dfrac{-5a^3}{3b^3} \cdot \dfrac{2b^2}{15a}$

Answers

1. a. $\dfrac{16y}{3x^2}$, **b.** $-\dfrac{2a^2}{9b}$

$$= \frac{-1 \cdot 7 \cdot 3 \cdot x^2 \cdot y \cdot y^4}{5 \cdot 2 \cdot 7 \cdot x^2 \cdot y}$$

$$= -\frac{3y^4}{10}$$

When multiplying rational expressions, it is usually best to first factor each numerator and denominator. This will help us when we apply the fundamental principle to write the product in lowest terms.

Example 2   Multiply: $\dfrac{x^2 + x}{3x} \cdot \dfrac{6}{5x + 5}$

Solution:

$$\frac{x^2 + x}{3x} \cdot \frac{6}{5x + 5} = \frac{x(x + 1)}{3x} \cdot \frac{2 \cdot 3}{5(x + 1)} \qquad \text{Factor numerators and denominators.}$$

$$= \frac{x(x + 1) \cdot 2 \cdot 3}{3x \cdot 5(x + 1)} \qquad \text{Multiply.}$$

$$= \frac{2}{5} \qquad \text{Apply the fundamental principle.}$$

The following steps may be used to multiply rational expressions.

> **TO MULTIPLY RATIONAL EXPRESSIONS**
>
> **Step 1.** Completely factor numerators and denominators.
> **Step 2.** Multiply numerators and multiply denominators.
> **Step 3.** Simplify or write the product in lowest terms by applying the fundamental principle to all common factors.

TRY THE CONCEPT CHECK IN THE MARGIN.

Example 3   Multiply: $\dfrac{3x + 3}{5x^2 - 5x} \cdot \dfrac{2x^2 + x - 3}{4x^2 - 9}$

Solution:

$$\frac{3x + 3}{5x^2 - 5x} \cdot \frac{2x^2 + x - 3}{4x^2 - 9} = \frac{3(x + 1)}{5x(x - 1)} \cdot \frac{(2x + 3)(x - 1)}{(2x - 3)(2x + 3)} \qquad \text{Factor.}$$

$$= \frac{3(x + 1)(2x + 3)(x - 1)}{5x(x - 1)(2x - 3)(2x + 3)} \qquad \text{Multiply.}$$

$$= \frac{3(x + 1)}{5x(2x - 3)} \qquad \text{Simplify.}$$

Practice Problem 2

Multiply: $\dfrac{6x + 6}{7} \cdot \dfrac{14}{x^2 - 1}$

✓ CONCEPT CHECK

Which of the following is a true statement?

a. $\dfrac{1}{3} \cdot \dfrac{1}{2} = \dfrac{1}{5}$ b. $\dfrac{2}{x} \cdot \dfrac{5}{x} = \dfrac{10}{x}$

c. $\dfrac{3}{x} \cdot \dfrac{1}{2} = \dfrac{3}{2x}$

d. $\dfrac{x}{7} \cdot \dfrac{x + 5}{4} = \dfrac{2x + 5}{28}$

Practice Problem 3

Multiply: $\dfrac{4x + 8}{7x^2 - 14x} \cdot \dfrac{3x^2 - 5x - 2}{9x^2 - 1}$

Answers

2. $\dfrac{12}{x - 1}$, **3.** $\dfrac{4(x + 2)}{7x(3x - 1)}$

✓ **Concept Check:** c

B DIVIDING RATIONAL EXPRESSIONS

We can divide by a rational expression in the same way we divide by a number fraction. Recall that to divide by a fraction, we multiply by its reciprocal.

> **HELPFUL HINT**
>
> Don't forget how to find reciprocals. The reciprocal of $\frac{a}{b}$ is $\frac{b}{a}$, $a \neq 0$, $b \neq 0$.

For example, to divide $\frac{3}{2}$ by $\frac{7}{8}$, we multiply $\frac{3}{2}$ by $\frac{8}{7}$.

$$\frac{3}{2} \div \frac{7}{8} = \frac{3}{2} \cdot \frac{8}{7} = \frac{3 \cdot 4 \cdot 2}{2 \cdot 7} = \frac{12}{7}$$

DIVIDING RATIONAL EXPRESSIONS

If $\frac{P}{Q}$ and $\frac{R}{S}$ are rational expressions and $\frac{R}{S}$ is not 0, then

$$\frac{P}{Q} \div \frac{R}{S} = \frac{P}{Q} \cdot \frac{S}{R} = \frac{PS}{QR}$$

To divide two rational expressions, multiply the first rational expression by the reciprocal of the second rational expression.

Example 4   Divide: $\dfrac{3x^3}{40} \div \dfrac{4x^3}{y^2}$

Solution:

$$\frac{3x^3}{40} \div \frac{4x^3}{y^2} = \frac{3x^3}{40} \cdot \frac{y^2}{4x^3} \quad \text{Multiply by the reciprocal of } \frac{4x^3}{y^2}.$$

$$= \frac{3}{160} \frac{x^3 y^2}{x^3}$$

$$= \frac{3y^2}{160} \quad \text{Simplify.}$$

Example 5   Divide: $\dfrac{(x-1)(x+2)}{10} \div \dfrac{2x+4}{5}$

Solution:

$$\frac{(x-1)(x+2)}{10} \div \frac{2x+4}{5} = \frac{(x-1)(x+2)}{10} \cdot \frac{5}{2x+4} \quad \text{Multiply by the reciprocal of } \frac{2x+4}{5}.$$

$$= \frac{(x-1)(x+2) \cdot 5}{5 \cdot 2 \cdot 2 \cdot (x+2)} \quad \text{Factor and multiply.}$$

$$= \frac{x-1}{4} \quad \text{Simplify.}$$

Practice Problem 4

Divide: $\dfrac{7x^2}{6} \div \dfrac{x}{2y}$

Practice Problem 5

Divide: $\dfrac{(2x+3)(x-4)}{6} \div \dfrac{3x-12}{2}$

Answers

4. $\dfrac{7xy}{3}$, **5.** $\dfrac{2x+3}{9}$

The following may be used to divide by a rational expression.

To Divide by a Rational Expression

Multiply by its reciprocal.

Practice Problem 6

Divide: $\dfrac{10x + 4}{x^2 - 4} \div \dfrac{5x^3 + 2x^2}{x + 2}$

Example 6

Divide: $\dfrac{6x + 2}{x^2 - 1} \div \dfrac{3x^2 + x}{x - 1}$

Solution:

$$\dfrac{6x + 2}{x^2 - 1} \div \dfrac{3x^2 + x}{x - 1} = \dfrac{6x + 2}{x^2 - 1} \cdot \dfrac{x - 1}{3x^2 + x} \qquad \text{Multiply by the reciprocal.}$$

$$= \dfrac{2(3x + 1)(x - 1)}{(x + 1)(x - 1) \cdot x(3x + 1)} \qquad \text{Factor and multiply.}$$

$$= \dfrac{2}{x(x + 1)} \qquad \text{Simplify.}$$

Practice Problem 7

Divide: $\dfrac{3x^2 - 10x + 8}{7x - 14} \div \dfrac{9x - 12}{21}$

Example 7

Divide: $\dfrac{2x^2 - 11x + 5}{5x - 25} \div \dfrac{4x - 2}{10}$

Solution:

$$\dfrac{2x^2 - 11x + 5}{5x - 25} \div \dfrac{4x - 2}{10} = \dfrac{2x^2 - 11x + 5}{5x - 25} \cdot \dfrac{10}{4x - 2} \qquad \text{Multiply by the reciprocal.}$$

$$= \dfrac{(2x - 1)(x - 5) \cdot 2 \cdot 5}{5(x - 5) \cdot 2(2x - 1)} \qquad \text{Factor and multiply.}$$

$$= \dfrac{1}{1} \text{ or } 1 \qquad \text{Simplify.}$$

C   Multiplying and Dividing Rational Expressions

Let's make sure that we understand the difference between multiplying and dividing rational expressions.

Rational Expressions	
Multiplication	Multiply the numerators and multiply the denominators.
Division	Multiply by the reciprocal of the divisor.

Practice Problem 8

Multiply or divide as indicated.

a. $\dfrac{x + 3}{x} \cdot \dfrac{7}{x + 3}$

b. $\dfrac{x + 3}{x} \div \dfrac{7}{x + 3}$

Example 8

Multiply or divide as indicated.

a. $\dfrac{x - 4}{5} \cdot \dfrac{x}{x - 4}$ b. $\dfrac{x - 4}{5} \div \dfrac{x}{x - 4}$

Solution:

a. $\dfrac{x - 4}{5} \cdot \dfrac{x}{x - 4} = \dfrac{(x - 4) \cdot x}{5 \cdot (x - 4)} = \dfrac{x}{5}$

b. $\dfrac{x - 4}{5} \div \dfrac{x}{x - 4} = \dfrac{x - 4}{5} \cdot \dfrac{x - 4}{x} = \dfrac{(x - 4)^2}{5x}$

Answers

6. $\dfrac{2}{x^2(x - 2)}$,   **7.** 1,   **8. a.** $\dfrac{7}{x}$,   **b.** $\dfrac{(x + 3)^2}{7x}$

D CONVERTING BETWEEN UNITS OF MEASURE

Now that we know how to multiply fractions and rational expressions, we can use this knowledge to help us convert between units of measure. To do so, we will use **unit fractions**. A unit fraction is a fraction that equals 1. For example, since 12 in. = 1 ft, we have the unit fractions

$$\frac{12 \text{ in.}}{1 \text{ ft}} = 1 \quad \text{and} \quad \frac{1 \text{ ft}}{12 \text{ in.}} = 1$$

Example 9 **Converting from Square Yards to Square Feet**

The largest casino in the world is the Foxwoods Resort Casino in Ledyard, CT. The gaming area for this casino is 21,444 *square yards*. Find the size of the gaming area in *square feet*. (*Source*: *The Guinness Book of Records*, 1996)

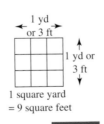

Solution: There are 9 square feet in 1 square yard.

Unit fraction

21,444 square yards = 21,444 s̶q̶.̶ ̶y̶d̶ · $\frac{9 \text{ sq. ft}}{1 \text{ s̶q̶.̶ ̶y̶d̶}}$

= 192,996 square feet

≈ 193,000 square feet

1 yd
or 3 ft
1 yd or
3 ft
1 square yard
= 9 square feet

HELPFUL HINT

When converting among units of measurement, if possible, write the unit fraction so that **the numerator contains units converting to** and **the denominator contains original units**.

Unit fraction

48 ft = $\frac{48 \text{ i̶n̶.̶}}{1} \cdot \frac{1 \text{ ft}}{12 \text{ i̶n̶.̶}}$ ← Units converting to
← Original units

= $\frac{48}{12}$ ft = 4 ft

Practice Problem 9

Convert 40,000 square feet to square yards.

Answer

9. 4444.44 sq. yards

Practice Problem 10

Carl Lewis holds the men's record for speed. He has been timed at 39.5 feet per second. Convert this to miles per hour. (*Source: The Guinness Book of Records*, 1996)

Example 10   Converting from Feet per Second to Miles per Hour

Florence Griffith Joyner holds the women's record for speed. She has been timed at 36 feet per second. Convert this to miles per hour. (*Source: The Guinness Book of Records*, 1996)

Solution: Recall that 1 mile = 5280 feet and 1 hour = 3600 seconds $(60 \cdot 60)$.

Unit fractions

$$36 \text{ feet/second} = \frac{36 \text{ feet}}{1 \text{ second}} \cdot \frac{3600 \text{ seconds}}{1 \text{ hour}} \cdot \frac{1 \text{ mile}}{5280 \text{ feet}}$$

$$= \frac{36 \cdot 3600}{5280} \text{ miles/hour}$$

$$\approx 24.5 \text{ miles/hour (rounded to the nearest tenth)}$$

Answer

10. 26.9 miles/hour

Name _____ Section _____ Date _____

MENTAL MATH

Find each product. See Example 1.

1. $\dfrac{2}{y} \cdot \dfrac{x}{3}$

2. $\dfrac{3x}{4} \cdot \dfrac{1}{y}$

3. $\dfrac{5}{7} \cdot \dfrac{y^2}{x^2}$

4. $\dfrac{x^5}{11} \cdot \dfrac{4}{z^3}$

5. $\dfrac{9}{x} \cdot \dfrac{x}{5}$

EXERCISE SET 6.2

ANSWERS

1. _____

2. _____

3. _____

4. _____

5. _____

6. _____

7. _____

8. _____

9. _____

10. _____

11. _____

12. _____

13. _____

14. _____

15. _____

16. _____

17. _____

18. _____

19. _____

20. _____

A *Find each product and simplify if possible. See Examples 1 through 3.*

1. $\dfrac{3x}{y^2} \cdot \dfrac{7y}{4x}$

2. $\dfrac{9x^2}{y} \cdot \dfrac{4y}{3x^2}$

3. $\dfrac{8x}{2} \cdot \dfrac{x^5}{4x^2}$

4. $\dfrac{6x^2}{10x^3} \cdot \dfrac{5x}{12}$

5. $-\dfrac{5a^2b}{30a^2b^2} \cdot b^3$

6. $\dfrac{x}{2x - 14} \cdot \dfrac{x^2 - 7x}{5}$

7. $\dfrac{4x - 24}{20x} \cdot \dfrac{5}{x - 6}$

8. $\dfrac{6x + 6}{5} \cdot \dfrac{10}{36x + 36}$

9. $\dfrac{x^2 + x}{8} \cdot \dfrac{16}{x + 1}$

10. $\dfrac{m^2 - n^2}{m + n} \cdot \dfrac{m}{m^2 - mn}$

B *Find each quotient and simplify. See Examples 4 through 7.*

11. $\dfrac{5x^7}{2x^5} \div \dfrac{10x}{4x^3}$

12. $\dfrac{9y^4}{6y} \div \dfrac{y^2}{3}$

13. $\dfrac{8x^2}{y^3} \div \dfrac{4x^2y^3}{6}$

14. $\dfrac{7a^2b}{3ab^2} \div \dfrac{21a^2b^2}{14ab}$

15. $\dfrac{(x - 6)(x + 4)}{4x} \div \dfrac{2x - 12}{8x^2}$

C *Multiply or divide as indicated. See Example 8.*

16. $\dfrac{5x - 10}{12} \div \dfrac{4x - 8}{8}$

17. $\dfrac{6x + 6}{5} \div \dfrac{3x + 3}{10}$

18. $\dfrac{x^2 + 5x}{8} \cdot \dfrac{9}{3x + 15}$

19. $\dfrac{3x^2 + 12x}{6} \cdot \dfrac{9}{2x + 8}$

20. $\dfrac{7}{6p^2 + q} \div \dfrac{14}{18p^2 + 3q}$

443

Problem Solving Notes

6.3 ADDING AND SUBTRACTING RATIONAL EXPRESSIONS WITH THE SAME DENOMINATOR AND LEAST COMMON DENOMINATOR

A ADDING AND SUBTRACTING RATIONAL EXPRESSIONS WITH THE SAME DENOMINATOR

Like multiplication and division, addition and subtraction of rational expressions is similar to addition and subtraction of rational numbers. In this section, we add and subtract rational expressions with a common denominator.

Add: $\frac{6}{5} + \frac{2}{5}$ | Add: $\frac{9}{x+2} + \frac{3}{x+2}$

Add the numerators and place the sum over the common denominator.

$$\frac{6}{5} + \frac{2}{5} = \frac{6+2}{5}$$ | $$\frac{9}{x+2} + \frac{3}{x+2} = \frac{9+3}{x+2}$$
$$= \frac{8}{5} \quad \text{Simplify.}$$ | $$= \frac{12}{x+2} \quad \text{Simplify.}$$

> **ADDING AND SUBTRACTING RATIONAL EXPRESSIONS WITH COMMON DENOMINATORS**
>
> If $\frac{P}{R}$ and $\frac{Q}{R}$ are rational expressions, then
>
> $$\frac{P}{R} + \frac{Q}{R} = \frac{P+Q}{R} \quad \text{and} \quad \frac{P}{R} - \frac{Q}{R} = \frac{P-Q}{R}$$
>
> To add or subtract rational expressions, add or subtract numerators and place the sum or difference over the common denominator.

Example 1   Add: $\frac{5m}{2n} + \frac{m}{2n}$

Solution: $\frac{5m}{2n} + \frac{m}{2n} = \frac{5m+m}{2n}$ Add the numerators.

$$= \frac{6m}{2n}$$ Simplify the numerator by combining like terms.

$$= \frac{3m}{n}$$ Simplify by applying the fundamental principle.

Example 2   Subtract: $\frac{2y}{2y-7} - \frac{7}{2y-7}$

Solution: $\frac{2y}{2y-7} - \frac{7}{2y-7} = \frac{2y-7}{2y-7}$ Subtract the numerators.

$$= \frac{1}{1} \text{ or } 1 \quad \text{Simplify.}$$

Example 3   Subtract: $\frac{3x^2 + 2x}{x-1} - \frac{10x-5}{x-1}$

Practice Problem 1

Add: $\frac{8x}{3y} + \frac{x}{3y}$

Practice Problem 2

Subtract: $\frac{3x}{3x-7} - \frac{7}{3x-7}$

Practice Problem 3

Subtract: $\frac{2x^2 + 5x}{x+2} - \frac{4x+6}{x+2}$

Answers

1. $\frac{3x}{y}$, **2.** 1, **3.** $2x - 3$

Solution:

$$\frac{3x^2 + 2x}{x - 1} - \frac{10x - 5}{x - 1} = \frac{3x^2 + 2x - (10x - 5)}{x - 1}$$ Subtract the numerators. Notice the parentheses.

$$= \frac{3x^2 + 2x - 10x + 5}{x - 1}$$ Use the distributive property.

$$= \frac{3x^2 - 8x + 5}{x - 1}$$ Combine like terms.

$$= \frac{(x - 1)(3x - 5)}{x - 1}$$ Factor.

$$= 3x - 5$$ Simplify.

HELPFUL HINT

Notice how the numerator $10x - 5$ has been subtracted in Example 3.

This $-$ sign applies to the entire numerator of $10x - 5$.   So parentheses are inserted here to indicate this.

$$\frac{3x^2 + 2x}{x - 1} - \frac{10x - 5}{x - 1} = \frac{3x^2 + 2x - (10x - 5)}{x - 1}$$

B   FINDING THE LEAST COMMON DENOMINATOR

Recall from Chapter R that to add and subtract fractions with different denominators, we first find a least common denominator (LCD). Then we write all fractions as equivalent fractions with the LCD.

For example, suppose we add $\frac{8}{3}$ and $\frac{2}{5}$. The LCD of denominators 3 and 5 is 15, since 15 is the smallest number that both 3 and 5 divide into evenly. So we rewrite each fraction so that its denominator is 15. (Notice how we apply the fundamental principle.)

$$\frac{8}{3} + \frac{2}{5} = \frac{8(5)}{3(5)} + \frac{2(3)}{5(3)} = \frac{40}{15} + \frac{6}{15} = \frac{40 + 6}{15} = \frac{46}{15}$$

To add or subtract rational expressions with different denominators, we also first find an LCD and then write all rational expressions as equivalent expressions with the LCD. The **least common denominator (LCD) of a list of rational expressions** is a polynomial of least degree whose factors include all the factors of the denominators in the list.

TO FIND THE LEAST COMMON DENOMINATOR (LCD)

Step 1. Factor each denominator completely.

Step 2. The least common denominator (LCD) is the product of all unique factors found in Step 1, each raised to a power equal to the greatest number of times that the factor appears in any one factored denominator.

Practice Problem 4

Find the LCD for each pair.

a. $\dfrac{2}{9}, \dfrac{7}{15}$ b. $\dfrac{5}{6x^3}, \dfrac{11}{8x^5}$

Answers

4. **a.** 45,   **b.** $24x^5$

Example 4   Find the LCD for each pair.

a. $\dfrac{1}{8}, \dfrac{3}{22}$ **b.** $\dfrac{7}{5x}, \dfrac{6}{15x^2}$

Solution: **a.** We start by finding the prime factorization of each denominator.

$$8 = 2 \cdot 2 \cdot 2 = 2^3 \quad \text{and} \quad 22 = 2 \cdot 11$$

Next we write the product of all the unique factors, each raised to a power equal to the greatest number of times that the factor appears.

> The greatest number of times that the factor 2 appears is 3.
> The greatest number of times that the factor 11 appears is 1.

$$\text{LCD} = 2^3 \cdot 11^1 = 8 \cdot 11 = 88$$

b. We factor each denominator.

$$5x = 5 \cdot x \quad \text{and} \quad 15x^2 = 3 \cdot 5 \cdot x^2$$

> The greatest number of times that the factor 5 appears is 1.
> The greatest number of times that the factor 3 appears is 1.
> The greatest number of times that the factor x appears is 2.

$$\text{LCD} = 3^1 \cdot 5^1 \cdot x^2 = 15x^2$$

Example 5   Find the LCD of $\dfrac{7x}{x+2}$ and $\dfrac{5x^2}{x-2}$.

Solution: The denominators $x + 2$ and $x - 2$ are completely factored already. The factor $x + 2$ appears once and the factor $x - 2$ appears once.

$$\text{LCD} = (x + 2)(x - 2)$$

Example 6   Find the LCD of $\dfrac{6m^2}{3m+15}$ and $\dfrac{2}{(m+5)^2}$.

Solution: We factor each denominator.

$$3m + 15 = 3(m + 5)$$
$$(m + 5)^2 = (m + 5)^2 \quad \text{\small This denominator is already factored.}$$

> The greatest number of times that the factor 3 appears is 1.
> The greatest number of times that the factor $m + 5$ appears *in any one denominator* is 2.

$$\text{LCD} = 3(m + 5)^2$$

TRY THE CONCEPT CHECK IN THE MARGIN.

Example 7   Find the LCD of $\dfrac{t-10}{t^2-t-6}$ and $\dfrac{t+5}{t^2+3t+2}$.

Solution:
$$t^2 - t - 6 = (t - 3)(t + 2)$$
$$t^2 + 3t + 2 = (t + 1)(t + 2)$$
$$\text{LCD} = (t - 3)(t + 2)(t + 1)$$

Practice Problem 5

Find the LCD of $\dfrac{3a}{a+5}$ and $\dfrac{7a}{a-5}$.

Practice Problem 6

Find the LCD of $\dfrac{7x^2}{(x-4)^2}$ and $\dfrac{5x}{3x-12}$.

✓ **CONCEPT CHECK**

Choose the correct LCD of $\dfrac{x}{(x+1)^2}$ and $\dfrac{5}{x+1}$.

a. $x + 1$ b. $(x + 1)^2$

c. $(x + 1)^3$ d. $5x(x + 1)^2$

Practice Problem 7

Find the LCD of $\dfrac{y+5}{y^2+2y-3}$ and $\dfrac{y+4}{y^2-3y+2}$.

Answers

5. $(a + 5)(a - 5)$,   **6.** $3(x - 4)^2$,
7. $(y + 3)(y - 2)(y - 1)$
✓ Concept Check: b

Practice Problem 8

Find the LCD of $\dfrac{6}{x-4}$ and $\dfrac{9}{4-x}$.

Example 8   Find the LCD of $\dfrac{2}{x-2}$ and $\dfrac{10}{2-x}$.

Solution: The denominators $x-2$ and $2-x$ are opposites. That is, $2-x = -1(x-2)$. We can use either $x-2$ or $2-x$ as the LCD.

$$\text{LCD} = x-2 \quad \text{or} \quad \text{LCD} = 2-x$$

C   WRITING EQUIVALENT RATIONAL EXPRESSIONS

Next we practice writing a rational expression as an equivalent rational expression with a given denominator. To do this, we apply the fundamental principle, which says that $\dfrac{PR}{QR} = \dfrac{P}{Q}$, or equivalently that $\dfrac{P}{Q} = \dfrac{PR}{QR}$. This can be seen by recalling that multiplying an expression by 1 produces an equivalent expression. In other words,

$$\frac{P}{Q} = \frac{P}{Q} \cdot 1 = \frac{P}{Q} \cdot \frac{R}{R} = \frac{PR}{QR}$$

Practice Problem 9

Write the rational expression as an equivalent rational expression with the given denominator.

$$\frac{2x}{5y} = \frac{}{20x^2y^2}$$

Example 9   Write the rational expression as an equivalent rational expression with the given denominator.

$$\frac{4b}{9a} = \frac{}{27a^2b}$$

Solution: We can ask ourselves: "What do we multiply $9a$ by to get $27a^2b$?" The answer is $3ab$, since $9a(3ab) = 27a^2b$. So we multiply the numerator and denominator by $3ab$.

$$\frac{4b}{9a} = \frac{4b(3ab)}{9a(3ab)} = \frac{12ab^2}{27a^2b}$$

Practice Problem 10

Write the rational expression as an equivalent rational expression with the given denominator.

$$\frac{3}{x^2-25} = \frac{}{(x+5)(x-5)(x-3)}$$

Example 10   Write the rational expression as an equivalent rational expression with the given denominator.

$$\frac{5}{x^2-4} = \frac{}{(x-2)(x+2)(x-4)}$$

Solution: First we factor the denominator x^2-4 as $(x-2)(x+2)$. If we multiply the original denominator $(x-2)(x+2)$ by $x-4$, the result is the new denominator $(x-2)(x+2)(x-4)$. Thus, we multiply the numerator and the denominator by $x-4$.

$$\frac{5}{x^2-4} = \frac{5}{(x-2)(x+2)} = \frac{5(x-4)}{(x-2)(x+2)(x-4)}$$

$$= \frac{5x-20}{(x-2)(x+2)(x-4)}$$

Answers

8. $(x-4)$ or $(4-x)$,   **9.** $\dfrac{8x^3y}{20x^2y^2}$,

10. $\dfrac{3x-9}{(x+5)(x-5)(x-3)}$

Name _____ **Section** _____ **Date** _____

MENTAL MATH

Perform each indicated operation.

1. $\dfrac{2}{3} + \dfrac{1}{3}$

2. $\dfrac{5}{11} + \dfrac{1}{11}$

3. $\dfrac{3x}{9} + \dfrac{4x}{9}$

4. $\dfrac{3y}{8} + \dfrac{2y}{8}$

5. $\dfrac{8}{9} - \dfrac{7}{9}$

6. $\dfrac{7y}{5} + \dfrac{10y}{5}$

EXERCISE SET 6.3

A *Add or subtract as indicated. Simplify the result if possible. See Examples 1 through 3.*

1. $\dfrac{a}{13} + \dfrac{9}{13}$

2. $\dfrac{x+1}{7} + \dfrac{6}{7}$

3. $\dfrac{4m}{3n} + \dfrac{5m}{3n}$

4. $\dfrac{3p}{2} + \dfrac{11p}{2}$

5. $\dfrac{4m}{m-6} - \dfrac{24}{m-6}$

6. $\dfrac{8y}{y-2} - \dfrac{16}{y-2}$

7. $\dfrac{9}{3+y} + \dfrac{y+1}{3+y}$

8. $\dfrac{9}{y+9} + \dfrac{y}{y+9}$

9. $\dfrac{5x+4}{x-1} - \dfrac{2x+7}{x-1}$

10. $\dfrac{x^2+9x}{x+7} - \dfrac{4x+14}{x+7}$

11. $\dfrac{a}{a^2+2a-15} - \dfrac{3}{a^2+2a-15}$

12. $\dfrac{3y}{y^2+3y-10} - \dfrac{6}{y^2+3y-10}$

13. $\dfrac{2x+3}{x^2-x-30} - \dfrac{x-2}{x^2-x-30}$

14. $\dfrac{3x-1}{x^2+5x-6} - \dfrac{2x-7}{x^2+5x-6}$

MENTAL MATH ANSWERS

1. _____
2. _____
3. _____
4. _____
5. _____
6. _____

ANSWERS

1. _____
2. _____
3. _____
4. _____
5. _____
6. _____
7. _____
8. _____
9. _____
10. _____
11. _____
12. _____
13. _____
14. _____

Problem Solving Notes

6.4 ADDING AND SUBTRACTING RATIONAL EXPRESSIONS WITH DIFFERENT DENOMINATORS

A ADDING AND SUBTRACTING RATIONAL EXPRESSIONS WITH DIFFERENT DENOMINATORS

In the previous section, we practiced all the skills we need to add and subtract rational expressions with different denominators. The steps are as follows:

> **TO ADD OR SUBTRACT RATIONAL EXPRESSIONS WITH DIFFERENT DENOMINATORS**
>
> **Step 1.** Find the LCD of the rational expressions.
> **Step 2.** Rewrite each rational expression as an equivalent expression whose denominator is the LCD found in Step 1.
> **Step 3.** Add or subtract numerators and write the sum or difference over the common denominator.
> **Step 4.** Simplify or write the rational expression in lowest terms.

Objective

A Add and subtract rational expressions with different denominators.

SSM   CD-ROM   Video
6.4

Example 1   Perform each indicated operation.

    a. $\dfrac{a}{4} - \dfrac{2a}{8}$ **b.** $\dfrac{3}{10x^2} + \dfrac{7}{25x}$

Solution:

a. First, we must find the LCD. Since $4 = 2^2$ and $8 = 2^3$, the LCD $= 2^3 = 8$. Next we write each fraction as an equivalent fraction with the denominator 8, then we subtract.

$$\dfrac{a}{4} - \dfrac{2a}{8} = \dfrac{a(2)}{4(2)} - \dfrac{2a}{8} = \dfrac{2a}{8} - \dfrac{2a}{8} = \dfrac{2a - 2a}{8} = \dfrac{0}{8} = 0$$

b. Since $10x^2 = 2 \cdot 5 \cdot x \cdot x$ and $25x = 5 \cdot 5 \cdot x$, the LCD $= 2 \cdot 5^2 \cdot x^2 = 50x^2$. We write each fraction as an equivalent fraction with a denominator of $50x^2$.

$$\dfrac{3}{10x^2} + \dfrac{7}{25x} = \dfrac{3(5)}{10x^2(5)} + \dfrac{7(2x)}{25x(2x)}$$

$$= \dfrac{15}{50x^2} + \dfrac{14x}{50x^2}$$

$$= \dfrac{15 + 14x}{50x^2} \quad \text{Add numerators. Write the sum over the common denominator.}$$

Example 2   Subtract: $\dfrac{6x}{x^2 - 4} - \dfrac{3}{x + 2}$

Solution: Since $x^2 - 4 = (x + 2)(x - 2)$, the LCD $= (x - 2)(x + 2)$. We write equivalent expressions with the LCD as denominators.

Practice Problem 1

Perform each indicated operation.

  a. $\dfrac{y}{5} - \dfrac{3y}{15}$ **b.** $\dfrac{5}{8x} + \dfrac{11}{10x^2}$

Practice Problem 2

Subtract: $\dfrac{10x}{x^2 - 9} - \dfrac{5}{x + 3}$

Answers

1. a. 0, **b.** $\dfrac{25x + 44}{40x^2}$, **2.** $\dfrac{5}{x - 3}$

$$\frac{6x}{x^2-4} - \frac{3}{x+2} = \frac{6x}{(x-2)(x+2)} - \frac{3(x-2)}{(x+2)(x-2)}$$

$$= \frac{6x - 3(x-2)}{(x+2)(x-2)} \qquad \text{Subtract numerators. Write the difference over the common denominator.}$$

$$= \frac{6x - 3x + 6}{(x+2)(x-2)} \qquad \text{Apply the distributive property in the numerator.}$$

$$= \frac{3x + 6}{(x+2)(x-2)} \qquad \text{Combine like terms in the numerator.}$$

Next we factor the numerator to see if this rational expression can be simplified.

$$= \frac{3(x+2)}{(x+2)(x-2)} \qquad \text{Factor.}$$

$$= \frac{3}{x-2} \qquad \text{Apply the fundamental principle to simplify.}$$

Practice Problem 3

Add: $\dfrac{5}{7x} + \dfrac{2}{x+1}$

Example 3   Add: $\dfrac{2}{3t} + \dfrac{5}{t+1}$

Solution: The LCD is $3t(t+1)$. We write each rational expression as an equivalent rational expression with a denominator of $3t(t+1)$.

$$\frac{2}{3t} + \frac{5}{t+1} = \frac{2(t+1)}{3t(t+1)} + \frac{5(3t)}{(t+1)(3t)}$$

$$= \frac{2(t+1) + 5(3t)}{3t(t+1)} \qquad \text{Add numerators. Write the sum over the common denominator.}$$

$$= \frac{2t + 2 + 15t}{3t(t+1)} \qquad \text{Apply the distributive property in the numerator.}$$

$$= \frac{17t + 2}{3t(t+1)} \qquad \text{Combine like terms in the numerator.}$$

Practice Problem 4

Subtract: $\dfrac{10}{x-6} - \dfrac{15}{6-x}$

Example 4   Subtract: $\dfrac{7}{x-3} - \dfrac{9}{3-x}$

Solution: To find a common denominator, we notice that $x - 3$ and $3 - x$ are opposites. That is, $3 - x = -(x-3)$. We write the denominator $3 - x$ as $-(x-3)$ and simplify.

$$\frac{7}{x-3} - \frac{9}{3-x} = \frac{7}{x-3} - \frac{9}{-(x-3)}$$

$$= \frac{7}{x-3} - \frac{-9}{x-3} \qquad \text{Apply } \frac{a}{-b} = \frac{-a}{b}.$$

$$= \frac{7 - (-9)}{x-3} \qquad \text{Subtract numerators. Write the difference over the common denominator.}$$

$$= \frac{16}{x-3}$$

Practice Problem 5

Add: $2 + \dfrac{x}{x+5}$

Example 5   Add: $1 + \dfrac{m}{m+1}$

Solution: Recall that 1 is the same as $\dfrac{1}{1}$. The LCD of $\dfrac{1}{1}$ and $\dfrac{m}{m+1}$ is $m + 1$.

Answers

3. $\dfrac{19x + 5}{7x(x+1)}$, **4.** $\dfrac{25}{x-6}$, **5.** $\dfrac{3x + 10}{x+5}$

$$1 + \frac{m}{m+1} = \frac{1}{1} + \frac{m}{m+1}$$ Write 1 as $\frac{1}{1}$.

$$= \frac{1(m+1)}{1(m+1)} + \frac{m}{m+1}$$ Multiply both the numerator and the denominator of $\frac{1}{1}$ by $m+1$.

$$= \frac{m+1+m}{m+1}$$ Add numerators. Write the sum over the common denominator.

$$= \frac{2m+1}{m+1}$$ Combine like terms in the numerator.

Example 6   Subtract: $\dfrac{3}{2x^2 + x} - \dfrac{2x}{6x + 3}$

Solution: First, we factor the denominators.

$$\frac{3}{2x^2 + x} - \frac{2x}{6x + 3} = \frac{3}{x(2x + 1)} - \frac{2x}{3(2x + 1)}$$

The LCD is $3x(2x + 1)$. We write equivalent expressions with denominators of $3x(2x + 1)$.

$$= \frac{3(3)}{x(2x + 1)(3)} - \frac{2x(x)}{3(2x + 1)(x)}$$

$$= \frac{9 - 2x^2}{3x(2x + 1)}$$ Subtract numerators. Write the difference over the common denominator.

Example 7   Add: $\dfrac{2x}{x^2 + 2x + 1} + \dfrac{x}{x^2 - 1}$

Solution: First we factor the denominators.

$$\frac{2x}{x^2 + 2x + 1} + \frac{x}{x^2 - 1} = \frac{2x}{(x + 1)(x + 1)} + \frac{x}{(x + 1)(x - 1)}$$

Now we write the rational expressions as equivalent expressions with denominators of $(x + 1)(x + 1)(x - 1)$, the LCD.

$$= \frac{2x(x - 1)}{(x + 1)(x + 1)(x - 1)} + \frac{x(x + 1)}{(x + 1)(x - 1)(x + 1)}$$

$$= \frac{2x(x - 1) + x(x + 1)}{(x + 1)^2(x - 1)}$$ Add numerators. Write the sum over the common denominator.

$$= \frac{2x^2 - 2x + x^2 + x}{(x + 1)^2(x - 1)}$$ Apply the distributive property in the numerator.

$$= \frac{3x^2 - x}{(x + 1)^2(x - 1)}   \quad \text{or} \quad   \frac{x(3x - 1)}{(x + 1)^2(x - 1)}$$

The numerator was factored as a last step to see if the rational expression could be simplified further. Since there are no factors common to the numerator and the denominator, we can't simplify further.

Practice Problem 6

Subtract: $\dfrac{4}{3x^2 + 2x} - \dfrac{3x}{12x + 8}$

Practice Problem 7

Add: $\dfrac{6x}{x^2 + 4x + 4} + \dfrac{x}{x^2 - 4}$

Answers

6. $\dfrac{16 - 3x^2}{4x(3x + 2)}$,   **7.** $\dfrac{x(7x - 10)}{(x + 2)^2(x - 2)}$

Focus On Business and Career

FAST-GROWING CAREERS

According to U.S. Bureau of Labor Statistics projections, the careers listed below will have the largest job growth into the next century.

Occupation	Employment [Numbers in thousands]		
	1994	**2005**	**Change**
1. Cashiers	3,005	3,567	+562
2. Janitors and cleaners, including maids and housekeeping cleaners	3,043	3,602	+559
3. Salespersons, retail	3,842	4,374	+532
4. Waiters and waitresses	1,847	2,326	+479
5. Registered nurses	1,906	2,379	+473
6. General managers and top executives	3,046	3,512	+466
7. Systems analysts	483	928	+445
8. Home health aides	420	848	+428
9. Guards	867	1,282	+415
10. Nursing aides, orderlies, and attendants	1,265	1,652	+387
11. Teachers, secondary	1,340	1,726	+386
12. Marketing and sales worker supervisors	2,293	2,673	+380
13. Teacher's aides and educational associates	932	1,296	+364
14. Receptionists and information clerks	1,019	1,337	+318
15. Truck drivers, light and heavy	2,565	2,837	+271
16. Secretaries, except legal and medical	2,842	3,109	+267
17. Clerical supervisors and managers	1,340	1,600	+261
18. Child care workers	757	1,005	+248
19. Maintenance repairers, general utility	1,273	1,505	+231
20. Teachers, elementary	1,419	1,639	+220

Source: Bureau of Labor Statistics, Office of Employment Projections, November 1995

What do all of these in-demand occupations have in common? They all require a knowledge of math! For some careers like cashiers, salespersons, waiters and waitresses, financial managers, and computer engineers, the ways math is used on the job may be obvious. For other occupations, the use of math may not be quite as obvious. However, tasks common to many jobs like filling in a time sheet, writing up an expense or mileage report, planning a budget, figuring a bill, ordering supplies, completing a packing list, and even making a work schedule all require math.

CRITICAL THINKING

Suppose that your college placement office is planning to publish an occupational handbook on math in popular occupations. Choose one of the occupations from the list above that interests you. Research the occupation. Then write a brief entry for the occupational handbook that describes how a person in that career would use math in his or her job. Include an example if possible.

EXERCISE SET 6.4

A *Perform each indicated operation. Simplify if possible. See Example 1.*

1. $\dfrac{4}{2x} + \dfrac{9}{3x}$

2. $\dfrac{15}{7a} + \dfrac{8}{6a}$

▭ **3.** $\dfrac{15a}{b} + \dfrac{6b}{5}$

4. $\dfrac{4c}{d} - \dfrac{8x}{5}$

5. $\dfrac{3}{x} + \dfrac{5}{2x^2}$

6. $\dfrac{3}{x + 2} - \dfrac{1}{x^2 - 4}$

7. $\dfrac{3}{4x} + \dfrac{8}{x - 2}$

8. $\dfrac{5}{y^2} - \dfrac{y}{2y + 1}$

▭ **9.** $\dfrac{6}{x - 3} + \dfrac{8}{3 - x}$

10. $\dfrac{9}{x - 3} + \dfrac{9}{3 - x}$

Perform each indicated operation. Simplify if possible. See Examples 4 through 7.

11. $\dfrac{5x}{x + 2} - \dfrac{3x - 4}{x + 2}$

12. $\dfrac{7x}{x - 3} - \dfrac{4x + 9}{x - 3}$

13. $\dfrac{3x^4}{x} - \dfrac{4x^2}{x^2}$

14. $\dfrac{5x}{6} + \dfrac{15x^2}{2}$

15. $\dfrac{1}{x + 3} - \dfrac{1}{(x + 3)^2}$

16. $\dfrac{5x}{(x - 2)^2} - \dfrac{3}{x - 2}$

17. $\dfrac{4}{5b} + \dfrac{1}{b - 1}$

18. $\dfrac{1}{y + 5} + \dfrac{2}{3y}$

19. $\dfrac{2}{m} + 1$

20. $\dfrac{6}{x} - 1$

ANSWERS

1. _____
2. _____
3. _____
4. _____
5. _____
6. _____
7. _____
8. _____
9. _____
10. _____
11. _____
12. _____
13. _____
14. _____
15. _____
16. _____
17. _____
18. _____
19. _____
20. _____

Problem Solving Notes

6.5 SOLVING EQUATIONS CONTAINING RATIONAL EXPRESSIONS

A SOLVING EQUATIONS CONTAINING RATIONAL EXPRESSIONS

In Chapter 2, we solved equations containing fractions. In this section, we continue the work we began in Chapter 2 by solving equations containing rational expressions. For example,

$$\frac{x}{5} + \frac{x+2}{9} = 8 \quad \text{and} \quad \frac{x+1}{9x-5} = \frac{2}{3x}$$

are equations containing rational expressions. To solve equations such as these, we use the multiplication property of equality to clear the equation of fractions by multiplying both sides of the equation by the LCD.

Objectives

A Solve equations containing rational expressions.

B Solve equations containing rational expressions for a specified variable.

SSM   CD-ROM   Video 6.5

Example 1   Solve: $\frac{x}{2} + \frac{8}{3} = \frac{1}{6}$

Solution: The LCD of denominators 2, 3, and 6 is 6, so we multiply both sides of the equation by 6.

$$6\left(\frac{x}{2} + \frac{8}{3}\right) = 6\left(\frac{1}{6}\right)$$

HELPFUL HINT

Make sure that *each* term is multiplied by the LCD.

$$6\left(\frac{x}{2}\right) + 6\left(\frac{8}{3}\right) = 6\left(\frac{1}{6}\right) \quad \text{Use the distributive property.}$$

$$3 \cdot x + 16 = 1 \quad \text{Multiply and simplify.}$$

$$3x = -15 \quad \text{Subtract 16 from both sides.}$$

$$x = -5 \quad \text{Divide both sides by 3.}$$

Check: To check, we replace x with -5 in the original equation.

$$\frac{-5}{2} + \frac{8}{3} \stackrel{?}{=} \frac{1}{6} \quad \text{Replace } x \text{ with } -5.$$

$$\frac{1}{6} = \frac{1}{6} \quad \text{True.}$$

This number checks, so the solution is -5.

Example 2   Solve: $\frac{t-4}{2} - \frac{t-3}{9} = \frac{5}{18}$

Solution: The LCD of denominators 2, 9, and 18 is 18, so we multiply both sides of the equation by 18.

$$18\left(\frac{t-4}{2} - \frac{t-3}{9}\right) = 18\left(\frac{5}{18}\right)$$

$$18\left(\frac{t-4}{2}\right) - 18\left(\frac{t-3}{9}\right) = 18\left(\frac{5}{18}\right) \quad \text{Use the distributive property.}$$

HELPFUL HINT

Multiply *each* term by 18.

$$9(t-4) - 2(t-3) = 5 \quad \text{Simplify. Use the distributive property.}$$

$$9t - 36 - 2t + 6 = 5 \quad$$

$$7t - 30 = 5 \quad \text{Combine like terms.}$$

$$7t = 35$$

$$t = 5 \quad \text{Solve for } t.$$

Practice Problem 1

Solve: $\frac{x}{4} + \frac{4}{5} = \frac{1}{20}$

Practice Problem 2

Solve: $\frac{x+2}{3} - \frac{x-1}{5} = \frac{1}{15}$

Answers

1. $x = -3$,   **2.** $x = -6$

Check:
$$\frac{t-4}{2} - \frac{t-3}{9} = \frac{5}{18}$$

$$\frac{5-4}{2} - \frac{5-3}{9} \stackrel{?}{=} \frac{5}{18} \quad \text{Replace } t \text{ with 5.}$$

$$\frac{1}{2} - \frac{2}{9} \stackrel{?}{=} \frac{5}{18} \quad \text{Simplify.}$$

$$\frac{5}{18} = \frac{5}{18} \quad \text{True.}$$

The solution is 5. ▬▬▬

Recall from Section 6.1 that a rational expression is defined for all real numbers except those that make the denominator of the expression 0. This means that if an equation contains *rational expressions with variables in the denominator*, we must be certain that the proposed solution does not make the denominator 0. If replacing the variable with the proposed solution makes the denominator 0, the rational expression is undefined and this proposed solution must be rejected.

Practice Problem 3

Solve: $2 + \dfrac{6}{x} = x + 7$

Example 3

Solve: $3 - \dfrac{6}{x} = x + 8$

Solution:

In this equation, 0 cannot be a solution because if x is 0, the rational expression $\dfrac{6}{x}$ is undefined. The LCD is x, so we multiply both sides of the equation by x.

$$x\left(3 - \frac{6}{x}\right) = x(x + 8)$$

> **HELPFUL HINT**
>
> Multiply *each* term by x.

$$x(3) - x\left(\frac{6}{x}\right) = x \cdot x + x \cdot 8 \quad \begin{array}{l}\text{Use the distributive}\\\text{property.}\end{array}$$

$$3x - 6 = x^2 + 8x \quad \text{Simplify.}$$

Now we write the quadratic equation in standard form and solve for x.

$$0 = x^2 + 5x + 6$$

$$0 = (x + 3)(x + 2) \quad \text{Factor.}$$

$$x + 3 = 0 \quad \text{or} \quad x + 2 = 0 \quad \begin{array}{l}\text{Set each factor equal}\\\text{to 0 and solve.}\end{array}$$

$$x = -3 \qquad\qquad x = -2$$

Notice that neither -3 nor -2 makes the denominator in the original equation equal to 0.

Check:

To check these solutions, we replace x in the original equation by -3, and then by -2.

If $x = -3$:

$$3 - \frac{6}{x} = x + 8$$

$$3 - \frac{6}{-3} \stackrel{?}{=} -3 + 8$$

$$3 - (-2) \stackrel{?}{=} 5$$

$$5 = 5 \quad \text{True.}$$

If $x = -2$:

$$3 - \frac{6}{x} = x + 8$$

$$3 - \frac{6}{-2} \stackrel{?}{=} -2 + 8$$

$$3 - (-3) \stackrel{?}{=} 6$$

$$6 = 6 \quad \text{True.}$$

Answer

3. $x = -6, x = 1$

Both -3 and -2 are solutions. ▬▬▬

The following steps may be used to solve an equation containing rational expressions.

To Solve an Equation Containing Rational Expressions

Step 1. Multiply both sides of the equation by the LCD of all rational expressions in the equation.

Step 2. Remove any grouping symbols and solve the resulting equation.

Step 3. Check the solution in the original equation.

Example 4 Solve: $\dfrac{4x}{x^2 - 25} + \dfrac{2}{x - 5} = \dfrac{1}{x + 5}$

Solution: The denominator $x^2 - 25$ factors as $(x + 5)(x - 5)$. The LCD is then $(x + 5)(x - 5)$, so we multiply both sides of the equation by this LCD.

Multiply by the LCD.

$$(x + 5)(x - 5)\left(\frac{4x}{x^2 - 25} + \frac{2}{x - 5}\right) = (x + 5)(x - 5)\left(\frac{1}{x + 5}\right)$$

$$(x + 5)(x - 5) \cdot \frac{4x}{x^2 - 25} + (x + 5)(x - 5) \cdot \frac{2}{x - 5}$$ Use the distributive property.

$$= (x + 5)(x - 5) \cdot \frac{1}{x + 5}$$

$$4x + 2(x + 5) = x - 5 \quad \text{Simplify.}$$

$$4x + 2x + 10 = x - 5 \quad \text{Use the distributive property.}$$

$$6x + 10 = x - 5 \quad \text{Combine like terms.}$$

$$5x = -15$$

$$x = -3 \quad \text{Divide both sides by 5.}$$

Check: Check by replacing x with -3 in the original equation. The solution is -3.

Example 5 Solve: $\dfrac{2x}{x - 4} = \dfrac{8}{x - 4} + 1$

Solution: Multiply both sides by the LCD, $x - 4$.

$$(x - 4)\left(\frac{2x}{x - 4}\right) = (x - 4)\left(\frac{8}{x - 4} + 1\right)$$ Multiply by the LCD.

$$(x - 4) \cdot \frac{2x}{x - 4} = (x - 4) \cdot \frac{8}{x - 4} + (x - 4) \cdot 1$$ Use the distributive property.

$$2x = 8 + (x - 4) \quad \text{Simplify.}$$

$$2x = 4 + x$$

$$x = 4$$

Notice that 4 makes the denominator 0 in the original equation. Therefore, 4 is *not* a solution and this equation has *no solution*.

Try the Concept Check in the margin.

HELPFUL HINT

As we can see from Example 5, it is important to check the proposed solution(s) in the original equation.

Practice Problem 4

Solve: $\dfrac{2}{x + 3} + \dfrac{3}{x - 3} = \dfrac{-2}{x^2 - 9}$

Practice Problem 5

Solve: $\dfrac{5x}{x - 1} = \dfrac{5}{x - 1} + 3$

✓ Concept Check

When can fractions be cleared by multiplying through by the LCD?

a. When adding or subtracting rational expressions

b. When solving an equation containing rational expressions

c. Both of these

d. Neither of these

Answers

4. $x = -1$,   **5.** No solution

✓ Concept Check: b

Practice Problem 6

Solve: $x - \dfrac{6}{x + 3} = \dfrac{2x}{x + 3} + 2$

Example 6

Solve: $x + \dfrac{14}{x - 2} = \dfrac{7x}{x - 2} + 1$

Solution: Notice the denominators in this equation. We can see that 2 can't be a solution. The LCD is $x - 2$, so we multiply both sides of the equation by $x - 2$.

$$(x - 2)\left(x + \frac{14}{x - 2}\right) = (x - 2)\left(\frac{7x}{x - 2} + 1\right)$$

$$(x - 2)(x) + (x - 2)\left(\frac{14}{x - 2}\right) = (x - 2)\left(\frac{7x}{x - 2}\right) + (x - 2)(1)$$

$$x^2 - 2x + 14 = 7x + x - 2 \qquad \text{Simplify.}$$

$$x^2 - 2x + 14 = 8x - 2 \qquad \text{Combine like terms.}$$

$$x^2 - 10x + 16 = 0 \qquad \text{Write the quadratic equation in standard form.}$$

$$(x - 8)(x - 2) = 0 \qquad \text{Factor.}$$

$$x - 8 = 0 \quad \text{or} \quad x - 2 = 0 \qquad \text{Set each factor equal to 0.}$$

$$x = 8 \qquad\qquad x = 2 \qquad \text{Solve.}$$

As we have already noted, 2 can't be a solution of the original equation. So we need only replace x with 8 in the original equation. We find that 8 is a solution; the only solution is 8.   ■

B   SOLVING EQUATIONS FOR A SPECIFIED VARIABLE

The last example in this section is an equation containing several variables, and we are directed to solve for one of the variables. The steps used in the preceding examples can be applied to solve equations for a specified variable as well.

Practice Problem 7

Solve $\dfrac{1}{a} + \dfrac{1}{b} = \dfrac{1}{x}$ for a.

Example 7

Solve $\dfrac{1}{a} + \dfrac{1}{b} = \dfrac{1}{x}$ for x

Solution: (This type of equation often models a work problem, as we shall see in the next section.) The LCD is abx, so we multiply both sides by abx.

$$abx\left(\frac{1}{a} + \frac{1}{b}\right) = abx\left(\frac{1}{x}\right)$$

$$abx\left(\frac{1}{a}\right) + abx\left(\frac{1}{b}\right) = abx \cdot \frac{1}{x}$$

$$bx + ax = ab \qquad \text{Simplify.}$$

$$x(b + a) = ab \qquad \text{Factor out } x \text{ from each term on the left side.}$$

$$\frac{x(b + a)}{b + a} = \frac{ab}{b + a} \qquad \text{Divide both sides by } b + a.$$

$$x = \frac{ab}{b + a} \qquad \text{Simplify.}$$

This equation is now solved for x.   ■

Answers

6. $x = 4$,   **7.** $a = \dfrac{bx}{b - x}$

Name _____ **Section** _____ **Date** _____

MENTAL MATH

Solve each equation for the variable.

1. $\dfrac{x}{5} = 2$ **2.** $\dfrac{x}{8} = 4$ **3.** $\dfrac{z}{6} = 6$ **4.** $\dfrac{y}{7} = 8$

EXERCISE SET 6.5

A *Solve each equation and check each solution. See Examples 1 and 2.*

1. $\dfrac{x}{5} + 3 = 9$ **2.** $\dfrac{x}{5} - 2 = 9$ **3.** $\dfrac{x}{2} + \dfrac{5x}{4} = \dfrac{x}{12}$

4. $\dfrac{x}{6} + \dfrac{4x}{3} = \dfrac{x}{18}$ **5.** $2 - \dfrac{8}{x} = 6$ **6.** $5 + \dfrac{4}{x} = 1$

7. $2 + \dfrac{10}{x} = x + 5$ **8.** $6 + \dfrac{5}{y} = y - \dfrac{2}{y}$ **9.** $\dfrac{a}{5} = \dfrac{a-3}{2}$

10. $\dfrac{b}{5} = \dfrac{b+2}{6}$ ▣ **11.** $\dfrac{x-3}{5} + \dfrac{x-2}{2} = \dfrac{1}{2}$

Solve each equation and check each answer. See Examples 3 through 6.

▣ **12.** $\dfrac{2}{y} + \dfrac{1}{2} = \dfrac{5}{2y}$ **13.** $\dfrac{6}{3y} + \dfrac{3}{y} = 1$ **14.** $\dfrac{11}{2x} + \dfrac{2}{3} = \dfrac{7}{2x}$

15. $\dfrac{5}{3} - \dfrac{3}{2x} = \dfrac{3}{2}$ ▣ **16.** $2 + \dfrac{3}{a-3} = \dfrac{a}{a-3}$

MENTAL MATH ANSWERS

1. _____
2. _____
3. _____
4. _____

ANSWERS

1. _____
2. _____
3. _____
4. _____
5. _____
6. _____
7. _____
8. _____
9. _____
10. _____
11. _____
12. _____
13. _____
14. _____
15. _____
16. _____

Problem Solving Notes

6.6 RATIONAL EQUATIONS AND PROBLEM SOLVING

A SOLVING PROBLEMS ABOUT NUMBERS

In this section, we solve problems that can be modeled by equations containing rational expressions. To solve these problems, we use the same problem-solving steps that were first introduced in Section 2.5. In our first example, our goal is to find an unknown number.

Example 1   Finding an Unknown Number

The quotient of a number and 6 minus $\frac{5}{3}$ is the quotient of the number and 2. Find the number.

Solution: **1.** UNDERSTAND. Read and reread the problem. Suppose that the unknown number is 2, then we see if the quotient of 2 and 6, or $\frac{2}{6}$, minus $\frac{5}{3}$ is equal to the quotient of 2 and 2, or $\frac{2}{2}$.

$$\frac{2}{6} - \frac{5}{3} = \frac{1}{3} - \frac{5}{3} = -\frac{4}{3}, \text{ not } \frac{2}{2}$$

Don't forget that the purpose of a proposed solution is to better understand the problem.

Let $x =$ the unknown number.

2. TRANSLATE.

In words:

the quotient of x and 6	minus	$\frac{5}{3}$	is	the quotient of x and 2
↓	↓	↓	↓	↓

Translate: $\dfrac{x}{6} \qquad - \qquad \dfrac{5}{3} \quad = \qquad \dfrac{x}{2}$

3. SOLVE. Here, we solve the equation $\frac{x}{6} - \frac{5}{3} = \frac{x}{2}$. We begin by multiplying both sides of the equation by the LCD 6.

$$6\left(\frac{x}{6} - \frac{5}{3}\right) = 6\left(\frac{x}{2}\right)$$

$$6\left(\frac{x}{6}\right) - 6\left(\frac{5}{3}\right) = 6\left(\frac{x}{2}\right) \qquad \text{Apply the distributive property.}$$

$$x - 10 = 3x \qquad \text{Simplify.}$$

$$-10 = 2x \qquad \text{Subtract } x \text{ from both sides.}$$

$$-\frac{10}{2} = \frac{2x}{2} \qquad \text{Divide both sides by 2.}$$

$$-5 = x \qquad \text{Simplify.}$$

Objectives

A Solve problems about numbers.
B Solve problems about work.
C Solve problems about distance, rate, and time.
D Solve problems about similar triangles.

SSM   CD-ROM   Video 6.6

Practice Problem 1

The quotient of a number and 2 minus $\frac{1}{3}$ is the quotient of the number and 6.

Answer
1. $x = 1$

4. INTERPRET.

Check: To check, we verify that "the quotient of -5 and 6 minus $\frac{5}{3}$ is the quotient of -5 and 2, or $-\frac{5}{6} - \frac{5}{3} = -\frac{5}{2}$.

State: The unknown number is -5.

B SOLVING PROBLEMS ABOUT WORK

The next example is often called a work problem. Work problems usually involve people or machines doing a certain task.

Example 2   Finding Work Rates

Sam Waterton and Frank Schaffer work in a plant that manufactures automobiles. Sam can complete a quality control tour of the plant in 3 hours while his assistant, Frank, needs 7 hours to complete the same job. The regional manager is coming to inspect the plant facilities, so both Sam and Frank are directed to complete a quality control tour together. How long will this take?

Solution:

1. UNDERSTAND. Read and reread the problem. The key idea here is the relationship between the **time** (hours) it takes to complete the job and the **part of the job** completed in 1 unit of time (hour). For example, if the **time** it takes Sam to complete the job is 3 hours, the **part of the job** he can complete in 1 hour is $\frac{1}{3}$.

Similarly, Frank can complete $\frac{1}{7}$ of the job in 1 hour.

Let $x =$ the **time** in hours it takes Sam and Frank to complete the job together.

Then $\frac{1}{x} =$ the **part of the job** they complete in 1 hour.

	Hours to Complete Total Job	Part of Job Completed in 1 Hour
Sam	3	$\frac{1}{3}$
Frank	7	$\frac{1}{7}$
Together	x	$\frac{1}{x}$

2. TRANSLATE.

In words:

part of job Sam completed in 1 hour	added to	part of job Frank completed in 1 hour	is equal to	part of job they completed together in 1 hour
↓	↓	↓	↓	↓

Translate:

$$\frac{1}{3} \quad + \quad \frac{1}{7} \quad = \quad \frac{1}{x}$$

Practice Problem 2

Andrew and Timothy Larson volunteer at a local recycling plant. Andrew can sort a batch of recyclables in 2 hours alone while his brother Timothy needs 3 hours to complete the same job. If they work together, how long will it take them to sort one batch?

Answer

2. $1\frac{1}{5}$ hours

3. SOLVE. Here, we solve the equation $\frac{1}{3} + \frac{1}{7} = \frac{1}{x}$. We begin by multiplying both sides of the equation by the LCD, $21x$.

$$21x\left(\frac{1}{3}\right) + 21x\left(\frac{1}{7}\right) = 21x\left(\frac{1}{x}\right)$$

$$7x + 3x = 21 \qquad \text{Simplify.}$$

$$10x = 21$$

$$x = \frac{21}{10} \quad \text{or} \quad 2\frac{1}{10} \text{ hours}$$

4. INTERPRET.

Check: Our proposed solution is $2\frac{1}{10}$ hours. This proposed solution is reasonable since $2\frac{1}{10}$ hours is more than half of Sam's time and less than half of Frank's time. Check this solution in the originally *stated* problem.

State: Sam and Frank can complete the quality control tour in $2\frac{1}{10}$ hours.

━━━━━

C Solving Problems about Distance, Rate, and Time

Next we look at a problem solved by the distance formula.

▣ **Example 3**   **Finding Speeds of Vehicles**

A car travels 180 miles in the same time that a truck travels 120 miles. If the car's speed is 20 miles per hour faster than the truck's, find the car's speed and the truck's speed.

Solution: **1.** UNDERSTAND. Read and reread the problem. Suppose that the truck's speed is 45 miles per hour. Then the car's speed is 20 miles per hour more, or 65 miles per hour.

We are given that the car travels 180 miles in the same time that the truck travels 120 miles. To find the time it takes the car to travel 180 miles, remember that since $d = rt$, we know that $\frac{d}{r} = t$.

Car's Time

$$t = \frac{d}{r} = \frac{180}{65} = 2\frac{50}{65} = 2\frac{10}{13} \text{ hours}$$

Truck's Time

$$t = \frac{d}{r} = \frac{120}{45} = 2\frac{30}{45} = 2\frac{2}{3} \text{ hours}$$

Since the times are not the same, our proposed solution is not correct. But we have a better understanding of the problem.

Practice Problem 3

A car travels 280 miles in the same time that a motorcycle travels 240 miles. If the car's speed is 10 miles per hour more than the motorcycle's, find the speed of the car and the speed of the motorcycle.

Answer

3. car: 70 mph; motorcycle: 60 mph

Let x = the speed of the truck.

Since the car's speed is 20 miles per hour faster than the truck's, then

$x + 20$ = the speed of the car

Use the formula $d = r \cdot t$ or **distance = rate · time**. Prepare a chart to organize the information in the problem.

	Distance	=	Rate	·	Time
Truck	120		x		$\dfrac{120}{x} \begin{matrix}\leftarrow \text{distance} \\ \leftarrow \text{rate}\end{matrix}$
Car	180		$x + 20$		$\dfrac{180}{x+20} \begin{matrix}\leftarrow \text{distance} \\ \leftarrow \text{rate}\end{matrix}$

> **HELPFUL HINT**
>
> If $d = r \cdot t$, then $t = \dfrac{d}{r}$ or *time* $= \dfrac{distance}{rate}$.

2. TRANSLATE. Since the car and the truck traveled the same amount of time, we have that

In words: car's time = truck's time

$$\text{Translate:} \qquad \frac{180}{x + 20} = \frac{120}{x}$$

3. SOLVE. We begin by multiplying both sides of the equation by the LCD, $x(x + 20)$, or cross multiplying.

$$\frac{180}{x + 20} = \frac{120}{x}$$

$$180x = 120(x + 20)$$

$$180x = 120x + 2400 \qquad \text{Use the distributive property.}$$

$$60x = 2400 \qquad \text{Subtract } 120x \text{ from both sides.}$$

$$x = 40 \qquad \text{Divide both sides by 60.}$$

4. INTERPRET. The speed of the truck is 40 miles per hour. The speed of the car must then be $x + 20$ or 60 miles per hour.

Check: Find the time it takes the car to travel 180 miles and the time it takes the truck to travel 120 miles.

Car's Time

$$t = \frac{d}{r} = \frac{180}{60} = 3 \text{ hours}$$

Truck's Time

$$t = \frac{d}{r} = \frac{120}{40} = 3 \text{ hours}$$

Since both travel the same amount of time, the proposed solution is correct.

State: The car's speed is 60 miles per hour and the truck's speed is 40 miles per hour. ▬▬▬

D SOLVING PROBLEMS ABOUT SIMILAR TRIANGLES

Similar triangles have the same shape but not necessarily the same size. In similar triangles, the measures of corresponding angles are equal, and corresponding sides are in proportion.

If triangle ABC and triangle XYZ shown are similar, then we know that the measure of angle A = the measure of angle X, the measure of angle B = the measure of angle Y, and the measure of angle C = the measure of angle Z. We also know that corresponding sides are in proportion: $\frac{a}{x} = \frac{b}{y} = \frac{c}{z}$.

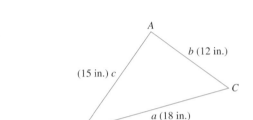

In this section, we will position similar triangles so that they have the same orientation.

To show that corresponding sides are in proportion for the triangles above, we write the ratios of the corresponding sides.

$$\frac{a}{x} = \frac{18}{6} = 3 \qquad \frac{b}{y} = \frac{12}{4} = 3 \qquad \frac{c}{z} = \frac{15}{5} = 3$$

Example 4   Finding the Length of a Side of a Triangle

If the following two triangles are similar, find the missing length x.

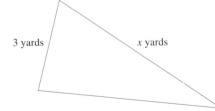

Solution: Since the triangles are similar, their corresponding sides are in proportion and we have

$$\frac{2}{3} = \frac{10}{x}$$

If the following two triangles are similar, find the missing length x.

To solve, we multiply both sides by the LCD, $3x$, or cross multiply.

$2x = 30$

$x = 15$ Divide both sides by 2.

The missing length is 15 yards.

EXERCISE SET 6.6

A *Solve. See Example 1.*

1. Three times the reciprocal of a number equals 9 times the reciprocal of 6. Find the number.

2. Twelve divided by the sum of a number and 2 equals the quotient of 4 and the difference of the number and 2. Find the number.

3. If twice a number added to 3 is divided by the number plus 1, the result is three halves. Find the number.

4. A number added to the product of 6 and the reciprocal of the number equals −5. Find the number.

5. Two divided by the difference of a number and 3, minus 4 divided by the number plus 3, equals 8 times the reciprocal of the difference of the number squared and 9. What is the number?

6. If 15 times the reciprocal of a number is added to the ratio of 9 times the number minus 7 and the number plus 2, the result is 9. What is the number?

7. One-fourth equals the quotient of a number and 8. Find the number.

8. Four times a number added to 5 is divided by 6. The result is $\frac{7}{2}$. Find the number.

B *Solve. See Example 2.*

9. Smith Engineering found that an experienced surveyor surveys a roadbed in 4 hours. An apprentice surveyor needs 5 hours to survey the same stretch of road. If the two work together, find how long it takes them to complete the job.

10. An experienced bricklayer constructs a small wall in 3 hours. The apprentice completes the job in 6 hours. Find how long it takes if they work together.

11. In 2 minutes, a conveyor belt moves 300 pounds of recyclable aluminum from the delivery truck to a storage area. A smaller belt moves the same quantity of cans the same distance in 6 minutes. If both belts are used, find how long it takes to move the cans to the storage area.

12. Find how long it takes the conveyor belts described in Exercise 11 to move 1200 pounds of cans. (*Hint*: Think of 1200 pounds as four 300-pound jobs.)

13. _____

13. Marcus and Tony work for Lombardo's Pipe and Concrete. Mr. Lombardo is preparing an estimate for a customer. He knows that Marcus lays a slab of concrete in 6 hours. Tony lays the same size slab in 4 hours. If both work on the job and the cost of labor is $45.00 per hour, decide what the labor estimate should be.

14. Mr. Dodson can paint his house by himself in 4 days. His son needs an additional day to complete the job if he works by himself. If they work together, find how long it takes to paint the house.

14. _____

15. _____

15. One custodian cleans a suite of offices in 3 hours. When a second worker is asked to join the regular custodian, the job takes only $1\frac{1}{2}$ hours. How long does it take the second worker to do the same job alone?

16. One person proofreads copy for a small newspaper in 4 hours. If a second proofreader is also employed, the job can be done in $2\frac{1}{2}$ hours. How long does it take for the second proofreader to do the same job alone?

16. _____

17. One pipe fills a storage pond in 20 hours. A second pipe fills the same pond in 15 hours. When a third pipe is added and all three are used to fill the pond, it takes only 6 hours. Find how long it takes the third pipe to do the job.

18. One pump fills a tank 3 times as fast as another pump. If the pumps work together, they fill the tank in 21 minutes. How long does it take for each pump to fill the tank?

17. _____

C *Solve. See Example 3.*

18. _____

19. A jogger begins her workout by jogging to the park, a distance of 3 miles. She then jogs home at the same speed but along a different route. This return trip is 9 miles and her time is one hour longer. Complete the accompanying chart and use it to find her jogging speed.

20. A marketing manager travels 1080 miles in a corporate jet and then an additional 240 miles by car. If the car ride takes one hour longer than the jet ride takes, and if the rate of the jet is 6 times the rate of the car, find the time the manager travels by jet and find the time the manager travels by car.

19. _____

	Distance	= Rate ·	Time
Trip to park	3		x
Return trip	9		$x + 1$

20. _____

EPIGRAM OF DIOPHANTES

One of the great algebraists of ancient times was a man named Diophantus. Little is known of his life other than that he lived and worked in Alexandria. Some historians believe he lived during the first century of the Christian era, about the time of Nero. The only clue to his personal life is the following epigram found in a collection called the Palatine Anthology.

God granted him youth for a sixth of his life and added a twelfth part to this. He clothed his cheeks in down. He lit him the light of wedlock after a seventh part, and five years after his marriage, He granted him a son. Alas, lateborn wretched child. After attaining the measure of half his father's life, cruel fate overtook him, thus leaving Diophantus during the last four years of his life only such consolation as the science of numbers. How old was Diophantus at his death?*

*From *The Nature and Growth of Modern Mathematics*, Edna Kramer, 1970, Fawcett Premier Books, Vol. 1, pages 107–108.

We are looking for Diophantus' age when he died, so let x represent that age. If we sum the parts of his life, we should get the total age.

Parts of his life
$$\begin{cases} \frac{1}{6} \cdot x + \frac{1}{12} \cdot x \text{ is the time of his youth.} \\ \frac{1}{7} \cdot x \text{ is the time between his youth and when he married.} \\ 5 \text{ years is the time between his marriage and the birth of his son.} \\ \frac{1}{2} \cdot x \text{ is the time Diophantus had with his son.} \\ 4 \text{ years is the time between his son's death and his own.} \end{cases}$$

The sum of these parts should equal Diophantus' age when he died.

$$\frac{1}{6} \cdot x + \frac{1}{12} \cdot x + \frac{1}{7} \cdot x + 5 + \frac{1}{2} \cdot x + 4 = x$$

CRITICAL THINKING

1. Solve the epigram.
2. How old was Diophantus when his son was born? How old was the son when he died?

3. Solve the following epigram:
   I was four when my mother packed my lunch and sent me off to school. Half my life was spent in school and another sixth was spent on a farm. Alas, hard times befell me. My crops and cattle fared poorly and my land was sold. I returned to school for 3 years and have spent one tenth of my life teaching. How old am I?

GROUP ACTIVITY

4. Write an epigram describing your life. Be sure that none of the time periods in your epigram overlap. Exchange epigrams with a partner to solve and check.

Problem Solving Notes

6.7   SIMPLIFYING COMPLEX FRACTIONS

Objectives

A Simplify complex fractions using method 1.

B Simplify complex fractions using method 2.

SSM   CD-ROM   Video
6.7

A rational expression whose numerator or denominator or both numerator and denominator contain fractions is called a **complex rational expression** or a **complex fraction**. Some examples are

$$\frac{4}{2-\frac{1}{2}} \qquad \frac{\frac{3}{2}}{\frac{4}{7}-x} \qquad \left.\frac{\frac{1}{x+2}}{x+2-\frac{1}{x}}\right\}$$

$\leftarrow$ Numerator of complex fraction
$\leftarrow$ Main fraction bar
$\leftarrow$ Denominator of complex fraction

Our goal in this section is to write complex fractions in simplest form. A complex fraction is in simplest form when it is in the form $\frac{P}{Q}$, where P and Q are polynomials that have no common factors.

A   SIMPLIFYING COMPLEX FRACTIONS—METHOD 1

In this section, two methods of simplifying complex fractions are represented. The first method presented uses the fact that the main fraction bar indicates division.

METHOD 1: TO SIMPLIFY A COMPLEX FRACTION

Step 1. Add or subtract fractions in the numerator or denominator so that the numerator is a single fraction and the denominator is a single fraction.

Step 2. Perform the indicated division by multiplying the numerator of the complex fraction by the reciprocal of the denominator of the complex fraction.

Step 3. Write the rational expression in lowest terms.

Example 1   Simplify the complex fraction $\dfrac{\frac{5}{8}}{\frac{2}{3}}$.

Solution: Since the numerator and denominator of the complex fraction are already single fractions, we proceed to step 2: perform the indicated division by multiplying the numerator $\frac{5}{8}$ by the reciprocal of the denominator $\frac{2}{3}$.

$$\frac{\frac{5}{8}}{\frac{2}{3}} = \frac{5}{8} \cdot \frac{3}{2} = \frac{15}{16}$$

The reciprocal of $\frac{2}{3}$ is $\frac{3}{2}$.

Example 2   Simplify: $\dfrac{\frac{2}{3}+\frac{1}{5}}{\frac{2}{3}-\frac{2}{9}}$

Solution: We simplify above and below the main fraction bar separately. First we add $\frac{2}{3}$ and $\frac{1}{5}$ to obtain a single

Practice Problem 1

Simplify the complex fraction $\dfrac{\frac{3}{7}}{\frac{5}{9}}$.

Practice Problem 2

Simplify: $\dfrac{\frac{3}{4}-\frac{2}{3}}{\frac{1}{2}+\frac{3}{8}}$

Answers

1. $\frac{27}{35}$,   **2.** $\frac{2}{21}$

fraction in the numerator. Then we subtract $\frac{2}{9}$ from $\frac{2}{3}$ to obtain a single fraction in the denominator.

$$\frac{\dfrac{2}{3} + \dfrac{1}{5}}{\dfrac{2}{3} - \dfrac{2}{9}} = \frac{\dfrac{2(5)}{3(5)} + \dfrac{1(3)}{5(3)}}{\dfrac{2(3)}{3(3)} - \dfrac{2}{9}}$$
 The LCD of the numerator's fractions is 15.
 The LCD of the denominator's fractions is 9.

$$= \frac{\dfrac{10}{15} + \dfrac{3}{15}}{\dfrac{6}{9} - \dfrac{2}{9}}$$
 Simplify.

$$= \frac{\dfrac{13}{15}}{\dfrac{4}{9}}$$
 Add the numerator's fractions.
 Subtract the denominator's fractions.

Next we perform the indicated division by multiplying the numerator of the complex fraction by the reciprocal of the denominator of the complex fraction.

$$\frac{\dfrac{13}{15}}{\dfrac{4}{9}} = \frac{13}{15} \cdot \frac{9}{4}$$
 The reciprocal of $\frac{4}{9}$ is $\frac{9}{4}$.

$$= \frac{13 \cdot 3 \cdot 3}{3 \cdot 5 \cdot 4} = \frac{39}{20}$$

Practice Problem 3

Simplify: $\dfrac{\dfrac{2}{5} - \dfrac{1}{x}}{\dfrac{x}{10} - \dfrac{1}{3}}$

Example 3   Simplify: $\dfrac{\dfrac{1}{z} - \dfrac{1}{2}}{\dfrac{1}{3} - \dfrac{z}{6}}$

Solution: Subtract to get a single fraction in the numerator and a single fraction in the denominator of the complex fraction.

$$\frac{\dfrac{1}{z} - \dfrac{1}{2}}{\dfrac{1}{3} - \dfrac{z}{6}} = \frac{\dfrac{2}{2z} - \dfrac{z}{2z}}{\dfrac{2}{6} - \dfrac{z}{6}}$$
 The LCD of the numerator's fractions is $2z$.
 The LCD of the denominator's fractions is 6.

$$= \frac{\dfrac{2 - z}{2z}}{\dfrac{2 - z}{6}}$$

$$= \frac{2 - z}{2z} \cdot \frac{6}{2 - z}$$
 Multiply by the reciprocal of $\frac{2 - z}{6}$.

$$= \frac{2 \cdot 3 \cdot (2 - z)}{2 \cdot z \cdot (2 - z)}$$
 Factor.

$$= \frac{3}{z}$$
 Write in lowest terms.

Answer

3. $\dfrac{6(2x - 5)}{x(3x - 10)}$

B SIMPLIFYING COMPLEX FRACTIONS—METHOD 2

Next we study a second method for simplifying complex fractions. In this method, we multiply the numerator and the denominator of the complex fraction by the LCD of all fractions in the complex fraction.

METHOD 2: TO SIMPLIFY A COMPLEX FRACTION

Step 1. Find the LCD of all the fractions in the complex fraction.

Step 2. Multiply both the numerator and the denominator of the complex fraction by the LCD from Step 1.

Step 3. Perform the indicated operations and write the result in lowest terms.

We use method 2 to rework Example 2.

Example 4 Simplify: $\dfrac{\dfrac{2}{3} + \dfrac{1}{5}}{\dfrac{2}{3} - \dfrac{2}{9}}$

Solution: The LCD of $\dfrac{2}{3}, \dfrac{1}{5}, \dfrac{2}{3}$ and $\dfrac{2}{9}$ is 45, so we multiply the numerator and the denominator of the complex fraction by 45. Then we perform the indicated operations, and write in lowest terms.

$$\frac{\dfrac{2}{3} + \dfrac{1}{5}}{\dfrac{2}{3} - \dfrac{2}{9}} = \frac{45\left(\dfrac{2}{3} + \dfrac{1}{5}\right)}{45\left(\dfrac{2}{3} - \dfrac{2}{9}\right)}$$

$$= \frac{45\left(\dfrac{2}{3}\right) + 45\left(\dfrac{1}{5}\right)}{45\left(\dfrac{2}{3}\right) - 45\left(\dfrac{2}{9}\right)} \quad \text{Apply the distributive property.}$$

$$= \frac{30 + 9}{30 - 10} = \frac{39}{20} \quad \text{Simplify.}$$

Practice Problem 4

Use method 2 to simplify the complex fraction in Practice Problem 2:

$$\frac{\dfrac{3}{4} - \dfrac{2}{3}}{\dfrac{1}{2} + \dfrac{3}{8}}$$

⌐ HELPFUL HINT

The same complex fraction was simplified using two different methods in Examples 2 and 4. Notice that each time the simplified result is the same.

Answer

4. $\dfrac{2}{21}$

Practice Problem 5

Simplify: $\dfrac{1 + \dfrac{x}{y}}{\dfrac{2x + 1}{y}}$

Practice Problem 6

Simplify: $\dfrac{\dfrac{5}{6y} + \dfrac{y}{x}}{\dfrac{y}{3} - x}$

Example 5   Simplify: $\dfrac{\dfrac{x + 1}{y}}{\dfrac{x}{y} + 2}$

Solution: The LCD of $\dfrac{x + 1}{y}$ and $\dfrac{x}{y}$ is y, so we multiply the numerator and the denominator of the complex fraction by y.

$$\dfrac{\dfrac{x + 1}{y}}{\dfrac{x}{y} + 2} = \dfrac{y\left(\dfrac{x + 1}{y}\right)}{y\left(\dfrac{x}{y} + 2\right)}$$

$$= \dfrac{y\left(\dfrac{x + 1}{y}\right)}{y\left(\dfrac{x}{y}\right) + y \cdot 2} \qquad \text{Apply the distributive property in the denominator.}$$

$$= \dfrac{x + 1}{x + 2y} \qquad \text{Simplify.}$$

Example 6   Simplify: $\dfrac{\dfrac{x}{y} + \dfrac{3}{2x}}{\dfrac{x}{2} + y}$

Solution: The LCD of $\dfrac{x}{y}, \dfrac{3}{2x}, \dfrac{x}{2},$ and $\dfrac{y}{1}$ is $2xy$, so we multiply both the numerator and the denominator of the complex fraction by $2xy$.

$$\dfrac{\dfrac{x}{y} + \dfrac{3}{2x}}{\dfrac{x}{2} + y} = \dfrac{2xy\left(\dfrac{x}{y} + \dfrac{3}{2x}\right)}{2xy\left(\dfrac{x}{2} + y\right)}$$

$$= \dfrac{2xy\left(\dfrac{x}{y}\right) + 2xy\left(\dfrac{3}{2x}\right)}{2xy\left(\dfrac{x}{2}\right) + 2xy(y)} \qquad \text{Apply the distributive property.}$$

$$= \dfrac{2x^2 + 3y}{x^2y + 2xy^2}$$

$$\text{or } \dfrac{2x^2 + 3y}{xy(x + 2y)}$$

Answers

5. $\dfrac{y + x}{2x + 1},$ **6.** $\dfrac{5x + 6y^2}{2yx(y - 3x)}$

EXERCISE SET 6.7

A **B** *Simplify each complex fraction. See Examples 1 through 6.*

1. $\dfrac{\frac{1}{2}}{\frac{3}{4}}$

2. $\dfrac{\frac{1}{8}}{-\frac{5}{12}}$

3. $\dfrac{-\frac{4x}{9}}{-\frac{2x}{3}}$

4. $\dfrac{-\frac{6y}{11}}{\frac{4y}{9}}$

5. $\dfrac{-\frac{5}{12x^2}}{\frac{25}{16x^3}}$

6. $\dfrac{-\frac{7}{8y}}{\frac{21}{4y}}$

7. $\dfrac{\frac{1}{3}}{\frac{1}{2}-\frac{1}{4}}$

8. $\dfrac{\frac{7}{10}-\frac{3}{5}}{\frac{1}{2}}$

9. $\dfrac{2+\frac{7}{10}}{1+\frac{3}{5}}$

10. $\dfrac{4-\frac{11}{12}}{5+\frac{1}{4}}$

11. $\dfrac{\frac{m}{n}-1}{\frac{m}{n}+1}$

12. $\dfrac{\frac{x}{2}+2}{\frac{x}{2}-2}$

13. $\dfrac{\frac{1}{5}-\frac{1}{x}}{\frac{7}{10}+\frac{1}{x^2}}$

14. $\dfrac{\frac{1}{y^2}+\frac{2}{3}}{\frac{1}{y}-\frac{5}{6}}$

15. $\dfrac{1+\frac{1}{y-2}}{y+\frac{1}{y-2}}$

16. $\dfrac{x-\frac{1}{2x+1}}{1-\frac{x}{2x+1}}$

17. $\dfrac{\frac{4y-8}{16}}{\frac{6y-12}{4}}$

18. $\dfrac{\frac{7y+21}{3}}{\frac{3y+9}{8}}$

19. $\dfrac{\frac{x}{y}+1}{\frac{x}{y}-1}$

20. $\dfrac{\frac{3}{5y}+8}{\frac{3}{5y}-8}$

1. _____
2. _____
3. _____
4. _____
5. _____
6. _____
7. _____
8. _____
9. _____
10. _____
11. _____
12. _____
13. _____
14. _____
15. _____
16. _____
17. _____
18. _____
19. _____
20. _____

CHAPTER 6 HIGHLIGHTS

DEFINITIONS AND CONCEPTS	EXAMPLES

SECTION 6.1 SIMPLIFYING RATIONAL EXPRESSIONS

A **rational expression** is an expression that can be written in the form $\dfrac{P}{Q}$, where P and Q are polynomials and Q does not equal 0.

$$\frac{7y^3}{4}, \frac{x^2 + 6x + 1}{x - 3}, \frac{-5}{s^3 + 8}$$

To find values for which a rational expression is undefined, find values for which the denominator is 0.

Find any values for which the expression $\dfrac{5y}{y^2 - 4y + 3}$ is undefined.

$$y^2 - 4y + 3 = 0 \quad \text{Set the denominator equal to 0.}$$
$$(y - 3)(y - 1) = 0 \quad \text{Factor.}$$
$$y - 3 = 0 \text{ or } y - 1 = 0 \quad \text{Set each factor equal to 0.}$$
$$y = 3 \qquad y = 1 \quad \text{Solve.}$$

The expression is undefined when y is 3 and when y is 1.

FUNDAMENTAL PRINCIPLE OF RATIONAL EXPRESSIONS

If P, Q, and R are polynomials, and Q and R are not 0, then

$$\frac{PR}{QR} = \frac{P}{Q}$$

By the fundamental principle,

$$\frac{(x - 3)(x + 1)}{x(x + 1)} = \frac{x - 3}{x}$$

as long as $x \neq 0$ and $x \neq -1$.

TO SIMPLIFY A RATIONAL EXPRESSION

Step 1. Factor the numerator and denominator.

Step 2. Apply the fundamental principle to divide out common factors.

Simplify: $\dfrac{4x + 20}{x^2 - 25}$

$$\frac{4x + 20}{x^2 - 25} = \frac{4(x + 5)}{(x + 5)(x - 5)} = \frac{4}{x - 5}$$

SECTION 6.2 MULTIPLYING AND DIVIDING RATIONAL EXPRESSIONS

TO MULTIPLY RATIONAL EXPRESSIONS

Step 1. Factor numerators and denominators.

Step 2. Multiply numerators and multiply denominators.

Step 3. Write the product in lowest terms.

$$\frac{P}{Q} \cdot \frac{R}{S} = \frac{PR}{QS}$$

Multiply: $\dfrac{4x + 4}{2x - 3} \cdot \dfrac{2x^2 + x - 6}{x^2 - 1}$

$$\frac{4x + 4}{2x - 3} \cdot \frac{2x^2 + x - 6}{x^2 - 1}$$

$$= \frac{4(x + 1)}{2x - 3} \cdot \frac{(2x - 3)(x + 2)}{(x + 1)(x - 1)}$$

$$= \frac{4(x + 1)(2x - 3)(x + 2)}{(2x - 3)(x + 1)(x - 1)}$$

$$= \frac{4(x + 2)}{x - 1}$$

SECTION 6.2 (CONTINUED)

To divide by a rational expression, multiply by the reciprocal.

$$\frac{P}{Q} \div \frac{R}{S} = \frac{P}{Q} \cdot \frac{S}{R} = \frac{PS}{QR}$$

Divide: $\dfrac{15x + 5}{3x^2 - 14x - 5} \div \dfrac{15}{3x - 12}$

$$\frac{15x + 5}{3x^2 - 14x - 5} \div \frac{15}{3x - 12}$$

$$= \frac{5(3x + 1)}{(3x + 1)(x - 5)} \cdot \frac{3(x - 4)}{3 \cdot 5}$$

$$= \frac{x - 4}{x - 5}$$

SECTION 6.3 ADDING AND SUBTRACTING RATIONAL EXPRESSIONS WITH THE SAME DENOMINATOR AND LEAST COMMON DENOMINATOR

To add or subtract rational expressions with the same denominator, add or subtract numerators, and place the sum or difference over the common denominator.

$$\frac{P}{R} + \frac{Q}{R} = \frac{P + Q}{R}$$

$$\frac{P}{R} - \frac{Q}{R} = \frac{P - Q}{R}$$

Perform each indicated operation.

$$\frac{5}{x + 1} + \frac{x}{x + 1} = \frac{5 + x}{x + 1}$$

$$\frac{2y + 7}{y^2 - 9} - \frac{y + 4}{y^2 - 9}$$

$$= \frac{(2y + 7) - (y + 4)}{y^2 - 9}$$

$$= \frac{2y + 7 - y - 4}{y^2 - 9}$$

$$= \frac{y + 3}{(y + 3)(y - 3)}$$

$$= \frac{1}{y - 3}$$

TO FIND THE LEAST COMMON DENOMINATOR (LCD)

Step 1. Factor the denominators.

Step 2. The LCD is the product of all unique factors, each raised to a power equal to the greatest number of times that it appears in any one factored denominator.

Find the LCD for

$$\frac{7x}{x^2 + 10x + 25} \quad \text{and} \quad \frac{11}{3x^2 + 15x}$$

$$x^2 + 10x + 25 = (x + 5)(x + 5)$$

$$3x^2 + 15x = 3x(x + 5)$$

$$\text{LCD} = 3x(x + 5)(x + 5) \text{ or}$$

$$3x(x + 5)^2$$

SECTION 6.4 ADDING AND SUBTRACTING RATIONAL EXPRESSIONS WITH DIFFERENT DENOMINATORS

TO ADD OR SUBTRACT RATIONAL EXPRESSIONS WITH DIFFERENT DENOMINATORS

Step 1. Find the LCD.

Perform the indicated operation.

$$\frac{9x + 3}{x^2 - 9} - \frac{5}{x - 3}$$

$$= \frac{9x + 3}{(x + 3)(x - 3)} - \frac{5}{x - 3}$$

LCD is $(x + 3)(x - 3)$.

SECTION 6.4 (CONTINUED)

Step 2. Rewrite each rational expression as an equivalent expression whose denominator is the LCD.

Step 3. Add or subtract numerators and place the sum or difference over the common denominator.

Step 4. Write the result in lowest terms.

$$= \frac{9x + 3}{(x + 3)(x - 3)} - \frac{5(x + 3)}{(x - 3)(x + 3)}$$

$$= \frac{9x + 3 - 5(x + 3)}{(x + 3)(x - 3)}$$

$$= \frac{9x + 3 - 5x - 15}{(x + 3)(x - 3)}$$

$$= \frac{4x - 12}{(x + 3)(x - 3)}$$

$$= \frac{4(x - 3)}{(x + 3)(x - 3)} = \frac{4}{x + 3}$$

SECTION 6.5 SOLVING EQUATIONS CONTAINING RATIONAL EXPRESSIONS

TO SOLVE AN EQUATION CONTAINING RATIONAL EXPRESSIONS

Step 1. Multiply both sides of the equation by the LCD of all rational expressions in the equation.

Step 2. Remove any grouping symbols and solve the resulting equation.

Step 3. Check the solution in the original equation.

Solve: $\dfrac{5x}{x + 2} + 3 = \dfrac{4x - 6}{x + 2}$ The LCD is $x + 2$.

$$(x + 2)\left(\frac{5x}{x + 2} + 3\right) = (x + 2)\left(\frac{4x - 6}{x + 2}\right)$$

$$(x + 2)\left(\frac{5x}{x + 2}\right) + (x + 2)(3)$$

$$= (x + 2)\left(\frac{4x - 6}{x + 2}\right)$$

$$5x + 3x + 6 = 4x - 6$$

$$4x = -12$$

$$x = -3$$

The solution checks; the solution is -3.

SECTION 6.6 RATIONAL EQUATIONS AND PROBLEM SOLVING

PROBLEM-SOLVING STEPS

1. UNDERSTAND. Read and reread the problem.

A small plane and a car leave Kansas City, Missouri, and head for Minneapolis, Minnesota, a distance of 450 miles. The speed of the plane is 3 times the speed of the car, and the plane arrives 6 hours ahead of the car. Find the speed of the car.

Let x = the speed of the car.
Then $3x$ = the speed of the plane.

Distance	= Rate	· Time
Car	450	x	$\dfrac{450}{x}\left(\dfrac{\text{distance}}{\text{rate}}\right)$
Plane	450	$3x$	$\dfrac{450}{3x}\left(\dfrac{\text{distance}}{\text{rate}}\right)$

2. TRANSLATE.

In words: $\underbrace{\text{plane's time}}$ + $\underbrace{\text{6 hours}}$ = $\underbrace{\text{car's time}}$

$\qquad\qquad\quad \downarrow \qquad\quad \downarrow \qquad\quad \downarrow$

Translate: $\quad \dfrac{450}{3x} \quad + \quad 6 \quad = \quad \dfrac{450}{x}$

3. SOLVE.

$$\dfrac{450}{3x} + 6 = \dfrac{450}{x}$$

$$3x\left(\dfrac{450}{3x}\right) + 3x(6) = 3x\left(\dfrac{450}{x}\right)$$
$$450 + 18x = 1350$$
$$18x = 900$$
$$x = 50$$

4. INTERPRET.

Check the solution by replacing x with 50 in the original equation. **State** the conclusion: The speed of the car is 50 miles per hour.

SECTION 6.7 SIMPLIFYING COMPLEX FRACTIONS

METHOD 1: TO SIMPLIFY A COMPLEX FRACTION

Step 1. Add or subtract fractions in the numerator and the denominator of the complex fraction.

Step 2. Perform the indicated division.

Step 3. Write the result in lowest terms.

Simplify:

$$\dfrac{\dfrac{1}{x} + 2}{\dfrac{1}{x} - \dfrac{1}{y}} = \dfrac{\dfrac{1}{x} + \dfrac{2x}{x}}{\dfrac{y}{xy} - \dfrac{x}{xy}}$$

$$= \dfrac{\dfrac{1 + 2x}{x}}{\dfrac{y - x}{xy}}$$

$$= \dfrac{1 + 2x}{x} \cdot \dfrac{xy}{y - x}$$

$$= \dfrac{y(1 + 2x)}{y - x}$$

SECTION 6.7 (CONTINUED)	
METHOD 2. TO SIMPLIFY A COMPLEX FRACTION **Step 1.** Find the LCD of all fractions in the complex fraction. **Step 2.** Multiply the numerator and the denominator of the complex fraction by the LCD. **Step 3.** Perform the indicated operations and write the result in lowest terms.	$$\dfrac{\dfrac{1}{x}+2}{\dfrac{1}{x}-\dfrac{1}{y}}=\dfrac{xy\left(\dfrac{1}{x}+2\right)}{xy\left(\dfrac{1}{x}-\dfrac{1}{y}\right)}$$ $$=\dfrac{xy\left(\dfrac{1}{x}\right)+xy(2)}{xy\left(\dfrac{1}{x}\right)-xy\left(\dfrac{1}{y}\right)}$$ $$=\dfrac{y+2xy}{y-x}\quad\text{or}\quad\dfrac{y(1+2x)}{y-x}$$

Focus On Business and Career

MORTGAGES

A loan for which the purpose is to buy a house or other property is called a **mortgage**. When you are thinking of getting a mortgage to buy a house, it is helpful to know how much your monthly mortgage payment will be. One way to calculate the monthly payment is to use the formula

$$P = \frac{\dfrac{Ar}{12}}{1 - \dfrac{1}{\left(1 + \dfrac{r}{12}\right)^{12t}}},$$

where $A =$ is the amount of the mortgage, $r =$ annual interest rate (written as a decimal), and $t =$ loan term in years. Try the exercises below.

CRITICAL THINKING

1. The average mortgage rate in the United States in 1996 was 7.93% (Source: National Association of Realtors®). Suppose you had borrowed $80,000 to buy a house in 1996. If your loan term is 30 years, calculate your monthly mortgage payment.

2. The average mortgage interest rate in the United States in 1989 was 10.11% (Source: National Association of Realtors®). Suppose you had borrowed $71,000 to buy a house in 1989. If your loan term is 20 years, calculate your monthly mortgage payment.

Another way to calculate a monthly mortgage payment is to use one of the many sites on the World Wide Web that offer an interactive mortgage calculator. For instance, by visiting the given World Wide Web address, you will be able to access the Fleet Bank Web site, or a related site, where you can calculate a monthly mortgage payment by entering the amount to be borrowed, the interest rate as a percent, and the term of the loan in years. Use the site below to solve the given exercises.

Go to http://www.prenhall.com/martin-gay

Internet Excursions

3. Suppose you would like to borrow $64,000 to buy a house. If the interest rate is 8.2% and you plan to take out a 25-year loan, what will be your monthly mortgage payment?
4. Suppose you would like to borrow $100,000 to buy a house. If the interest rate is 7.5% and you plan to take out a 20-year loan, what will be your monthly mortgage payment?

Graphs and Functions

In this chapter, we continue our investigation of graphs of equations in two variables, which began in Chapter 3. These equations and their graphs lead to the notion of relation and to the notion of function, perhaps the single most important and useful concept in all of mathematics.

At the beginning of the 20th century, there were approximately 237,600 students enrolled in the 977 institutions of higher education in the United States. At that time, only 19% of bachelor's degree recipients were women. By the year 2000, the projected 3800 colleges and universities in the United States will have an estimated 14,800,000 students. Roughly 56% of bachelor's degree recipients are expected to be women. The phenomenal growth of colleges and universities can also be seen in the average tuition costs at these institutions of higher learning. For instance, the average annual tuition at a private four-year college or university has increased from $1809 in 1970 to $13,664 in 1998, an increase of about 655%!

Problem Solving Notes

7.1 THE SLOPE-INTERCEPT FORM

A GRAPHING A LINE USING SLOPE AND *Y*-INTERCEPT

From Section 3.4, we know that when a linear equation is solved for y, the coefficient of x is the slope of the line. For example, the slope of the line whose equation is $y = 3x + 1$ is 3. In this equation, $y = 3x + 1$, what does 1 represent? To find out, let $x = 0$ and watch what happens.

$$y = 3x + 1$$
$$y = 3 \cdot 0 + 1 \quad \text{Let } x = 0.$$
$$y = 1$$

We now have the ordered pair $(0, 1)$, which means that 1 is the y-intercept.

This is true in general. To see this, let $x = 0$ and solve for y in $y = mx + b$.

$$y = m \cdot 0 + b \quad \text{Let } x = 0.$$
$$y = b$$

We obtain the ordered pair $(0, b)$, which means that b is the y-intercept.

The form $y = mx + b$ is appropriately called the **slope-intercept form** of a linear equation.

slope   y-intercept

SLOPE-INTERCEPT FORM

When a linear equation in two variables is written in slope-intercept form,
$$y = mx + b$$
then m is the slope of the line and b is the y-intercept of the line.

We can use the slope-intercept form to graph a linear equation.

Example 1   Graph: $y = \dfrac{1}{4}x - 3$

Solution: Recall that the slope of the graph of $y = \dfrac{1}{4}x - 3$ is $\dfrac{1}{4}$ and the y-intercept is -3. To graph the line, we first plot the y-intercept point $(0, -3)$. To find another point on the line, we recall that slope is $\dfrac{\text{rise}}{\text{run}} = \dfrac{1}{4}$. Another point may then be plotted by starting at $(0, -3)$, rising 1 unit up, and then running 4 units to the right. We are now at the point $(4, -2)$. The graph of $y = \dfrac{1}{4}x - 3$ is the line through points $(0, -3)$ and $(4, -2)$, as shown.

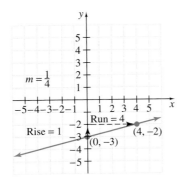

Objectives

A Graph a line using its slope and y-intercept.

B Use the slope-intercept form to write an equation of the line.

C Interpret the slope-intercept form in an application.

SSM   CD-ROM   Video
7.1

Practice Problem 1

Graph: $y = \dfrac{2}{3}x + 1$

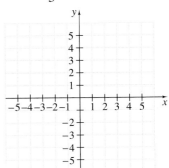

Answer

1.

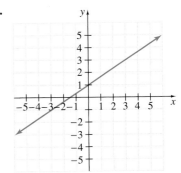

Practice Problem 2

Graph: $3x + y = -2$

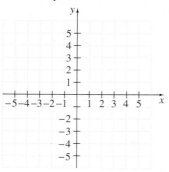

Practice Problem 3

Write an equation of the line with slope $\frac{2}{3}$ and y-intercept 1.

✓ CONCEPT CHECK

What is wrong with the following equation of a line with y-intercept 4 and slope 2?

$y = 4x + 2$

Answers

2.

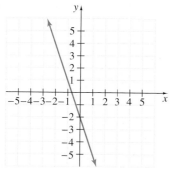

3. $y = \frac{2}{3}x + 1$

✓ **Concept Check:**   The y-intercept and slope were switched. It should be $y = 2x + 4$.

Example 2

Graph: $2x + y = 3$

Solution: First, we solve the equation for y to write it in slope-intercept form. In slope-intercept form, the equation is $y = -2x + 3$. Next we plot the y-intercept point, $(0, 3)$. To find another point on the line, we use the slope -2, which can be written as $\dfrac{\text{rise}}{\text{run}} = \dfrac{-2}{1}$. We start at $(0, 3)$ and move vertically 2 units down, since the numerator of the slope is -2; then we move horizontally 1 unit to the right since the denominator of the slope is 1. We arrive at the point $(1, 1)$. The line through $(1, 1)$ and $(0, 3)$ will have the required slope of -2.

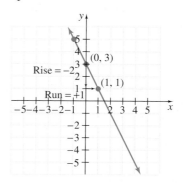

B USING THE SLOPE-INTERCEPT FORM TO WRITE AN EQUATION

Given the slope and y-intercept of a line, we may write its equation as well as graph the line.

Example 3

Write an equation of the line with y-intercept -3 and slope of $\frac{1}{4}$.

Solution: We let $m = \dfrac{1}{4}$ and $b = -3$, and write the equation in slope-intercept form, $y = mx + b$.

$y = mx + b$

$y = \dfrac{1}{4}x + (-3)$ Let $m = \frac{1}{4}$ and $b = -3$.

$y = \dfrac{1}{4}x - 3$ Simplify.

Notice that the graph of this equation has slope $\dfrac{1}{4}$ and y-intercept -3, as desired.

TRY THE CONCEPT CHECK IN THE MARGIN.

C INTERPRETING THE SLOPE-INTERCEPT FORM

On the next page is a graph of an adult one-day pass price for Disney World over time. Often, businesses depend on linear equations that "closely fit" data to model the data and predict future trends. For example, by a method called least squares regression, the linear equation $y = 1.462x + 29.35$ approximates the data shown, where x is the number of years since 1988 and y is the ticket price for that year.

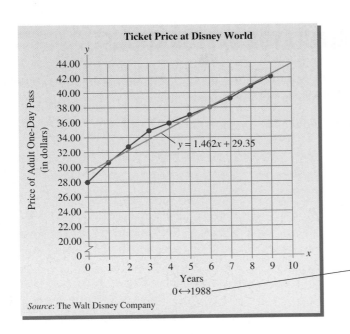

Ticket Price at Disney World

$y = 1.462x + 29.35$

Source: The Walt Disney Company

Example 4   Predicting Future Prices

The adult one-day pass price y for Disney World is given by
$$y = 1.462x + 29.35$$

where x is the number of years since 1988.

a. Use this equation to predict the ticket price for the year 2002.
b. What does the slope of this equation mean?
c. What does the y-intercept of this equation mean?

Solution:

a. To predict the price of a pass in 2002, we need to find y when x is 14. (Since year 1988 corresponds to $x = 0$, year 2002 corresponds to $x = 2002 - 1988 = 14$.

$$y = 1.462x + 29.35$$
$$= 1.462(14) + 29.35 \quad \text{Let } x = 14.$$
$$= 49.818$$

We predict that in the year 2002, the price of an adult one-day pass to Disney World will be about $49.82.

b. The slope of $y = 1.462x + 29.35$ is 1.462. We can think of this number as $\dfrac{\text{rise}}{\text{run}}$ or $\dfrac{1.462}{1}$. This means that the ticket price increases on the average by $1.462 every 1 year.

c. The y-intercept of $y = 1.462x + 29.35$ is 29.35. Notice that it corresponds to the point of the graph $(0, 29.35)$.
↑   ↖
year   price

This means that at year $x = 0$ or 1988, the ticket price was $29.35.

Practice Problem 4

The yearly average income y of an American woman with some high school education but no diploma is given by the equation
$$y = 356.5x + 8912.2,$$
where x is the number of years since 1991. (*Source*: Based on data from the U.S. Bureau of the Census, 1991–1996)

a. Predict the income for the year 2001.
b. What does the slope of this equation mean?
c. What does the y-intercept of this equation mean?

Answers
4. a. $12,477.20,   **b.** The yearly average income increases by $356.50 every year,   **c.** At year $x = 0$, or 1991, the yearly average income was $8912.20.

GRAPHING CALCULATOR EXPLORATIONS

You may have noticed by now that to use the $\boxed{Y=}$ key on a grapher to graph an equation, the equation must be solved for y.

Graph each equation by first solving the equation for y.

1. $x = 3.5y$

2. $-2.7y = x$

3. $5.78x + 2.31y = 10.98$

4. $-7.22x + 3.89y = 12.57$

5. $y - x = 3.78$

6. $3y - 5x = 6x - 4$

7. $y - 5.6x = 7.7x + 1.5$

8. $y + 2.6x = -3.2$

MENTAL MATH

Find the slope and the y-intercept of each line.

1. $y = -4x + 12$

2. $y = \frac{2}{3}x - \frac{7}{2}$

3. $y = 5x$

4. $y = -x$

5. $y = \frac{1}{2}x + 6$

EXERCISE SET 7.1

A *Graph each line passing through the given point with the given slope. See Examples 1 and 2.*

1. Through $(1, 3)$ with slope $\frac{3}{2}$

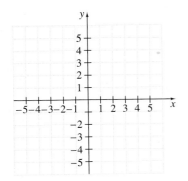

2. Through $(-2, -4)$ with slope $\frac{2}{5}$

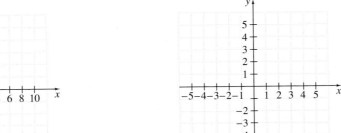

3. Through $(0, 0)$ with slope 5

4. Through $(-5, 2)$ with slope 2

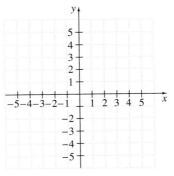

5. Through $(0, 7)$ with slope -1

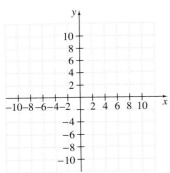

6. Through $(3, 0)$ with slope -3

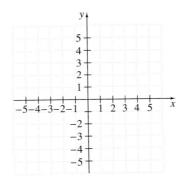

MENTAL MATH ANSWERS

1. _____

2. _____

3. _____

4. _____

5. _____

ANSWERS

1. see graph

2. see graph

3. see graph

4. see graph

5. see graph

6. see graph

Graph each linear equation using the slope and y-intercept. See Examples 1 and 2.

7. $y = -2x$ **8.** $y = 2x$ **9.** $y = -2x + 3$ **10.** $y = 2x + 6$

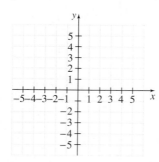 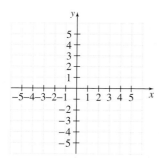 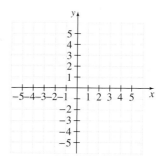 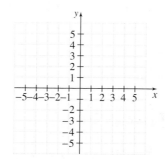

11. $y = \dfrac{1}{2}x$

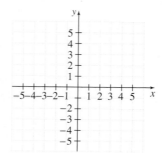

B *Use the slope-intercept form of a linear equation to write the equation of each line with the given slope and y-intercept. See Example 3.*

12. Slope -1; y-intercept 1 **13.** Slope $\dfrac{1}{2}$; y-intercept -6

14. Slope 2; y-intercept $\dfrac{3}{4}$ **15.** Slope -3; y-intercept $-\dfrac{1}{5}$

12. _____

13. _____

14. _____

15. _____

7.2   MORE EQUATIONS OF LINES

A   USING THE POINT-SLOPE FORM TO WRITE AN EQUATION

When the slope of a line and a point on the line are known, the equation of the line can also be found. To do this, we use the slope formula to write the slope of a line that passes through points (x, y), and (x_1, y_1). We have

$$m = \frac{y - y_1}{x - x_1}$$

We multiply both sides of this equation by $x - x_1$ to obtain

$$y - y_1 = m(x - x_1)$$

This form is called the **point-slope form** of the equation of a line.

POINT-SLOPE FORM OF THE EQUATION OF A LINE

The **point-slope form** of the equation of a line is

$$y - \underset{\text{point}}{y_1} = \overset{\overset{\text{slope}}{\downarrow}}{m}(x - x_1)$$

where m is the slope of the line and (x_1, y_1) is a point on the line.

Objectives

A Use the point-slope form to write the equation of a line.

B Write equations of vertical and horizontal lines.

C Write equations of parallel and perpendicular lines.

D Use the point-slope form in real-world applications.

SSM   CD-ROM   Video
7.2

Example 1   Write an equation of the line with slope -3 containing the point $(1, -5)$.

Solution: Because we know the slope and a point on the line, we use the point-slope form with $m = -3$ and $(x_1, y_1) = (1, -5)$.

$$y - y_1 = m(x - x_1) \quad \text{Point-slope form}$$
$$y - (-5) = -3(x - 1) \quad \text{Let } m = -3 \text{ and } (x_1, y_1) = (1, -5).$$
$$y + 5 = -3x + 3 \quad \text{Use the distributive property.}$$
$$y = -3x - 2$$

The equation is $y = -3x - 2$.

Practice Problem 1

Write an equation of the line with slope -2 containing the point $(2, -4)$. Write the equation in slope-intercept form, $y = mx + b$.

Example 2   Write an equation of the line through points $(4, 0)$ and $(-4, -5)$.

Solution: First we find the slope of the line.

$$m = \frac{-5 - 0}{-4 - 4} = \frac{-5}{-8} = \frac{5}{8}$$

Next we make use of the point-slope form. We replace (x_1, y_1) by either $(4, 0)$ or $(-4, -5)$ in the point-slope equation. We will choose the point $(4, 0)$. The line through $(4, 0)$ with slope $\frac{5}{8}$ is

Practice Problem 2

Write an equation of the line through points $(3, 0)$ and $(-2, 4)$. Write the equation in slope-intercept form, $y = mx + b$.

Answers

1. $y = -2x$,   **2.** $y = -\frac{4}{5}x + \frac{12}{5}$

$$y - y_1 = m(x - x_1) \qquad \text{Point-slope form}$$

$$y - 0 = \frac{5}{8}(x - 4) \qquad \text{Let } m = \frac{5}{8} \text{ and } (x_1, y_1) = (4, 0).$$

$$y = \frac{5}{8}x - \frac{5}{8} \cdot 4 \qquad \text{Use the distributive property.}$$

$$y = \frac{5}{8}x - \frac{5}{2} \qquad \text{Simplify.}$$

The equation is $y = \frac{5}{8}x - \frac{5}{2}$. If we choose to use the point $(-4, -5)$, we have $y - (-5) = \frac{5}{8}[x - (-4)]$, which also simplifies to $y = \frac{5}{8}x - \frac{5}{2}$.

> **⌐HELPFUL HINT**
>
> If two points of a line are given, either one may be used with the slope-intercept form to write an equation of the line.

B WRITING EQUATIONS OF VERTICAL AND HORIZONTAL LINES

A few special types of linear equations are those whose graphs are vertical and horizontal lines.

Practice Problem 3

Write an equation of the horizontal line containing the point $(-1, 6)$.

Example 3   Write an equation of the horizontal line containing the point $(2, 3)$.

Solution: Recall from Section 3.1, that a horizontal line has an equation of the form $y = b$. Since the line contains the point $(2, 3)$, the equation is $y = 3$.

Practice Problem 4

Write an equation of the line containing the point $(4, 7)$ with undefined slope.

Example 4   Write an equation of the line containing the point $(2, 3)$ with undefined slope.

Solution: Since the line has undefined slope, the line must be vertical. A vertical line has an equation of the form $x = c$, and since the line contains the point $(2, 3)$, the equation is $x = 2$.

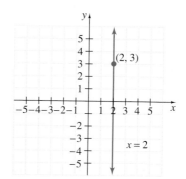

C WRITING EQUATIONS OF PARALLEL AND PERPENDICULAR LINES

Next, we write equations of parallel and perpendicular lines.

Example 5   Write an equation of the line containing the point $(4, 4)$ and parallel to the line $2x + y = -6$.

Solution: Because the line we want to find is *parallel* to the line $2x + y = -6$, the two lines must have equal slopes. So we first find the slope of $2x + y = -6$ by solving it for y to write it in the form $y = mx + b$. Here $y = -2x - 6$ so the slope is -2.

Now we use the point-slope form to write the equation of a line through $(4, 4)$ with slope -2.

$$y - y_1 = m(x - x_1)$$
$$y - 4 = -2(x - 4) \quad \text{Let } m = -2, x_1 = 4, \text{ and } y_1 = 4.$$
$$y - 4 = -2x + 8 \quad \text{Use the distributive property.}$$
$$y = -2x + 12$$

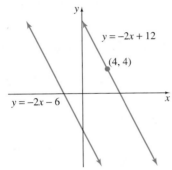

The equation, $y = -2x - 6$, and the new equation, $y = -2x + 12$, have the same slope but different y-intercepts so their graphs are parallel. Also, the graph of $y = -2x + 12$ contains the point $(4, 4)$, as desired.

Example 6   Write an equation of the line containing the point $(-2, 1)$ and perpendicular to the line $3x + 5y = 4$.

Solution: First we find the slope of $3x + 5y = 4$ by solving it for y.

$$5y = -3x + 4$$
$$y = -\frac{3}{5}x + \frac{4}{5}$$

The slope of the given line is $-\frac{3}{5}$. A line perpendicular to this line will have a slope that is the negative reciprocal of $-\frac{3}{5}$, or $\frac{5}{3}$. We use the point-slope form to write an equation of a new line through $(-2, 1)$ with slope $\frac{5}{3}$.

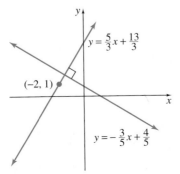

Practice Problem 5

Write an equation of the line containing the point $(-1, 2)$ and parallel to the line $3x + y = 5$. Write the equation in the form $y = mx + b$.

Practice Problem 6

Write an equation of the line containing the point $(3, 4)$ and perpendicular to the line $2x + 4y = 5$. Write the equation in standard form.

Answers

5. $y = -3x - 1$, **6.** $2x - y = 2$

$$y - 1 = \frac{5}{3}[x - (-2)]$$

$$y - 1 = \frac{5}{3}(x + 2) \qquad \text{Simplify.}$$

$$y - 1 = \frac{5}{3}x + \frac{10}{3} \qquad \text{Use the distributive property.}$$

$$y = \frac{5}{3}x + \frac{13}{3} \qquad \text{Add 1 to both sides.}$$

The equation $y = -\frac{3}{5}x + \frac{4}{5}$ and the new equation $y = \frac{5}{3}x + \frac{13}{3}$ have negative reciprocal slopes so their graphs are perpendicular. Also, the graph of $y = \frac{5}{3}x + \frac{13}{3}$ contains the point $(-2, 1)$, as desired.

D USING THE POINT-SLOPE FORM IN APPLICATIONS

The point-slope form of an equation is very useful for solving real-world problems.

Example 7   Predicting Sales

Southern Star Realty is an established real estate company that has enjoyed constant growth in sales since 1990. In 1992 the company sold 200 houses, and in 1997 the company sold 275 houses. Use these figures to predict the number of houses this company will sell in the year 2002.

Solution:

1. UNDERSTAND. Read and reread the problem. Then let

$x =$ the number of years after 1990 and

$y =$ the number of houses sold in the year corresponding to x.

The information provided then gives the ordered pairs $(2, 200)$ and $(7, 275)$. To better visualize the sales of Southern Star Realty, we graph the linear equation that passes through the points $(2, 200)$ and $(7, 275)$.

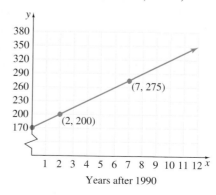

Years after 1990

2. TRANSLATE. We write a linear equation that passes through the points $(2, 200)$ and $(7, 275)$. To do so, we first find the slope of the line.

Practice Problem 7

Southwest Regional is an established office product maintenance company that has enjoyed constant growth in new maintenance contracts since 1985. In 1990, the company obtained 15 new contracts and in 1997, the company obtained 36 new contracts. Use these figures to predict the number of new contracts this company can expect in 2004.

Answer

7. 57 new contracts

$$m = \frac{275 - 200}{7 - 2} = \frac{75}{5} = 15$$

Then, using the point-slope form to write the equation, we have

$$y - y_1 = m(x - x_1)$$
$$y - 200 = 15(x - 2) \quad \text{Let } m = 15 \text{ and } (x_1, y_1) = (2, 200).$$
$$y - 200 = 15x - 30 \quad \text{Multiply.}$$
$$y = 15x + 170 \quad \text{Add 200 to both sides.}$$

3. SOLVE. To predict the number of houses sold in the year 2002, we use $y = 15x + 170$ and complete the ordered pair $(12, \quad)$, since $2002 - 1990 = 12$.

$$y = 15(12) + 170 \quad \text{Let } x = 12.$$
$$y = 350$$

4. INTERPRET.

Check: Verify that the point $(12, 350)$ is a point on the line graphed in step 1.

State: Southern Star Realty should expect to sell 350 houses in the year 2002.

GRAPHING CALCULATOR EXPLORATIONS

Many graphing calculators have a TRACE feature. This feature allows you to trace along a graph and see the corresponding x- and y-coordinates appear on the screen. Use this feature for the following exercises.

Graph each equation and then use the TRACE feature to complete each ordered pair solution. (Many times the tracer will not show an exact x- or y-value asked for. In each case, trace as closely as you can to the given x- or y-coordinate and approximate the other, unknown, coordinate to one decimal place.)

1. $y = 2.3x + 6.7$
$x = 5.1, y = ?$

2. $y = -4.8x + 2.9$
$x = -1.8, y = ?$

3. $y = -5.9x - 1.6$
$x = ?, y = 7.2$

4. $y = 0.4x - 8.6$
$x = ?, y = -4.4$

5. $y = 5.2x - 3.3$
$x = 2.3, y = ?$
$x = ?, y = 36$

6. $y = -6.2x - 8.3$
$x = 3.2, y = ?$
$x = ?, y = 12$

Focus On Business and Career

LINEAR MODELING

As we saw in Section 3.3, businesses often depend on equations that "closely fit" data. To *model* the data means to find an equation that describes the relationship between the paired data of two variables, such as time in years and profit. A model that accurately summarizes the relationship between two variables can be used to replace a potentially lengthy listing of the raw data. An accurate model might also be used to predict future trends by answering questions such as "If the trend seen in our company's performance in the last several years continues, what level of profit can we reasonably expect in 3 years?"

There are several ways to find a linear equation that models a set of data. If only two ordered pair data points are involved, an exact equation that contains both points can be found using the methods of Section 3.4. When more than two ordered pair data points are involved, it may be impossible to find a linear equation that contains all of the data points. In this case, the graph of the **best fit equation** should have a majority of the plotted ordered pair data points on the graph or close to it. In statistics, a technique called least squares regression is used to determine an equation that best fits a set of data. Various graphing utilities have built-in capabilities for finding an equation (called a regression equation) that best fits a set of ordered pair data points. Regression capabilities are often found with a graphing utility's statistics features.* A best fit equation can also be estimated using an algebraic method, which is outlined in the Group Activity below. In either case, a useful first step when finding a linear equation that models a set of data is creating a scatter diagram of the ordered pair data points to verify that a linear equation is an appropriate model.

GROUP ACTIVITY

Coca-Cola Company is the world's largest producer of soft drinks and juices. The table shows Coca-Cola's net profit (in billions of dollars) for the years 1993–1997. Use the table

Year	1993	1994	1995	1996	1997
Net Profit (in billions of dollars)	2.2	2.6	3.0	3.5	4.1

(*Source:* The Coca-Cola Company)

along with your answers to the questions below to find a linear function $f(x)$ that represents net profit (in billions of dollars) as a function of the number of years after 1993.

1. Create a scatter diagram of the paired data given in the table. Does a linear model seem appropriate for the data?

2. Use a straight edge to draw on your graph what appears to be the line that "best fits" the data you plotted.

3. Estimate the coordinates of two points that fall on your best fit line. Use these points to find a linear function $f(x)$ for the line.

4. Use your linear function to find $f(8)$, and interpret its meaning in context.

5. Compare your group's linear function with other groups' functions. Are they the same or different? Explain why.

6. (Optional) Enter the data from the table into a graphing utility and use the linear regression feature to find a linear function that models the data. Compare this function with the one you found in Question 3. How are they alike or different?

7. (Optional) Using corporation annual reports or articles from magazines or newspapers, search for a set of business-related data that could be modeled with a linear function. Explain how modeling this data could be useful to a business. Then find the best fit equation for the data.

*To find out more about using a graphing utility to find a regression equation, consult the user's manual for your graphing utility.

Name _____ Section _____ Date _____

MENTAL MATH

Find the slope and a point of the graph of each equation.

1. $y - 4 = -2(x - 1)$ **2.** $y - 6 = -3(x - 4)$ **3.** $y - 0 = \frac{1}{4}(x - 2)$

4. $y - 1 = -\frac{2}{3}(x - 0)$ **5.** $y + 2 = 5(x - 3)$

EXERCISE SET 7.2

A *Write an equation of each line with the given slope and containing the given point. Write the equation in the form $y = mx + b$. See Example 1.*

1. Slope 3; through $(1, 2)$

2. Slope 4; through $(5, 1)$

3. Slope -2; through $(1, -3)$

4. Slope -4; through $(2, -4)$

5. Slope $\frac{1}{2}$; through $(-6, 2)$

Write an equation of the line passing through the given points. Write the equation in the form $y = mx + b$. See Example 2.

6. $(2, 0)$ and $(4, 6)$

7. $(3, 0)$ and $(7, 8)$

8. $(-2, 5)$ and $(-6, 13)$

9. $(7, -4)$ and $(2, 6)$

10. $(-2, -4)$ and $(-4, -3)$

B *Write an equation of each line. See Examples 3 and 4.*

11. Vertical; through $(2, 6)$

12. Slope 0; through $(-2, -4)$

13. Horizontal; through $(-3, 1)$

14. Vertical; through $(4, 7)$

15. Undefined slope; through $(0, 5)$

C *Write an equation of each line. Write the equation in the form $y = mx + b$. See Examples 5 and 6.*

16. Through $(3, 8)$; parallel to $y = 4x - 2$

17. Through $(1, 5)$; parallel to $y = 3x - 4$

18. Through $(2, -5)$; perpendicular to $y = -2x - 6$

19. Through $(-4, 8)$; perpendicular to $y = -4x - 1$

20. Through $(-2, -3)$; parallel to $3x + 2y = 5$

ANSWERS

1. _____

2. _____

3. _____

4. _____

5. _____

6. _____

7. _____

8. _____

9. _____

10. _____

11. _____

12. _____

13. _____

14. _____

15. _____

16. _____

17. _____

18. _____

19. _____

20. _____

Problem Solving Notes

7.3 INTRODUCTION TO FUNCTIONS

A DEFINING RELATION, DOMAIN, AND RANGE

Equations in two variables, such as $y = 2x + 1$, describe **relations** between x-values and y-values. For example, if $x = 1$, then this equation describes how to find the y-value related to $x = 1$. In words, the equation $y = 2x + 1$ says that twice the x-value increased by 1 gives the corresponding y-value. The x-value of 1 corresponds to the y-value of $2(1) + 1 = 3$ for this equation, and we have the ordered pair $(1, 3)$.

There are other ways of describing relations or correspondences between two numbers or, in general, a first set (sometimes called the set of *inputs*) and a second set (sometimes called the set of *outputs*). For example,

First Set: Input	Correspondence	Second Set: Output
People in a certain city	Each person's age	The set of nonnegative integers

A few examples of ordered pairs from this relation might be (Ana, 4); (Bob, 36); (Trey, 21); and so on.

Below are just a few other ways of describing relations between two sets and the ordered pairs that they generate.

First Set: **Second Set:**
Input   **Output**
Correspondence

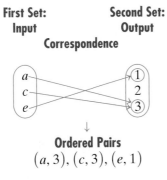

↓
Ordered Pairs
$(a, 3), (c, 3), (e, 1)$

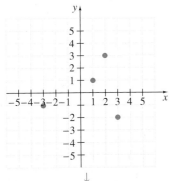

↓
Ordered Pairs
$(-3, -1), (1, 1), (2, 3), (3, -2)$

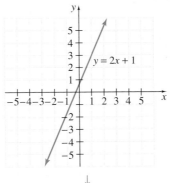

↓
Some Ordered Pairs
$(1, 3), (0, 1)$, and so on

RELATION, DOMAIN, AND RANGE

A **relation** is a set of ordered pairs.
The **domain** of the relation is the set of all first components of the ordered pairs.
The **range** of the relation is the set of all second components of the ordered pairs.

For example, the domain for our middle relation above is $\{a, c, e\}$ and the range is $\{1, 3\}$. Notice that the range does not include the element 2 of the second set. This is because no element of the first set is assigned to this element. If a relation is defined in terms of *x*- and *y*-values, we will agree that the domain corresponds to *x*-values and that the range corresponds to *y*-values.

Practice Problems 1–3

Determine the domain and range of each relation.

1. $\{(1, 6), (2, 8), (0, 3), (0, -2)\}$

2.

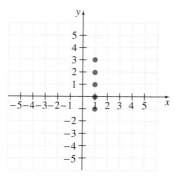

3.

Input: States	Output: Number of Representatives

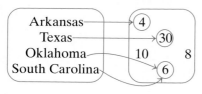

Arkansas ——— 4
Texas ———— 30
Oklahoma ——— 10   8
South Carolina ——— 6

◼ Examples

Determine the domain and range of each relation.

1. $\{(2, 3), (2, 4), (0, -1), (3, -1)\}$
   The domain is the set of all first coordinates of the ordered pairs, $\{2, 0, 3\}$.
   The range is the set of all second coordinates, $\{3, 4, -1\}$.

2.

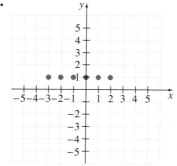

The relation is $\{(-3, 1), (-2, 1), (-1, 1), (0, 1), (1, 1), (2, 1)\}$.
The domain is $\{-3, -2, -1, 0, 1, 2\}$.
The range is $\{1\}$.

3.

Input: Cities	Output: Population (in thousands)

Erie ——— 109
Miami ——— 359   200
Escondido ——— 117   52
Waco ——— 182
Gary ——— 104

The domain is the first set, $\{$ Erie, Escondido, Gary, Miami, Waco$\}$.
The range is the numbers in the second set that correspond to elements in the first set, $\{104, 109, 117, 359\}$.

B IDENTIFYING FUNCTIONS

Now we consider a special kind of relation called a function.

> **FUNCTION**
>
> A **function** is a relation in which each first component in the ordered pairs corresponds to *exactly one* second component.

Answers

1. domain: $\{1, 2, 0\}$, range: $\{6, 8, 3, -2\}$,
2. domain: $\{1\}$, range: $\{-1, 0, 1, 2, 3\}$,
3. domain: $\{$Arkansas, Texas, Oklahoma, South Carolina$\}$, range: $\{4, 30, 6\}$

HELPFUL HINT

A function is a special type of relation, so all functions are relations, but not all relations are functions.

Examples

Determine whether each relation is also a function.

4. $\{(-2, 5), (2, 7), (-3, 5), (9, 9)\}$

Although the ordered pairs $(-2, 5)$ and $(-3, 5)$ have the same y-value, each x-value is assigned to only one y-value, so this set of ordered pairs is a function.

5.

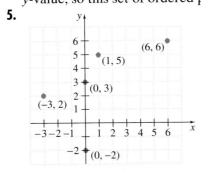

The x-value 0 is assigned to two y-values, -2 and 3, in this graph so this relation is not a function.

6.

Input	Correspondence	Output
People in a certain city	Each person's age	The set of nonnegative integers

This relation is a function because although two different people may have the same age, each person has only one age. This means that each element in the first set is assigned to only one element in the second set. ━━━

TRY THE CONCEPT CHECK IN THE MARGIN.

We will call an equation such as $y = 2x + 1$ a relation since this equation defines a set of ordered pair solutions.

Example 7

Determine whether the relation $y = 2x + 1$ is also a function.

Solution:

The relation $y = 2x + 1$ is a function if each x-value corresponds to just one y-value. For each x-value substituted in the equation $y = 2x + 1$, the multiplication and addition performed gives a single result, so only one y-value will be associated with each x-value. Thus, $y = 2x + 1$ is a function. ━━━

Example 8

Determine whether the relation $x = y^2$ is also a function.

Solution:

In $x = y^2$, if $y = 3$, then $x = 9$. Also, if $y = -3$, then $x = 9$. In other words, we have the ordered pairs $(9, 3)$ and $(9, -3)$. Since the x-value 9 corresponds to two y-values, 3 and -3, $x = y^2$ is not a function. ━━━

Practice Problems 4–6

Determine whether each relation is also a function.

4. $\{(-3, 7), (1, 7), (2, 2)\}$

5.

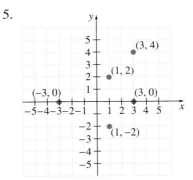

6.

Input	Correspondence	Output
People in a certain state	County/ Parish that a person lives in	Counties of that state

✓ CONCEPT CHECK

Explain why a function can contain both the ordered pairs $(1, 3)$ and $(2, 3)$ but not both $(3, 1)$ and $(3, 2)$.

Practice Problem 7

Determine whether the relation $y = 3x + 2$ is also a function.

Practice Problem 8

Determine whether the relation $x = y^2 + 1$ is also a function.

Answers

4. function,   **5.** not a function,   **6.** function
7. yes,   **8.** no

✓ Concept Check: Two different ordered pairs can have the same y-value, but not the same x-value in a function.

Practice Problems 9–13

Use the vertical line test to determine which are graphs of functions.

9.

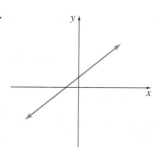

10.

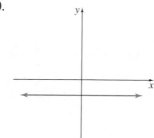

11.

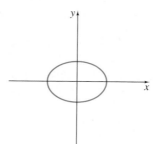

12.

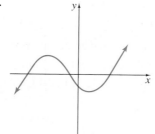

13.

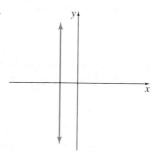

Answers

9. function,   **10.** function,   **11.** not a function,
12. function,   **13.** not a function

C USING THE VERTICAL LINE TEST

As we have seen, not all relations are functions. Consider the graphs of $y = 2x + 1$ and $x = y^2$ shown next. On the graph of $y = 2x + 1$, notice that each x-value corresponds to only one y-value. Recall from Example 7 that $y = 2x + 1$ is a function.

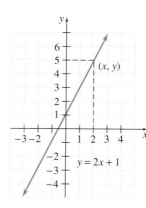
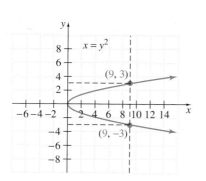

On the graph of $x = y^2$, the x-value 9, for example, corresponds to two y-values, 3 and -3, as shown by the vertical line. Recall from Example 8 that $x = y^2$ is not a function.

Graphs can be used to help determine whether a relation is also a function by the following vertical line test.

VERTICAL LINE TEST

If no vertical line can be drawn so that it intersects a graph more than once, the graph is the graph of a function.

Examples   Use the vertical line test to determine which are graphs of functions.

9.

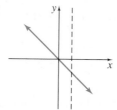

This is the graph of a function since no vertical line will intersect this graph more than once.

10.

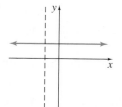

This is the graph of a function.

11.

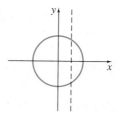

This is not the graph of a function. Note that vertical lines can be drawn that intersect the graph in two points.

12.

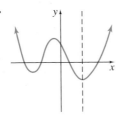

This is the graph
of a function.

13.

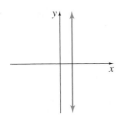

This is not the graph
of a function. A vertical line
can be drawn that intersects
this line at every point.

TRY THE CONCEPT CHECK IN THE MARGIN.

D **FINDING DOMAIN AND RANGE FROM A GRAPH**

Next we practice finding the domain and range of a relation from its graph.

Examples   Find the domain and range of each relation.

Solutions:

14.

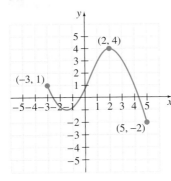

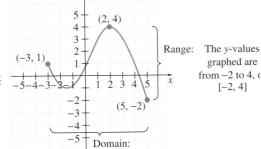

Range: The *y*-values
graphed are
from −2 to 4, or
[−2, 4]

Domain:
The *x*-values graphed are
from −3 to 5, or
[−3, 5]

15.

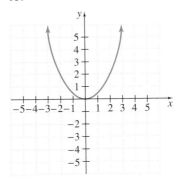

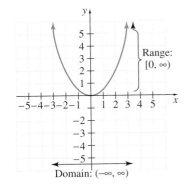

Range:
[0, ∞)

Domain: (−∞, ∞)

✓ **CONCEPT CHECK**

Determine which equations represent
functions. Explain your answer.
a. $y = 14$
b. $x = -5$
c. $x + y = 6$

Practice Problems 14–17

Find the domain and range of each
relation.

14.

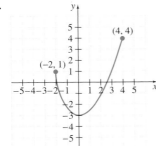

15.

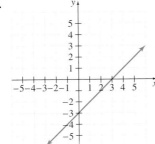

16.

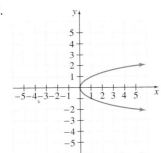

17.

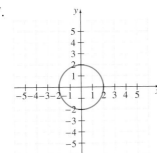

Answers
14. domain: $[-2, 4]$, range: $[-3, 4]$,
15. domain: $[0, \infty)$, range: $(-\infty, \infty)$,
16. domain: $(-\infty, \infty)$, range: $(-\infty, \infty)$,
17. domain: $[-2, 2]$, range: $[-2, 2]$
✓ **Concept Check:**   a, c

16.

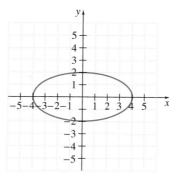 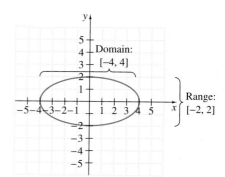

17.

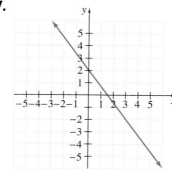 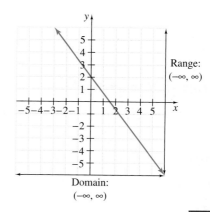

E   USING FUNCTION NOTATION

Many times letters such as f, g, and h are used to name functions. To denote that y is a function of x, we can write

$$y = f(x)$$

This means that **y is a function of x** or that y *depends on x*. For this reason, y is called the **dependent variable** and x the **independent variable**. The notation $f(x)$ is read "f of x" and is called **function notation**.

For example, to use function notation with the function $y = 4x + 3$, we write $f(x) = 4x + 3$. The notation $f(1)$ means to replace x with 1 and find the resulting y or function value. Since

$$f(x) = 4x + 3$$

then

$$f(1) = 4(1) + 3 = 7$$

This means that when $x = 1$, y or $f(x) = 7$. The corresponding ordered pair is $(1, 7)$. Here, the input is 1 and the output is $f(1)$ or 7. Now let's find $f(2)$, $f(0)$, and $f(-1)$.

$$
\begin{aligned}
f(x) &= 4x + 3 \\
f(2) &= 4(2) + 3 \\
&= 8 + 3 \\
&= 11
\end{aligned}
\qquad
\begin{aligned}
f(x) &= 4x + 3 \\
f(0) &= 4(0) + 3 \\
&= 0 + 3 \\
&= 3
\end{aligned}
\qquad
\begin{aligned}
f(x) &= 4x + 3 \\
f(-1) &= 4(-1) + 3 \\
&= -4 + 3 \\
&= -1
\end{aligned}
$$

Ordered Pairs:

$$(2, 11) \qquad\qquad (0, 3) \qquad\qquad (-1, -1)$$

HELPFUL HINT

Note that $f(x)$ is a special symbol in mathematics used to denote a function. The symbol $f(x)$ is read "f of x." It does *not* mean $f \cdot x$ (f times x).

Examples

Find each function value.

18. If $g(x) = 3x - 2$, find $g(1)$.

$$g(1) = 3(1) - 2 = 1$$

19. If $g(x) = 3x - 2$, find $g(0)$.

$$g(0) = 3(0) - 2 = -2$$

20. If $f(x) = 7x^2 - 3x + 1$, find $f(1)$.

$$f(1) = 7(1)^2 - 3(1) + 1 = 5$$

21. If $f(x) = 7x^2 - 3x + 1$, find $f(-2)$.

$$f(-2) = 7(-2)^2 - 3(-2) + 1 = 35$$

TRY THE CONCEPT CHECK IN THE MARGIN.

F GRAPHING LINEAR FUNCTIONS

Recall that the graph of a linear equation in two variables is a line, and a line that is not vertical will always pass the vertical line test. Thus, *all linear equations are functions except those whose graph is a vertical line*. We call such functions **linear functions**.

> **LINEAR FUNCTION**
>
> A **linear function** is a function that can be written in the form
>
> $$f(x) = mx + b$$

Example 22

Graph the function $f(x) = 2x + 1$.

Solution: Since $y = f(x)$, we could replace $f(x)$ with y and graph as usual. The graph of $y = 2x + 1$ has slope 2 and y-intercept 1. Its graph is shown.

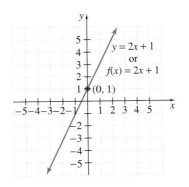

Practice Problems 18–21

Find each function value.

18. If $g(x) = 4x + 5$, find $g(0)$.

19. If $g(x) = 4x + 5$, find $g(-5)$.

20. If $f(x) = 3x^2 - x + 2$, find $f(2)$.

21. If $f(x) = 3x^2 - x + 2$, find $f(-1)$.

✓ CONCEPT CHECK

Suppose $y = f(x)$ and we are told that $f(3) = 9$. Which is not true?
a. When $x = 3$, $y = 9$.
b. A possible function is $f(x) = x^2$.
c. A point on the graph of the function is $(3, 9)$.
d. A possible function is $f(x) = 2x + 4$.

Practice Problem 22

Graph the function $f(x) = 3x - 2$.

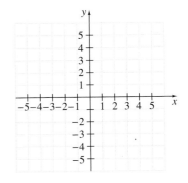

Answers

18. $g(0) = 5$, **19.** $g(-5) = -15$,
20. $f(2) = 12$, **21.** $f(-1) = 6$
22.

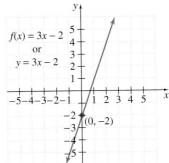

✓ Concept Check: d

Focus On Mathematical Connections

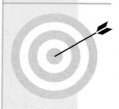

PERPENDICULAR BISECTORS

A **perpendicular bisector** is a line that is perpendicular to a given line segment and divides the segment into two equal lengths. A perpendicular bisector crosses the line segment at the point that is located exactly halfway between the two endpoints of the line segment. That point is called the **midpoint** of the line segment. If a line segment has the endpoints (x_1, y_1) and (x_2, y_2), then the midpoint of this line segment is the point with coordinates $\left(\dfrac{x_1 + x_2}{2}, \dfrac{y_1 + y_2}{2} \right)$. An example of a line segment and its perpendicular bisector is shown in the figure.

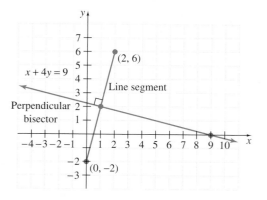

To find the equation of a line segment's perpendicular bisector, follow these steps:

Step 1. Find the midpoint of the line segment.
Step 2. Find the slope of the line segment.
Step 3. Find the slope of a line that is perpendicular to the line segment.
Step 4. Use the midpoint and the slope of the perpendicular line to find the equation of the perpendicular bisector.

CRITICAL THINKING

Use the steps given above and what you have learned in this chapter to find the equation of the perpendicular bisector of each line segment whose endpoints are given.

1. $(3, -1); (-5, 1)$

2. $(-6, -3); (-8, -1)$

3. $(-2, 6); (-22, -4)$

4. $(5, 8); (7, 2)$

5. $(2, 3); (-4, 7)$

6. $(-6, 8); (-4, -2)$

7.4 POLYNOMIAL AND RATIONAL FUNCTIONS

A EVALUATING POLYNOMIAL FUNCTIONS

At times it is convenient to use function notation to represent polynomials. For example, we may write $P(x)$ to represent the polynomial $3x^2 - 2x - 5$. In symbols, we would write

$$P(x) = 3x^2 - 2x - 5$$

This function is called a **polynomial function** because the expression $3x^2 - 2x - 5$ is a polynomial.

> ⌐**HELPFUL HINT**
>
> Recall that the symbol $P(x)$ **does not mean** P times x. It is a special symbol used to denote a function.

Examples If $P(x) = 3x^2 - 2x - 5$, find each function value.

1. $P(1) = 3(1)^2 - 2(1) - 5 = -4$ Let $x = 1$ in the function $P(x)$.
2. $P(-2) = 3(-2)^2 - 2(-2) - 5 = 11$ Let $x = -2$ in the function $P(x)$.

Many real-world phenomena are modeled by polynomial functions. If the polynomial function model is given, we can often find the solution of a problem by evaluating the function at a certain value.

Example 3   Finding the Height of an Object

The world's highest bridge, Royal Gorge suspension bridge in Colorado, is 1053 feet above the Arkansas River. An object is dropped from the top of this bridge. Neglecting air resistance, the height of the object at time t seconds is given by the polynomial function $P(t) = -16t^2 + 1053$. Find the height of the object when $t = 1$ second and when $t = 8$ seconds.

Solution: To find the height of the object at 1 second, we find $P(1)$.

$$P(t) = -16t^2 + 1053$$
$$P(1) = -16(1)^2 + 1053$$
$$P(1) = 1037$$

When $t = 1$ second, the height of the object is 1037 feet.

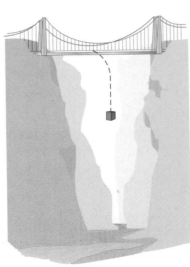

Objectives

A Evaluate polynomial functions.
B Evaluate rational functions.

SSM   CD-ROM   Video 7.4

Practice Problems 1–2

If $P(x) = 5x^2 - 3x + 7$, find each function value.

1. $P(2)$
2. $P(-1)$

Practice Problem 3

Use the polynomial function in Example 3 to find the height of the object when $t = 3$ seconds and $t = 7$ seconds.

Answers

1. $P(2) = 21$, **2.** $P(-1) = 15$, **3.** At 3 seconds, height is 909 feet; at 7 seconds, height is 269 feet

To find the height of the object at 8 seconds, we find $P(8)$.

$$P(t) = -16t^2 + 1037$$
$$P(8) = -16(8)^2 + 1037$$
$$P(8) = -1024 + 1037$$
$$P(8) = 13$$

When $t = 8$ seconds, the height of the object is 13 feet. Notice that as time t increases, the height of the object decreases. ▬▬▬

B   EVALUATING RATIONAL FUNCTIONS

Functions can also represent rational expressions. For example, we call the function $f(x) = \dfrac{x^2 + 2}{x - 3}$ a **rational function** since $\dfrac{x^2 + 2}{x - 3}$ is a rational expression in one variable.

The domain of a rational function such as $f(x) = \dfrac{x^2 + 2}{x - 3}$ is the set of all possible replacement values for x. In other words, since the rational expression $\dfrac{x^2 + 2}{x - 3}$ is not defined when $x = 3$, we say that the domain of $f(x) = \dfrac{x^2 + 2}{x - 3}$ is all real numbers except 3. We can write the domain as:

$$\{x \mid x \text{ is a real number and } x \neq 3\}$$

Practice Problem 4

A company's cost per book for printing x particular books is given by the rational function $C(x) = \dfrac{0.8x + 5000}{x}$. Find the cost per book for printing:

a. 100 books

b. 1000 books

Example 4   Finding Unit Cost

For the ICL Production Company, the rational function $C(x) = \dfrac{2.6x + 10,000}{x}$ describes the company's cost per disc of pressing x compact discs. Find the cost per disc for pressing:

a. 100 compact discs
b. 1000 compact discs

Solution: **a.** $C(100) = \dfrac{2.6(100) + 10,000}{100} = \dfrac{10,260}{100} = 102.6$

The cost per disc for pressing 100 compact discs is $102.60.

b. $C(1000) = \dfrac{2.6(1000) + 10,000}{1000} = \dfrac{12,600}{1000} = 12.6$

The cost per disc for pressing 1000 compact discs is $12.60. Notice that as more compact discs are produced, the cost per disc decreases. ▬▬▬

Answers

4. a. $50.80,   **b.** $5.80

GRAPHING CALCULATOR EXPLORATIONS

Recall that since the rational expression $\dfrac{7x - 2}{(x - 2)(x + 5)}$ is not defined when $x = 2$ or when $x = -5$, we

say that the domain of the rational function $f(x) = \dfrac{7x - 2}{(x - 2)(x + 5)}$ is all real numbers except 2 and -5.
This domain can be written as $\{x \mid x \text{ is a real number and } x \neq 2, x \neq -5\}$. This means that the graph of $f(x)$
should not cross the vertical lines $x = 2$ and $x = -5$. The graph of $f(x)$ in *connected* mode follows. In connected mode the grapher tries to connect all dots of the graph so that the result is a smooth curve. This is what has happened in the graph. Notice that the graph appears to contain vertical lines at $x = 2$ and at $x = -5$. We know that this cannot happen because the function is not defined at $x = 2$ and at $x = -5$. We also know that this cannot happen because the graph of this function would not pass the vertical line test.

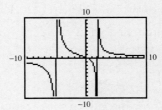

If we graph $f(x)$ in *dot* mode, the graph appears as follows. In dot mode the grapher will not connect dots with a smooth curve. Notice that the vertical lines have disappeared, and we have a better picture of the graph. The graph, however, actually appears more like the hand-drawn graph to its right. By using a Table feature, a Calculate Value feature, or by tracing, we can see that the function is not defined at $x = 2$ and at $x = -5$.

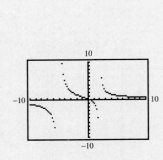   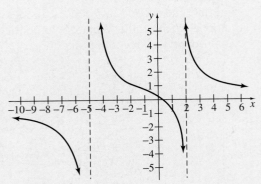

Find the domain of each rational function. Then graph each rational function and use the graph to confirm the domain.

1. $f(x) = \dfrac{x + 1}{x^2 - 4}$

2. $g(x) = \dfrac{5x}{x^2 - 9}$

3. $h(x) = \dfrac{x^2}{2x^2 + 7x - 4}$

4. $f(x) = \dfrac{3x + 2}{4x^2 - 19x - 5}$

Focus On Mathematical Connections

FINITE DIFFERENCES

When polynomial functions are evaluated at successive integer values, a list of values called a **sequence** is generated. The differences between successive pairs of numbers in such a sequence have special properties. Let's investigate these properties, beginning with a first-degree polynomial function, the linear function.

Notice in the table below on the left that *first differences* are the differences between the successive pairs of numbers in the original sequence. Find the first differences for any other linear function and fill in the table on the right. What do you notice? (*Note:* You may wish to try several different linear functions.)

x	Original Sequence $f(x) = 3x + 4$	First Differences
8	28	
7	25	3
6	22	3
5	19	3
4	16	3
3	13	3
2	10	3
1	7	3

x	Original Sequence $f(x) =$	First Differences

Now let's look at differences for a second-degree polynomial. Notice in the table below on the left that *second differences* are the differences between successive pairs of first differences. Find first and second differences for any other second-degree polynomial function and fill in the table on the right. What do you notice? (*Note:* You may wish to try several different second-degree polynomial functions.)

x	Original Sequence $f(x) = 2x^2 - 3x + 4$	First Differences	Second Differences
8	108		
7	81	27	
6	58	23	4
5	39	19	4
4	24	15	4
3	13	11	4
2	6	7	4
1	3	3	4

x	Original Sequence $f(x) =$	First Differences	Second Differences

CRITICAL THINKING

1. As you might guess, third differences are the differences between successive pairs of second differences. Find the first, second, and third differences for any two third-degree polynomial functions. What do you notice?

2. What would you expect to be true about the differences for a fourth-degree polynomial function?

3. What would you expect to be true about the differences for an nth-degree polynomial function?

EXERCISE SET 7.4

A *If* $P(x) = x^2 + x + 1$ *and* $Q(x) = 5x^2 - 1$, *find each function value. See Examples 1 and 2.*

1. $P(7)$ **2.** $Q(4)$ **3.** $Q(-10)$

4. $P(-4)$ **5.** $P(0)$ **6.** $Q(0)$

If $P(x) = x^3 + 2x - 3$ *and* $Q(x) = 7x + 5$, *find each function value. See Examples 1 and 2.*

7. $P(2)$ **8.** $Q(6)$ **9.** $Q(-3)$

10. $P(-1)$

B *If* $f(x) = \dfrac{x + 8}{2x - 1}$ *and* $g(x) = \dfrac{x - 2}{x - 5}$, *find each function value. See Example 4.*

11. $f(2)$ **12.** $g(10)$ **13.** $g(0)$

14. $f(0)$ **15.** $f(-1)$ **16.** $g(-5)$

If $f(x) = \dfrac{x^2 + 5}{x}$ *and* $g(x) = \dfrac{x^2 + 2x}{x + 3}$, *find each function value. See Example 4.*

17. $f(1)$ **18.** $g(3)$ **19.** $g(-6)$

20. $f(-2)$

1. _____
2. _____
3. _____
4. _____
5. _____
6. _____
7. _____
8. _____
9. _____
10. _____
11. _____
12. _____
13. _____
14. _____
15. _____
16. _____
17. _____
18. _____
19. _____
20. _____

Problem Solving Notes

7.5   AN INTRODUCTION TO GRAPHING POLYNOMIAL FUNCTIONS

A   ANALYZING GRAPHS OF POLYNOMIAL FUNCTIONS

Some polynomial functions are given special names according to their degree. For example,

$f(x) = 2x - 6$ is called a **linear function**; its **degree is one**.
$f(x) = 5x^2 - x + 3$ is called a **quadratic function**; its **degree is two**.
$f(x) = 7x^3 + 3x^2 - 1$ is called a **cubic function**; its **degree is three**.
$f(x) = -8x^4 - 3x^3 + 2x^2 + 20$ is called a **quartic function**; its **degree is four**.

All the above functions are also polynomial functions.

Example 1   Given the graph of the function $g(x)$ below to the left:

a. Find the domain and the range of the function.
b. List the x- and y-intercept points.
c. Find the coordinates of the point with the greatest y-value.
d. Find the coordinates of the point with the least y-value.

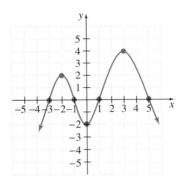
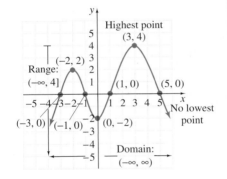

Solution: **a.** The domain is the set of all real numbers, or in interval notation, $(-\infty, \infty)$. The range is $(-\infty, 4]$.
b. The x-intercept points are $(-3, 0), (-1, 0), (1, 0)$, and $(5, 0)$. The y-intercept point is $(0, -2)$.
c. The point with the greatest y-value corresponds to the "highest" point. This is the point with coordinates $(3, 4)$. (This means that for all real number values for x, the greatest y-value, or $g(x)$ value, is 4.)
d. The point with the least y-value corresponds to the "lowest" point. This graph contains no "lowest" point, so there is no point with the least y-value.

The graph of any polynomial function (linear, quadratic, cubic, and so on) can be sketched by plotting a sufficient number of ordered pairs that satisfy the function and connecting them to form a smooth curve. The graphs of all polynomial functions will pass the vertical line test since they are graphs of functions.

Practice Problem 1

Given the graph of the function $f(x)$:

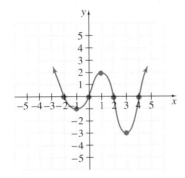

a. Find the domain and the range of the function.
b. List the x- and y-intercept points.
c. Find the coordinates of the point with the greatest y-value.
d. Find the coordinates of the point with the least y-value.

B GRAPHING QUADRATIC FUNCTIONS

Since we know how to graph linear functions (see Section 7.3), we will now graph quadratic functions and discuss special characteristics of their graphs.

> QUADRATIC FUNCTION
>
> A quadratic function is a function that can be written in the form
>
> $$f(x) = ax^2 + bx + c$$
>
> where a, b, and c are real numbers and $a \neq 0$.

We know that an equation of the form $f(x) = ax^2 + bx + c$ may be written as $y = ax^2 + bx + c$. Thus, both $f(x) = ax^2 + bx + c$ and $y = ax^2 + bx + c$ define quadratic functions as long as a is not 0.

Practice Problem 2

Graph the function $f(x) = -x^2$.

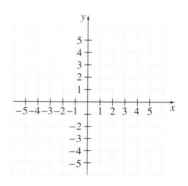

Example 2   Graph the function $f(x) = x^2$ by plotting points.

Solution: This function is not linear, and its graph is not a line. We begin by finding ordered pair solutions. Then we plot the points and draw a smooth curve through them.

If $x = -3$, then $f(-3) = (-3)^2$, or 9.
If $x = -2$, then $f(-2) = (-2)^2$, or 4.
If $x = -1$, then $f(-1) = (-1)^2$, or 1.
If $x = 0$, then $f(0) = 0^2$, or 0.
If $x = 1$, then $f(1) = 1^2$, or 1.
If $x = 2$, then $f(2) = 2^2$, or 4.
If $x = 3$, then $f(3) = 3^2$, or 9.

x	$y = f(x)$
-3	9
-2	4
-1	1
0	0
1	1
2	4
3	9

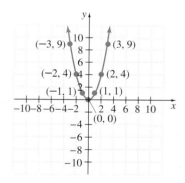

Notice that the graph of Example 2 passes the vertical line test, as it should since it is a function. This curve is called a **parabola**. The highest point on a parabola that opens downward or the lowest point on a parabola that opens upward is called the **vertex** of the parabola. The vertex of this parabola is $(0, 0)$, the lowest point on the graph. If we fold the graph along the y-axis, we can see that the two sides of the graph coincide. This means that this curve is symmetric about the y-axis, and the y-axis, or the line $x = 0$, is called the **axis of symmetry**. The graph of every quadratic function is a parabola and has an axis of symmetry: the vertical line that passes through the vertex of the parabola.

Answer

2.

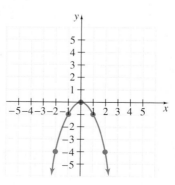

Example 3   Graph the quadratic function $f(x) = -x^2 + 2x - 3$ by plotting points.

Solution: To graph, we choose values for x and find corresponding $f(x)$ or y-values. Then we plot the points and draw a smooth curve through them.

x	$y = f(x)$
-2	-11
-1	-6
0	-3
1	-2
2	-3
3	-6

$f(x) = -x^2 + 2x - 3$

The vertex of this parabola is $(1, -2)$, the highest point on the graph. The vertical line $x = 1$ is the axis of symmetry. Recall that to find the x-intercepts of a graph, we let $y = 0$. Using function notation, this is the same as letting $f(x) = 0$. Since this graph has no x-intercepts, it means that $0 = -x^2 + 2x - 3$ has no real number solutions.

Notice that the parabola $f(x) = -x^2 + 2x - 3$ opens downward, whereas $f(x) = x^2$ opens upward. When the equation of a quadratic function is written in the form $f(x) = ax^2 + bx + c$, the coefficient of the squared variable, a, determines whether the parabola opens downward or upward. If $a > 0$, the parabola opens upward, and if $a < 0$, the parabola opens downward.

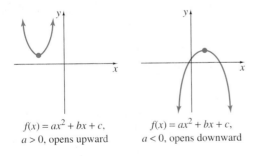

$f(x) = ax^2 + bx + c$,
$a > 0$, opens upward

$f(x) = ax^2 + bx + c$,
$a < 0$, opens downward

C FINDING THE VERTEX OF A PARABOLA

In both $f(x) = x^2$ and $f(x) = -x^2 + 2x - 3$, the vertex happens to be one of the points we chose to plot. Since this is not always the case, and since plotting the vertex allows us to draw the graph quickly, we need a consistent method for finding the vertex. One method is to use the following formula, which we shall derive in Chapter 10.

Practice Problem 3

Graph the quadratic function $f(x) = -x^2 - 2x - 3$.

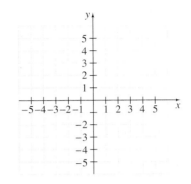

Answer

3.

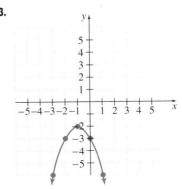

VERTEX FORMULA

The graph of $f(x) = ax^2 + bx + c, a \neq 0$, is a parabola with vertex

$$\left(\frac{-b}{2a}, f\left(\frac{-b}{2a} \right) \right)$$

We can also find the x- and y-intercepts of a parabola to aid in graphing. Recall that x-intercepts of the graph of any equation may be found by letting $y = 0$ or $f(x) = 0$ in the equation and solving for x. Also, y-intercepts may be found by letting $x = 0$ in the equation and solving for y or $f(x)$.

Example 4   Graph $f(x) = x^2 + 2x - 3$. Find the vertex and any intercepts.

Solution: To find the vertex, we use the vertex formula. For the function $f(x) = x^2 + 2x - 3$, $a = 1$ and $b = 2$. Thus,

$$x = \frac{-b}{2a} = \frac{-2}{2(1)} = -1 \qquad f(-1) = (-1)^2 + 2(-1) - 3 \quad \text{Find } f(-1).$$
$$= 1 - 2 - 3$$
$$= -4$$

The vertex is $(-1, -4)$, and since $a = 1$ is greater than 0, this parabola opens upward. Graph the vertex and notice that this parabola will have two x-intercepts because its vertex lies below the x-axis and it opens upward. To find the x-intercepts, we let y or $f(x) = 0$ and solve for x.

To find the y-intercept, let $x = 0$.

$$f(x) = x^2 + 2x - 3 \qquad\qquad f(x) = x^2 + 2x - 3$$
$$0 = x^2 + 2x - 3 \qquad\qquad f(0) = 0^2 + 2(0) - 3$$
$$0 = (x + 3)(x - 1) \qquad\qquad f(0) = -3$$
$$x + 3 = 0 \quad \text{or} \quad x - 1 = 0$$
$$x = -3 \qquad\qquad x = 1$$

The x-intercepts are -3 and 1 and the corresponding points are $(-3, 0)$ and $(1, 0)$. The y-intercept is -3 and the corresponding point is $(0, -3)$.

Now we plot these points and connect them with a smooth curve.

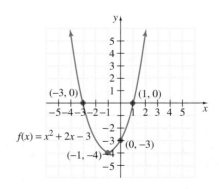

Practice Problem 4

Graph $f(x) = x^2 - 2x - 3$. Find the vertex and any intercepts.

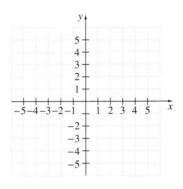

Answer

4.

┌─ **HELPFUL HINT** ─────────────────────────────────┐

Not all graphs of parabolas have x-intercepts. To see this, first plot the vertex of the parabola and decide whether the parabola opens upward or downward. Then use this information to decide whether the graph of the parabola has x-intercepts.

└──┘

Example 5   Graph $f(x) = 3x^2 - 12x + 13$. Find the vertex and any intercepts.

Solution: To find the vertex, we use the vertex formula. For the function $y = 3x^2 - 12x + 13$, $a = 3$ and $b = -12$. Thus

$$x = \frac{-b}{2a} = \frac{-(-12)}{2(3)} = \frac{12}{6} = 2 \quad f(2) = 3(2)^2 - 12(2) + 13 \quad \text{Find } f(2).$$
$$= 3(4) - 24 + 13$$
$$= 1$$

The vertex is $(2, 1)$. Also, this parabola opens upward since $a = 3$ is greater than 0. Graph the vertex and notice that this parabola has no x-intercepts: Its vertex lies above the x-axis, and it opens upward.

To find the y-intercept, let $x = 0$.

$$f(0) = 3(0)^2 - 12(0) + 13$$
$$= 0 - 0 + 13$$
$$= 13$$

The y-intercept is 13. Use this information along with symmetry of a parabola to sketch the graph of $f(x) = 3x^2 - 12x + 13$.

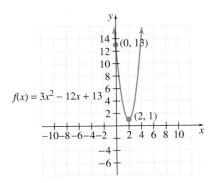

In Section 10.5 we study the graphing of quadratic functions further.

D GRAPHING CUBIC FUNCTIONS

To sketch the graph of a cubic function, we again plot points and then connect the points with a smooth curve. The general shapes of cubic graphs are given below.

Practice Problem 5

Graph $f(x) = 2x^2 + 4x + 4$. Find the vertex and any intercepts.

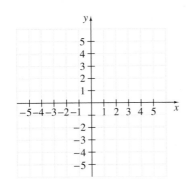

Answer

5.

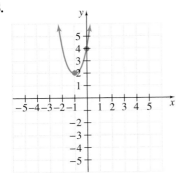

Graph of a Cubic Function
(Degree 3)

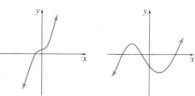

Coefficient of x^3
is a positive number.

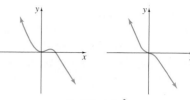

Coefficient of x^3
is a negative number.

Practice Problem 6

Graph $f(x) = x^3 - 9x$. Find any intercepts.

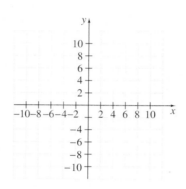

Example 6   Graph $f(x) = x^3 - 4x$. Find any intercepts.

Solution: To find x-intercepts, we let y or $f(x) = 0$ and solve for x.

$$f(x) = x^3 - 4x$$
$$0 = x^3 - 4x \quad\quad \text{Let } f(x) = 0.$$
$$0 = x(x^2 - 4)$$
$$0 = x(x + 2)(x - 2) \quad\quad \text{Factor.}$$
$$x = 0 \quad \text{or} \quad x + 2 = 0 \quad \text{or} \quad x - 2 = 0 \quad \begin{array}{l}\text{Set each factor}\\ \text{equal to 0.}\end{array}$$
$$x = 0 \quad\quad\quad\quad\quad x = -2 \quad\quad\quad x = 2 \quad \text{Solve.}$$

This graph has three x-intercepts. They are 0, -2, and 2. To find the y-intercept, we let $x = 0$.

$$f(0) = 0^3 - 4(0) = 0$$

Next let's select some x-values and find their corresponding $f(x)$ or y-values.

$$f(x) = x^3 - 4x$$
$$f(-3) = (-3)^3 - 4(-3) = -27 + 12 = -15$$
$$f(-1) = (-1)^3 - 4(-1) = -1 + 4 = 3$$
$$f(1) = 1^3 - 4(1) = 1 - 4 = -3$$
$$f(3) = 3^3 - 4(3) = 27 - 12 = 15$$

x	$f(x)$
-3	-15
-1	3
1	-3
3	15

Finally, we plot the intercepts and points and connect them with a smooth curve.

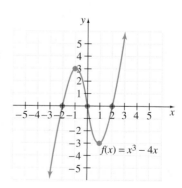

Answer

6.

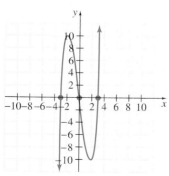

⌐ **HELPFUL HINT**

When a graph has an x-intercept of 0, notice that the y-intercept will also be 0.

HELPFUL HINT

If you are unsure about the graph of a function, plot more points.

Example 7   Graph $f(x) = -x^3$. Find any intercepts.

Solution: To find x-intercepts, we let y or $f(x) = 0$ and solve for x.

$$f(x) = -x^3$$
$$0 = -x^3$$
$$0 = x$$

The only x-intercept is 0. This means that the y-intercept is 0 also.

   Next we choose some x-values and find their corresponding y-values.

$$f(x) = -x^3$$
$$f(-2) = -(-2)^3 = 8$$
$$f(-1) = -(-1)^3 = 1$$
$$f(1) = -(1)^3 = -1$$
$$f(2) = -2^3 = -8$$

x	$f(x)$
-2	8
-1	1
1	-1
2	-8

Now we plot the points and sketch the graph of $f(x) = -x^3$.

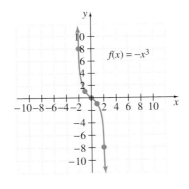

Practice Problem 7

Graph $f(x) = 2x^3$. Find any intercepts.

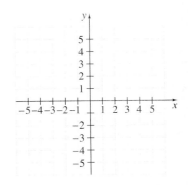

Answer

7.

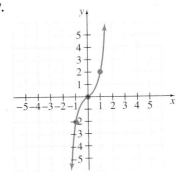

GRAPHING CALCULATOR EXPLORATIONS

We can use a grapher to approximate real number solutions of any quadratic equation in standard form, whether the associated polynomial is factorable or not. For example, let's solve the quadratic equation $x^2 - 2x - 4 = 0$. The solutions of this equation will be the x-intercepts of the graph of the function $f(x) = x^2 - 2x - 4$. (Recall that to find x-intercepts, we let $f(x) = 0$, or $y = 0$.) When we use a standard window, the graph of this function looks like this:

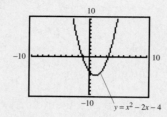

The graph appears to have one x-intercept between -2 and -1 and one between 3 and 4. To find the x-intercept between 3 and 4 to the nearest hundredth, we can use a Root feature, a Zoom feature, which magnifies a portion of the graph around the cursor, or we can redefine our window. If we redefine our window to

Xmin = 2	Ymin = -1
Xmax = 5	Ymax = 1
Xscl = 1	Yscl = 1

the resulting screen is

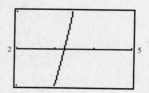

By using the Trace feature, we can now see that one of the intercepts is between 3.21 and 3.25. To approximate to the nearest hundredth, Zoom again or redefine the window to

Xmin = 3.2	Ymin = -0.1
Xmax = 3.3	Ymax = 0.1
Xscl = 1	Yscl = 1

If we use the Trace feature again, we see that, to the nearest thousandth, the x-intercept is 3.236. By repeating this process, we can approximate the other x-intercept to be -1.236.

To check, find $f(3.236)$ and $f(-1.236)$. Both of these values should be close to 0. (They will not be exactly 0 since we approximated these solutions.)

$$f(3.236) = -0.000304 \quad \text{and} \quad f(-1.236) = -0.000304$$

Solve each of these quadratic equations by graphing a related function and approximating the x-intercepts to the nearest thousandth.

1. $x^2 + 3x - 2 = 0$ **2.** $5x^2 - 7x + 1 = 0$

3. $2.3x^2 - 4.4x - 5.6 = 0$ **4.** $0.2x^2 + 6.2x + 2.1 = 0$

5. $0.09x^2 - 0.13x - 0.08 = 0$ **6.** $x^2 + 0.08x - 0.01 = 0$

EXERCISE SET 7.5

A *For the graph of each function* $f(x)$ *answer the following. See Example 1.*
  a. *Find the domain and the range of the function.*
  b. *List the x- and y-intercept points.*
  c. *Find the coordinates of the point with the greatest y-value.*
  d. *Find the coordinates of the point with the least y-value.*

1.

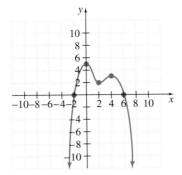

2.

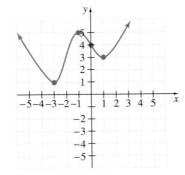

3. The graph in Example 4 of this section

4. The graph in Example 5 of this section

5. The graph in Example 6 of this section

6. The graph in Example 7 of this section

ANSWERS

1. a. _____
   b. _____
   c. _____
   d. _____

2. a. _____
   b. _____
   c. _____
   d. _____

3. a. _____
   b. _____
   c. _____
   d. _____

4. a. _____
   b. _____
   c. _____
   d. _____

5. a. _____
   b. _____
   c. _____
   d. _____

6. a. _____
   b. _____
   c. _____
   d. _____

B *Graph each quadratic function by plotting points. See Examples 2 and 3.*

7. $f(x) = 2x^2$

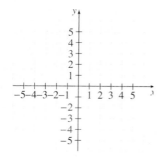

8. $f(x) = -3x^2$

9. $f(x) = x^2 + 1$

10. $f(x) = x^2 - 2$

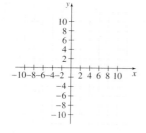

11. $f(x) = -x^2$

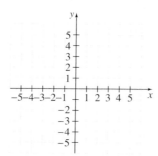

12. $f(x) = \frac{1}{2}x^2$

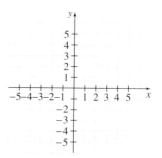

C *Graph each quadratic function. Find and label the vertex and intercepts. See Examples 4 and 5.*

13. $f(x) = x^2 + 8x + 7$ **14.** $f(x) = x^2 + 6x + 5$ **15.** $f(x) = x^2 - 2x - 24$ **16.** $f(x) = x^2 - 12x + 35$

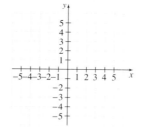
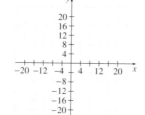
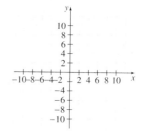

17. $f(x) = 2x^2 - 6x$

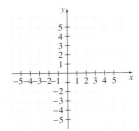

18. $f(x) = -3x^2 + 6x$

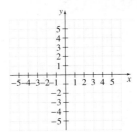

19. $f(x) = x^2 + 1$

20. $f(x) = x^2 + 4$

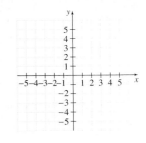

CHAPTER 7 HIGHLIGHTS

DEFINITIONS AND CONCEPTS	EXAMPLES

SECTION 7.1   THE SLOPE-INTERCEPT FORM

We can use the slope-intercept form to write an equation of a line given its slope and y-intercept.

Write an equation of the line with y-intercept -1 and slope $\frac{2}{3}$.

$$y = mx + b$$
$$y = \frac{2}{3}x - 1$$

SECTION 7.2   MORE EQUATIONS OF LINES

The **point-slope form** of the equation of a line is

$$y - y_1 = m(x - x_1)$$

where m is the slope of the line and (x_1, y_1) is a point on the line.

Find an equation of the line with slope 2 containing the point $(1, -4)$. Write the equation in standard form: $Ax + By = C$.

$$y - y_1 = m(x - x_1)$$
$$y - (-4) = 2(x - 1)$$
$$y + 4 = 2x - 2$$
$$-2x + y = -6 \quad \text{Standard form}$$

SECTION 7.3   INTRODUCTION TO FUNCTIONS

A **relation** is a set of ordered pairs. The **domain** of the relation is the set of all first components of the ordered pairs. The **range** of the relation is the set of all second components of the ordered pairs.

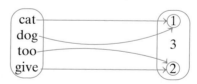

Domain: {cat, dog, too, give}
Range: {1, 2}

A **function** is a relation in which each element of the first set corresponds to exactly one element of the second set.

The previous relation is a function. Each word contains one exact number of vowels.

VERTICAL LINE TEST

If no vertical line can be drawn so that it intersects a graph more than once, the graph is the graph of a function.

Find the domain and the range of the relation. Also determine whether the relation is a function.

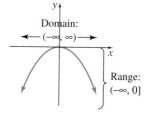

By the vertical line test, this is the graph of a function.

SECTION 7.3	**(CONTINUED)**

The symbol $f(x)$ means **function of x** and is called **function notation**.

A **linear function** is a function that can be written in the form

$$f(x) = mx + b$$

To graph a linear function, use the slope and y-intercept.

If $f(x) = 2x^2 - 5$, find $f(-3)$.

$$f(-3) = 2(-3)^2 - 5 = 2(9) - 5 = 13$$

$$f(x) = -3, \; g(x) = 5x, \; h(x) = -\frac{1}{3}x - 7$$

Graph: $f(x) = -2x$
(or $y = -2x + 0$)

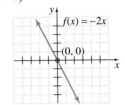

The slope is $\dfrac{2}{-1}$.

y-intercept is 0, or the point $(0, 0)$.

SECTION 7.4	**POLYNOMIAL AND RATIONAL FUNCTIONS**

A function P is a **polynomial function** if $P(x)$ is a polynomial.

A **rational function** is a function described by a rational expression.

For the polynomial function

$$P(x) = -x^2 + 6x - 12$$

find $P(-2)$.

$$P(-2) = -(-2)^2 + 6(-2) - 12 = -28$$

$$f(x) = \frac{2x - 6}{7}, \qquad h(t) = \frac{t^2 - 3t + 5}{t - 1}$$

SECTION 7.5	**AN INTRODUCTION TO GRAPHING POLYNOMIAL FUNCTIONS**

TO GRAPH A POLYNOMIAL FUNCTION

Find and plot x- and y-intercepts and a sufficient number of ordered pair solutions. Then connect the plotted points with a smooth curve.

Graph: $f(x) = x^3 + 2x^2 - 3x$

$$0 = x^3 + 2x^2 - 3x$$
$$0 = x(x - 1)(x + 3)$$
$$x = 0 \quad \text{or} \quad x = 1 \quad \text{or} \quad x = -3$$

The x-intercept points are $(0, 0)$, $(1, 0)$, and $(-3, 0)$.

$$f(0) = 0^3 + 2 \cdot 0^2 - 3 \cdot 0 = 0$$

The y-intercept point is $(0, 0)$.

x	$f(x)$
-4	-20
-2	6
-1	4
$\frac{1}{2}$	$-\frac{7}{8}$
2	10

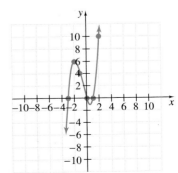

SECTION 7.5 (CONTINUED)	

A **quadratic function** is a function that can be written in the form

$$f(x) = ax^2 + bx + c, \quad a \neq 0$$

The graph of this quadratic function is a **parabola** with **vertex** $\left(\frac{-b}{2a}, f\left(\frac{-b}{2a} \right) \right)$.

Find the vertex of the graph of the quadratic function

$$f(x) = 2x^2 - 8x + 1$$

Here $a = 2$ and $b = -8$.

$$\frac{-b}{2a} = \frac{-(-8)}{2 \cdot 2} = 2$$

$$f(2) = 2 \cdot 2^2 - 8 \cdot (2) + 1 = -7$$

The vertex has coordinates $(2, -7)$.

Focus On History

CARTESIAN COORDINATE SYSTEM

The French mathematician and philosopher René Descartes (1596–1650) is generally credited with devising the rectangular coordinate system that we use in mathematics today. It is said that Descartes thought of describing the location of a point in a plane using a fixed frame of reference while watching a fly crawl on his ceiling as he laid in bed one morning meditating. He incorporated this idea of defining a point's position in the plane by giving its distances, x and y, to two fixed axes in his text *La Géométrie*.

Although Descartes is credited with the concept of the rectangular coordinate system, nowhere in his written works does an example of the modern gridlike coordinate system appear. He also never referred to a point's location as we do today with (x, y)-notation giving an ordered pair of coordinates. In fact, Descartes never even used the term *coordinate*! Instead, his basic ideas were expanded upon by later mathematicians. The Dutch mathematician Frans van Schooten (1615–1660) is credited with making Descartes' concept of a coordinate system widely accepted with his text, *Geometria a Renato Des Cartes* (*Geometry by René Descartes*). The German mathematician Gottfried Wilhelm Leibniz (1646–1716) later contributed the terms *abscissa* (the x-axis in the modern rectangular coordinate system), *ordinate* (the y-axis), and *coordinate* to the development of the rectangular coordinate system.

Focus On the Real World

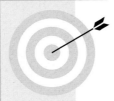

CENTERS OF MASS

The **center of mass**, also known as **center of gravity**, of an object is the point at which the mass of the object may be considered to be concentrated. For a two-dimensional object or surface, such as a flat board, the center of mass can be described as the point on which the surface would balance.

The idea of center of mass is an important one in many disciplines, especially physics and its applications. The following list describes situations in which an object's center of mass is important.

▲ Geographers are sometimes concerned with pinpointing the *geographic center* of a county, state, country, or continent. A geographic center is actually the center of mass of a geographic region if it is considered as a two-dimensional surface. The geographic center of the 48 contiguous United States is near Lebanon, Kansas. The geographic center of the North American continent is 6 miles west of Balta, North Dakota.

▲ A top-loading washing machine is designed so its center of mass is located within its agitator post. During the spin cycle, the washer tub spins around its center of mass. If clothes aren't carefully distributed within the tub, they can bunch up and throw off the center of mass. This causes the machine to vibrate, sometimes jumping or shaking wildly. Some washing machine models will stop operating when the loads become "unbalanced" in this way.

▲ Single-hulled boats are normally designed so that their centers of mass are below the water line. This provides a boat with stability. Otherwise, with a center of mass above the water line, the boat would have a tendency to tip over in the water.

▲ Most small airplanes must be carefully loaded so as not to affect the location of the airplane's center of mass. The center of mass of an airplane must be near the center of the wings. Pilots of small aircraft usually try to balance the weight of their cargo around the airplane's center of mass.

GROUP ACTIVITY

Attach a piece of graph paper to a piece of cardboard. Cut out a triangle and label the vertices with their coordinates. Lay the triangle on a horizontal table top with the graph paper face down. Slide the triangle toward the edge of the table until it is balanced on the edge, just about to tip over the side of the table. Firmly hold the triangle in place while another group member uses the straight edge of the table to draw a line on the graph paper side of the triangle marking the position of the table edge. Rotate the triangle a quarter turn and rebalance the triangle on the edge of the table. Draw a second line on the graph paper side of the triangle marking the position of the table edge.

1. The point where the two lines drawn on the triangle intersect is the center of mass of the triangle. Find the coordinates of this point.

2. Verify that the point you have located is roughly the center of mass of the triangle by balancing it at this point on the tip of a pencil or pen.

3. List the coordinates of the vertices of your triangle. What is the relationship between the coordinates of the center of mass and the coordinates of the triangle's vertices? (*Hint:* You may find it helpful to examine the sum of the x-coordinates and the sum of the y-coordinates of the vertices of the triangle.)

4. Test your observation in Question 3. Cut out another triangle. Label its vertices and, using your observation from Question 3, predict the location of the center of mass of the triangle. Use the balancing procedure to find the center of mass. How close was your prediction?

Systems of Equations and Inequalities

In this chapter, two or more equations in two or more variables are solved simultaneously. Such a collection of equations is called a **system of equations**. Systems of equations are good mathematical models for many real-world problems because these problems may involve several related patterns. We will study various methods for solving systems of equations and will conclude with a look at systems of inequalities.

Lightning, most often produced during thunderstorms, is a rapid discharge of high-current electricity into the atmosphere. At any given moment around the world, there are about 2000 thunderstorms in progress producing approximately 100 lightning flashes per second. In the United States, lightning causes an average of 75 fatalities per year. An estimated 5% of all residential insurance claims in the United States are due to lightning damage, totaling more than $1 billion per year. In addition, roughly 30% of all power outages in the United States are lightning related. Because of lightning's potentially destructive nature, meteorologists track lightning activity by recording and plotting the positions of lightning strikes.

Problem Solving Notes

8.1 Solving Systems of Linear Equations by Graphing

A **system of linear equations** consists of two or more linear equations. In this section, we focus on solving systems of linear equations containing two equations in two variables. Examples of such linear systems are

$$\begin{cases} 3x - 3y = 0 \\ x = 2y \end{cases} \quad \begin{cases} x - y = 0 \\ 2x + y = 10 \end{cases} \quad \begin{cases} y = 7x - 1 \\ y = 4 \end{cases}$$

A Deciding Whether an Ordered Pair is a Solution

A **solution** of a system of two equations in two variables is an ordered pair of numbers that is a solution of both equations in the system.

Example 1 Determine whether $(12, 6)$ is a solution of the system

$$\begin{cases} 2x - 3y = 6 \\ x = 2y \end{cases}$$

Solution: To determine whether $(12, 6)$ is a solution of the system, we replace x with 12 and y with 6 in both equations.

$2x - 3y = 6$	First equation	$x = 2y$	Second equation
$2(12) - 3(6) \stackrel{?}{=} 6$	Let $x = 12$ and $y = 6$.	$12 \stackrel{?}{=} 2(6)$	Let $x = 12$ and $y = 6$.
$24 - 18 \stackrel{?}{=} 6$	Simplify.	$12 = 12$	True.
$6 = 6$	True.		

Since $(12, 6)$ is a solution of both equations, it is a solution of the system.

Example 2 Determine whether $(-1, 2)$ is a solution of the system

$$\begin{cases} x + 2y = 3 \\ 4x - y = 6 \end{cases}$$

Solution: We replace x with -1 and y with 2 in both equations.

$x + 2y = 3$	First equation	$4x - y = 6$	Second equation
$-1 + 2(2) \stackrel{?}{=} 3$	Let $x = -1$ and $y = 2$.	$4(-1) - 2 \stackrel{?}{=} 6$	Let $x = -1$ and $y = 2$.
$-1 + 4 \stackrel{?}{=} 3$	Simplify.	$-4 - 2 \stackrel{?}{=} 6$	Simplify.
$3 = 3$	True.	$-6 = 6$	False.

$(-1, 2)$ is not a solution of the second equation, $4x - y = 6$, so it is not a solution of the system.

B Solving Systems of Equations by Graphing

Since a solution of a system of two equations in two variables is a solution common to both equations, it is also a point common to the graphs of both equations. Let's practice finding solutions of both equations in a system—that is, solutions of a system—by graphing and identifying points of intersection.

Practice Problem 1

Determine whether $(3, 9)$ is a solution of the system

$$\begin{cases} 5x - 2y = -3 \\ y = 3x \end{cases}$$

Practice Problem 2

Determine whether $(3, -2)$ is a solution of the system

$$\begin{cases} 2x - y = 8 \\ x + 3y = 4 \end{cases}$$

Answers

1. $(3, 9)$ is a solution of the system,   **2.** $(3, -2)$ is not a solution of the system

Practice Problem 3

Solve the system of equations by graphing

$$\begin{cases} -3x + y = -10 \\ x - y = 6 \end{cases}$$

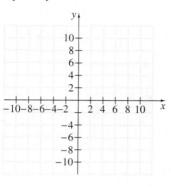

Example 3

Solve the system of equations by graphing.

$$\begin{cases} -x + 3y = 10 \\ x + y = 2 \end{cases}$$

Solution: On a single set of axes, graph each linear equation.

$-x + 3y = 10$

x	y
0	$\frac{10}{3}$
-4	2
2	4

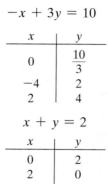

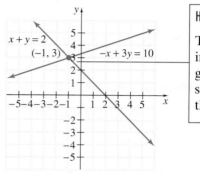

> **HELPFUL HINT**
>
> The point of intersection gives the solution of the system.

$x + y = 2$

x	y
0	2
2	0
1	1

The two lines appear to intersect at the point $(-1, 3)$. To check, we replace x with -1 and y with 3 in both equations.

$$\begin{aligned} -x + 3y &= 10 && \text{First equation} \\ -(-1) + 3(3) &\stackrel{?}{=} 10 && \text{Let } x = -1 \\ & && \text{and } y = 3. \\ 1 + 9 &\stackrel{?}{=} 10 && \text{Simplify.} \\ 10 &= 10 && \text{True.} \end{aligned}$$

$$\begin{aligned} x + y &= 2 && \text{Second equation} \\ -1 + 3 &\stackrel{?}{=} 2 && \text{Let } x = -1 \\ & && \text{and } y = 3. \\ 2 &= 2 && \text{True.} \end{aligned}$$

$(-1, 3)$ checks, so it is the solution of the system.

> **HELPFUL HINT**
>
> Neatly drawn graphs can help when "guessing" the solution of a system of linear equations by graphing.

Practice Problem 4

Solve the system of equations by graphing.

$$\begin{cases} x + 3y = -1 \\ y = 1 \end{cases}$$

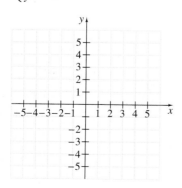

Example 4

Solve the system of equations by graphing.

$$\begin{cases} 2x + 3y = -2 \\ x = 2 \end{cases}$$

Solution: We graph each linear equation on a single set of axes.

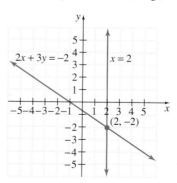

The two lines appear to intersect at the point $(2, -2)$. To determine whether $(2, -2)$ is the solution, we replace x with 2 and y with -2 in both equations.

Answers

3. please see page 536, **4.** please see page 536

$$2x + 3y = -2 \quad \text{First equation}$$
$$2(2) + 3(-2) \stackrel{?}{=} -2 \quad \text{Let } x = 2 \text{ and } y = -2.$$
$$4 + (-6) \stackrel{?}{=} -2 \quad \text{Simplify.}$$
$$-2 = -2 \quad \text{True.}$$

$$x = 2 \quad \text{Second equation}$$
$$2 = 2 \quad \text{Let } x = 2.$$
$$2 = 2 \quad \text{True.}$$

Since a true statement results in both equations, $(2, -2)$ is the solution of the system.

C IDENTIFYING SPECIAL SYSTEMS OF LINEAR EQUATIONS

Not all systems of linear equations have a single solution. Some systems have no solution and some have an infinite number of solutions.

▣ Example 5

Solve the system of equations by graphing.

$$\begin{cases} 2x + y = 7 \\ 2y = -4x \end{cases}$$

Solution: We graph the two equations in the system. The equations in slope-intercept form are $y = -2x + 7$ and $y = -2x$. Notice from the equations that the lines have the same slope, -2, and different y-intercepts. This means that the lines are parallel.

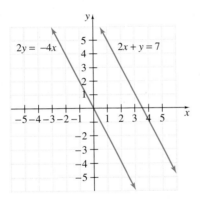

Since the lines are parallel, they do not intersect. This means that the system has *no solution*.

▣ Example 6

Solve the system of equations by graphing.

$$\begin{cases} x - y = 3 \\ -x + y = -3 \end{cases}$$

Solution: We graph each equation.

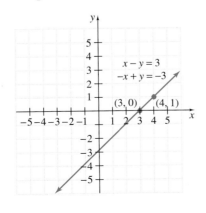

Practice Problem 5

Solve the system of equations by graphing.

$$\begin{cases} 3x - y = 6 \\ 6x = 2y \end{cases}$$

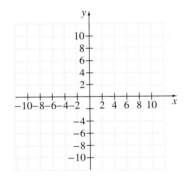

Practice Problem 6

Solve the system of equations by graphing.

$$\begin{cases} 3x + 4y = 12 \\ 9x + 12y = 36 \end{cases}$$

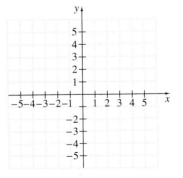

Answers

5. please see page 536,   **6.** please see page 536.

Answers

3. $(2, -4)$

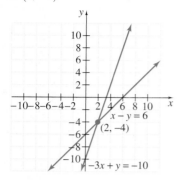

4. $(-4, 1)$

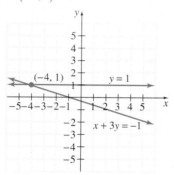

5. no solution

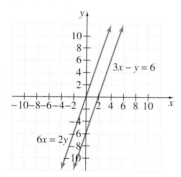

6. infinite number of solutions

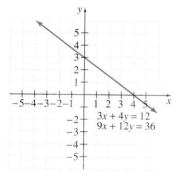

The graphs of the equations are the same line. To see this, notice that if both sides of the first equation in the system are multiplied by -1, the result is the second equation.

$$x - y = 3 \qquad \text{First equation}$$
$$-1(x - y) = -1(3) \qquad \text{Multiply both sides by } -1.$$
$$-x + y = -3 \qquad \text{Simplify. This is the second equation.}$$

This means that the system has an infinite number of solutions. Any ordered pair that is a solution of one equation is a solution of the other and is then a solution of the system.

——————

Examples 5 and 6 are special cases of systems of linear equations. A system that has no solution is said to be an **inconsistent system**. If the graphs of the two equations of a system are identical—we call the equations **dependent equations**.

As we have seen, three different situations can occur when graphing the two lines associated with the equations in a linear system. These situations are shown in the figures.

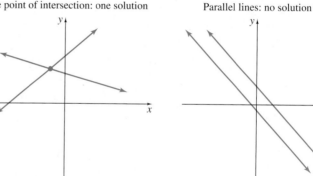

One point of intersection: one solution

Parallel lines: no solution

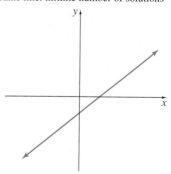

Same line: infinite number of solutions

GRAPHING CALCULATOR EXPLORATIONS

A graphing calculator may be used to approximate solutions of systems of equations. For example, to approximate the solution of the system

$$\begin{cases} y = -3.14x - 1.35 \\ y = 4.88x + 5.25, \end{cases}$$

first graph each equation on the same set of axes. Then use the intersect feature of your calculator to approximate the point of intersection.

 The approximate point of intersection is $(-0.82, 1.23)$.

Solve each system of equations. Approximate the solutions to two decimal places.

1. $\begin{cases} y = -2.68x + 1.21 \\ y = 5.22x - 1.68 \end{cases}$

2. $\begin{cases} y = 4.25x + 3.89 \\ y = -1.88x + 3.21 \end{cases}$

3. $\begin{cases} 4.3x - 2.9y = 5.6 \\ 8.1x + 7.6y = -14.1 \end{cases}$

4. $\begin{cases} -3.6x - 8.6y = 10 \\ -4.5x + 9.6y = -7.7 \end{cases}$

Problem Solving Notes

8.2 SOLVING SYSTEMS OF LINEAR EQUATIONS BY SUBSTITUTION

A USING THE SUBSTITUTION METHOD

You may have suspected by now that graphing alone is not an accurate way to solve a system of linear equations. For example, a solution of $\left(\frac{1}{2}, \frac{2}{9}\right)$ is unlikely to be read correctly from a graph. In this section, we discuss a second, more accurate method for solving systems of equations. This method is called the **substitution method** and is introduced in the next example.

Example 1   Solve the system:

$$\begin{cases} 2x + y = 10 & \text{First equation} \\ x = y + 2 & \text{Second equation} \end{cases}$$

Solution: The second equation in this system is $x = y + 2$. This tells us that x and $y + 2$ have the same value. This means that we may substitute $y + 2$ for x in the first equation.

$$2x + y = 10 \quad \text{First equation}$$

$$2(y + 2) + y = 10 \quad \text{Substitute } y + 2 \text{ for } x \text{ since } x = y + 2.$$

Notice that this equation now has one variable, y. Let's now solve this equation for y.

HELPFUL HINT

Don't forget the distributive property.

$$2(y + 2) + y = 10$$
$$2y + 4 + y = 10 \quad \text{Use the distributive property.}$$
$$3y + 4 = 10 \quad \text{Combine like terms.}$$
$$3y = 6 \quad \text{Subtract 4 from both sides.}$$
$$y = 2 \quad \text{Divide both sides by 3.}$$

Now we know that the y-value of the ordered pair solution of the system is 2. To find the corresponding x-value, we replace y with 2 in the equation $x = y + 2$ and solve for x.

$$x = y + 2$$
$$x = 2 + 2 \quad \text{Let } y = 2.$$
$$x = 4$$

The solution of the system is the ordered pair $(4, 2)$. Since an ordered pair solution must satisfy both linear equations in the system, we could have chosen the equation $2x + y = 10$ to find the corresponding x-value. The resulting x-value is the same.

Check: We check to see that $(4, 2)$ satisfies both equations of the original system.

First Equation

$$2x + y = 10$$
$$2(4) + 2 \stackrel{?}{=} 10$$
$$10 = 10 \quad \text{True.}$$

Second Equation

$$x = y + 2$$
$$4 \stackrel{?}{=} 2 + 2 \quad \text{Let } x = 4 \text{ and } y = 2.$$
$$4 = 4 \quad \text{True.}$$

Objective

A Use the substitution method to solve a system of linear equations.

SSM   CD-ROM   Video 8.2

Practice Problem 1

Use the substitution method to solve the system:

$$\begin{cases} 2x + 3y = 13 \\ x = y + 4 \end{cases}$$

Answer

1. $(5, 1)$

The solution of the system is $(4, 2)$.

A graph of the two equations shows the two lines intersecting at the point $(4, 2)$.

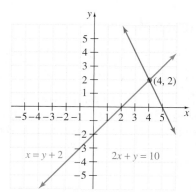

Practice Problem 2

Use the substitution method to solve the system:

$$\begin{cases} 4x - y = 2 \\ y = 5x \end{cases}$$

Example 2

Solve the system:

$$\begin{cases} 5x - y = -2 \\ y = 3x \end{cases}$$

Solution: The second equation is solved for y in terms of x. We substitute $3x$ for y in the first equation.

$$5x - y = -2 \quad \text{First equation}$$
$$\downarrow$$
$$5x - (3x) = -2$$

Now we solve for x.

$$5x - 3x = -2$$
$$2x = -2 \quad \text{Combine like terms.}$$
$$x = -1 \quad \text{Divide both sides by 2.}$$

The x-value of the ordered pair solution is -1. To find the corresponding y-value, we replace x with -1 in the equation $y = 3x$.

$$y = 3x$$
$$y = 3(-1) \quad \text{Let } x = -1.$$
$$y = -3$$

Check to see that the solution of the system is $(-1, -3)$.

To solve a system of equations by substitution, we first need an equation solved for one of its variables.

Practice Problem 3

Solve the system:

$$\begin{cases} 3x + y = 5 \\ 3x - 2y = -7 \end{cases}$$

Example 3

Solve the system:

$$\begin{cases} x + 2y = 7 \\ 2x + 2y = 13 \end{cases}$$

Solution: We choose one of the equations and solve for x or y. We will solve the first equation for x by subtracting $2y$ from both sides.

$$x + 2y = 7 \quad \text{First equation}$$
$$x = 7 - 2y \quad \text{Subtract } 2y \text{ from both sides.}$$

Answers

2. $(-2, -10)$,   **3.** $\left(\frac{1}{3}, 4\right)$

Since $x = 7 - 2y$, we now substitute $7 - 2y$ for x in the second equation and solve for y.

$$2x + 2y = 13 \qquad \text{Second equation}$$

HELPFUL HINT
Don't forget to insert parentheses.

$$2(7 - 2y) + 2y = 13 \qquad \text{Let } x = 7 - 2y.$$

$$14 - 4y + 2y = 13 \qquad \text{Apply the distributive property.}$$

$$14 - 2y = 13 \qquad \text{Simplify.}$$

$$-2y = -1 \qquad \text{Subtract 14 from both sides.}$$

$$y = \frac{1}{2} \qquad \text{Divide both sides by } -2.$$

To find x, we let $y = \frac{1}{2}$ in the equation $x = 7 - 2y$.

$$x = 7 - 2y$$

$$x = 7 - 2\left(\frac{1}{2}\right) \qquad \text{Let } y = \frac{1}{2}.$$

$$x = 7 - 1$$

$$x = 6$$

Check the solution in both equations of the original system. The solution is $\left(6, \frac{1}{2}\right)$.

The following steps summarize how to solve a system of equations by the substitution method.

TO SOLVE A SYSTEM OF TWO LINEAR EQUATIONS BY THE SUBSTITUTION METHOD

Step 1. Solve one of the equations for one of its variables.

Step 2. Substitute the expression for the variable found in Step 1 into the other equation.

Step 3. Solve the equation from Step 2 to find the value of one variable.

Step 4. Substitute the value found in Step 3 in any equation containing both variables to find the value of the other variable.

Step 5. Check the proposed solution in the original system.

TRY THE CONCEPT CHECK IN THE MARGIN.

Example 4   Solve the system:

$$\begin{cases} 7x - 3y = -14 \\ -3x + y = 6 \end{cases}$$

Solution: To avoid introducing fractions, we will solve the second equation for y.

$$-3x + y = 6 \qquad \text{Second equation}$$

$$y = 3x + 6$$

Next, we substitute $3x + 6$ for y in the first equation.

✓ **CONCEPT CHECK**

As you solve the system
$$\begin{cases} 2x + y = -5 \\ x - y = 5 \end{cases}$$
you find that $y = -5$. Is this the solution of the system?

Practice Problem 4

Solve the system:

$$\begin{cases} 5x - 2y = 6 \\ -3x + y = -3 \end{cases}$$

Answers

4. $(0, -3)$

✓ **Concept Check:** No, the solution will be an ordered pair.

$$7x - 3y = -14 \quad \text{First equation}$$
$$7x - 3(3x + 6) = -14 \quad \text{Let } y = 3x + 6.$$
$$7x - 9x - 18 = -14 \quad \text{Use the distributive property.}$$
$$-2x - 18 = -14 \quad \text{Simplify.}$$
$$-2x = 4 \quad \text{Add 18 to both sides.}$$
$$\frac{-2x}{-2} = \frac{4}{-2} \quad \text{Divide both sides by } -2.$$
$$x = -2$$

To find the corresponding y-value, we substitute -2 for x in the equation $y = 3x + 6$. Then $y = 3(-2) + 6$ or $y = 0$. The solution of the system is $(-2, 0)$. Check this solution in both equations of the system.

HELPFUL HINT

When solving a system of equations by the substitution method, begin by solving an equation for one of its variables. If possible, solve for a variable that has a coefficient of 1 or −1 to avoid working with time-consuming fractions.

Example 5   Solve the system: $\begin{cases} \dfrac{1}{2}x - y = 3 \\ x = 6 + 2y \end{cases}$

Solution: The second equation is already solved for x in terms of y. Thus we substitute $6 + 2y$ for x in the first equation and solve for y.

$$\frac{1}{2}x - y = 3 \quad \text{First equation}$$

$$\frac{1}{2}(6 + 2y) - y = 3 \quad \text{Let } x = 6 + 2y.$$
$$3 + y - y = 3 \quad \text{Apply the distributive property.}$$
$$3 = 3 \quad \text{Simplify.}$$

Arriving at a true statement such as $3 = 3$ indicates that the two linear equations in the original system are equivalent. This means that their graphs are identical, as shown in the figure. There is an infinite number of solutions to the system, and any solution of one equation is also a solution of the other.

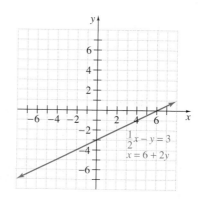

Practice Problem 5

Solve the system:

$$\begin{cases} -x + 3y = 6 \\ y = \dfrac{1}{3}x + 2 \end{cases}$$

Answer

5. infinite number of solutions

Example 6   Solve the system:

$$\begin{cases} 6x + 12y = 5 \\ -4x - 8y = 0 \end{cases}$$

Solution: We choose the second equation and solve for y.

$$-4x - 8y = 0 \qquad \text{Second equation}$$

$$-8y = 4x \qquad \text{Add } 4x \text{ to both sides.}$$

$$\frac{-8y}{-8} = \frac{4x}{-8} \qquad \text{Divide both sides by } -8.$$

$$y = -\frac{1}{2}x \quad \text{Simplify.}$$

Now we replace y with $-\frac{1}{2}x$ in the first equation.

$$6x + 12y = 5 \qquad \text{First equation}$$

$$6x + 12\left(-\frac{1}{2}x\right) = 5 \qquad \text{Let } y = -\frac{1}{2}x.$$

$$6x + (-6x) = 5 \qquad \text{Simplify.}$$

$$0 = 5 \qquad \text{Combine like terms.}$$

The false statement $0 = 5$ indicates that this system has no solution. The graph of the linear equations in the system is a pair of parallel lines, as shown in the figure.

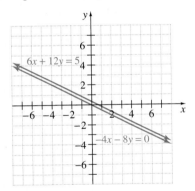

TRY THE CONCEPT CHECK IN THE MARGIN.

Practice Problem 6

Solve the system:

$$\begin{cases} 2x - 3y = 6 \\ -4x + 6y = -12 \end{cases}$$

✓ **CONCEPT CHECK**

Describe how the graphs of the equations in a system appear if the system has

a. no solution
b. one solution
c. an infinite number of solutions

Answers

6. infinite number of solutions

✓ **Concept Check**

a. parallel lines
b. intersect at one point
c. identical graphs

Focus On Business and Career

BREAK-EVEN POINT

When a business sells a new product, it generally does not start making a profit right away. There are usually many expenses associated with producing a new product. These expenses might include an advertising blitz to introduce the product to the public. These start-up expenses might also include the cost of market research and product development or any brand-new equipment needed to manufacture the product. Start-up costs like these are generally called *fixed costs* because they don't depend on the number of items manufactured. Expenses that depend on the number of items manufactured, such as the cost of materials and shipping, are called *variable costs*. The total cost of manufacturing the new product is given by the cost equation: Total cost = Fixed costs + Variable costs.

For instance, suppose a greeting card company is launching a new line of greeting cards. The company spent $7000 doing product research and development for the new line and spent $15,000 on advertising the new line. The company did not need to buy any new equipment to manufacture the cards, but the paper and ink needed to make each card will cost $0.20 per card. The total cost y in dollars for manufacturing x cards is $y = 22{,}000 + 0.20x$.

Once a business sets a price for the new product, the company can find the product's expected *revenue*. Revenue is the amount of money the company takes in from the sales of its product. The revenue from selling a product is given by the revenue equation: Revenue = Price per item × Number of items sold.

For instance, suppose that the card company plans to sell its new cards for $1.50 each. The revenue y, in dollars, that the company can expect to receive from the sales of x cards is $y = 1.50x$.

If the total cost and revenue equations are graphed on the same coordinate system, the graphs should intersect. The point of intersection is where total cost equals revenue and is called the *break-even point*. The break-even point gives the number of items x that must be sold for the company to recover its expenses. If fewer than this number of items is sold, the company loses money. If more than this number of items is sold, the company makes a profit. In the case of the greeting card company, approximately 16,923 cards must be sold for the company to break even on this new card line. The total cost and revenue of producing and selling 16,923 cards is the same. It is approximately $25,385.

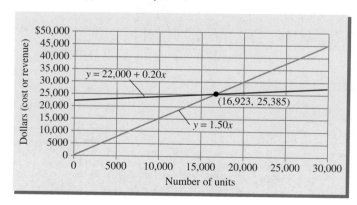

GROUP ACTIVITY

Suppose your group is starting a small business near your campus.
 a. Choose a business and decide what campus-related product or service you will provide.
 b. Research the fixed costs of starting up such a business.
 c. Research the variable costs of producing such a product or providing such a service.
 d. Decide how much you would charge per unit of your product or service.
 e. Find a system of equations for the total cost and revenue of your product or service.
 f. How many units of your product or service must be sold before your business will break even?

EXERCISE SET 8.2

A *Solve each system of equations by the substitution method. See Examples 1 and 2.*

1. $\begin{cases} x + y = 3 \\ x = 2y \end{cases}$

2. $\begin{cases} x + y = 20 \\ x = 3y \end{cases}$

3. $\begin{cases} x + y = 6 \\ y = -3x \end{cases}$

4. $\begin{cases} x + y = 6 \\ y = -4x \end{cases}$

5. $\begin{cases} 3x + 2y = 16 \\ x = 3y - 2 \end{cases}$

6. $\begin{cases} 2x + 3y = 18 \\ x = 2y - 5 \end{cases}$

7. $\begin{cases} 3x - 4y = 10 \\ x = 2y \end{cases}$

8. $\begin{cases} 3x - 4y = 10 \\ y = 2x \end{cases}$

9. $\begin{cases} y = 3x + 1 \\ 4y - 8x = 12 \end{cases}$

10. $\begin{cases} y = 2x + 3 \\ 5y - 7x = 18 \end{cases}$

11. $\begin{cases} y = 2x + 9 \\ y = 7x + 10 \end{cases}$

12. $\begin{cases} y = 5x - 3 \\ y = 8x + 4 \end{cases}$

Solve each system of equations by the substitution method. See Examples 3 through 6.

13. $\begin{cases} x + 2y = 6 \\ 2x + 3y = 8 \end{cases}$

14. $\begin{cases} x + 3y = -5 \\ 2x + 2y = 6 \end{cases}$

15. $\begin{cases} 2x - 5y = 1 \\ 3x + y = -7 \end{cases}$

16. $\begin{cases} 4x + 2y = 5 \\ 2x + y = -4 \end{cases}$

17. $\begin{cases} 2y = x + 2 \\ 6x - 12y = 0 \end{cases}$

18. $\begin{cases} 3y = x + 6 \\ 4x + 12y = 0 \end{cases}$

19. $\begin{cases} 4x + y = 11 \\ 2x + 5y = 1 \end{cases}$

20. $\begin{cases} 3x + y = -14 \\ 4x + 3y = -22 \end{cases}$

21. $\begin{cases} 2x - 3y = -9 \\ 3x = y + 4 \end{cases}$

22. $\begin{cases} 8x - 3y = -4 \\ 7x = y + 3 \end{cases}$

23. $\begin{cases} 6x - 3y = 5 \\ x + 2y = 0 \end{cases}$

24. $\begin{cases} 10x - 5y = -21 \\ x + 3y = 0 \end{cases}$

25. $\begin{cases} 3x - y = 1 \\ 2x - 3y = 10 \end{cases}$

ANSWERS

1. _____
2. _____
3. _____
4. _____
5. _____
6. _____
7. _____
8. _____
9. _____
10. _____
11. _____
12. _____
13. _____
14. _____
15. _____
16. _____
17. _____
18. _____
19. _____
20. _____
21. _____
22. _____
23. _____
24. _____
25. _____

Problem Solving Notes

8.3 SOLVING SYSTEMS OF LINEAR EQUATIONS BY ADDITION

A USING THE ADDITION METHOD

We have seen that substitution is an accurate method for solving a system of linear equations. Another accurate method is the **addition** or **elimination method**. The addition method is based on the addition property of equality: Adding equal quantities to both sides of an equation does not change the solution of the equation. In symbols,

if $A = B$ and $C = D$, then $A + C = B + D$

Example 1   Solve the system: $\begin{cases} x + y = 7 \\ x - y = 5 \end{cases}$

Solution: Since the left side of each equation is equal to its right side, we are adding equal quantities when we add the left sides of the equations together and the right sides of the equations together. This adding eliminates the variable y and gives us an equation in one variable, x. We can then solve for x.

$$\begin{array}{ll} x + y = 7 & \text{First equation} \\ \underline{x - y = 5} & \text{Second equation} \\ 2x \quad\;\; = 12 & \text{Add the equations to eliminate } y. \\ \quad x = 6 & \text{Divide both sides by 2.} \end{array}$$

The x-value of the solution is 6. To find the corresponding y-value, we let $x = 6$ in either equation of the system. We will use the first equation.

$$\begin{array}{ll} x + y = 7 & \text{First equation} \\ 6 + y = 7 & \text{Let } x = 6. \\ \quad\;\; y = 1 & \text{Solve for } y. \end{array}$$

The solution is $(6, 1)$.

Check: Check the solution in both equations.

First Equation

$$\begin{array}{ll} x + y = 7 & \\ 6 + 1 \stackrel{?}{=} 7 & \text{Let } x = 6 \text{ and } y = 1. \\ \quad\;\; 7 = 7 & \text{True.} \end{array}$$

Second Equation

$$\begin{array}{ll} x - y = 5 & \\ 6 - 1 \stackrel{?}{=} 5 & \text{Let } x = 6 \text{ and } y = 1. \\ \quad\;\; 5 = 5 & \text{True.} \end{array}$$

Thus, the solution of the system is $(6, 1)$ and the graphs of the two equations intersect at the point $(6, 1)$, as shown.

Objective

A Use the addition method to solve a system of linear equations.

SSM   CD-ROM   Video
8.3

Practice Problem 1

Use the addition method to solve the system: $\begin{cases} x + y = 13 \\ x - y = 5 \end{cases}$

Answer

1. $(9, 4)$

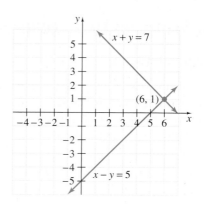

Practice Problem 2

Solve the system: $\begin{cases} 2x - y = -6 \\ -x + 4y = 17 \end{cases}$

Example 2   Solve the system: $\begin{cases} -2x + y = 2 \\ -x + 3y = -4 \end{cases}$

Solution: If we simply add these two equations, the result is still an equation in two variables. However, our goal is to eliminate one of the variables so that we have an equation in the other variable. To do this, notice what happens if we multiply *both sides* of the first equation by -3. We are allowed to do this by the multiplication property of equality. Then the system

$$\begin{cases} -3(-2x + y) = -3(2) \\ -x + 3y = -4 \end{cases} \quad \text{simplifies to} \quad \begin{cases} 6x - 3y = -6 \\ -x + 3y = -4 \end{cases}$$

When we add the resulting equations, the y variable is eliminated.

$$\begin{array}{rl} 6x - 3y = -6 & \\ \underline{-x + 3y = -4} & \\ 5x \phantom{{}-3y} = -10 & \text{Add.} \\ x = -2 & \text{Divide both sides by 5.} \end{array}$$

To find the corresponding y-value, we let $x = -2$ in any of the preceding equations containing both variables. We use the first equation of the original system.

$$\begin{array}{rl} -2x + y = 2 & \text{First equation} \\ -2(-2) + y = 2 & \text{Let } x = -2. \\ 4 + y = 2 & \\ y = -2 & \end{array}$$

Check the ordered pair $(-2, -2)$ in both equations of the *original* system. The solution is $(-2, -2)$. ▬

In Example 2, the decision to multiply the first equation by -3 was no accident. **To eliminate a variable** when adding two equations, **the coeffi-**

cient of the variable in one equation must be the opposite of its coefficient in the other equation.

> **HELPFUL HINT**
>
> Be sure to multiply *both sides* of an equation by a chosen number when solving by the addition method. A common mistake is to multiply only the side containing the variables.

Example 3   Solve the system: $\begin{cases} 2x - y = 7 \\ 8x - 4y = 1 \end{cases}$

Practice Problem 3

Solve the system: $\begin{cases} x - 3y = -2 \\ -3x + 9y = 5 \end{cases}$

Solution: When we multiply both sides of the first equation by -4, the resulting coefficient of x is -8. This is the opposite of 8, the coefficient of x in the second equation. Then the system

> **HELPFUL HINT**
>
> Don't forget to multiply both sides by -4.

$\begin{cases} -4(2x - y) = -4(7) \\ 8x - 4y = 1 \end{cases}$ simplifies to

$\begin{cases} -8x + 4y = -28 \\ \underline{8x - 4y = 1} \\ 0 = -27 \end{cases}$ Add the equations.

When we add the equations, both variables are eliminated and we have $0 = -27$, a false statement. This means that the system has no solution. The equations, if graphed, are parallel lines.   ▬▬▬

Example 4   Solve the system: $\begin{cases} 3x - 2y = 2 \\ -9x + 6y = -6 \end{cases}$

Practice Problem 4

Solve the system: $\begin{cases} 2x + 5y = 1 \\ -4x - 10y = -2 \end{cases}$

Solution: First we multiply both sides of the first equation by 3, then we add the resulting equations.

$\begin{cases} 3(3x - 2y) = 3 \cdot 2 \\ -9x + 6y = -6 \end{cases}$ simplifies to $\begin{cases} 9x - 6y = 6 \\ \underline{-9x + 6y = -6} \\ 0 = 0 \end{cases}$ Add the equations.

Both variables are eliminated and we have $0 = 0$, a true statement. This means that the system has an infinite number of solutions.   ▬▬▬

Answers

3. no solution,   **4.** infinite number of solutions

✓ CONCEPT CHECK

Suppose you are solving the system

$$\begin{cases} 3x + 8y = -5 \\ 2x - 4y = 3 \end{cases}$$

You decide to use the addition method by multiplying both sides of the second equation by 2. In which of the following was the multiplication performed correctly? Explain.

a. $4x - 8y = 3$
b. $4x - 8y = 6$

Practice Problem 5

Solve the system: $\begin{cases} 4x + 5y = 14 \\ 3x - 2y = -1 \end{cases}$

✓ CONCEPT CHECK

Suppose you are solving the system

$$\begin{cases} -4x + 7y = 6 \\ x + 2y = 5 \end{cases}$$

by the addition method.

a. What step(s) should you take if you wish to eliminate x when adding the equations?

b. What step(s) should you take if you wish to eliminate y when adding the equations?

Answer

5. $(1, 2)$

✓ **Concept Check:** b

a. Multiply the second equation by 4.
b. Possible answer: Multiply the first equation by -2 and the second equation by 7.

TRY THE CONCEPT CHECK IN THE MARGIN.

Example 5   Solve the system: $\begin{cases} 3x + 4y = 13 \\ 5x - 9y = 6 \end{cases}$

Solution: We can eliminate the variable y by multiplying the first equation by 9 and the second equation by 4. Then we add the resulting equations.

$$\begin{cases} 9(3x + 4y) = 9(13) \\ 4(5x - 9y) = 4(6) \end{cases} \text{ simplifies to}$$

$$\begin{cases} 27x + 36y = 117 \\ \underline{20x - 36y = 24} \end{cases}$$ Add the equations.

$$47x = 141$$

$$x = 3 \quad \text{Solve for } x.$$

To find the corresponding y-value, we let $x = 3$ in any equation in this example containing two variables. Doing so in any of these equations will give $y = 1$. Check to see that $(3, 1)$ satisfies each equation in the original system. The solution is $(3, 1)$.

If we had decided to eliminate x instead of y in Example 5, the first equation could have been multiplied by 5 and the second by -3. Try solving the original system this way to check that the solution is $(3, 1)$.

The following steps summarize how to solve a system of linear equations by the addition method.

TO SOLVE A SYSTEM OF TWO LINEAR EQUATIONS BY THE ADDITION METHOD

Step 1. Rewrite each equation in standard from $Ax + By = C$.

Step 2. If necessary, multiply one or both equations by a nonzero number so that the coefficients of a chosen variable in the system are opposites.

Step 3. Add the equations.

Step 4. Find the value of one variable by solving the resulting equation from Step 3.

Step 5. Find the value of the second variable by substituting the value found in Step 4 into either of the original equations.

Step 6. Check the proposed solution in the orignal system.

TRY THE CONCEPT CHECK IN THE MARGIN.

Example 6   Solve the system: $\begin{cases} -x - \dfrac{y}{2} = \dfrac{5}{2} \\ -\dfrac{x}{2} + \dfrac{y}{4} = 0 \end{cases}$

Practice Problem 6

Solve the system: $\begin{cases} -\dfrac{x}{3} + y = \dfrac{4}{3} \\ \dfrac{x}{2} - \dfrac{5}{2}y = -\dfrac{1}{2} \end{cases}$

Solution: We begin by clearing each equation of fractions. To do so, we multiply both sides of the first equation by the LCD 2 and both sides of the second equation by the LCD 4. Then the system

$\begin{cases} 2\left(-x - \dfrac{y}{2}\right) = 2\left(\dfrac{5}{2}\right) \\ 4\left(-\dfrac{x}{2} + \dfrac{y}{4}\right) = 4(0) \end{cases}$ simplifies to $\begin{cases} -2x - y = 5 \\ -2x + y = 0 \end{cases}$

Now we add the resulting equations in the simplified system.

$\begin{array}{r} -2x - y = 5 \\ -2x + y = 0 \\ \hline -4x \quad\quad = 5 \end{array}$ Add.

$$x = -\dfrac{5}{4}$$

To find y, we could replace x with $-\dfrac{5}{4}$ in one of the equations with two variables. Instead, let's go back to the simplified system and multiply by appropriate factors to eliminate the variable x and solve for y. To do this, we multiply the first equation in the simplified system by -1. Then the system

$\begin{cases} -1(-2x - y) = -1(5) \\ -2x + y = 0 \end{cases}$ simplifies to $\begin{array}{r} 2x + y = -5 \\ -2x + y = 0 \\ \hline 2y = -5 \end{array}$ Add.

$$y = -\dfrac{5}{2}$$

Check the ordered pair $\left(-\dfrac{5}{4}, -\dfrac{5}{2}\right)$ in both equations of the original system. The solution is $\left(-\dfrac{5}{4}, -\dfrac{5}{2}\right)$. ▬

Answer

6. $\left(-\dfrac{17}{2}, -\dfrac{3}{2}\right)$

Focus On History

The oldest known arithmetic book is a Chinese textbook called *Nine Chapters on the Mathematical Art*. No one knows for sure who wrote this text or when it was first written. Experts believe that it was a collection of works written by many different people. It was probably written over the course of several centuries. Even though no one knows the original date of the *Nine Chapters*, we do know that it existed in 213 BC. In that year, all of the original copies of the *Nine Chapters*, along with many other books, were burned when the first emperor of the Qin Dynasty (221–206 BC) tried to erase all traces of previous rulers and dynasties.

The Qin Emperor was not quite successful in destroying all of the *Nine Chapters*. Pieces of the text were found and many Chinese mathematicians filled in the missing material. In 263 AD, the Chinese mathematician Liu Hui wrote a summary of the *Nine Chapters*, adding his own solutions to its problems. Liu Hui's version was studied in China for over a thousand years. At one point, the Chinese government even adopted *Nine Chapters* as the official study aid for university students to use when preparing for civil service exams.

The *Nine Chapters* is a guide to everyday math in ancient China. It contains a total of 246 problems covering widely encountered problems like field measurement, rice exchange, fair taxation, and construction. It includes the earliest known use of negative numbers and shows the first development of solving systems of linear equations. The following problem appears in Chapter 7, "Excess and Deficiency," of the *Nine Chapters*. (Note: A *wen* is a unit of currency.)

> A certain number of people are purchasing some chickens together. If each person contributes 9 wen, there is an excess of 11 wen. If each person contributes just 6 wen, there is a deficiency of 16 wen. Find the number of people and the total price of the chickens. (Adapted from *The History of Mathematics: An Introduction*, second edition, David M. Burton, 1991, Wm. C. Brown Publishers, p. 164)

CRITICAL THINKING

The information in the excess/deficiency problem from *Nine Chapters* can be translated into two equations in two variables. Let c represent the total price of the chickens, and let x represent the number of people pooling their money to buy the chickens. In this situation, an excess of 11 wen can be interpreted as 11 more than the price of the chickens. A deficiency of 16 wen can be interpreted as 16 less than the price of the chickens.

1. Use what you have learned so far in this book about translating sentences into equations to write two equations in two variables for the excess/deficiency problem.
2. Solve the problem from *Nine Chapters* by solving the system of equations you wrote in Question 1. How many people pooled their money? What was the price of the chickens?
3. Write a modern-day excess/deficiency problem of your own.

8.4   SYSTEMS OF LINEAR EQUATIONS AND PROBLEM SOLVING

A USING A SYSTEM OF EQUATIONS FOR PROBLEM SOLVING

Many of the word problems solved earlier using one-variable equations can also be solved using two equations in two variables. We use the same problem-solving steps that have been used throughout this text. The only difference is that two variables are assigned to represent the two unknown quantities and that the problem is translated into two equations.

PROBLEM-SOLVING STEPS

1. UNDERSTAND the problem. During this step, become comfortable with the problem. Some ways of doing this are to

   Read and reread the problem.
   Choose two variables to represent the two unknowns.
   Construct a drawing.
   Propose a solution and check. Pay careful attention to how you check your proposed solution. This will help when writing an equation to model the problem.

2. TRANSLATE the problem into two equations.
3. SOLVE the system of equations.
4. INTERPRET the results: *Check* the proposed solution in the stated problem and *state* your conclusion.

▣ Example 1   **Finding Unknown Numbers**

Find two numbers whose sum is 37 and whose difference is 21.

Solution: **1.** UNDERSTAND. Read and reread the problem. Suppose that one number is 20. If their sum is 37, the other number is 17 because $20 + 17 = 37$. Is their difference 21? No; $20 - 17 = 3$. Our proposed solution is incorrect, but we now have a better understanding of the problem.

Since we are looking for two numbers, we let

x = first number

y = second number

2. TRANSLATE. Since we have assigned two variables to this problem, we translate our problem into two equations.

In words:	two numbers whose sum	is	37
	↓	↓	↓
Translate:	$x + y$	$=$	37

In words:	two numbers whose difference	is	21
	↓	↓	↓
Translate:	$x - y$	$=$	21

Objective

A Use a system of equations to solve problems.

SSM   CD-ROM   Video
8.4

Practice Problem 1

Find two numbers whose sum is 50 and whose difference is 22.

Answer

1. The numbers are 36 and 14.

3. SOLVE. Now we solve the system

$$\begin{cases} x + y = 37 \\ x - y = 21 \end{cases}$$

Notice that the coefficients of the variable y are opposites. Let's then solve by the addition method and begin by adding the equations.

$$\begin{array}{r} x + y = 37 \\ \underline{x - y = 21} \\ 2x = 58 \end{array} \quad \text{Add the equations.}$$

$$x = \frac{58}{2} = 29 \quad \text{Divide both sides by 2.}$$

Now we let $x = 29$ in the first equation to find y.

$$\begin{array}{ll} x + y = 37 & \text{First equation} \\ 29 + y = 37 & \\ y = 37 - 29 = 8 \end{array}$$

4. INTERPRET. The solution of the system is $(29, 8)$.

Check: Notice that the sum of 29 and 8 is $29 + 8 = 37$, the required sum. Their difference is $29 - 8 = 21$, the required difference.

State: The numbers are 29 and 8. ■

Practice Problem 2

Admission prices at a local weekend fair were $5 for children and $7 for adults. The total money collected was $3379, and 587 people attended the fair. How many children and how many adults attended the fair?

Example 2 Solving a Problem about Prices

The Barnum and Bailey Circus is in town. Admission for 4 adults and 2 children is $22, while admission for 2 adults and 3 children is $16.

a. What is the price of an adult's ticket?
b. What is the price of a child's ticket?
c. A special rate of $60 is charged for groups of 20 persons. Should a group of 4 adults and 16 children use the group rate? Why or why not?

Solution: **1.** UNDERSTAND. Read and reread the problem and guess a solution. Let's suppose that the price of an adult's ticket is $5 and the price of a child's ticket is $4. To check our proposed solution, let's see if admission for 4 adults and 2 children is $22. Admission for 4 adults is 4($5) or $20 and admission for 2 children is 2($4) or $8. This gives a total admission of $20 + $8 = $28, not the required $22. Again though, we have accomplished the purpose of this process: We have a better understanding of the problem. To continue, we let

A = the price of an adult's ticket

C = the price of a child's ticket

2. TRANSLATE. We translate the problem into two equations using both variables.

In words:	admission for 4 adults	and	admission for 2 children	is	$22
	↓	↓	↓	↓	↓
Translate:	$4A$	$+$	$2C$	$=$	22

In words:	admission for 2 adults	and	admission for 3 children	is	$16
	↓	↓	↓	↓	↓
Translate:	$2A$	$+$	$3C$	$=$	16

3. SOLVE. We solve the system

$$\begin{cases} 4A + 2C = 22 \\ 2A + 3C = 16 \end{cases}$$

Since both equations are written in standard form, we solve by the addition method. First we multiply the second equation by -2 to eliminate the variable A. Then the system

$$\begin{cases} 4A + 2C = 22 \\ -2(2A + 3C) = -2(16) \end{cases} \quad \text{simplifies to} \quad \begin{cases} 4A + 2C = 22 \\ \underline{-4A - 6C = -32} \\ -4C = -10 \end{cases}$$

Add the equations.

$$-4C = -10$$

$$C = \frac{5}{2} = 2.5 \text{ or } \$2.50, \text{ the children's ticket price.}$$

To find A, we replace C with 2.5 in the first equation.

$$4A + 2C = 22 \qquad \text{First equation}$$
$$4A + 2(2.5) = 22 \qquad \text{Let } C = 2.5.$$
$$4A + 5 = 22$$
$$4A = 17$$
$$A = \frac{17}{4} = \begin{array}{l} 4.25 \text{ or } \$4.25, \\ \text{the adult's ticket price.} \end{array}$$

4. INTERPRET.

Check: Notice that 4 adults and 2 children will pay
$4(\$4.25) + 2(\$2.50) = \$17 + \$5 = \$22$, the required amount. Also, the price for 2 adults and 3 children is $2(\$4.25) + 3(\$2.50) = \$8.50 + \$7.50 = \$16$, the required amount.

State: Answer the three original questions.

a. Since $A = 4.25$, the price of an adult's ticket is $4.25.
b. Since $C = 2.5$, the price of a child's ticket is $2.50.
c. The regular admission price for 4 adults and 16 children is

$$4(\$4.25) + 16(\$2.50) = \$17.00 + \$40.00$$
$$= \$57.00$$

This is $3 less than the special group rate of $60, so they should *not* request the group rate.

Practice Problem 3

Two cars are 440 miles apart and traveling toward each other. They meet in 3 hours. If one car's speed is 10 miles per hour faster than the other car's speed, find the speed of each car.

	r	$\cdot$ t	$= d$
Faster car			
Slower car			

Example 3   Finding Rates

Albert and Louis live 15 miles away from each other. They decide to meet one day by walking toward one another. After 2 hours they meet. If Louis walks one mile per hour faster than Albert, find both walking speeds.

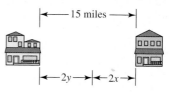

Solution

1. UNDERSTAND. Read and reread the problem. Let's propose a solution and use the formula $d = r \cdot t$ to check. Suppose that Louis' rate is 4 miles per hour. Since Louis' rate is 1 mile per hour faster, Albert's rate is 3 miles per hour. To check, see if they can walk a total of 15 miles in 2 hours. Louis' distance is rate $\cdot$ time $= 4(2) = 8$ miles and Albert's distance is rate $\cdot$ time $= 3(2) = 6$ miles. Their total distance is 8 miles $+$ 6 miles $=$ 14 miles, not the required 15 miles. Now that we have a better understanding of the problem, let's model it with a system of equations.

First, we let

$x =$ Albert's rate in miles per hour

$y =$ Louis' rate in miles per hour

Now we use the facts stated in the problem and the formula $d = rt$ to fill in the following chart.

	r	$\cdot$ t	$= d$
Albert	x	2	$2x$
Louis	y	2	$2y$

2. TRANSLATE. We translate the problem into two equations using both variables.

In words:	Albert's distance	$+$	Louis' distance	$=$	15
	$\downarrow$		$\downarrow$		$\downarrow$
Translate:	$2x$	$+$	$2y$	$=$	15

In words:	Louis' rate	is	1 mile per hour faster than Albert's
	$\downarrow$	$\downarrow$	$\downarrow$
Translate:	y	$=$	$x + 1$

3. SOLVE. The system of equations we are solving is

$$\begin{cases} 2x + 2y = 15 \\ y = x + 1 \end{cases}$$

Let's use substitution to solve the system since the second equation is solved for y.

Answer

3. One car's speed is $68\frac{1}{3}$ mph and the other car's speed is $78\frac{1}{3}$ mph.

$$2x + 2y = 15 \quad \text{First equation}$$

$$2x + 2(x + 1) = 15 \quad \text{Replace } y \text{ with } x + 1.$$

$$2x + 2x + 2 = 15$$

$$4x = 13$$

$$x = \frac{13}{4} = 3.25$$

$$y = x + 1 = 3.25 + 1 = 4.25$$

4. INTERPRET. Albert's proposed rate is 3.25 miles per hour and Louis' proposed rate is 4.25 miles per hour.

Check: Use the formula $d = rt$ and find that in 2 hours, Albert's distance is $(3.25)(2)$ miles or 6.5 miles. In 2 hours, Louis' distance is $(4.25)(2)$ miles or 8.5 miles. The total distance walked is 6.5 miles $+$ 8.5 miles or 15 miles, the given distance.

State: Albert walks at a rate of 3.25 miles per hour and Louis walks at a rate of 4.25 miles per hour.

Example 4   Finding Amounts of Solutions

Eric Daly, a chemistry teaching assistant, needs 10 liters of a 20% saline solution (salt water) for his 2 p.m. laboratory class. Unfortunately, the only mixtures on hand are a 5% saline solution and a 25% saline solution. How much of each solution should he mix to produce the 20% solution?

Solution: **1.** UNDERSTAND. Read and reread the problem. Suppose that we need 4 liters of the 5% solution. Then we need $10 - 4 = 6$ liters of the 25% solution. To see if this gives us 10 liters of a 20% saline solution, let's find the amount of pure salt in each solution.

	concentration rate	$\times$	amount of solution	$=$	amount of pure salt
	$\downarrow$		$\downarrow$		$\downarrow$
5% solution:	0.05	$\times$	4 liters	$=$	0.2 liters
25% solution:	0.25	$\times$	6 liters	$=$	1.5 liters
20% solution:	0.20	$\times$	10 liters	$=$	2 liters

Since 0.2 liters $+$ 1.5 liters $=$ 1.7 liters, not 2 liters, our proposed solution is incorrect. But we have gained some insight into how to model and check this problem.

We let

$x =$ number of liters of 5% solution

$y =$ number of liters of 25% solution

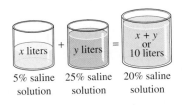

5% saline solution + 25% saline solution = 20% saline solution

Now we use a table to organize the given data.

	Concentration Rate	Liters of Solution	Liters of Pure Salt
First solution	5%	x	$0.05x$
Second solution	25%	y	$0.25y$
Mixture needed	20%	10	$(0.20)(10)$

2. TRANSLATE. We translate into two equations using both variables.

In words: liters of 5% solution   +   liters of 25% solution   =   10

$$\downarrow \qquad\qquad \downarrow \qquad\qquad \downarrow$$

Translate: $x \qquad + \qquad y \qquad = \qquad 10$

In words: salt in 5% solution   +   salt in 25% solution   =   salt in mixture

$$\downarrow \qquad\qquad \downarrow \qquad\qquad \downarrow$$

Translate: $0.05x \qquad + \qquad 0.25y \qquad = \qquad (0.20)(10)$

3. SOLVE. Here we solve the system

$$\begin{cases} x + y = 10 \\ 0.05x + 0.25y = 2 \end{cases}$$

To solve by the addition method, we first multiply the first equation by -25 and the second equation by 100. Then the system

$$\begin{cases} -25(x + y) = -25(10) \\ 100(0.05x + 0.25y) = 100(2) \end{cases}$$

simplifies to

$$\begin{cases} -25x - 25y = -250 \\ \underline{\quad 5x + 25y = 200} \\ -20x \qquad\quad = -50 \quad \text{Add.} \\ \qquad\quad x = 2.5 \end{cases}$$

To find y, we let $x = 2.5$ in the first equation of the original system.

$$x + y = 10$$
$$2.5 + y = 10 \quad \text{Let } x = 2.5.$$
$$y = 7.5$$

4. INTERPRET. Thus, we propose that Eric needs to mix 2.5 liters of 5% saline solution with 7.5 liters of 25% saline solution.

Check: Notice that $2.5 + 7.5 = 10$, the required number of liters. Also, the sum of the liters of salt in the two solutions equals the liters of salt in the required mixture:

$$0.05(2.5) + 0.25(7.5) = 0.20(10)$$
$$0.125 + 1.875 = 2$$

State: Eric needs 2.5 liters of the 5% saline solution and 7.5 liters of the 25% solution.

TRY THE CONCEPT CHECK IN THE MARGIN.

✓ **CONCEPT CHECK**

Suppose you mix an amount of 30% acid solution with an amount of 50% acid solution. Which of the following acid strengths would be possible for the resulting acid mixture?
a. 22% b. 44% c. 63%

✓ Concept Check: b

Focus On Business and Career

When you finish your present course of study, you will probably look for a job. When your job search has paid off and you receive a job offer, how will you decide whether to take the job? How do you decide between two or more job offers? These decisions are an important part of the job search and may not be easy to make. To evaluate the job offer, you should consider the nature of the work involved in the job, the type of company or organization that has offered the job, and the salary and benefits offered by the employer. You may also need to compare the compensation packages of two or more job offers. The following hints on assessing a job's compensation package were included in an article by Max Carry in the *Winter, 1990–91 Occupational Outlook Handbook Quarterly*.

Most companies will not talk about pay until they have decided to hire you. In order to know if their offer is reasonable, you need a rough estimate of what the job should pay. You may have to go to several sources for this information. Talk to friends who recently were hired in similar jobs. Ask your teachers and the staff in the college placement office about starting pay for graduates with your qualifications. Scan the help-wanted ads in newspapers. Check the library or your school's career center for salary surveys, such as the College Council Salary Survey and the Bureau of Labor Statistics wage surveys. If you are considering the salary and benefits for a job in another geographic area, make allowances for differences in the cost of living. Use the research to come up with a base salary range for yourself, the top being the best you can hope to get and the bottom being the least you will take. When negotiating, aim for the top of your estimated salary range, but be prepared to settle for less. Entry level salaries sometimes are not negotiable, particularly in government agencies. If you are not pleased with an offer, however, what harm could come from asking for more?

An employer cannot be specific about the amount of pay if it includes commissions and bonuses. The way the pay plan works, however, should be explained. The employer also should be able to tell you what most people in the job are earning. You should also learn the organization's policy regarding overtime. Depending on the job, you may or may not be exempt from laws requiring the employer to compensate you for overtime. Find out how many hours you will be expected to work each week and whether evening and weekend work is required or expected. Will you receive overtime pay or time off for working more than the specified number of hours in a week? Also take into account that the starting salary is just that, the start. Your salary should be reviewed on a regular basis—many organizations do it every 12 months. If the employer is pleased with your performance, how much can you expect to make after one year? Two years? Three years and so on?

Don't think of salary as the only compensation you will receive. Consider benefits. What on the surface looks like a great salary could be accompanied by little else. Do you know the value of the employer's contribution to your benefits? According to a 1989 Bureau of Labor Statistics study, for each dollar employers in private industry spent on straight-time wages and salaries, they contributed on average another 38 cents to employee benefits, including contributions required by law. Benefits can add a lot to your base pay. Your benefit package probably will consist of health insurance, life insurance, a pension plan, and paid vacations, holidays, and sick leave. It also may include items as diverse as profit sharing, moving expenses, parking space, a company car, and on-site day care. Do you know exactly what the benefit package includes and how much of the cost must be borne by you? Depending on your circumstances, you might want to increase or decrease particular benefits.

CRITICAL THINKING

1. Suppose you have been searching for a position as an electronics sales associate. You have received two job offers. The first job pays a monthly salary of $1500 per month plus a commission of 4% on all sales made. The second pays a monthly salary of $1800 per month plus a commission of 2% on all sales made. At what level of monthly sales would the jobs pay the same amount? Based only on the given information about the jobs, which job would you choose? Why? What other information would you want to have about the jobs before making a decision?

2. Suppose you have been searching for an entry-level bookkeeping position. You have received two job offers. The first company offers you a starting hourly wage of $6.50 per hour and says that each year entry-level workers receive a raise of $0.75 per hour. The second company offers you a starting hourly wage of $7.50 per hour and says that you can expect a $0.50 per hour raise each year. After how many years will the two jobs pay the same hourly wage? Based only on the given information about the jobs, which job would you choose? Why? What other information would you want to have about the jobs before making a decision?

Problem Solving Notes

8.5 SOLVING SYSTEMS OF LINEAR EQUATIONS IN THREE VARIABLES

Objectives

A Solve a system of three linear equations in three variables.

B Solve problems that can be modeled by a system of three linear equations

SSM   CD-ROM   Video
8.5

In this section, we solve systems of three linear equations in three variables. We call the equation $3x - y + z = -15$, for example, a **linear equation in three variables** since there are three variables and each variable is raised only to the power 1. A solution of this equation is an **ordered triple** (x, y, z) that makes the equation a true statement.

For example, the ordered triple $(2, 0, -21)$ is a solution of $3x - y + z = -15$ since replacing x with 2, y with 0, and z with -21 yields the true statement

$$3(2) - 0 + (-21) = -15$$

The graph of this equation is a plane in three-dimensional space, just as the graph of a linear equation in two variables is a line in two-dimensional space.

Although we will not discuss the techniques for graphing equations in three variables, visualizing the possible patterns of intersecting planes gives us insight into the possible patterns of solutions of a system of three three-variable linear equations. There are four possible patterns.

1. Three planes have a single point in common. This point represents the single solution of the system. This system is **consistent**.

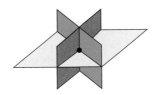

2. Three planes intersect at no point common to all three. This system has no solution. A few ways that this can occur are shown. This system is **inconsistent**.

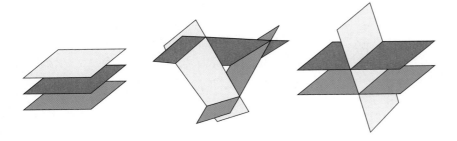

3. Three planes intersect at all the points of a single line. The system has infinitely many solutions. This system is **consistent**.

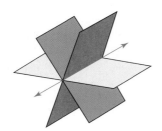

SOLVING A SYSTEM OF THREE LINEAR EQUATIONS BY THE ELIMINATION METHOD

Step 1. Write each equation in standard form,
$Ax + By + Cz = D$.

Step 2. Choose a pair of equations and use the equations to eliminate a variable.

Step 3. Choose any other pair of equations and eliminate the *same variable* as in Step 2.

Step 4. Two equations in two variables should be obtained from Step 2 and Step 3. Use methods from Section 4.1 to solve this system for both variables.

Step 5. To solve for the third variable, substitute the values of the variables found in Step 4 into any of the original equations containing the third variable.

Step 6. Check the ordered triple solution in *all three* original equations.

Practice Problem 1

Solve the system:

$$\begin{cases} 2x - y + 3z = 13 \\ x + y - z = -2 \\ 3x + 2y + 2z = 13 \end{cases}$$

4. Three planes coincide at all points on the plane. The system is consistent, and the equations are **dependent**.

A SOLVING A SYSTEM OF THREE LINEAR EQUATIONS IN THREE VARIABLES

Just as with systems of two equations in two variables, we can use the elimination method to solve a system of three equations in three variables. To do so, we eliminate a variable and obtain a system of two equations in two variables. Then we use the methods we learned in the previous section to solve the system of two equations. See the box in the margin for steps.

Example 1   Solve the system:

$$\begin{cases} 3x - y + z = -15 & \text{Equation (1)} \\ x + 2y - z = 1 & \text{Equation (2)} \\ 2x + 3y - 2z = 0 & \text{Equation (3)} \end{cases}$$

Solution: We add equations (1) and (2) to eliminate z.

$$\begin{array}{r} 3x - y + z = -15 \\ \underline{x + 2y - z = \quad 1} \\ 4x + y \quad\quad = -14 \quad \text{Equation (4)} \end{array}$$

Next we add two *other* equations and *eliminate z again*. To do so, we multiply both sides of equation (1) by 2 and add this resulting equation to equation (3). Then

$$\begin{cases} 2(3x - y + z) = 2(-15) \\ 2x + 3y - 2z = 0 \end{cases} \begin{array}{c} \text{simplifies} \\ \text{to} \end{array} \begin{cases} 6x - 2y + 2z = -30 \\ \underline{2x + 3y - 2z = \quad 0} \\ 8x + y \quad\quad = -30 \end{cases}$$
$$\text{Equation (5)}$$

Now we solve equations (4) and (5) for x and y. To solve by elimination, we multiply both sides of equation (4) by -1 and add this resulting equation to equation (5). Then

$$\begin{cases} -1(4x + y) = -1(-14) \\ 8x + y = -30 \end{cases} \begin{array}{c} \text{simplifies} \\ \text{to} \end{array} \begin{cases} -4x - y = 14 \\ \underline{8x + y = -30} \quad \text{Add the} \\ 4x \quad\quad = -16 \quad \text{equations.} \\ \quad\quad x = -4 \quad \text{Solve for } x. \end{cases}$$

We now replace x with -4 in equation (4) or (5).

$$\begin{array}{rl} 4x + y = -14 & \text{Equation (4)} \\ 4(-4) + y = -14 & \text{Let } x = -4. \\ y = 2 & \text{Solve for } y. \end{array}$$

Finally, we replace x with -4 and y with 2 in equation (1), (2), or (3).

$$\begin{array}{rl} x + 2y - z = 1 & \text{Equation (2)} \\ -4 + 2(2) - z = 1 & \text{Let } x = -4 \text{ and } y = 2. \\ -4 + 4 - z = 1 & \\ -z = 1 & \\ z = -1 & \end{array}$$

The ordered triple solution is $(-4, 2, -1)$. To check, let $x = -4$, $y = 2$, and $z = -1$ in *all three* original equations of the system.

Equation (1)

$$3x - y + z = -15$$
$$3(-4) - 2 + (-1) = -15$$
$$-12 - 2 - 1 = -15$$
$$-15 = -15$$
True.

Equation (2)

$$x + 2y - z = 1$$
$$-4 + 2(2) - (-1) = 1$$
$$-4 + 4 + 1 = 1$$
$$1 = 1$$
True.

Equation (3)

$$2x + 3y - 2z = 0$$
$$2(-4) + 3(2) - 2(-1) = 0$$
$$-8 + 6 + 2 = 0$$
$$0 = 0$$
True.

All three statements are true, so the ordered triple solution is $(-4, 2, -1)$.

Example 2   Solve the system:

$$\begin{cases} 2x - 4y + 8z = 2 & (1) \\ -x - 3y + z = 11 & (2) \\ x - 2y + 4z = 0 & (3) \end{cases}$$

Solution: When we add equations (2) and (3) to eliminate x, the new equation is

$$-5y + 5z = 11 \quad (4)$$

To eliminate x again, we multiply both sides of equation (2) by 2 and add the resulting equation to equation (1). Then

$$\begin{cases} 2x - 4y + 8z = 2 \\ 2(-x - 3y + z) = 2(11) \end{cases} \quad \begin{array}{c} \text{simplifies} \\ \text{to} \end{array} \quad \begin{cases} 2x - 4y + 8z = 2 \\ \underline{-2x - 6y + 2z = 22} \\ -10y + 10z = 24 \quad (5) \end{cases}$$

Next we solve for y and z using equations (4) and (5). To do so, we multiply both sides of equation (4) by -2 and add the resulting equation to equation (5).

$$\begin{cases} -2(-5y + 5z) = -2(11) \\ -10y + 10z = 24 \end{cases} \quad \begin{array}{c} \text{simplifies} \\ \text{to} \end{array} \quad \begin{cases} 10y - 10z = -22 \\ \underline{-10y + 10z = 24} \\ 0 = 2 \quad \text{False.} \end{cases}$$

Since the statement is false, this system is inconsistent and has no solution. The solution set is the empty set $\{\ \}$ or $\varnothing$.

TRY THE CONCEPT CHECK IN THE MARGIN.

Example 3   Solve the system:

$$\begin{cases} 2x + 4y = 1 & (1) \\ 4x - 4z = -1 & (2) \\ y - 4z = -3 & (3) \end{cases}$$

Solution: Notice that equation (2) has no term containing the variable y. Let us eliminate y using equations (1) and (3). We multiply both sides of equation (3) by -4 and add the resulting equation to equation (1). Then

$$\begin{cases} 2x + 4y = 1 \\ -4(y - 4z) = -4(-3) \end{cases} \quad \begin{matrix} \text{simplifies} \\ \text{to} \end{matrix} \quad \begin{cases} 2x + 4y = 1 \\ \underline{\quad -4y + 16z = 12} \\ 2x \quad\quad + 16z = 13 \end{cases} \quad (4)$$

Next we solve for z using equations (4) and (2). We multiply both sides of equation (4) by -2 and add the resulting equation to equation (2).

$$\begin{cases} -2(2x + 16z) = -2(13) \\ 4x - 4z = -1 \end{cases} \quad \begin{matrix} \text{simplifies} \\ \text{to} \end{matrix} \quad \begin{cases} -4x - 32z = -26 \\ \underline{\quad 4x - 4z = -1} \\ -36z = -27 \\ z = \dfrac{3}{4} \end{cases}$$

Now we replace z with $\dfrac{3}{4}$ in equation (3) and solve for y.

$$y - 4\left(\frac{3}{4}\right) = -3 \quad \text{Let } z = \frac{3}{4} \text{ in equation (3).}$$
$$y - 3 = -3$$
$$y = 0$$

Finally, we replace y with 0 in equation (1) and solve for x.

$$2x + 4(0) = 1 \quad \text{Let } y = 0 \text{ in equation (1).}$$
$$2x = 1$$
$$x = \frac{1}{2}$$

The ordered triple solution is $\left(\dfrac{1}{2}, 0, \dfrac{3}{4}\right)$. Check to see that this solution satisfies *all three* equations of the system.

Practice Problem 4

Solve the system:

$$\begin{cases} x - 3y + 4z = 2 \\ -2x + 6y - 8z = -4 \\ \dfrac{1}{2}x - \dfrac{3}{2}y + 2z = 1 \end{cases}$$

Example 4   Solve the system:

$$\begin{cases} x - 5y - 2z = 6 & (1) \\ -2x + 10y + 4z = -12 & (2) \\ \dfrac{1}{2}x - \dfrac{5}{2}y - z = 3 & (3) \end{cases}$$

Solution: We multiply both sides of equation (3) by 2 to eliminate fractions, and we multiply both sides of equation (2) by $-\dfrac{1}{2}$ so that the coefficient of x is 1. The resulting system is then

$$\begin{cases} x - 5y - 2z = 6 & (1) \\ x - 5y - 2z = 6 & \text{Multiply (2) by } -\dfrac{1}{2}. \\ x - 5y - 2z = 6 & \text{Multiply (3) by 2.} \end{cases}$$

All three resulting equations are identical, and therefore equations (1), (2), and (3) are all equivalent. There are infinitely many solutions of this system. The equations are dependent. The solution set can be written as $\{(x, y, z) \mid x - 5y - 2z = 6\}$.

B SOLVING PROBLEMS MODELED BY SYSTEMS OF THREE EQUATIONS

To introduce problem solving by writing a system of three linear equations in three variables, we solve a problem about triangles.

Example 5 Finding Angle Measures

The measure of the largest angle of a triangle is 80° more than the measure of the smallest angle, and the measure of the remaining angle is 10° more than the measure of the smallest angle. Find the measure of each angle.

Solution: **1.** UNDERSTAND.   Read and reread the problem. Recall that the sum of the measures of the angles of a triangle is 180°. Then guess a solution. If the smallest angle measures 20°, the measure of the largest angle is 80° more, or 20° + 80° = 100°. The measure of the remaining angle is 10° more than the measure of the smallest angle, or 20° + 10° = 30°. The sum of these three angles is 20° + 100° + 30° = 150°, not the required 180°. We now know that the measure of the smallest angle is greater than 20°.

To model this problem we will let

x = degree measure of the smallest angle
y = degree measure of the largest angle
z = degree measure of the remaining angle

2. TRANSLATE.   We translate the given information into three equations.

In words: | the sum of the measures | = | 180 |

Translate: $x + y + z = 180$

In words: | the largest angle | is | 80 more than the smallest angle |

Translate: $y = x + 80$

In words: | the remaining angle | is | 10 more than the smallest angle |

Translate: $z = x + 10$

3. SOLVE.   We solve the system

$$\begin{cases} x + y + z = 180 \\ y = x + 80 \\ z = x + 10 \end{cases}$$

Since y and z are both expressed in terms of x, we will solve using the subsitution method. We substitute $y = x + 80$ and $z = x + 10$ in the first equation. Then

Practice Problem 5

The measure of the largest angle of a triangle is 90° more than the measure of the smallest angle, and the measure of the remaining angle is 30° more than the measure of the smallest angle. Find the measure of each angle.

Answer

5. 20°, 50°, 110°

$$x + y + z = 180 \quad \text{First equation}$$

$$x + \overbrace{(x + 80)} + \overbrace{(x + 10)} = 180 \quad \text{Let } y = x + 80 \text{ and } z = x + 10.$$

$$3x + 90 = 180$$

$$3x = 90$$

$$x = 30$$

Then $y = x + 80 = 30 + 80 = 110$, and $z = x + 10 = 30 + 10 = 40$. The ordered triple solution is $(30, 110, 40)$.

4. INTERPRET.

Check: Notice that $30° + 40° + 110° = 180°$. Also, the measure of the largest angle, $110°$, is $80°$ more than the measure of the smallest angle, $30°$. The measure of the remaining angle, $40°$, is $10°$ more than the measure of the smallest angle, $30°$.

State: The angles measure $30°$, $110°$, and $40°$.   ▬▬▬

8.6   SOLVING SYSTEMS OF EQUATIONS USING MATRICES

By now, you may have noticed that the solution of a system of equations depends on the coefficients of the equations in the system and not on the variables. In this section, we introduce how to solve a system of equations using a **matrix**.

A   USING MATRICES TO SOLVE A SYSTEM OF TWO EQUATIONS

A **matrix** (plural: **matrices**) is a rectangular array of numbers. The following are examples of matrices.

$$\begin{bmatrix} 1 & 0 \\ 0 & 1 \end{bmatrix} \quad \begin{bmatrix} 2 & 1 & 3 & -1 \\ 0 & -1 & 4 & 5 \\ -6 & 2 & 1 & 0 \end{bmatrix} \quad \begin{bmatrix} a & b & c \\ d & e & f \end{bmatrix}$$

The numbers aligned horizontally in a matrix are in the same **row**. The numbers aligned vertically are in the same **column**.

$$\begin{array}{c} \text{row 1} \rightarrow \\ \text{row 2} \rightarrow \end{array} \begin{bmatrix} 2 & 1 & 0 \\ -1 & 6 & 2 \end{bmatrix}$$

column 1
column 2
column 3

This matrix has 2 rows and 3 columns. It is called a 2×3 (read "two by three") matrix.

To see the relationship between systems of equations and matrices, study the example below.

System of Equations

$$\begin{cases} 2x - 3y = 6 & \text{Equation 1} \\ x + y = 0 & \text{Equation 2} \end{cases}$$

Corresponding Matrix

$$\begin{bmatrix} 2 & -3 & | & 6 \\ 1 & 1 & | & 0 \end{bmatrix} \begin{array}{l} \text{Row 1} \\ \text{Row 2} \end{array}$$

Notice that the rows of the matrix correspond to the equations in the system. The coefficients of each variable are placed to the left of a vertical dashed line. The constants are placed to the right. Each of these numbers in the matrix is called an **element**.

The method of solving systems by matrices is to write this matrix as an equivalent matrix from which we easily identify the solution. Two matrices are equivalent if they represent systems that have the same solution set. The following **row operations** can be performed on matrices, and the result is an equivalent matrix.

ELEMENTARY ROW OPERATIONS

1. Any two rows in a matrix may be interchanged.
2. The elements of any row may be multiplied (or divided) by the same nonzero number.
3. The elements of any row may be multiplied (or divided) by a nonzero number and added to their corresponding elements in any other row.

HELPFUL HINT

Notice that these *row* operations are the same operations that we can perform on *equations* in a system.

Objectives

A Use matrices to solve a system of two equations.

B Use matrices to solve a system of three equations.

SSM   CD-ROM   Video
8.6

Practice Problem 1

Use matrices to solve the system:
$$\begin{cases} x + 2y = -4 \\ 2x - 3y = 13 \end{cases}$$

Example 1

Use matrices to solve the system:
$$\begin{cases} x + 3y = 5 \\ 2x - y = -4 \end{cases}$$

Solution: The corresponding matrix is $\begin{bmatrix} 1 & 3 & | & 5 \\ 2 & -1 & | & -4 \end{bmatrix}$. We use elementary row operations to write an equivalent matrix that looks like $\begin{bmatrix} 1 & a & | & b \\ 0 & 1 & | & c \end{bmatrix}$.

For the matrix given, the element in the first row, first column is already 1, as desired. Next we write an equivalent matrix with a 0 below the 1. To do this, we multiply row 1 by -2 and add to row 2. *We will change only row 2.*

$$\begin{bmatrix} 1 & 3 & | & 5 \\ -2(1) + 2 & -2(3) + (-1) & | & -2(5) + (-4) \end{bmatrix}$$ simplifies to

row 1 row 2   row 1 row 2   row 1 row 2
element element element element element element

$$\begin{bmatrix} 1 & 3 & | & 5 \\ 0 & -7 & | & -14 \end{bmatrix}$$

Now we change the -7 to a 1 by use of an elementary row operation. We divide row 2 by -7, then

$$\begin{bmatrix} 1 & 3 & | & 5 \\ \dfrac{0}{-7} & \dfrac{-7}{-7} & | & \dfrac{-14}{-7} \end{bmatrix}$$ simplifies to $$\begin{bmatrix} 1 & 3 & | & 5 \\ 0 & 1 & | & 2 \end{bmatrix}$$

This last matrix corresponds to the system
$$\begin{cases} x + 3y = 5 \\ y = 2 \end{cases}$$

To find x, we let $y = 2$ in the first equation, $x + 3y = 5$.

$x + 3y = 5$ First equation
$x + 3(2) = 5$ Let $y = 2$.
$x = -1$

The ordered pair solution is $(-1, 2)$. Check to see that this ordered pair satisfies both equations. ▬▬▬

Practice Problem 2

Use matrices to solve the system:
$$\begin{cases} -3x + y = 0 \\ -6x + 2y = 2 \end{cases}$$

Example 2

Use matrices to solve the system:
$$\begin{cases} 2x - y = 3 \\ 4x - 2y = 5 \end{cases}$$

Solution: The corresponding matrix is $\begin{bmatrix} 2 & -1 & | & 3 \\ 4 & -2 & | & 5 \end{bmatrix}$. To get 1 in the row 1, column 1 position, we divide the elements of row 1 by 2.

$$\begin{bmatrix} \dfrac{2}{2} & -\dfrac{1}{2} & | & \dfrac{3}{2} \\ 4 & -2 & | & 5 \end{bmatrix}$$ simplifies to $$\begin{bmatrix} 1 & -\dfrac{1}{2} & | & \dfrac{3}{2} \\ 4 & -2 & | & 5 \end{bmatrix}$$

Answers

1. $(2, -3)$, **2.** no solution

To get 0 under the 1, we multiply the elements of row 1 by -4 and add the new elements to the elements of row 2.

$$\begin{bmatrix} 1 & -\dfrac{1}{2} & \vdots & \dfrac{3}{2} \\ -4(1) + 4 & -4\left(-\dfrac{1}{2}\right) - 2 & \vdots & -4\left(\dfrac{3}{2}\right) + 5 \end{bmatrix} \quad \text{simplifies to}$$

$$\begin{bmatrix} 1 & -\dfrac{1}{2} & \vdots & \dfrac{3}{2} \\ 0 & 0 & \vdots & -1 \end{bmatrix}$$

The corresponding system is $\begin{cases} x - \dfrac{1}{2}y = \dfrac{3}{2} \\ 0 = -1 \end{cases}$. The equation $0 = -1$ is false for all y or x values; hence the system is inconsistent and has no solution.

TRY THE CONCEPT CHECK IN THE MARGIN.

B   USING MATRICES TO SOLVE A SYSTEM OF THREE EQUATIONS

To solve a system of three equations in three variables using matrices, we will write the corresponding matrix in the form

$$\begin{bmatrix} 1 & a & b & \vdots & d \\ 0 & 1 & c & \vdots & e \\ 0 & 0 & 1 & \vdots & f \end{bmatrix}$$

Example 3   Use matrices to solve the system:

$$\begin{cases} x + 2y + z = 2 \\ -2x - y + 2z = 5 \\ x + 3y - 2z = -8 \end{cases}$$

Solution: The corresponding matrix is $\begin{bmatrix} 1 & 2 & 1 & \vdots & 2 \\ -2 & -1 & 2 & \vdots & 5 \\ 1 & 3 & -2 & \vdots & -8 \end{bmatrix}$.

Our goal is to write an equivalent matrix with 1s along the diagonal (see the numbers in red) and 0s below the 1s. The element in row 1, column 1 is already 1. Next we get 0s for each element in the rest of column 1. To do this, first we multiply the elements of row 1 by 2 and add the new elements to row 2. Also, we multiply the elements of row 1 by -1 and add the new elements to the elements of row 3. We *do not change row 1.* Then

$$\begin{bmatrix} 1 & 2 & 1 & \vdots & 2 \\ 2(1) - 2 & 2(2) - 1 & 2(1) + 2 & \vdots & 2(2) + 5 \\ -1(1) + 1 & -1(2) + 3 & -1(1) - 2 & \vdots & -1(2) - 8 \end{bmatrix} \quad \text{simplifies to}$$

$$\begin{bmatrix} 1 & 2 & 1 & \vdots & 2 \\ 0 & 3 & 4 & \vdots & 9 \\ 0 & 1 & -3 & \vdots & -10 \end{bmatrix}$$

✓ **CONCEPT CHECK**

Consider the system
$$\begin{cases} 2x - 3y = 8 \\ x + 5y = -3 \end{cases}$$
What is wrong with its corresponding matrix shown below?
$$\begin{bmatrix} 2 & 3 & \vdots & 8 \\ 0 & 5 & \vdots & 3 \end{bmatrix}$$

Practice Problem 3

Use matrices to solve the system:

$$\begin{cases} x + 3y + z = 5 \\ -3x + y - 3z = 5 \\ x + 2y - 2z = 9 \end{cases}$$

Answers

3. $(1, 2, -2)$

✓ **Concept Check:**   Its matrix is $\begin{bmatrix} 2 & -3 & \vdots & 8 \\ 1 & 5 & \vdots & -3 \end{bmatrix}$.

We continue down the diagonal and use elementary row operations to get 1 where the element 3 is now. To do this, we interchange rows 2 and 3.

$$\begin{bmatrix} 1 & 2 & 1 & \vdots & 2 \\ 0 & 3 & 4 & \vdots & 9 \\ 0 & 1 & -3 & \vdots & -10 \end{bmatrix} \text{ is equivalent to } \begin{bmatrix} 1 & 2 & 1 & \vdots & 2 \\ 0 & 1 & -3 & \vdots & -10 \\ 0 & 3 & 4 & \vdots & 9 \end{bmatrix}$$

Next we want the new row 3, column 2 element to be 0. We multiply the elements of row 2 by -3 and add the result to the elements of row 3.

$$\begin{bmatrix} 1 & 2 & 1 & \vdots & 2 \\ 0 & 1 & -3 & \vdots & -10 \\ -3(0) + 0 & -3(1) + 3 & -3(-3) + 4 & \vdots & -3(-10) + 9 \end{bmatrix} \text{ simplifies to}$$

$$\begin{bmatrix} 1 & 2 & 1 & \vdots & 2 \\ 0 & 1 & -3 & \vdots & -10 \\ 0 & 0 & 13 & \vdots & 39 \end{bmatrix}$$

Finally, we divide the elements of row 3 by 13 so that the final diagonal element is 1.

$$\begin{bmatrix} 1 & 2 & 1 & \vdots & 2 \\ 0 & 1 & -3 & \vdots & -10 \\ \frac{0}{13} & \frac{0}{13} & \frac{13}{13} & \vdots & \frac{39}{13} \end{bmatrix} \text{ simplifies to } \begin{bmatrix} 1 & 2 & 1 & \vdots & 2 \\ 0 & 1 & -3 & \vdots & -10 \\ 0 & 0 & 1 & \vdots & 3 \end{bmatrix}$$

This matrix corresponds to the system

$$\begin{cases} x + 2y + z = 2 \\ \quad\quad y - 3z = -10 \\ \quad\quad\quad\quad z = 3 \end{cases}$$

We identify the z-coordinate of the solution as 3. Next we replace z with 3 in the second equation and solve for y.

$$\begin{aligned} y - 3z &= -10 \quad \text{Second equation} \\ y - 3(3) &= -10 \quad \text{Let } z = 3. \\ y &= -1 \end{aligned}$$

To find x, we let $z = 3$ and $y = -1$ in the first equation.

$$\begin{aligned} x + 2y + z &= 2 \quad \text{First equation} \\ x + 2(-1) + 3 &= 2 \quad \text{Let } z = 3 \text{ and } y = -1. \\ x &= 1 \end{aligned}$$

The ordered triple solution is $(1, -1, 3)$. Check to see that it satisfies all three equations in the original system.

8.7   SOLVING SYSTEMS OF EQUATIONS USING DETERMINANTS

We have solved systems of two linear equations in two variables in four different ways: graphically, by substitution, by elimination, and by matrices. Now we analyze another method called **Cramer's rule**.

A   EVALUATING 2 × 2 DETERMINANTS

Recall that a matrix is a rectangular array of numbers. If a matrix has the same number of rows and columns, it is called a **square matrix**. Examples of square matrices are

$$\begin{bmatrix} 1 & 6 \\ 5 & 2 \end{bmatrix} \qquad \begin{bmatrix} 2 & 4 & 1 \\ 0 & 5 & 2 \\ 3 & 6 & 9 \end{bmatrix}$$

A **determinant** is a real number associated with a square matrix. The determinant of a square matrix is denoted by placing vertical bars about the array of numbers. Thus,

The determinant of the square matrix $\begin{bmatrix} 1 & 6 \\ 5 & 2 \end{bmatrix}$ is $\begin{vmatrix} 1 & 6 \\ 5 & 2 \end{vmatrix}$.

The determinant of the square matrix $\begin{bmatrix} 2 & 4 & 1 \\ 0 & 5 & 2 \\ 3 & 6 & 9 \end{bmatrix}$ is $\begin{vmatrix} 2 & 4 & 1 \\ 0 & 5 & 2 \\ 3 & 6 & 9 \end{vmatrix}$.

We define the determinant of a 2 × 2 matrix first. (Recall that 2 × 2 is read "two by two." It means that the matrix has 2 rows and 2 columns.)

DETERMINANT OF A 2 × 2 MATRIX

$$\begin{vmatrix} a & b \\ c & d \end{vmatrix} = ad - bc$$

Example 1   Evaluate each determinant.

a. $\begin{vmatrix} -1 & 2 \\ 3 & -4 \end{vmatrix}$ **b.** $\begin{vmatrix} 2 & 0 \\ 7 & -5 \end{vmatrix}$

Solution: First we identify the values of a, b, c, and d. Then we perform the evaluation.

a. Here $a = -1$, $b = 2$, $c = 3$, and $d = -4$.

$$\begin{vmatrix} -1 & 2 \\ 3 & -4 \end{vmatrix} = ad - bc = (-1)(-4) - (2)(3) = -2$$

b. In this example, $a = 2$, $b = 0$, $c = 7$, and $d = -5$.

$$\begin{vmatrix} 2 & 0 \\ 7 & -5 \end{vmatrix} = ad - bc = 2(-5) - (0)(7) = -10$$

Objectives

A Define and evaluate a 2 × 2 determinant.

B Use Cramer's rule to solve a system of two linear equations in two variables.

C Define and evaluate a 3 × 3 determinant.

D Use Cramer's rule to solve a linear system of three equations in three variables.

SSM   CD-ROM   Video
8.7

Practice Problem 1

Evaluate each determinant.

a. $\begin{vmatrix} -3 & 6 \\ 2 & 1 \end{vmatrix}$

b. $\begin{vmatrix} 4 & 5 \\ 0 & -5 \end{vmatrix}$

Answers

1. a. -15,   **b.** -20

B USING CRAMER'S RULE TO SOLVE A SYSTEM OF TWO LINEAR EQUATIONS

To develop Cramer's rule, we solve the system $\begin{cases} ax + by = h \\ cx + dy = k \end{cases}$ using elimination. First, we eliminate y by multiplying both sides of the first equation by d and both sides of the second equation by $-b$ so that the coefficients of y are opposites. The result is that

$$\begin{cases} d(ax + by) = d \cdot h \\ -b(cx + dy) = -b \cdot k \end{cases} \quad \text{simplifies to} \quad \begin{cases} adx + bdy = hd \\ -bcx - bdy = -kb \end{cases}$$

We now add the two equations and solve for x.

$$\begin{array}{rl}
adx + bdy &= hd \\
-bcx - bdy &= -kb \\
\hline
adx - bcx &= hd - kb \qquad \text{Add the equations.}\\
(ad - bc)x &= hd - kb \\
x &= \dfrac{hd - kb}{ad - bc} \qquad \text{Solve for } x.
\end{array}$$

When we replace x with $\dfrac{hd - kb}{ad - bc}$ in the equation $ax + by = h$ and solve for y, we find that $y = \dfrac{ak - ch}{ad - bc}$.

Notice that the numerator of the value of x is the determinant of

$$\begin{vmatrix} h & b \\ k & d \end{vmatrix} = hd - kb$$

Also, the numerator of the value of y is the determinant of

$$\begin{vmatrix} a & h \\ c & k \end{vmatrix} = ak - hc$$

Finally, the denominators of the values of x and y are the same and are the determinant of

$$\begin{vmatrix} a & b \\ c & d \end{vmatrix} = ad - bc$$

This means that the values of x and y can be written in determinant notation:

$$x = \frac{\begin{vmatrix} h & b \\ k & d \end{vmatrix}}{\begin{vmatrix} a & b \\ c & d \end{vmatrix}} \quad \text{and} \quad y = \frac{\begin{vmatrix} a & h \\ c & k \end{vmatrix}}{\begin{vmatrix} a & b \\ c & d \end{vmatrix}}$$

For convenience, we label the determinants D, D_x, and D_y.

$$\begin{vmatrix} a & b \\ c & d \end{vmatrix} = D \qquad \begin{vmatrix} h & b \\ k & d \end{vmatrix} = D_x \qquad \begin{vmatrix} a & h \\ c & k \end{vmatrix} = D_y$$

x-column replaced by constants y-column replaced by constants

These determinant formulas for the coordinates of the solution of a system are known as **Cramer's rule**.

CRAMER'S RULE FOR TWO LINEAR EQUATIONS IN TWO VARIABLES

The solution of the system $\begin{cases} ax + by = h \\ cx + dy = k \end{cases}$ is given by

$$x = \frac{\begin{vmatrix} h & b \\ k & d \end{vmatrix}}{\begin{vmatrix} a & b \\ c & d \end{vmatrix}} = \frac{D_x}{D} \qquad y = \frac{\begin{vmatrix} a & h \\ c & k \end{vmatrix}}{\begin{vmatrix} a & b \\ c & d \end{vmatrix}} = \frac{D_y}{D}$$

as long as $D = ad - bc$ is not 0.

When $D = 0$, the system is either inconsistent or the equations are dependent. When this happens, we need to use another method to see which is the case.

Example 2   Use Cramer's rule to solve the system:

$$\begin{cases} 3x + 4y = -7 \\ x - 2y = -9 \end{cases}$$

Solution: First we find D, D_x, and D_y.

$$\begin{array}{ccc} a & b & h \\ \downarrow & \downarrow & \downarrow \end{array}$$
$$\begin{cases} 3x + 4y = -7 \\ x - 2y = -9 \end{cases}$$
$$\begin{array}{ccc} \uparrow & \uparrow & \uparrow \\ c & d & k \end{array}$$

$$D = \begin{vmatrix} a & b \\ c & d \end{vmatrix} = \begin{vmatrix} 3 & 4 \\ 1 & -2 \end{vmatrix} = 3(-2) - 4(1) = -10$$

$$D_x = \begin{vmatrix} h & b \\ k & d \end{vmatrix} = \begin{vmatrix} -7 & 4 \\ -9 & -2 \end{vmatrix} = (-7)(-2) - 4(-9) = 50$$

$$D_y = \begin{vmatrix} a & h \\ c & k \end{vmatrix} = \begin{vmatrix} 3 & -7 \\ 1 & -9 \end{vmatrix} = 3(-9) - (-7)(1) = -20$$

Then $x = \dfrac{D_x}{D} = \dfrac{50}{-10} = -5$ and $y = \dfrac{D_y}{D} = \dfrac{-20}{-10} = 2$.
The ordered pair solution is $(-5, 2)$.

As always, check the solution in both original equations.

Example 3   Use Cramer's rule to solve the system:

$$\begin{cases} 5x + y = 5 \\ -7x - 2y = -7 \end{cases}$$

Solution: First we find D, D_x, and D_y.

$$D = \begin{vmatrix} 5 & 1 \\ -7 & -2 \end{vmatrix} = 5(-2) - (-7)(1) = -3$$

$$D_x = \begin{vmatrix} 5 & 1 \\ -7 & -2 \end{vmatrix} = 5(-2) - (-7)(1) = -3$$

$$D_y = \begin{vmatrix} 5 & 5 \\ -7 & -7 \end{vmatrix} = 5(-7) - 5(-7) = 0$$

Practice Problem 2

Use Cramer's rule to solve the system.

$$\begin{cases} x - y = -4 \\ 2x + 3y = 2 \end{cases}$$

Practice Problem 3

Use Cramer's rule to solve the system.

$$\begin{cases} 4x + y = 3 \\ 2x - 3y = -9 \end{cases}$$

Answers

2. $(-2, 2)$,   **3.** $(0, 3)$

Then

$$x = \frac{D_x}{D} = \frac{-3}{-3} = 1 \qquad y = \frac{D_y}{D} = \frac{0}{-3} = 0$$

The ordered pair solution is $(1, 0)$.

C EVALUATING 3×3 DETERMINANTS

A 3×3 determinant can be used to solve a system of three equations in three variables. The determinant of a 3×3 matrix, however, is considerably more complex than a 2×2 one.

DETERMINANT OF A 3×3 MATRIX

$$\begin{vmatrix} a_1 & b_1 & c_1 \\ a_2 & b_2 & c_2 \\ a_3 & b_3 & c_3 \end{vmatrix} = a_1 \cdot \begin{vmatrix} b_2 & c_2 \\ b_3 & c_3 \end{vmatrix} - a_2 \cdot \begin{vmatrix} b_1 & c_1 \\ b_3 & c_3 \end{vmatrix} + a_3 \cdot \begin{vmatrix} b_1 & c_1 \\ b_2 & c_2 \end{vmatrix}$$

Notice that the determinant of a 3×3 matrix is related to the determinants of three 2×2 matrices. Each determinant of these 2×2 matrices is called a **minor**, and every element of a 3×3 matrix has a minor associated with it. For example, the minor of c_2 is the determinant of the 2×2 matrix found by deleting the row and column containing c_2.

$$\begin{matrix} a_1 & b_1 & c_1 \\ a_2 & b_2 & c_2 \\ a_3 & b_3 & c_3 \end{matrix} \qquad \text{The minor of } c_2 \text{ is} \qquad \begin{vmatrix} a_1 & b_1 \\ a_3 & b_3 \end{vmatrix}$$

Also, the minor of element a_1 is the determinant of the 2×2 matrix that has no row or column containing a_1.

$$\begin{matrix} a_1 & b_1 & c_1 \\ a_2 & b_2 & c_2 \\ a_3 & b_3 & c_3 \end{matrix} \qquad \text{The minor of } a_1 \text{ is} \qquad \begin{vmatrix} b_2 & c_2 \\ b_3 & c_3 \end{vmatrix}$$

So the determinant of a 3×3 matrix can be written as

$$a_1 \cdot (\text{minor of } a_1) - a_2 \cdot (\text{minor of } a_2) + a_3 \cdot (\text{minor of } a_3)$$

Finding the determinant by using minors of elements in the first column is called **expanding** by the minors of the first column. *The value of a determinant can be found by expanding by the minors of any row or column.* The following **array of signs** is helpful in determining whether to add or subtract the product of an element and its minor.

$$\begin{matrix} + & - & + \\ - & + & - \\ + & - & + \end{matrix}$$

If an element is in a position marked $+$, we add. If marked $-$, we subtract.

TRY THE CONCEPT CHECK IN THE MARGIN.

Example 4   Evaluate by expanding by the minors of the given row or column.

$$\begin{vmatrix} 0 & 5 & 1 \\ 1 & 3 & -1 \\ -2 & 2 & 4 \end{vmatrix}$$

a. First column   **b.** Second row

✓ **CONCEPT CHECK**

Suppose you are interested in finding the determinant of a 4×4 matrix. Study the pattern shown in the array of signs for a 3×3 matrix. Use the pattern to expand the array of signs for use with a 4×4 matrix.

Practice Problem 4

Evaluate by expanding by the minors of the given row or column.

a. First column   b. Third row

$$\begin{vmatrix} 2 & 0 & 1 \\ -1 & 3 & 2 \\ 5 & 1 & 4 \end{vmatrix}$$

Answers

4. a. 4,   **b.** 4

✓ **Concept Check:** $\begin{matrix} + & - & + & - \\ - & + & - & + \\ + & - & + & - \\ - & + & - & + \end{matrix}$

Solution: **a.** The elements of the first column are 0, 1, and -2. The first column of the array of signs is $+$, $-$, $+$.

$$\begin{vmatrix} 0 & 5 & 1 \\ 1 & 3 & -1 \\ -2 & 2 & 4 \end{vmatrix} = 0 \cdot \begin{vmatrix} 3 & -1 \\ 2 & 4 \end{vmatrix} - 1 \cdot \begin{vmatrix} 5 & 1 \\ 2 & 4 \end{vmatrix} + (-2) \cdot \begin{vmatrix} 5 & 1 \\ 3 & -1 \end{vmatrix}$$

$$= 0(12 - (-2)) - 1(20 - 2) + (-2)(-5 - 3)$$
$$= 0 - 18 + 16 = -2$$

b. The elements of the second row are 1, 3, and -1. This time, the signs begin with $-$ and again alternate.

$$\begin{vmatrix} 0 & 5 & 1 \\ 1 & 3 & -1 \\ -2 & 2 & 4 \end{vmatrix} = -1 \cdot \begin{vmatrix} 5 & 1 \\ 2 & 4 \end{vmatrix} + 3 \cdot \begin{vmatrix} 0 & 1 \\ -2 & 4 \end{vmatrix} - (-1) \cdot \begin{vmatrix} 0 & 5 \\ -2 & 2 \end{vmatrix}$$

$$= -1(20 - 2) + 3(0 - (-2)) - (-1)(0 - (-10))$$
$$= -18 + 6 + 10 = -2$$

Notice that the determinant of the 3×3 matrix is the same regardless of the row or column you select to expand by.

TRY THE CONCEPT CHECK IN THE MARGIN.

D **USING CRAMER'S RULE TO SOLVE A SYSTEM OF THREE LINEAR EQUATIONS**

A system of three equations in three variables may be solved with Cramer's rule also. Using the elimination process to solve a system with unknown constants as coefficients leads to the following.

CRAMER'S RULE FOR THREE EQUATIONS IN THREE VARIABLES

The solution of the system $\begin{cases} a_1x + b_1y + c_1z = k_1 \\ a_2x + b_2y + c_2z = k_2 \\ a_3x + b_3y + c_3z = k_3 \end{cases}$ is given by

$$x = \frac{D_x}{D} \qquad y = \frac{D_y}{D} \qquad \text{and} \qquad z = \frac{D_z}{D}$$

where

$$D = \begin{vmatrix} a_1 & b_1 & c_1 \\ a_2 & b_2 & c_2 \\ a_3 & b_3 & c_3 \end{vmatrix} \quad D_x = \begin{vmatrix} k_1 & b_1 & c_1 \\ k_2 & b_2 & c_2 \\ k_3 & b_3 & c_3 \end{vmatrix}$$

$$D_y = \begin{vmatrix} a_1 & k_1 & c_1 \\ a_2 & k_2 & c_2 \\ a_3 & k_3 & c_3 \end{vmatrix} \quad D_z = \begin{vmatrix} a_1 & b_1 & k_1 \\ a_2 & b_2 & k_2 \\ a_3 & b_3 & k_3 \end{vmatrix}$$

as long as D is not 0.

Example 5   Use Cramer's rule to solve the system:

$$\begin{cases} x - 2y + z = 4 \\ 3x + y - 2z = 3 \\ 5x + 5y + 3z = -8 \end{cases}$$

✓ CONCEPT CHECK

Why would expanding by minors of the second row be a good choice for

the determinant $\begin{vmatrix} 3 & 4 & -2 \\ 5 & 0 & 0 \\ 6 & -3 & 7 \end{vmatrix}$?

Practice Problem 5

Use Cramer's rule to solve the system:

$$\begin{cases} x + 2y - z = 3 \\ 2x - 3y + z = -9 \\ -x + y - 2z = 0 \end{cases}$$

Answers

5. $(-1, 3, 2)$

✓ Concept Check: Two elements of the second row are 0, which makes calculations easier.

Solution: First we find D, D_x, D_y, and D_z. Beginning with D, we expand by the minors of the first column.

$$D = \begin{vmatrix} 1 & -2 & 1 \\ 3 & 1 & -2 \\ 5 & 5 & 3 \end{vmatrix} = 1 \cdot \begin{vmatrix} 1 & -2 \\ 5 & 3 \end{vmatrix} - 3 \cdot \begin{vmatrix} -2 & 1 \\ 5 & 3 \end{vmatrix} + 5 \cdot \begin{vmatrix} -2 & 1 \\ 1 & -2 \end{vmatrix}$$

$$= 1(3 - (-10)) - 3(-6 - 5) + 5(4 - 1)$$

$$= 13 + 33 + 15 = 61$$

$$D_x = \begin{vmatrix} 4 & -2 & 1 \\ 3 & 1 & -2 \\ -8 & 5 & 3 \end{vmatrix} = 4 \cdot \begin{vmatrix} 1 & -2 \\ 5 & 3 \end{vmatrix} - 3 \cdot \begin{vmatrix} -2 & 1 \\ 5 & 3 \end{vmatrix} + (-8) \cdot \begin{vmatrix} -2 & 1 \\ 1 & -2 \end{vmatrix}$$

$$= 4(3 - (-10)) - 3(-6 - 5) + (-8)(4 - 1)$$

$$= 52 + 33 - 24 = 61$$

$$D_y = \begin{vmatrix} 1 & 4 & 1 \\ 3 & 3 & -2 \\ 5 & -8 & 3 \end{vmatrix} = 1 \cdot \begin{vmatrix} 3 & -2 \\ -8 & 3 \end{vmatrix} - 3 \cdot \begin{vmatrix} 4 & 1 \\ -8 & 3 \end{vmatrix} + 5 \cdot \begin{vmatrix} 4 & 1 \\ 3 & -2 \end{vmatrix}$$

$$= 1(9 - 16) - 3(12 - (-8)) + 5(-8 - 3)$$

$$= -7 - 60 - 55 = -122$$

$$D_z = \begin{vmatrix} 1 & -2 & 4 \\ 3 & 1 & 3 \\ 5 & 5 & -8 \end{vmatrix} = 1 \cdot \begin{vmatrix} 1 & 3 \\ 5 & -8 \end{vmatrix} - 3 \cdot \begin{vmatrix} -2 & 4 \\ 5 & -8 \end{vmatrix} + 5 \cdot \begin{vmatrix} -2 & 4 \\ 1 & 3 \end{vmatrix}$$

$$= 1(-8 - 15) - 3(16 - 20) + 5(-6 - 4)$$

$$= -23 + 12 - 50 = -61$$

From these determinants, we calculate the solution:

$$x = \frac{D_x}{D} = \frac{61}{61} = 1 \qquad y = \frac{D_y}{D} = \frac{-122}{61} = -2 \qquad z = \frac{D_z}{D} = \frac{-61}{61} = -1$$

The ordered triple solution is $(1, -2, -1)$. Check this solution by verifying that it satisfies each equation of the system.

8.8 SYSTEMS OF LINEAR INEQUALITIES

A GRAPHING SYSTEMS OF LINEAR INEQUALITIES

In Section 3.5 we solved linear inequalities in two variables. Just as two linear equations make a system of linear equations, two linear inequalities make a **system of linear inequalities**. Systems of inequalities are very important in a process called linear programming. Many businesses use linear programming to find the most profitable way to use limited resources such as employees, machines, or buildings.

A **solution of a system of linear inequalities** is an ordered pair that satisfies each inequality in the system. The set of all such ordered pairs is the solution set of the system. Graphing this set gives us a picture of the solution set. We can graph a system of inequalities by graphing each inequality in the system and identifying the region of overlap.

GRAPHING THE SOLUTIONS OF A SYSTEM OF LINEAR INEQUALITIES

Step 1. Graph each inequality in the system on the same set of axes.

Step 2. The solutions of the system are the points common to the graphs of all the inequalities in the system.

Example 1   Graph the solutions of the system $\begin{cases} 3x \geq y \\ x + 2y \leq 8 \end{cases}$

Solution: We begin by graphing each inequality on the *same* set of axes. The graph of the solutions of the system is the region contained in the graphs of both inequalities. In other words, it is their intersection.

First let's graph $3x \geq y$. The boundary line is the graph of $3x = y$. We sketch a solid boundary line since the inequality $3x \geq y$ means $3x > y$ or $3x = y$. The test point $(1, 0)$ satisfies the inequality, so we shade the half-plane that includes $(1, 0)$.

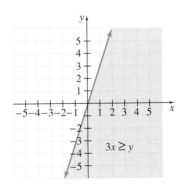   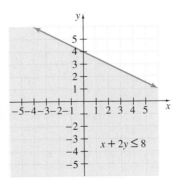

Next we sketch a solid boundary line $x + 2y = 8$ on the same set of axes. The test point $(0, 0)$ satisfies the inequality $x + 2y \leq 8$, so we shade the half-plane that includes $(0, 0)$. (For clarity, the graph of $x + 2y \leq 8$ is shown here on a separate set of axes.)

An ordered pair solution of the system must satisfy both inequalities. These solutions are points that lie in

Objective

A Graph a system of linear inequalities.

SSM   CD-ROM   Video
8.8

Practice Problem 1

Graph the solutions of the system:

$\begin{cases} 2x \leq y \\ x + 4y \geq 4 \end{cases}$

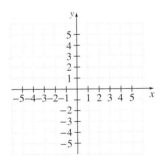

Answer

1.

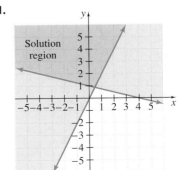

both shaded regions. The solution of the system is the darkest shaded region. This solution includes parts of both boundary lines.

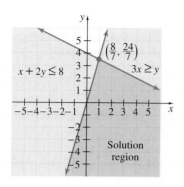

In linear programming, it is sometimes necessary to find the coordinates of the **corner point**: the point at which the two boundary lines intersect. To find the point of intersection for the system of Example 1, we solve the related linear system

$$\begin{cases} 3x = y \\ x + 2y = 8 \end{cases}$$

using either the subsitution or the elimination method. The lines intersect at $\left(\dfrac{8}{7}, \dfrac{24}{7} \right)$, the corner point of the graph.

Practice Problem 2

Graph the solutions of the system:

$$\begin{cases} -x + y < 3 \\ y < 1 \\ 2x + y > -2 \end{cases}$$

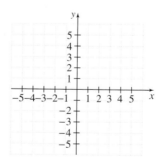

Answer

2.

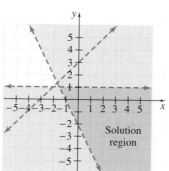

Example 2   Graph the solutions of the system: $\begin{cases} x - y < 2 \\ x + 2y > -1 \\ y < 2 \end{cases}$

Solution: First we graph all three inequalities on the same set of axes. All boundary lines are dashed lines since the inequality symbols are $<$ and $>$. The solution of the system is the region shown by the darkest shading. In this example, the boundary lines are *not* a part of the solution.

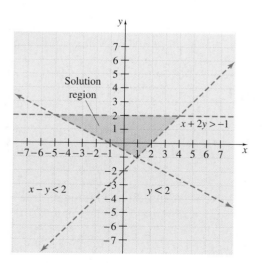

TRY THE CONCEPT CHECK IN THE MARGIN.

Example 3   Graph the solutions of the system $\begin{cases} -3x + 4y \le 12 \\ x \le 3 \\ x \ge 0 \\ y \ge 0 \end{cases}$

     Solution: We graph the inequalities on the same set of axes. The intersection of the inequalities is the solution region. It is the only region shaded in this graph and includes portions of all four boundary lines.

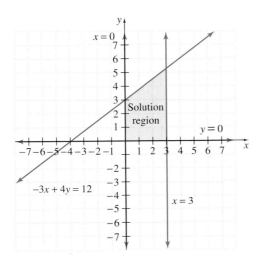

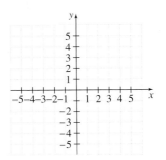

✓ CONCEPT CHECK

Describe the solution of the system of inequalities:

$$\begin{cases} x \le 2 \\ x \ge 2 \end{cases}$$

Practice Problem 3

Graph the solutions of the system:

$$\begin{cases} 2x - 3y \le 6 \\ y \ge 0 \\ y \le 4 \\ x \ge 0 \end{cases}$$

Answers

3.

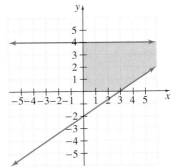

✓ Concept Check: It is the line $x = 2$.

Focus On the Real World

ANOTHER MATHEMATICAL MODEL

Sometimes mathematical models other than linear models are appropriate for data. Suppose that an equation of the form $y = ax^2 + bx + c$ is an appropriate model for the ordered pairs (x_1, y_1), (x_2, y_2), and (x_3, y_3). Then it is necessary to find the values of a, b, and c such that the given ordered pairs are solutions of the equation $y = ax^2 + bx + c$. To do so, substitute each ordered pair into the equation. Each time, the result is an equation in three unknowns: a, b, and c. Solving the resulting system of three linear equation in three unknowns will give the required values of a, b, and c.

GROUP ACTIVITY

1. The table gives the average annual fatalities from lightning in each of the years listed.
   a. Write the data as ordered pairs of the form (x, y) where y is the number of lightning fatalities in the year x ($x = 0$ represents 1900).
   b. Find the values of a, b, and c such that the equation $y = ax^2 + bx + c$ models this data.
   c. Verify that the model you found in part (b) gives each of the ordered pair solutions from part (a).
   d. According to the model, what was the average annual number of fatalities from lightning in 1955?

AVERAGE ANNUAL LIGHTNING FATALITIES	
Year	Fatalities
1940	337
1950	184
1960	133

(*Source:* National Weather Service)

2. The table gives the world production of red meat (in millions of metric tons) for each of the years listed.
   a. Write the data as ordered pairs of the form (x, y) where y is the red meat production (in millions of metric tons) in the year x ($x = 0$ represents 1990).
   b. Find the values of a, b, and c such that the equation $y = ax^2 + bx + c$ models this data.
   c. According to the model, what is the world production of red meat in 1999?

WORLD PRODUCTION OF RED MEAT	
Year	Millions of metric tons
1993	119.3
1996	135.5
1998	140.1

(*Source:* Economic Research Service, U.S. Department of Agriculture)

3. **a.** Make up an equation of the form $y = ax^2 + bx + c$.
   b. Find three ordered pair solutions of the equation.
   c. Without revealing your equation from part (a), exchange lists of ordered pair solutions with another group.
   d. Use the method described above to find the values of a, b, and c such that the equation $y = ax^2 + bx + c$ has the ordered pair solutions you received from the other group.
   e. Check with the other group to see if your equation from part (d) is the correct one.

Focus On the Real World

LINEAR MODELING

In Chapter 3, we learned several ways to find a linear model when given either two ordered pairs or an ordered pair and slope. Another way to find a linear model of the form $y = mx + b$ for two ordered pairs (x_1, y_1) and (x_2, y_2) is to solve the following system of linear equations for m and b:

$$\begin{cases} y_1 = mx_1 + b \\ y_2 = mx_2 + b \end{cases}$$

For example, suppose a researcher wishes to find a linear model for the number of traffic accidents involving teenagers. The researcher locates statistics stating that there were 5215 teenage deaths from motor vehicle accidents in the United States in 1992. By 1996, this number had increased to 5805 teenage deaths in motor vehicle accidents. (*Source:* Insurance Institute for Highway Safety)

This data gives two ordered pairs: (1992, 5215) and (1996, 5805). Alternatively, the ordered pairs could be written as (2, 5215) and (6, 5805), where the *x*-coordinate represents the number of years after 1990. (Adjusting data given as years in this way often simplifies calculations.) By substituting the coordinates of the second set of ordered pairs into the general linear system, we obtain the system

$$\begin{cases} 5215 = 2m + b \\ 5805 = 6m + b \end{cases}$$

The solution of this system is $m = 147.5$ and $b = 4920$. We can use these values to write the model the researcher wished to find: $y = 147.5x + 4920$, where y is the number of teenage deaths in motor vehicle accidents x years after 1990.

CHAPTER 8 HIGHLIGHTS

DEFINITIONS AND CONCEPTS	EXAMPLES

SECTION 8.1   SOLVING SYSTEMS OF LINEAR EQUATIONS BY GRAPHING

A **system of linear equations** consists of two or more linear equations.

A **solution** of a system of two equations in two variables is an ordered pair of numbers that is a solution of both equations in the system.

$$\begin{cases} 2x + y = 6 \\ x = -3y \end{cases} \quad \begin{cases} -3x + 5y = 10 \\ x - 4y = -2 \end{cases}$$

Determine whether $(-1, 3)$ is a solution of the system.

$$\begin{cases} 2x - y = -5 \\ x = 3y - 10 \end{cases}$$

Replace x with -1 and y with 3 in both equations.

$$2x - y = -5$$
$$2(-1) - 3 \stackrel{?}{=} -5$$
$$-5 = -5 \quad \text{True.}$$

$$x = 3y - 10$$
$$-1 \stackrel{?}{=} 3(3) - 10$$
$$-1 = -1 \quad \text{True.}$$

$(-1, 3)$ is a solution of the system.

Graphically, a solution of a system is a point common to the graphs of both equations.

Solve by graphing: $\begin{cases} 3x - 2y = -3 \\ x + y = 4 \end{cases}$

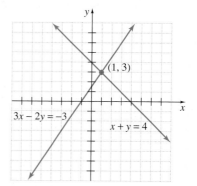

Three different situations can occur when graphing the two lines associated with the equations in a linear system.

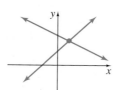

One point of intersection; one solution

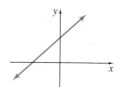

Same line; infinite number of solutions

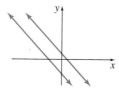

Parallel lines; no solution

SECTION 8.2   SOLVING SYSTEMS OF LINEAR EQUATIONS BY SUBSTITUTION

To SOLVE A SYSTEM OF LINEAR EQUATIONS BY THE SUBSTITUTION METHOD

Step 1. Solve one equation for a variable.
Step 2. Substitute the expression for the variable into the other equation.
Step 3. Solve the equation from Step 2 to find the value of one variable.
Step 4. Substitute the value from Step 3 in either original equation to find the value of the other variable.
Step 5. Check the solution in both equations.

Solve by substitution.

$$\begin{cases} 3x + 2y = 1 \\ x = y - 3 \end{cases}$$

Substitute $y - 3$ for x in the first equation

$$3x + 2y = 1$$
$$3(y - 3) + 2y = 1$$
$$3y - 9 + 2y = 1$$
$$5y = 10$$
$$y = 2 \qquad \text{Divide by 5.}$$

To find x, substitute 2 for y in $x = y - 3$ so that $x = 2 - 3$ or -1. The solution $(-1, 2)$ checks.

SECTION 8.3   SOLVING SYSTEMS OF LINEAR EQUATIONS BY ADDITION

TO SOLVE A SYSTEM OF LINEAR EQUATIONS BY THE ADDITION METHOD

Step 1. Rewrite each equation in standard form $Ax + By = C$.
Step 2. Multiply one or both equations by a nonzero number so that the coefficients of a variable are opposites.
Step 3. Add the equations.
Step 4. Find the value of one variable by solving the resulting equation.
Step 5. Substitute the value from Step 4 into either original equation to find the value of the other variable.
Step 6. Check the solution in both equations.

If solving a system of linear equations by substitution or addition yields a true statement such as $-2 = -2$, then the graphs of the equations in the system are identical and there is an infinite number of solutions of the system.

If solving a system of linear equations yields a false statement such as $0 = 3$, the graphs of the equations in the system are parallel lines and the system has no solution.

Solve by addition

$$\begin{cases} x - 2y = 8 \\ 3x + y = -4 \end{cases}$$

Multiply both sides of the first equation by -3.

$$\begin{cases} -3x + 6y = -24 \\ \underline{3x + y = -4} \end{cases}$$
$$7y = -28 \qquad \text{Add.}$$
$$y = -4 \qquad \text{Divide by 7.}$$

To find x, let $y = -4$ in an original equation.

$$x - 2(-4) = 8 \qquad \text{First equation}$$
$$x + 8 = 8$$
$$x = 0$$

The solution $(0, -4)$ checks.

Solve: $\begin{cases} 2x - 6y = -2 \\ x = 3y - 1 \end{cases}$

Substitute $3y - 1$ for x in the first equation.

$$2(3y - 1) - 6y = -2$$
$$6y - 2 - 6y = -2$$
$$-2 = -2 \qquad \text{True.}$$

The system has an infinite number of solutions.

Solve: $\begin{cases} 5x - 2y = 6 \\ \underline{-5x + 2y = -3} \\ 0 = 3 \end{cases} \qquad \text{False.}$

The system has no solution.

SECTION 8.4   SYSTEMS OF LINEAR EQUATIONS AND PROBLEM SOLVING

PROBLEM-SOLVING STEPS

1. UNDERSTAND. Read and reread the problem.

Two angles are supplementary if their sum is 180°. The larger of two supplementary angles is three times the smaller, decreased by twelve. Find the measure of each angle. Let

$$x = \text{measure of smaller angle}$$
$$y = \text{measure of larger angle}$$

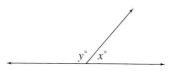

2. TRANSLATE.

In words:

the sum of supplementary angles	is	180°
↓	↓	↓

Translate: $x + y \quad = \quad 180$

In words:

larger angle	is	3 times smaller	decreased by	12
↓	↓	↓	↓	↓

Translate: $y \quad = \quad 3x \quad - \quad 12$

3. SOLVE.

Solve the system

$$\begin{cases} x + y = 180 \\ y = 3x - 12 \end{cases}$$

Use the substitution method and replace y with $3x - 12$ in the first equation.

$$x + y = 180$$
$$x + (3x - 12) = 180$$
$$4x = 192$$
$$x = 48$$

4. INTERPRET.

Since $y = 3x - 12$, then $y = 3 \cdot 48 - 12$ or 132.

The solution checks. The smaller angle measures 48° and the larger angle measures 132°.

SECTION 8.5   SOLVING SYSTEMS OF LINEAR EQUATIONS IN THREE VARIABLES

A **solution** of an equation in three variables x, y and z is an **ordered triple** (x, y, z) that makes the equation a true statement.

Verify that $(-2, 1, 3)$ is a solution of $2x + 3y - 2z = -7$.
Replace x with -2, y with 1, and z with 3.

$$2(-2) + 3(1) - 2(3) = -7$$
$$-4 + 3 - 6 = -7$$
$$-7 = -7 \quad \text{True.}$$

$(-2, 1, 3)$ is a solution.

SECTION 8.5 (CONTINUED)

SOLVING A SYSTEM OF THREE LINEAR EQUATIONS BY THE ELIMINATION METHOD

Step 1. Write each equation in standard form, $Ax + By + Cz = D$.

Step 2. Choose a pair of equations and use the equations to eliminate a variable.

Step 3. Choose any other pair of equations and eliminate the same variable.

Step 4. Solve the system of two equations in two variables from Steps 2 and 3.

Step 5. Solve for the third variable by substituting the values of the variables from Step 4 into any of the original equations.

Step 6. Check the solution in all three original equations.

Solve:

$$\begin{cases} 2x + y - z = 0 & (1) \\ x - y - 2z = -6 & (2) \\ -3x - 2y + 3z = -22 & (3) \end{cases}$$

1. Each equation is written in standard form.

2.
$$\begin{aligned} 2x + y - z &= 0 \quad (1) \\ x - y - 2z &= -6 \quad (2) \\ \hline 3x \qquad - 3z &= -6 \quad (4) \quad \text{Add.} \end{aligned}$$

3. Eliminate y from equations (1) and (3) also.

$$\begin{aligned} 4x + 2y - 2z &= 0 \qquad \qquad \text{Multiply equation} \\ -3x - 2y + 3z &= -22 \quad (3) \quad \text{(1) by 2.} \\ \hline x \qquad + z &= -22 \quad (5) \quad \text{Add.} \end{aligned}$$

4. Solve.

$$\begin{cases} 3x - 3z = -6 & (4) \\ x + z = -22 & (5) \end{cases}$$

$$\begin{aligned} x - z &= -2 \qquad \text{Divide equation (4) by 3.} \\ x + z &= -22 \quad (5) \\ \hline 2x \quad &= -24 \\ x \quad &= -12 \end{aligned}$$

To find z, use equation (5).

$$\begin{aligned} x + z &= -22 \\ -12 + z &= -22 \\ z &= -10 \end{aligned}$$

5. To find y, use equation (1).

$$\begin{aligned} 2x + y - z &= 0 \\ 2(-12) + y - (-10) &= 0 \\ -24 + y + 10 &= 0 \\ y &= 14 \end{aligned}$$

The solution $(-12, 14, -10)$ checks.

SECTION 8.6 SOLVING SYSTEMS OF EQUATIONS USING MATRICES

A **matrix** is a rectangular array of numbers.

$$\begin{bmatrix} -7 & 0 & 3 \\ 1 & 2 & 4 \end{bmatrix} \qquad \begin{bmatrix} a & b & c \\ d & e & f \\ g & h & i \end{bmatrix}$$

The **corresponding matrix of the system** is obtained by writing a matrix composed of the coefficients of the variables and the constants of the system.

The corresponding matrix of the system

$$\begin{cases} x - y = 1 \\ 2x + y = 11 \end{cases} \text{ is } \begin{bmatrix} 1 & -1 & \vdots & 1 \\ 2 & 1 & \vdots & 11 \end{bmatrix}$$

SECTION 8.6 (CONTINUED)	

The following **row operations** can be performed on matrices, and the result is an equivalent matrix.

Elementary row operations:

1. Interchange any two rows.
2. Multiply (or divide) the elements of one row by the same nonzero number.
3. Multiply (or divide) the elements of one row by the same nonzero number and add to its corresponding elements in any other row.

Use matrices to solve: $\begin{cases} x - y = 1 \\ 2x + y = 11 \end{cases}$

The corresponding matrix is

$$\left[\begin{array}{cc|c} 1 & -1 & 1 \\ 2 & 1 & 11 \end{array}\right]$$

Use row operations to write an equivalent matrix with 1s along the diagonal and 0s below each 1 in the diagonal. Multiply row 1 by -2 and add to row 2. Change row 2 only.

$$\left[\begin{array}{cc|c} 1 & -1 & 1 \\ -2(1) + 2 & -2(-1) + 1 & -2(1) + 11 \end{array}\right]$$

simplifies to $\left[\begin{array}{cc|c} 1 & -1 & 1 \\ 0 & 3 & 9 \end{array}\right]$

Divide row 2 by 3.

$$\left[\begin{array}{cc|c} 1 & -1 & 1 \\ \dfrac{0}{3} & \dfrac{3}{3} & \dfrac{9}{3} \end{array}\right] \quad \text{simplifies to} \quad \left[\begin{array}{cc|c} 1 & -1 & 1 \\ 0 & 1 & 3 \end{array}\right]$$

This matrix corresponds to the system

$$\begin{cases} x - y = 1 \\ y = 3 \end{cases}$$

Let $y = 3$ in the first equation.

$$x - 3 = 1$$
$$x = 4$$

The ordered pair solution is $(4, 3)$.

SECTION 8.7 SOLVING SYSTEMS OF EQUATIONS USING DETERMINANTS	

A **square matrix** is a matrix with the same number of rows and columns.

$$\left[\begin{array}{cc} -2 & 1 \\ 6 & 8 \end{array}\right] \qquad \left[\begin{array}{ccc} 4 & -1 & 6 \\ 0 & 2 & 5 \\ 1 & 1 & 2 \end{array}\right]$$

A **determinant** is a real number associated with a square matrix. To denote the determinant, place vertical bars about the array of numbers.

The determinant of a 2×2 matrix is

$$\begin{vmatrix} a & b \\ c & d \end{vmatrix} = ad - bc$$

The determinant of $\left[\begin{array}{cc} -2 & 1 \\ 6 & 8 \end{array}\right]$ is $\begin{vmatrix} -2 & 1 \\ 6 & 8 \end{vmatrix}$.

$$\begin{vmatrix} -2 & 1 \\ 6 & 8 \end{vmatrix} = -2 \cdot 8 - 1 \cdot 6 = -22$$

SECTION 8.7 (CONTINUED)

CRAMER'S RULE FOR TWO LINEAR EQUATIONS IN TWO VARIABLES

The solution of the system $\begin{cases} ax + by = h \\ cx + dy = k \end{cases}$ is given by

$$x = \frac{\begin{vmatrix} h & b \\ k & d \end{vmatrix}}{\begin{vmatrix} a & b \\ c & d \end{vmatrix}} = \frac{D_x}{D} \qquad y = \frac{\begin{vmatrix} a & h \\ c & k \end{vmatrix}}{\begin{vmatrix} a & b \\ c & d \end{vmatrix}} = \frac{D_y}{D}$$

as long as $D = ad - bc$ is not 0.

DETERMINANT OF A 3 × 3 MATRIX

$$\begin{vmatrix} a_1 & b_1 & c_1 \\ a_2 & b_2 & c_2 \\ a_3 & b_3 & c_3 \end{vmatrix} = a_1 \cdot \begin{vmatrix} b_2 & c_2 \\ b_3 & c_3 \end{vmatrix} - a_2 \cdot$$

$$\begin{vmatrix} b_1 & c_1 \\ b_3 & c_3 \end{vmatrix} + a_3 \cdot \begin{vmatrix} b_1 & c_1 \\ b_2 & c_2 \end{vmatrix}$$

Each 2 × 2 matrix above is called a **minor**.

CRAMER'S RULE FOR THREE EQUATIONS IN THREE VARIABLES

The solution of the system $\begin{cases} a_1x + b_1y + c_1z = k_1 \\ a_2x + b_2y + c_2z = k_2 \\ a_3x + b_3y + c_3z = k_3 \end{cases}$ is given by

$$x = \frac{D_x}{D} \quad y = \frac{D_y}{D} \quad \text{and} \quad z = \frac{D_z}{D}$$

where

$$D = \begin{vmatrix} a_1 & b_1 & c_1 \\ a_2 & b_2 & c_2 \\ a_3 & b_3 & c_3 \end{vmatrix} \qquad D_x = \begin{vmatrix} k_1 & b_1 & c_1 \\ k_2 & b_2 & c_2 \\ k_3 & b_3 & c_3 \end{vmatrix}$$

$$D_y = \begin{vmatrix} a_1 & k_1 & c_1 \\ a_2 & k_2 & c_2 \\ a_3 & k_3 & c_3 \end{vmatrix} \qquad D_z = \begin{vmatrix} a_1 & b_1 & k_1 \\ a_2 & b_2 & k_2 \\ a_3 & b_3 & k_3 \end{vmatrix}$$

as long as D is not 0.

Use Cramer's rule to solve

$$\begin{cases} 3x + 2y = 8 \\ 2x - y = -11 \end{cases}$$

$$D = \begin{vmatrix} 3 & 2 \\ 2 & -1 \end{vmatrix} = 3(-1) - 2(2) = -7$$

$$D_x = \begin{vmatrix} 8 & 2 \\ -11 & -1 \end{vmatrix} = 8(-1) - 2(-11) = 14$$

$$D_y = \begin{vmatrix} 3 & 8 \\ 2 & -11 \end{vmatrix} = 3(-11) - 8(2) = -49$$

$$x = \frac{D_x}{D} = \frac{14}{-7} = -2 \qquad y = \frac{D_y}{D} = \frac{-49}{-7} = 7$$

The ordered pair solution is $(-2, 7)$.

$$\begin{vmatrix} 0 & 2 & -1 \\ 5 & 3 & 0 \\ 2 & -2 & 4 \end{vmatrix} = 0 \begin{vmatrix} 3 & 0 \\ -2 & 4 \end{vmatrix} - 2 \begin{vmatrix} 5 & 0 \\ 2 & 4 \end{vmatrix}$$

$$+ (-1) \begin{vmatrix} 5 & 3 \\ 2 & -2 \end{vmatrix}$$

$$= 0(12 - 0) - 2(20 - 0)$$
$$- 1(-10 - 6)$$
$$= 0 - 40 + 16 = -24$$

Use Cramer's rule to solve

$$\begin{cases} 3y + 2z = 8 \\ x + y + z = 3 \\ 2x - y + z = 2 \end{cases}$$

$$D = \begin{vmatrix} 0 & 3 & 2 \\ 1 & 1 & 1 \\ 2 & -1 & 1 \end{vmatrix} = -3$$

$$D_x = \begin{vmatrix} 8 & 3 & 2 \\ 3 & 1 & 1 \\ 2 & -1 & 1 \end{vmatrix} = 3$$

$$D_y = \begin{vmatrix} 0 & 8 & 2 \\ 1 & 3 & 1 \\ 2 & 2 & 1 \end{vmatrix} = 0$$

$$D_z = \begin{vmatrix} 0 & 3 & 8 \\ 1 & 1 & 3 \\ 2 & -1 & 2 \end{vmatrix} = -12$$

$$x = \frac{D_x}{D} = \frac{3}{-3} = -1 \qquad y = \frac{D_y}{D} = \frac{0}{-3} = 0$$

$$z = \frac{D_z}{D} = \frac{-12}{-3} = 4$$

The ordered triple solution is $(-1, 0, 4)$.

SECTION 8.8   SYSTEMS OF LINEAR INEQUALITIES

A system of linear inequalities consists of two or more linear inequalities.

To graph a system of inequalities, graph each inequality in the system. The overlapping region is the solution of the system.

$$\begin{cases} x - y \geq 3 \\ \ y \leq -2x \end{cases}$$

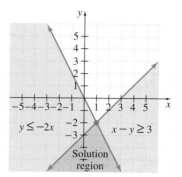

Rational Exponents, Radicals, and Complex Numbers

In this chapter, we define radical notation, and then introduce rational exponents. As the name implies, rational exponents are exponents that are rational numbers. We present an interpretation of rational exponents that is consistent with the meaning and rules already established for integer exponents, and we present two forms of notation for roots: radical and exponent. We conclude this chapter with complex numbers, a natural extension of the real number system.

Mount Vesuvius is the only active volcano on the European continent. Although the Romans thought the volcano was extinct, Vesuvius erupted violently in 79 A.D. The eruption buried the cities of Pompeii, Herculaneum, and Stabiae under up to 60 feet of ash and mud, killing approximately 16,000 people. Although Pompeii was completely engulfed, the city was far from destroyed. The blanket of mud and ash perfectly preserved much of the city in a snapshot of daily life of the ancient Romans. Pompeii lay undisturbed for over 1500 years until the first excavations were made and its archaeological significance was proven. The 79 A.D. eruption of Vesuvius seemed to be the volcano's renewal. It has erupted with varying degrees of violence more than 30 times since Pompeii's burial, most recently in 1944.

Problem Solving Notes

9.1   RADICAL EXPRESSIONS

A FINDING SQUARE ROOTS

The opposite of squaring a number is taking the **square root** of a number. For example, since the square of 4, or 4^2, is 16, we say that a square root of 16 is 4.

SQUARE ROOT

The number b is a **square root** of a if $b^2 = a$.

Examples   Find the square roots of each number.

1. 25   Since $5^2 = 25$ and $(-5)^2 = 25$, the square roots of 25 are 5 and -5.

2. 49   Since $7^2 = 49$ and $(-7)^2 = 49$, the square roots of 49 are 7 and -7.

3. -4 There is no real number whose square is -4. The number -4 has no real number square root.

The notation $\sqrt{a}$ is used to denote the **positive**, or **principal**, **square root** of a nonnegative number a. We then have in symbols that

$$\sqrt{16} = 4$$

We denote the **negative square root** with the **negative radical sign**:

$$-\sqrt{16} = -4$$

An expression containing a radical sign is called a **radical expression**. An expression within, or "under," a radical sign is called a **radicand**.

radical expression: $\overset{\text{radical sign}}{\underset{\text{radicand}}{\sqrt{a}}}$

PRINCIPAL AND NEGATIVE SQUARE ROOTS

The **principal square root** of a nonnegative number a is its nonnegative square root. The principal square root is written as $\sqrt{a}$. The **negative square root** of a is written as $-\sqrt{a}$.

Examples   Find each square root. Assume that all variables represent nonnegative real numbers.

4. $\sqrt{36} = 6$ because $6^2 = 36$.

5. $\sqrt{0} = 0$ because $0^2 = 0$.

6. $\sqrt{\dfrac{4}{49}} = \dfrac{2}{7}$ because $\left(\dfrac{2}{7}\right)^2 = \dfrac{4}{49}$.

7. $\sqrt{0.25} = 0.5$ because $(0.5)^2 = 0.25$.

8. $\sqrt{x^6} = x^3$ because $(x^3)^2 = x^6$.

9. $\sqrt{9x^{10}} = 3x^5$ because $(3x^5)^2 = 9x^{10}$.

10. $-\sqrt{81} = -9$. The negative in front of the radical indicates the negative square root of 81.

11. $\sqrt{-81}$ is not a real number.

Objectives

A Find square roots.

B Approximate roots using a calculator.

C Find cube roots.

D Find nth roots.

E Find $\sqrt[n]{a^n}$ when a is any real number.

F Find function values of radical functions.

SSM   CD-ROM   Video 9.1

Practice Problems 1–3

Find the square roots of each number.

1. 36

2. 81

3. -16

Practice Problems 4–11

Find each square root. Assume that all variables represent nonnegative real numbers.

4. $\sqrt{25}$ 5. $\sqrt{0}$

6. $\sqrt{\dfrac{9}{25}}$ 7. $\sqrt{0.36}$

8. $\sqrt{x^{10}}$ 9. $\sqrt{36x^6}$

10. $-\sqrt{25}$ 11. $\sqrt{-25}$

Answers

1. $6, -6$, **2.** $9, -9$, **3.** no real number square root, **4.** 5, **5.** 0, **6.** $\dfrac{3}{5}$, **7.** 0.6, **8.** x^5, **9.** $6x^3$, **10.** -5, **11.** not a real number

HELPFUL HINT

Don't forget that the square root of a negative number is not a real number. For example,

$$\sqrt{-9} \quad \text{it not a real number}$$

because there is no real number that when multiplied by itself would give a product of -9. In Section 9.7, we will see what kind of a number $\sqrt{-9}$ is.

Practice Problem 12

Use a calculator or the appendix to approximate $\sqrt{30}$. Round the approximation to 3 decimal places and check to see that your approximation is reasonable.

Practice Problems 13–17

Find each cubic root.

13. $\sqrt[3]{0}$

14. $\sqrt[3]{-8}$

15. $\sqrt[3]{\dfrac{1}{64}}$

16. $\sqrt[3]{x^9}$

17. $\sqrt[3]{-64x^6}$

Answers

12. 5.477, **13.** 0, **14.** -2, **15.** $\dfrac{1}{4}$, **16.** x^3,
17. $-4x^2$

B APPROXIMATING ROOTS

Numbers such as 1, 4, 9, and 25 are called **perfect squares**, since $1 = 1^2$, $4 = 2^2$, $9 = 3^2$, and $25 = 5^2$. Square roots of perfect square radicands simplify to rational numbers. What happens when we try to simplify a root such as $\sqrt{3}$? Since 3 is not a perfect square, $\sqrt{3}$ is not a rational number. It is called an **irrational number**, and we can find a decimal **approximation** of it. To find decimal approximations, we can use the table in the appendix or a calculator. For example, an approximation for $\sqrt{3}$ is

$$\sqrt{3} \approx 1.732$$
$$\uparrow$$
approximation symbol

To see if the approximation is reasonable, notice that since

$$1 < 3 < 4, \text{ then}$$
$$\sqrt{1} < \sqrt{3} < \sqrt{4}, \text{ or}$$
$$1 < \sqrt{3} < 2.$$

We found $\sqrt{3} \approx 1.732$, a number between 1 and 2, so our result is reasonable.

Example 12 Use a calculator or the appendix to approximate $\sqrt{20}$. Round the approximation to 3 decimal places and check to see that your approximation is reasonable.

$$\sqrt{20} \approx 4.472$$

Is this reasonable? Since $16 < 20 < 25$, then $\sqrt{16} < \sqrt{20} < \sqrt{25}$, or $4 < \sqrt{20} < 5$. The approximation is between 4 and 5 and is thus reasonable. ▬▬▬

C FINDING CUBE ROOTS

Finding roots can be extended to other roots such as cube roots. For example, since $2^3 = 8$, we call 2 the **cube root** of 8. In symbols, we write

$$\sqrt[3]{8} = 2$$

CUBE ROOT

The **cube root** of a real number a is written as $\sqrt[3]{a}$, and

$$\sqrt[3]{a} = b \text{ only if } b^3 = a$$

From this definition, we have

$$\sqrt[3]{64} = 4 \text{ since } 4^3 = 64$$
$$\sqrt[3]{-27} = -3 \text{ since } (-3)^3 = -27$$
$$\sqrt[3]{x^3} = x \text{ since } x^3 = x^3$$

Notice that, unlike with square roots, *it is possible to have a negative radicand when finding a cube root.* This is so because the *cube* of a negative number is a negative number. Therefore, the *cube root* of a negative number is a negative number.

Examples Find each cube root.

13. $\sqrt[3]{1} = 1$ because $1^3 = 1$.

14. $\sqrt[3]{-64} = -4$ because $(-4)^3 = -64$.

15. $\sqrt[3]{\dfrac{8}{125}} = \dfrac{2}{5}$ because $\left(\dfrac{2}{5}\right)^3 = \dfrac{8}{125}$.

16. $\sqrt[3]{x^6} = x^2$ because $(x^2)^3 = x^6$.

17. $\sqrt[3]{-8x^9} = -2x^3$ because $(-2x^3)^3 = -8x^9$.   ▬▬▬

D FINDING nTH ROOTS

Just as we can raise a real number to powers other than 2 or 3, we can find roots other than square roots and cube roots. In fact, we can find the **nth root** of a number, where n is any natural number. In symbols, the nth root of a is written as $\sqrt[n]{a}$, where n is called the **index**. The index 2 is usually omitted for square roots.

HELPFUL HINT

If the index is even, such as $\sqrt{\ }$, $\sqrt[4]{\ }$, $\sqrt[6]{\ }$, and so on, the radicand must be nonnegative for the root to be a real number. For example,

$\sqrt[4]{16} = 2$, but $\sqrt[4]{-16}$ is not a real number,

$\sqrt[6]{64} = 2$, but $\sqrt[6]{-64}$ is not a real number.

If the index is odd, such as $\sqrt[3]{\ }$, $\sqrt[5]{\ }$, and so on, the radicand may be any real number. For example,

$\sqrt[3]{64} = 4$ and $\sqrt[3]{-64} = -4$,

$\sqrt[5]{32} = 2$ and $\sqrt[5]{-32} = -2$.

TRY THE CONCEPT CHECK IN THE MARGIN.

Examples   Find each root.

18. $\sqrt[4]{81} = 3$ because $3^4 = 81$ and 3 is positive.

19. $\sqrt[5]{-243} = -3$ because $(-3)^5 = -243$.

20. $-\sqrt{25} = -5$ because -5 is the opposite of $\sqrt{25}$.

21. $\sqrt[4]{-81}$ is not a real number. There is no real number that, when raised to the fourth power, is -81.

22. $\sqrt[3]{64x^3} = 4x$ because $(4x)^3 = 64x^3$.   ▬▬▬

E FINDING $\sqrt[n]{a^n}$ WHEN a IS ANY REAL NUMBER

Recall that the notation $\sqrt{a^2}$ indicates the positive square root of a^2 only. For example,

$\sqrt{(-5)^2} = \sqrt{25} = 5$

When variables are present in the radicand and it is *unclear whether the variable represents a positive number or a negative number*, absolute value bars are sometimes needed to ensure that the result is a positive number. For example,

$\sqrt{x^2} = |x|$

This ensures that the result is positive. This same situation may occur when the index is any *even* positive integer. When the index is any *odd* positive integer, absolute value bars are not necessary.

✓ **CONCEPT CHECK**

Which one is not a real number?

a. $\sqrt[3]{-15}$

b. $\sqrt[4]{-15}$

c. $\sqrt[5]{-15}$

d. $\sqrt{(-15)^2}$

Practice Problems 18–22

Find each root.

18. $\sqrt[4]{16}$ 19. $\sqrt[5]{-32}$

20. $-\sqrt{36}$ 21. $\sqrt[4]{-16}$

22. $\sqrt[3]{8x^6}$

Answers

18. 2,   **19.** -2,   **20.** -6,   **21.** not a real number,   **22.** $2x^2$

✓ **Concept Check:**   b

FINDING $\sqrt[n]{a^n}$

If n is an *even* positive integer, then $\sqrt[n]{a^n} = |a|$.
If n is an *odd* positive integer, then $\sqrt[n]{a^n} = a$.

Practice Problems 23–29

Simplify. Assume that the variables represent any real number.

23. $\sqrt{(-5)^2}$

24. $\sqrt{x^6}$

25. $\sqrt[4]{(x + 6)^4}$

26. $\sqrt[3]{(-3)^3}$

27. $\sqrt[5]{(7x - 1)^5}$

28. $\sqrt{36x^2}$

29. $\sqrt{x^2 + 6x + 9}$

Examples   Simplify. Assume that the variables represent any real number.

23. $\sqrt{(-3)^2} = |-3| = 3$ When the index is even, the absolute value bars ensure that the result is not negative.

24. $\sqrt{x^2} = |x|$

25. $\sqrt[4]{(x - 2)^4} = |x - 2|$

26. $\sqrt[3]{(-5)^3} = -5$ Absolute value bars are not needed when the index is odd.

27. $\sqrt[5]{(2x - 7)^5} = 2x - 7$

28. $\sqrt{25x^2} = 5|x|$

29. $\sqrt{x^2 + 2x + 1} = \sqrt{(x + 1)^2} = |x + 1|$ ▬▬▬

F FINDING FUNCTION VALUES

Functions of the form

$$f(x) = \sqrt[n]{x}$$

are called **radical functions**. Recall that the domain of a function in x is the set of all possible replacement values of x. This means that if n is even, the domain is the set of all nonnegative numbers, or $\{x | x \geq 0\}$. If n is odd, the domain is the set of all real numbers. Keep this in mind as we find function values. In Chapter 9, we will graph these functions.

Practice Problems 30–33

If $f(x) = \sqrt{x + 2}$ and $g(x) = \sqrt[3]{x - 1}$, find each function value.

30. $f(7)$

31. $g(9)$

32. $f(0)$

33. $g(10)$

Examples   If $f(x) = \sqrt{x - 4}$ and $g(x) = \sqrt[3]{x + 2}$, find each function value.

30. $f(8) = \sqrt{8 - 4} = \sqrt{4} = 2$

31. $f(6) = \sqrt{6 - 4} = \sqrt{2}$

32. $g(-1) = \sqrt[3]{-1 + 2} = \sqrt[3]{1} = 1$

33. $g(1) = \sqrt[3]{1 + 2} = \sqrt[3]{3}$ ▬▬▬

> **HELPFUL HINT**
>
> Notice that for the function $f(x) = \sqrt{x - 4}$, the domain includes all real numbers that make the radicand ≥ 0. To see what numbers these are, solve $x - 4 \geq 0$ and find that $x \geq 4$. The domain is $\{x | x \geq 4\}$.
> The domain of the cube root function $g(x) = \sqrt[3]{x + 2}$ is the set of real numbers.

Answers

23. 5, **24.** $|x^3|$, **25.** $|x + 6|$, **26.** -3,
27. $7x - 1$, **28.** $6|x|$, **29.** $|x + 3|$,
30. 3, **31.** 2, **32.** $\sqrt{2}$, **33.** $\sqrt[3]{9}$

EXERCISE SET 9.1

A *Find the square roots of each number. See Examples 1 through 3.*

1. 4

2. 9

Find each square root. Assume that all variables represent nonnegative real numbers. See Examples 4 through 11.

3. $\sqrt{100}$

4. $\sqrt{400}$

5. $\sqrt{\dfrac{1}{4}}$

6. $\sqrt{\dfrac{9}{25}}$

7. $\sqrt{0.0001}$

8. $\sqrt{0.04}$

B *Use a calculator or the appendix to approximate each square root to 3 decimal places. Check to see that each approximation is reasonable. See Example 12.*

9. $\sqrt{7}$

10. $\sqrt{11}$

C *Find each cube root. See Examples 13 through 17.*

11. $\sqrt[3]{64}$

12. $\sqrt[3]{27}$

13. $\sqrt[3]{\dfrac{1}{8}}$

14. $\sqrt[3]{\dfrac{27}{64}}$

15. $\sqrt[3]{-1}$

16. $\sqrt[3]{-125}$

D *Find each root. Assume that all variables represent nonnegative real numbers. See Examples 18 through 22.*

17. $-\sqrt[4]{16}$

18. $\sqrt[5]{-243}$

19. $\sqrt[4]{-16}$

20. $\sqrt{-16}$

21. $\sqrt[5]{-32}$

F *If $f(x) = \sqrt{2x + 3}$ and $g(x) = \sqrt[3]{x - 8}$, find each function value. See Examples 30 through 33.*

22. $f(0)$

23. $g(0)$

24. $g(7)$

25. $f(-1)$

26. $g(-19)$

1. _____

2. _____

3. _____

4. _____

5. _____

6. _____

7. _____

8. _____

9. _____

10. _____

11. _____

12. _____

13. _____

14. _____

15. _____

16. _____

17. _____

18. _____

19. _____

20. _____

21. _____

22. _____

23. _____

24. _____

25. _____

26. _____

Problem Solving Notes

9.2 RATIONAL EXPONENTS

A UNDERSTANDING $a^{1/n}$

So far in this text, we have not defined expressions with rational exponents such as $3^{1/2}$, $x^{2/3}$, and $-9^{-1/4}$. We will define these expressions so that the rules for exponents shall apply to these rational exponents as well.

Suppose that $x = 5^{1/3}$. Then

$$x^3 = \left(5^{1/3}\right)^3 = 5^{1/3 \cdot 3} = 5^1 \text{ or } 5$$
$$\underset{\substack{\text{using rules} \uparrow \\ \text{for exponents}}}{}$$

Since $x^3 = 5$, then x is the number whose cube is 5, or $x = \sqrt[3]{5}$. Notice that we also know that $x = 5^{1/3}$. This means that

$$5^{1/3} = \sqrt[3]{5}$$

DEFINITION OF $a^{1/n}$

If n is a positive integer greater than 1 and $\sqrt[n]{a}$ is a real number, then

$$a^{1/n} = \sqrt[n]{a}$$

Notice that the denominator of the rational exponent corresponds to the index of the radical.

Examples Use radical notation to write each expression. Simplify if possible.

1. $4^{1/2} = \sqrt{4} = 2$
2. $64^{1/3} = \sqrt[3]{64} = 4$
3. $x^{1/4} = \sqrt[4]{x}$
4. $-9^{1/2} = -\sqrt{9} = -3$
5. $\left(81x^8\right)^{1/4} = \sqrt[4]{81x^8} = 3x^2$
6. $5y^{1/3} = 5\sqrt[3]{y}$

━━━━━━

B UNDERSTANDING $a^{m/n}$

As we expand our use of exponents to include $\dfrac{m}{n}$, we define their meaning so that rules for exponents still hold true. For example, by properties of exponents,

$$8^{2/3} = \left(8^{1/3}\right)^2 = \left(\sqrt[3]{8}\right)^2 \quad \text{or}$$
$$8^{2/3} = \left(8^2\right)^{1/3} = \sqrt[3]{8^2}$$

DEFINITION OF $a^{m/n}$

If m and n are positive integers greater than 1 with $\dfrac{m}{n}$ in lowest terms, then

$$a^{m/n} = \sqrt[n]{a^m} = \left(\sqrt[n]{a}\right)^m$$

as long as $\sqrt[n]{a}$ is a real number.

Objectives

A Understand the meaning of $a^{1/n}$.
B Understand the meaning of $a^{m/n}$.
C Understand the meaning of $a^{-m/n}$.
D Use rules for exponents to simplify expressions that contain rational exponents.
E Use rational exponents to simplify radical expressions.

SSM CD-ROM Video 9.2

Practice Problems 1–6

Use radical notation to write each expression. Simplify if possible.

1. $25^{1/2}$

2. $27^{1/3}$

3. $x^{1/5}$

4. $-25^{1/2}$

5. $\left(-27y^6\right)^{1/3}$

6. $7x^{1/5}$

Answers

1. 5, **2.** 3, **3.** $\sqrt[5]{x}$, **4.** -5, **5.** $-3y^2$,
6. $7\sqrt[5]{x}$

Notice that the denominator n of the rational exponent corresponds to the index of the radical. The numerator m of the rational exponent indicates that the base is to be raised to the mth power. This means that

$$8^{2/3} = \sqrt[3]{8^2} = \sqrt[3]{64} = 4 \qquad \text{or}$$
$$8^{2/3} = \left(\sqrt[3]{8}\right)^2 = 2^2 = 4$$

> **HELPFUL HINT**
>
> Most of the time, $\left(\sqrt[n]{a}\right)^m$ will be easier to calculate than $\sqrt[n]{a^m}$.

Practice Problems 7–11

Use radical notation to write each expression. Simplify if possible.

7. $9^{3/2}$

8. $-256^{3/4}$

9. $(-32)^{2/5}$

10. $\left(\dfrac{1}{4}\right)^{3/2}$

11. $(2x + 1)^{2/7}$

Examples Use radical notation to write each expression. Simplify if possible.

7. $4^{3/2} = \left(\sqrt{4}\right)^3 = 2^3 = 8$

8. $-16^{3/4} = -\left(\sqrt[4]{16}\right)^3 = -(2)^3 = -8$

9. $(-27)^{2/3} = \left(\sqrt[3]{-27}\right)^2 = (-3)^2 = 9$

10. $\left(\dfrac{1}{9}\right)^{3/2} = \left(\sqrt{\dfrac{1}{9}}\right)^3 = \left(\dfrac{1}{3}\right)^3 = \dfrac{1}{27}$

11. $(4x - 1)^{3/5} = \sqrt[5]{(4x - 1)^3}$

> **HELPFUL HINT**
>
> The *denominator* of a rational exponent is the index of the corresponding radical. For example, $x^{1/5} = \sqrt[5]{x}$ and $z^{2/3} = \sqrt[3]{z^2}$, or $z^{2/3} = \left(\sqrt[3]{z}\right)^2$.

C UNDERSTANDING $a^{-m/n}$

The rational exponents we have given meaning to exclude negative rational numbers. To complete the set of definitions, we define $a^{-m/n}$.

> **DEFINITION OF $a^{-m/n}$**
>
> $$a^{-m/n} = \dfrac{1}{a^{m/n}}$$
>
> as long as $a^{m/n}$ is a nonzero real number.

Practice Problems 12–13

Write each expression with a positive exponent. Then simplify.

12. $27^{-2/3}$

13. $-256^{-3/4}$

Examples Write each expression with a positive exponent. Then simplify.

12. $16^{-3/4} = \dfrac{1}{16^{3/4}} = \dfrac{1}{\left(\sqrt[4]{16}\right)^3} = \dfrac{1}{2^3} = \dfrac{1}{8}$

13. $(-27)^{-2/3} = \dfrac{1}{(-27)^{2/3}} = \dfrac{1}{\left(\sqrt[3]{-27}\right)^2} = \dfrac{1}{(-3)^2} = \dfrac{1}{9}$

Answers

7. 27, **8.** -64, **9.** 4, **10.** $\dfrac{1}{8}$,

11. $\sqrt[7]{(2x + 1)^2}$, **12.** $\dfrac{1}{9}$, **13.** $-\dfrac{1}{64}$

HELPFUL HINT

If an expression contains a negative rational exponent, you may want to first write the expression with a positive exponent, then interpret the rational exponent. Notice that the sign of the base is not affected by the sign of its exponent. For example,

$$9^{-3/2} = \frac{1}{9^{3/2}} = \frac{1}{(\sqrt{9})^3} = \frac{1}{27}$$

Also,

$$(-27)^{-1/3} = \frac{1}{(-27)^{1/3}} = -\frac{1}{3}$$

TRY THE CONCEPT CHECK IN THE MARGIN.

D USING RULES FOR EXPONENTS

It can be shown that the properties of integer exponents hold for rational exponents. By using these properties and definitions, we can now simplify expressions that contain rational exponents. These rules are repeated here for review.

SUMMARY OF EXPONENT RULES

If m and n are rational numbers, and a, b, and c are numbers for which the expressions below exist, then

Product rule for exponents:	$a^m \cdot a^n = a^{m+n}$
Power rule for exponents:	$(a^m)^n = a^{m \cdot n}$
Power rules for products and quotients:	$(ab)^n = a^n b^n$ and
	$\left(\dfrac{a}{c}\right)^n = \dfrac{a^n}{c^n}, c \neq 0$
Quotient rule for exponents:	$\dfrac{a^m}{a^n} = a^{m-n}, a \neq 0$
Zero exponent:	$a^0 = 1, a \neq 0$
Negative exponent:	$a^{-n} = \dfrac{1}{a^n}, a \neq 0$

Examples   Use the properties of exponents to simplify.

14. $x^{1/2}x^{1/3} = x^{1/2+1/3} = x^{3/6+2/6} = x^{5/6}$ Use the product rule.

15. $\dfrac{7^{1/3}}{7^{4/3}} = 7^{1/3-4/3} = 7^{-3/3} = 7^{-1} = \dfrac{1}{7}$ Use the quotient rule.

16. $\dfrac{(2x^{2/5})^5}{x^2} = \dfrac{2^5(x^{2/5})^5}{x^2}$ Use the power rule.

$\qquad = \dfrac{32x^2}{x^2}$ Simplify.

$\qquad = 32x^{2-2}$ Use the quotient rule.

$\qquad = 32x^0$ Simplify.

$\qquad = 32 \cdot 1 \quad$ or $\quad 32$ Substitute 1 for x^0.

✓ **CONCEPT CHECK**

Which one is correct?

a. $-8^{2/3} = \dfrac{1}{4}$

b. $8^{-2/3} = -\dfrac{1}{4}$

c. $8^{-2/3} = -4$

d. $-8^{-2/3} = -\dfrac{1}{4}$

Practice Problems 14–16

Use the properties of exponents to simplify.

14. $x^{1/3}x^{1/4}$

15. $\dfrac{9^{2/5}}{9^{12/5}}$

16. $\dfrac{(3x^{2/3})^3}{x^2}$

Answers

14. $x^{7/12}$,   **15.** $\dfrac{1}{81}$,   **16.** 27

✓ Concept Check: d

E USING RATIONAL EXPONENTS TO SIMPLIFY RADICAL EXPRESSIONS

Some radical expressions are easier to simplify when we first write them with rational exponents. We can simplify some radical expressions by first writing the expression with rational exponents. Use properties of exponents to simplify, and then convert back to radical notation.

Practice Problems 17–19

Use rational exponents to simplify. Assume that all variables represent positive numbers.

17. $\sqrt[10]{y^5}$

18. $\sqrt[4]{9}$

19. $\sqrt[9]{a^6 b^3}$

Examples Use rational exponents to simplify. Assume that all variables represent positive numbers.

17. $\sqrt[8]{x^4} = x^{4/8}$ Write with rational exponents.

$\phantom{\sqrt[8]{x^4}} = x^{1/2}$ Simplify the exponent.

$\phantom{\sqrt[8]{x^4}} = \sqrt{x}$ Write with radical notation.

18. $\sqrt[6]{25} = 25^{1/6}$ Write with rational exponents.

$\phantom{\sqrt[6]{25}} = \left(5^2\right)^{1/6}$ Write 25 as 5^2.

$\phantom{\sqrt[6]{25}} = 5^{2/6}$ Use the power rule.

$\phantom{\sqrt[6]{25}} = 5^{1/3}$ Simplify the exponent.

$\phantom{\sqrt[6]{25}} = \sqrt[3]{5}$ Write with radical notation.

19. $\sqrt[6]{r^2 s^4} = \left(r^2 s^4\right)^{1/6}$ Write with rational exponents.

$\phantom{\sqrt[6]{r^2 s^4}} = r^{2/6} s^{4/6}$ Use the power rule.

$\phantom{\sqrt[6]{r^2 s^4}} = r^{1/3} s^{2/3}$ Simplify the exponents.

$\phantom{\sqrt[6]{r^2 s^4}} = \left(r s^2\right)^{1/3}$ Use $a^n b^n = (ab)^n$.

$\phantom{\sqrt[6]{r^2 s^4}} = \sqrt[3]{r s^2}$ Write with radical notation.

Practice Problems 20–22

Use rational exponents to write as a single radical.

20. $\sqrt{y} \cdot \sqrt[3]{y}$

21. $\dfrac{\sqrt[3]{x}}{\sqrt[4]{x}}$

22. $\sqrt{5} \cdot \sqrt[3]{2}$

Examples Use rational exponents to write as a single radical.

20. $\sqrt{x} \cdot \sqrt[4]{x} = x^{1/2} \cdot x^{1/4} = x^{1/2 + 1/4}$

$\phantom{\sqrt{x} \cdot \sqrt[4]{x}} = x^{3/4} = \sqrt[4]{x^3}$

21. $\dfrac{\sqrt{x}}{\sqrt[3]{x}} = \dfrac{x^{1/2}}{x^{1/3}} = x^{1/2 - 1/3} = x^{3/6 - 2/6}$

$\phantom{\dfrac{\sqrt{x}}{\sqrt[3]{x}}} = x^{1/6} = \sqrt[6]{x}$

22. $\sqrt[3]{3} \cdot \sqrt{2} = 3^{1/3} \cdot 2^{1/2}$ Write with rational exponents.

$\phantom{\sqrt[3]{3} \cdot \sqrt{2}} = 3^{2/6} \cdot 2^{3/6}$ Write the exponents so that they have the same denominator.

$\phantom{\sqrt[3]{3} \cdot \sqrt{2}} = \left(3^2 \cdot 2^3\right)^{1/6}$ Use $a^n b^n = (ab)^n$.

$\phantom{\sqrt[3]{3} \cdot \sqrt{2}} = \sqrt[6]{3^2 \cdot 2^3}$ Write with radical notation.

$\phantom{\sqrt[3]{3} \cdot \sqrt{2}} = \sqrt[6]{72}$ Multiply $3^2 \cdot 2^3$.

9.3 SIMPLIFYING RADICAL EXPRESSIONS

A USING THE PRODUCT RULE

It is possible to simplify some radicals that do not evaluate to rational numbers. To do so, we use a product rule and a quotient rule for radicals. To discover the product rule, notice the following pattern:

$$\sqrt{9} \cdot \sqrt{4} = 3 \cdot 2 = 6$$
$$\sqrt{9 \cdot 4} = \sqrt{36} = 6$$

Since both expressions simplify to 6, it is true that

$$\sqrt{9} \cdot \sqrt{4} = \sqrt{9 \cdot 4}$$

This pattern suggests the following product rule for exponents.

> **PRODUCT RULE FOR RADICALS**
>
> If $\sqrt[n]{a}$ and $\sqrt[n]{b}$ are real numbers, then
>
> $$\sqrt[n]{a} \cdot \sqrt[n]{b} = \sqrt[n]{ab}$$

Notice that the product rule is the relationship $a^{1/n} \cdot b^{1/n} = (ab)^{1/n}$ stated in radical notation.

Examples   Use the product rule to multiply.

1. $\sqrt{3} \cdot \sqrt{5} = \sqrt{3 \cdot 5} = \sqrt{15}$
2. $\sqrt{21} \cdot \sqrt{x} = \sqrt{21x}$
3. $\sqrt[3]{4} \cdot \sqrt[3]{2} = \sqrt[3]{4 \cdot 2} = \sqrt[3]{8} = 2$
4. $\sqrt[4]{5} \cdot \sqrt[4]{2x^3} = \sqrt[4]{5 \cdot 2x^3} = \sqrt[4]{10x^3}$
5. $\sqrt{\dfrac{2}{a}} \cdot \sqrt{\dfrac{b}{3}} = \sqrt{\dfrac{2}{a} \cdot \dfrac{b}{3}} = \sqrt{\dfrac{2b}{3a}}$

B USING THE QUOTIENT RULE

To discover the quotient rule for radicals, notice the following pattern:

$$\sqrt{\dfrac{4}{9}} = \dfrac{2}{3}$$

$$\dfrac{\sqrt{4}}{\sqrt{9}} = \dfrac{2}{3}$$

Since both expressions simplify to $\dfrac{2}{3}$, it is true that

$$\sqrt{\dfrac{4}{9}} = \dfrac{\sqrt{4}}{\sqrt{9}}$$

This pattern suggests the following quotient rule for radicals.

> **QUOTIENT RULE FOR RADICALS**
>
> If $\sqrt[n]{a}$ and $\sqrt[n]{b}$ are real numbers and $\sqrt[n]{b}$ is not zero, then
>
> $$\sqrt[n]{\dfrac{a}{b}} = \dfrac{\sqrt[n]{a}}{\sqrt[n]{b}}$$

Objectives

A Use the product rule for radicals.
B Use the quotient rule for radicals.
C Simplify radicals.

SSM   CD-ROM   Video 9.3

Practice Problems 1–5

Use the product rule to multiply.

1. $\sqrt{2} \cdot \sqrt{7}$
2. $\sqrt{17} \cdot \sqrt{y}$
3. $\sqrt[3]{2} \cdot \sqrt[3]{32}$
4. $\sqrt[4]{6} \cdot \sqrt[4]{3x^2}$
5. $\sqrt{\dfrac{3}{x}} \cdot \sqrt{\dfrac{y}{2}}$

Answers

1. $\sqrt{14}$, 2. $\sqrt{17y}$, 3. 4, 4. $\sqrt[4]{18x^2}$,

5. $\sqrt{\dfrac{3y}{2x}}$

Notice that the quotient rule is the relationship $\left(\dfrac{a}{b}\right)^{1/n} = \dfrac{a^{1/n}}{b^{1/n}}$ stated in radical notation. We can use the quotient rule to simplify radical expressions by reading the rule from left to right or to divide radicals by reading the rule from right to left.

For example,

$$\sqrt{\frac{x}{16}} = \frac{\sqrt{x}}{\sqrt{16}} = \frac{\sqrt{x}}{4} \qquad \text{Using } \sqrt[n]{\frac{a}{b}} = \frac{\sqrt[n]{a}}{\sqrt[n]{b}}$$

$$\frac{\sqrt{75}}{\sqrt{3}} = \sqrt{\frac{75}{3}} = \sqrt{25} = 5 \qquad \text{Using } \frac{\sqrt[n]{a}}{\sqrt[n]{b}} = \sqrt[n]{\frac{a}{b}}$$

Note: *For the remainder of this chapter, we will assume that variables represent positive real numbers. If this is so, we need not insert absolute value bars when we simplify even roots.*

Practice Problems 6–9

Use the quotient rule to simplify.

6. $\sqrt{\dfrac{9}{25}}$

7. $\sqrt{\dfrac{y}{36}}$

8. $\sqrt[3]{\dfrac{27}{64}}$

9. $\sqrt[5]{\dfrac{7}{32x^5}}$

Examples Use the quotient rule to simplify.

6. $\sqrt{\dfrac{25}{49}} = \dfrac{\sqrt{25}}{\sqrt{49}} = \dfrac{5}{7}$

7. $\sqrt{\dfrac{x}{9}} = \dfrac{\sqrt{x}}{\sqrt{9}} = \dfrac{\sqrt{x}}{3}$

8. $\sqrt[3]{\dfrac{8}{27}} = \dfrac{\sqrt[3]{8}}{\sqrt[3]{27}} = \dfrac{2}{3}$

9. $\sqrt[4]{\dfrac{3}{16y^4}} = \dfrac{\sqrt[4]{3}}{\sqrt[4]{16y^4}} = \dfrac{\sqrt[4]{3}}{2y}$

C SIMPLIFYING RADICALS

Both the product and quotient rules can be used to simplify a radical. If the product rule is read from right to left, we have that $\sqrt[n]{ab} = \sqrt[n]{a} \cdot \sqrt[n]{b}$. We use this to simplify the following radicals.

Practice Problem 10

Simplify: $\sqrt{18}$

Example 10 Simplify: $\sqrt{50}$

Solution: We factor 50 such that one factor is the largest perfect square that divides 50. The largest perfect square factor of 50 is 25, so we write 50 as $25 \cdot 2$ and use the product rule for radicals to simplify.

$$\sqrt{50} = \sqrt{25 \cdot 2} = \sqrt{25} \cdot \sqrt{2} = 5\sqrt{2}$$

⌐ The largest perfect square factor of 50.

> **HELPFUL HINT**
>
> Don't forget that, for example, $5\sqrt{2}$ means $5 \cdot \sqrt{2}$.

Practice Problems 11–13

Simplify:

11. $\sqrt[3]{40}$

12. $\sqrt{14}$

13. $\sqrt[4]{162}$

Examples Simplify.

11. $\sqrt[3]{24} = \sqrt[3]{8 \cdot 3} = \sqrt[3]{8} \cdot \sqrt[3]{3} = 2\sqrt[3]{3}$

⌐ The largest perfect cube factor of 24.

12. $\sqrt{26}$ The largest perfect square factor of 26 is 1, so $\sqrt{26}$ cannot be simplified further.

13. $\sqrt[4]{32} = \sqrt[4]{16 \cdot 2} = \sqrt[4]{16} \cdot \sqrt[4]{2} = 2\sqrt[4]{2}$

⌐ The largest 4th power factor of 32.

Answers

6. $\dfrac{3}{5}$, 7. $\dfrac{\sqrt{y}}{6}$, 8. $\dfrac{3}{4}$, 9. $\dfrac{\sqrt[5]{7}}{2x}$, 10. $3\sqrt{2}$,

11. $2\sqrt[3]{5}$, 12. $\sqrt{14}$, 13. $3\sqrt[4]{2}$

After simplifying a radical such as a square root, always check the radicand to see that it contains no other perfect square factors. It may, if the largest perfect square factor of the radicand was not originally recognized. For example,

$$\sqrt{200} = \sqrt{4 \cdot 50} = \sqrt{4} \cdot \sqrt{50} = 2\sqrt{50}$$

Notice that the radicand 50 still contains the perfect square factor 25. This is because 4 is not the largest perfect square factor of 200. We continue as follows:

$$2\sqrt{50} = 2\sqrt{25 \cdot 2} = 2 \cdot \sqrt{25} \cdot \sqrt{2} = 2 \cdot 5 \cdot \sqrt{2} = 10\sqrt{2}$$

The radical is now simplified since 2 contains no perfect square factors (other than 1).

HELPFUL HINT

To help you recognize largest perfect power factors of a radicand, it will help if you are familiar with some perfect powers. A few are listed below

Perfect Squares	1, 4, 9, 16, 25, 36, 49, 64, 81, 100, 121, 144
	1^2 2^2 3^2 4^2 5^2 6^2 7^2 8^2 9^2 10^2 11^2 12^2
Perfect Cubes	1, 8, 27, 64, 125
	1^3 2^3 3^3 4^3 5^3
Perfect 4th powers	1, 16, 81, 256
	1^4 2^4 3^4 4^4

HELPFUL HINT

We say that a radical of the form $\sqrt[n]{a}$ is simplified when the radicand a contains no factors that are perfect nth powers (other than 1 or -1).

Examples   Simplify.

14. $\sqrt{25x^3} = \sqrt{25 \cdot x^2 \cdot x}$ Find the largest perfect square factor.

$= \sqrt{25 \cdot x^2} \cdot \sqrt{x}$ Use the product rule.

$= 5x\sqrt{x}$ Simplify.

15. $\sqrt[3]{54x^6y^8} = \sqrt[3]{27 \cdot 2 \cdot x^6 \cdot y^6 \cdot y^2}$ Factor the radicand and identify perfect cube factors.

$= \sqrt[3]{27 \cdot x^6 \cdot y^6 \cdot 2y^2}$

$= \sqrt[3]{27 \cdot x^6 \cdot y^6} \cdot \sqrt[3]{2y^2}$ Use the product rule.

$= 3x^2y^2\sqrt[3]{2y^2}$ Simplify.

16. $\sqrt[4]{81z^{11}} = \sqrt[4]{81 \cdot z^8 \cdot z^3}$ Factor the radicand and identify perfect fourth power factors.

$= \sqrt[4]{81 \cdot z^8} \cdot \sqrt[4]{z^3}$ Use the product rule.

$= 3z^2\sqrt[4]{z^3}$ Simplify.

Examples   Use the quotient rule to divide. Then simplify if possible.

17. $\dfrac{\sqrt{20}}{\sqrt{5}} = \sqrt{\dfrac{20}{5}}$ Use the quotient rule.

$= \sqrt{4}$ Simplify.

$= 2$ Simplify.

18. $\dfrac{\sqrt{50x}}{2\sqrt{2}} = \dfrac{1}{2} \cdot \sqrt{\dfrac{50x}{2}}$ Use the quotient rule.

Practice Problems 14–16

Simplify.

14. $\sqrt{49a^5}$

15. $\sqrt[3]{24x^9y^7}$

16. $\sqrt[4]{16z^9}$

Practice Problems 17–20

Use the quotient rule to divide. Then simplify if possible.

17. $\dfrac{\sqrt{75}}{\sqrt{3}}$ **18.** $\dfrac{\sqrt{80y}}{3\sqrt{5}}$

19. $\dfrac{5\sqrt[3]{162x^8}}{\sqrt[3]{3x^2}}$ **20.** $\dfrac{3\sqrt[4]{243x^9y^6}}{\sqrt[4]{x^{-3}y}}$

Answers

14. $7a^2\sqrt{a}$, **15.** $2x^3y^2\sqrt[3]{3y}$, **16.** $2z^2\sqrt[4]{z}$,

17. 5, **18.** $\dfrac{4}{3}\sqrt{y}$, **19.** $15x^2\sqrt[3]{2}$, **20.** $9x^3y\sqrt[4]{3y}$

$$= \frac{1}{2} \cdot \sqrt{25x} \qquad \text{Simplify.}$$

$$= \frac{1}{2} \cdot \sqrt{25} \cdot \sqrt{x} \qquad \text{Factor } 25x.$$

$$= \frac{1}{2} \cdot 5 \cdot \sqrt{x} \qquad \text{Simplify.}$$

$$= \frac{5}{2}\sqrt{x}$$

19. $\dfrac{7\sqrt[3]{48y^4}}{\sqrt[3]{2y}} = 7\sqrt[3]{\dfrac{48y^4}{2y}} = 7\sqrt[3]{24y^3} = 7\sqrt[3]{8 \cdot y^3 \cdot 3}$

$$= 7\sqrt[3]{8 \cdot y^3} \cdot \sqrt[3]{3} = 7 \cdot 2y\sqrt[3]{3} = 14y\sqrt[3]{3}$$

20. $\dfrac{2\sqrt[4]{32a^8b^6}}{\sqrt[4]{a^{-1}b^2}} = 2\sqrt[4]{\dfrac{32a^8b^6}{a^{-1}b^2}} = 2\sqrt[4]{32a^9b^4} = 2\sqrt[4]{16 \cdot a^8 \cdot b^4 \cdot 2 \cdot a}$

$$= 2\sqrt[4]{16 \cdot a^8 \cdot b^4} \cdot \sqrt[4]{2 \cdot a} = 2 \cdot 2a^2b \cdot \sqrt[4]{2a} = 4a^2b\sqrt[4]{2a}$$

✓ CONCEPT CHECK

Find and correct the error:

$$\frac{\sqrt[3]{27}}{\sqrt{9}} = \sqrt[3]{\frac{27}{9}} = \sqrt[3]{3}$$

TRY THE CONCEPT CHECK IN THE MARGIN.

Focus On History

DEVELOPMENT OF THE RADICAL SYMBOL

The first mathematician to use the symbol we use today to denote a square root was Christoff Rudolff (1499–1545). In 1525, Rudolff wrote and published the first German algebra text, *Die Coss*. In it, he used $\sqrt{\ }$ to represent a square root, the symbol $\mathcal{W}$ to represent a cube root, and the symbol $\mathcal{WW}$ to represent a fourth root, It was another 100 years before the square root symbol was extended with an overbar called a *vinculum*, $\sqrt{\overline{\ \ }}$, to indicate the inclusion of several terms under the radical symbol. This innovation was introduced by René Descartes (1596–1650) in 1637 in his text *La Géométrie*. The modern use of a numeral as part of a radical sign to indicate the index of the radical for higher roots did not appear until 1690 when this notation was used by French mathematician Michel Rolle (1652–1719) in his text *Traité d'Algébre*.

Answer

✓ Concept Check: $\dfrac{\sqrt[3]{27}}{\sqrt{9}} = \dfrac{3}{3} = 1$

EXERCISE SET 9.3

A *Use the product rule to multiply. See Examples 1 through 5.*

1. $\sqrt{7} \cdot \sqrt{2}$

2. $\sqrt{11} \cdot \sqrt{10}$

3. $\sqrt[4]{8} \cdot \sqrt[4]{2}$

4. $\sqrt[4]{27} \cdot \sqrt[4]{3}$

5. $\sqrt[3]{4} \cdot \sqrt[3]{9}$

6. $\sqrt[3]{10} \cdot \sqrt[3]{5}$

7. $\sqrt{2} \cdot \sqrt{3x}$

8. $\sqrt{3y} \cdot \sqrt{5x}$

9. $\sqrt{\dfrac{7}{x}} \cdot \sqrt{\dfrac{2}{y}}$

10. $\sqrt{\dfrac{6}{m}} \cdot \sqrt{\dfrac{n}{5}}$

11. $\sqrt[4]{4x^3} \cdot \sqrt[4]{5}$

12. $\sqrt[4]{ab^2} \cdot \sqrt[4]{27ab}$

B *Use the quotient rule to simplify. See Examples 6 through 9.*

13. $\sqrt{\dfrac{6}{49}}$

14. $\sqrt{\dfrac{8}{81}}$

15. $\sqrt{\dfrac{2}{49}}$

16. $\sqrt{\dfrac{5}{121}}$

17. $\sqrt[4]{\dfrac{x^3}{16}}$

18. $\sqrt[4]{\dfrac{y}{81x^4}}$

19. $\sqrt[3]{\dfrac{4}{27}}$

20. $\sqrt[3]{\dfrac{3}{64}}$

21. $\sqrt[4]{\dfrac{8}{x^8}}$

22. $\sqrt[4]{\dfrac{a^3}{81}}$

23. $\sqrt[3]{\dfrac{2x}{81y^{12}}}$

24. $\sqrt[3]{\dfrac{3}{8x^6}}$

25. $\sqrt{\dfrac{x^2y}{100}}$

ANSWERS

1. _____
2. _____
3. _____
4. _____
5. _____
6. _____
7. _____
8. _____
9. _____
10. _____
11. _____
12. _____
13. _____
14. _____
15. _____
16. _____
17. _____
18. _____
19. _____
20. _____
21. _____
22. _____
23. _____
24. _____
25. _____

Problem Solving Notes

9.4 ADDING, SUBTRACTING, AND MULTIPLYING RADICAL EXPRESSIONS

A ADDING OR SUBTRACTING RADICAL EXPRESSIONS

We have learned that the sum or difference of like terms can be simplified. To simplify these sums or differences, we use the distributive property. For example,

$$2x + 3x = (2 + 3)x = 5x$$

The distributive property can also be used to add **like radicals**.

LIKE RADICALS

Radicals with the same index and the same radicand are like radicals. For example,

$$2\sqrt{7} + 3\sqrt{7} = (2 + 3)\sqrt{7} = 5\sqrt{7}$$

HELPFUL HINT

The expression

$$5\sqrt{7} - 3\sqrt{6}$$

does not contain like radicals and cannot be simplified further.

Examples   Add or subtract as indicated.

1. $4\sqrt{11} + 8\sqrt{11} = (4 + 8)\sqrt{11} = 12\sqrt{11}$

2. $5\sqrt[3]{3x} - 7\sqrt[3]{3x} = (5 - 7)\sqrt[3]{3x} = -2\sqrt[3]{3x}$

3. $2\sqrt{7} + 2\sqrt[3]{7}$ This expression cannot be simplified since $2\sqrt{7}$ and $2\sqrt[3]{7}$ do not contain like radicals.

TRY THE CONCEPT CHECK IN THE MARGIN.

When adding or subtracting radicals, always check first to see whether any radicals can be simplified.

Examples   Add or subtract as indicated.

4. $\sqrt{20} + 2\sqrt{45} = \sqrt{4 \cdot 5} + 2\sqrt{9 \cdot 5}$ Factor 20 and 45.

$\qquad = \sqrt{4} \cdot \sqrt{5} + 2 \cdot \sqrt{9} \cdot \sqrt{5}$ Use the product rule.

$\qquad = 2 \cdot \sqrt{5} + 2 \cdot 3 \cdot \sqrt{5}$ Simplify $\sqrt{4}$ and $\sqrt{9}$.

$\qquad = 2\sqrt{5} + 6\sqrt{5}$ Add like radicals.

$\qquad = 8\sqrt{5}$

5. $\sqrt[3]{54} - 5\sqrt[3]{16} + \sqrt[3]{2}$

$\qquad = \sqrt[3]{27} \cdot \sqrt[3]{2} - 5 \cdot \sqrt[3]{8} \cdot \sqrt[3]{2} + \sqrt[3]{2}$ Factor and use the product rule.

$\qquad = 3 \cdot \sqrt[3]{2} - 5 \cdot 2 \cdot \sqrt[3]{2} + \sqrt[3]{2}$ Simplify $\sqrt[3]{27}$ and $\sqrt[3]{8}$.

$\qquad = 3\sqrt[3]{2} - 10\sqrt[3]{2} + \sqrt[3]{2}$ Write $5 \cdot 2$ as 10.

$\qquad = -6\sqrt[3]{2}$ Combine like radicals.

Objectives

A Add or subtract radical expressions.

B Multiply radical expressions.

SSM   CD-ROM   Video 9.4

Practice Problems 1–3

Add or subtract as indicated.

1. $5\sqrt{15} + 2\sqrt{15}$

2. $9\sqrt[3]{2y} - 15\sqrt[3]{2y}$

3. $6\sqrt{10} - 3\sqrt[3]{10}$

✓ CONCEPT CHECK

True or false?
$$\sqrt{a} + \sqrt{b} = \sqrt{a + b}$$
Explain.

Practice Problems 4–8

Add or subtract as indicated.

4. $\sqrt{50} + 5\sqrt{18}$

5. $\sqrt[3]{24} - 4\sqrt[3]{192} + \sqrt[3]{3}$

6. $\sqrt{20x} - 6\sqrt{16x} + \sqrt{45x}$

7. $\sqrt[4]{32} + \sqrt{32}$

8. $\sqrt[3]{8y^5} + \sqrt[3]{27y^5}$

Answers

1. $7\sqrt{15}$,   **2.** $-6\sqrt[3]{2y}$,   **3.** $6\sqrt{10} - 3\sqrt[3]{10}$,
4. $20\sqrt{2}$,   **5.** $-13\sqrt[3]{3}$,   **6.** $5\sqrt{5x} - 24\sqrt{x}$,
7. $2\sqrt[4]{2} + 4\sqrt{2}$,   **8.** $5y\sqrt[3]{y^2}$

✓ **Concept Check:**   false; answers may vary

6. $\sqrt{27x} - 2\sqrt{9x} + \sqrt{72x}$

$= \sqrt{9} \cdot \sqrt{3x} - 2 \cdot \sqrt{9} \cdot \sqrt{x} + \sqrt{36} \cdot \sqrt{2x}$ Factor and use the product rule.

$= 3 \cdot \sqrt{3x} - 2 \cdot 3 \cdot \sqrt{x} + 6 \cdot \sqrt{2x}$ Simplify $\sqrt{9}$ and $\sqrt{36}$.

$= 3\sqrt{3x} - 6\sqrt{x} + 6\sqrt{2x}$ Write $2 \cdot 3$ as 6.

> **HELPFUL HINT**
>
> None of these terms contain like radicals. We can simplify no further.

7. $\sqrt[3]{98} + \sqrt{98} = \sqrt[3]{98} + \sqrt{49} \cdot \sqrt{2}$ Factor and use the product rule.

$= \sqrt[3]{98} + 7\sqrt{2}$ No further simplification is possible.

8. $\sqrt[3]{48y^4} + \sqrt[3]{6y^4} = \sqrt[3]{8y^3} \cdot \sqrt[3]{6y} + \sqrt[3]{y^3} \cdot \sqrt[3]{6y}$ Factor and use the product rule.

$= 2y\sqrt[3]{6y} + y\sqrt[3]{6y}$ Simplify $\sqrt[3]{8y^3}$ and $\sqrt[3]{y^3}$.

$= 3y\sqrt[3]{6y}$ Combine like radicals.

Practice Problems 9–10

Add or subtract as indicated.

9. $\dfrac{\sqrt{75}}{9} - \dfrac{\sqrt{3}}{2}$

10. $\sqrt[3]{\dfrac{5x}{27}} + 4\sqrt[3]{5x}$

Examples   Add or subtract as indicated.

9. $\dfrac{\sqrt{45}}{4} - \dfrac{\sqrt{5}}{3} = \dfrac{3\sqrt{5}}{4} - \dfrac{\sqrt{5}}{3}$ To subtract, notice that the LCD is 12.

$= \dfrac{3\sqrt{5} \cdot 3}{4 \cdot 3} - \dfrac{\sqrt{5} \cdot 4}{3 \cdot 4}$ Write each expression as an equivalent expression with a denominator of 12.

$= \dfrac{9\sqrt{5}}{12} - \dfrac{4\sqrt{5}}{12}$ Multiply factors in the numerators and the denominators.

$= \dfrac{5\sqrt{5}}{12}$ Subtract.

10. $\sqrt[3]{\dfrac{7x}{8}} + 2\sqrt[3]{7x} = \dfrac{\sqrt[3]{7x}}{\sqrt[3]{8}} + 2\sqrt[3]{7x}$ Use the quotient rule for radicals.

$= \dfrac{\sqrt[3]{7x}}{2} + 2\sqrt[3]{7x}$ Simplify.

$= \dfrac{\sqrt[3]{7x}}{2} + \dfrac{2\sqrt[3]{7x} \cdot 2}{2}$ Write each expression as an equivalent expression with a denominator of 2.

$= \dfrac{\sqrt[3]{7x}}{2} + \dfrac{4\sqrt[3]{7x}}{2}$

$= \dfrac{5\sqrt[3]{7x}}{2}$ Add.

B   MULTIPLYING RADICAL EXPRESSIONS

We can multiply radical expressions by using many of the same properties used to multiply polynomial expressions. For instance, to multiply $\sqrt{2}(\sqrt{6} - 3\sqrt{2})$, we use the distributive property and multiply $\sqrt{2}$ by each term inside the parentheses.

$\sqrt{2}(\sqrt{6} - 3\sqrt{2}) = \sqrt{2}(\sqrt{6}) - \sqrt{2}(3\sqrt{2})$ Use the distributive property.

$= \sqrt{2 \cdot 6} - 3\sqrt{2 \cdot 2}$

$= \sqrt{2 \cdot 2 \cdot 3} - 3 \cdot 2$ Use the product rule for radicals.

$= 2\sqrt{3} - 6$

Answers

9. $\dfrac{19\sqrt{3}}{18}$,   **10.** $\dfrac{13\sqrt[3]{5x}}{3}$

Example 11

Multiply: $\sqrt{3}(5 + \sqrt{30})$

Solution:

$$\sqrt{3}(5 + \sqrt{30}) = \sqrt{3}(5) + \sqrt{3}(\sqrt{30})$$
$$= 5\sqrt{3} + \sqrt{3 \cdot 30}$$
$$= 5\sqrt{3} + \sqrt{3 \cdot 3 \cdot 10}$$
$$= 5\sqrt{3} + 3\sqrt{10}$$

Examples   Multiply.

12. $(\sqrt{5} - \sqrt{6})(\sqrt{7} + 1) = \overset{\text{First}}{\sqrt{5} \cdot \sqrt{7}} + \overset{\text{Outer}}{\sqrt{5} \cdot 1} - \overset{\text{Inner}}{\sqrt{6} \cdot \sqrt{7}} - \overset{\text{Last}}{\sqrt{6} \cdot 1}$

$$= \sqrt{35} + \sqrt{5} - \sqrt{42} - \sqrt{6} \quad \text{Using the FOIL order, simplify.}$$

13. $(\sqrt{2x} + 5)(\sqrt{2x} - 5) = (\sqrt{2x})^2 - 5^2$ Multiply the sum and difference of two terms:

$$= 2x - 25 \quad (a + b)(a - b) = a^2 - b^2.$$

14. $(\sqrt{3} - 1)^2 = (\sqrt{3})^2 - 2 \cdot \sqrt{3} \cdot 1 + 1^2$ Square the binomial:

$$= 3 - 2\sqrt{3} + 1 \quad (a - b)^2 = a^2 - 2ab + b^2.$$
$$= 4 - 2\sqrt{3}$$

Square the binomial: $(a + b)^2 = a^2 + 2ab + b^2$

15. $(\sqrt{x - 3} + 5)^2 = (\sqrt{x - 3})^2 + 2 \cdot \sqrt{x - 3} \cdot 5 + 5^2$

$$= x - 3 + 10\sqrt{x - 3} + 25 \quad \text{Simplify.}$$
$$= x + 22 + 10\sqrt{x - 3} \quad \text{Combine like terms.}$$

Practice Problem 11

Multiply: $\sqrt{2}(6 + \sqrt{10})$

Practice Problems 12–15

Multiply.

12. $(\sqrt{3} - \sqrt{5})(\sqrt{2} + 7)$

13. $(\sqrt{5y} + 2)(\sqrt{5y} - 2)$

14. $(\sqrt{3} - 7)^2$

15. $(\sqrt{x + 1} + 2)^2$

Focus On the Real World

DIFFUSION

Diffusion is the spontaneous movement of the molecules of a substance from a region of higher concentration to a region of lower concentration until a uniform concentration throughout the region is reached. For example, if a drop of food coloring is added to a glass of water, the molecules of the coloring are diffused so that the entire glass of water is colored evenly without any kind of stirring. Diffusion is also mostly responsible for the spread of the smell of baking brownies throughout a house.

Diffusion is used or seen in important aspects of many disciplines. The following list describes situations in which diffusion plays a role.

▲ In the commercial production of sugar, sugar can be extracted from sugar cane through a diffusion process.

▲ Solid-state diffusion plays a role in the manufacturing process of silicon computer chips.

▲ In biology, the diffusion phenomenon allows water molecules, nutrient molecules, and dissolved gas molecules (such as oxygen and carbon dioxide) to pass through the semipermeable membranes of cell walls.

▲ During a human pregnancy, the fetus is nourished from the mother's blood supply via diffusion through the placenta. Waste materials from the fetus are also diffused through the placenta to be carried away by the mother's circulatory system.

▲ The medical treatment known as kidney dialysis, in which waste materials are removed from the blood of a patient without kidney function, is made possible by diffusion.

▲ A diffusion process is widely used to separate the uranium isotope U-235, which can be used as a fuel in nuclear power plants, from the uranium isotope U-238, which cannot be used to create nuclear energy.

In chemistry, Graham's law states that the diffusion rate of a substance in its gaseous state is inversely proportional to the square root of its molecular weight. Another useful property of diffusion is that the distance a material diffuses over time is directly proportional to the square root of the time.

CRITICAL THINKING

1. Write an equation for the relationship described by Graham's law. Be sure to define the variables and constants that you use.

2. According to Graham's law, which molecule will diffuse more rapidly: a molecule with a molecular weight of 58.4 or a molecule with a molecular weight of 180.2? Explain your reasoning.

3. Write an equation for the relationship between the distance that a material diffuses and time. Again, be sure to define the variables and constants that you use.

4. Suppose it takes sugar 1 week to diffuse a distance of 1 cm from its starting point in a particular liquid. How long will it take the sugar to diffuse a total of 3 cm from its starting point in the liquid?

9.6 RADICAL EQUATIONS AND PROBLEM SOLVING

A SOLVING EQUATIONS THAT CONTAIN RADICAL EXPRESSIONS

In this section, we present techniques to solve equations containing radical expressions such as

$$\sqrt{2x - 3} = 9$$

We use the power rule to help us solve these radical equations.

> **POWER RULE**
>
> If both sides of an equation are raised to the same power, *all* solutions of the original equation are *among* the solutions of the new equation.

This property *does not* say that raising both sides of an equation to a power yields an equivalent equation. A solution of the new equation *may or may not* be a solution of the original equation. Thus, *each solution of the new equation must be checked* to make sure it is a solution of the original equation. Recall that a proposed solution that is not a solution of the original equation is called an extraneous solution.

Example 1   Solve: $\sqrt{2x - 3} = 9$

Solution: We use the power rule to square both sides of the equation to eliminate the radical.

$$\sqrt{2x - 3} = 9$$
$$(\sqrt{2x - 3})^2 = 9^2$$
$$2x - 3 = 81$$
$$2x = 84$$
$$x = 42$$

Now we check the solution in the original equation.

Check:
$$\sqrt{2x - 3} = 9$$
$$\sqrt{2(42) - 3} \stackrel{?}{=} 9 \quad \text{Let } x = 42.$$
$$\sqrt{84 - 3} \stackrel{?}{=} 9$$
$$\sqrt{81} \stackrel{?}{=} 9$$
$$9 = 9 \quad \text{True.}$$

The solution checks, so we conclude that the solution set is $\{42\}$. ■

To solve a radical equation, first isolate a radical on one side of the equation.

Example 2   Solve: $\sqrt{-10x - 1} + 3x = 0$

Solution: First, isolate the radical on one side of the equation. To do this, we subtract $3x$ from both sides.

$$\sqrt{-10x - 1} + 3x = 0$$
$$\sqrt{-10x - 1} + 3x - 3x = 0 - 3x$$
$$\sqrt{-10x - 1} = -3x$$

Objectives

A Solve equations that contain radical expressions.

B Use the Pythagorean theorem to model problems.

SSM   CD-ROM   Video 9.6

Practice Problem 1

Solve: $\sqrt{3x - 2} = 5$

Practice Problem 2

Solve: $\sqrt{9x - 2} - 2x = 0$

Answers

1. $\{9\}$,   **2.** $\left\{\frac{1}{4}, 2\right\}$

Next we use the power rule to eliminate the radical.

$$(\sqrt{-10x - 1})^2 = (-3x)^2$$
$$-10x - 1 = 9x^2$$

Since this is a quadratic equation, we can set the equation equal to 0 and try to solve by factoring.

$$9x^2 + 10x + 1 = 0$$
$$(9x + 1)(x + 1) = 0 \quad \text{Factor.}$$
$$9x + 1 = 0 \quad \text{or} \quad x + 1 = 0 \quad \text{Set each factor equal to 0.}$$
$$x = -\frac{1}{9} \qquad\qquad x = -1$$

Check: Let $x = -\frac{1}{9}$.   Let $x = -1$.

$$\sqrt{-10x - 1} + 3x = 0 \qquad\qquad \sqrt{-10x - 1} + 3x = 0$$

$$\sqrt{-10\left(-\frac{1}{9}\right) - 1} + 3\left(-\frac{1}{9}\right) \stackrel{?}{=} 0 \qquad \sqrt{-10(-1) - 1} + 3(-1) \stackrel{?}{=} 0$$

$$\sqrt{\frac{10}{9} - \frac{9}{9}} - \frac{3}{9} \stackrel{?}{=} 0 \qquad\qquad \sqrt{10 - 1} - 3 \stackrel{?}{=} 0$$

$$\sqrt{\frac{1}{9}} - \frac{1}{3} \stackrel{?}{=} 0 \qquad\qquad\qquad \sqrt{9} - 3 \stackrel{?}{=} 0$$

$$\frac{1}{3} - \frac{1}{3} = 0 \quad \text{True.} \qquad\qquad\qquad 3 - 3 = 0$$
$$\text{True.}$$

Both solutions check. The solution set is $\left\{-\frac{1}{9}, -1\right\}$.

The following steps may be used to solve a radical equation.

Solving a Radical Equation

Step 1.   Isolate one radical on one side of the equation.

Step 2.   Raise each side of the equation to a power equal to the index of the radical and simplify.

Step 3.   If the equation still contains a radical term, repeat Steps 1 and 2. If not, solve the equation.

Step 4.   Check all proposed solutions in the original equation.

Practice Problem 3

Solve: $\sqrt[3]{x - 5} + 2 = 1$

Example 3   Solve: $\sqrt[3]{x + 1} + 5 = 3$

Solution: First we isolate the radical by subtracting 5 from both sides of the equation.

$$\sqrt[3]{x + 1} + 5 = 3$$
$$\sqrt[3]{x + 1} = -2$$

Next we raise both sides of the equation to the third power to eliminate the radical.

Answer

3. $\{4\}$

$$(\sqrt[3]{x + 1})^3 = (-2)^3$$
$$x + 1 = -8$$
$$x = -9$$

The solution checks in the original equation, so the solution set is $\{-9\}$.

Example 4   Solve: $\sqrt{4 - x} = x - 2$

Solution:
$$\sqrt{4 - x} = x - 2$$
$$(\sqrt{4 - x})^2 = (x - 2)^2$$
$$4 - x = x^2 - 4x + 4$$
$$x^2 - 3x = 0 \quad \text{Write the quadratic equation in standard form.}$$
$$x(x - 3) = 0 \quad \text{Factor.}$$
$$x = 0 \text{ or } x - 3 = 0 \quad \text{Set each factor equal to 0.}$$
$$x = 3$$

Check:

$\sqrt{4 - x} = x - 2$	$\sqrt{4 - x} = x - 2$
$\sqrt{4 - 0} \stackrel{?}{=} 0 - 2$ Let $x = 0$.	$\sqrt{4 - 3} \stackrel{?}{=} 3 - 2$ Let $x = 3$.
$2 = -2$ False.	$1 = 1$ True.

The proposed solution 3 checks, but 0 does not. Since 0 is an extraneous solution, the solution set is $\{3\}$.

HELPFUL HINT

In Example 4, notice that $(x - 2)^2 = x^2 - 4x + 4$. Make sure binomials are squared correctly.

TRY THE CONCEPT CHECK IN THE MARGIN.

Example 5   Solve: $\sqrt{2x + 5} + \sqrt{2x} = 3$

Solution: We get one radical alone by subtracting $\sqrt{2x}$ from both sides.

$$\sqrt{2x + 5} + \sqrt{2x} = 3$$
$$\sqrt{2x + 5} = 3 - \sqrt{2x}$$

Now we use the power rule to begin eliminating the radicals. First we square both sides.

$$(\sqrt{2x + 5})^2 = (3 - \sqrt{2x})^2$$
$$2x + 5 = 9 - 6\sqrt{2x} + 2x \quad \begin{array}{l}\text{Multiply:}\\ (3 - \sqrt{2x})(3 - \sqrt{2x}).\end{array}$$

There is still a radical in the equation, so we get the radical alone again. Then we square both sides.

Practice Problem 4

Solve: $\sqrt{9 + x} = x + 3$

✓ CONCEPT CHECK

How can you immediately tell that the equation $\sqrt{2y + 3} = -4$ has no real solution?

Practice Problem 5

Solve: $\sqrt{3x + 1} + \sqrt{3x} = 2$

Answers

4. $\{0\}$,   **5.** $\left\{\dfrac{3}{16}\right\}$

✓ Concept Check: answers may vary

$$2x + 5 = 9 - 6\sqrt{2x} + 2x$$

$$6\sqrt{2x} = 4$$ Get the radical alone.

$$(6\sqrt{2x})^2 = 4^2$$ Square both sides of the equation to eliminate the radical.

$$36(2x) = 16$$

$$72x = 16$$ Multiply.

$$x = \frac{16}{72}$$ Solve.

$$x = \frac{2}{9}$$ Simplify.

The proposed solution, $\frac{2}{9}$ checks in the original equation.

The solution set is $\left\{\frac{2}{9}\right\}$.

HELPFUL HINT

Make sure expressions are squared correctly. In Example 5, we squared $(3 - \sqrt{2x})$ as

$$(3 - \sqrt{2x})^2 = (3 - \sqrt{2x})(3 - \sqrt{2x})$$

$$= 3 \cdot 3 - 3\sqrt{2x} - 3\sqrt{2x} + \sqrt{2x} \cdot \sqrt{2x}$$

$$= 9 - 6\sqrt{2x} + 2x$$

✓ CONCEPT CHECK

What is wrong with the following solution?

$$\sqrt{2x + 5} + \sqrt{4 - x} = 8$$

$$(\sqrt{2x + 5} + \sqrt{4 - x})^2 = 8^2$$

$$(2x + 5) + (4 - x) = 64$$

$$x + 9 = 64$$

$$x = 55$$

TRY THE CONCEPT CHECK IN THE MARGIN.

B USING THE PYTHAGOREAN THEOREM

Recall that the Pythagorean theorem states that in a right triangle, the length of the hypotenuse squared equals the sum of the lengths of each of the legs squared.

> **PYTHAGOREAN THEOREM**
>
> If a and b are the lengths of the legs of a right triangle and c is the length of the hypotenuse, then $a^2 + b^2 = c^2$.
>
>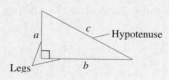

Practice Problem 6

Find the length of the unknown leg of the right triangle.

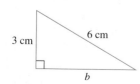

Example 6

Find the length of the unknown leg of the right triangle.

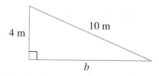

Solution: In the formula $a^2 + b^2 = c^2$, c is the hypotenuse. Here, $c = 10$, the length of the hypotenuse, and $a = 4$. We solve for b. Then $a^2 + b^2 = c^2$ becomes

Answers

6. $3\sqrt{3}$ cm

✓ **Concept Check:**   answers may vary

$$4^2 + b^2 = 10^2$$
$$16 + b^2 = 100$$
$$b^2 = 84 \quad \text{Subtract 16 from both sides.}$$

Since b is a length and thus is positive, we have that

$$b = \sqrt{84} = \sqrt{4 \cdot 21} = 2\sqrt{21}$$

The unknown leg of the triangle is $2\sqrt{21}$ meters long.

Example 7   Calculating Placement of a Wire

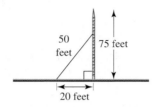

A 50-foot supporting wire is to be attached to a 75-foot antenna. Because of surrounding buildings, sidewalks, and roadways, the wire must be anchored exactly 20 feet from the base of the antenna.

a. How high from the base of the antenna is the wire attached?

b. Local regulations require that a supporting wire be attached at a height no less than $\frac{3}{5}$ of the total height of the antenna. From part (a), have local regulations been met?

Solution:

1. UNDERSTAND. Read and reread the problem. From the diagram we notice that a right triangle is formed with hypotenuse 50 feet and one leg 20 feet. Let $x =$ the height from the base of the antenna to the attached wire.

2. TRANSLATE. Use the Pythagorean theorem.

$$(a)^2 + (b)^2 = (c)^2$$
$$(20)^2 + x^2 = (50)^2 \quad a = 20, c = 50$$

3. SOLVE.
$$(20)^2 + x^2 = (50)^2$$
$$400 + x^2 = 2500$$
$$x^2 = 2100 \quad \text{Subtract 400 from both sides.}$$
$$x = \sqrt{2100}$$
$$= 10\sqrt{21}$$

4. INTERPRET. *Check* the work and *state* the solution.

a. The wire is attached exactly $10\sqrt{21}$ feet from the base of the pole, or approximately 45.8 feet.

b. The supporting wire must be attached at a height no less than $\frac{3}{5}$ of the total height of the antenna. This height is $\frac{3}{5}(75 \text{ feet})$, or 45 feet. Since we know from part (a) that the wire is to be attached at a height of approximately 45.8 feet, local regulations have been met.

Practice Problem 7

A furniture upholsterer wishes to cut a strip from a piece of fabric that is 45 inches by 45 inches. The strip must be cut on the bias of the fabric. What is the longest strip that can be cut? Give an exact answer and a two decimal place approximation.

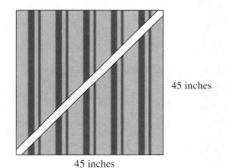

45 inches

45 inches

Answer

7. $45\sqrt{2}$ in. ≈ 63.64 in.

GRAPHING CALCULATOR EXPLORATIONS

We can use a grapher to solve radical equations. For example, to use a grapher to approximate the solutions of the equation solved in Example 4, we graph the following:

$$Y_1 = \sqrt{4 - x} \qquad \text{and} \qquad Y_2 = x - 2$$

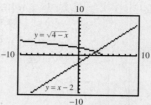

The x-value of the point of intersection is the solution. Use the Intersect feature or the Zoom and Trace features of your grapher to see that the solution is 3.

Use a grapher to solve each radical equation. Round all solutions to the nearest hundredth.

1. $\sqrt{x + 7} = x$

2. $\sqrt{3x + 5} = 2x$

3. $\sqrt{2x + 1} = \sqrt{2x + 2}$

4. $\sqrt{10x - 1} = \sqrt{-10x + 10} - 1$

5. $1.2x = \sqrt{3.1x + 5}$

6. $\sqrt{1.9x^2 - 2.2} = -0.8x + 3$

9.7 COMPLEX NUMBERS

A WRITING NUMBERS IN THE FORM *bi*

Our work with radical expressions has excluded expressions such as $\sqrt{-16}$ because $\sqrt{-16}$ is not a real number; there is no real number whose square is -16. In this section, we discuss a number system that includes roots of negative numbers. This number system is the **complex number system**, and it includes the set of real numbers as a subset. The complex number system allows us to solve equations such as $x^2 + 1 = 0$ that have no real number solutions. The set of complex numbers includes the **imaginary unit**.

IMAGINARY UNIT

The **imaginary unit**, written i, is the number whose square is -1. That is,

$$i^2 = -1 \quad \text{and} \quad i = \sqrt{-1}$$

To write the square root of a negative number in terms of i, we use the property that if a is a positive number, then

$$\sqrt{-a} = \sqrt{-1} \cdot \sqrt{a}$$
$$= i \cdot \sqrt{a}$$

Using i, we can write $\sqrt{-16}$ as

$$\sqrt{-16} = \sqrt{-1 \cdot 16} = \sqrt{-1} \cdot \sqrt{16} = i \cdot 4 \text{ or } 4i$$

Examples   Write using *i* notation.

1. $\sqrt{-36} = \sqrt{-1 \cdot 36} = \sqrt{-1} \cdot \sqrt{36} = i \cdot 6 \text{ or } 6i$

2. $\sqrt{-5} = \sqrt{-1(5)} = \sqrt{-1} \cdot \sqrt{5} = i\sqrt{5}$. Since $\sqrt{5}i$ can easily be confused with $\sqrt{5i}$, we write $\sqrt{5}i$ as $i\sqrt{5}$.

3. $-\sqrt{-20} = -\sqrt{-1 \cdot 20} = -\sqrt{-1} \cdot \sqrt{4 \cdot 5} = -i \cdot 2\sqrt{5} = -2i\sqrt{5}$

The product rule for radicals does not necessarily hold true for imaginary numbers. *To multiply square roots of negative numbers, first we write each number in terms of the imaginary unit i.* For example, to multiply $\sqrt{-4}$ and $\sqrt{-9}$, we first write each number in the form bi:

$$\sqrt{-4}\sqrt{-9} = 2i(3i) = 6i^2 = 6(-1) = -6$$

We will also use this method to simplify quotients of square roots of negative numbers.

Examples   Multiply or divide as indicated.

4. $\sqrt{-3} \cdot \sqrt{-5} = i\sqrt{3}(i\sqrt{5}) = i^2\sqrt{15} = -1\sqrt{15} = -\sqrt{15}$

5. $\sqrt{-36} \cdot \sqrt{-1} = 6i(i) = 6i^2 = 6(-1) = -6$

6. $\sqrt{8} \cdot \sqrt{-2} = 2\sqrt{2}(i\sqrt{2}) = 2i(\sqrt{2}\sqrt{2}) = 2i(2) = 4i$

7. $\dfrac{\sqrt{-125}}{\sqrt{5}} = \dfrac{i\sqrt{125}}{\sqrt{5}} = i\sqrt{25} = 5i$

Now that we have practiced working with the imaginary unit, we define complex numbers.

Objectives

A Write square roots of negative numbers in the form *bi*.

B Add or subtract complex numbers.

C Multiply complex numbers.

D Divide complex numbers.

E Raise *i* to powers.

SSM   CD-ROM   Video 9.7

Practice Problems 1–3

Write using *i* notation.

1. $\sqrt{-25}$

2. $\sqrt{-3}$

3. $-\sqrt{-50}$

Practice Problems 4–7

Multiply or divide as indicated.

4. $\sqrt{-2} \cdot \sqrt{-7}$

5. $\sqrt{-25} \cdot \sqrt{-1}$

6. $\sqrt{27} \cdot \sqrt{-3}$

7. $\dfrac{\sqrt{-8}}{\sqrt{2}}$

Answers

1. $5i$, **2.** $i\sqrt{3}$, **3.** $-5i\sqrt{2}$, **4.** $-\sqrt{14}$, **5.** -5, **6.** $9i$, **7.** $2i$

> **COMPLEX NUMBERS**
>
> A **complex number** is a number that can be written in the form $a + bi$, where a and b are real numbers.

Notice that the set of real numbers is a subset of the complex numbers since any real number can be written in the form of a complex number. For example,

$$16 = 16 + 0i$$

In general, a complex number $a + bi$ is a real number if $b = 0$. Also, a complex number is called an **imaginary number** if $a = 0$. For example,

$$3i = 0 + 3i \quad \text{and} \quad i\sqrt{7} = 0 + i\sqrt{7}$$

are imaginary numbers.

The following diagram shows the relationship between complex numbers and their subsets.

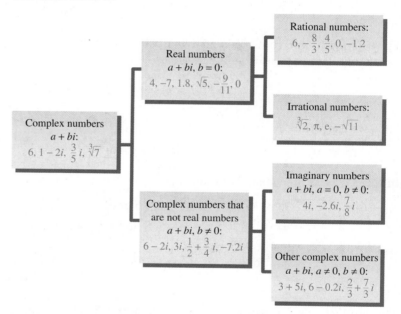

TRY THE CONCEPT CHECK IN THE MARGIN.

B ADDING OR SUBTRACTING COMPLEX NUMBERS

Two complex numbers $a + bi$ and $c + di$ are equal if and only if $a = c$ and $b = d$. Complex numbers can be added or subtracted by adding or subtracting their real parts and then adding or subtracting their imaginary parts.

> **SUM OR DIFFERENCE OF COMPLEX NUMBERS**
>
> If $a + bi$ and $c + di$ are complex numbers, then their sum is
>
> $$(a + bi) + (c + di) = (a + c) + (b + d)i$$
>
> Their difference is
>
> $$(a + bi) - (c + di) = a + bi - c - di = (a - c) + (b - d)i$$

Examples   Add or subtract as indicated.

8. $(2 + 3i) + (-3 + 2i) = (2 - 3) + (3 + 2)i = -1 + 5i$

9. $5i - (1 - i) = 5i - 1 + i$
$$= -1 + (5 + 1)i$$
$$= -1 + 6i$$

10. $(-3 - 7i) - (-6) = -3 - 7i + 6$
$$= (-3 + 6) - 7i$$
$$= 3 - 7i$$

Practice Problems 8–10

Add or subtract as indicated.

8. $(5 + 2i) + (4 - 3i)$

9. $6i - (2 - i)$

10. $(-2 - 4i) - (-3)$

C MULTIPLYING COMPLEX NUMBERS

To multiply two complex numbers of the form $a + bi$, we multiply as though they are binomials. Then use the relationship $i^2 = -1$ to simplify.

Examples   Multiply.

11. $-7i \cdot 3i = -21i^2$
$$= -21(-1) \quad \text{Replace } i^2 \text{ with } -1.$$
$$= 21$$

12. $3i(2 - i) = 3i \cdot 2 - 3i \cdot i \quad \text{Use the distributive property.}$
$$= 6i - 3i^2 \quad\quad \text{Multiply.}$$
$$= 6i - 3(-1) \quad\quad \text{Replace } i^2 \text{ with } -1.$$
$$= 6i + 3$$
$$= 3 + 6i$$

Use the FOIL method. (First, Outer, Inner, Last)

13. $(2 - 5i)(4 + i) = 2(4) + 2(i) - 5i(4) - 5i(i)$
$$\quad\quad\quad\quad\quad\quad F \quad\quad O \quad\quad I \quad\quad L$$
$$= 8 + 2i - 20i - 5i^2$$
$$= 8 - 18i - 5(-1) \quad i^2 = -1.$$
$$= 8 - 18i + 5$$
$$= 13 - 18i$$

14. $(2 - i)^2 = (2 - i)(2 - i)$
$$= 2(2) - 2(i) - 2(i) + i^2$$
$$= 4 - 4i + (-1) \quad\quad i^2 = -1.$$
$$= 3 - 4i$$

15. $(7 + 3i)(7 - 3i) = 7(7) - 7(3i) + 3i(7) - 3i(3i)$
$$= 49 - 21i + 21i - 9i^2$$
$$= 49 - 9(-1) \quad i^2 = -1.$$
$$= 49 + 9$$
$$= 58$$

Practice Problems 11–15

Multiply.

11. $-5i \cdot 3i$

12. $-2i(6 - 2i)$

13. $(3 - 4i)(6 + i)$

14. $(1 - 2i)^2$

15. $(6 + 5i)(6 - 5i)$

Notice that if you add, subtract, or multiply two complex numbers, the result is a complex number.

D DIVIDING COMPLEX NUMBERS

From Example 15, notice that the product of $7 + 3i$ and $7 - 3i$ is a real number. These two complex numbers are called **complex conjugates** of one another. In general, we have the following definition.

Answers

8. $9 - i$,   **9.** $-2 + 7i$,   **10.** $1 - 4i$,
11. 15,   **12.** $-4 - 12i$,   **13.** $22 - 21i$,
14. $-3 - 4i$,   **15.** 61

> **COMPLEX CONJUGATES**
>
> The complex numbers $(a + bi)$ and $(a - bi)$ are called **complex conjugates** of each other, and
>
> $$(a + bi)(a - bi) = a^2 + b^2$$

To see that the product of a complex number $a + bi$ and its conjugate $a - bi$ is the real number $a^2 + b^2$, we multiply:

$$(a + bi)(a - bi) = a^2 - abi + abi - b^2 i^2$$
$$= a^2 - b^2(-1)$$
$$= a^2 + b^2$$

We will use complex conjugates to divide by a complex number.

Practice Problem 16

Divide and write in the form $a + bi$:
$$\frac{3 + i}{2 - 3i}$$

Example 16   Divide and write in the form $a + bi$: $\dfrac{2 + i}{1 - i}$

Solution: We multiply the numerator and the denominator by the complex conjugate of $1 - i$ to eliminate the imaginary number in the denominator.

$$\frac{2 + i}{1 - i} = \frac{(2 + i)(1 + i)}{(1 - i)(1 + i)}$$
$$= \frac{2(1) + 2(i) + 1(i) + i^2}{1^2 - i^2}$$
$$= \frac{2 + 3i - 1}{1 + 1}$$
$$= \frac{1 + 3i}{2} = \frac{1}{2} + \frac{3}{2}i$$

Practice Problem 17

Divide and write in the form $a + bi$: $\dfrac{6}{5i}$

Example 17   Divide and write in the form $a + bi$: $\dfrac{7}{3i}$

Solution: We multiply the numerator and the denominator by the conjugate of $3i$. Note that $3i = 0 + 3i$, so its conjugate is $0 - 3i$ or $-3i$.

$$\frac{7}{3i} = \frac{7(-3i)}{(3i)(-3i)} = \frac{-21i}{-9i^2} = \frac{-21i}{-9(-1)} = \frac{-21i}{9} = \frac{-7i}{3} = -\frac{7}{3}i$$

E   FINDING POWERS OF *i*

We can use the fact that $i^2 = -1$ to simplify i^3 and i^4.

$$i^3 = i^2 \cdot i = (-1)i = -i$$
$$i^4 = i^2 \cdot i^2 = (-1) \cdot (-1) = 1$$

We continue this process and use the fact that $i^4 = 1$ and $i^2 = -1$ to simplify i^5 and i^6.

$$i^5 = i^4 \cdot i = 1 \cdot i = i$$
$$i^6 = i^4 \cdot i^2 = 1 \cdot (-1) = -1$$

Answers

16. $\dfrac{3}{13} + \dfrac{11}{13}i$,   **17.** $-\dfrac{6}{5}i$

If we continue finding powers of i, we generate the following pattern. Notice that the values $i, -1, -i$, and 1 repeat as i is raised to higher and higher powers.

$i^1 = i$ $\quad$ $i^5 = i$ $\quad$ $i^9 = i$

$i^2 = -1$ $\quad$ $i^6 = -1$ $\quad$ $i^{10} = -1$

$i^3 = -i$ $\quad$ $i^7 = -i$ $\quad$ $i^{11} = -i$

$i^4 = 1$ $\quad$ $i^8 = 1$ $\quad$ $i^{12} = 1$

This pattern allows us to find other powers of i. To do so, we will use the fact that $i^4 = 1$ and rewrite a power of i in terms of i^4.

For example, $i^{22} = i^{20} \cdot i^2 = (i^4)^5 \cdot i^2 = 1^5 \cdot (-1) = 1 \cdot (-1) = -1$.

Examples   Find each power of i.

18. $i^7 = i^4 \cdot i^3 = 1(-i) = -i$

19. $i^{20} = (i^4)^5 = 1^5 = 1$

20. $i^{46} = i^{44} \cdot i^2 = (i^4)^{11} \cdot i^2 = 1^{11}(-1) = -1$

21. $i^{-12} = \dfrac{1}{i^{12}} = \dfrac{1}{(i^4)^3} = \dfrac{1}{(1)^3} = \dfrac{1}{1} = 1$

────────

Practice Problems 18–21

Find the powers of i.

18. i^{11}

19. i^{40}

20. i^{50}

21. i^{-10}

Focus On History

HERON OF ALEXANDRIA

Heron (also Hero) was a Greek mathematician and engineer. He lived and worked in Alexandria, Egypt, around 75 A.D. During his prolific work life, Heron developed a rotary steam engine called an aeolipile, a surveying tool called a dioptra, as well as a wind organ and a fire engine. As an engineer, he must have had the need to approximate square roots because he described an iterative method for doing so in his work *Metrica*. Heron's method for approximating a square root can be summarized as follows:

Suppose that x is not a perfect square and a^2 is the nearest perfect square to x. For a rough estimate of the value of $\sqrt{x}$, find the value of $y_1 = \frac{1}{2}\left(a + \frac{x}{a}\right)$. This estimate can be improved by calculating a second estimate using the first estimate y_1 in place of a: $y_2 = \frac{1}{2}\left(y_1 + \frac{x}{y_1}\right)$. Repeating this process several times will give more and more accurate estimates of $\sqrt{x}$.

CRITICAL THINKING

1. **a.** Which perfect square is closest to 80?
   b. Use Heron's method for approximating square roots to calculate the first estimate of the square root of 80.
   c. Use the first estimate of the square root of 80 to find a more refined second estimate.
   d. Use a calculator to find the actual value of the square root of 80. List all digits shown on your calculator's display.
   e. Compare the actual value from part (d) to the values of the first and second estimates. What do you notice?
   f. How many iterations of this process are necessary to get an estimate that differs no more than one digit from the actual value recorded in part (d)?

2. Repeat Question 1 for finding an estimate of the square root of 30.

3. Repeat Question 1 for finding an estimate of the square root of 4572.

4. Why would this iterative method have been important to people of Heron's era? Would you say that this method is as important today? Why or why not?

CHAPTER 9 HIGHLIGHTS

DEFINITIONS AND CONCEPTS	EXAMPLES

SECTION 9.1   RADICAL EXPRESSIONS

The **positive**, or **principal**, **square root** of a nonnegative number a is written as $\sqrt{a}$.

$$\sqrt{a} = b \text{ only if } b^2 = a \text{ and } b \geq 0$$

The **negative square root** of a is written as $-\sqrt{a}$.

The **cube root** of a real number a is written as $\sqrt[3]{a}$.

$$\sqrt[3]{a} = b \text{ only if } b^3 = a$$

If n is an even positive integer, then $\sqrt[n]{a^n} = |a|$.
If n is an odd positive integer, then $\sqrt[n]{a^n} = a$.

A **radical function** in x is a function defined by an expression containing a root of x.

—

$$\sqrt{36} = 6 \qquad \sqrt{\frac{9}{100}} = \frac{3}{10}$$

$$-\sqrt{36} = -6 \quad \sqrt{0.04} = 0.2$$

$$\sqrt[3]{27} = 3 \qquad \sqrt[3]{-\frac{1}{8}} = -\frac{1}{2}$$
$$\sqrt[3]{y^6} = y^2 \quad \sqrt[3]{64x^9} = 4x^3$$
$$\sqrt{(-3)^2} = |-3| = 3$$
$$\sqrt[3]{(-7)^3} = -7$$

If $f(x) = \sqrt{x} + 2$,
$$f(1) = \sqrt{1} + 2 = 1 + 2 = 3$$
$$f(3) = \sqrt{3} + 2 \approx 3.73$$

SECTION 9.2   RATIONAL EXPONENTS

$a^{1/n} = \sqrt[n]{a}$ if $\sqrt[n]{a}$ is a real number.

If m and n are positive integers greater than 1 with $\frac{m}{n}$ in lowest terms and $\sqrt[n]{a}$ is a real number, then

$$a^{m/n} = \left(a^{1/n}\right)^m = \left(\sqrt[n]{a}\right)^m$$

$a^{-m/n} = \dfrac{1}{a^{m/n}}$ as long as $a^{m/n}$ is a nonzero number.

Exponent rules are true for rational exponents.

—

$$81^{1/2} = \sqrt{81} = 9$$
$$(-8x^3)^{1/3} = \sqrt[3]{-8x^3} = -2x$$

$$4^{5/2} = \left(\sqrt{4}\right)^5 = 2^5 = 32$$
$$27^{2/3} = \left(\sqrt[3]{27}\right)^2 = 3^2 = 9$$

$$16^{-3/4} = \frac{1}{16^{3/4}} = \frac{1}{\left(\sqrt[4]{16}\right)^3} = \frac{1}{2^3} = \frac{1}{8}$$

$$x^{2/3} \cdot x^{-5/6} = x^{2/3 - 5/6} = x^{-1/6} = \frac{1}{x^{1/6}}$$

$$\left(8^{14}\right)^{1/7} = 8^2 = 64$$

$$\frac{a^{4/5}}{a^{-2/5}} = a^{4/5 - (-2/5)} = a^{6/5}$$

SECTION 9.3   SIMPLIFYING RADICAL EXPRESSIONS

PRODUCT AND QUOTIENT RULES

If $\sqrt[n]{a}$ and $\sqrt[n]{b}$ are real numbers,

$$\sqrt[n]{a} \cdot \sqrt[n]{b} = \sqrt[n]{a \cdot b}$$
$$\frac{\sqrt[n]{a}}{\sqrt[n]{b}} = \sqrt[n]{\frac{a}{b}}, \text{ provided } \sqrt[n]{b} \neq 0$$

A radical of the form $\sqrt[n]{a}$ is **simplified** when a contains no factors that are perfect nth powers.

—

Multiply or divide as indicated:
$$\sqrt{11} \cdot \sqrt{3} = \sqrt{33}$$

$$\frac{\sqrt[3]{40x}}{\sqrt[3]{5x}} = \sqrt[3]{8} = 2$$

$$\sqrt{40} = \sqrt{4 \cdot 10} = 2\sqrt{10}$$
$$\sqrt{36x^5} = \sqrt{36x^4 \cdot x} = 6x^2\sqrt{x}$$
$$\sqrt[3]{24x^7y^3} = \sqrt[3]{8x^6y^3 \cdot 3x} = 2x^2y\sqrt[3]{3x}$$

SECTION 9.4 ADDING, SUBTRACTING, AND MULTIPLYING RADICAL EXPRESSIONS

Radicals with the same index and the same radicand are **like radicals**.

The distributive property can be used to add like radicals.

Radical expressions are multiplied by using many of the same properties used to multiply polynomials.

$$5\sqrt{6} + 2\sqrt{6} = (5 + 2)\sqrt{6} = 7\sqrt{6}$$

$$-\sqrt[3]{3x} - 10\sqrt[3]{3x} + 3\sqrt[3]{10x}$$
$$= (-1 - 10)\sqrt[3]{3x} + 3\sqrt[3]{10x}$$
$$= -11\sqrt[3]{3x} + 3\sqrt[3]{10x}$$

Multiply:
$$\left(\sqrt{5} - \sqrt{2x}\right)\left(\sqrt{2} + \sqrt{2x}\right)$$
$$= \sqrt{10} + \sqrt{10x} - \sqrt{4x} - 2x$$
$$= \sqrt{10} + \sqrt{10x} - 2\sqrt{x} - 2x$$
$$\left(2\sqrt{3} - \sqrt{8x}\right)\left(2\sqrt{3} + \sqrt{8x}\right)$$
$$= 4(3) - 8x = 12 - 8x$$

SECTION 9.5 RATIONALIZING NUMERATORS AND DENOMINATORS OF RADICAL EXPRESSIONS

The **conjugate** of $a + b$ is $a - b$.

The process of writing the denominator of a radical expression without a radical is called **rationalizing the denominator**.

The conjugate of $\sqrt{7} + \sqrt{3}$ is $\sqrt{7} - \sqrt{3}$.

Rationalize each denominator:
$$\frac{\sqrt{5}}{\sqrt{3}} = \frac{\sqrt{5} \cdot \sqrt{3}}{\sqrt{3} \cdot \sqrt{3}} = \frac{\sqrt{15}}{3}$$
$$\frac{6}{\sqrt{7} + \sqrt{3}} = \frac{6(\sqrt{7} - \sqrt{3})}{(\sqrt{7} + \sqrt{3})(\sqrt{7} - \sqrt{3})}$$
$$= \frac{6(\sqrt{7} - \sqrt{3})}{7 - 3}$$
$$= \frac{6(\sqrt{7} - \sqrt{3})}{4} = \frac{3(\sqrt{7} - \sqrt{3})}{2}$$

The process of writing the numerator of a radical expression without a radical is called **rationalizing the numerator**.

Rationalize each numerator:
$$\frac{\sqrt[3]{9}}{\sqrt[3]{5}} = \frac{\sqrt[3]{9} \cdot \sqrt[3]{3}}{\sqrt[3]{5} \cdot \sqrt[3]{3}} = \frac{\sqrt[3]{27}}{\sqrt[3]{15}} = \frac{3}{\sqrt[3]{15}}$$
$$\frac{\sqrt{9} + \sqrt{3x}}{12} = \frac{(\sqrt{9} + \sqrt{3x})(\sqrt{9} - \sqrt{3x})}{12(\sqrt{9} - \sqrt{3x})}$$
$$= \frac{9 - 3x}{12(\sqrt{9} - \sqrt{3x})}$$
$$= \frac{3(3 - x)}{3 \cdot 4(3 - \sqrt{3x})} = \frac{3 - x}{4(3 - \sqrt{3x})}$$

SECTION 9.6 RADICAL EQUATIONS AND PROBLEM SOLVING

TO SOLVE A RADICAL EQUATION

Step 1. Write the equation so that one radical is by itself on one side of the equation.

Step 2. Raise each side of the equation to a power equal to the index of the radical and simplify.

Step 3. If the equation still contains a radical, repeat Steps 1 and 2. If not, solve the equation.

Step 4. Check all proposed solutions in the original equation.

Solve: $x = \sqrt{4x + 9} + 3$

1. $x - 3 = \sqrt{4x + 9}$

2. $(x - 3)^2 = (\sqrt{4x + 9})^2$
   $x^2 - 6x + 9 = 4x + 9$

3. $x^2 - 10x = 0$
   $x(x - 10) = 0$
   $x = 0$ or $x = 10$

4. The proposed solution 10 checks, but 0 does not. The solution is $\{10\}$.

SECTION 9.7   COMPLEX NUMBERS

A **complex number** is a number that can be written in the form $a + bi$, where a and b are real numbers.

$$i^2 = -1 \text{ and } i = \sqrt{-1}$$

Simplify: $\sqrt{-9}$

$$\sqrt{-9} = \sqrt{-1 \cdot 9} = \sqrt{-1} \cdot \sqrt{9} = i \cdot 3 \text{, or } 3i$$

Complex Numbers	Written in Form $a + bi$
12	$12 + 0i$
$-5i$	$0 + (-5)i$
$-2 - 3i$	$-2 + (-3)i$

Multiply.

$$\sqrt{-3} \cdot \sqrt{-7} = i\sqrt{3} \cdot i\sqrt{7}$$
$$= i^2\sqrt{21}$$
$$= -\sqrt{21}$$

To add or subtract complex numbers, add or subtract their real parts and then add or subtract their imaginary parts.

Perform each indicated operation.

$$(-3 + 2i) - (7 - 4i) = -3 + 2i - 7 + 4i$$
$$= -10 + 6i$$

To multiply complex numbers, multiply as though they are binomials.

$$(-7 - 2i)(6 + i) = -42 - 7i - 12i - 2i^2$$
$$= -42 - 19i - 2(-1)$$
$$= -42 - 19i + 2$$
$$= -40 - 19i$$

The complex numbers $(a + bi)$ and $(a - bi)$ are called **complex conjugates**.

The complex conjugate of

$$(3 + 6i) \text{ is } (3 - 6i).$$

Their product is a real number:

$$(3 - 6i)(3 + 6i) = 9 - 36i^2$$
$$= 9 - 36(-1) = 9 + 36 = 45$$

To divide complex numbers, multiply the numerator and the denominator by the conjugate of the denominator.

Divide: $\dfrac{4}{2 - i} = \dfrac{4(2 + i)}{(2 - i)(2 + i)}$

$$= \frac{4(2 + i)}{4 - i^2}$$
$$= \frac{4(2 + i)}{5}$$
$$= \frac{8 + 4i}{5} = \frac{8}{5} + \frac{4}{5}i$$

Quadratic Equations and Functions

An important part of algebra is learning to model and solve problems. Often, the model of a problem is a quadratic equation or a function containing a second-degree polynomial. In this chapter, we continue the work from Chapter 5, solving quadratic equations in one variable by factoring. Two other methods of solving quadratic equations are analyzed in this chapter, with methods of solving nonlinear inequalities in one variable and the graphs of quadratic functions.

The surface of the Earth is heated by the sun and then slowly radiated into outer space. Sometimes, certain gases in the atmosphere reflect the heat radiation back to Earth, preventing it from escaping. The gradual warming of the atmosphere is known as the greenhouse effect. This effect is compounded by the increase of certain gases (greenhouse gases) such as carbon dioxide, methane, and nitrous oxide. Although these gases occur naturally and are needed to keep the surface of the Earth at a temperature that is hospitable to life, the recent buildup of these gases is due primarily to human activities. According to the Natural Resources Defense Council, carbon dioxide concentrations have increased by 30% globally over the past century.

Problem Solving Notes

10.1 SOLVING QUADRATIC EQUATIONS BY COMPLETING THE SQUARE

A USING THE SQUARE ROOT PROPERTY

In Chapter 5, we solved quadratic equations by factoring. Recall that a **quadratic**, or **second-degree**, **equation** is an equation that can be written in the form $ax^2 + bx + c = 0$, where a, b, and c are real numbers and a is not 0. To solve a quadratic equation such as $x^2 = 9$ by factoring, we use the zero-factor theorem. To use the zero-factor theorem, the equation must first be written in standard form, $ax^2 + bx + c = 0$.

$$x^2 = 9$$
$$x^2 - 9 = 0 \quad \text{Subtract 9 from both sides to write in standard form.}$$
$$(x + 3)(x - 3) = 0 \quad \text{Factor.}$$
$$x + 3 = 0 \text{ or } x - 3 = 0 \quad \text{Set each factor equal to 0.}$$
$$x = -3 \qquad x = 3 \quad \text{Solve.}$$

The solution set is $\{-3, 3\}$, the positive and negative square roots of 9.

Not all quadratic equations can be solved by factoring, so we need to explore other methods. Notice that the solutions of the equation $x^2 = 9$ are two numbers whose square is 9:

$$3^2 = 9 \quad \text{and} \quad (-3)^2 = 9$$

Thus, we can solve the equation $x^2 = 9$ by taking the square root of both sides. Be sure to include both $\sqrt{9}$ and $-\sqrt{9}$ as solutions since both $\sqrt{9}$ and $-\sqrt{9}$ are numbers whose square is 9.

$$x^2 = 9$$
$$x = \pm\sqrt{9} \quad \text{The notation } \pm\sqrt{9} \text{ (read as plus or minus } \sqrt{9}\text{) indicates the pair of}$$
$$\qquad \qquad \text{numbers } +\sqrt{9} \text{ and } -\sqrt{9}.$$
$$x = \pm 3$$

This illustrates the square root property.

HELPFUL HINT

The notation ± 3, for example, is read as "plus or minus 3." It is a shorthand notation for the pair of numbers $+3$ and -3.

SQUARE ROOT PROPERTY

If b is a real number and if $a^2 = b$, then $a = \pm\sqrt{b}$.

Example 1   Use the square root property to solve $x^2 = 50$.

Solution: $$x^2 = 50$$
$$x = \pm\sqrt{50} \quad \text{Use the square root property.}$$
$$x = \pm 5\sqrt{2} \quad \text{Simplify the radical.}$$

Objectives

A Use the square root property to solve quadratic equations.

B Write perfect square trinomials.

C Solve quadratic equations by completing the square.

D Use quadratic equations to solve problems.

SSM   CD-ROM   Video 10.1

Practice Problem 1

Use the square root property to solve $x^2 = 20$.

Answer

1. $\{2\sqrt{5}, -2\sqrt{5}\}$

Check:

Let $x = 5\sqrt{2}$.	Let $x = -5\sqrt{2}$.
$x^2 = 50$	$x^2 = 50$
$(5\sqrt{2})^2 \stackrel{?}{=} 50$	$(-5\sqrt{2})^2 \stackrel{?}{=} 50$
$25 \cdot 2 \stackrel{?}{=} 50$	$25 \cdot 2 \stackrel{?}{=} 50$
$50 = 50$ True.	$50 = 50$ True.

The solution set is $\{5\sqrt{2}, -5\sqrt{2}\}$.

Practice Problem 2

Use the square root property to solve $5x^2 = 55$.

Example 2 Use the square root property to solve $2x^2 = 14$.

Solution: First we get the squared variable alone on one side of the equation.

$2x^2 = 14$

$x^2 = 7$ Divide both sides by 2.

$x = \pm\sqrt{7}$ Use the square root property.

Check:

Let $x = \sqrt{7}$.	Let $x = -\sqrt{7}$.
$2x^2 = 14$	$2x^2 = 14$
$2(\sqrt{7})^2 \stackrel{?}{=} 14$	$2(-\sqrt{7})^2 \stackrel{?}{=} 14$
$2 \cdot 7 \stackrel{?}{=} 14$	$2 \cdot 7 \stackrel{?}{=} 14$
$14 = 14$ True.	$14 = 14$ True.

The solution set is $\{\sqrt{7}, -\sqrt{7}\}$.

Practice Problem 3

Use the square root property to solve $(x + 2)^2 = 18$.

Example 3 Use the square root property to solve $(x + 1)^2 = 12$.

Solution:

$(x + 1)^2 = 12$

$x + 1 = \pm\sqrt{12}$ Use the square root property.

$x + 1 = \pm 2\sqrt{3}$ Simplify the radical.

$x = -1 \pm 2\sqrt{3}$ Subtract 1 from both sides.

Check: Below is a check for $-1 + 2\sqrt{3}$. The check for $-1 - 2\sqrt{3}$ is almost the same and is left for you to do on your own.

$(x + 1)^2 = 12$

$(-1 + 2\sqrt{3} + 1)^2 \stackrel{?}{=} 12$

$(2\sqrt{3})^2 \stackrel{?}{=} 12$

$4 \cdot 3 \stackrel{?}{=} 12$

$12 = 12$ True.

The solution set is $\{-1 + 2\sqrt{3}, -1 - 2\sqrt{3}\}$.

Practice Problem 4

Use the square root property to solve $(3x - 1)^2 = -4$.

Example 4 Use the square root property to solve $(2x - 5)^2 = -16$.

Solution:

$(2x - 5)^2 = -16$

$2x - 5 = \pm\sqrt{-16}$ Use the square root property.

$2x - 5 = \pm 4i$ Simplify the radical.

$2x = 5 \pm 4i$ Add 5 to both sides.

$x = \dfrac{5 \pm 4i}{2}$ Divide both sides by 2.

Check each proposed solution in the original equation to see that the solution set is $\left\{\dfrac{5 + 4i}{2}, \dfrac{5 - 4i}{2}\right\}$.

✓ CONCEPT CHECK

How do you know just by looking that $(x - 2)^2 = -4$ has complex solutions?

Answers

2. $\{\sqrt{11}, -\sqrt{11}\}$,

3. $\{-2 + 3\sqrt{2}, -2 - 3\sqrt{2}\}$,

4. $\left\{\dfrac{1 - 2i}{3}, \dfrac{1 + 2i}{3}\right\}$

✓ Concept Check: answers may vary

TRY THE CONCEPT CHECK IN THE MARGIN.

B WRITING PERFECT SQUARE TRINOMIALS

Notice from Examples 3 and 4 that, if we write a quadratic equation so that one side is the square of a binomial, we can solve by using the square root property. To write the square of a binomial, we write perfect square trinomials. Recall that a perfect square trinomial is a trinomial that can be factored into two identical binomial factors, that is, as a binomial squared.

Perfect Square Trinomials	Factored Form
$x^2 + 8x + 16$	$(x + 4)^2$
$x^2 - 6x + 9$	$(x - 3)^2$
$x^2 + 3x + \dfrac{9}{4}$	$\left(x + \dfrac{3}{2}\right)^2$

Notice that for each perfect square trinomial, *the constant term of the trinomial is the square of half the coefficient of the x-term.* For example,

$$x^2 + 8x + 16 \qquad\qquad x^2 - 6x + 9$$

$$\frac{1}{2}(8) = 4 \text{ and } 4^2 = 16 \qquad \frac{1}{2}(-6) = -3 \text{ and } (-3)^2 = 9$$

Example 5   Add the proper constant to $x^2 + 6x$ so that the result is a perfect square trinomial. Then factor.

Solution: We add the square of half the coefficient of x.

$$x^2 + 6x + 9 \qquad = (x + 3)^2 \quad \text{In factored form}$$

$$\frac{1}{2}(6) = 3 \text{ and } 3^2 = 9$$

_____

Example 6   Add the proper constant to $x^2 - 3x$ so that the result is a perfect square trinomial. Then factor.

Solution: We add the square of half the coefficient of x.

$$x^2 - 3x + \frac{9}{4} \qquad = \left(x - \frac{3}{2}\right)^2 \quad \text{In factored form}$$

$$\frac{1}{2}(-3) = -\frac{3}{2} \text{ and } \left(-\frac{3}{2}\right)^2 = \frac{9}{4}$$

_____

C SOLVING BY COMPLETING THE SQUARE

The process of writing a quadratic equation so that one side is a perfect square trinomial is called **completing the square**. We will use this process in the next examples.

Example 7   Solve $p^2 + 2p = 4$ by completing the square.

Solution: First we add the square of half the coefficient of p to both sides so that the resulting trinomial will be a perfect square trinomial. The coefficient of p is 2.

$$\frac{1}{2}(2) = 1 \qquad \text{and} \qquad 1^2 = 1$$

Practice Problem 5

Add the proper constant to $x^2 + 12x$ so that the result is a perfect square trinomial. Then factor.

Practice Problem 6

Add the proper constant to $y^2 - 5y$ so that the result is a perfect square trinomial. Then factor.

Practice Problem 7

Solve $x^2 + 8x = 1$ by completing the square.

Answers

5. $x^2 + 12x + 36 = (x + 6)^2$,

6. $y^2 - 5y + \dfrac{25}{4} = \left(x - \dfrac{5}{2}\right)^2$,

7. $\{-4 - \sqrt{17}, -4 + \sqrt{17}\}$

Now we add 1 to both sides of the original equation.

$$p^2 + 2p = 4$$

$$p^2 + 2p + 1 = 4 + 1 \qquad \text{Add 1 to both sides.}$$

$$(p + 1)^2 = 5 \qquad \text{Factor the trinomial; simplify the right side.}$$

We may now use the square root property and solve for p.

$$p + 1 = \pm\sqrt{5} \qquad \text{Use the square root property.}$$

$$p = -1 \pm \sqrt{5} \qquad \text{Subtract 1 from both sides.}$$

Notice that there are two solutions: $-1 + \sqrt{5}$ and $-1 - \sqrt{5}$. The solution set is $\{-1 + \sqrt{5}, -1 - \sqrt{5}\}$.

Practice Problem 8

Solve $y^2 - 5y + 2 = 0$ by completing the square.

Example 8 Solve $m^2 - 7m - 1 = 0$ by completing the square.

Solution: First we add 1 to both sides of the equation so that the left side has no constant term. We can then add the constant term on both sides that will make the left side a perfect square trinomial.

$$m^2 - 7m - 1 = 0$$

$$m^2 - 7m = 1$$

Now we find the constant term that makes the left side a perfect square trinomial by squaring half the coefficient of m. We add this constant to both sides of the equation.

$$\frac{1}{2}(-7) = -\frac{7}{2} \qquad \text{and} \qquad \left(-\frac{7}{2}\right)^2 = \frac{49}{4}$$

$$m^2 - 7m + \frac{49}{4} = 1 + \frac{49}{4} \qquad \text{Add } \frac{49}{4} \text{ to both sides of the equation.}$$

$$\left(m - \frac{7}{2}\right)^2 = \frac{53}{4} \qquad \text{Factor the perfect square trinomial and simplify the right side.}$$

$$m - \frac{7}{2} = \pm\sqrt{\frac{53}{4}} \qquad \text{Use the square root property.}$$

$$m = \frac{7}{2} \pm \frac{\sqrt{53}}{2} \qquad \text{Add } \frac{7}{2} \text{ to both sides and simplify } \sqrt{\frac{53}{4}}.$$

$$m = \frac{7 \pm \sqrt{53}}{2} \qquad \text{Simplify.}$$

The solution set is $\left\{\dfrac{7 + \sqrt{53}}{2}, \dfrac{7 - \sqrt{53}}{2}\right\}$.

The following steps may be used to solve a quadratic equation such as $ax^2 + bx + c = 0$ by completing the square. This method may be used whether or not the polynomial $ax^2 + bx + c$ is factorable.

Answer

8. $\left\{\dfrac{5 - \sqrt{17}}{2}, \dfrac{5 + \sqrt{17}}{2}\right\}$

SOLVING A QUADRATIC EQUATION IN x BY COMPLETING THE SQUARE

Step 1. If the coefficient of x^2 is 1, go to Step 2. Otherwide, divide both sides of the equation by the coefficient of x^2.

Step 2. Get all variable terms alone on one side of the equation.

Step 3. Complete the square for the resulting binomial by adding the square of half of the coefficient of x to both sides of the equation.

Step 4. Factor the resulting perfect square trinomial and write it as the square of a binomial.

Step 5. Use the square root property to solve for x.

Example 9   Solve $4x^2 - 24x + 41 = 0$ by completing the square.

Solution: First we divide both sides of the equation by 4 so that the coefficient of x^2 is 1.

$$4x^2 - 24x + 41 = 0$$

Step 1. $x^2 - 6x + \dfrac{41}{4} = 0$ Divide both sides of the equation by 4.

Step 2. $x^2 - 6x = -\dfrac{41}{4}$ Subtract $\frac{41}{4}$ from both sides.

Since $\dfrac{1}{2}(-6) = -3$ and $(-3)^2 = 9$, we add 9 to both sides of the equation.

Step 3. $x^2 - 6x + 9 = -\dfrac{41}{4} + 9$ Add 9 to both sides.

Step 4. $(x - 3)^2 = -\dfrac{41}{4} + \dfrac{36}{4}$ Factor the perfect square trinomial.

$(x - 3)^2 = -\dfrac{5}{4}$

Step 5. $x - 3 = \pm\sqrt{-\dfrac{5}{4}}$ Use the square root property.

$x - 3 = \pm\dfrac{i\sqrt{5}}{2}$ Simplify the radical.

$x = 3 \pm \dfrac{i\sqrt{5}}{2}$ Add 3 to both sides.

$= \dfrac{6}{2} \pm \dfrac{i\sqrt{5}}{2}$ Find a common denominator.

$= \dfrac{6 \pm i\sqrt{5}}{2}$ Simplify.

The solution set is $\left\{\dfrac{6 + i\sqrt{5}}{2}, \dfrac{6 - i\sqrt{5}}{2}\right\}$.

Practice Problem 9

Solve $2x^2 - 2x + 7 = 0$ by completing the square.

Answer

9. $\left\{\dfrac{1 + i\sqrt{13}}{2}, \dfrac{1 - i\sqrt{13}}{2}\right\}$

D SOLVING PROBLEMS MODELED BY QUADRATIC EQUATIONS

Recall the **simple interest** formula $I = Prt$, where I is the interest earned, P is the principal, r is the rate of interest, and t is time. If \$100 is invested at a simple interest rate of 5% annually, at the end of 3 years the total interest I earned is

$$I = P \cdot r \cdot t$$

or

$$I = 100 \cdot 0.05 \cdot 3 = \$15$$

and the new principal is

$$\$100 + \$15 = \$115$$

Most of the time, the interest computed on money borrowed or money deposited is **compound interest**. Compound interest, unlike simple interest, is computed on original principal *and* on interest already earned. To see the difference between simple interest and compound interest, suppose that \$100 is invested at a rate of 5% compounded annually. To find the total amount of money at the end of 3 years, we calculate as follows:

$$I \quad = \quad P \cdot \quad r \cdot t$$

First year:	Interest = \$100 · 0.05 · 1 = \$5.00
	New principal = \$100.00 + \$5.00 = \$105.00
Second year:	Interest = \$105.00 · 0.05 · 1 = \$5.25
	New principal = \$105.00 + \$5.25 = \$110.25
Third year:	Interest = \$110.25 · 0.05 · 1 ≈ \$5.51
	New principal = \$110.25 + \$5.51 = \$115.76

At the end of the third year, the total compound interest earned is \$15.76, whereas the total simple interest earned is \$15.

It is tedious to calculate compound interest as we did above, so we use a compound interest formula. The formula for calculating the total amount of money when interest is compounded annually is

$$A = P(1 + r)^t$$

where P is the original investment, r is the interest rate per compounding period, and t is the number of periods. For example, the amount of money A at the end of 3 years if \$100 is invested at 5% compounded annually is

$$A = \$100(1 + 0.05)^3 \approx \$100(1.1576) = \$115.76$$

as we previously calculated.

Practice Problem 10

Use the formula from Example 10 to find the interest rate r if \$1600 compounded annually grows to \$1764 in 2 years.

Example 10   Finding Interest Rates

Find the interest rate r if \$2000 compounded annually grows to \$2420 in 2 years.

Solution: **1.** UNDERSTAND the problem. For this example, make sure that you understand the formula for compounding interest annually.

2. TRANSLATE. We substitute the given values into the formula:

$$A = P(1 + r)^t$$

$$2420 = 2000(1 + r)^2 \quad \text{Let } A = 2420, P = 2000, \text{ and } t = 2.$$

Answer

10. 5%

3. SOLVE. Solve the equation for r.

$$2420 = 2000(1 + r)^2$$

$$\frac{2420}{2000} = (1 + r)^2 \qquad \text{Divide both sides by 2000.}$$

$$\frac{121}{100} = (1 + r)^2 \qquad \text{Simplify the fraction.}$$

$$\pm\sqrt{\frac{121}{100}} = 1 + r \qquad \text{Use the square root property.}$$

$$\pm\frac{11}{10} = 1 + r \qquad \text{Simplify.}$$

$$-1 \pm \frac{11}{10} = r$$

$$-\frac{10}{10} \pm \frac{11}{10} = r$$

$$\frac{1}{10} = r \ \text{ or } \ -\frac{21}{10} = r$$

4. INTERPRET. The rate cannot be negative, so we reject $-\frac{21}{10}$.

Check: $\frac{1}{10} = 0.10 = 10\%$ per year. If we invest \$2000 at 10% compounded annually, in 2 years the amount in the account would be $2000(1 + 0.10)^2 = 2420$ dollars, the desired amount.

State: The interest rate is 10% compounded annually. ▬▬▬

GRAPHING CALCULATOR EXPLORATIONS

In Section 7.5, we showed how we can use a grapher to approximate real number solutions of a quadratic equation written in standard form. We can also use a grapher to solve a quadratic equation when it is not written in standard form. For example, to solve $(x + 1)^2 = 12$, the quadratic equation in Example 3, we graph the following on the same set of axes. We use $\text{Xmin} = -10$, $\text{Xmax} = 10$, $\text{Ymin} = -13$, and $\text{Ymax} = 13$.

$$Y_1 = (x + 1)^2 \text{ and } Y_2 = 12$$

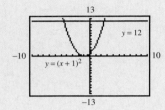

Use the Intersect feature or the Zoom and Trace features to locate the points of intersection of the graphs. The x-values of these points are the solutions of $(x + 1)^2 = 12$. The solutions, rounded to two decimal points, are 2.46 and -4.46.

Check to see that these numbers are approximations of the exact solutions $-1 \pm 2\sqrt{3}$.

Use a grapher to solve each quadratic equation. Round all solutions to the nearest hundredth.

1. $x(x - 5) = 8$ **2.** $x(x + 2) = 5$

3. $x^2 + 0.5x = 0.3x + 1$ **4.** $x^2 - 2.6x = -2.2x + 3$

5. Use a grapher and solve $(2x - 5)^2 = -16$, Example 4 in this section, using the window

$\text{Xmin} = -20$
$\text{Xmax} = 20$
$\text{Xscl} = 1$
$\text{Ymin} = -20$
$\text{Ymax} = 20$
$\text{Yscl} = 1$

Explain the results. Compare your results with the solution found in Example 4.

6. What are the advantages and disadvantages of using a grapher to solve quadratic equations?

EXERCISE SET 10.1

A *Use the square root property to solve each equation. See Examples 1 through 4.*

1. $x^2 = 16$

2. $x^2 = 49$

3. $x^2 - 7 = 0$

4. $x^2 - 11 = 0$

5. $x^2 = 18$

6. $y^2 = 20$

7. $3z^2 - 30 = 0$

8. $2x^2 = 4$

9. $(x + 5)^2 = 9$

10. $(y - 3)^2 = 4$

11. $(z - 6)^2 = 18$

12. $(y + 4)^2 = 27$

13. $(2x - 3)^2 = 8$

14. $(4x + 9)^2 = 6$

15. $x^2 + 9 = 0$

16. $x^2 + 4 = 0$

17. $x^2 - 6 = 0$

18. $y^2 - 10 = 0$

19. $2z^2 + 16 = 0$

20. $3p^2 + 36 = 0$

21. $(x - 1)^2 = -16$

22. $(y + 2)^2 = -25$

23. $(z + 7)^2 = 5$

24. $(x + 10)^2 = 11$

25. $(x + 3)^2 = -8$

1. _____
2. _____
3. _____
4. _____
5. _____
6. _____
7. _____
8. _____
9. _____
10. _____
11. _____
12. _____
13. _____
14. _____
15. _____
16. _____
17. _____
18. _____
19. _____
20. _____
21. _____
22. _____
23. _____
24. _____
25. _____

Focus On Business and Career

FLOWCHARTS

We saw in the Focus On Business and Career feature in Chapter 6 that three of the top 10 fastest-growing jobs into the 21st century are computer related. A useful skill in computer-related careers is *flowcharting*. A **flowchart** is a diagram showing a sequence of procedures used to complete a task. Flowcharts are commonly used in computer programming to help a programmer plan the steps and commands needed to write a program. Flowcharts are also used in other types of careers, such as manufacturing or finance, to describe the sequence of events needed in a certain process.

A flowchart usually uses the following symbols to represent certain types of actions.

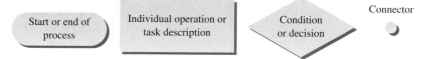

For instance, suppose we want to write a flowchart for the process of computing a household's monthly electric bill. The flowchart might look something like the one here.

CRITICAL THINKING

1. Using the given flowchart as a guide, describe in words this utility company's pricing structure for household electricity usage.

2. Make a flowchart for the process of computing the human-equivalent age for the age of a dog if 1 dog year is equivalent to 7 human years.

3. Make a flowchart for the process of determining the number and type of solutions of a quadratic equation of the form $ax^2 + bx + c = 0$ using the discriminant.

4. (Optional) Write a programmable calculator program for your graphing calculator that determines the number and type of solutions of a quadratic equation of the form $ax^2 + bx + c = 0$.

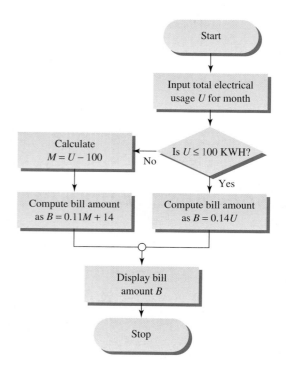

644

10.2 SOLVING QUADRATIC EQUATIONS BY THE QUADRATIC FORMULA

A SOLVING EQUATIONS BY USING THE QUADRATIC FORMULA

Any quadratic equation can be solved by completing the square. Since the same sequence of steps is repeated each time we complete the square, let's complete the square for a general quadratic equation, $ax^2 + bx + c = 0$. By doing so, we find a pattern for the solutions of a quadratic equation known as the **quadratic formula**.

Recall that to complete the square for an equation such as $ax^2 + bx + c = 0$, $a \neq 0$, we first divide both sides by the coefficient of x^2.

$$ax^2 + bx + c = 0$$

$$x^2 + \frac{b}{a}x + \frac{c}{a} = 0 \qquad \text{Divide both sides by } a, \text{ the coefficient of } x^2.$$

$$x^2 + \frac{b}{a}x = -\frac{c}{a} \qquad \text{Subtract the constant } \frac{c}{a} \text{ from both sides.}$$

Next we find the square of half $\frac{b}{a}$, the coefficient of x.

$$\frac{1}{2}\left(\frac{b}{a}\right) = \frac{b}{2a} \qquad \text{and} \qquad \left(\frac{b}{2a}\right)^2 = \frac{b^2}{4a^2}$$

Now we add this result to both sides of the equation.

$$x^2 + \frac{b}{a}x + \frac{b^2}{4a^2} = -\frac{c}{a} + \frac{b^2}{4a^2} \qquad \text{Add } \frac{b^2}{4a^2} \text{ to both sides.}$$

$$x^2 + \frac{b}{a}x + \frac{b^2}{4a^2} = \frac{-c \cdot 4a}{a \cdot 4a} + \frac{b^2}{4a^2} \qquad \text{Find a common denominator on the right side.}$$

$$x^2 + \frac{b}{a}x + \frac{b^2}{4a^2} = \frac{b^2 - 4ac}{4a^2} \qquad \text{Simplify the right side.}$$

$$\left(x + \frac{b}{2a}\right)^2 = \frac{b^2 - 4ac}{4a^2} \qquad \text{Factor the perfect square trinomial on the left side.}$$

$$x + \frac{b}{2a} = \pm\sqrt{\frac{b^2 - 4ac}{4a^2}} \qquad \text{Use the square root property.}$$

$$x + \frac{b}{2a} = \pm\frac{\sqrt{b^2 - 4ac}}{2a} \qquad \text{Simplify the radical.}$$

$$x = -\frac{b}{2a} \pm \frac{\sqrt{b^2 - 4ac}}{2a} \qquad \text{Subtract } \frac{b}{2a} \text{ from both sides.}$$

$$x = \frac{-b \pm \sqrt{b^2 - 4ac}}{2a} \qquad \text{Simplify.}$$

The resulting equation identifies the solutions of the general quadratic equation in standard form and is called the quadratic formula. It can be used to solve any equation written in standard form $ax^2 + bx + c = 0$ as long as a is not 0.

QUADRATIC FORMULA

A quadratic equation written in the form $ax^2 + bx + c = 0$ has the solutions

$$x = \frac{-b \pm \sqrt{b^2 - 4ac}}{2a}$$

Practice Problem 1

Solve: $2x^2 + 9x + 10 = 0$

Example 1

Solve: $3x^2 + 16x + 5 = 0$

Solution: This equation is in standard form with $a = 3$, $b = 16$, and $c = 5$. We substitute these values into the quadratic formula.

$$x = \frac{-b \pm \sqrt{b^2 - 4ac}}{2a} \qquad \text{Quadratic formula}$$

$$= \frac{-16 \pm \sqrt{16^2 - 4(3)(5)}}{2(3)} \qquad \text{Let } a = 3, b = 16, \text{ and } c = 5.$$

$$= \frac{-16 \pm \sqrt{256 - 60}}{6}$$

$$= \frac{-16 \pm \sqrt{196}}{6} = \frac{-16 \pm 14}{6}$$

$$x = \frac{-16 + 14}{6} = -\frac{1}{3} \quad \text{or} \quad x = \frac{-16 - 14}{6} = -\frac{30}{6} = -5$$

The solution set is $\left\{-\frac{1}{3}, -5\right\}$.

> **HELPFUL HINT**
>
> To replace a, b, and c correctly in the quadratic formula, write the quadratic equation in standard form $ax^2 + bx + c = 0$.

Practice Problem 2

Solve: $2x^2 - 6x - 1 = 0$

Example 2

Solve: $2x^2 - 4x = 3$

Solution: First we write the equation in standard form by subtracting 3 from both sides.

$$2x^2 - 4x - 3 = 0$$

Now $a = 2$, $b = -4$, and $c = -3$. We substitute these values into the quadratic formula.

$$x = \frac{-b \pm \sqrt{b^2 - 4ac}}{2a}$$

$$= \frac{-(-4) \pm \sqrt{(-4)^2 - 4(2)(-3)}}{2(2)}$$

$$= \frac{4 \pm \sqrt{16 + 24}}{4}$$

$$= \frac{4 \pm \sqrt{40}}{4} = \frac{4 \pm 2\sqrt{10}}{4}$$

$$= \frac{2(2 \pm \sqrt{10})}{2 \cdot 2} = \frac{2 \pm \sqrt{10}}{2}$$

The solution set is $\left\{\dfrac{2 + \sqrt{10}}{2}, \dfrac{2 - \sqrt{10}}{2}\right\}$.

TRY THE CONCEPT CHECK IN THE MARGIN.

✓ CONCEPT CHECK

For the quadratic equation $x^2 = 7$, which substitution is correct?
a. $a = 1$, $b = 0$, and $c = -7$
b. $a = 1$, $b = 0$, and $c = 7$
c. $a = 0$, $b = 0$, and $c = 7$
d. $a = 1$, $b = 1$, and $c = -7$

Answers

1. $\left\{-\frac{5}{2}, -2\right\}$, **2.** $\left\{\dfrac{3 + \sqrt{11}}{2}, \dfrac{3 - \sqrt{11}}{2}\right\}$

✓ Concept Check: a

Example 3   Solve: $\dfrac{1}{4}m^2 - m + \dfrac{1}{2} = 0$

Solution: We could use the quadratic formula with $a = \dfrac{1}{4}$, $b = -1$,

and $c = \dfrac{1}{2}$. Instead, let's find a simpler, equivalent standard form equation whose coefficients are not fractions.

First we multiply both sides of the equation by 4 to clear the fractions.

$$4\left(\frac{1}{4}m^2 - m + \frac{1}{2}\right) = 4 \cdot 0$$

$$m^2 - 4m + 2 = 0 \qquad \text{Simplify.}$$

Now we can substitute $a = 1$, $b = -4$, and $c = 2$ into the quadratic formula and simplify.

$$m = \frac{-(-4) \pm \sqrt{(-4)^2 - 4(1)(2)}}{2(1)}$$

$$= \frac{4 \pm \sqrt{16 - 8}}{2}$$

$$= \frac{4 \pm \sqrt{8}}{2} = \frac{4 \pm 2\sqrt{2}}{2} = \frac{2(2 \pm \sqrt{2})}{2} = 2 \pm \sqrt{2}$$

The solution set is $\{2 + \sqrt{2}, 2 - \sqrt{2}\}$.   ▬▬

Example 4   Solve: $p = -3p^2 - 3$

Solution: The equation in standard form is $3p^2 + p + 3 = 0$. Thus, $a = 3$, $b = 1$, and $c = 3$ in the quadratic formula.

$$p = \frac{-1 \pm \sqrt{1^2 - 4(3)(3)}}{2(3)} = \frac{-1 \pm \sqrt{1 - 36}}{6}$$

$$= \frac{-1 \pm \sqrt{-35}}{6} = \frac{-1 \pm i\sqrt{35}}{6}$$

The solution set is $\left\{\dfrac{-1 + i\sqrt{35}}{6}, \dfrac{-1 - i\sqrt{35}}{6}\right\}$.   ▬▬

TRY THE CONCEPT CHECK IN THE MARGIN.

Practice Problem 3

Solve: $\dfrac{1}{6}x^2 - \dfrac{1}{2}x - 1 = 0$

Practice Problem 4

Solve: $x = -4x^2 - 4$

✓ CONCEPT CHECK

What is the first step in solving $-3x^2 = 5x - 4$ using the quadratic formula?

Answers

3. $\left\{\dfrac{3 + \sqrt{33}}{2}, \dfrac{3 - \sqrt{33}}{2}\right\}$,

4. $\left\{\dfrac{-1 - 3i\sqrt{7}}{8}, \dfrac{-1 + 3i\sqrt{7}}{8}\right\}$

✓ Concept Check   answers may vary

B USING THE DISCRIMINANT

In the quadratic formula $x = \dfrac{-b \pm \sqrt{b^2 - 4ac}}{2a}$, the radicand $b^2 - 4ac$ is called the **discriminant** because when we know its value, we can **discriminate** among the possible number and type of solutions of a quadratic equation. Possible values of the discriminant and their meanings are summarized next.

DISCRIMINANT

The following table corresponds the discriminant $b^2 - 4ac$ of a quadratic equation of the form $ax^2 + bx + c = 0$ with the number and type of solutions of the equation.

$b^2 - 4ac$	Number and Type of Solutions
Positive	Two real solutions
Zero	One real solution
Negative	Two complex but not real solutions

Practice Problem 5

Use the discriminant to determine the number and type of solutions of $x^2 + 4x + 4 = 0$.

Example 5 Use the discriminant to determine the number and type of solutions of $x^2 + 2x + 1$.

Solution: In $x^2 + 2x + 1 = 0$, $a = 1$, $b = 2$, and $c = 1$. Thus,

$$b^2 - 4ac = 2^2 - 4(1)(1) = 0$$

Since $b^2 - 4ac = 0$, this quadratic equation has one real solution.

Practice Problem 6

Use the discriminant to determine the number and type of solutions of $5x^2 + 7 = 0$.

Example 6 Use the discriminant to determine the number and type of solutions of $3x^2 + 2 = 0$.

Solution: In this equation, $a = 3$, $b = 0$, and $c = 2$. Then $b^2 - 4ac = 0^2 - 4(3)(2) = -24$. Since $b^2 - 4ac$ is negative, this quadratic equation has two complex but not real solutions.

Practice Problem 7

Use the discriminant to determine the number and type of solutions of $3x^2 - 2x - 2 = 0$.

Example 7 Use the discriminant to determine the number and type of solutions of $2x^2 - 7x - 4 = 0$.

Solution: In this equation, $a = 2$, $b = -7$, and $c = -4$. Then

$$b^2 - 4ac = (-7)^2 - 4(2)(-4) = 81$$

Since $b^2 - 4ac$ is positive, this quadratic equation has two real solutions.

C SOLVING PROBLEMS MODELED BY QUADRATIC EQUATIONS

The quadratic formula is useful in solving problems that are modeled by quadratic equations.

Answers

5. 1 real solution, **6.** 2 complex but not real solutions, **7.** 2 real solutions

Example 8   Calculating Distance Saved

At a local university, students often leave the sidewalk and cut across the lawn to save walking distance. Given the diagram below of a favorite place to cut across the lawn, approximate how many feet of walking distance a student saves by cutting across the lawn instead of walking on the sidewalk.

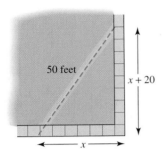

Practice Problem 8

Given the diagram below, approximate to the nearest foot how many feet of walking distance a person saves by cutting across the lawn instead of walking on the sidewalk.

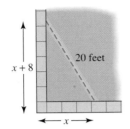

Solution: **1.** UNDERSTAND. Read and reread the problem. You may want to review the Pythagorean theorem.

2. TRANSLATE. By the Pythagorean theorem, we have

In words: $(\text{leg})^2 + (\text{leg})^2 = (\text{hypotenuse})^2$

Translate: $x^2 + (x + 20)^2 = 50^2$

3. SOLVE. Use the quadratic formula to solve.

$x^2 + x^2 + 40x + 400 = 2500$ Square $(x + 20)$ and 50.

$2x^2 + 40x - 2100 = 0$ Set the equation equal to 0.

$x^2 + 20x - 1050 = 0$ Divide by 2.

Here, $a = 1, b = 20, c = -1050$. By the quadratic formula,

$$x = \frac{-20 \pm \sqrt{20^2 - 4(1)(-1050)}}{2 \cdot 1}$$

$$= \frac{-20 \pm \sqrt{400 + 4200}}{2} = \frac{-20 \pm \sqrt{4600}}{2}$$

$$= \frac{-20 \pm \sqrt{100 \cdot 46}}{2} = \frac{-20 \pm 10\sqrt{46}}{2}$$

$$= -10 \pm 5\sqrt{46}$$ Simplify.

Check: **4.** INTERPRET. Check your calculations in the quadratic formula. The length of a side of a triangle can't be negative, so we reject $-10 - 5\sqrt{46}$. Since $-10 + 5\sqrt{46} \approx 24$ feet, the walking distance along the sidewalk is

$$x + (x + 20) \approx 24 + (24 + 20) = 68 \text{ feet.}$$

State: A person saves $68 - 50$ or 18 feet of walking distance by cutting across the lawn.

Answer

8. 9 feet

Practice Problem 9

Use the equation given in Example 9 to find how long after the object is thrown it will be 100 feet from the ground. Round to the nearest tenth of a second.

Example 9   An object is thrown upward from the top of a 200-foot cliff with a velocity of 12 feet per second. The height h of the object after t seconds is

$$h = -16t^2 + 12t + 200$$

How long after the object is thrown will it strike the ground? Round to the nearest tenth of a second.

200 feet

Solution:

1. UNDERSTAND. Read and reread the problem.
2. TRANSLATE. Since we want to know when the object strikes the ground, we want to know when the height $h = 0$, or

$$0 = -16t^2 + 12t + 200$$

3. SOLVE. First, divide both sides of the equation by -4.

$$0 = 4t^2 - 3t - 50 \quad \text{Divide both sides by } -4.$$

Here, $a = 4$, $b = -3$, and $c = -50$. By the quadratic formula,

$$t = \frac{-(-3) \pm \sqrt{(-3)^2 - 4(4)(-50)}}{2 \cdot 4}$$

$$= \frac{3 \pm \sqrt{9 + 800}}{8}$$

$$= \frac{3 \pm \sqrt{809}}{8}$$

Check:

4. INTERPRET. Check your calculations in the quadratic formula. Since the time won't be negative, we reject the proposed solution $\dfrac{3 - \sqrt{809}}{8}$.

State:

The time it takes for the object to strike the ground is exactly $\dfrac{3 + \sqrt{809}}{8}$ seconds ≈ 3.9 seconds.

Name _____ Section _____ Date _____

MENTAL MATH

Identify the values of a, b, and c in each quadratic equation.

1. $x^2 + 3x + 1 = 0$ **2.** $2x^2 - 5x - 7 = 0$ **3.** $7x^2 - 4 = 0$

4. $x^2 + 9 = 0$ **5.** $6x^2 - x = 0$

EXERCISE SET 10.2

A Use the quadratic formula to solve each equation. See Examples 1 through 4.

1. $m^2 + 5m - 6 = 0$ **2.** $p^2 + 11p - 12 = 0$ **3.** $2y = 5y^2 - 3$

4. $5x^2 - 3 = 14x$ **5.** $x^2 - 6x + 9 = 0$ **6.** $y^2 + 10y + 25 = 0$

7. $x^2 + 7x + 4 = 0$ **8.** $y^2 + 5y + 3 = 0$ **9.** $8m^2 - 2m = 7$

10. $11n^2 - 9n = 1$ **11.** $3m^2 - 7m = 3$ **12.** $x^2 - 13 = 5x$

13. $\frac{1}{2}x^2 - x - 1 = 0$ **14.** $\frac{1}{6}x^2 + x + \frac{1}{3} = 0$ **15.** $\frac{2}{5}y^2 + \frac{1}{5}y = \frac{3}{5}$

16. $\frac{1}{8}x^2 + x = \frac{5}{2}$ **17.** $\frac{1}{3}y^2 - y - \frac{1}{6} = 0$ **18.** $\frac{1}{2}y^2 = y + \frac{1}{2}$

19. $10y^2 + 10y + 3 = 0$ **20.** $3y^2 + 6y + 5 = 0$

651

Problem Solving Notes

10.3 SOLVING EQUATIONS BY USING QUADRATIC METHODS

A SOLVING EQUATIONS THAT ARE QUADRATIC IN FORM

In this section, we discuss various types of equations that can be solved in part by using the methods for solving quadratic equations.

Once each equation is simplified, you may want to use these steps when deciding what method to use to solve the quadratic equation.

SOLVING A QUADRATIC EQUATION

Step 1. If the equation is in the form $(ax + b)^2 = c$, use the square root property and solve. If not, go to Step 2.

Step 2. Write the equation in standard form: $ax^2 + bx + c = 0$.

Step 3. Try to solve the equation by the factoring method. If not possible, go to Step 4.

Step 4. Solve the equation by the quadratic formula.

The first example is a radical equation that becomes a quadratic equation once we square both sides.

Example 1   Solve: $x - \sqrt{x} - 6 = 0$

Solution: Recall that to solve a radical equation, first get the radical alone on one side of the equation. Then square both sides.

$$x - 6 = \sqrt{x} \quad \text{Add } \sqrt{x} \text{ to both sides.}$$
$$x^2 - 12x + 36 = x \quad \text{Square both sides.}$$
$$x^2 - 13x + 36 = 0 \quad \text{Set the equation equal to 0.}$$
$$(x - 9)(x - 4) = 0$$
$$x - 9 = 0 \quad \text{or} \quad x - 4 = 0$$
$$x = 9 \qquad\qquad x = 4$$

Check:

Let $x = 9$

$$x - \sqrt{x} - 6 = 0$$
$$9 - \sqrt{9} - 6 \overset{?}{=} 0$$
$$9 - 3 - 6 \overset{?}{=} 0$$
$$0 = 0 \quad \text{True}$$

Let $x = 4$

$$x - \sqrt{x} - 6 = 0$$
$$4 - \sqrt{4} - 6 \overset{?}{=} 0$$
$$4 - 2 - 6 \overset{?}{=} 0$$
$$-4 = 0 \quad \text{False}$$

The solution set is $\{9\}$.

Example 2   Solve: $\dfrac{3x}{x - 2} - \dfrac{x + 1}{x} = \dfrac{6}{x(x - 2)}$

Solution: In this equation, x cannot be either 2 or 0, because these values cause denominators to equal zero. To solve for x, we first multiply both sides of the equation by $x(x - 2)$ to clear the fractions. By the distributive property, this means that we multiply each term by $x(x - 2)$.

Objectives

A Solve various equations that are quadratic in form.

B Solve problems that lead to quadratic equations.

SSM   CD-ROM   Video
10.3

Practice Problem 1

Solve: $x - \sqrt{x - 1} - 3 = 0$

Practice Problem 2

Solve: $\dfrac{2x}{x - 1} - \dfrac{x + 2}{x} = \dfrac{5}{x(x - 1)}$

Answers

1. $\{5\}$, **2.** $\left\{\dfrac{1 + \sqrt{13}}{2}, \dfrac{1 - \sqrt{13}}{2}\right\}$

$$x(x-2)\left(\frac{3x}{x-2}\right) - x(x-2)\left(\frac{x+1}{x}\right) = x(x-2)\left[\frac{6}{x(x-2)}\right]$$

$$3x^2 - (x-2)(x+1) = 6 \quad \text{Simplify.}$$

$$3x^2 - (x^2 - x - 2) = 6 \quad \text{Multiply.}$$

$$3x^2 - x^2 + x + 2 = 6$$

$$2x^2 + x - 4 = 0 \quad \text{Simplify.}$$

This equation cannot be factored using integers, so we solve by the quadratic formula.

$$x = \frac{-1 \pm \sqrt{1^2 - 4(2)(-4)}}{2 \cdot 2} \quad \text{Let } a = 2, b = 1, \text{ and } c = -4 \text{ in the quadratic formula.}$$

$$= \frac{-1 \pm \sqrt{1 + 32}}{4} \quad \text{Simplify.}$$

$$= \frac{-1 \pm \sqrt{33}}{4}$$

Neither proposed solution will make the denominators 0.

The solution set is $\left\{\frac{-1 + \sqrt{33}}{4}, \frac{-1 - \sqrt{33}}{4}\right\}$. ■

Practice Problem 3

Solve: $x^4 - 5x^2 - 36 = 0$

✓ CONCEPT CHECK

a. *True or False?* The maximum number of solutions that a quadratic equation can have is 2.
b. *True or False?* The maximum number of solutions that an equation in quadratic form can have is 2.

Practice Problem 4

Solve: $(x+4)^2 - (x+4) - 6 = 0$

Example 3 Solve: $p^4 - 3p^2 - 4 = 0$

Solution: First we factor the trinomial.

$$p^4 - 3p^2 - 4 = 0$$

$$(p^2 - 4)(p^2 + 1) = 0 \quad \text{Factor.}$$

$$(p - 2)(p + 2)(p^2 + 1) = 0 \quad \text{Factor further.}$$

$$p - 2 = 0 \quad \text{or} \quad p + 2 = 0 \quad \text{or} \quad p^2 + 1 = 0 \quad \text{Set each factor equal to 0 and solve.}$$

$$p = 2 \qquad\qquad p = -2 \qquad\qquad p^2 = -1$$

$$p = \pm\sqrt{-1} = \pm i$$

The solution set is $\{2, -2, i, -i\}$. ■

TRY THE CONCEPT CHECK IN THE MARGIN.

Example 4 Solve: $(x - 3)^2 - 3(x - 3) - 4 = 0$

Solution: Notice that the quantity $(x - 3)$ is repeated in this equation. Sometimes it is helpful to substitute a variable (in this case other than x) for the repeated quantity. We will let $y = x - 3$. Then

becomes

$$(x - 3)^2 - 3(x - 3) - 4 = 0$$

$$y^2 - 3y - 4 = 0 \quad \text{Let } x - 3 = y.$$

$$(y - 4)(y + 1) = 0 \quad \text{Factor.}$$

To solve, we use the zero factor property.

$$y - 4 = 0 \quad \text{or} \quad y + 1 = 0 \quad \text{Set each factor equal to 0.}$$

$$y = 4 \qquad\qquad y = -1 \quad \text{Solve.}$$

To find values of x, we substitute back. That is, we substitute $x - 3$ for y.

$$x - 3 = 4 \quad \text{or} \quad x - 3 = -1$$
$$x = 7 \qquad\qquad x = 2$$

> **HELPFUL HINT**
>
> When using substitution, don't forget to substitute back to the original variable.

Both 2 and 7 check. The solution is $\{2, 7\}$.

Example 5   Solve: $x^{2/3} - 5x^{1/3} + 6 = 0$

Solution: The key to solving this equation is recognizing that $x^{2/3} = \left(x^{1/3}\right)^2$. We replace $x^{1/3}$ with m so that

$$\left(x^{1/3}\right)^2 - 5x^{1/3} + 6 = 0$$

becomes

$$m^2 - 5m + 6 = 0$$

Now we solve by factoring.

$$m^2 - 5m + 6 = 0$$
$$(m - 3)(m - 2) = 0 \qquad \text{Factor.}$$
$$m - 3 = 0 \quad \text{or} \quad m - 2 = 0 \qquad \text{Set each factor equal to 0.}$$
$$m = 3 \qquad\qquad m = 2$$

Since $m = x^{1/3}$, we have

$$x^{1/3} = 3 \qquad\qquad \text{or} \quad x^{1/3} = 2$$
$$x = 3^3 = 27 \quad \text{or} \qquad x = 2^3 = 8$$

Both 8 and 27 check. The solution set is $\{8, 27\}$.

> **HELPFUL HINT**
>
> Example 3 can be solved using substitution also. Think of $p^4 - 3p^2 - 4 = 0$ as
>
> $$\left(p^2\right)^2 - 3p^2 - 4 = 0 \qquad \text{Then let } x = p^2, \text{ and}$$
> $$\downarrow \qquad \swarrow \qquad\qquad \text{solve and substitute back. The solution set will}$$
> $$x^2 - 3x - 4 = 0 \qquad\quad \text{be the same.}$$

B SOLVING PROBLEMS THAT LEAD TO QUADRATIC EQUATIONS

The next example is a work problem. This problem is modeled by a rational equation that simplifies to a quadratic equation.

Example 6   Finding Work Time

Together, an experienced typist and an apprentice typist can process a document in 6 hours. Alone, the experienced typist can process the document 2 hours faster than the apprentice typist can. Find the time in which each person can process the document alone.

Practice Problem 5

Solve: $x^{2/3} - 7x^{1/3} + 10 = 0$

Practice Problem 6

Together, Karen and Doug Lewis can clean a strip of beach in 5 hours. Alone, Karen can clean the strip of beach one hour faster than Doug. Find the time that each person can clean the strip of beach alone. Give an exact answer and a one decimal place approximation.

Answers

5. $\{8, 125\}$, **6.** Doug, $\dfrac{11 + \sqrt{101}}{2} \approx 10.5$

hours; Karen, $\dfrac{9 + \sqrt{101}}{2} \approx 9.5$ hours

Solution:

1. UNDERSTAND. Read and reread the problem. The key idea here is the relationship between the *time* (hours) it takes to complete the job and the *part of the job* completed in one unit of time (hour). For example, because they can complete the job together in 6 hours, the *part of the job* they can complete in 1 hour is $\frac{1}{6}$. Let

$x =$ the *time* in hours it takes the apprentice typist to complete the job alone

$x - 2 =$ the time in hours it takes the experienced typist to complete the job alone

We can summarize in a chart the information discussed.

	Total Hours to Complete Job	Part of Job Completed in 1 Hour
Apprentice Typist	x	$\frac{1}{x}$
Experienced Typist	$x - 2$	$\frac{1}{x - 2}$
Together	6	$\frac{1}{6}$

2. TRANSLATE.

In words:

part of job completed by apprentice typist in 1 hour	added to	part of job completed by experienced typist in 1 hour	is equal to	part of job completed together in 1 hour
↓	↓	↓	↓	↓

Translate:
$$\frac{1}{x} \quad + \quad \frac{1}{x - 2} \quad = \quad \frac{1}{6}$$

3. SOLVE.

$$\frac{1}{x} + \frac{1}{x - 2} = \frac{1}{6}$$

$$6x(x - 2)\left(\frac{1}{x} + \frac{1}{x - 2}\right) = 6x(x - 2) \cdot \frac{1}{6} \quad \text{Multiply both sides by the LCD, } 6x(x-2).$$

$$6x(x - 2) \cdot \frac{1}{x} + 6x(x - 2) \cdot \frac{1}{x - 2} = 6x(x - 2) \cdot \frac{1}{6} \quad \text{Use the distributive property.}$$

$$6(x - 2) + 6x = x(x - 2)$$

$$6x - 12 + 6x = x^2 - 2x$$

$$0 = x^2 - 14x + 12$$

Now we can substitute $a = 1$, $b = -14$, and $c = 12$ into the quadratic formula and simplify.

$$x = \frac{-(-14) \pm \sqrt{(-14)^2 - 4(1)(12)}}{2(1)} = \frac{14 \pm \sqrt{148}}{2}$$

Using a calculator or a square root table, we see that $\sqrt{148} \approx 12.2$ rounded to one decimal place. Thus,

$$x \approx \frac{14 \pm 12.2}{2}$$

$$x \approx \frac{14 + 12.2}{2} = 13.1 \quad \text{or} \quad x \approx \frac{14 - 12.2}{2} = 0.9$$

4. INTERPRET.

Check: If the apprentice typist completes the job alone in 0.9 hours, the experienced typist completes the job alone in $x - 2 = 0.9 - 2 = -1.1$ hours. Since this is not possible, we reject the solution of 0.9. The approximate solution is thus 13.1 hours.

State: The apprentice typist can complete the job alone in approximately 13.1 hours, and the experienced typist can complete the job alone in approximately $x - 2 = 13.1 - 2 = 11.1$ hours.

Example 7 Beach and Fargo are about 400 miles apart. A salesperson travels from Fargo to Beach one day at a certain speed. She returns to Fargo the next day and drives 10 mph faster. Her total travel time was $14\frac{2}{3}$ hours. Find her speed to Beach and the return speed to Fargo.

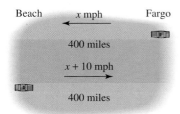

Beach   *x* mph   Fargo

400 miles

x + 10 mph

400 miles

Solution: **1.** UNDERSTAND. Read and reread the problem. Let

$x =$ the speed to Beach, so

$x + 10 =$ the return speed to Fargo.

Then organize the given information in a table.

distance	=	rate	·	time
To Beach	400		x	$\dfrac{400}{x}$
Return to Fargo	400		$x + 10$	$\dfrac{400}{x + 10}$

2. TRANSLATE.

In words:

time to Beach	+	return time to Fargo	=	$14\frac{2}{3}$ hours

Translate: $\dfrac{400}{x} + \dfrac{400}{x + 10} = \dfrac{44}{3}$

Practice Problem 7

A family drives 500 miles to the beach for a vacation. The return trip was made at a speed that was 10 mph faster. The total traveling time was $18\frac{1}{3}$ hours. Find the speed to the beach and the return speed.

Answer

7. 50 mph to the beach; 60 mph returning

3. SOLVE.

$$\frac{400}{x} + \frac{400}{x+10} = \frac{44}{3}$$

Divide both sides by 4.

$$\frac{100}{x} + \frac{100}{x+10} = \frac{11}{3}$$

$$3x(x+10)\left(\frac{100}{x} + \frac{100}{x+10}\right) = 3x(x+10) \cdot \frac{11}{3}$$

Multiply both sides by the LCD, $3x(x+10)$.

$$3x(x+10) \cdot \frac{100}{x} + 3x(x+10) \cdot \frac{100}{x+10} = 3x(x+10) \cdot \frac{11}{3}$$

Use the distributive property.

$$3(x+10) \cdot 100 + 3x \cdot 100 = x(x+10) \cdot 11$$

$$300x + 3000 + 300x = 11x^2 + 110x$$

$$0 = 11x^2 - 490x - 3000$$

Set equation equal to 0.

$$0 = (11x + 60)(x - 50)$$

$11x + 60 = 0$ or $x - 50 = 0$ Set each factor equal to 0.   Factor.

$x = -\dfrac{60}{11}$ or $-5\dfrac{5}{11}$ $x = 50$

4. INTERPRET.

Check: The speed is not negative, so it's not $-5\dfrac{5}{11}$. The number 50 does check.

State: The speed to Beach was 50 mph and her return speed to Fargo was 60 mph.   ▬▬▬

THE EVOLUTION OF SOLVING QUADRATIC EQUATIONS

The ancient Babylonians (circa 2000 B.C.) are sometimes credited with being the first to solve quadratic equations. This is only partially true because the Babylonians had no concept of an equation. However, what they did develop was a method for completing the square to apply to problems that today would be solved with a quadratic equation. The Babylonians only recognized positive solutions to such problems and did not acknowledge the existence of negative solutions at all.

Babylonian mathematical knowledge influenced much of the ancient world, most notably Hindu Indians. The Hindus were the first culture to denote debts in everyday business affairs with negative numbers. With this level of comfort with negative numbers, the Indian mathematician Brahmagupta (598–665 A.D.) extended the Babylonian methods and was the first to recognize negative solutions to quadratic equations. Later Hindu mathematicians noted that every positive number has two square roots: a positive square root and a negative square root. Hindus allowed irrational solutions (quite an innovation in the ancient world!) to quadratic equations and were the first to realize that quadratic equations could have 0, 1, or 2 real number solutions. They did not, however, acknowledge complex numbers and, therefore, could not solve equations with solutions requiring the square root of a negative number.

Complex numbers were finally developed by European mathematicians during the 17th and 18th centuries. Up until that time, what we would today consider complex solutions to quadratic equations were routinely ignored by mathematicians.

Problem Solving Notes

10.4 NONLINEAR INEQUALITIES IN ONE VARIABLE

A SOLVING POLYNOMIAL INEQUALITIES

Just as we can solve linear inequalities in one variable, so we can also solve quadratic inequalities in one variable. A **quadratic inequality** is an inequality that can be written so that one side is a quadratic expression and the other side is 0. Here are examples of quadratic inequalities in one variable. Each is written in **standard form**.

$$x^2 - 10x + 7 \le 0 \qquad 3x^2 + 2x - 6 > 0$$
$$2x^2 + 9x - 2 < 0 \qquad x^2 - 3x + 11 \ge 0$$

A solution of a quadratic inequality in one variable is a value of the variable that makes the inequality a true statement.

The value of an expression such as $x^2 - 3x - 10$ will sometimes be positive, sometimes negative, and sometimes 0, depending on the value substituted for x. To solve the inequality $x^2 - 3x - 10 < 0$, we are looking for all values of x that make the expression $x^2 - 3x - 10$ **less than 0**, or **negative**. To understand how we find these values, we'll study the graph of the quadratic function $y = x^2 - 3x - 10$.

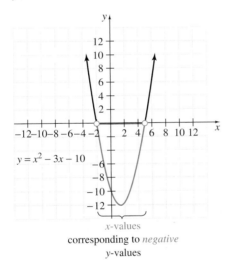

$y = x^2 - 3x - 10$

x-values
corresponding to *negative*
y-values

Notice that the x-values for which y or $x^2 - 3x - 10$ is positive are separated from the x values for which y or $x^2 - 3x - 10$ is negative by the values for which y or $x^2 - 3x - 10$ is 0, the x-intercepts. Thus, the solution set of $x^2 - 3x - 10 < 0$ consists of all real numbers from -2 to 5, or in interval notation, $(-2, 5)$.

It is not necessary to graph $y = x^2 - 3x - 10$ to solve the related inequality $x^2 - 3x - 10 < 0$. Instead, we can draw a number line representing the x-axis and keep the following in mind: *A region on the number line for which the value of $x^2 - 3x - 10$ is positive is separated from a region on the number line for which the value of $x^2 - 3x - 10$ is negative by a value for which the expression is 0.*

Let's find these values for which the expression is 0 by solving the related equation:

$$x^2 - 3x - 10 = 0$$
$$(x - 5)(x + 2) = 0 \qquad \text{Factor.}$$
$$x - 5 = 0 \quad \text{or} \quad x + 2 = 0 \qquad \text{Set each factor equal to 0.}$$
$$x = 5 \qquad\qquad x = -2 \qquad \text{Solve.}$$

These two numbers -2 and 5, divide the number line into three regions. We will call the regions A, B, and C. These regions are important because, if the value of $x^2 - 3x - 10$ is negative when a number from a region is substituted for x, then $x^2 - 3x - 10$ is negative when any number in that region is substituted for x. The same is true if the value of $x^2 - 3x - 10$ is positive for a particular value of x in a region.

To see whether the inequality $x^2 - 3x - 10 < 0$ is true or false in each region, we choose a test point from each region and substitute its value for x in the inequality $x^2 - 3x - 10 < 0$. If the resulting inequality is true, the region containing the test point is a solution region.

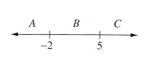

Region	Test Point Value	$(x - 5)(x + 2) < 0$	Result
A	-3	$(-8)(-1) < 0$	False.
B	0	$(-5)(2) < 0$	True.
C	6	$(1)(8) < 0$	False.

The values in region B satisfy the inequality. The numbers -2 and 5 are not included in the solution set since the inequality symbol is $<$. The solution set is $(-2, 5)$, and its graph is shown.

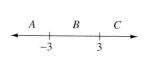

Practice Problem 1

Solve: $(x - 2)(x + 4) > 0$

SOLVING A POLYNOMIAL INEQUALITY

Step 1. Write the inequality in standard form and then solve the related equation.

Step 2. Separate the number line into regions with the solutions from Step 1.

Step 3. For each region, choose a test point and determine whether its value satisfies the *original inequality*.

Step 4. The solution set includes the regions whose test point value is a solution. If the inequality symbol is $\leq$ or $\geq$, the values from Step 1 are solutions; if $<$ or $>$, they are not.

✓ CONCEPT CHECK

When choosing a test point in Step 4, why would the solutions from Step 2 not make good choices for test points?

Answer

1. $(-\infty, -4) \cup (2, \infty)$

✓ **Concept Check:** The solutions found in Step 2 have a value of zero in the original inequality.

Example 1   Solve: $(x + 3)(x - 3) > 0$

Solution: First we solve the related equation $(x + 3)(x - 3) = 0$.

$$(x + 3)(x - 3) = 0$$
$$x + 3 = 0 \quad \text{or} \quad x - 3 = 0$$
$$x = -3 \qquad\qquad x = 3$$

The two numbers -3 and 3 separate the number line into three regions, A, B, and C.

Now we substitute the value of a test point from each region. If the test value satisfies the inequality, every value in the region containing the test value is a solution.

Region	Test Point Value	$(x + 3)(x - 3) > 0$	Result
A	-4	$(-1)(-7) > 0$	True.
B	0	$(3)(-3) > 0$	False.
C	4	$(7)(1) > 0$	True.

The points in regions A and C satisfy the inequality. The numbers -3 and 3 are not included in the solution since the inequality symbol is $>$. The solution set is $(-\infty, -3) \cup (3, \infty)$, and its graph is shown.

The steps in the margin may be used to solve a polynomial inequality.

TRY THE CONCEPT CHECK IN THE MARGIN.

Example 2

Solve: $x^2 - 4x \leq 0$

Solution: First we solve the related equation $x^2 - 4x = 0$.

$$x^2 - 4x = 0$$
$$x(x - 4) = 0$$
$$x = 0 \text{ or } x = 4$$

The numbers 0 and 4 separate the number line into three regions, A, B and C.

Check a test value in each region in the original inequality. Values in region B satisfy the inequality. The numbers 0 and 4 are included in the solution since the inequality symbol is $\leq$. The solution set is $[0, 4]$, and its graph is shown.

Example 3

Solve: $(x + 2)(x - 1)(x - 5) \leq 0$

Solution: First we solve $(x + 2)(x - 1)(x - 5) = 0$. By inspection, we see that the solutions are $-2, 1,$ and 5. They separate the number line into four regions, A, B, C, and D. Next we check test points from each region.

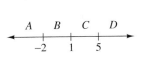

Region	Test Point Value	$(x + 2)(x - 1)$ $(x - 5) \leq 0$	Result
A	-3	$(-1)(-4)(-8) \leq 0$	True.
B	0	$(2)(-1)(-5) \leq 0$	False.
C	2	$(4)(1)(-3) \leq 0$	True.
D	6	$(8)(5)(1) \leq 0$	False.

The solution set is $(-\infty, -2] \cup [1, 5]$, and its graph is shown. We include the numbers $-2, 1$ and 5 because the inequality symbol is $\leq$.

B   SOLVING RATIONAL INEQUALITIES

Inequalities containing rational expressions with variables in the denominator are solved by using a similar procedure.

Example 4

Solve: $\dfrac{x + 2}{x - 3} \leq 0$

Solution: First we find all values that make the denominator equal to 0. To do this, we solve $x - 3 = 0$, or $x = 3$.

Next, we solve the related equation $\dfrac{x + 2}{x - 3} = 0$.

Practice Problem 2

Solve: $x^2 - 6x \leq 0$

Practice Problem 3

Solve: $(x - 2)(x + 1)(x + 5) \leq 0$

Practice Problem 4

Solve: $\dfrac{x - 3}{x + 5} \leq 0$

Answers

2. $[0, 6]$,   **3.** $(-\infty, -5] \cup [-1, 2]$,   **4.** $(-5, 3]$

SOLVING A RATIONAL INEQUALITY

Step 1. Solve for values that make all denominators 0.

Step 2. Solve the related equation.

Step 3. Separate the number line into regions with the solutions from Steps 1 and 2.

Step 4. For each region, choose a test point and determine whether its value satisfies the *original inequality*.

Step 5. The solution set includes the regions whose test point value is a solution. Check whether to include values from Step 2. Be sure *not* to include values that make any denominator 0.

$$\frac{x + 2}{x - 3} = 0 \quad \text{Multiply both sides by the LCD, } x - 3.$$

$$x + 2 = 0$$

$$x = -2$$

Now we place these numbers on a number line and proceed as before, checking test point values in the original inequality.

Choose −3 from region A.

$$\frac{x + 2}{x - 3} \leq 0$$

$$\frac{-3 + 2}{-3 - 3} \leq 0$$

$$\frac{-1}{-6} \leq 0$$

$$\frac{1}{6} \leq 0 \quad \text{False.}$$

Choose 0 from region B.

$$\frac{x + 2}{x - 3} \leq 0$$

$$\frac{0 + 2}{0 - 3} \leq 0$$

$$-\frac{2}{3} \leq 0 \quad \text{True.}$$

Choose 4 from region C.

$$\frac{x + 2}{x - 3} \leq 0$$

$$\frac{4 + 2}{4 - 3} \leq 0$$

$$6 \leq 0 \quad \text{False.}$$

The solution set is $[-2, 3)$. This interval includes -2 because -2 satisfies the original inequality. This interval does not include 3, because 3 would make the denominator 0.

The steps in the margin may be used to solve a rational inequality with variables in the denominator.

Practice Problem 5

Solve: $\dfrac{3}{x - 2} < 2$

Example 5 Solve: $\dfrac{5}{x + 1} < -2$

Solution: First we find values for x that make the denominator equal to 0.

$$x + 1 = 0$$

$$x = -1$$

Next we solve $\dfrac{5}{x + 1} = -2$.

$$(x + 1) \cdot \frac{5}{x + 1} = (x + 1) \cdot -2 \quad \text{Multiply both sides by the LCD, } x + 1.$$

$$5 = -2x - 2 \quad \text{Simplify.}$$

$$7 = -2x$$

$$-\frac{7}{2} = x$$

We use these two solutions to divide a number line into three regions and choose test points. Only a test point value from region B satisfies the *original inequality*. The solution set is $\left(-\dfrac{7}{2}, -1\right)$, and its graph is shown.

10.5 QUADRATIC FUNCTIONS AND THEIR GRAPHS

A GRAPHING $f(x) = x^2 + k$

We graphed the quadratic function $f(x) = x^2$ in Section 7.5. In that section, we discovered that the graph of a quadratic function is a parabola opening upward or downward. Now, as we continue our study, we will discover more details about quadratic functions and their graphs.

First, let's recall the definition of a quadratic function.

> **QUADRATIC FUNCTION**
>
> A quadratic function is a function that can be written in the form $f(x) = ax^2 + bx + c$, where a, b, and c are real numbers and $a \neq 0$.

Notice that equations of the form $y = ax^2 + bx + c$, where $a \neq 0$, also define quadratic functions since y is a function of x or $y = f(x)$.

Recall that if $a > 0$, the parabola opens upward and if $a < 0$, the parabola opens downward. Also, the vertex of a parabola is the lowest point if the parabola opens upward and the highest point if the parabola opens downward. The axis of symmetry is the vertical line that passes through the vertex.

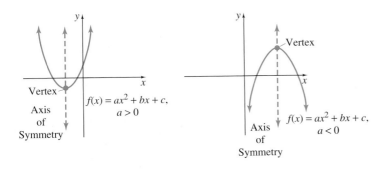

Example 1 Graph $f(x) = x^2$ and $g(x) = x^2 + 3$ on the same set of axes.

Solution: First we construct a table of values for $f(x)$ amd plot the points. Notice that for each x-value, the corresponding value of $g(x)$ must be 3 more than the corresponding value of $f(x)$ since $f(x) = x^2$ and $g(x) = x^2 + 3$. In other words, the graph of $g(x) = x^2 + 3$ is the same as the graph of $f(x) = x^2$ shifted upward 3 units. The axis of symmetry for both graphs is the y-axis.

Objectives

 A Graph quadratic functions of the form $f(x) = x^2 + k$.

 B Graph quadratic functions of the form $f(x) = (x - h)^2$.

C Graph quadratic functions of the form $f(x) = (x - h)^2 + k$.

D Graph quadratic functions of the form $f(x) = ax^2$.

 E Graph quadratic functions of the form $f(x) = a(x - h)^2 + k$.

SSM   CD-ROM   Video
10.5

Practice Problem 1

Graph $f(x) = x^2$ and $g(x) = x^2 + 4$ on the same set of axes.

Answer

1.

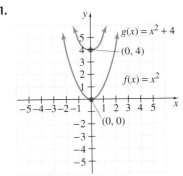

Practice Problems 2–3

Graph each function.

2. $F(x) = x^2 + 1$

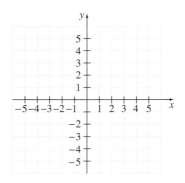

3. $g(x) = x^2 - 2$

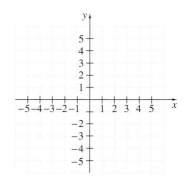

Answers

2.

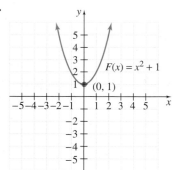

$F(x) = x^2 + 1$
$(0, 1)$

3.

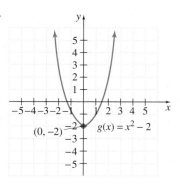

$(0, -2)$ $g(x) = x^2 - 2$

x	$f(x) = x^2$	$g(x) = x^2 + 3$
-2	4	7
-1	1	4
0	0	3
1	1	4
2	4	7

Each y-value
is increased
by 3.

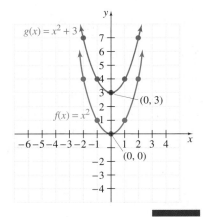

In general, we have the following properties.

GRAPHING THE PARABOLA DEFINED BY $f(x) = x^2 + k$

If k is positive, the graph of $f(x) = x^2 + k$ is the graph of $y = x^2$ shifted upward k units.

If k is negative, the graph of $f(x) = x^2 + k$ is the graph of $y = x^2$ shifted downward $|k|$ units.

The vertex is $(0, k)$, and the axis of symmetry is the y-axis.

Examples   Graph each function.

2. $F(x) = x^2 + 2$

The graph of $F(x) = x^2 + 2$ is obtained by shifting the graph of $y = x^2$ upward 2 units.

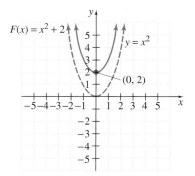

3. $g(x) = x^2 - 3$

The graph of $g(x) = x^2 - 3$ is obtained by shifting the graph of $y = x^2$ downward 3 units.

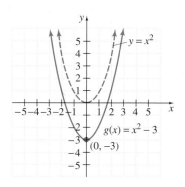

B GRAPHING $f(x) = (x - h)^2$

Now we will graph functions of the form $f(x) = (x - h)^2$.

Example 4   Graph $f(x) = x^2$ and $g(x) = (x - 2)^2$ on the same set of axes.

Solution: By plotting points, we see that for each x-value, the corresponding value of $g(x)$ is the same as the value of $f(x)$ when the x-value is increased by 2. Thus, the graph of $g(x) = (x - 2)^2$ is the graph of $f(x) = x^2$ shifted to the right 2 units. The axis of symmetry for the graph of $g(x) = (x - 2)^2$ is also shifted 2 units to the right and is the line $x = 2$.

Practice Problem 4

Graph $f(x) = x^2$ and $g(x) = (x - 1)^2$ on the same set of axes.

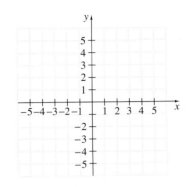

x	$f(x) = x^2$	x	$g(x) = (x - 2)^2$
-2	4	0	4
-1	1	1	1
0	0	2	0
1	1	3	1
2	4	4	4

Each x-value increased by 2 corresponds to same y-value.

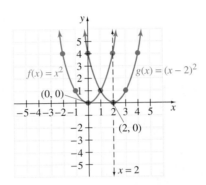

In general, we have the following properties.

GRAPHING THE PARABOLA DEFINED BY $f(x) = (x - h)^2$

If h is positive, the graph of $f(x) = (x - h)^2$ is the graph of $y = x^2$ shifted to the right h units.

If h is negative, the graph of $f(x) = (x - h)^2$ is the graph of $y = x^2$ shifted to the left $|h|$ units.

The vertex is $(h, 0)$, and the axis of symmetry is the vertical line $x = h$.

Examples   Graph each function.

5. $G(x) = (x - 3)^2$

The graph of $G(x) = (x - 3)^2$ is obtained by shifting the graph of $y = x^2$ to the right 3 units.

Answer

4.

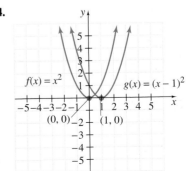

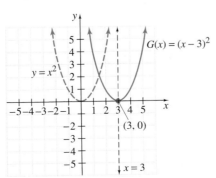

Practice Problems 5–6

Graph each function.

5. $G(x) = (x - 4)^2$

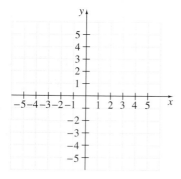

6. $F(x) = (x + 2)^2$

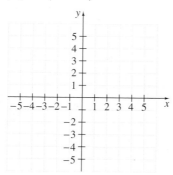

Answers

5.

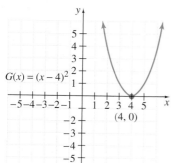

6.

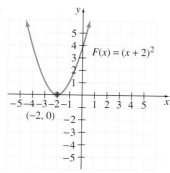

---

6. $F(x) = (x + 1)^2$

The equation $F(x) = (x + 1)^2$ can be written as $F(x) = [x - (-1)]^2$. The graph of $F(x) = [x - (-1)]^2$ is obtained by shifting the graph of $y = x^2$ to the left 1 unit.

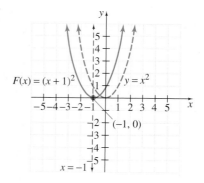

C GRAPHING $f(x) = (x - h)^2 + k$

As we will see in graphing functions of the form $f(x) = (x - h)^2 + k$, it is possible to combine vertical and horizontal shifts.

> **GRAPHING THE PARABOLA DEFINED BY $f(x) = (x - h)^2 + k$**
>
> The parabola has the same shape as $y = x^2$.
> The vertex is (h, k), and the axis of symmetry is the vertical line $x = h$.

Example 7   Graph: $F(x) = (x - 3)^2 + 1$

Solution: The graph of $F(x) = (x - 3)^2 + 1$ is the graph of $y = x^2$ shifted 3 units to the right and 1 unit up. The vertex is then $(3, 1)$, and the axis of symmetry is $x = 3$. A few ordered pair solutions are plotted to aid in graphing.

x	$F(x) = (x - 3)^2 + 1$
1	5
2	2
4	2
5	5

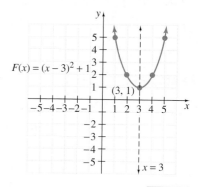

D GRAPHING $f(x) = ax^2$

Next, we discover the change in the shape of the graph when the coefficient of x^2 is not 1.

Example 8   Graph $f(x) = x^2$, $g(x) = 3x^2$, and $h(x) = \frac{1}{2}x^2$ on the same set of axes.

Solution: Comparing the table of values, we see that for each x-value, the corresponding value of $g(x)$ is triple the corresponding value of $f(x)$. Similarly, the value of $h(x)$ is half the value of $f(x)$.

x	$f(x) = x^2$	x	$g(x) = 3x^2$
-2	4	-2	12
-1	1	-1	3
0	0	0	0
1	1	1	3
2	4	2	12

x	$h(x) = \frac{1}{2}x^2$
-2	2
-1	$\frac{1}{2}$
0	0
1	$\frac{1}{2}$
2	2

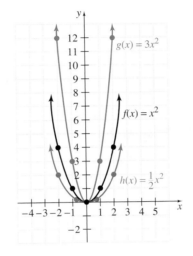

The result is that the graph of $g(x) = 3x^2$ is narrower than the graph of $f(x) = x^2$ and the graph of $h(x) = \frac{1}{2}x^2$ is wider. The vertex for each graph is $(0, 0)$, and the axis of symmetry is the y-axis.

GRAPHING THE PARABOLA DEFINED BY $f(x) = ax^2$

If a is positive, the parabola opens upward, and if a is negative, the parabola opens downward.

If $|a| > 1$, the graph of the parabola is narrower than the graph of $y = x^2$.

If $|a| < 1$, the graph of the parabola is wider than the graph of $y = x^2$.

Example 9   Graph: $f(x) = -2x^2$

Solution: Because $a = -2$, a negative value, this parabola opens downward. Since $|-2| = 2$ and $2 > 1$, the parabola is narrower than the graph of $y = x^2$. The vertex is $(0, 0)$, and the axis of symmetry is the y-axis. We verify this by plotting a few points.

Graph: $F(x) = (x - 2)^2 + 3$

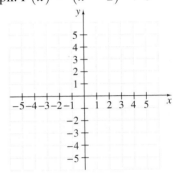

Graph $f(x) = x^2$, $g(x) = 2x^2$, and $h(x) = \frac{1}{3}x^2$ on the same set of axes.

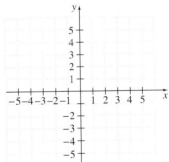

Answers

7.

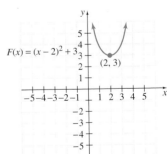

8.

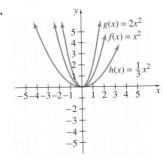

Practice Problem 9

Graph: $f(x) = -3x^2$

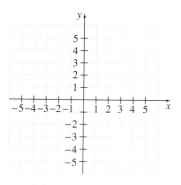

Practice Problem 10

Graph: $f(x) = 2(x + 3)^2 - 4$. Find the vertex and axis of symmetry.

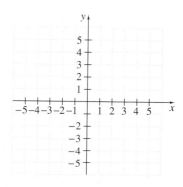

Answers

9.

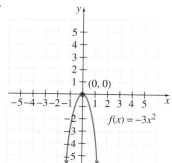

10.

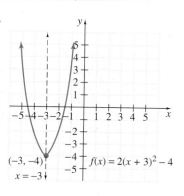

x	$f(x) = -2x^2$
-2	-8
-1	-2
0	0
1	-2
2	-8

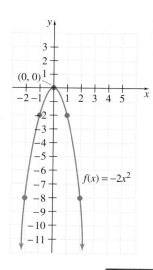

E GRAPHING $f(x) = a(x - h)^2 + k$

Now we will see the shape of the graph of a quadratic function of the form $f(x) = a(x - h)^2 + k$.

Example 10 Graph: $g(x) = \frac{1}{2}(x + 2)^2 + 5$. Find the vertex and the axis of symmetry.

Solution: The function $g(x) = \frac{1}{2}(x + 2)^2 + 5$ may be written as $g(x) = \frac{1}{2}[x - (-2)]^2 + 5$. Thus, this graph is the same as the graph of $y = x^2$ shifted 2 units to the left and 5 units up, and it is wider because a is $\frac{1}{2}$. The vertex is $(-2, 5)$, and the axis of symmetry is $x = -2$. We plot a few points to verify.

x	$g(x) = \frac{1}{2}(x + 2)^2 + 5$
-4	7
-3	$5\frac{1}{2}$
-2	5
-1	$5\frac{1}{2}$
0	7

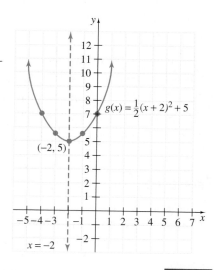

In general, the following holds.

GRAPH OF A QUADRATIC FUNCTION

The graph of a quadratic function written in the form
$f(x) = a(x - h)^2 + k$ is a parabola with vertex (h, k). If $a > 0$, the parabola opens upward, and if $a < 0$, the parabola opens downward. The axis of symmetry is the line whose equation is $x = h$.

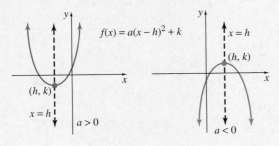

TRY THE CONCEPT CHECK IN THE MARGIN.

✓ **CONCEPT CHECK**

Which description of the graph of $f(x) = -0.35(x + 3)^2 - 4$ is correct?

a. The graph opens downward and has its vertex at $(-3, 4)$.
b. The graph opens upward and has its vertex at $(-3, 4)$.
c. The graph opens downward and has its vertex at $(-3, -4)$.
d. The graph is narrower than the graph of $y = x^2$.

GRAPHING CALCULATOR EXPLORATIONS

Use a grapher to graph the first function of each pair that follows. Then use its graph to predict the graph of the second function. Check your prediction by graphing both on the same set of axes.

1. $F(x) = \sqrt{x}$; $G(x) = \sqrt{x} + 1$

2. $g(x) = x^3$; $H(x) = x^3 - 2$

3. $H(x) = |x|$; $f(x) = |x - 5|$

4. $h(x) = x^3 + 2$; $g(x) = (x - 3)^3 + 2$

5. $f(x) = |x + 4|$; $F(x) = |x + 4| + 3$

6. $G(x) = \sqrt{x} - 2$; $g(x) = \sqrt{x - 4} - 2$

Focus On Business and Career

FINANCIAL RATIOS

A financial ratio is a number found with a rational expression that tells something about a company's activities. Such ratios allow a comparison between the financial positions of two companies, even if the values of the companies' financial data are very different. Here are some common financial ratios:

▲ The **current ratio** gauges a company's ability to pay its short-term debts. It is given by the formula

$$\text{current ratio} = \frac{\text{total current assets}}{\text{total current liabilities}}$$

The higher the value of this ratio, the better able the company is to pay off its short-term debts.

▲ The **total asset turnover ratio** gauges how effectively a company is using all of its resources to generate sales of its products and services. It is given by the formula

$$\text{total asset turnover ratio} = \frac{\text{sales}}{\text{total assets}}$$

The higher the value of this ratio, the more effective the company is at utilizing its resources for sales generation.

▲ The **gross profit margin ratio** gauges how effectively the company is making pricing decisions as well as controlling production costs. It is given by the formula

$$\text{gross profit margin ratio} = \frac{\text{sales} - \text{cost of sales}}{\text{sales}}$$

The higher the value of this ratio, the better the company is doing with regards to controlling costs and pricing products.

▲ The **price-to-earnings (P/E) ratio** gauges the stock market's view of a company with respect to risk. It is given by the formula

$$\text{P/E ratio} = \frac{\text{stock market price per share}}{\text{earnings per share}}$$

A company with low risk will generally have a high P/E ratio. A high P/E ratio also translates into better growth potential for the company's earnings.

For all of these ratios, a higher-than-industry-average ratio is generally considered to be a sign of good financial health.

ADDITIONAL DEFINITIONS

▲ **Assets**—things of value that are owned by a company. *Current assets* include cash and assets that can be converted into cash quickly. *Total assets* are all things of value, including property and equipment, owned by the company. Current assets and total assets may be found on a company's consolidated balance sheet or statement of financial position in an annual report.

▲ **Liabilities**—what a company owes to creditors. *Current liabilities* include any debts expected to come due within the next year. Current liabilities may be found on a company's consolidated balance sheet or statement of financial position in an annual report.

▲ **Sales**—the total amount of money collected by a company from the sales of its goods or services. Sales (or sometimes noted as "net sales") may be found on a company's consolidated statement of income/earnings/operations in an annual report.

▲ **Cost of sales**—a company's cost of inventory actually sold to customers. This is also sometimes referred to as "cost of goods/merchandise sold." Cost of sales may be found on a company's consolidated statement of income/earnings/operations in an annual report.

▲ **Earnings per share**—the value of a company's earnings available for each share of common stock held by stockholders. Earnings per share (or sometimes noted as net income per share) may be found on a company's consolidated statement of income/earnings/operations in an annual report.

▲ **Stock market price per share**—the current price of a company's share of stock as given on one of the major stock markets. Current share prices may be found in newspapers or on the World-Wide Web.

GROUP ACTIVITY

Locate annual reports for two companies involved in similar industries. Using the information and definitions given above, compute these four financial ratios for each company. Then compare the companies' ratios and discuss what the ratios indicate about the two companies. Which company do you think is in better overall financial health? Why?

Problem Solving Notes

10.6 FURTHER GRAPHING OF QUADRATIC FUNCTIONS

A WRITING QUADRATIC FUNCTIONS IN THE FORM $y = a(x - h)^2 + k$

We know that the graph of a quadratic function is a parabola. If a quadratic function is written in the form

$$f(x) = a(x - h)^2 + k$$

we can easily find the vertex (h, k) and graph the parabola. To write a quadratic function in this form, we need to complete the square. (See Section 10.1 for a review of completing the square.)

Example 1   Graph: $f(x) = x^2 - 4x - 12$. Find the vertex and any intercepts.

Solution: The graph of this quadratic function is a parabola. To find the vertex of the parabola, we complete the square on the binomial $x^2 - 4x$. To simplify our work, we let $f(x) = y$.

$$y = x^2 - 4x - 12 \quad \text{Let } f(x) = y.$$
$$y + 12 = x^2 - 4x \quad \text{Add 12 to both sides to get the } x\text{-variable terms alone.}$$

Now we add the square of half of -4 to both sides.

$$\frac{1}{2}(-4) = -2 \quad \text{and} \quad (-2)^2 = 4$$

$$y + 12 + 4 = x^2 - 4x + 4 \quad \text{Add 4 to both sides.}$$
$$y + 16 = (x - 2)^2 \quad \text{Factor the trinomial.}$$
$$y = (x - 2)^2 - 16 \quad \text{Subtract 16 from both sides.}$$
$$f(x) = (x - 2)^2 - 16 \quad \text{Replace } y \text{ with } f(x).$$

From this equation, we can see that the vertex of the parabola is $(2, -16)$, a point in quadrant IV, and the axis of symmetry is the line $x = 2$.

Notice that $a = 1$. Since $a > 0$, the parabola opens upward. This parabola opening upward with vertex $(2, -16)$ will have two x-intercepts.

To find them, we let $f(x)$ or $y = 0$.

$$0 = x^2 - 4x - 12$$
$$0 = (x - 6)(x + 2)$$
$$0 = x - 6 \quad \text{or} \quad 0 = x + 2$$
$$6 = x \qquad\qquad -2 = x$$

The two x-intercepts are 6 and -2. To find the y-intercept, we let $x = 0$.

$$f(0) = 0^2 - 4 \cdot 0 - 12 = -12$$

The y-intercept is -12. The sketch of $f(x) = x^2 - 4x - 12$ is shown.

Objectives

A Write quadratic functions in the form $y = a(x - h)^2 + k$.

B Derive a formula for finding the vertex of a parabola.

C Find the minimum or maximum value of a quadratic function.

SSM   CD-ROM   Video
10.6

Practice Problem 1

Graph: $f(x) = x^2 - 4x - 5$. Find the vertex and any intercepts.

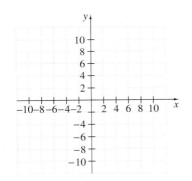

Answer

1. vertex: $(2, -9)$; x-intercepts: $-1, 5$; y-intercept: -5

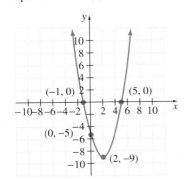

⌐HELPFUL HINT

Parabola Opens Upward
Vertex in I or II: no x-intercepts
Vertex in III or IV: 2 x-intercepts

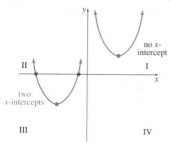

Parabola Opens Downward
Vertex in I or II: 2 x-intercepts
Vertex in III or IV: no x-intercepts.

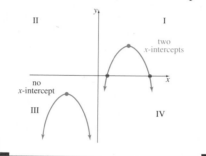

Practice Problem 2

Graph: $f(x) = 2x^2 + 2x + 5$. Find the vertex and any intercepts.

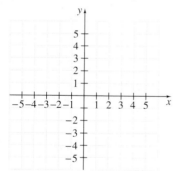

Answer

2. vertex: $\left(-\frac{1}{2}, \frac{9}{2}\right)$; y-intercept: 5

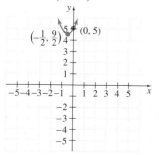

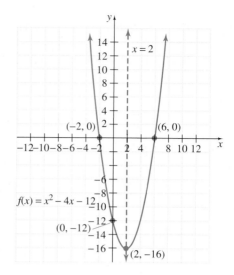

$f(x) = x^2 - 4x - 12$

Example 2   Graph: $f(x) = 3x^2 + 3x + 1$. Find the vertex and any intercepts.

Solution: We replace $f(x)$ with y and complete the square on x to write the equation in the form $y = a(x - h)^2 + k$.

$$y = 3x^2 + 3x + 1 \qquad \text{Replace } f(x) \text{ with } y.$$

$$y - 1 = 3x^2 + 3x \qquad \text{Get the } x\text{-variable terms alone.}$$

Next we factor 3 from the terms $3x^2 + 3x$ so that the coefficient of x^2 is 1.

$$y - 1 = 3(x^2 + x) \qquad \text{Factor out 3.}$$

The coefficient of x is 1. Then $\frac{1}{2}(1) = \frac{1}{2}$ and $\left(\frac{1}{2}\right)^2 = \frac{1}{4}$.

Since we are adding $\frac{1}{4}$ inside the parentheses, we are really adding $3\left(\frac{1}{4}\right)$, so we *must* add $3\left(\frac{1}{4}\right)$ to the left side.

$$y - 1 + 3\left(\frac{1}{4}\right) = 3\left(x^2 + x + \frac{1}{4}\right)$$

$$y - \frac{1}{4} = 3\left(x + \frac{1}{2}\right)^2 \qquad \begin{array}{l}\text{Simplify the left side and}\\ \text{factor the right side.}\end{array}$$

$$y = 3\left(x + \frac{1}{2}\right)^2 + \frac{1}{4} \qquad \text{Add } \frac{1}{4} \text{ to both sides.}$$

$$f(x) = 3\left(x + \frac{1}{2}\right)^2 + \frac{1}{4} \qquad \text{Replace } y \text{ with } f(x).$$

Then $a = 3$, $h = -\frac{1}{2}$, and $k = \frac{1}{4}$. This means that the parabola opens upward with vertex $\left(-\frac{1}{2}, \frac{1}{4}\right)$ and that the axis of symmetry is the line $x = -\frac{1}{2}$.

To find the y-intercept, we let $x = 0$. Then

$$f(0) = 3(0)^2 + 3(0) + 1 = 1$$

This parabola has no x-intercepts since the vertex is in the second quadrant and it opens upward. We use the vertex, axis of symmetry, and y-intercept to graph the parabola.

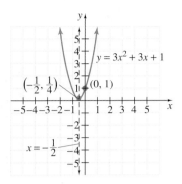

Example 3

Graph: $f(x) = -x^2 - 2x + 3$. Find the vertex and any intercepts.

Solution:

We write $f(x)$ in the form $a(x - h)^2 + k$ by completing the square. First we replace $f(x)$ with y.

$$f(x) = -x^2 - 2x + 3$$
$$y = -x^2 - 2x + 3$$
$$y - 3 = -x^2 - 2x \qquad \text{Subtract 3 from both sides to get the } x\text{-variable terms alone.}$$
$$y - 3 = -1(x^2 + 2x) \qquad \text{Factor } -1 \text{ from the terms } -x^2 - 2x.$$

The coefficient of x is 2. Then $\frac{1}{2}(2) = 1$ and $1^2 = 1$. We add 1 to the right side inside the parentheses and add $-1(1)$ to the left side.

$$y - 3 - 1(1) = -1(x^2 + 2x + 1)$$
$$y - 4 = -1(x + 1)^2 \qquad \text{Simplify the left side and factor the right side.}$$
$$y = -1(x + 1)^2 + 4 \qquad \text{Add 4 to both sides.}$$
$$f(x) = -1(x + 1)^2 + 4 \qquad \text{Replace } y \text{ with } f(x).$$

Since $a = -1$, the parabola opens downward with vertex $(-1, 4)$ and axis of symmetry $x = -1$.

To find the y-intercept, we let $x = 0$ and solve for y. Then

$$f(0) = -0^2 - 2(0) + 3 = 3$$

Thus, 3 is the y-intercept.

To find the x-intercepts, we let y or $f(x) = 0$ and solve for x.

$$f(x) = -x^2 - 2x + 3$$
$$0 = -x^2 - 2x + 3 \qquad \text{Let } f(x) = 0.$$

Now we divide both sides by -1 so that the coefficient of x^2 is 1:

Practice Problem 3

Graph: $f(x) = -x^2 - 2x + 8$. Find the vertex and any intercepts.

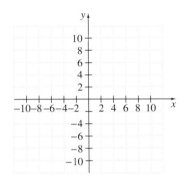

HELPFUL HINT

This can be written as $f(x) = -1[x - (-1)]^2 + 4$. Notice that the vertex is $(-1, 4)$.

Answer

3. vertex: $(-1, 9)$; x-intercepts: $-4, 2$; y-intercept: 8

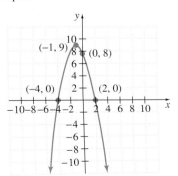

$$\frac{0}{-1} = \frac{-x^2}{-1} - \frac{2x}{-1} + \frac{3}{-1} \quad \text{Divide both sides by } -1.$$

$$0 = x^2 + 2x - 3 \qquad \text{Simplify.}$$

$$0 = (x + 3)(x - 1) \qquad \text{Factor.}$$

$$x + 3 = 0 \quad \text{or} \quad x - 1 = 0 \quad \text{Set each factor equal to 0.}$$

$$x = -3 \qquad\qquad x = 1 \quad \text{Solve.}$$

The x-intercepts are -3 and 1. We use these points to graph the parabola.

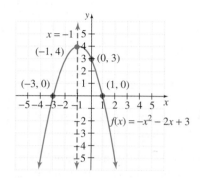

B DERIVING A FORMULA FOR FINDING THE VERTEX

Recall from Section 7.5 that we introduced a formula for finding the vertex of a parabola. Now that we have practiced completing the square, we will show that the x-coordinate of the vertex of the graph of $f(x)$ or $y = ax^2 + bx + c$ can be found by the formula $x = \frac{-b}{2a}$. To do so, we complete the square on x and write the equation in the form $y = (x - h)^2 + k$.

First we get the x-variable terms alone by subtracting c from both sides.

$$y = ax^2 + bx + c$$

$$y - c = ax^2 + bx$$

$$y - c = a\left(x^2 + \frac{b}{a}x\right) \quad \text{Factor } a \text{ from the terms } ax^2 + bx.$$

Now we add the square of half of $\frac{b}{a}$, or $\left(\frac{b}{2a}\right)^2 = \frac{b^2}{4a^2}$, to the right side inside the parentheses. Because of the factor a, what we really added is $a\left(\frac{b^2}{4a^2}\right)$ and this must be added to the left side.

$$y - c + a\left(\frac{b^2}{4a^2}\right) = a\left(x^2 + \frac{b}{a}x + \frac{b^2}{4a^2}\right)$$

$$y - c + \frac{b^2}{4a} = a\left(x + \frac{b}{2a}\right)^2 \quad \begin{array}{l}\text{Simplify the left side and} \\ \text{factor the right side.}\end{array}$$

$$y = a\left(x + \frac{b}{2a}\right)^2 + c - \frac{b^2}{4a} \quad \begin{array}{l}\text{Add } c \text{ to both sides and} \\ \text{subtract } \frac{b^2}{4a} \text{ from both sides.}\end{array}$$

Compare this form with $f(x)$ or $y = a(x - h)^2 + k$ and see that h is $\frac{-b}{2a}$, which means that the x-coordinate of the vertex of the graph of $f(x) = ax^2 + bx + c$ is $\frac{-b}{2a}$.

Let's use this vertex formula in the margin to find the vertex of the parabola we graphed in Example 1.

Example 4   Find the vertex of the graph of $f(x) = x^2 - 4x - 12$.

Solution: In the quadratic function $f(x) = x^2 - 4x - 12$, notice that $a = 1$, $b = -4$, and $c = -12$. Then

$$\frac{-b}{2a} = \frac{-(-4)}{2(1)} = 2$$

The x-value of the vertex is 2. To find the corresponding $f(x)$ or y-value, find $f(2)$. Then

$$f(2) = 2^2 - 4(2) - 12 = 4 - 8 - 12 = -16$$

The vertex is $(2, -16)$. These results agree with our findings in Example 1.

> **VERTEX FORMULA**
>
> The graph of
> $f(x) = ax^2 + bx + c$, when
> $a \neq 0$, is a parabola with vertex
> $$\left(\frac{-b}{2a}, f\left(\frac{-b}{2a}\right)\right)$$

Practice Problem 4

Find the vertex of the graph of $f(x) = x^2 - 4x - 5$. Compare your result with the result of Practice Problem 1.

C FINDING MINIMUM AND MAXIMUM VALUES

The quadratic function whose graph is a parabola that opens upward has a minimum value, and the quadratic function whose graph is a parabola that opens downward has a maximum value. The $f(x)$ or y-value of the vertex is the minimum or maximum value of the function.

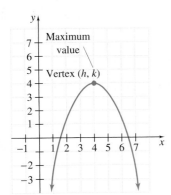

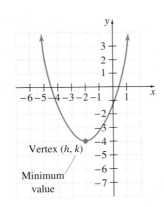

Recall from Section 8.2 that the discriminant, $b^2 - 4ac$, tells us how many solutions the quadratic equation $0 = ax^2 + bx + c$ has. It also tells us how many x-intercepts the graph of a quadratic equation $y = ax^2 + bx + c$ has.

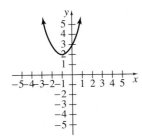

$y = x^2 + 2x + 3$
$b^2 - 4ac < 0$
No x-intercepts

$y = x^2 - 2x + 1$
$b^2 - 4ac = 0$
One x-intercept

$y = x^2 - 2x - 3$
$b^2 - 4ac > 0$
Two x-intercepts

Answer

4. $(2, -9)$

✓ CONCEPT CHECK

Without making any calculations, tell whether the graph of $f(x) = 7 - x - 0.3x^2$ has a maximum value or a minimum value. Explain your reasoning.

Practice Problem 5

An object is thrown upward from the top of a 100-foot cliff. Its height above ground after t seconds is given by the function $f(t) = -16t^2 + 10t + 100$. Find the maximum height of the object and the number of seconds it took for the object to reach its maximum height.

TRY THE CONCEPT CHECK IN THE MARGIN.

Example 5 Finding Maximum Height

A rock is thrown upward from the ground. Its height in feet above ground after t seconds is given by the function $f(t) = -16t^2 + 20t$. Find the maximum height of the rock and the number of seconds it took for the rock to reach its maximum height.

Solution:

1. UNDERSTAND. The maximum height of the rock is the largest value of $f(t)$. Since the function $f(t) = -16t^2 + 20t$ is a quadratic function, its graph is a parabola. It opens downward since $-16 < 0$. Thus, the maximum value of $f(t)$ is the $f(t)$ or y-value of the vertex of its graph.
2. TRANSLATE. To find the vertex (h, k), notice that for $f(t) = -16t^2 + 20t$, $a = -16$, $b = 20$, and $c = 0$. We will use these values and the vertex formula

$$\left(\frac{-b}{2a}, f\left(\frac{-b}{2a}\right)\right)$$

3. SOLVE. $h = \dfrac{-b}{2a} = \dfrac{-20}{-32} = \dfrac{5}{8}$

$$f\left(\frac{5}{8}\right) = -16\left(\frac{5}{8}\right)^2 + 20\left(\frac{5}{8}\right) = -16\left(\frac{25}{64}\right) + \frac{25}{2} = -\frac{25}{4} + \frac{50}{4} = \frac{25}{4}$$

4. INTERPRET. The graph of $f(t)$ is a parabola opening downward with vertex $\left(\frac{5}{8}, \frac{25}{4}\right)$. This means that the rock's maximum height is $\frac{25}{4}$ feet, or $6\frac{1}{4}$ feet, which was reached in $\frac{5}{8}$ second.

Answers

5. Maximum height: $101\frac{9}{16}$ feet in $\frac{5}{16}$ seconds

✓ **Concept Check:** $f(x)$ has a maximum value since it opens downward.

CHAPTER 10 HIGHLIGHTS

DEFINITIONS AND CONCEPTS	EXAMPLES

SECTION 10.1   SOLVING QUADRATIC EQUATIONS BY COMPLETING THE SQUARE

SQUARE ROOT PROPERTY

If b is a real number and if $a^2 = b$, then $a = \pm\sqrt{b}$.

Solve: $(x + 3)^2 = 14$

$$x + 3 = \pm\sqrt{14}$$
$$x = -3 \pm \sqrt{14}$$

TO SOLVE A QUADRATIC EQUATION IN x BY COMPLETING THE SQUARE

Step 1. If the coefficient of x^2 is not 1, divide both sides of the equation by the coefficient of x^2.

Step 2. Get the variable terms alone.

Step 3. Complete the square by adding the square of half of the coefficient of x to both sides.

Step 4. Write the resulting trinomial as the square of a binomial.

Step 5. Use the square root property.

Solve: $3x^2 - 12x - 18 = 0$

1. $x^2 - 4x - 6 = 0$

2. $\qquad x^2 - 4x = 6$

3. $\frac{1}{2}(-4) = -2$ and $(-2)^2 = 4$
    $$x^2 - 4x + 4 = 6 + 4$$

4. $(x - 2)^2 = 10$

5. $x - 2 = \pm\sqrt{10}$
    $$x = 2 \pm \sqrt{10}$$

SECTION 10.2   SOLVING QUADRATIC EQUATIONS BY THE QUADRATIC FORMULA

QUADRATIC FORMULA

A quadratic equation written in the form $ax^2 + bx + c = 0$ has solutions

$$x = \frac{-b \pm \sqrt{b^2 - 4ac}}{2a}$$

Solve: $x^2 - x - 3 = 0$

$$a = 1, b = -1, c = -3$$
$$x = \frac{-(-1) \pm \sqrt{(-1)^2 - 4(1)(-3)}}{2 \cdot 1}$$
$$x = \frac{1 \pm \sqrt{13}}{2}$$

SECTION 10.3   SOLVING EQUATIONS BY USING QUADRATIC METHODS

Substitution is often helpful in solving an equation that contains a repeated variable expression.

Solve: $(2x + 1)^2 - 5(2x + 1) + 6 = 0$

Let $m = 2x + 1$. Then

$$m^2 - 5m + 6 = 0 \qquad \text{Let } m = 2x + 1.$$
$$(m - 3)(m - 2) = 0$$

$$m = 3 \quad \text{or} \quad m = 2$$
$$2x + 1 = 3 \qquad 2x + 1 = 2 \qquad \text{Substitute}$$
$$x = 1 \qquad\qquad x = \frac{1}{2} \qquad \text{back.}$$

SECTION 10.4 NONLINEAR INEQUALITIES IN ONE VARIABLE

TO SOLVE A POLYNOMIAL INEQUALITY

Step 1. Write the inequality in standard form.

Step 2. Solve the related equation.

Step 3. Use solutions from Step 2 to separate the number line into regions.

Step 4. Use a test point to determine whether values in each region satisfy the original inequality.

Step 5. Write the solution set as the union of regions whose test point values are solutions.

TO SOLVE A RATIONAL INEQUALITY

Step 1. Solve for values that make all denominators 0.

Step 2. Solve the related equation.

Step 3. Use solutions from Steps 1 and 2 to separate the number line into regions.

Step 4. Use a test point to determine whether values in each region satisfy the original inequality.

Step 5. Write the solution set as the union of regions whose test point value is a solution.

Solve: $x^2 \geq 6x$

1. $x^2 - 6x \geq 0$

2. $x^2 - 6x = 0$

$x(x - 6) = 0$

$x = 0$ or $x = 6$

3.

4.

Region	Test Point Value	$x^2 \geq 6x$	Result
A	-2	$(-2)^2 \geq 6(-2)$	True.
B	1	$1^2 \geq 6(1)$	False.
C	7	$7^2 \geq 6(7)$	True.

5.

The solution set is $(-\infty, 0] \cup [6, \infty)$.

Solve: $\dfrac{6}{x - 1} < -2$

1. $x - 1 = 0$ Set denominator equal to 0.

$x = 1$

2. $\dfrac{6}{x - 1} = -2$

$6 = -2(x - 1)$ Multiply by $(x - 1)$.

$6 = -2x + 2$

$4 = -2x$

$-2 = x$

3.

4. Only a test value from region B satisfies the original inequality.

5.

The solution set is $(-2, 1)$.

SECTION 10.5   QUADRATIC FUNCTIONS AND THEIR GRAPHS

GRAPH OF A QUADRATIC FUNCTION

The graph of a quadratic function written in the form $f(x) = a(x - h)^2 + k$ is a parabola with vertex (h, k). If $a > 0$, the parabola opens upward; if $a < 0$, the parabola opens downward. The axis of symmetry is the line whose equation is $x = h$.

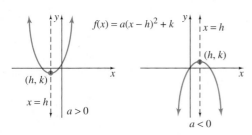

Graph: $g(x) = 3(x - 1)^2 + 4$

The graph is a parabola with vertex $(1, 4)$ and axis of symmetry $x = 1$. Since $a = 3$ is positive, the graph opens upward.

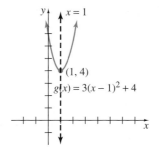

SECTION 10.6   FURTHER GRAPHING OF QUADRATIC FUNCTIONS

The graph of $f(x) = ax^2 + bx + c$, $a \neq 0$, is a parabola with vertex

$$\left(\frac{-b}{2a}, f\left(\frac{-b}{2a} \right) \right).$$

Graph: $f(x) = x^2 - 2x - 8$. Find the vertex and x- and y-intercepts.

$$\frac{-b}{2a} = \frac{-(-2)}{2 \cdot 1} = 1$$
$$f(1) = 1^2 - 2(1) - 8 = -9$$

The vertex is $(1, -9)$.

$$0 = x^2 - 2x - 8$$
$$0 = (x - 4)(x + 2)$$
$$x = 4 \text{ or } x = -2$$

The x-intercept points are $(4, 0)$ and $(-2, 0)$.

$$f(0) = 0^2 - 2 \cdot 0 - 8 = -8$$

The y-intercept point is $(0, -8)$.

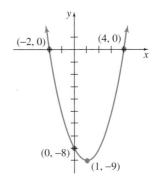

Exponential and Logarithmic Functions

In this chapter, we discuss two closely related functions: exponential and logarithmic functions. These functions are vital in applications in economics, finance, engineering, the sciences, education, and other fields. Models of tumor growth and learning curves are two examples of the uses of exponential and logarithmic functions.

An earthquake is a series of vibrations in the crust of the earth. The size, or magnitude, of an earthquake is measured on the Richter scale. The magnitude of an earthquake can range broadly, from barely detectable (2.5 or less on the Richter scale) to massively destructive (7.0 or greater on the Richter scale). According to the United States Geological Survey, earthquakes are an everyday occurrence. In 1997, there were a total of 20,824 earthquakes around the world, or an average of 57 earthquakes per day. However, most of these were minor tremors with magnitudes of 3.9 or less, and many could only be detected by seismographs. Only 0.09% of the earthquakes occurring during 1997 could be classified as major earthquakes (7.0 or greater on the Richter scale). Even so, earthquakes were responsible for 2907 deaths that year.

Problem Solving Notes

11.1   THE ALGEBRA OF FUNCTIONS

A   ADDING, SUBTRACTING, MULTIPLYING, AND DIVIDING FUNCTIONS

As we have seen in earlier chapters, it is possible to add, subtract, multiply, and divide functions. Although we have not stated it as such, the sums, differences, products, and quotients of functions are themselves functions. For example, if $f(x) = 3x$ and $g(x) = x + 1$, their product, $f(x) \cdot g(x) = 3x(x + 1) = 3x^2 + 3x$, is a new function. We can use the notation $(f \cdot g)(x)$ to denote this new function. Finding the sum, difference, product, and quotient of functions to generate new functions is called the **algebra of functions**.

> **ALGEBRA OF FUNCTIONS**
>
> Let f and g be functions. New functions from f and g are defined as follows:
>
> Sum $(f + g)(x) = f(x) + g(x)$
> Difference $(f - g)(x) = f(x) - g(x)$
> Product $(f \cdot g)(x) = f(x) \cdot g(x)$
> Quotient $\left(\dfrac{f}{g}\right)(x) = \dfrac{f(x)}{g(x)}$

Example 1   If $f(x) = x - 1$ and $g(x) = 2x - 3$, find:

  a. $(f + g)(x)$
  b. $(f - g)(x)$
  c. $(f \cdot g)(x)$
  d. $\left(\dfrac{f}{g}\right)(x)$

Solution: Use the algebra of functions and replace $f(x)$ by $x - 1$ and $g(x)$ by $2x - 3$. Then we simplify.

  a. $(f + g)(x) = f(x) + g(x)$
           $= (x - 1) + (2x - 3)$
           $= 3x - 4$
  b. $(f - g)(x) = f(x) - g(x)$
           $= (x - 1) - (2x - 3)$
           $= x - 1 - 2x + 3$
           $= -x + 2$
  c. $(f \cdot g)(x) = f(x) \cdot g(x)$
           $= (x - 1)(2x - 3)$
           $= 2x^2 - 5x + 3$
  d. $\left(\dfrac{f}{g}\right)(x) = \dfrac{f(x)}{g(x)} = \dfrac{x - 1}{2x - 3}$, where $x \neq \dfrac{3}{2}$ ▬▬▬

There is an interesting but not surprising relationship between the graphs of functions and the graph of their sum, difference, product, and quotient. For example, the graph of $(f + g)(x)$ can be found by adding the graph of $f(x)$ to the graph of $g(x)$. We add two graphs by adding corresponding y-values.

Objectives

A Add, subtract, multiply, and divide functions.

B Compose functions.

SSM   CD-ROM   Video
11.1

Practice Problem 1

If $f(x) = x + 3$ and $g(x) = 3x - 1$, find:

a. $(f + g)(x)$
b. $(f - g)(x)$
c. $(f \cdot g)(x)$
d. $\left(\dfrac{f}{g}\right)(x)$

Answers

1. a. $4x + 2$, **b.** $-2x + 4$, **c.** $3x^2 + 8x - 3$,
d. $\dfrac{x + 3}{3x - 1}$ where $x \neq \dfrac{1}{3}$

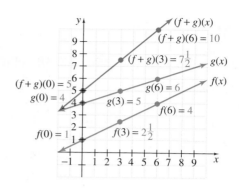

B COMPOSITION OF FUNCTIONS

Another way to combine functions is called **function composition**. To understand this new way of combining functions, study the tables below. They show degrees Fahrenheit converted to equivalent degrees Celsius, and then degrees Celsius converted to equivalent degrees Kelvin. (The Kelvin scale is a temperature scale devised by Lord Kelvin in 1848.)

x = Degrees Fahrenheit (Input)	−31	−13	32	68	149	212
$C(x)$ = Degrees Celsius (Output)	−35	−25	0	20	65	100

C = Degrees Celsius (Input)	−35	−25	0	20	65	100
$K(C)$ = Kelvins (Output)	238.15	248.15	273.15	293.15	338.15	373.15

Suppose that we want a table that shows a direct conversion from degrees Fahrenheit to kelvins. In other words, suppose that a table is needed that shows kelvins as a function of degrees Fahrenheit. This can easily be done because in the tables, the output of the first table is the same as the input of the second table. The new table is as follows.

x = Degrees Fahrenheit (Input)	−31	−13	32	68	149	212
$K(C(x))$ = Kelvins (Output)	238.15	248.15	273.15	293.15	338.15	373.15

Since the output of the first table is used as the input of the second table, we write the new function as $K(C(x))$. The new function is formed from the composition of the other two functions. The mathematical symbol for this composition is $(K \circ C)(x)$. Thus, $(K \circ C)(x) = K(C(x))$.

It is possible to find an equation for the composition of the two functions $C(x)$ and $K(x)$. In other words, we can find a function that converts degrees Fahrenheit directly to kelvins. The function $C(x) = \dfrac{5}{9}(x - 32)$ converts degrees Fahrenheit to degrees Celsius, and the function $K(C) = C + 273.15$ converts degrees Celsius to kelvins. Thus,

$$(K \circ C)(x) = K(C(x)) = K\left(\frac{5}{9}(x - 32)\right) = \frac{5}{9}(x - 32) + 273.15$$

In general, the notation **$f(g(x))$** means "f composed with g" and can be written as **$(f \circ g)(x)$**. Also $g(f(x))$, or $(g \circ f)(x)$, means "g composed with f."

COMPOSITE FUNCTIONS

The composition of functions f and g is

$$(f \circ g)(x) = f(g(x))$$

HELPFUL HINT

$(f \circ g)(x)$ does not mean the same as $(f \cdot g)(x)$.

$$(f \circ g)(x) = f(g(x)) \text{ while } (f \cdot g)(x) = f(x) \cdot g(x)$$

Example 2   If $f(x) = x^2$ and $g(x) = x + 3$, find each composition.

a. $(f \circ g)(2)$ and $(g \circ f)(2)$
b. $(f \circ g)(x)$ and $(g \circ f)(x)$

Solution:

a. $(f \circ g)(2) = f(g(2))$
$= f(5)$ Since $g(x) = x + 3$, then $g(2) = 2 + 3 = 5$.
$= 5^2 = 25$
$(g \circ f)(2) = g(f(2))$
$= g(4)$ Since $f(x) = x^2$, then $f(2) = 2^2 = 4$.
$= 4 + 3 = 7$

b. $(f \circ g)(x) = f(g(x))$
$= f(x + 3)$ Replace $g(x)$ with $x + 3$.
$= (x + 3)^2$ $f(x + 3) = (x + 3)^2$
$= x^2 + 6x + 9$ Square $(x + 3)$.
$(g \circ f)(x) = g(f(x))$
$= g(x^2)$ Replace $f(x)$ with x^2.
$= x^2 + 3$ $g(x^2) = x^2 + 3$

Example 3   If $f(x) = |x|$ and $g(x) = x - 2$, find each composition.

a. $(f \circ g)(x)$
b. $(g \circ f)(x)$

Solution: **a.** $(f \circ g)(x) = f(g(x)) = f(x - 2) = |x - 2|$
b. $(g \circ f)(x) = g(f(x)) = g(|x|) = |x| - 2$

HELPFUL HINT

In Examples 2 and 3, notice that $(g \circ f)(x) \neq (f \circ g)(x)$. In general, $(g \circ f)(x)$ *may* or *may not* equal $(f \circ g)(x)$.

Example 4   If $f(x) = 5x$, $g(x) = x - 2$, and $h(x) = \sqrt{x}$, write each function as a composition with f, g, or h.

a. $F(x) = \sqrt{x - 2}$
b. $G(x) = 5x - 2$

Practice Problem 2

If $f(x) = x^2$ and $g(x) = 2x + 1$, find each composition.

a. $(f \circ g)(3)$ and $(g \circ f)(3)$
b. $(f \circ g)(x)$ and $(g \circ f)(x)$

Practice Problem 3

If $f(x) = \sqrt{x}$ and $g(x) = x + 1$, find each composition.

a. $(f \circ g)(x)$
b. $(g \circ f)(x)$

Practice Problem 4

If $f(x) = 2x$, $g(x) = x + 5$, and $h(x) = |x|$, write each function as a composition of f, g, or h.

a. $F(x) = |x + 5|$
b. $G(x) = 2x + 5$

Answers

2. a. 49; 19, **b.** $4x^2 + 4x + 1$; $2x^2 + 1$,
3. a. $\sqrt{x + 1}$, **b.** $\sqrt{x} + 1$, **4. a.** $(h \circ g)(x)$,
b. $(g \circ f)(x)$

Solution: **a.** Notice the order in which the function F operates on an input value x. First, 2 is subtracted from x, and then the square root of that result is taken. This means that $F = h \circ g$. To check, we find $h \circ g$:

$$(h \circ g)(x) = h(g(x)) = h(x - 2) = \sqrt{x - 2}$$

b. Notice the order in which the function G operates on an input value x. First, x is multiplied by 5, and then 2 is subtracted from the result. This means that $G = g \circ f$. To check, we find $g \circ f$:

$$(g \circ f)(x) = g(f(x)) = g(5x) = 5x - 2$$

GRAPHING CALCULATOR EXPLORATIONS

If $f(x) = \frac{1}{2}x + 2$ and $g(x) = \frac{1}{3}x^2 + 4$, then

$$(f + g)(x) = f(x) + g(x)$$

$$= \left(\frac{1}{2}x + 2\right) + \left(\frac{1}{3}x^2 + 4\right)$$

$$= \frac{1}{3}x^2 + \frac{1}{2}x + 6.$$

To visualize this addition of functions with a grapher, graph

$$Y_1 = \frac{1}{2}x + 2, \qquad Y_2 = \frac{1}{3}x^2 + 4, \qquad Y_3 = \frac{1}{3}x^2 + \frac{1}{2}x + 6$$

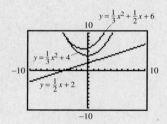

Use a TABLE feature to verify that for a given x value, $Y_1 + Y_2 = Y_3$. For example, verify that when $x = 0$, $Y_1 = 2$, $Y_2 = 4$ and $Y_3 = 2 + 4 = 6$.

11.2   INVERSE FUNCTIONS

In the next section, we begin a study of two new functions: exponential and logarithmic functions. As we learn more about these functions, we will discover that they share a special relation to each other; they are inverses of each other.

Before we study these functions, we need to learn about inverses. We begin by defining one-to-one functions.

A   DETERMINING WHETHER A FUNCTION IS ONE-TO-ONE

Study the following table.

Degrees Fahrenheit (Input)	−31	−13	32	68	149	212
Degrees Celsius (Output)	−35	−25	0	20	65	100

Recall that since each degrees Fahrenheit (input) corresponds to exactly one degrees Celsius (output), this table of inputs and outputs does describe a function. Also notice that each output corresponds to a different input. This type of function is given a special name—a one-to-one function.

Does the set $f = \{(0, 1), (2, 2), (-3, 5), (7, 6)\}$ describe a one-to-one function? It is a function since each x-value corresponds to a unique y-value. For this particular function f, each y-value corresponds to a unique x-value. Thus, this function is also a **one-to-one function**.

ONE-TO-ONE FUNCTION

For a **one-to-one function**, each x-value (input) corresponds to only one y-value (output) and each y-value (output) corresponds to only one x-value (input).

Examples   Determine whether each function described is one-to-one.

1. $f\{(6, 2), (5, 4), (-1, 0), (7, 3)\}$

The function f is one-to-one since each y-value corresponds to only one x-value.

2. $g = \{(3, 9), (-4, 2), (-3, 9), (0, 0)\}$

The function g is not one-to-one because the y-value 9 in $(3, 9)$ and $(-3, 9)$ corresponds to two different x-values.

3. $h = \{(1, 1), (2, 2), (10, 10), (-5, -5)\}$

The function h is one-to-one since each y-value corresponds to only one x-value.

4.

Mineral (Input)	Talc	Gypsum	Diamond	Topaz	Stibnite
Hardness on the Mohs Scale (Output)	1	2	10	8	2

Objectives

A Determine whether a function is a one-to-one function.

B Use the horizontal line test to decide whether a function is a one-to-one function.

C Find the inverse of a function.

D Find the equation of the inverse of a function.

E Graph functions and their inverses.

SSM   CD-ROM   Video 11.2

Practice Problems 1–5

Determine whether each function described is one-to-one.

1. $f = \{(7, 3), (-1, 1), (5, 0), (4, -2)\}$

2. $g = \{(-3, 2), (6, 3), (2, 14), (-6, 2)\}$

3. $h = \{(0, 0), (1, 2), (3, 4), (5, 6)\}$

4.

State (Input)	Colo-rado	Missis-sippi	Nevada	New Mexico	Utah
Number of Colleges and Universities (Output)	9	44	13	44	21

Source: The Chronicle of Higher Education, Vol. XLV, No. 1, August 28, 1998.

5.

<!-- graph with plotted points on coordinate plane -->

Answers

1. one-to-one, **2.** not one-to-one, **3.** one-to-one, **4.** not one-to-one, **5.** not one-to-one

This table does not describe a one-to-one function since the output 2 corresponds to two different inputs, gypsum and stibnite.

5.

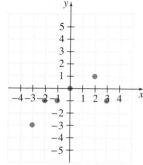

This graph does not describe a one-to-one function since the y-value -1 corresponds to three different x-values, -2, -1, and 3, as shown to the right.

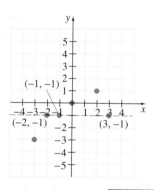

B USING THE HORIZONTAL LINE TEST

Recall that we recognize the graph of a function when it passes the vertical line test. Since every x-value of the function corresponds to exactly one y-value, each vertical line intersects the function's graph at most once. The graph shown next, for instance, is the graph of a function.

Is this function a *one-to-one* function? The answer is no. To see why not, notice that the y-value of the ordered pair $(-3, 3)$, for example, is the same as the y-value of the ordered pair $(3, 3)$. This function is therefore not one-to-one.

To test whether a graph is the graph of a one-to-one function, we can apply the vertical line test to see if it is a function, and then apply a similar **horizontal line test** to see if it is a one-to-one function.

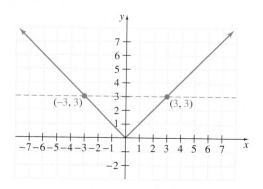

Example 6   Use the vertical and horizontal line tests to determine whether each graph is the graph of a one-to-one function.

a.

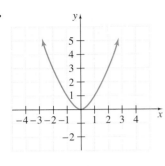

b.

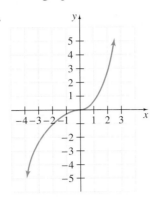

c.

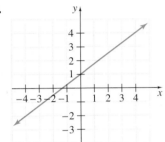

d.

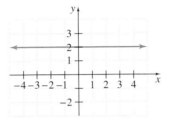

e.

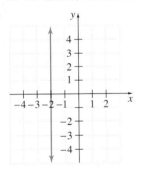

Solution: Graphs **a**, **b**, **c**, and **d** all pass the vertical line test, so only these graphs are graphs of functions. But, of these, only **b** and **c** pass the horizontal line test, so only **b** and **c** are graphs of one-to-one functions.

HELPFUL HINT

All linear equations are one-to-one functions except those whose graphs are horizontal or vertical lines. A vertical line does not pass the vertical line test and hence is not the graph of a function. A horizontal line is the graph of a function but does not pass the horizontal line test and hence is not the graph of a one-to-one function.

C FINDING THE INVERSE OF A FUNCTION

One-to-one functions are special in that their graphs pass the vertical and horizontal line tests. They are special, too, in another sense: We can find its

Practice Problem 6

Use the vertical and horizontal line tests to determine whether each graph is the graph of a one-to-one function.

a.

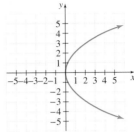

b.

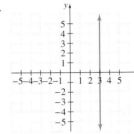

c.

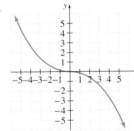

d.

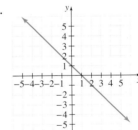

e.

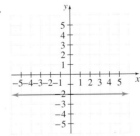

Answers

6. a. not one-to-one,   **b.** not one-to-one,
c. one-to-one,   **d.** one-to-one,   **e.** not one-to-one

inverse function for any one-to-one function by switching the coordinates of the ordered pairs of the function, or the inputs and the outputs. For example, the inverse of the one-to-one function

Degrees Fahrenheit (Input)	−31	−13	32	68	149	212
Degrees Celsius (Output)	−35	−25	0	20	65	100

is the function

Degrees Celsius (Input)	−35	−25	0	20	65	100
Degrees Fahrenheit (Output)	−31	−13	32	68	149	212

Notice that the ordered pair $(-31, -35)$ of the function, for example, becomes the ordered pair $(-35, -31)$ of its inverse.

Also, the inverse of the one-to-one function $f = \{(2, -3), (5, 10), (9, 1)\}$ is $\{(-3, 2), (10, 5), (1, 9)\}$. For a function f, we use the notation f^{-1}, read "f inverse," to denote its inverse function. Notice that since the coordinates of each ordered pair have been switched, the domain (set of inputs) of f is the range (set of outputs) of f^{-1}, and the range of f is the domain of f^{-1}. See the definition of inverse function in the margin.

Practice Problem 7

Find the inverse of the one-to-one function: $f = \{(2, -4), (-1, 13), (0, 0), (-7, -8)\}$

✓ **CONCEPT CHECK**

Suppose that $f(x)$ is a one-to-one function. If the ordered pair $(1, 5)$ belongs to f, name one point that we know must belong to the inverse function f^{-1}.

HELPFUL HINT

The symbol f^{-1} is the single symbol used to denote the inverse of the function f. It is read as "f inverse." This symbol *does not* mean $\dfrac{1}{f}$.

Practice Problem 8

Find the equation of the inverse of $f(x) = x - 6$.

Answers

7. $f^{-1} = \{(-4, 2), (13, -1), (0, 0), (-8, -7)\}$,
8. $f^{-1}(x) = x + 6$
✓ **Concept Check:** $(5, 1)$ belongs to f^{-1}

Example 7 Find the inverse of the one-to-one function:

$$f = \{(0, 1), (-2, 7), (3, -6), (4, 4)\}$$

Solution: $f^{-1} = \{(1, 0), (7, -2), (-6, 3), (4, 4)\}$

Switch coordinates of each ordered pair.

TRY THE CONCEPT CHECK IN THE MARGIN.

D FINDING THE EQUATION OF THE INVERSE OF A FUNCTION

If a one-to-one function f is defined as a set of ordered pairs, we can find f^{-1} by interchanging the x- and y-coordinates of the ordered pairs. If a one-to-one function f is given in the form of an equation, we can find the equation of f^{-1} by using a similar procedure.

FINDING AN EQUATION OF THE INVERSE OF A ONE-TO-ONE FUNCTION $f(x)$

Step 1. Replace $f(x)$ with y.
Step 2. Interchange x with y.
Step 3. Solve the equation for y.
Step 4. Replace y with the notation $f^{-1}(x)$.

Example 8 Find an equation of the inverse of $f(x) = x + 3$.

Solution: $f(x) = x + 3$

Step 1. $y = x + 3$ Replace $f(x)$ with y.

Step 2. $x = y + 3$ Interchange x and y.

Step 3. $x - 3 = y$ Solve for y.

Step 4. $f^{-1}(x) = x - 3$ Replace y with $f^{-1}(x)$.

The inverse of $f(x) = x + 3$ is $f^{-1}(x) = x - 3$. Notice that, for example,

$$f(1) = 1 + 3 = 4 \quad \text{and} \quad f^{-1}(4) = 4 - 3 = 1$$

Ordered pair: $(1, 4)$ Ordered pair: $(4, 1)$

The coordinates are switched, as expected.

Example 9   Find the equation of the inverse of $f(x) = 3x - 5$. Graph f and f^{-1} on the same set of axes.

Solution: $f(x) = 3x - 5$

Step 1. $y = 3x - 5$ Replace $f(x)$ with y.

Step 2. $x = 3y - 5$ Interchange x and y.

Step 3. $3y = x + 5$ Solve for y.

$$y = \frac{x + 5}{3}$$

Step 4. $f^{-1}(x) = \dfrac{x + 5}{3}$ Replace y with $f^{-1}(x)$.

Now we graph $f(x)$ and $f^{-1}(x)$ on the same set of axes. Both $f(x) = 3x - 5$ and $f^{-1}(x) = \dfrac{x + 5}{3}$ are linear functions, so each graph is a line.

$$f(x) = 3x - 5 \qquad\qquad f^{-1}(x) = \frac{x + 5}{3}$$

x	$y = f(x)$
1	-2
0	-5
$\frac{5}{3}$	0

x	$y = f^{-1}(x)$
-2	1
-5	0
0	$\frac{5}{3}$

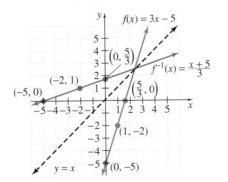

E GRAPHING INVERSE FUNCTIONS

Notice that the graphs of f and f^{-1} in Example 9 are mirror images of each other, and the "mirror" is the dashed line $y = x$. This is true for every function and its inverse. For this reason, we say that *the graphs of f and f^{-1} are symmetric about the line $y = x$.*

Practice Problem 9

Find an equation of the inverse of $f(x) = 2x + 3$. Graph f and f^{-1} on the same set of axes.

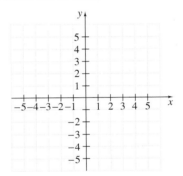

Answers

9. $f^{-1}(x) = \dfrac{x - 3}{2}$

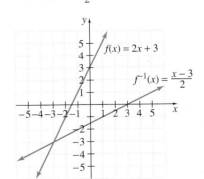

To see why this happens, study the graph of a few ordered pairs and their switched coordinates.

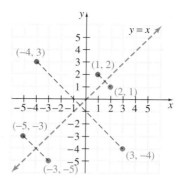

Practice Problem 10

Graph the inverse of each function.

a.

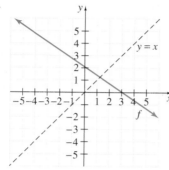

b.

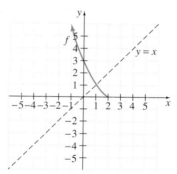

Example 10 Graph the inverse of each function.

Solution: The function is graphed in blue and the inverse is graphed in red.

a.

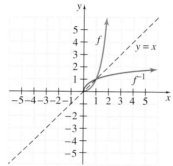

b.

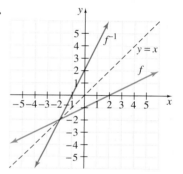

Answers

10. a.

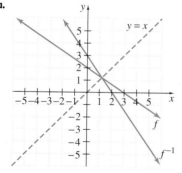

b.

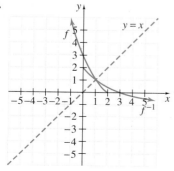

GRAPHING CALCULATOR EXPLORATIONS

A grapher can be used to visualize functions and their inverses. Recall that the graph of a function f and its inverse f^{-1} are mirror images of each other across the line $y = x$. To see this for the function $f(x) = 3x + 2$, use a square window and graph

the given function: $Y_1 = 3x + 2$

its inverse: $Y_2 = \dfrac{x - 2}{3}$

and the line: $Y_3 = x$

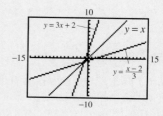

11.3 EXPONENTIAL FUNCTIONS

Objectives

A Graph exponential functions.

B Solve equations of the form $b^x = b^y$.

C Solve problems modeled by exponential equations.

SSM   CD-ROM   Video
11.3

In earlier chapters, we gave meaning to exponential expressions such as 2^x, where x is a rational number. For example,

$$2^3 = 2 \cdot 2 \cdot 2 \qquad \text{Three factors; each factor is 2}$$
$$2^{3/2} = (2^{1/2})^3 = \sqrt{2} \cdot \sqrt{2} \cdot \sqrt{2} \qquad \text{Three factors; each factor is } \sqrt{2}$$

When x is an irrational number (for example, $\sqrt{3}$), what meaning can we give to $2^{\sqrt{3}}$?

It is beyond the scope of this book to give precise meaning to 2^x if x is irrational. We can confirm your intuition and say that $2^{\sqrt{3}}$ is a real number, and since $1 < \sqrt{3} < 2$, then $2^1 < 2^{\sqrt{3}} < 2^2$. We can also use a calculator and approximate $2^{\sqrt{3}}$: $2^{\sqrt{3}} \approx 3.321997$. In fact, as long as the base b is positive, b^x is a real number for all real numbers x. Finally, the rules of exponents apply whether x is rational or irrational, as long as b is positive.

In this section, we are interested in functions of the form $f(x) = b^x$, where $b > 0$. A function of this form is called an **exponential function**.

EXPONENTIAL FUNCTION

A function of the form

$$f(x) = b^x$$

is called an **exponential function** if $b > 0$, b is not 1, and x is a real number.

A GRAPHING EXPONENTIAL FUNCTIONS

Now let's practice graphing exponential functions.

Example 1   Graph the exponential functions defined by $f(x) = 2^x$ and $g(x) = 3^x$ on the same set of axes.

Solution: To graph these functions, we find some ordered pair solutions, plot the points, and connect them with a smooth curve.

$f(x) = 2^x$

x	0	1	2	3	-1	-2
$f(x)$	1	2	4	8	$\frac{1}{2}$	$\frac{1}{4}$

$g(x) = 3^x$

x	0	1	2	3	-1	-2
$g(x)$	1	3	9	27	$\frac{1}{3}$	$\frac{1}{9}$

Practice Problem 1

Graph the exponential function $f(x) = 4^x$.

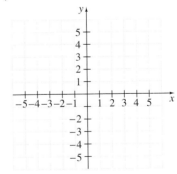

Answer

1.

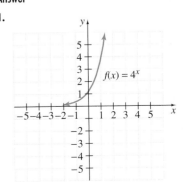

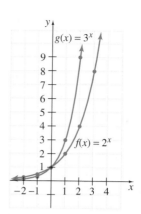

A number of things should be noted about the two graphs of exponential functions in Example 1. First, the graphs show that $f(x) = 2^x$ and $g(x) = 3^x$ are one-to-one functions since each graph passes the vertical and horizontal line tests. The y-intercept of each graph is 1, but neither graph has an x-intercept. From the graph, we can also see that the domain of each function is all real numbers and that the range is $(0, \infty)$. We can also see that as x-values are increasing, y-values are increasing also.

Practice Problem 2

Graph the exponential function $f(x) = \left(\dfrac{1}{5}\right)^x$.

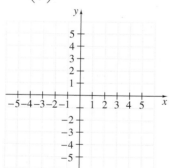

Example 2   Graph the exponential functions $y = \left(\dfrac{1}{2}\right)^x$ and $y = \left(\dfrac{1}{3}\right)^x$ on the same set of axes.

Solution: As before, we find some ordered pair solutions, plot the points, and connect them with a smooth curve.

$y = \left(\dfrac{1}{2}\right)^x$

x	0	1	2	3	-1	-2
y	1	$\dfrac{1}{2}$	$\dfrac{1}{4}$	$\dfrac{1}{8}$	2	4

$y = \left(\dfrac{1}{3}\right)^x$

x	0	1	2	3	-1	-2
y	1	$\dfrac{1}{3}$	$\dfrac{1}{9}$	$\dfrac{1}{27}$	3	9

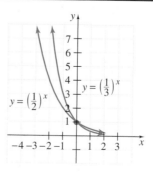

Answer

2.

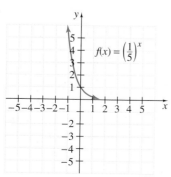

Each function in Example 2 again is a one-to-one function. The y-intercept of both is 1. The domain is the set of all real numbers, and the range is $(0, \infty)$.

Notice the difference between the graphs of Example 1 and the graphs of Example 2. An exponential function is always increasing if the base is greater than 1. When the base is between 0 and 1, the graph is always decreasing. The following figures summarize these characteristics of exponential functions.

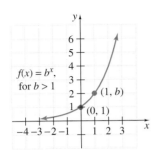 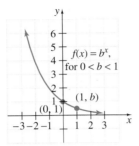

Example 3   Graph the exponential function $f(x) = 3^{x+2}$.

Solution: As before, we find and plot a few ordered pair solutions. Then we connect the points with a smooth curve.

$y = 3^{x+2}$

x	0	-1	-2	-3	-4
y	9	3	1	$\frac{1}{3}$	$\frac{1}{9}$

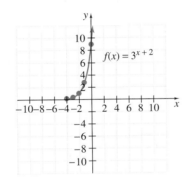

TRY THE CONCEPT CHECK IN THE MARGIN.

B   SOLVING EQUATIONS OF THE FORM $b^x = b^y$

We have seen that an exponential function $y = b^x$ is a one-to-one function. Another way of stating this fact is a property that we can use to solve exponential equations.

> **UNIQUENESS OF b^x**
>
> Let $b > 0$ and $b \neq 1$. Then $b^x = b^y$ is equivalent to $x = y$.

Example 4   Solve: $2^x = 16$

Solution: We write 16 as a power of 2 and then use the uniqueness of b^x to solve.

$2^x = 16$

$2^x = 2^4$

Practice Problem 3

Graph the exponential function $f(x) = 2^{x-1}$.

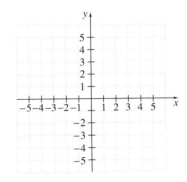

✓ CONCEPT CHECK

Which functions are exponential functions?

a. $f(x) = x^3$ b. $g(x) = \left(\dfrac{2}{3}\right)^x$

c. $h(x) = 5^{x-2}$ d. $w(x) = (2x)^2$

Answers

3.

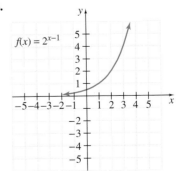

✓ Concept Check: b and c are exponential

Practice Problem 4

Solve: $5^x = 125$

Practice Problem 5

Solve: $4^x = 8$

Practice Problem 6

Solve: $9^{x-1} = 27^x$

Since the bases are the same and are nonnegative, by the uniqueness of b^x we then have that the exponents are equal. Thus,

$$x = 4$$

To check, we replace x with 4 in the original equation. The solution set is $\{4\}$.

Example 5   Solve: $9^x = 27$

Solution: Since both 9 and 27 are powers of 3, we can use the uniqueness of b^x.

$$9^x = 27$$
$$(3^2)^x = 3^3 \quad \text{Write 9 and 27 as powers of 3.}$$
$$3^{2x} = 3^3$$
$$2x = 3 \quad \text{Use the uniqueness of } b^x.$$
$$x = \frac{3}{2} \quad \text{Divide both sides by 2.}$$

To check, we replace x with $\frac{3}{2}$ in the original equation. The solution set is $\left\{\frac{3}{2}\right\}$.

Example 6   Solve: $4^{x+3} = 8^x$

Solution: We write both 4 and 8 as powers of 2, and then use the uniqueness of b^x.

$$4^{x+3} = 8^x$$
$$(2^2)^{x+3} = (2^3)^x$$
$$2^{2x+6} = 2^{3x}$$
$$2x + 6 = 3x \quad \text{Use the uniqueness of } b^x.$$
$$6 = x \quad \text{Subtract } 2x \text{ from both sides.}$$

Check to see that the solution set is $\{6\}$.

There is one major problem with the preceding technique. Often the two sides of an equation, $4 = 3^x$ for example, cannot easily be written as powers of a common base. We explore how to solve such an equation with the help of **logarithms** later.

C SOLVING PROBLEMS MODELED BY EXPONENTIAL EQUATIONS

The bar graph on the next page shows the increase in the number of cellular phone users. Notice that the graph of the exponential function $y = 3.638 \, (1.443)^x$ approximates the heights of the bars. This is just one example of how the world abounds with patterns that can be modeled by exponential functions. To make these applications realistic, we use numbers that warrant a calculator. Another application of an exponential function has to do with interest rates on loans.

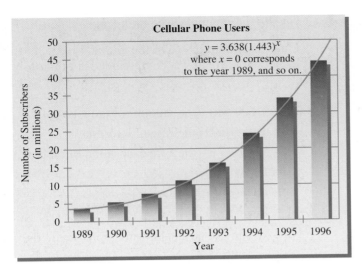

Cellular Phone Users

$y = 3.638(1.443)^x$
where $x = 0$ corresponds
to the year 1989, and so on.

Number of Subscribers (in millions)

Year

The exponential function defined by $A = P\left(1 + \dfrac{r}{n}\right)^{nt}$ models the pattern relating the dollars A accrued (or owed) after P dollars are invested (or loaned) at an annual rate of interest r compounded n times each year for t years. This function is known as the compound interest formula.

Example 7   Using the Compound Interest Formula

Find the amount owed at the end of 5 years if $1600 is loaned at a rate of 9% compounded monthly.

Solution: We use the formula $A = P\left(1 + \dfrac{r}{n}\right)^{nt}$, with the following values:

$P = \$1600$ (the amount of the loan)

$r = 9\% = 0.09$ (the annual rate of interest)

$n = 12$ (the number of times interest is compounded each year)

$t = 5$ (the duration of the loan, in years)

$A = P\left(1 + \dfrac{r}{n}\right)^{nt}$ Compound interest formula

$\quad = 1600\left(1 + \dfrac{0.09}{12}\right)^{12(5)}$ Substitute known values.

$\quad = 1600(1.0075)^{60}$

To approximate A, use the $\boxed{y^x}$ or $\boxed{\wedge}$ key on your calculator.

$$\boxed{2505.0896}$$

Thus, the amount A owed is approximately $2505.09.

When interest is compounded continuously, the formula $A = Pe^{rt}$ is used, where r is the annual interest rate and interest is compounded continuously for t years. The number e used in the formula is an irrational number approximately equal to 2.7183.

Practice Problem 7

As a result of the Chernobyl nuclear accident, radioactive debris was carried through the atmosphere. One immediate concern was the impact that debris had on the milk supply. The percent y of radioactive material in raw milk after t days is estimated by $y = 100(2.7)^{-0.1t}$. Estimate the expected percent of radioactive material in the milk after 30 days.

Answer

7. approximately 5.08%

Practice Problem 8

Find the amount owed at the end of 3 years if $1200 is loaned at a rate of 8% compounded continuously.

Example 8   Finding the Amount Owed on a Loan

Find the amount owed at the end of 5 years if $1600 is loaned at a rate of 9% compounded continuously.

Solution: We use the formula $A = Pe^{rt}$, where

$P = \$1600$ (the amount of the loan)

$r = 9\% = 0.09$ (the rate of interest)

$t = 5$ (the 5-year duration of the loan)

$A = Pe^{rt}$

$\quad = 1600e^{0.09(5)}$ Substitute known values.

$\quad = 1600e^{0.45}$

Now we can use a calculator to approximate the solution.

$A \approx 2509.30$

The total amount of money owed is approximately $2509.30.

Answer

8. $1525.50

GRAPHING CALCULATOR EXPLORATIONS

We can use a graphing calculator and its TRACE feature to solve Practice Problem 7 graphically.

To estimate the expected percent of radioactive material in the milk after 30 days, enter $Y_1 = 100(2.7)^{-0.1x}$. The graph does not appear on a standard viewing window, so we need to determine an appropriate viewing window. Because it doesn't make sense to look at radioactivity *before* the Chernobyl nuclear accident, we use Xmin = 0. We are interested in finding the percent of radioactive material in the milk when $x = 30$, so we choose Xmax = 35 to leave enough space to see the graph at $x = 30$. Because the values of y are percents, it seems appropriate that $0 \leq y \leq 100$. (We also use Xscl = 1 and Yscl = 10.) Now we graph the function.

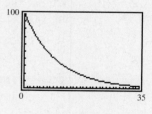

We can use the TRACE feature to obtain an approximation of the expected percent of radioactive material in the milk when $x = 30$. (A TABLE feature may also be used to approximate the percent.) To obtain a better approximation, let's use the ZOOM feature several times to zoom in near $x = 30$.

The percent of radioactive material in the milk 30 days after the Chernobyl accident was 5.08%, accurate to two decimal places.

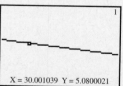

Use a grapher to find each percent. Approximate your solutions so that they are accurate to two decimal places.

1. Estimate the expected percent of radioactive material in the milk 2 days after the Chernobyl nuclear accident.

2. Estimate the expected percent of radioactive material in the milk 10 days after the Chernobyl nuclear accident.

3. Estimate the expected percent of radioactive material in the milk 15 days after the Chernobyl nuclear accident.

4. Estimate the expected percent of radioactive material in the milk 25 days after the Chernobyl nuclear accident.

EXERCISE SET 11.3

A *Graph each exponential function. See Examples 1 through 3.*

1. $y = 4^x$

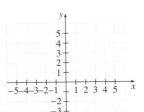

2. $y = 5^x$

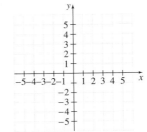

3. $y = 1 + 2^x$

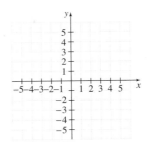

4. $y = 3^x - 1$

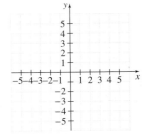

5. $y = \left(\dfrac{1}{4}\right)^x$

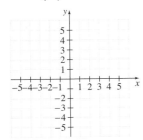

6. $y = \left(\dfrac{1}{5}\right)^x$

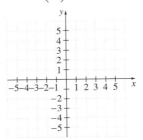

7. $y = \left(\dfrac{1}{2}\right)^x - 2$

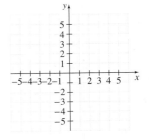

8. $y = \left(\dfrac{1}{3}\right)^x + 2$

9. $y = -2^x$

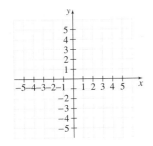

10. $y = -3^x$

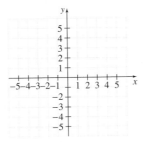

Problem Solving Notes

11.4   LOGARITHMIC FUNCTIONS

A   USING LOGARITHMIC NOTATION

Since the exponential function $f(x) = 2^x$ is a one-to-one function, it has an inverse. We can create a table of values for f^{-1} by switching the coordinates in the accompanying table of values for $f(x) = 2^x$.

x	$y = f(x)$	x	$y = f^{-1}(x)$
-3	$\frac{1}{8}$	$\frac{1}{8}$	-3
-2	$\frac{1}{4}$	$\frac{1}{4}$	-2
-1	$\frac{1}{2}$	$\frac{1}{2}$	-1
0	1	1	0
1	2	2	1
2	4	4	2
3	8	8	3

The graphs of $f(x)$ and its inverse are shown in the margin. Notice that the graphs of f and f^{-1} are symmetric about the line $y = x$, as expected.

Now we would like to be able to write an equation for f^{-1}. To do so, we follow the steps for finding the equation of an inverse:

$$f(x) = 2^x$$

Step 1. Replace $f(x)$ by y. $y = 2^x$

Step 2. Interchange x and y. $x = 2^y$

Step 3. Solve for y.

At this point, we are stuck. To solve this equation for y, a new notation, the **logarithmic notation**, is needed.

The symbol $\log_b x$ means "the power to which b is raised to produce a result of x." In other words,

$$\log_b x = y \text{ means } b^y = x$$

We say that $\log_b x$ is "the logarithm of x to the base b" or "the log of x to the base b." See the logarithmic definition in the margin.

Before returning to the function $x = 2^y$ and solving it for y in terms of x, let's practice using the new notation $\log_b x$.

It is important to be able to write exponential equations with logarithmic notation, and vice versa. The following table shows examples of both forms.

Logarithmic Equation	Corresponding Exponential Equation
$\log_3 9 = 2$	$3^2 = 9$
$\log_6 1 = 0$	$6^0 = 1$
$\log_2 8 = 3$	$2^3 = 8$
$\log_4 \frac{1}{16} = -2$	$4^{-2} = \frac{1}{16}$
$\log_8 2 = \frac{1}{3}$	$8^{1/3} = 2$

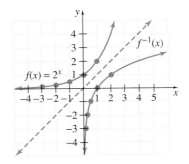

$f(x) = 2^x$ $f^{-1}(x)$

LOGARITHMIC DEFINITION

If $b > 0$, and $b \neq 1$, then

$$y = \log_b x \text{ means } x = b^y$$

for every $x > 0$ and every real number y.

HELPFUL HINT

Notice that a *logarithm* is an *exponent*. In other words, $\log_3 9$ is the *power* that we raise 3 to in order to get 9.

Practice Problems 1–3

Write as an exponential equation.

1. $\log_7 49 = 2$ 2. $\log_8 \dfrac{1}{8} = -1$

3. $\log_3 \sqrt{3} = \dfrac{1}{2}$

Practice Problems 4–6

Write as a logarithmic equation.

4. $3^4 = 81$ 5. $2^{-3} = \dfrac{1}{8}$

6. $7^{1/3} = \sqrt[3]{7}$

Practice Problem 7

Find the value of each logarithmic expression.

a. $\log_2 8$ b. $\log_3 \dfrac{1}{9}$

c. $\log_{25} 5$

Practice Problem 8

Solve: $\log_2 x = 4$

Practice Problem 9

Solve: $\log_x 9 = 2$

Practice Problem 10

Solve: $\log_2 1 = x$

PROPERTIES OF LOGARITHMS

If b is a real number, $b > 0$ and $b \neq 1$, then
1. $\log_b 1 = 0$
2. $\log_b b^x = x$
3. $b^{\log_b x} = x$

Answers

1. $7^2 = 49$,   2. $8^{-1} = \dfrac{1}{8}$,   3. $3^{1/2} = \sqrt{3}$,

4. $\log_3 81 = 4$,   5. $\log_2 \dfrac{1}{8} = -3$,

6. $\log_7 \sqrt[3]{7} = \dfrac{1}{3}$,   7. a. 3,   b. -2,   c. $\dfrac{1}{2}$

8. $\{16\}$,   9. $\{3\}$,   10. $\{0\}$

Examples   Write as an exponential equation.

1. $\log_5 25 = 2$ means $5^2 = 25$

2. $\log_6 \dfrac{1}{6} = -1$ means $6^{-1} = \dfrac{1}{6}$

3. $\log_2 \sqrt{2} = \dfrac{1}{2}$ means $2^{1/2} = \sqrt{2}$

Examples   Write as a logarithmic equation.

4. $9^3 = 729$ means $\log_9 729 = 3$

5. $6^{-2} = \dfrac{1}{36}$ means $\log_6 \dfrac{1}{36} = -2$

6. $5^{1/3} = \sqrt[3]{5}$ means $\log_5 \sqrt[3]{5} = \dfrac{1}{3}$

Example 7   Find the value of each logarithmic expression.

a. $\log_4 16$

b. $\log_{10} \dfrac{1}{10}$

c. $\log_9 3$

Solution: a. $\log_4 16 = 2$ because $4^2 = 16$.

b. $\log_{10} \dfrac{1}{10} = -1$ because $10^{-1} = \dfrac{1}{10}$.

c. $\log_9 3 = \dfrac{1}{2}$ because $9^{1/2} = \sqrt{9} = 3$.

B SOLVING LOGARITHMIC EQUATIONS

The ability to interchange the logarithmic and exponential forms of a statement is often the key to solving logarithmic equations.

Example 8   Solve: $\log_5 x = 3$

Solution: $\log_5 x = 3$

$5^3 = x$ Write as an exponential equation.

$125 = x$

The solution set is $\{125\}$.

Example 9   Solve: $\log_x 25 = 2$

Solution: $\log_x 25 = 2$

$x^2 = 25$ Write as an exponential equation.

$x = 5$

Even though $(-5)^2 = 25$, the base b of a logarithm must be positive. The solution set is $\{5\}$.

Example 10   Solve: $\log_3 1 = x$

Solution: $\log_3 1 = x$

$3^x = 1$ Write as an exponential equation.

$3^x = 3^0$ Write 1 as 3^0.

$x = 0$ Use the uniqueness of b^x.

The solution set is $\{0\}$.

In Example 10, we illustrated an important property of logarithms. That is, $\log_b 1$ is always 0. This property as well as two important others are given in the margin on page 706.

To see that $\log_b b^x = x$, change the logarithmic form to exponential form. Then, $\log_b b^x = x$ means $b^x = b^x$. In exponential form, the statement is true, so in logarithmic form, the statement is also true.

Example 11   Simplify.

a. $\log_3 3^2$ **b.** $\log_7 7^{-1}$
c. $5^{\log_5 3}$ **d.** $2^{\log_2 6}$

Solution: **a.** From property 2, $\log_3 3^2 = 2$.
b. From property 2, $\log_7 7^{-1} = -1$.
c. From property 3, $5^{\log_5 3} = 3$.
d. From property 3, $2^{\log_2 6} = 6$.

C GRAPHING LOGARITHMIC FUNCTIONS

Let us now return to the function $f(x) = 2^x$ and write an equation for its inverse, $f^{-1}(x)$. Recall our earlier work.

$$f(x) = 2^x$$

Step 1. Replace $f(x)$ by y. $y = 2^x$
Step 2. Interchange x and y. $x = 2^y$

Having gained proficiency with the notation $\log_b x$, we can now complete the steps for writing the inverse equation.

Step 3. Solve for y. $y = \log_2 x$
Step 4. Replace y with $f^{-1}(x)$. $f^{-1}(x) = \log_2 x$

Thus, $f^{-1}(x) = \log_2 x$ defines a function that is the inverse function of the function $f(x) = 2^x$. The function $f^{-1}(x)$ or $y = \log_2 x$ is called a **logarithmic function**.

TRY THE CONCEPT CHECK IN THE MARGIN.

We can explore logarithmic functions by graphing them.

Example 12   Graph the logarithmic function $y = \log_2 x$.

Solution: First we write the equation with exponential notation as $2^y = x$. Then we find some ordered pair solutions that satisfy this equation. Finally, we plot the points and connect them with a smooth curve. The domain of this function is $(0, \infty)$, and the range is all real numbers.
Since $x = 2^y$ is solved for x, we choose y-values and compute corresponding x-values.

If $y = 0$, $x = 2^0 = 1$
If $y = 1$, $x = 2^1 = 2$
If $y = 2$, $x = 2^2 = 4$
If $y = -1$, $x = 2^{-1} = \frac{1}{2}$

$x = 2^y$	y
1	0
2	1
4	2
$\frac{1}{2}$	-1

Practice Problem 12

Graph the logarithmic function $y = \log_4 x$.

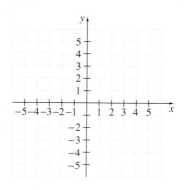

Practice Problem 13

Graph the logarithmic function $f(x) = \log_{1/2} x$.

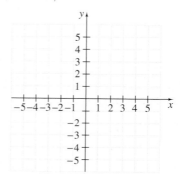

Answers

12.

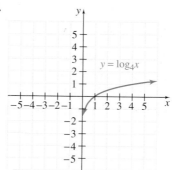

13.

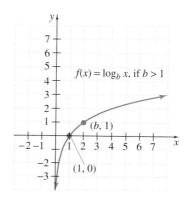

Example 13   Graph the logarithmic function $f(x) = \log_{1/3} x$.

Solution: We can replace $f(x)$ with y, and write the result with exponential notation.

$$f(x) = \log_{1/3} x$$
$$y = \log_{1/3} x \qquad \text{Replace } f(x) \text{ with } y.$$
$$\left(\frac{1}{3}\right)^y = x \qquad \text{Write in exponential form.}$$

Now we can find ordered pair solutions that satisfy $\left(\frac{1}{3}\right)^y = x$, plot these points, and connect them with a smooth curve.

If $y = 0$, $x = \left(\frac{1}{3}\right)^0 = 1$

If $y = 1$, $x = \left(\frac{1}{3}\right)^1 = \frac{1}{3}$

If $y = -1$, $x = \left(\frac{1}{3}\right)^{-1} = 3$

If $y = -2$, $x = \left(\frac{1}{3}\right)^{-2} = 9$

$x = \left(\frac{1}{3}\right)^y$	y
1	0
$\frac{1}{3}$	1
3	−1
9	−2

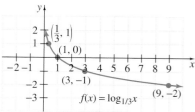

The domain of this function is $(0, \infty)$, and the range is the set of all real numbers.

The following figures summarize characteristics of logarithmic functions.

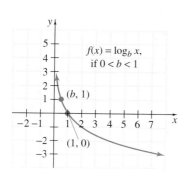

EXERCISE SET 11.4

A *Write each as an exponential equation. See Examples 1 through 3.*

1. $\log_6 36 = 2$ **2.** $\log_2 32 = 5$ **3.** $\log_3 \frac{1}{27} = -3$

4. $\log_5 \frac{1}{25} = -2$ **5.** $\log_{10} 1000 = 3$

Write each as a logarithmic equation. See Examples 4 through 6.

6. $2^4 = 16$ **7.** $5^3 = 125$ **8.** $10^2 = 100$

9. $10^4 = 1000$ **10.** $e^3 = x$

Find the value of each logarithmic expression. See Example 7.

11. $\log_2 8$ **12.** $\log_3 9$ **13.** $\log_3 \frac{1}{9}$

14. $\log_2 \frac{1}{32}$ **15.** $\log_{25} 5$

B *Solve. See Examples 8 through 10.*

16. $\log_3 9 = x$ **17.** $\log_2 8 = x$ **18.** $\log_3 x = 4$

19. $\log_2 x = 3$ **20.** $\log_x 49 = 2$

ANSWERS

1. _____
2. _____
3. _____
4. _____
5. _____
6. _____
7. _____
8. _____
9. _____
10. _____
11. _____
12. _____
13. _____
14. _____
15. _____
16. _____
17. _____
18. _____
19. _____
20. _____

C *Graph each logarithmic function. See Examples 12 and 13.*

21. $y = \log_3 x$

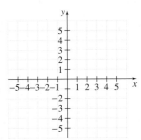

22. $y = \log_2 x$

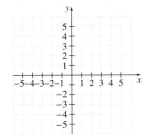

23. $f(x) = \log_{1/4} x$

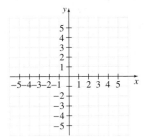

24. $f(x) = \log_{1/2} x$

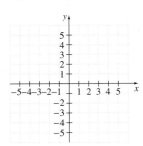

25. $f(x) = \log_5 x$

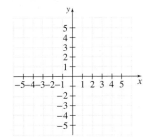

11.5 PROPERTIES OF LOGARITHMS

In the previous section we explored some basic properties of logarithms. We now introduce and explore additional properties. Because a logarithm is an exponent, logarithmic properties are just restatements of exponential properties.

A USING THE PRODUCT PROPERTY

The first of these properties is called the **product property of logarithms** because it deals with the logarithm of a product.

> **PRODUCT PROPERTY OF LOGARITHMS**
>
> If x, y, and b are positive real numbers and $b \neq 1$, then
>
> $$\log_b xy = \log_b x + \log_b y$$

To prove this, we let $\log_b x = M$ and $\log_b y = N$. Now we write each logarithm with exponential notation.

$$\log_b x = M \quad \text{is equivalent to} \quad b^M = x$$
$$\log_b y = N \quad \text{is equivalent to} \quad b^N = y$$

When we multiply the left sides and the right sides of the exponential equations, we have that

$$xy = (b^M)(b^N) = b^{M+N}$$

If we write the equation $xy = b^{M+N}$ in equivalent logarithmic form, we have

$$\log_b xy = M + N$$

But since $M = \log_b x$ and $N = \log_b y$, we can write

$$\log_b xy = \log_b x + \log_b y \quad \text{Let } M = \log_b x \text{ and } N = \log_b y.$$

In other words, the logarithm of a product is the sum of the logarithms of the factors. This property is sometimes used to simplify logarithmic expressions.

Example 1   Write as a single logarithm: $\log_{11} 10 + \log_{11} 3$

   Solution: $\log_{11} 10 + \log_{11} 3 = \log_{11} (10 \cdot 3)$ Use the product property.
$$= \log_{11} 30 \quad \rule{2cm}{2pt}$$

Example 2   Write as a single logarithm: $\log_2 (x + 2) + \log_2 x$

   Solution:

$$\log_2 (x + 2) + \log_2 x = \log_2 [(x + 2) \cdot x] = \log_2 (x^2 + 2x) \quad \rule{2cm}{2pt}$$

B USING THE QUOTIENT PROPERTY

The second property is the **quotient property of logarithms**.

Practice Problem 1

Write as a single logarithm:
$\log_2 7 + \log_2 5$

Practice Problem 2

Write as a single logarithm:
$\log_3 x + \log_3 (x - 9)$

Answers

1. $\log_2 35$,   **2.** $\log_3 (x^2 - 9x)$

✓ CONCEPT CHECK

Which of the following is the correct way to rewrite $\log_5 \dfrac{7}{2}$?

a. $\log_5 7 - \log_5 2$

b. $\log_5 (7 - 2)$

c. $\dfrac{\log_5 7}{\log_5 2}$

d. $\log_5 14$

Practice Problem 3

Write as a single logarithm:
$\log_7 40 - \log_7 8$

Practice Problem 4

Write as a single logarithm:
$\log_3 (x^3 + 4) - \log_3 (x^2 + 2)$

> **QUOTIENT PROPERTY OF LOGARITHMS**
>
> If x, y, and b are positive real numbers and $b \neq 1$, then
>
> $$\log_b \frac{x}{y} = \log_b x - \log_b y$$

The proof of the quotient property of logarithms is similar to the proof of the product property. Notice that the quotient property says that the logarithm of a quotient is the difference of the logarithms of the dividend and divisor.

TRY THE CONCEPT CHECK IN THE MARGIN.

Example 3   Write as a single logarithm: $\log_{10} 27 - \log_{10} 3$

Solution: $\log_{10} 27 - \log_{10} 3 = \log_{10} \dfrac{27}{3}$ Use the quotient property.

$$= \log_{10} 9$$

Example 4   Write as a single logarithm: $\log_3 (x^2 + 5) - \log_3 (x^2 + 1)$

Solution:

$\log_3 (x^2 + 5) - \log_3 (x^2 + 1) = \log_3 \dfrac{x^2 + 5}{x^2 + 1}$ Use the quotient property.

C USING THE POWER PROPERTY

The third and final property we introduce is the **power property of logarithms**.

> **POWER PROPERTY OF LOGARITHMS**
>
> If x and b are positive real numbers, $b \neq 1$, and r is a real number, then
>
> $$\log_b x^r = r \log_b x$$

Practice Problems 5–6

Use the power property to rewrite each expression.

5. $\log_3 x^5$ 6. $\log_7 \sqrt[3]{4}$

Examples   Use the power property to rewrite each expression.

5. $\log_5 x^3 = 3 \log_5 x$

6. $\log_4 \sqrt{2} = \log_4 2^{1/2} = \dfrac{1}{2} \log_4 2$

Practice Problems 7–8

Write as a single logarithm.

7. $3 \log_4 2 + 2 \log_4 5$

8. $5 \log_2 (2x - 1) - \log_2 x$

D USING MORE THAN ONE PROPERTY

Many times we must use more than one property of logarithms to simplify a logarithmic expression.

Examples   Write as a single logarithm.

7. $2 \log_5 3 + 3 \log_5 2 = \log_5 3^2 + \log_5 2^3$ Use the power property.

$$= \log_5 9 + \log_5 8$$

$$= \log_5 (9 \cdot 8)$$ Use the product property.

$$= \log_5 72$$

Answers

3. $\log_7 5$,   **4.** $\log_3 \dfrac{x^3 + 4}{x^2 + 2}$,   **5.** $5 \log_3 x$,

6. $\dfrac{1}{3} \log_7 4$,   **7.** $\log_4 200$, 8,   **8.** $\log_2 \dfrac{(2x - 1)^5}{x}$

✓ Concept Check: a

8. $3 \log_9 x - \log_9 (x + 1) = \log_9 x^3 - \log_9 (x + 1)$ Use the power property.

$$= \log_9 \frac{x^3}{x + 1}$$ Use the quotient property.

Examples

Write each expression as sums or differences of logarithms.

9. $\log_3 \dfrac{5 \cdot 7}{4} = \log_3 (5 \cdot 7) - \log_3 4$ Use the quotient property.

$$= \log_3 5 + \log_3 7 - \log_3 4$$ Use the product property.

10. $\log_2 \dfrac{x^5}{y^2} = \log_2 (x^5) - \log_2 (y^2)$ Use the quotient property.

$$= 5 \log_2 x - 2 \log_2 y$$ Use the power property.

> ### HELPFUL HINT
>
> Notice that we are not able to simplify further a logarithmic expression such as $\log_5 (2x - 1)$. None of the basic properties gives a way to write the logarithm of a difference in some equivalent form.

TRY THE CONCEPT CHECK IN THE MARGIN.

Examples

If $\log_b 2 = 0.43$ and $\log_b 3 = 0.68$, use the properties of logarithms to evaluate each expression.

11. $\log_b 6 = \log_b (2 \cdot 3)$ Write 6 as $2 \cdot 3$.

$$= \log_b 2 + \log_b 3$$ Use the product property.

$$= 0.43 + 0.68$$ Substitute given values.

$$= 1.11$$ Simplify.

12. $\log_b 9 = \log_b 3^2$ Write 9 as 3^2.

$$= 2 \log_b 3$$ Use the power property.

$$= 2(0.68)$$ Substitute the given value.

$$= 1.36$$ Simplify.

13. $\log_b \sqrt{2} = \log_b 2^{1/2}$ Write $\sqrt{2}$ as $2^{1/2}$.

$$= \frac{1}{2} \log_b 2$$ Use the power property.

$$= \frac{1}{2} (0.43)$$ Substitute the given value.

$$= 0.215$$ Simplify.

Practice Problems 9–10

Write each expression as sums or differences of logarithms.

9. $\log_7 \dfrac{6 \cdot 2}{5}$

10. $\log_3 \dfrac{x^4}{y^3}$

✓ CONCEPT CHECK

What is wrong with the following?

$$\log_{10} (x^2 + 5) = \log_{10} x^2 + \log_{10} 5$$
$$= 2 \log_{10} x + \log_{10} 5$$

Use a numerical example to demonstrate that the result is incorrect.

Practice Problems 11–13

If $\log_b 4 = 0.86$ and $\log_b 7 = 1.21$, use the properties of logarithms to evaluate each expression.

11. $\log_b 28$

12. $\log_b 49$

13. $\log_b \sqrt[3]{4}$

Answers

9. $\log_7 6 + \log_7 2 - \log_7 5$,
10. $4 \log_3 x - 3 \log_3 y$,   **11.** 2.07,   **12.** 2.42,
13. 0.286

✓ **Concept Check:** The properties do not give any way to simplify the logarithm of a sum; answers may vary

Focus On the Real World

SOUND INTENSITY

The decibel (dB) measures sound intensity, or the relative loudness or strength of a sound. One decibel is the smallest difference in sound levels that is detectable by humans. The decibel is a logarithmic unit. This means that for approximately every 3-decibel increase in sound intensity, the relative loudness of the sound is doubled. For example, a 35 dB sound is twice as loud as a 32 dB sound.

In the modern world, noise pollution has increasingly become a concern. Sustained exposure to high sound intensities can lead to hearing loss. Regular exposure to 90 dB sounds can eventually lead to loss of hearing. Sounds of 130 dB and more can cause permanent loss of hearing instantaneously.

The relative loudness of a sound D in decibels is given by the equation

$$D = 10 \log_{10} \frac{I}{10^{-16}}$$

where I is the intensity of a sound given in watts per square centimeter. Some sound intensities of common noises are listed in the table in order of increasing sound intensity.

SOME SOUND INTENSITIES OF COMMON NOISES	
Noise	Intensity (watts/cm^2)
Whispering	10^{-15}
Rustling leaves	$10^{-14.2}$
Normal conversation	10^{-13}
Background noise in a quiet residence	$10^{-12.2}$
Typewriter	10^{-11}
Air conditioning	10^{-10}
Freight train at 50 feet	$10^{-8.5}$
Vacuum cleaner	10^{-8}
Nearby thunder	10^{-7}
Air hammer	$10^{-6.5}$
Jet plane at takeoff	10^{-6}
Threshold of pain	10^{-4}

GROUP ACTIVITY

1. Work together to create a table of the relative loudness (in decibels) of the sounds listed in the table.

2. Research the loudness of other common noises. Add these sounds and their decibel levels to your table. Be sure to list the sounds in order of increasing sound intensity.

11.6   COMMON LOGARITHMS, NATURAL LOGARITHMS, AND CHANGE OF BASE

In this section we look closely at two particular logarithmic bases. These two logarithmic bases are used so frequently that logarithms to their bases are given special names. **Common logarithms** are logarithms to base 10. **Natural logarithms** are logarithms to base e, which we introduced in Section 11.3. The work in this section is based on the use of the calculator which has both the common "log" [LOG] and the natural "log" [LN] keys.

A   APPROXIMATING COMMON LOGARITHMS

Logarithms to base 10—common logarithms—are used frequently because our number system is a base 10 decimal system. The notation $\log x$ means the same as $\log_{10} x$.

> **COMMON LOGARITHMS**
>
> $\log x$ means $\log_{10} x$

Example 1

Use a calculator to approximate $\log 7$ to four decimal places.

Solution: Press the following sequence of keys:

[7] [LOG]   or   [LOG] [7] [ENTER]

To four decimal places,

$\log 7 \approx 0.8451$

B   EVALUATING COMMON LOGARITHMS OF POWERS OF 10

To evaluate the common log of a power of 10, a calculator is not needed. According to the property of logarithms,

$\log_b b^x = x$

It follows that if b is replaced with 10, we have

$\log 10^x = x$

> ⌐ **HELPFUL HINT**
>
>   The base of this logarithm is understood to be 10.

Examples

Find the exact value of each logarithm.

2. $\log 10 = \log 10^1 = 1$

3. $\log \dfrac{1}{10} = \log 10^{-1} = -1$

4. $\log 1000 = \log 10^3 = 3$

5. $\log \sqrt{10} = \log 10^{1/2} = \dfrac{1}{2}$

As we will soon see, equations containing common logs are useful models of many natural phenomena.

Objectives

A Identify common logarithms and approximate them with a calculator.

B Evaluate common logarithms of powers of 10.

C Identify natural logarithms and approximate them with a calculator.

D Evaluate natural logarithms of powers of e.

E Use the change of base formula.

SSM   CD-ROM   Video
                          11.6

Practice Problem 1

Use a calculator to aproximate $\log 21$ to four decimal places.

Practice Problems 2–5

Find the exact value of each logarithm.

2. $\log 100$

3. $\log \dfrac{1}{100}$

4. $\log 10,000$

5. $\log \sqrt[3]{10}$

Answers

1. 1.3222,   **2.** 2,   **3.** −2,   **4.** 4,   **5.** $\dfrac{1}{3}$

Practice Problem 6

Solve: $\log x = 2.9$. Give an exact solution, and then approximate the solution to four decimal places.

Example 6   Solve: $\log x = 1.2$. Give an exact solution, and then approximate the solution to four decimal places.

Solution: Remember that the base of a common log is understood to be 10.

> HELPFUL HINT
>
> The understood base is 10.

$$\log x = 1.2$$
$$10^{1.2} = x \qquad \text{Write with exponential notation.}$$

The exact solution is $10^{1.2}$. To four decimal places, $x \approx 15.8489$.

C   APPROXIMATING NATURAL LOGARITHMS

Natural logarithms are also frequently used, especially to describe natural events; hence the label "natural logarithm." Natural logarithms are logarithms to the base e, which is a constant approximately equal to 2.7183. The notation $\log_e x$ is usually abbreviated to $\ln x$. (The abbreviation ln is read "el en.")

> NATURAL LOGARITHMS
>
> $\ln x$ means $\log_e x$

Practice Problem 7

Use a calculator to approximate $\ln 11$ to four decimal places.

Example 7   Use a calculator to approximate $\ln 8$ to four decimal places.

Solution: Press the following sequence of keys:

| 8 | ln | or | ln | 8 | ENTER |

To four decimal places,

$\ln 8 \approx 2.0794$

D   EVALUATING NATURAL LOGARITHMS OF POWERS OF e

As a result of the property $\log_b b^x = x$, we know that $\log_e e^x = x$, or $\ln e^x = x$.

Practice Problems 8–9

Find the exact value of each natural logarithm.

8. $\ln e^5$ 9. $\ln \sqrt{e}$

Examples   Find the exact value of each natural logarithm.

8. $\ln e^3 = 3$

9. $\ln \sqrt[5]{e} = \ln e^{1/5} = \dfrac{1}{5}$

Practice Problem 10

Solve: $\ln 7x = 10$. Give an exact solution, and then approximate the solution to four decimal places.

Example 10   Solve: $\ln 3x = 5$. Give an exact solution, and then approximate the solution to four decimal places.

Solution: Remember that the base of a natural logarithm is understood to be e.

> HELPFUL HINT
>
> The understood base is e.

$$\ln 3x = 5$$
$$e^5 = 3x \qquad \text{Write with exponential notation.}$$
$$\frac{e^5}{3} = x \qquad \text{Solve for } x.$$

The exact solution is $\dfrac{e^5}{3}$. To four decimal places, $x \approx 49.4711$.

Answers

6. $x = 10^{2.9}$; $x \approx 794.3282$,   **7.** 2.3979,

8. 5,   **9.** $\dfrac{1}{2}$,   **10.** $x = \dfrac{e^{10}}{7}$; $x \approx 3146.6380$

E USING THE CHANGE OF BASE FORMULA

Calculators are handy tools for approximating natural and common logarithms. Unfortunately, some calculators cannot be used to approximate logarithms to bases other than e or 10—at least not directly. In such cases, we use the change of base formula.

CHANGE OF BASE

If a, b, and c are positive real numbers and neither b nor c is 1, then

$$\log_b a = \frac{\log_c a}{\log_c b}$$

Example 11   Approximate $\log_5 3$ to four decimal places.

Solution: We use the change of base property to write $\log_5 3$ as a quotient of logarithms to base 10.

$$\log_5 3 = \frac{\log 3}{\log 5} \qquad \text{Use the change of base property.}$$

$$\approx \frac{0.4771213}{0.69897} \qquad \text{Approximate the logarithms by calculator.}$$

$$\approx 0.6826063 \qquad \text{Simplify by calculator.}$$

To four decimal places, $\log_5 3 \approx 0.6826$.

TRY THE CONCEPT CHECK IN THE MARGIN.

Practice Problem 11

Approximate $\log_7 5$ to four decimal places.

✓ CONCEPT CHECK

If a graphing calculator cannot directly evaluate logarithms to base 5, describe how you could use the graphing calculator to graph the function $f(x) = \log_5 x$.

Answer

11. 0.8271

✓ Concept Check: $f(x) = \dfrac{\log x}{\log 5}$

Focus On History

THE INVENTION OF LOGARITHMS

Logarithms were the invention of John Napier (1550–1617), a Scottish land owner and theologian. Napier was also fascinated by mathematics and made it his hobby. Over a period of 20 years in his spare time, he developed his theory of logarithms, which were explained in his Latin text *Mirifici Logarithmorum Canonis Descriptio* (A Description of an Admirable Table of Logarithms), published in 1614. He hoped that his discovery would help to simplify the many time-consuming calculations required in astronomy. In fact, Napier's logarithms revolutionized astronomy and many other advanced mathematical fields by replacing "the multiplications, divisions, square and cubical extractions of great numbers, which besides the tedious expense of time are for the most part subject to many slippery errors" with related numbers that can be easily added and subtracted instead. His discovery was a great time-saving device. Some historians suggest that the use of logarithms to simplify calculations enabled German astronomer Johannes Kepler to develop his three laws of planetary motion, which in turn helped English physicist Sir Isaac Newton develop his theory of gravitation. Two hundred years after Napier's discovery, the French mathematician Pierre de Laplace wrote that logarithms, "by shortening the labors, doubled the life of the astronomer."

Napier's original logarithm tables had several flaws: They did not actually use a particular logarithmic base per se and log 1 was not defined to be equal to 0. An English mathematician, Henry Briggs, read Napier's Latin text soon after it was published and was very impressed by his ideas. Briggs wrote to Napier, asking to meet in person to discuss his wonderful discovery and to offer several improvements. The two mathematicians met in the summer of 1615. Briggs suggested redefining logarithms to base 10 and defining log 1 = 0. Napier had also thought of using base 10 but hadn't been well enough to start a new set of tables. He asked Briggs to undertake the construction of a new set of base 10 tables. And so it was that the first table of common logarithms was constructed by Briggs over the next two years. Napier died in 1617 before Briggs was able to complete his new tables.

CRITICAL THINKING

Locate a table of common logarithms and describe how to use it. Give several examples. Explain why a table of common logarithms would have been invaluable to many calculations before the invention of the hand-held calculator.

EXERCISE SET 11.6

A **C** *Use a calculator to approximate each logarithm to four decimal places. See Examples 1 and 7.*

1. log 8   **2.** log 6   **3.** log 2.31   **4.** log 4.86

5. ln 2   **6.** ln 3   **7.** ln 0.0716   **8.** ln 0.0032

9. log 12.6   **10.** log 25.9

B **D** *Find the exact value of each logarithm. See Examples 2 through 5, 8, and 9.*

11. log 100   **12.** log 10,000   **13.** $\log \dfrac{1}{1000}$ **14.** $\log \dfrac{1}{10}$

15. $\ln e^2$ **16.** $\ln e^4$ **17.** $\ln \sqrt[4]{e}$ **18.** $\ln \sqrt[5]{e}$

19. $\log 10^3$ **20.** $\ln e^5$

ANSWERS

1. _____

2. _____

3. _____

4. _____

5. _____

6. _____

7. _____

8. _____

9. _____

10. _____

11. _____

12. _____

13. _____

14. _____

15. _____

16. _____

17. _____

18. _____

19. _____

20. _____

Problem Solving Notes

11.7   EXPONENTIAL AND LOGARITHMIC EQUATIONS AND PROBLEM SOLVING

A   SOLVING EXPONENTIAL EQUATIONS

In Section 11.3 we solved exponential equations such as $2^x = 16$ by writing 16 as a power of 2 and using the uniqueness of b^x:

$$2^x = 16$$
$$2^x = 2^4 \quad \text{Write 16 as } 2^4.$$
$$x = 4 \quad \text{Use the uniqueness of } b^x.$$

To solve an equation such as $3^x = 7$, we use the fact that $f(x) = \log_b x$ is a one-to-one function. Another way of stating this fact is as a property of equality.

> **LOGARITHM PROPERTY OF EQUALITY**
>
> Let a, b, and c be real numbers such that $\log_b a$ and $\log_b c$ are real numbers and b is not 1. Then
>
> $$\log_b a = \log_b c \quad \text{is equivalent to} \quad a = c$$

Example 1   Solve: $3^x = 7$. Give an exact answer and a four-decimal-place approximation.

Solution: We use the logarithm property of equality and take the logarithm of both sides. For this example, we use the common logarithm.

$$3^x = 7$$
$$\log 3^x = \log 7 \quad \text{Take the common log of both sides.}$$
$$x \log 3 = \log 7 \quad \text{Use the power property of logarithms.}$$
$$x = \frac{\log 7}{\log 3} \quad \text{Divide both sides by log 3.}$$

The exact solution is $\dfrac{\log 7}{\log 3}$. To approximate to four decimal places, we have

$$\frac{\log 7}{\log 3} \approx \frac{0.845098}{0.4771213} \approx 1.7712$$

The solution set is $\left\{ \dfrac{\log 7}{\log 3} \right\}$, or approximately $\{1.7712\}$.

B   SOLVING LOGARITHMIC EQUATIONS

By applying the appropriate properties of logarithms, we can solve a broad variety of logarithmic equations.

Example 2   Solve: $\log_4 (x - 2) = 2$

Solution: Notice that $x - 2$ must be positive, so x must be greater than 2. With this in mind, we first write the equation with exponential notation.

Objectives

A Solve exponential equations.

B Solve logarithmic equations.

C Solve problems that can be modeled by exponential and logarithmic equations.

SSM   CD-ROM   Video
11.7

Practice Problem 1

Solve: $2^x = 5$. Give an exact answer and a four-decimal-place aproximation.

Practice Problem 2

Solve: $\log_3 (x + 5) = 2$

Answers

1. $\left\{ \dfrac{\log 5}{\log 2} \right\}$; $\{2.3219\}$, **2.** $\{4\}$

$$\log_4 (x - 2) = 2$$
$$4^2 = x - 2$$
$$16 = x - 2$$
$$18 = x \qquad \text{Add 2 to both sides.}$$

To check, we replace x with 18 in the original equation.

$$\log_4 (x - 2) = 2$$
$$\log_4 (18 - 2) \overset{?}{=} 2 \qquad \text{Let } x = 18.$$
$$\log_4 16 \overset{?}{=} 2$$
$$4^2 = 16 \qquad \text{True.}$$

The solution set is $\{18\}$.

Practice Problem 3

Solve: $\log_6 x + \log_6 (x + 1) = 1$

Example 3

Solution:

Solve: $\log_2 x + \log_2 (x - 1) = 1$

Notice that $x - 1$ must be positive, so x must be greater than 1. We use the product property on the left side of the equation.

$$\log_2 x + \log_2 (x - 1) = 1$$
$$\log_2 [x(x - 1)] = 1 \qquad \text{Use the product property.}$$
$$\log_2 (x^2 - x) = 1$$

Next we write the equation with exponential notation and solve for x.

$$2^1 = x^2 - x$$
$$0 = x^2 - x - 2 \qquad \text{Subtract 2 from both sides.}$$
$$0 = (x - 2)(x + 1) \qquad \text{Factor.}$$
$$0 = x - 2 \text{ or } 0 = x + 1 \qquad \text{Set each factor equal to 0.}$$
$$2 = x \qquad -1 = x$$

Recall that -1 cannot be a solution because x must be greater than 1. If we forgot this, we would still reject -1 after checking. To see this, we replace x with -1 in the original equation.

$$\log_2 x + \log_2 (x - 1) = 1$$
$$\log_2 (-1) + \log_2 (-1 - 1) \overset{?}{=} 1 \qquad \text{Let } x = -1.$$

Because the logarithm of a negative number is undefined, -1 is rejected. Check to see that the solution set is $\{2\}$.

Practice Problem 4

Solve: $\log (x - 1) - \log x = 1$

Example 4

Solution:

Solve: $\log (x + 2) - \log x = 2$

We use the quotient property of logarithms on the left side of the equation.

$$\log (x + 2) - \log x = 2$$
$$\log \frac{x + 2}{x} = 2 \qquad \text{Use the quotient property.}$$
$$10^2 = \frac{x + 2}{x} \qquad \text{Write using exponential notation.}$$
$$100 = \frac{x + 2}{x}$$

Answers

3. $\{2\}$, **4.** $\varnothing$

$$100x = x + 2 \qquad \text{Multiply both sides by } x.$$
$$99x = 2 \qquad \text{Subtract } x \text{ from both sides.}$$
$$x = \frac{2}{99} \qquad \text{Divide both sides by 99.}$$

Check to see that the solution set is $\left\{\frac{2}{99}\right\}$. ▬▬▬

C SOLVING PROBLEMS MODELED BY EXPONENTIAL AND LOGARITHMIC EQUATIONS

Logarithmic and exponential functions are used in a variety of scientific, technical, and business settings. A few examples follow.

Example 5   Estimating Population Size

The population size y of a community of lemmings varies according to the relationship $y = y_0 e^{0.15t}$. In this formula, t is time in months, and y_0 is the initial population at time 0. Estimate the population after 6 months if there were originally 5000 lemmings.

Solution: We substitute 5000 for y_0 and 6 for t.

$$y = y_0 e^{0.15t}$$
$$= 5000 e^{0.15(6)} \qquad \text{Let } t = 6 \text{ and } y_0 = 5000.$$
$$= 5000 e^{0.9} \qquad \text{Multiply.}$$

Using a calculator, we find that $y \approx 12{,}298.016$. In 6 months the population will be approximately 12,300 lemmings. ▬▬▬

Example 6   Doubling an Investment

How long does it take an investment of $2000 to double if it is invested at 5% interest compounded quarterly?

The necessary formula is $A = P\left(1 + \dfrac{r}{n}\right)^{nt}$, where A is the accrued amount, P is the principal invested, r is the annual rate of interest, n is the number of compounding periods per year, and t is the number of years.

Solution: We are given that $P = \$2000$ and $r = 5\% = 0.05$. Compounding quarterly means 4 times a year, so $n = 4$. The investment is to double, so A must be $4000. We substitute these values and solve for t.

$$A = P\left(1 + \frac{r}{n}\right)^{nt}$$
$$4000 = 2000\left(1 + \frac{0.05}{4}\right)^{4t} \qquad \text{Substitute known values.}$$
$$4000 = 2000(1.0125)^{4t} \qquad \text{Simplify } 1 + \frac{0.05}{4}.$$
$$2 = (1.0125)^{4t} \qquad \text{Divide both sides by 2000.}$$
$$\log 2 = \log 1.0125^{4t} \qquad \text{Take the logarithm of both sides.}$$
$$\log 2 = 4t(\log 1.0125) \qquad \text{Use the power property.}$$

Practice Problem 5

Use the equation in Example 5 to estimate the lemming population in 8 months.

Practice Problem 6

How long does it take an investment of $1000 to double if it is invested at 6% compounded quarterly?

Answers

5. approximately 16,601 lemmings,   **6.** $11\frac{3}{4}$ years

$$\frac{\log 2}{4 \log 1.0125} = t$$ 　Divide both sides by 4 log 1.0125.

$$13.949408 \approx t$$ 　Approximate by calculator.

It takes 14 years for the money to double in value.

GRAPHING CALCULATOR EXPLORATIONS

Use a grapher to find how long it takes an investment of $1500 to triple if it is invested at 8% interest compounded monthly.

First, let $P = \$1500$, $r = 0.08$, and $n = 12$ (for 12 months) in the formula

$$A = P\left(1 + \frac{r}{n}\right)^{nt}$$

Notice that when the investment has tripled, the accrued amount A is $4500. Thus,

$$4500 = 1500\left(1 + \frac{0.08}{12}\right)^{12t}$$

Determine an appropriate viewing window and enter and graph the equations

$$Y_1 = 1500\left(1 + \frac{0.08}{12}\right)^{12x}$$

and

$$Y_2 = 4500$$

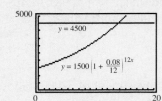

The point of intersection of the two curves is the solution. The x-coordinate tells how long it takes for the investment to triple.

Use a TRACE feature or an INTERSECT feature to approximate the coordinates of the point of intersection of the two curves. It takes approximately 13.78 years, or 13 years and 10 months, for the investment to triple in value to $4500.

Use this graphical solution method to solve each problem. Round each answer to the nearest hundredth.

1. Find how long it takes an investment of $5000 to grow to $6000 if it is invested at 5% interest compounded quarterly.

2. Find how long it takes an investment of $1000 to double if it is invested at 4.5% interest compounded daily. (Use 365 days in a year.)

3. Find how long it takes an investment of $10,000 to quadruple if it is invested at 6% interest compounded monthly.

4. Find how long it takes $500 to grow to $800 if it is invested at 4% interest compounded semiannually.

CHAPTER 11 HIGHLIGHTS

DEFINITIONS AND CONCEPTS	EXAMPLES

SECTION 11.1　THE ALGEBRA OF FUNCTIONS

ALGEBRA OF FUNCTIONS

Sum $(f + g)(x) = f(x) + g(x)$

Difference $(f - g)(x) = f(x) - g(x)$

Product $(f \cdot g)(x) = f(x) \cdot g(x)$

Quotient $\left(\dfrac{f}{g}\right)(x) = \dfrac{f(x)}{g(x)}$

If $f(x) = 7x$ and $g(x) = x^2 + 1$,

$(f + g)(x) = f(x) + g(x) = 7x + x^2 + 1$

$(f - g)(x) = f(x) - g(x) = 7x - (x^2 + 1)$
$= 7x - x^2 - 1$

$(f \cdot g)(x) = f(x) \cdot g(x) = 7x(x^2 + 1)$
$= 7x^3 + 7x^2$

$\left(\dfrac{f}{g}\right)(x) = \dfrac{f(x)}{g(x)} = \dfrac{7x}{x^2 + 1}$

COMPOSITE FUNCTIONS

The notation $(f \circ g)(x)$ means "f composed with g."

$(f \circ g)(x) = f(g(x))$
$(g \circ f)(x) = g(f(x))$

If $f(x) = x^2 + 1$ and $g(x) = x - 5$, find $(f \circ g)(x)$.

$(f \circ g)(x) = f(g(x))$
$= f(x - 5)$
$= (x - 5)^2 + 1$
$= x^2 - 10x + 26$

SECTION 11.2　INVERSE FUNCTIONS

If f is a function, then f is a **one-to-one function** only if each y-value (output) corresponds to only one x-value (input).

HORIZONTAL LINE TEST

If every horizontal line intersects the graph of a function at most once, then the function is a one-to-one function.

Determine whether each graph is a one-to-one function.

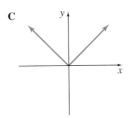

Graphs A and C pass the vertical line test, so only these are graphs of functions. Of graphs A and C, only graph A passes the horizontal line test, so only graph A is the graph of a one-to-one function.

SECTION 11.2   (CONTINUED)

The **inverse** of a one-to-one function f is the one-to-one function f^{-1} that is the set of all ordered pairs (b, a) such that (a, b) belongs to f.

TO FIND THE INVERSE OF A ONE-TO-ONE FUNCTION $f(x)$

Step 1.   Replace $f(x)$ with y.
Step 2.   Interchange x and y.
Step 3.   Solve for y.
Step 4.   Replace y with $f^{-1}(x)$.

Find the inverse of $f(x) = 2x + 7$.

$$y = 2x + 7 \quad \text{Replace } f(x) \text{ with } y.$$
$$x = 2y + 7 \quad \text{Interchange } x \text{ and } y.$$
$$2y = x - 7 \quad \text{Solve for } y.$$
$$y = \frac{x - 7}{2}$$
$$f^{-1}(x) = \frac{x - 7}{2} \quad \text{Replace } y \text{ with } f^{-1}(x).$$

The inverse of $f(x) = 2x + 7$ is $f^{-1}(x) = \dfrac{x - 7}{2}$.

SECTION 11.3   EXPONENTIAL FUNCTIONS

A function of the form $f(x) = b^x$ is an **exponential function**, where $b > 0$, $b \neq 1$, and x is a real number.

UNIQUENESS OF b^x

If $b > 0$ and $b \neq 1$, then $b^x = b^y$ is equivalent to $x = y$.

COMPOUND INTEREST FORMULA

$$A = P\left(1 + \frac{r}{n}\right)^{nt}$$

where r is the annual interest rate for P dollars compounded n times a year for t years.

CONTINUOUSLY COMPOUNDED INTEREST FORMULA

$$A = Pe^{rt}$$

where r is the annual interest rate for P dollars invested for t years.

Graph the exponential function $y = 4^x$.

x	y
-2	$\frac{1}{16}$
-1	$\frac{1}{4}$
0	1
1	4
2	16

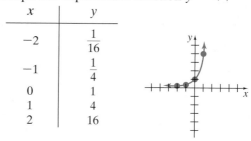

Solve: $2^{x+5} = 8$

$$2^{x+5} = 2^3 \quad \text{Write 8 as } 2^3.$$
$$x + 5 = 3 \quad \text{Use the uniqueness of } b^x.$$
$$x = -2 \quad \text{Subtract 5 from both sides.}$$

Find the amount in an account at the end of 3 years if $1000 is invested at an interest rate of 4% compounded continuously.
Here, $t = 3$ years, $P = \$1000$, and $r = 0.04$.

$$A = Pe^{rt}$$
$$= 1000e^{0.04(3)}$$
$$\approx \$1127.50$$

SECTION 11.4   LOGARITHMIC FUNCTIONS

LOGARITHMIC DEFINITION

If $b > 0$ and $b \neq 1$, then

$$y = \log_b x \quad \text{means} \quad x = b^y$$

for any positive number x and real number y.

PROPERTIES OF LOGARITHMS

If b is a real number, $b > 0$ and $b \neq 1$, then

$$\log_b 1 = 0, \quad \log_b b^x = x, \quad b^{\log_b x} = x$$

LOGARITHMIC FORM	CORRESPONDING EXPONENTIAL STATEMENT
$\log_5 25 = 2$	$5^2 = 25$
$\log_9 3 = \frac{1}{2}$	$9^{1/2} = 3$

$$\log_5 1 = 0, \quad \log_7 7^2 = 2, \quad 3^{\log_3 6} = 6$$

SECTION 11.4   (CONTINUED)

LOGARITHMIC FUNCTION

If $b > 0$ and $b \neq 1$, then a **logarithmic function** is a function that can be defined as

$$f(x) = \log_b x$$

The domain of f is the set of positive real numbers, and the range of f is the set of real numbers.

Graph: $y = \log_3 x$

Write $y = \log_3 x$ as $3^y = x$. Plot the ordered pair solutions listed in the table, and connect them with a smooth curve.

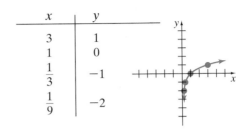

x	y
3	1
1	0
$\frac{1}{3}$	-1
$\frac{1}{9}$	-2

SECTION 11.5   PROPERTIES OF LOGARITHMS

Let x, y, and b be positive numbers and $b \neq 1$.

PRODUCT PROPERTY

$$\log_b xy = \log_b x + \log_b y$$

QUOTIENT PROPERTY

$$\log_b \frac{x}{y} = \log_b x - \log_b y$$

POWER PROPERTY

$$\log_b x^r = r \log_b x$$

Write as a single logarithm:

$2 \log_5 6 + \log_5 x - \log_5 (y + 2)$

$= \log_5 6^2 + \log_5 x - \log_5 (y + 2)$ Power property

$= \log_5 36 \cdot x - \log_5 (y + 2)$ Product property

$= \log_5 \dfrac{36x}{y + 2}$ Quotient property

SECTION 11.6   COMMON LOGARITHMS, NATURAL LOGARITHMS, AND CHANGE OF BASE

COMMON LOGARITHMS

$\log x$ means $\log_{10} x$

NATURAL LOGARITHMS

$\ln x$ means $\log_e x$

CHANGE OF BASE

$$\log_b a = \frac{\log_c a}{\log_c b}$$

$\log 5 = \log_{10} 5 \approx 0.69897$

$\ln 7 = \log_e 7 \approx 1.94591$

$\log_8 12 = \dfrac{\log 12}{\log 8}$

SECTION 11.7   EXPONENTIAL AND LOGARITHMIC EQUATIONS AND PROBLEM SOLVING

LOGARITHM PROPERTY OF EQUALITY

Let $\log_b a$ and $\log_b c$ be real numbers and $b \neq 1$. Then

$\log_b a = \log_b c$ is equivalent to $a = c$

Solve: $2^x = 5$

$\log 2^x = \log 5$ Log property of equality

$x \log 2 = \log 5$ Power property

$x = \dfrac{\log 5}{\log 2}$ Divide both sides by log 2.

$x \approx 2.3219$ Use a calculator.

Geometry

The word *geometry* is formed from the Greek words *geo*, meaning earth, and *metron*, meaning measure. Geometry literally means to measure the earth. In this chapter we learn about various geometric figures and their properties, such as perimeter, area, and volume. Knowledge of geometry can help us solve practical problems in real-life situations. For instance, knowing certain measures of a circular swimming pool allows us to calculate how much water it can hold.

Just outside of Cairo, Egypt, is a famous plateau called Giza. This is the home of the Great Pyramids, the only surviving entry on the list of the Seven Wonders of the Ancient World. There are three pyramids at Giza. The largest, and oldest, was built as the tomb of the Pharaoh Khufu around 2550 B.C. This pyramid is made from over 2,300,000 blocks of stone weighing a total of 6.5 million tons. It took about 30 years to build this monument. Prior to the 20th century, Khufu's pyramid was the tallest building in the world. Khufu's son Khafre is responsible for the second-largest pyramid at Giza during his rule as pharaoh between 2520 to 2494 B.C. The smallest of the pyramids at Giza is credited to Menkaure, believed to be the son of Khafre and grandson of Khufu.

Problem Solving Notes

A.1   LINES AND ANGLES

A   IDENTIFY LINES, LINE SEGMENTS, RAYS, AND ANGLES

Let's begin with a review of two important concepts—plane and space.

A **plane** is a flat surface that extends indefinitely. Surfaces like a plane are a classroom floor or a blackboard or whiteboard.

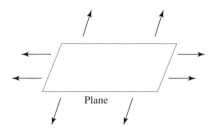

Plane

Space extends in all directions indefinitely. Examples of *objects* in space are houses, grains of salt, bushes, your Basic College Mathematics textbook, and you.

The most basic concept of geometry is the idea of a point in space. A **point** has no length, no width, and no height, but it does have location. We will represent a point by a dot, and we will label points with letters.

Point *P*

A **line** is a set of points extending indefinitely in two directions. A line has no width or height, but it does have length. We can name a line by any two of its points. A **line segment** is a piece of a line with two endpoints.

Line *AB* or $\overleftrightarrow{AB}$ Line segment *AB* or $\overline{AB}$

A **ray** is a part of a line with one endpoint. A ray extends indefinitely in one direction. An **angle** is made up of two rays that share the same endpoint. The common endpoint is called the **vertex**.

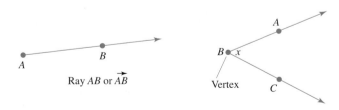

Ray *AB* or $\overrightarrow{AB}$ Vertex

The angle in the figure above can be named

$$\angle ABC \qquad \angle CBA \qquad \angle B \qquad \text{or} \qquad \angle x$$

    ↑   ↑

The vertex is the
middle point.

Rays *BA* and *BC* are **sides** of the angle.

Practice Problem 1

Identify each figure as a line, a ray, a line segment, or an angle. Then name the figure using the given points.

a.

b.

c.

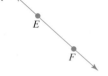

d.

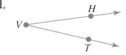

Practice Problem 2

Use the figure for Example 2 to list other ways to name ∠z.

Example 1

Identify each figure as a line, a ray, a line segment, or an angle. Then name the figure using the given points.

a.

b.

c.

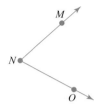

d.

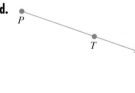

Solution: Figure (a) extends indefinitely in two directions. It is line *CD* or $\overleftrightarrow{CD}$.

Figure (b) has two endpoints. It is line segment *EF* or $\overline{EF}$.

Figure (c) has two rays with a common endpoint. It is ∠*MNO*, ∠*ONM*, or ∠*N*.

Figure (d) is part of a line with one endpoint. It is ray *PT* or $\overrightarrow{PT}$.

Example 2

List other ways to name ∠*y*.

Solution: Two other ways to name ∠*y* are ∠*QTR* and ∠*RTQ*. We may not use the vertex alone to name this angle because three different angles have *T* as their vertex.

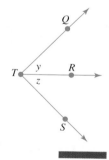

B   CLASSIFYING ANGLES

An angle can be measured in **degrees**. The symbol for degrees is a small, raised circle, °. There are 360° in a full revolution, or full circle.

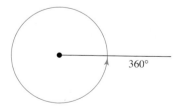

$\frac{1}{2}$ of a revolution measures $\frac{1}{2}\left(360°\right) = 180°$. An angle that measures 180° is called a **straight angle**.

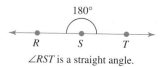

∠*RST* is a straight angle.

$\frac{1}{4}$ of a revolution measures $\frac{1}{4}(360°) = 90°$. An angle that measures $90°$ is called a **right angle**. The symbol ⌐ is used to denote a right angle.

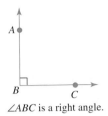

∠ABC is a right angle.

An angle whose measure is between $0°$ and $90°$ is called an **acute angle**.

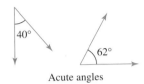

Acute angles

An angle whose measure is between $90°$ and $180°$ is called an **obtuse angle**.

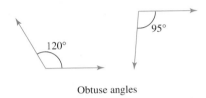

Obtuse angles

Example 3 Classify each angle as acute, right, obtuse, or straight.

a.   b.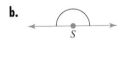

c.   d.

Solution:
  a. ∠R is a right angle, denoted by ⌐.
  b. ∠S is a straight angle.
  c. ∠T is an acute angle. It measures between $0°$ and $90°$.
  d. ∠Q is an obtuse angle. It measures between $90°$ and $180°$.

C IDENTIFYING COMPLEMENTARY AND SUPPLEMENTARY ANGLES

Two angles that have a sum of $90°$ are called **complementary angles**. We say that each angle is the **complement** of the other.

  ∠R and ∠S are complementary angles because
  $\downarrow$ $\downarrow$
  $60° + 30° = 90°$

 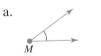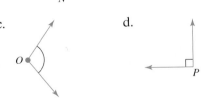

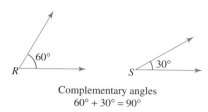

Complementary angles
$60° + 30° = 90°$

Two angles that have a sum of 180° are called **supplementary angles**. We say that each angle is the **supplement** of the other.

$\angle M$ and $\angle N$ are supplementary angles because

$125° + 55° = 180°$

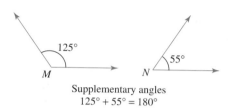

Supplementary angles
$125° + 55° = 180°$

Practice Problem 4

Find the complement of a 36° angle.

Practice Problem 5

Find the supplement of an 88° angle.

✓ Concept Check

True or false? The supplement of a 48° angle is 42°. Explain.

Practice Problem 6

Find the measure of $\angle y$.

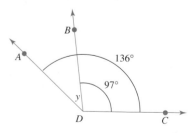

Example 4   Find the complement of a 48° angle.

Solution: The complement of an angle that measures 48° is an angle that measures $90° - 48° = 42°$.

Example 5   Find the supplement of a 107° angle.

Solution: The supplement of an angle that measures 107° is an angle that measures $180° - 107° = 73°$.

TRY THE CONCEPT CHECK IN THE MARGIN.

D   FINDING MEASURES OF ANGLES

Measures of angles can be added or subtracted to find measures of related angles.

Example 6   Find the measure of $\angle x$.

Solution: $\angle x = 87° - 52° = 35°$

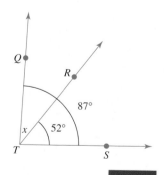

Two lines in a plane can be either parallel or intersecting. **Parallel lines** never meet. **Intersecting lines** meet at a point. The symbol $\parallel$ is used to indicate "is parallel to." For example, in the figure $p \parallel q$.

Answers

4. 54°, **5.** 92°, **6.** 39°

✓ **Concept Check:**   False; the complement of a 48° angle is 42°, the supplement of a 48° angle is 132°.

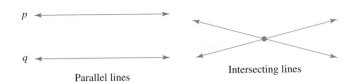

Parallel lines   Intersecting lines

Some intersecting lines are perpendicular. Two lines are **perpendicular** if they form right angles when they intersect. The symbol ⊥ is used to denote "is perpendicular to." For example, in the figure $n \perp m$.

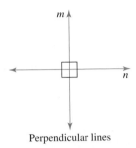

Perpendicular lines

When two lines intersect, four angles are formed. Two of these angles that are opposite each other are called **vertical angles**. Vertical angles have the same measure. Two angles that share a common side are called **adjacent angles**. Adjacent angles formed by intersecting lines are supplementary. That is, they have a sum of 180°.

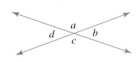

Vertical angles:
∠ a and ∠ c
∠ d and ∠ b

Adjacent angles:
∠ a and ∠ b
∠ b and ∠ c
∠ c and ∠ d
∠ d and ∠ a

Example 7   Find the measure of ∠x, ∠y, and ∠z if the measure of ∠t is 42°.

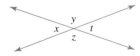

Solution: Since ∠t and ∠x are vertical angles, they have the same measure, so ∠x measures 42°.

Since ∠t and ∠y are adjacent angles, their measures have a sum of 180°. So ∠y measures $180° - 42° = 138°$.

Since ∠y and ∠z are vertical angles, they have the same measure. So ∠z measures 138°.

Practice Problem 7

Find the measure of ∠a, ∠b, and ∠c.

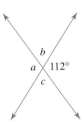

Answer

7. ∠$a = 112°$; ∠$b = 68°$; ∠$c = 68°$

A line that intersects two or more lines at different points is called a **transversal**. Line *l* is a transversal that intersects lines *m* and *n*. The eight angles formed have special names. Some of these names are:

Corresponding Angles: ∠*a* and ∠*e*, ∠*c* and ∠*g*, ∠*b* and ∠*f*, ∠*d* and ∠*h*

Alternate Interior Angles: ∠*c* and ∠*f*, ∠*d* and ∠*e*

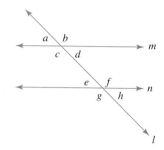

When two lines cut by a transversal are *parallel*, the following are true:

> **PARALLEL LINES CUT BY A TRANSVERSAL**
>
> If two parallel lines are cut by a transversal, then the measures of **corresponding angles are equal** and **alternate interior angles are equal**.

Example 8   Given that *m* ‖ *n* and that the measure of ∠*w* is 100°, find the measures of ∠*x*, ∠*y*, and ∠*z*.

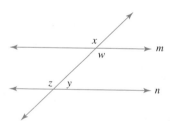

Solution:

The measure of ∠*x* = 100°.   ∠*x* and ∠*w* are vertical angles.

The measure of ∠*z* = 100°.   ∠*x* and ∠*z* are corresponding angles.

The measure of ∠*y* = 180° − 100° = 80°.   ∠*z* and ∠*y* are supplementary angles.

Practice Problem 8

Given that *m* ‖ *n* and that the measure of ∠*w* = 40°, find the measures of ∠*x*, ∠*y*, and ∠*z*.

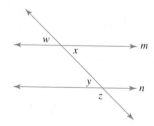

Answer

8. ∠*x* = 40°; ∠*y* = 40°; ∠*z* = 140°

A.2   PLANE FIGURES AND SOLIDS

In order to prepare for the sections ahead in this chapter, we first review plane figures and solids.

A   IDENTIFYING PLANE FIGURES

Recall from Section A.1 that a **plane** is a flat surface that extends indefinitely.

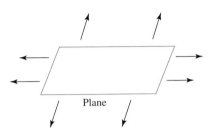

A **plane figure** is a figure that lies on a plane. Plane figures, like planes, have length and width but no thickness or depth.

A **polygon** is a closed plane figure that basically consists of three or more line segments that meet at their endpoints.

A **regular polygon** is one whose sides are all the same length and whose angles are the same measure.

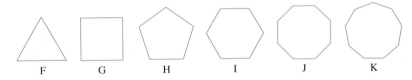

A polygon is named according to the number of its sides.

Some triangles and quadrilaterals are given special names, so let's study these special polygons further. We will begin with triangles. The sum of the measures of the angles of a triangle is 180°.

Example 1   Find the measure of ∠a.

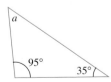

Solution: Since the sum of the measures of the three angles is 180°, we have

measure of $\angle a = 180° - 95° - 35° = 50°$

To check, see that $95° + 35° + 50° = 180°$.   ▬▬▬

We can classify triangles according to the lengths of their sides. (We will use tick marks to denote sides and angles in a figure that are equal.)

Objectives

A Identify plane figures.
B Identify solid figures.

Number of Sides	Name	Figure Examples
3	Triangle	A, F
4	Quadrilateral	B, E, G
5	Pentagon	H
6	Hexagon	I
7	Heptagon	C
8	Octagon	J
9	Nonagon	K
10	Decagon	D

Practice Problem 1

Find the measure of ∠x.

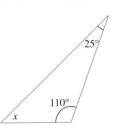

Answer

1. 45°

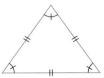
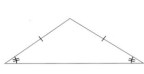

Equilateral triangle
All three sides are
the same length. Also,
all three angles have
the same measure.

Isoceles triangle
Two sides are the
same length. Also,
the angles opposite
the equal sides
have equal measure.

Scalene triangle
No sides are the
same length. No
angles are the
same measure.

One other important type of triangle is a right triangle. A **right triangle** is a triangle with a right angle. The side opposite the right angle is called the **hypotenuse** and the other two sides are called **legs**.

Practice Problem 2

Find the measure of $\angle y$.

HELPFUL HINT

From the previous example, can you see that in a right triangle, the sum of the other two acute angles is 90°? This is because

$$90° + 90° = 180°$$

↑   ↑   ↑

right
angle's
measure

sum of
other
two
angles'
measures

sum of
angles'
measures

Example 2

Find the measure of $\angle b$.

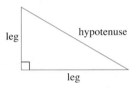

Solution: We know that the measure of the right angle, ∟, is 90°. Since the sum of the measures of the angles is 180°, we have

measure of $\angle b = 180° - 90° - 30° = 60°$ ▬▬▬

Now we review some special quadrilaterals. A **parallelogram** is a special quadrilateral with opposite sides parallel and equal in length.

A **rectangle** is a special **parallelogram** that has four right angles.

A **square** is a special **rectangle** that has all four sides equal in length.

Answer

2. 65°

A **rhombus** is a special **parallelogram** that has all four sides equal in length.

A **trapezoid** is a quadrilateral with exactly one pair of opposite sides parallel.

parallel
sides

TRY THE CONCEPT CHECK IN THE MARGIN.

In addition to triangles and quadrilaterals, circles are common plane figures. A **circle** is a plane figure that consists of all points that are the same fixed distance from a point c. The point c is called the **center** of the circle. The **radius** of a circle is the distance from the center of the circle to any point of the circle. The **diameter** of a circle is the distance across the circle passing through the center. The diameter is twice the radius and the radius is half the diameter.

radius
center
diameter

$$\text{diameter} = 2 \cdot \text{radius} \qquad \text{radius} = \frac{\text{diameter}}{2}$$

$$d = 2 \cdot r \qquad r = \frac{d}{2}$$

Example 3   Find the diameter of the circle.

Solution: The diameter is twice the radius.

$d = 2 \cdot r$

$d = 2 \cdot 5\,\text{cm} = 10\,\text{cm}$

The diameter is 10 centimeters.

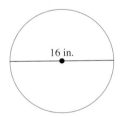

5 cm

B   IDENTIFYING SOLID FIGURES

Recall from Section A.1 that space extends in all directions indefinitely.

A **solid** is a figure that lies in space. Solids have length, width, and height or depth.

A **rectangular solid** is a solid that consists of six sides, or faces, all of which are rectangles.

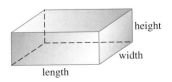

height
width
length

✓ CONCEPT CHECK

True or false? All quadrilaterals are parallelograms. Explain.

Practice Problem 3

Find the radius of the circle.

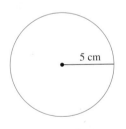

16 in.

Answers

3. 8 in.

✓ Concept Check: false

A **cube** is a rectangular solid whose six sides are squares.

A **pyramid** is shown below.

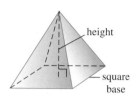

A **sphere** consists of all points in space that are the same distance from a point *c*. The point *c* is called the **center** of the sphere. The **radius** of a sphere is the distance from the center to any point of the sphere. The **diameter** of a sphere is the distance across the sphere passing through the center.

The radius and diameter of a sphere are related in the same way as the radius and diameter of a circle.

$$d = 2 \cdot r \quad \text{or} \quad r = \frac{d}{2}$$

Practice Problem 4

Find the diameter of the sphere.

Example 4 Find the radius of the sphere.

Solution: The radius is half the diameter.

$$r = \frac{d}{2}$$

$$r = \frac{36 \text{ feet}}{2} = 18 \text{ feet}$$

The radius is 18 feet.

The **cylinders** we will study have bases that are in the shape of circles and are perpendicular to their height.

The **cones** we will study have bases that are circles and are perpendicular to their height.

Answer

4. 14 mi

A.3 PERIMETER

A USING FORMULAS TO FIND PERIMETERS

Recall from Section 1.2 that the perimeter of a polygon is the distance around the polygon. This means that the perimeter of a polygon is the sum of the lengths of its sides.

Example 1 Find the perimeter of the rectangle below.

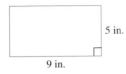

5 in.

9 in.

Solution: perimeter = 9 inches + 9 inches + 5 inches + 5 inches
= 28 inches

Notice that the perimeter of the rectangle in Example 1 can be written as $2 \cdot (9 \text{ inches}) + 2 \cdot (5 \text{ inches})$.

length width

In general, we can say that the perimeter of a rectangle is always

$$2 \cdot \text{length} + 2 \cdot \text{width}$$

As we have just seen, the perimeter of some special figures such as rectangles form patterns. These patterns are given as **formulas**. The formula for the perimeter of a rectangle is shown next.

PERIMETER OF A RECTANGLE

perimeter = 2 · length + 2 · width

In symbols, this can be written as

$$P = 2 \cdot l + 2 \cdot w$$

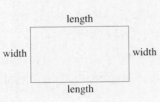

length

width   width

length

Example 2 Find the perimeter of a rectangle with a length of 11 inches and a width of 3 inches.

11 in.

3 in.

Solution: We will use the formula for perimeter and replace the letters by their known lengths.

$$P = 2 \cdot l + 2 \cdot w$$
$$= 2 \cdot 11 \text{ in.} + 2 \cdot 3 \text{ in.} \quad \text{Replace } l \text{ with 11 in. and } w \text{ with 3 in.}$$
$$= 22 \text{ in.} + 6 \text{ in.}$$
$$= 28 \text{ in.}$$

The perimeter is 28 inches.

Practice Problem 1

Find the perimeter of the rectangular lot shown below.

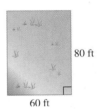

80 ft

60 ft

Practice Problem 2

Find the perimeter of a rectangle with a length of 22 cm and a width of 10 cm.

Answers

1. 280 ft, **2.** 64 cm

Recall that a square is a special rectangle with all four sides the same length. The formula for the perimeter of a square is shown next.

PERIMETER OF A SQUARE

Perimeter = side + side + side + side
             = 4 · side

In symbols,

$$P = 4 \cdot s$$

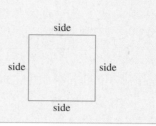

Practice Problem 3

How much fencing is needed to enclose a square field 50 yd on a side?

50 yd

Example 3   Finding Perimeter of a Table Top

Find the perimeter of a square table top if each side is 5 feet long.

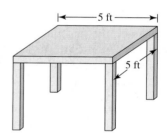

5 ft

5 ft

Solution: The formula for the perimeter of a square is $P = 4 \cdot s$. We will use this formula and replace s by 5 feet.

$$P = 4 \cdot s$$
$$= 4 \cdot 5 \text{ feet}$$
$$= 20 \text{ feet}$$

The perimeter of the square table top is 20 feet. ▬▬▬

The formula for the perimeter of a triangle with sides of length a, b, and c is given next.

PERIMETER OF A TRIANGLE

Perimeter = side a + side b + side c

In symbols,

$$P = a + b + c$$

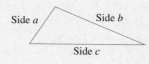

Side a Side b

Side c

Answer

3. 200 yd

Example 4 Find the perimeter of a triangle if the sides are 3 inches, 7 inches, and 6 inches.

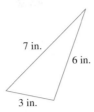

Solution: The formula is $P = a + b + c$, where a, b, and c are the lengths of the sides. Thus,

$P = a + b + c$

$\quad = 3 \text{ in.} + 7 \text{ in.} + 6 \text{ in.}$

$\quad = 16 \text{ in.}$

The perimeter of the triangle is 16 inches.

Recall that to find the perimeter of other polygons, we find the sum of the lengths of their sides.

Example 5 Find the perimeter of the trapezoid shown.

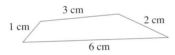

Solution: To find the perimeter, we find the sum of the lengths of its sides.

perimeter $= 3 \text{ cm} + 2 \text{ cm} + 5 \text{ cm} + 2 \text{ cm} = 12 \text{ cm}$

The perimeter is 12 centimeters.

Example 6 **Finding the Perimeter of a Room**

Find the perimeter of the room shown below.

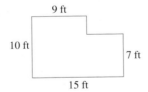

Solution: To find the perimeter of the room, we first need to find the lengths of all sides of the room.

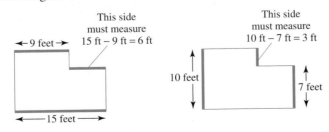

Now that we know the measures of all sides of the room, we can add the measures to find the perimeter.

Practice Problem 4

Find the perimeter of a triangle if the sides are 5 cm, 9 cm, and 7 cm in length.

Practice Problem 5

Find the perimeter of the trapezoid shown.

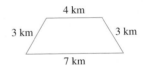

Practice Problem 6

Find the perimeter of the room shown.

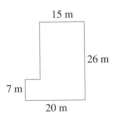

Answers
4. 21 cm, **5.** 17 km, **6.** 92 m

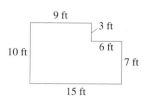

perimeter = 10 ft + 9 ft + 3 ft + 6 ft + 7 ft + 15 ft

= 50 ft

The perimeter of the room is 50 feet.

Practice Problem 7

A rectangular lot measures 60 ft by 120 ft. Find the cost to install fencing around the lot if the cost of fencing is $1.90 per foot.

Example 7   Calculating the Cost of Baseboard

A rectangular room measures 10 feet by 12 feet. Find the cost to install new baseboard around the room if the cost of the baseboard is $0.66 per foot.

Solution: First we find the perimeter of the room.

$$P = 2 \cdot l + 2 \cdot w$$
$$= 2 \cdot 12 \text{ feet} + 2 \cdot 10 \text{ feet}$$
Replace *l* with 12 feet and *w* with 10 feet.
$$= 24 \text{ feet} + 20 \text{ feet}$$
$$= 44 \text{ feet}$$

The cost for the baseboard is

cost = 0.66 · 44 = 29.04

The cost of the baseboard is $29.04.

B   USING FORMULAS TO FIND CIRCUMFERENCES

Recall from Section 4.4 that the distance around a circle is given a special name called the **circumference**. This distance depends on the radius or the diameter of the circle.

The formulas for circumference are shown next.

CIRCUMFERENCE OF A CIRCLE

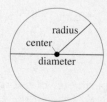

Circumference = 2 · π · radius   or   Circumference = π · diameter
In symbols,

$$C = 2 \cdot \pi \cdot r \quad \text{or} \quad C = \pi \cdot d$$

where $\pi \approx 3.14$ or $\pi \approx \dfrac{22}{7}$.

Answer

7. $684

To better understand circumference and π(pi), try the following experiment. Take any can and measure its circumference and its diameter.

The can in the figure above has a circumference of 23.5 centimeters and a diameter of 7.5 centimeters. Now divide the circumference by the diameter.

$$\frac{\text{circumference}}{\text{diameter}} = \frac{23.5 \text{ cm}}{7.5 \text{ cm}} \approx 3.1$$

Try this with other sizes of cylinders and circles—you should always get a number close to 3.1. The exact ratio of circumference to diameter is π. (Recall that $\pi \approx 3.14$ or $\pi \approx \frac{22}{7}$.)

Example 8   Mary Catherine Dooley plans to install a border of new tiling around the circumference of her circular spa. If her spa has a diameter of 14 feet, find its circumference.

Solution: Because we are given the diameter, we use the formula $C = \pi \cdot d$.

$C = \pi \cdot d$

$\quad = \pi \cdot 14 \text{ ft}$ Replace *d* with 14 feet.

$\quad = 14 \pi \text{ ft}$

The circumference of the spa is *exactly* 14π feet. By replacing π with the *approximation* 3.14, we find that the circumference is *approximately* 14 feet $\cdot$ 3.14 = 43.96 feet.

TRY THE CONCEPT CHECK IN THE MARGIN.

Practice Problem 8

An irrigation device waters a circular region with a diameter of 20 yd. What is the circumference of the watered region?

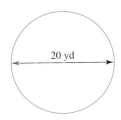

✓ CONCEPT CHECK

The distance around which figure is greater: a square with side length 5 in. or a circle with radius 3 in.?

Answers

8. 62.8 yd

✓ Concept Check: a square with length 5 in.

Focus On History

THE PYTHAGOREAN THEOREM

Pythagoras was a Greek teacher who lived from about 580 B.C. to 501 B.C. He founded his school at Croton in southern Italy. The school was a tightly knit community of eager young students with the motto "All is number." Pythagoras himself taught by lecturing on the subjects of arithmetic, music, geometry, and astronomy—all based on mathematics. In fact, Pythagoras and his students held numbers to be sacred and tried to find mathematical order in all parts of life and nature. Pythagoras' followers, known as Pythagoreans, were sworn to not reveal any of the Master's teachings to outsiders. It is for this reason that very little of his life or teachings are known today—Pythagoreans were forbidden from recording any of the Master's teachings in written form, and Pythagoras left none of his own writings.

Traditionally, the so-called Pythagorean theorem is attributed to Pythagoras himself. However, some scholars question whether Pythagoras ever gave any rigorous proof of the theorem at all! There is ample evidence that several ancient cultures had knowledge of this important theorem and used its results in practical matters well before the time of Pythagoras.

▲ A Babylonian clay tablet (#322 in the Plimpton collection at Columbia University) dating from 1900 B.C. gives numerical evidence that the Babylonians were well aware of what we today call the Pythagorean theorem.

▲ The ancient Chinese knew of a very simple geometric proof of the Pythagorean theorem, which can be seen in the *Arithmetic Classics of the Gnomon and the Circular Paths of Heaven*, dating from about 600 B.C.

▲ The ancient Egyptians used the Pythagorean theorem to form right angles when they needed to measure a plot of land. They put equally spaced knots in pieces of rope so they could form a triangle that had a right angle by stretching the rope out on the ground. They used equally spaced knots to ensure that the lengths of the measuring ropes were in the ratio 3:4:5.

CRITICAL THINKING

In the Egyptian use of the Pythagorean theorem, if three pieces of rope were used and one piece had a total of 10 knots equally spaced and another piece had a total of 13 knots with the same equal spacing, how many knots with the same equal spacing would a third piece of rope need so that the three pieces form a right triangle with the first two pieces as the legs of the triangle? Draw a figure and explain your answer. (Assume that each piece of rope was knotted at both ends.)

EXERCISE SET A.3

A *Find the perimeter of each figure. See Examples 1 through 6.*

1.

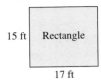

15 ft | Rectangle

17 ft

2.

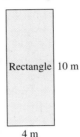

Rectangle | 10 m

4 m

3.

Square

9 cm

4.

Square

46 mi

5.

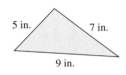

5 in.   7 in.

9 in.

6.

4 units   10 units

8 units

7.

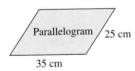

Parallelogram / 25 cm

35 cm

8.

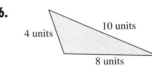

Parallelogram

3 yd

2 yd

9.

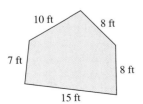

10 ft   8 ft

7 ft   8 ft

15 ft

10.

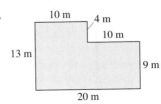

10 m   4 m

10 m

13 m   9 m

20 m

Find the perimeter of each figure. See Example 6.

11.

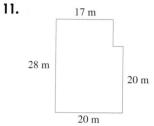

17 m
28 m
20 m
20 m

12.

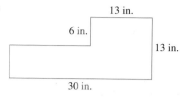

13 in.
6 in.
13 in.
30 in.

13.

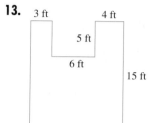

3 ft
4 ft
5 ft
6 ft
15 ft

14.

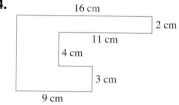

16 cm
2 cm
11 cm
4 cm
3 cm
9 cm

15.

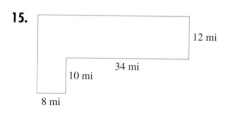

12 mi
34 mi
10 mi
8 mi

16.

22 km
12 km
5 km
6 km

B *Find the circumference of each circle. Give the exact circumference and then an approximation. Use* $\pi \approx 3.14$. *See Example 8.*

17.

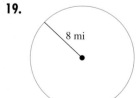

17 cm

18.

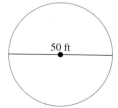

6 in.

19.

8 mi

20.

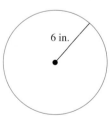

50 ft

11. _____

12. _____

13. _____

14. _____

15. _____

16. _____

17. _____

18. _____

19. _____

20. _____

A.4 AREA

A FINDING AREA

Recall that area measures the amount of surface of a region. Thus far, we know how to find the area of a rectangle and a square. These formulas, as well as formulas for finding the areas of other common geometic figures, are given next.

AREA FORMULAS OF COMMON GEOMETRIC FIGURES

Geometric Figure	**Area Formula**

RECTANGLE

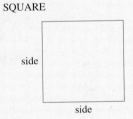

Area of a rectangle:
Area = **l**ength · **w**idth
$A = lw$

SQUARE

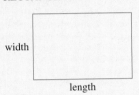

Area of a square:
Area = **s**ide · **s**ide
$A = s \cdot s = s^2$

TRIANGLE

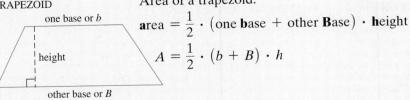

Area of a triangle:
Area = $\frac{1}{2}$ · **b**ase · **h**eight

$A = \frac{1}{2} \cdot b \cdot h$

PARALLELOGRAM

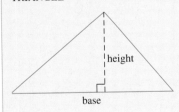

Area of a parallelogram:
Area = **b**ase · **h**eight
$A = b \cdot h$

TRAPEZOID

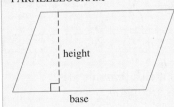
one base or b
height
other base or B

Area of a trapezoid:

area = $\frac{1}{2}$ · (one **b**ase + other **B**ase) · **h**eight

$A = \frac{1}{2} \cdot (b + B) \cdot h$

Use these formulas for the following examples.

┌─ **HELPFUL HINT**

Area is always measured in square units.
└─

Practice Problem 1

Find the area of the triangle.

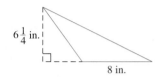

Example 1

Find the area of the triangle.

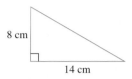

Solution: $A = \dfrac{1}{2} \cdot b \cdot h$

$= \dfrac{1}{2} \cdot 14 \text{ cm} \cdot 8 \text{ cm}$

$= \dfrac{\overset{1}{\cancel{2}} \cdot 7 \cdot 8}{\underset{1}{\cancel{2}}}$ square centimeters

$= 56$ square centimeters

The area is 56 square centimeters.

Practice Problem 2

Find the area of the square.

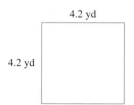

Example 2

Find the area of the parallelogram.

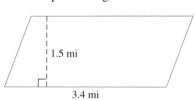

Solution: $A = b \cdot h$

$= 3.4 \text{ miles} \cdot 1.5 \text{ miles}$

$= 5.1$ square miles

The area is 5.1 square miles.

┌─ **HELPFUL HINT**

When finding the area of figures, be sure all measurements are changed to the same unit before calculations are made.
└─

Practice Problem 3

Find the area of the figure.

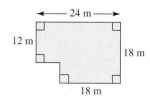

Example 3

Find the area of the figure.

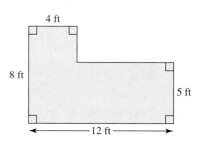

Answers

1. 25 sq. in.,   **2.** 17.64 sq. yd,   **3.** 396 sq. m

Solution: Split the figure into two rectangles. To find the area of the figure, we find the sum of the areas of the two rectangles.

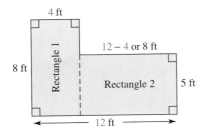

Area of Rectangle 1 $= l \cdot w$

$\qquad\qquad\qquad = 8 \text{ feet} \cdot 4 \text{ feet}$

$\qquad\qquad\qquad = 32 \text{ square feet}$

Notice that the length of Rectangle 2 is 12 feet $-$ 4 feet or 8 feet.

Area of Rectangle 2 $= l \cdot w$

$\qquad\qquad\qquad = 8 \text{ feet} \cdot 5 \text{ feet}$

$\qquad\qquad\qquad = 40 \text{ square feet}$

Area of the Figure $=$ Area of Rectangle 1 $+$ Area of Rectangle 2

$\qquad\qquad\qquad = 32 \text{ square feet} + 40 \text{ square feet}$

$\qquad\qquad\qquad = 72 \text{ square feet}$

HELPFUL HINT

The figure in Example 3 can also be split into two rectangles as shown.

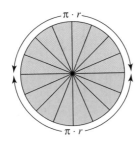

To better understand the formula for area of a circle, try the following. Cut a circle into many pieces as shown.

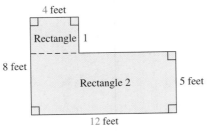

The circumference of a circle is $2 \cdot \pi \cdot r$. This means that the circumference of half a circle is half of $2 \cdot \pi \cdot r$, or $\pi \cdot r$.

Then unfold the two halves of the circle and place them together as shown.

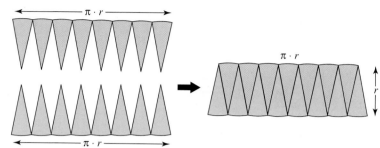

The figure on the right is almost a parallelogram with a base of $\pi \cdot r$ and a height of r. The area is

$$A = \boxed{\text{base}} \cdot \boxed{\text{height}}$$
$$= (\pi \cdot r) \cdot r$$
$$= \pi \cdot r^2$$

This is the formula for area of a circle.

AREA FORMULA OF A CIRCLE

Circle

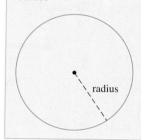

radius

Area of a circle:
Area $= \pi \cdot (\text{radius})^2$
$A = \pi \cdot r^2$

(A fraction approximation for π is $\dfrac{22}{7}$.)

(A decimal approximation for π is 3.14.)

Practice Problem 4

Find the area of the given circle. Find the exact area and an approximation. Use 3.14 as an approximation for π.

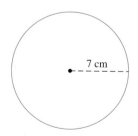

7 cm

✓ CONCEPT CHECK

Use estimation to decide which figure would have a larger area: a circle of diameter 10 in. or a square 10 in. long on each side.

Answers

4. 49π sq. cm ≈ 153.86 sq. cm

✓ **Concept Check:** a square 10 in. long on each side

Example 4 Find the area of a circle with a radius of 3 feet. Find the exact area and an approximation. Use 3.14 as an approximation for π.

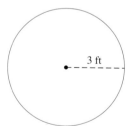

3 ft

Solution: We let $r = 3$ feet and use the formula
$$A = \pi \cdot r^2$$
$$= \pi \cdot (3 \text{ feet})^2$$
$$= 9 \cdot \pi \text{ square feet}$$

To approximate this area, we substitute 3.14 for π.

$9 \cdot \pi$ square feet $\approx 9 \cdot 3.14$ square feet
$= 28.26$ square feet

The *exact* area of the circle is 9π square feet, which is *approximately* 28.26 square feet.

TRY THE CONCEPT CHECK IN THE MARGIN.

Name _____ Section _____ Date _____

EXERCISE SET A.4

A *Find the area of the geometric figure. If the figure is a circle, give an exact area and then use the given **approximation** for π to approximate the area. See Examples 1 through 4.*

1.
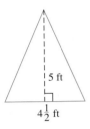
2 m | Rectangle
3.5 m

2.
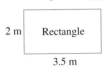
2.75 ft | Rectangle
7 ft

3.
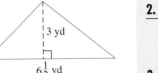
3 yd
$6\frac{1}{2}$ yd

4.
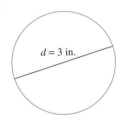
5 ft
$4\frac{1}{2}$ ft

5.
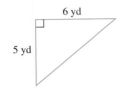
6 yd
5 yd

6.
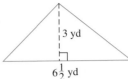
5 ft   7 ft

7. Use 3.14 for π.

$d = 3$ in.

8. Use $\frac{22}{7}$ for π.
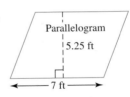
$r = 2$ cm

9.
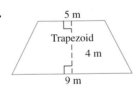
Parallelogram
5.25 ft
7 ft

10.
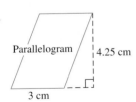
Parallelogram | 4.25 cm
3 cm

11.

5 m
Trapezoid
4 m
9 m

12.
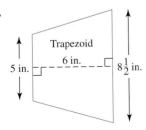
Trapezoid
6 in.
5 in. $8\frac{1}{2}$ in.

13.
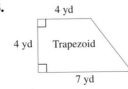
4 yd
4 yd | Trapezoid
7 yd

14. _____

15. _____

16. _____

17. _____

18. _____

19. _____

20. _____

21. _____

22. _____

14.

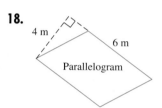

10 ft
3 ft
Trapezoid
5 ft

15.

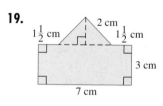

7 ft
Parallelogram
$5\frac{1}{4}$ ft

16.

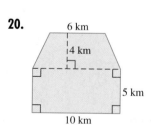

Parallelogram
$4\frac{1}{4}$ cm
3 cm

17.

7 ft (in image)
$4\frac{1}{2}$ in.
Parallelogram
5 in.

18.

4 m
6 m
Parallelogram

19.

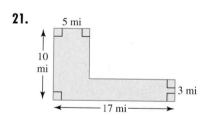

2 cm
$1\frac{1}{2}$ cm $1\frac{1}{2}$ cm
3 cm
7 cm

20.

6 km
4 km
5 km
10 km

21.

5 mi
10 mi
3 mi
17 mi

22.

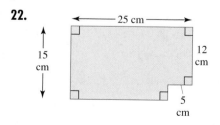

25 cm
15 cm
12 cm
5 cm

23.

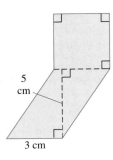

5
cm

3 cm

24.

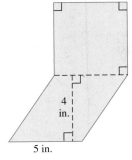

4
in.

5 in.

25. Use $\frac{22}{7}$ for π.

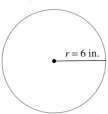

$r = 6$ in.

24. . _____

25. _____

Problem Solving Notes

A.5 VOLUME

A FINDING VOLUME

Volume is a measure of the space of a region. The volume of a box or can, for example, is the amount of space inside. Volume can be used to describe the amount of juice in a pitcher or the amount of concrete needed to pour a foundation for a house.

The volume of a solid is the number of **cubic units** in the solid. A cubic centimeter and a cubic inch are illustrated.

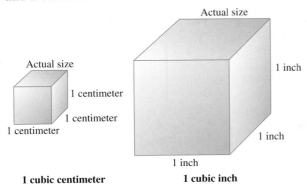

Actual size

1 inch

1 inch

1 inch

Actual size

1 centimeter

1 centimeter

1 centimeter

1 cubic centimeter   **1 cubic inch**

Formulas for finding the volumes of some common solids are given next.

VOLUME FORMULAS OF COMMON SOLIDS

Solid

Volume Formulas

Rectangular Solid

height

length

width

Volume of a rectangular solid:
Volume = length · width · height
$$V = l \cdot w \cdot h$$

CUBE

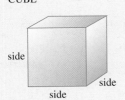

side

side

side

Volume of a cube:
Volume = side · side · side
$$V = s^3$$

SPHERE

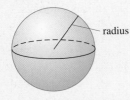

radius

Volume of a sphere:
$$\textbf{Volume} = \frac{4}{3} \cdot \pi \cdot (\textbf{radius})^3$$

$$V = \frac{4}{3} \cdot \pi \cdot r^3$$

CIRCULAR CYLINDER

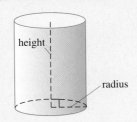

height

radius

Volume of a circular cylinder:
$$\textbf{Volume} = \pi \cdot (\textbf{radius})^2 \cdot \textbf{height}$$

$$V = \pi \cdot r^2 \cdot h$$

Ojective

A Find the volume of solids.

VOLUME FORMULAS OF COMMON SOLIDS

Solid

CONE

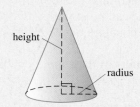

height

radius

Volume Formulas

Volume of a cone:
Volume
$$= \frac{1}{3} \cdot \pi \cdot (\mathbf{radius})^2 \cdot \mathbf{height}$$
$$V = \frac{1}{3} \cdot \pi \cdot r^2 \cdot h$$

SQUARE-BASED PYRAMID

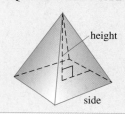

height

side

Volume of a square-based pyramid:
$$\mathbf{V}\text{olume} = \frac{1}{3} \cdot (\mathbf{side})^2 \cdot \mathbf{height}$$
$$V = \frac{1}{3} \cdot s^2 \cdot h$$

Practice Problem 1

Draw a diagram and find the volume of a rectangular box that is 5 ft long, 2 ft wide, and 4 ft deep.

✔ CONCEPT CHECK

Juan is calculating the volume of the following rectangular solid. Find the error in his calculation.

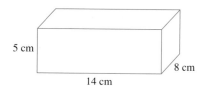

5 cm

14 cm

8 cm

$$\text{Volume} = l + w + h$$
$$= 14 + 8 + 5$$
$$= 27 \text{ cubic cm}$$

Practice Problem 2

Draw a diagram and approximate the volume of a ball of radius $\frac{1}{2}$ cm. Use $\frac{22}{7}$ for π.

Answers

1. 40 cu. ft,   **2.** $\frac{11}{21}$ cu. cm

✔ Concept Check

$$\text{Volume} = l \cdot w \cdot h$$
$$= 14 \cdot 8 \cdot 5$$
$$= 560 \text{ cu. cm}$$

HELPFUL HINT

Volume is always measured in cubic units.

Example 1   Find the volume of a rectangular box that is 12 inches long, 6 inches wide, and 3 inches high.

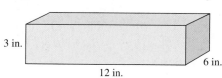

3 in.

12 in.

6 in.

Solution: $V = l \cdot w \cdot h$

$V = 12 \text{ inches} \cdot 6 \text{ inches} \cdot 3 \text{ inches} = 216 \text{ cubic inches}$

The volume of the rectangular box is 216 cubic inches.

TRY THE CONCEPT CHECK IN THE MARGIN.

Example 2   Approximate the volume of a ball of radius 3 inches.

Use the approximation $\frac{22}{7}$ for π.

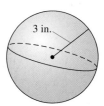

3 in.

Solution: $V = \dfrac{4}{3} \cdot \pi \cdot r^3$

$\approx \dfrac{4}{3} \cdot \dfrac{22}{7} (3 \text{ inches})^3$

$= \dfrac{4}{3} \cdot \dfrac{22}{7} \cdot 27 \text{ cubic inches}$

$= \dfrac{4 \cdot 22 \cdot \overset{1}{\cancel{3}} \cdot 9}{\underset{1}{\cancel{3}} \cdot 7} \text{ cubic inches}$

$= \dfrac{792}{7}$ or $113\dfrac{1}{7}$ cubic inches

The volume is *approximately* $113\dfrac{1}{7}$ cubic inches. ▬▬▬

Example 3 Approximate the volume of a can that has a $3\dfrac{1}{2}$-inch radius and a height of 6 inches. Use $\dfrac{22}{7}$ for π.

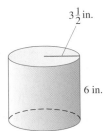

$3\dfrac{1}{2}$ in.

6 in.

Solution: Using the formula for a circular cylinder, we have

$V = \pi \cdot r^2 \cdot h$

$\qquad\qquad 3\dfrac{1}{2} = \dfrac{7}{2}$

$= \pi \cdot \left(\dfrac{7}{2} \text{ inches}\right)^2 \cdot 6 \text{ inches}$

or approximately

$\approx \dfrac{22}{7} \cdot \dfrac{49}{4} \cdot 6 \text{ cubic inches}$

$= 231 \text{ cubic inches}$

The volume is approximately 231 cubic inches. ▬▬▬

Practice Problem 3

Approximate the volume of a cylinder of radius 5 in. and height 7 in. Use 3.14 for π.

Answer

3. 549.5 cu. in.

Focus On the Real World

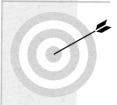

There are nearly 4 million miles of streets and roads in the United States. With streets, roads, and highways comes the need for traffic control, guidance, warning, and regulation. Road signs perform many of these tasks. Just in our routine travels, we see a wide variety of road signs every day. Think how many road signs must exist on the 4 million miles of roads in the United States. Have you ever wondered how much signs like these cost?

The cost of a road sign generally depends on the type of sign. Costs for several types of signs and sign posts are listed in the table. Examples of various types of signs are shown below.

ROAD SIGN COSTS	
Type of Sign	Cost
Regulatory, warning, marker	$15–$18 per square foot
Large guide	$20–$25 per square foot
Type of Post	Cost
U-channel	$125–$200 each
Square tube	$10–$15 per foot
Steel breakaway posts	$15–$25 per foot

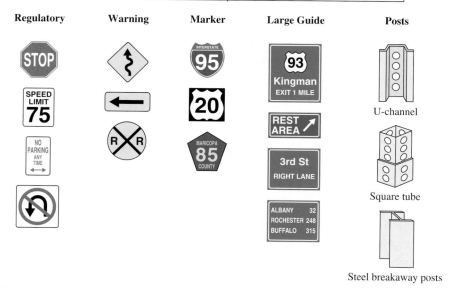

The cost of a sign is based on its area. For diamond, square, or rectangular signs, the area is found by multiplying the length (in feet) times the width (in feet). Then the area is multiplied by the cost per square foot. For signs with irregular shapes, costs are generally figured *as if* the sign were a rectangle, multiplying the height and width at the tallest and widest parts of the sign.

GROUP ACTIVITY

Locate four different kinds of road signs on or near your campus. Measure the dimensions of each sign, including the height of the post on which it is mounted. Using the cost data given in the table, find the minimum and maximum cost of each sign, including its post. Summarize your results in a table, and include a sketch of each sign.

Name _____ **Section** _____ **Date** _____

EXERCISE SET A.5

A *Find the volume of each solid. See Examples 1 through 3. Use $\frac{22}{7}$ for π.*

1.

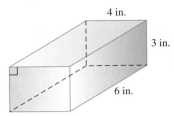

4 in.

3 in.

6 in.

2.

3 mi

3.

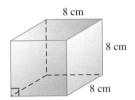

8 cm

8 cm

8 cm

4.

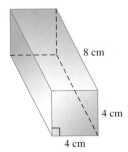

8 cm

4 cm

4 cm

5.

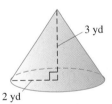

3 yd

2 yd

6.

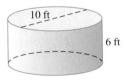

10 ft

6 ft

7.

10 in.

8.

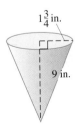

$1\frac{3}{4}$ in.

9 in.

9. _____

9.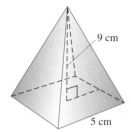

9 cm

5 cm

10. _____

10.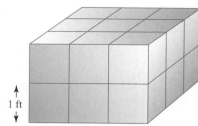

1 ft

11. _____

Solve.

11. Find the volume of a cube with edges of $1\frac{1}{3}$ inches.

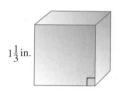

$1\frac{1}{3}$ in.

12. A water storage tank is in the shape of a cone with the pointed end down. If the radius is 14 ft and the depth of the tank is 15 ft, approximate the volume of the tank in cubic feet. Use $\frac{22}{7}$ for π.

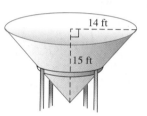

14 ft

15 ft

12. _____

13. _____

13. Find the volume of a rectangular box 2 ft by 1.4 ft by 3 ft.

14. Find the volume of a box in the shape of a cube that is 5 ft on each side.

14. _____

15. Find the volume of a pyramid with a square base 5 in. on a side and a height of 1.3 in.

15. _____

A.6   Square Roots and the Pythagorean Theorem

A Finding Square Roots

The square of a number is the number times itself. For example,

The square of 5 is 25 because 5^2 or $5 \cdot 5 = 25$.
The square of 3 is 9 because 3^2 or $3 \cdot 3 = 9$.
The square of 10 is 10^2 or $10 \cdot 10 = 100$.

The reverse process of squaring is finding a **square root**. For example,

A square root of 9 is 3 because $3^2 = 9$.
A square root of 25 is 5 because $5^2 = 25$.
A square root of 100 is 10 because $10^2 = 100$.

We use the symbol $\sqrt{}$, called a **radical sign**, to name square roots. For example,

$\sqrt{9} = 3$ because $3^2 = 9$.
$\sqrt{25} = 5$ because $5^2 = 25$.

SQUARE ROOT OF A NUMBER

A square root of a number a is a number b whose square is a. We use the radical sign $\sqrt{}$ to name square roots.

Example 1   Find each square root.

    a. $\sqrt{49}$ **b.** $\sqrt{36}$ **c.** $\sqrt{1}$ **d.** $\sqrt{81}$

Solution: **a.** $\sqrt{49} = 7$ because $7^2 = 49$.
    b. $\sqrt{36} = 6$ because $6^2 = 36$.
    c. $\sqrt{1} = 1$ because $1^2 = 1$.
    d. $\sqrt{81} = 9$ because $9^2 = 81$.

Example 2   Find: $\sqrt{\dfrac{1}{36}}$

Solution: $\sqrt{\dfrac{1}{36}} = \dfrac{1}{6}$ because $\dfrac{1}{6} \cdot \dfrac{1}{6} = \dfrac{1}{36}$.

Example 3   Find: $\sqrt{\dfrac{4}{25}}$

Solution: $\sqrt{\dfrac{4}{25}} = \dfrac{2}{5}$ because $\dfrac{2}{5} \cdot \dfrac{2}{5} = \dfrac{4}{25}$.

B Approximating Square Roots

Thus far, we have found square roots of perfect squares. Numbers like $\dfrac{1}{4}$, 36, $\dfrac{4}{25}$, and 1 are called **perfect squares** because their square root is a whole number or a fraction. A square root such as $\sqrt{5}$ cannot be written as a whole number or a fraction since 5 is not a perfect square.

Although $\sqrt{5}$ cannot be written as a whole number or a fraction, it can be approximated by estimating, by using a table, or by using a calculator.

Objectives

A Find the square root of a number.
B Approximate square roots.
C Use the Pythagorean theorem.

Practice Problem 1

Find each square root.

a. $\sqrt{100}$

b. $\sqrt{64}$

c. $\sqrt{121}$

d. $\sqrt{0}$

Practice Problem 2

Find: $\sqrt{\dfrac{1}{4}}$

Practice Problem 3

Find: $\sqrt{\dfrac{9}{16}}$

Answers

1. a. 10, **b.** 8, **c.** 11, **d.** 0, **2.** $\dfrac{1}{2}$, **3.** $\dfrac{3}{4}$

Practice Problem 4

Use a calculator to approximate the square root of 11 to the nearest thousandth.

Practice Problem 5

Approximate $\sqrt{29}$ to the nearest thousandth.

Example 4   Use a calculator to approximate the square root of 43 to the nearest thousandth.

Solution: $\sqrt{43} \approx 6.557$

HELPFUL HINT

$\sqrt{43}$ is *approximately* 6.557. This means that if we multiply 6.557 by 6.557, the product is *close* to 43.

$6.557 \times 6.557 = 42.994249$

Example 5   Approximate $\sqrt{32}$ to the nearest thousandth.

Solution: $\sqrt{32} \approx 5.657$

C   USING THE PYTHAGOREAN THEOREM

One important application of square roots has to do with right triangles. Recall that a **right triangle** is a triangle in which one of the angles is a right angle, or measures 90° (degrees). The **hypotenuse** of a right triangle is the side opposite the right angle. The **legs** of a right triangle are the other two sides. These are shown in the following figure. The right angle in the triangle is indicated by the small square drawn in that angle.

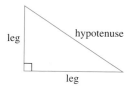

The following theorem is true for all right triangles.

PYTHAGOREAN THEOREM

In any **right triangle**,

$$(\text{leg})^2 + (\text{other leg})^2 = (\text{hypotenuse})^2$$

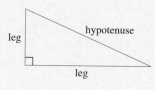

Using the Pythagorean theorem, we can use one of the following formulas to find an unknown length of a right triangle.

FINDING AN UNKNOWN LENGTH OF A RIGHT TRIANGLE

$$\text{hypotenuse} = \sqrt{(\text{leg})^2 + (\text{other leg})^2}$$

or

$$\text{leg} = \sqrt{(\text{hypotenuse})^2 - (\text{other leg})^2}$$

Answers

4. 3.317,   **5.** 5.385

Example 6 Find the length of the hypotenuse of the given right triangle.

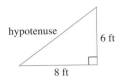

Solution: Since we are finding the hypotenuse, we use the formula

$$\text{hypotenuse} = \sqrt{(\text{leg})^2 + (\text{other leg})^2}$$

Putting the known values into the formula, we have

$$\begin{aligned} \text{hypotenuse} &= \sqrt{(6)^2 + (8)^2} \quad \text{Legs are 6 feet and 8 feet.} \\ &= \sqrt{36 + 64} \\ &= \sqrt{100} \\ &= 10 \end{aligned}$$

The hypotenuse is 10 feet long.

Example 7 Approximate the length of the hypotenuse of the given right triangle. Round the length to the nearest whole unit.

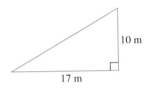

Solution:
$$\begin{aligned} \text{hypotenuse} &= \sqrt{(\text{leg})^2 + (\text{other leg})^2} \\ &= \sqrt{(17)^2 + (10)^2} \quad \text{The legs are 10 meters} \\ &\qquad\qquad\qquad\qquad \text{and 17 meters.} \\ &= \sqrt{289 + 100} \\ &= \sqrt{389} \\ &\approx 20 \quad \text{From a calculator} \end{aligned}$$

The hypotenuse is exactly $\sqrt{389}$ meters, which is approximately 20 meters.

Example 8 Find the length of the leg in the given right triangle. Give the exact length and a two-decimal-place approximation.

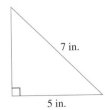

Practice Problem 6

Find the length of the hypotenuse of the given right triangle.

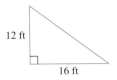

Practice Problem 7

Approximate the length of the hypotenuse of the given right triangle. Round to the nearest whole unit.

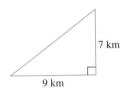

Practice Problem 8

Find the length of the leg in the given right triangle. Give the exact length and a two-decimal-place approximation.

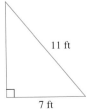

Answers

6. 20 ft, **7.** 11 km, **8.** $\sqrt{72}$ ft $\approx$ 8.49 ft

Solution: Notice that the hypotenuse measures 7 inches and that the length of one leg measures 5 inches. Since we are looking for the length of the other leg, we use the formula

$$\text{leg} = \sqrt{(\text{hypotenuse})^2 - (\text{other leg})^2}$$

Putting the known values into the formula, we have

$$\text{leg} = \sqrt{(7)^2 - (5)^2}$$ The hypotenuse is 7 inches and the other leg is 5 inches.
$$= \sqrt{49 - 25}$$
$$= \sqrt{24}$$
$$\approx 4.90$$ From a calculator

The length of the leg is exactly $\sqrt{24}$ inches, which is approximately 4.90 inches.

TRY THE CONCEPT CHECK IN THE MARGIN.

Example 9   Finding the Diagonal Length of a City Block

A standard city block is a square that measures 300 feet on a side. Find the length of the diagonal of a city block rounded to the nearest whole foot.

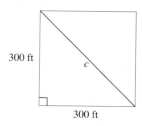

300 ft

c

300 ft

Solution: The diagonal is the hypotenuse of a right triangle, so we use the formula

$$\text{hypotenuse} = \sqrt{(\text{leg})^2 + (\text{other leg})^2}$$

Putting the known values into the formula, we have

$$\text{hypotenuse} = \sqrt{(300)^2 + (300)^2}$$ The legs are both 300 feet.
$$= \sqrt{90,000 + 90,000}$$
$$= \sqrt{180,000}$$
$$\approx 424$$ From a calculator

The length of the diagonal is approximately 424 feet.

✓ **CONCEPT CHECK**

The following lists are the lengths of the sides of two triangles. Which set forms a right triangle?
a. 8, 15, 17
b. 24, 30, 40

Practice Problem 9

A football field is a rectangle measuring 100 yards by 53 yards. Draw a diagram and find the length of the diagonal of a football field to the nearest yard.

Answers

9. 113 yd

✓ **Concept Check:**   **a.** $8^2 + 15^2 = 17^2$

CALCULATOR EXPLORATIONS
FINDING SQUARE ROOTS

To simplify or approximate square roots using a calculator, locate the key marked $\boxed{\sqrt{}}$.

To simplify $\sqrt{64}$, for example, press the keys

$\boxed{64}$ $\boxed{\sqrt{}}$ or $\boxed{\sqrt{}}$ $\boxed{64}$

The display should read $\boxed{8}$. Then

$\sqrt{64} = 8$

To *approximate* $\sqrt{10}$, press the keys

$\boxed{10}$ $\boxed{\sqrt{}}$ or $\boxed{\sqrt{}}$ $\boxed{10}$

The display should read $\boxed{3.16227766}$. This is an *approximation* for $\sqrt{10}$. A three-decimal-place approximation is:

$\sqrt{10} \approx 3.162$

Is this answer reasonable? Since 10 is between perfect squares 9 and 16, $\sqrt{10}$ is between $\sqrt{9} = 3$ and $\sqrt{16} = 4$. Our answer is reasonable since 3.162 is between 3 and 4.

Simplify.

1. $\sqrt{1024}$

2. $\sqrt{676}$

Approximate each square root. Round each answer to the nearest thousandth.

3. $\sqrt{15}$

4. $\sqrt{19}$

5. $\sqrt{97}$

6. $\sqrt{56}$

Focus On Business and Career

PRODUCT PACKAGING

Suppose you have just developed a new product that you would like to market. You will need to think about who would like to buy it, where and how it should be sold, for how much it will sell, how to package it, and other pressing concerns. Although all of these items are important to think through, many package designers believe that the packaging in which a product is sold is at least as important as the product itself.

Product packaging contains the product, keeping it from leaking out, keeping it fresh if perishable, providing protective cushioning against breakage, and keeping all the pieces together as a bundle. Product packaging also provides a way to give information about the product: what it is, how to use it, for whom it is designed, how it is beneficial or advantageous, who to contact if more information is needed, what other products are necessary to use with the product, etc. Product packaging must be pleasing and eyecatching to the product's audience. It must be capable of selling its contents without further assistance.

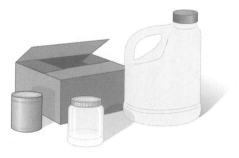

CRITICAL THINKING

1. How can a knowledge of geometry be helpful in the packaging design process?

2. Design two different packages for the same product that have roughly the same volume. Does one package "look" larger than the other? How could this be useful to a package designer?

Name _____ **Section** _____ **Date** _____

EXERCISE SET A.6

C *Find the unknown length of each right triangle. If necessary, approximate the length to the nearest thousandth. See Examples 6 through 8.*

1.

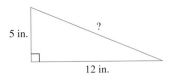

5 in.

?

12 in.

2.

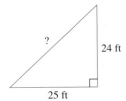

?

24 ft

25 ft

3.

12 cm

10 cm

4.

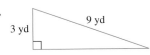

3 yd

9 yd

Sketch each right triangle and find the length of the side not given. If necessary, approximate the length to the nearest thousandth. See Examples 6 through 8.

5. leg = 3, leg = 4

6. leg = 9, leg = 12

7. leg = 6, hypotenuse = 10

8. leg = 48, hypotenuse = 53

9. leg = 10, leg = 14

10. leg = 32, leg = 19

11. leg = 2, leg = 16

12. leg = 27, leg = 36

13. leg = 5, hypotenuse = 13

14. leg = 45, hypotenuse = 117

15. leg = 35, leg = 28

16. leg = 30, leg = 15

17. leg = 30, leg = 30

18. leg = 110, leg = 132

19. hypotenuse = 2, leg = 1

20. hypotenuse = 7, leg = 6

ANSWERS

1. _____

2. _____

3. _____

4. _____

5. _____

6. _____

7. _____

8. _____

9. _____

10. _____

11. _____

12. _____

13. _____

14. _____

15. _____

16. _____

17. _____

18. _____

19. _____

20. _____

769

Problem Solving Notes

A.7   CONGRUENT AND SIMILAR TRIANGLES

A   DECIDING WHETHER TRIANGLES ARE CONGRUENT

Two triangles are **congruent** if they have the same shape and the same size. In congruent triangles, the measures of corresponding angles are equal and the lengths of corresponding sides are equal. The following triangles are congruent.

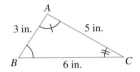   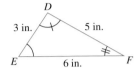

Since these triangles are congruent, the measures of corresponding angles are equal.

Angles with Equal Measure: $\angle A$ and $\angle D$, $\angle B$ and $\angle E$, $\angle C$ and $\angle F$

Also, the lengths of corresponding sides are equal.

Equal Corresponding Sides: $\overline{AB}$ and $\overline{DE}$, $\overline{BC}$ and $\overline{EF}$, $\overline{CA}$ and $\overline{FD}$

Any one of the following may be used to determine whether two triangles are congruent.

CONGRUENT TRIANGLES

Angle-Side-Angle (ASA)

If the measures of two angles of a triangle equal the measures of two angles of another triangle and the lengths of the sides between each pair of angles are equal, the triangles are congruent.

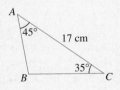   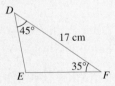

For example, these two triangles are congruent by Angle-Side-Angle.

Side-Side-Side (SSS)

If the lengths of the three sides of a triangle equal the lengths of the corresponding sides of another triangle, the triangles are congruent.

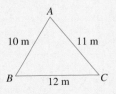   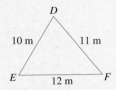

For example, these two triangles are congruent by Side-Side-Side.

Objectives

A Decide whether two triangles are congruent.

B Find the ratio of corresponding sides in similar triangles.

C Find unknown lengths of sides in similar triangles.

CONGRUENT TRIANGLES, CONTINUED

Side-Angle-Side (SAS)

If the lengths of two sides of a triangle equal the lengths of corresponding sides of another triangle, and the measures of the angles between each pair of sides are equal, the triangles are congruent.

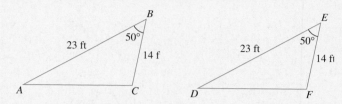

For example, these two triangles are congruent by Side-Angle-Side.

Practice Problem 1

Determine whether triangle *MNO* is congruent to triangle *RTS*.

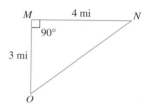

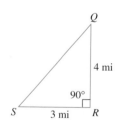

Example 1   Determine whether triangle *ABC* is congruent to triangle *DEF*.

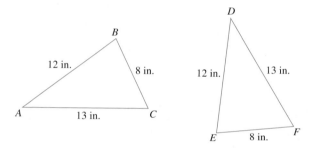

Solution: Since the lengths of all three sides of triangle *ABC* equal the lengths of all three sides of triangle *DEF*, the triangles are congruent. ▬▬▬

In Example 1, notice that once we know that the two triangles are congruent, we know that all three corresponding angles are congruent.

B FINDING THE RATIOS OF CORRESPONDING SIDES IN SIMILAR TRIANGLES

Two triangles are **similar** if they have the same shape but not necessarily the same size. In similar triangles, the measures of corresponding angles are equal and corresponding sides are in proportion. The following triangles are similar.

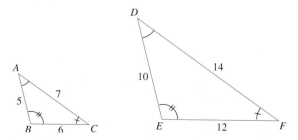

Since these triangles are similar, the measures of corresponding angles are equal.

Answer

1. congruent

Angles with Equal Measure: $\angle A$ and $\angle D$, $\angle B$ and $\angle E$, $\angle C$ and $\angle F$

Also, the lengths of corresponding sides are in proportion.

Sides in Proportion: $\dfrac{AB}{DE} = \dfrac{5}{10} = \dfrac{1}{2}, \dfrac{BC}{EF} = \dfrac{6}{12} = \dfrac{1}{2}, \dfrac{CA}{FD} = \dfrac{7}{14} = \dfrac{1}{2}$

The ratio of corresponding sides is $\dfrac{1}{2}$.

Example 2   Find the ratio of corresponding sides for the similar triangles *ABC* and *DEF*.

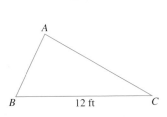
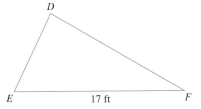

Solution: We are given the lengths of two corresponding sides. Their ratio is

$$\frac{12\ \cancel{\text{feet}}}{17\ \cancel{\text{feet}}} = \frac{12}{17}$$

C FINDING UNKNOWN LENGTHS OF SIDES IN SIMILAR TRIANGLES

Because the ratios of lengths of corresponding sides are equal, we can use proportions to find unknown lengths in similar triangles.

Example 3   Given that the triangles are similar, find the missing length *n*.

Solution: Since the triangles are similar, corresponding sides are in proportion. Thus, the ratio of 2 to 3 is the same as the ratio of 10 to *n*, or

$$\frac{2}{3} = \frac{10}{n}$$

To find the unknown length *n*, we set cross products equal.

$$\frac{2}{3} \;=\; \frac{10}{n}$$

$2 \cdot n = 30$ Set cross products equal.

$n = \dfrac{30}{2}$

$n = 15$

The missing length is 15 units.

Practice Problem 2

Find the ratio of corresponding sides for the similar triangles *QRS* and *XYZ*.

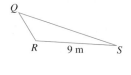
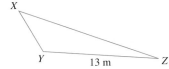

Practice Problem 3

Given that the triangles are similar, find the missing length *n*.

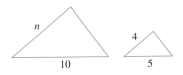

Answers

2. $\dfrac{9}{13}$,   **3.** $n = 8$

✓ CONCEPT CHECK

The following two triangles are similar. Which vertices of the first triangle appear to correspond to which vertices of the second triangle.

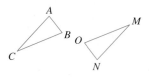

Practice Problem 4

Tammy Shultz, a firefighter, needs to estimate the height of a burning building. She estimates the length of her shadow to be 8 feet long and the length of the building's shadow to be 60 feet long. Find the height of the building if she is 5 feet tall.

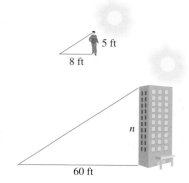

TRY THE CONCEPT CHECK IN THE MARGIN.

Many applications involve a diagram containing similar triangles. Surveyors, astronomers, and many other professionals use ratios of similar triangles continually in their work.

Example 4   Finding the Height of a Tree

Mel Rose is a 6-foot-tall park ranger who needs to know the height of a particular tree. He notices that when the shadow of the tree is 69 feet long, his own shadow is 9 feet long. Find the height of the tree.

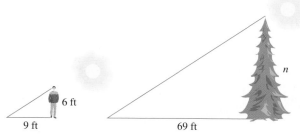

Solution:

1. **UNDERSTAND.** Read and reread the problem. Notice that the triangle formed by the sun's rays, Mel, and his shadow is similar to the triangle formed by the sun's rays, the tree, and its shadow.

2. **TRANSLATE.** Write a proportion from the similar triangles formed.

$$\begin{array}{l} \text{Mel's height} \rightarrow \\ \text{height of tree} \rightarrow \end{array} \dfrac{6}{n} = \dfrac{9}{69} \begin{array}{l} \leftarrow \text{length of Mel's shadow} \\ \leftarrow \text{length of tree's shadow} \end{array}$$

$$\text{or } \frac{6}{n} = \frac{3}{23} \text{ (ratio in lowest terms)}$$

3. **SOLVE** for n.

$$\frac{6}{n} = \frac{3}{23}$$

$$6 \cdot 23 = n \cdot 3 \quad \text{Set cross products equal.}$$
$$138 = n \cdot 3$$
$$\frac{138}{3} = n$$
$$46 = n$$

4. **INTERPRET.** *Check* to see that replacing n with 46 in the proportion makes the proportion true. *State* your conclusion: The height of the tree is 46 feet. ■

Answers

4. 37.5 ft

✓ **Concept Check:**   *A* corresponds to *O*; *B* corresponds to *N*; *C* corresponds to *M*

APPENDIX A HIGHLIGHTS

DEFINITIONS AND CONCEPTS	EXAMPLES

SECTION A.1   LINES AND ANGLES

A **line** is a set of points extending indefinitely in two directions. A line has no width or height, but it does have length. We can name a line by any two of its points.

Line AB or $\overleftrightarrow{AB}$

A **line segment** is a piece of a line with two endpoints.

Line segment AB or $\overline{AB}$

A **ray** is a part of a line with one endpoint. A ray extends indefinitely in one direction.

Ray AB or $\overrightarrow{AB}$

An **angle** is made up of two rays that share the same endpoint. The common endpoint is called the **vertex**.

An angle that measures 180° is called a **straight angle**.

$\angle RST$ is a straight angle.

An angle that measures 90° is called a **right angle**. The symbol $\llcorner$ is used to denote a right angle.

$\angle ABC$ is a right angle.

An angle whose measure is between 0° and 90° is called an **acute angle**.

Acute angles

An angle whose measure is between 90° and 180° is called an **obtuse angle**.

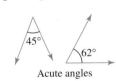

Obtuse angles

Two angles that have a sum of 90° are called **complementary angles**. We say that each angle is the **complement** of the other.

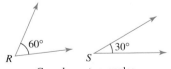

Complementary angles
60° + 30° = 90°

Two angles that have a sum of 180° are called **supplementary angles**. We say that each angle is the **supplement** of the other

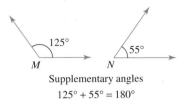

Supplementary angles
125° + 55° = 180°

When two lines intersect, four angles are formed. Two of these angles that are opposite each other are called **vertical angles**. Vertical angles have the same measure.

Two of these angles that share a common side are called **adjacent angles**. Adjacent angles formed in intersecting lines are supplementary.

Vertical angles:
∠a and ∠c
∠d and ∠b
Adjacent angles:
∠a and ∠b
∠b and ∠c
∠c and ∠d
∠d and ∠a

A line that intersects two or more lines at different points is called a **transversal**. Line *l* is a transversal that intersects lines *m* and *n*. The eight angles formed have special names. Some of these names are:

　　Corresponding Angles: ∠a and ∠e, ∠c and ∠g, ∠b and ∠f, ∠d and ∠h
　　Alternate Interior Angles: ∠c and ∠f, ∠d and ∠e

PARALLEL LINES CUT BY A TRANSVERSAL

If two parallel lines are cut by a transversal, then the measures of **corresponding angles are equal** and the measures of **alternate interior angles are equal**.

SECTION A.2   PLANE FIGURES AND SOLIDS

The **sum of the measures** of the angles of a triangle is 180°.

Find the measure of ∠x.

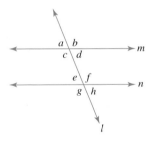

The measure of ∠x = 180° − 85° − 45° = 50°

A **right triangle** is a triangle with a right angle. The side opposite the right angle is called the **hypotenuse** and the other two sides are called **legs**.

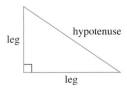

SECTION A.2 **(CONTINUED)**	

For a circle or a sphere:

$$\text{diameter} = 2 \cdot \text{radius}$$

$$d = 2 \cdot r$$

$$\text{radius} = \frac{\text{diameter}}{2}$$

$$r = \frac{d}{2}$$

Find the diameter of the circle.

$$d = 2 \cdot r$$
$$= 2 \cdot 6 \text{ feet} = 12 \text{ feet}$$

SECTION A.3 **PERIMETER**	

PERIMETER FORMULAS

Rectangle:

$$P = 2 \cdot l + 2 \cdot w$$

Square:

$$P = 4 \cdot s$$

Triangle:

$$P = a + b + c$$

Circumference of a Circle:

$$C = 2 \cdot \pi \cdot r \quad \text{or} \quad C = \pi \cdot d$$

where $\pi \approx 3.14 \quad$ or $\quad \pi \approx \frac{22}{7}$

Find the perimeter of the rectangle.

$$P = 2 \cdot l + 2 \cdot w$$
$$= 2 \cdot 28 \text{ m} + 2 \cdot 15 \text{ m}$$
$$= 56 \text{ m} + 30 \text{ m}$$
$$= 86 \text{ m}$$

The perimeter is 86 meters.

SECTION A.4 **AREA**	

AREA FORMULAS

Rectangle:

$$A = l \cdot w$$

Square:

$$A = s^2$$

Triangle:

$$A = \frac{1}{2} \cdot b \cdot h$$

Parallelogram:

$$A = b \cdot h$$

Trapezoid:

$$A = \frac{1}{2} \cdot (b + B) \cdot h$$

Circle:

$$A = \pi \cdot r^2$$

Find the area of the square.

$$A = s^2$$
$$= (8 \text{ cm})^2$$
$$= 64 \text{ square centimeters}$$

The area of the square is 64 square centimeters.

SECTION A.5   VOLUME

VOLUME FORMULAS

Rectangular Solid:

$$V = l \cdot w \cdot h$$

Cube:

$$V = s^3$$

Sphere:

$$V = \frac{4}{3} \cdot \pi \cdot r^3$$

Right Circular Cylinder:

$$V = \pi \cdot r^2 \cdot h$$

Cone:

$$V = \frac{1}{3} \cdot \pi \cdot r^2 \cdot h$$

Square-Based Pyramid:

$$V = \frac{1}{3} \cdot s^2 \cdot h$$

Find the volume of the sphere. Use $\frac{22}{7}$ for π.

$$V = \frac{4}{3} \cdot \pi \cdot r^3$$

$$\approx \frac{4}{3} \cdot \frac{22}{7} \cdot (2 \text{ inches})^3$$

$$= \frac{4 \cdot 22 \cdot 8}{3 \cdot 7} \text{ cubic inches}$$

$$= \frac{704}{21} \quad \text{or} \quad 33\frac{11}{21} \text{ cubic inches}$$

SECTION A.6   SQUARE ROOTS AND THE PYTHAGOREAN THEOREM

SQUARE ROOT OF A NUMBER

A **square root** of a number a is a number b whose square is a. We use the radical sign $\sqrt{}$ to name square roots.

PYTHAGOREAN THEOREM

$$(\text{leg})^2 + (\text{other leg})^2 = (\text{hypotenuse})^2$$

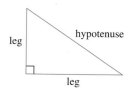

TO FIND AN UNKNOWN LENGTH OF A RIGHT TRIANGLE

$$\text{hypotenuse} = \sqrt{(\text{leg})^2 + (\text{other leg})^2}$$

$$\text{leg} = \sqrt{(\text{hypotenuse})^2 - (\text{other leg})^2}$$

$$\sqrt{9} = 3, \ \sqrt{100} = 10, \ \sqrt{1} = 1$$

Find the hypotenuse of the given triangle.

3 in.   hypotenuse

8 in.

$$\text{hypotenuse} = \sqrt{(\text{leg})^2 + (\text{other leg})^2}$$

$$= \sqrt{(3)^2 + (8)^2} \quad \text{The legs are}$$
$$\text{3 and 8 inches.}$$

$$= \sqrt{9 + 64}$$
$$= \sqrt{73} \text{ inches}$$
$$\approx 8.5 \text{ inches}$$

SECTION A.7   CONGRUENT AND SIMILAR TRIANGLES

Congruent triangles have the same shape and the same size. Corresponding angles are equal and corresponding sides are equal.

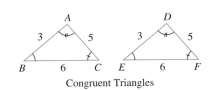

Congruent Triangles

Similar triangles have exactly the same shape but not necessarily the same size. Corresponding angles are equal and the ratios of the lengths of corresponding sides are equal.

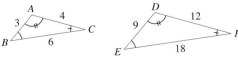

$$\frac{AB}{DE} = \frac{3}{9} = \frac{1}{3}, \quad \frac{BC}{EF} = \frac{6}{18} = \frac{1}{3},$$

$$\frac{CA}{FD} = \frac{4}{12} = \frac{1}{3}$$

PART VI

English
Handbook
and Grammar
Reference

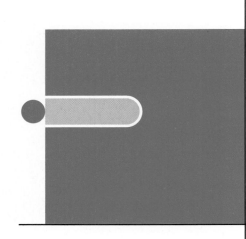

The Writer's FAQs
Muriel Harris

BRIEF CONTENTS

HOW TO USE THIS BOOK

Keep this book nearby to help you find answers to your writing questions in the quickest, easiest manner possible.

- **Before you start**
  Read the "Writing" part for reminders, advice, and writing tips,

- **As you write a draft of your paper**
  For sentence structure:
  Check "Sentence Choices" for suggestions for general sentence construction, clarity, word choice, and smooth flow.
  For research papers and finding information:
  The "Research" section offers advice on choosing a topic, finding information, using the library, evaluating and integrating sources into your paper, and avoiding plagiarism.

- **As an ESL student looking for information on using English**
  If you are a multilingual speaker (ESL=English as a Second Language), the whole book is useful as a guide for all writers. But questions that don't come up for native speakers are in the "Multilingual Speakers (ESL)" part and in ESL HINT boxes.

- **As you finish up and are editing and polishing a draft**
  When you are ready to check a point of grammar and mechanics, use "Sentence Grammar," "Punctuation," and "Mechanics" for the frequently used rules of grammar, spelling, capitalization, and other matters that indicate to readers that you're a literate user of the English language and that your ideas have merit. But what do you do if, like many writers, you can tell when something *isn't* right, but aren't sure what *is* right? Read an because this book is designed to help.

HOW DO YOU FIND WHAT YOU'RE LOOKING FOR?

- **If you know the term (or want to check on a term)**

1. Go to the Index at the back of the book.
2. Go to the Glossary of Usage and Glossary of Terms near the end of the book.

- **If you know the general topic (such as using commas)**

1. Find the topic in the Brief Contents inside the front cover.
2. Read the descriptions listed here and then do the following:
   a. Browse in the Brief Contents.
   b. Look at the questions at the beginning of that part.
   c. Turn to the list of questions inside the back cover.

WRITING Suggestions on writing concerns, linking sentences and paragraphs, writing introductions and conclusions, and so on.

SENTENCE CHOICES Information on writing clear, effective sentences that don't overuse the passive, and are varied, concise, and nonsexist in word choices.

SENTENCE GRAMMAR Grammar rules to help you avoid errors such as fragments, comma splices, and so on.

PUNCTUATION Guidelines for punctuation marks as well as useful diagram indicating the punctuation patterns for sentences.

MECHANICS Guidelines for capitalization, use of italics, numbers, abbreviations, and spelling.

MULTILINGUAL SPEAKERS Help with questions about English that ESL students are likely to have.

RESEARCH PAPERS Advice and guidelines for moving through the research paper process.

GLOSSARY OF USAGE Questions about word choices (Do I use "accept" or "except"?) and whether a certain word is acceptable in standard English (Can I write "and etc."?).

GLOSSARY OF TERMS Definitions of grammatical terms (such as "linking verb" or "reflexive pronoun") and charts to illustrate terms such as "personal pronouns" and "sentence diagram."

- **If you have a question**

1. If it's about one of those sticky word choices (for example: Is it "affect" or "effect"? Should I write "it's" or "its"?), go to the Glossary of Usage near the and of the book.
2. If the question is one that writers frequently have (the kind where you don't quite know the terminology to use), go to the inside back cover of the book and see if your question is like one of those FAQs (Frequently Asked Questions).

You'll also find HINT boxes throughout the book offering advice to avoid various problems writers encounter. Other user-friendly aids are lists and diagrams to help explain and clarify. As the term "user-friendly" implies, I hope this book is easy to use and becomes a writing friend that you keep nearby as you write.

Muriel Harris

USER'S GUIDE

YOUR QUESTION	SECTION
Writing	
What should I look for when I revise?	1
What goes into introductions and conclusions?	1
How useful is a spell checker?	1,28
Sentence Choices	
Are phrases like "It is a fact that . . ." OK to use?	2
How can I make my writing less choppy?	4,9
What is passive voice?	7
Can I start sentences with "And" or "But"?	9
Sentence Grammar	
What's wrong with the following?	
—I like catastrophe movies. Such as *Titanic*.	11
Which is correct? —Between you and (I/me)	14
Are these phrases correct? —"real bad"	
—"talk loud"	15
Can I write the following? —She is so happy.	15
Punctuation	
What are some basic patterns for punctuating sentences?	18
Am I overusing commas?	19
How do I punctuate —dates and addresses?	19
—quotations?	19,22
What's the difference between "its" and "it's"?	20
Are these apostrophes correct? —his' car	14,20
—the melon's are ripe	20
How do I show that I've left out some words in a quotation?	23
Which is correct?	
—well-known speaker/well known speaker	23
Mechanics	
Which is correct?	
—spring semester (or) Spring semester	24
—"The Simpsons" (or) The Simpsons	25
—April 1 (or) April first	26
—six million (or) 6,000,000	27
—The data is (or) The data are	28
What does "e.g." mean, and how do I use it?	27
Do I write "your" or "you're" and "whose" or "who's"?	28
Multilingual Speakers (ESL)	
How is academic writing in American English different from academic writing in my country?	29
What are the meanings of English verb tenses?	30
Which is correct?	
—She enjoys (to drink/drinking) coffee.	30
—two furnitures (or) some furniture	31
When do I use "the," "a," and "an"?	31
Do I write "in Tuesday" or "on Tuesday"?	33

YOUR QUESTION	SECTION
Research	
How do I start a library search for information?	36
How do I know if a source is reliable?	37
What are summarizing, paraphrasing, and quoting?	38
When do I cite a source?	38
Glossary of Usage	
How do I distinguish between "accept" or "except"?	Usage
Is it acceptable to write "ain't" or "alot" or "and etc."?	Usage
What's the difference between similar words such as "anyone" and "any one" or between "awful" and "awfully"?	Usage
When do I use words such as "that" and "which" or "who" and "whom"?	Usage
Glossary of Grammatical Terms	
How do I define terms such as "adjective" or "case" or "linking verb"?	**Grammatical Terms**
Where do I look for charts to illustrate terms such as "personal pronouns" or "sentence diagram" or "verb conjugations"?	**Grammatical Terms**

I WRITING

The section in this part of the book offers help with general writing processes and answers the following questions:

YOUR QUESTION   **SECTION**

- When I start planning or when I'm drafting my paper, what questions should I ask myself?   1
- What are HOCs (Higher Order Concerns), and why are they important?   1
- Why is thinking about the purpose and audience for an assignment important?   1
- What's a thesis statement, and how can I tell when I've phrased it appropriately?   1
- How can I make my paper flow?   1
- What should I put in the introduction and conclusion?   1
- When the paper is close to being done, what should I check for?   1
- What are LOCs (Later Order Concerns), and why are they important?   1
- How do I proofread the paper before I hand it in?   1
- What are some techniques for proofreading?   1
- How do I proofread for spelling?   1
- Should I depend on the spell checker on my computer?   1
- If I plan and write my paper on a computer, are there some strategies that I can use on a computer to help?   1

1 CHECKLIST FOR EFFECTIVE PAPERS

HOCs (Higher Order Concerns)

Listed here are the Higher Order Concerns (HOCs) that help make your writing effective:

Purpose

Be sure your purpose fits the assignment. If you are asked to persuade your readers, that is different than writing to explain or summarize. If you are asked to describe, your purpose is to help your readers envision your subject. Read the assignment carefully, and note the verbs carefully. Are you asked to compare two or more things? *Analyze? State your opinion?*

You can clarify your purpose by answering the questions, "Why am I writing this? What am I trying to accomplish?" For example, do you want your readers to take some action? Accept your view? Understand something they didn't know before? Share some experience of yours?

Audience

Think about your readers. Check to see that they are the appropriate audience for your assignment and purpose. Think about what they already know and what they most likely want to know or don't need to know. Do you need to add any information or background summary to help them understand your topic?

If you are arguing or persuading, are you writing to those who are likely to disagree? If so, how can you convince them? Is there some common ground, some aspects of the argument you share with them, that will help in getting those who disagree to consider what you are writing?

Thesis statement

Your thesis is the main idea or subject of the paper. You should be able to summarize it briefly in a sentence or two, and your paper should state this clearly for the reader. Think of the the-

sis as a promise that you will discuss this topic—a contract you will fulfill. When you read over your draft, check to see you have kept all parts of your promise.

Organization

As you look over your draft, note the central idea of each paragraph (the topic sentence) and ask yourself if each paragraph contributes to the larger thesis in some way. Make an outline of the topic sentences and look over that organization to see if it is logical. You want to avoid gaps or jumps in the development of your thesis that might confuse the reader.

Development

Be sure you have enough details, examples, specifics, supporting evidence, and information to support your thesis. You may need to delete irrelevant material or add material that will strengthen your thesis and help you achieve your purpose.

Paragraph length

Paragraphs are the large building blocks of the paper. As you look over your draft, check to see that the paragraphs are of the same approximate length on the page. If you have a paragraph that takes most of a page, followed by a paragraph that has only a few sentences, you may need to make the paragraphs more equal in length.

Transitions between sentences and paragraphs

Transitions connect or knit sentences and paragraphs together into a smooth whole. Like road signals, they indicate where the writing is heading and keep your reader following along easily. Check to see that you've supplied the needed connectors to indicate how your writing is moving forward.

Introductions and conclusions

The introduction brings the reader into your world, builds interest in your subject, and announces the thesis or topic of the paper. Sometimes, writers write the introduction after revising the rest of the paper and clarifying their topic through their writing and revising.

The conclusion of the paper signals that the end is approaching and helps the reader to put the whole paper in perspective. You can either look backward and offer a conclusion that summarizes the content or look forward and offer advice, suggestions, or actions the reader can take, based on what you have presented.

LOCs (Later Order Concerns)

When you have finished your major revisions and checked the HOCs (page 0), look more closely at words, sentences, and punctuation for problem areas that detract from your credibility as a writer. As you make your own list of problems to check for, consider whether you need to check for these common problems:

fragments   (see section 11)

subject-verb agreement   (see section 13)

HINT

LOCS (PROOFREADING CHECKLIST)

After your draft is well on the way to being completed, check for the Later Order Concerns (LOCs) as you edit and proofread.

verb endings   (see section 13)

verb tenses   (see section 13)

comma splices and fused sentences   (see section 12)

misplaced or omitted apostrophes   (see section 20)

pronoun reference   (see section 14)

omitted words   (see section 34)

omitted commas   (see section 19)

unnecessary commas   (see section 19)

spelling errors   (see section 28)

Strategies for checking on HOCs and LOCs

Find writing strategies that are effective for you, such as the following:

- Have someone (such as a tutor in your writing center) read your paper aloud as you listen and look at it, or read the paper aloud yourself. You'll see problems that are not as evident when you read silently.

- Put the draft away for a while so that when you read, the paper is not as fresh in your mind. To revise effectively, having some distance from the paper helps because you can more easily identify readers' concerns.

- Try to put yourself in the place of your intended readers and think about what they would want to know what they might object to in your arguments, what counterarguments they would make, what questions they would have. Ask yourself if your paper responds adequately to these considerations.

- For proofreading you need to help your eyes slow down and see each word. (Readers tend to see whole groups of words at once.) Try sliding a card down the page as you read because that permits your eyes and ears to work together.

- Proofread for spelling by reading backward, either from the end of the paper to the beginning or from the right side of the line to the left. Then you are not focusing on the meaning of the sentences and can notice smaller matters such as word choice, punctuation, and spelling. Computer spell checkers catch some, but not all, spelling errors.

- Don't depend on computerized grammar checkers. They may help slightly, but they cannot analyze language well enough to check completely for grammar problems, and the options suggested are sometimes not appropriate.

- Draw up a personal list of problem areas and keep those in mind as you reread your draft.

Make a personal checklist here:

```
┌─────────────────────────────────────────┐
│  Areas I should check for:               │
│                                          │
│                                          │
│                                          │
│                                          │
│                                          │
│                                          │
│                                          │
│                                          │
│                                          │
│                                          │
└─────────────────────────────────────────┘
```

Strategies for using computers

Word processing on a computer can help as you write and as you check your paper. Try the following strategies to see which are helpful for you:

- Copy the topic sentences from each paragraph and put them in an outline onscreen. If you have any questions about the organization, cut and paste and see if other arrangements are more effective.

- Be sure to use a spell checker because it is especially useful in catching typos. But remember that spell checkers can't find all spelling problems. They cannot, for example, distinguish between "it's" and "its" or "here" and "hear" to see if you have used the right form of these sound-alike words.

- As you write, highlight in some way (such as boldface) problem areas or phrases that you have questions about. Then you can find them later and reconsider them.

- If you think it might be better to delete a chunk of text, cut and paste it to a now file or to the end of the paper while you see if the paper is better without it. By putting it in a separate file, you can save it for later use or retrieve it if you decide you need it.

- If a fresh idea pops into your mind as you're writing but probably belongs elsewhere in the paper, write that in a separate file. (Some word processing programs permit you to make notes to yourself as you write that are not visible in the main text.)

- Many writers need to print out a hard copy of the paper as it develops to get a better sense of the whole paper.

- To check on paragraph length, switch to page or print view so you can see a whole page on the monitor. See if the paragraphs look about the same length.

- Working with a copy of your file, hit the return key after each period so each sentence looks like a separate paragraph. If all the sentences are approximately the same length, consider varying your sentences more (see section 4). If most of the sentences begin the same way (with the subject of the main clause), think about using different sentence patterns.

II   SENTENCE CHOICES

The sections in this part of the book discuss choices you make as you write. In most cases, there is no right or wrong answer, but you want to choose carefully so your writing is clear, concise, and smooth.

YOUR QUESTION	SECTION
• Why shouldn't I use negatives such as "no" and "not" in my writing?	2
• Why is "don't want no paper" wrong?	2
• What are some other negatives I should watch out for?	2
• What's the best order for including information in a sentence? Why?	2
• Should I write "the consideration of" or "they consider" or "the preparation of" or "they prepare"? Which is better and why?	2
• I sometimes start sentences with "It is . . ." or "There is the problem that. . . ." How can I make such sentences more effective?	2

2 CLARITY

Positive instead of negative

Put information in the positive because negative statements are harder to understand than positive ones.

Unclear negative: Less attention is paid to commercials that lack human interest stories.

Revised: People pay more attention to commercials with human interest stories.

Negatives can also make the writer seem evasive or unsure.

Evasive negative: Congresswoman Petros is not often heard to favor raising the minimum wage.

Revised: Congresswoman Petros prefers keeping the minimum wage at its present level.

Double negatives

Use only one negative at a time in your sentences. Double negatives are grammatically incorrect and may be difficult to understand.

Double negative: They don't want no phone calls.
Revised: They don't want any phone calls.

> **HINT**
> ### AVOIDING NEGATIVE WORDS
> Watch out for negative words such as the following:
>
> | hardly | no place | nothing |
> | neither | nobody | nowhere |
> | no one | none | scarcely |
>
> They hardly had ~~no~~ *any* popcorn left.

Known information to new or unknown information

Start your sentences with information that is known or generally familiar to your reader before you introduce new or unknown material.

Familiar ⟶ Unfamiliar

Familiar to new: When I visit my grandmother, she often has an old book from her childhood days to show me.

(This sentence should be easy to understand.)

New to familiar: An old book from her childhood days is something my grandmother often shows me when I visit.

(This sentence takes longer to understand and is less clear.)

Verbs instead of nouns

Actions expressed as verbs are more easily understood and usually more concise than actions named as nouns.

Unnecessary noun forms: Pay raises are a motivation improvement.

Revised: Pay raises improve motivation.

Intended subject as sentence subject

The real subject or doer of the action in the verb should be the grammatical subject of the sentence. Sometimes the real subject can get buried in prepositional phrases or other less noticeable places.

> **HINT**
> ### USING VERBS INSTEAD OF NOUNS
> Try rereading your sentences to see which nouns could be changed to verbs.
>
> | Some noun forms | Verbs to use instead |
> | The determination of | They determine |
> | The approval of | They approve |
> | The preparation of | They prepare |
> | The utilization of | They use |
> | The analysis of | They analyze |

Buried subject: It was the preference of the instructor to begin each lecture with a quiz.

(The grammatical subject here is it. Who begins each lecture? The instructor.)

Revised: The instructor preferred to begin each lecture with a quiz.

3 CONCISENESS

To be concise, eliminate the following:

- what your readers do not need to know
- what your readers already know
- whatever doesn't further the purpose of your paper

Sometimes writers are wordy when they are tempted to include everything they know about a subject, add a description of how they found their information (to impress readers with how hard they've worked to get the information), or add words they think will make their writing sound more formal or academic.

Strategies to eliminate unnecessary words:

- **Avoid repetition.** Some phrases, such as the following, say the same thing twice:

first beginning	9 A.M. in the morning
circular in shape	true facts
return again	really and truly
green in color	each and every

- **Avoid fillers.** Some phrases, such as the following, add little or nothing to your meaning:

in view of the fact that	due to the fact that
I am going to discuss	there are (or) is
the topic that I will explain here	

~~I am going to discuss~~ *T*the cloning of human

beings, ~~which~~ is a subject that raises many difficult ethical questions.

- **Combine sentences.** When the same nouns appear in two sentences, combine the sentences.

Global warming is a critically important topic/

and
~~Global warming~~ has been the subject of recent

specials, government regulations, and conferences.

- **Eliminate** *who, which,* **and** *that.*

The marking pen ~~that was~~ on my desk is gone.

- **Turn phrases and clauses into adjectives and adverbs.**

The football player who was graceful = the graceful football player

The building built out of cement = the cement building

The entrance to the station = the station entrance

- **Remove excess nouns and change to verbs whenever possible.**

agreed
He ~~made the statement that he was in agreement with~~

~~the concept~~ that inflation could be controlled.

- Use active rather than passive. (See section 7, page 00.)

research department
The ~~figures were~~ checked by the research

department.

4 VARIETY

A series of short sentences or sentences with the same subject-verb word order can be monotonous and sound choppy. Try these strategies for adding variety.

- **Combine short, choppy sentences.** Connect two sentences into one longer sentence with one subject and two verbs, or a comma and coordinating conjunction, or a semicolon, (see sections 12, 13, 18, and 19).

The school band performed at the local Apple

, and they
Festival. ~~They~~ were a great success.

Tuck a phrase, clause, or sentence inside a related sentence.

The school band performed at the local Apple

, who
Festival. ~~They~~ were a great success.

- **Rearrange sentence order.** Often, a series of sentences that sound choppy all have a subject-verb-object order. You can make one sentence depend on another or add or change phrases and clauses to break up the monotonous sound.

Choppy: The reporter asked each candidate the same question. He wanted to compare their campaign promises. They all evaded his questions. He wrote a story about their lack of answers.

Revised: Because the reporter wanted to compare the candidates' campaign promises, he asked each one the same question. Hearing them evade his questions, he wrote a story about their lack of answers.

5 VOICE (FORMAL AND INFORMAL)

In writing, an appropriate voice is one that fits the level of formality of your paper and your subject. Just as you don't wear a suit or dress when you go on a picnic or jeans to a formal dinner, you should match your word choices to the type of paper you are writing.

Formal documents such as research papers, reports, and applications avoid slang but may include some technical language, or jargon, appropriate to the field and the intended readers. Such documents are normally written in the third person, using "he" and "they."

Informal documents such as e-mail and letters to friends, informal essays, and some memos may include more informal word choices (for example, "kids" instead of "children") and frequent contractions, and they are normally written in the first person, using "I."

Compare these recommendations:

Informal: Be sure to see *Titanic*. I saw it last week, and it's great!

Formal: The laminate is the recommended choice for this product because test results show that it holds up well under stress and heat.

Slang

Slang words may be shared by a small group or may be generally known. Some slang enters the general vocabulary, such as "cab" or "yuppie," and some eventually disappears or becomes outdated, such as "far out" or "BMOC" (Big Man on Campus). It is usually too informal for most written work.

Jargon

Jargon words are specialized terms used by those in the same field or profession to refer quickly to complex concepts. For someone who is knowledgeable about computers, the terms "gigabytes" and "bit maps" are useful technical terms when writing to someone else in that field. Such shorthand vocabulary should only be used when you are sure your readers will be familiar with the words.

The term "jargon" is also applied to pompous language that is inflated and unnecessarily formal. The result is wordy prose that is hard to read and makes the writer sound pretentious.

Pompous: She was inordinately predisposed to render her perspective on all matters of national and international import.

Revised: She frequently offered her opinion on world affairs.

6 MIXED CONSTRUCTIONS

Mixed constructions are caused by mismatches when fitting parts of a sentence together. A writer can start off in one direction and then switch to another, causing grammar or logic problems in the sentence.

Mixed: For groups who want to reduce violence on television, students carrying knives to school are acting out what they see on the television screen.

Revised: Groups who want to reduce violence on television claim that students carrying knives to school are acting out what they see on the television screen.

Dangling modifiers

Some mixed constructions are caused by dangling modifiers—phrases or clauses that should modify the subject but don't.

Dangling: After <u>finishing</u> her degree, the <u>search</u> for a job began.

(This sentence says that the search, the subject of the sentence, finished her degree.)

Revised: After <u>finishing</u> her degree, <u>she</u> began the search for a job.

Mismatched subjects and predicates

Sometimes the subject and predicate don't match or fit together.

Mismatched: <u>Driver education</u> in high schools <u>assumes</u> that parents can pay the costs involved.

(<u>Driver education</u>, the subject, can't make assumptions about anything.)

Revised: High school administrators assume that parents can pay the costs involved for driver education programs.

7 ACTIVE AND PASSIVE VERBS

An active verb expresses the action completed by the subject. A passive verb expresses action done to the subject.

The active voice is usually more direct, clearer, and more concise than passive. However, sometimes the passive is a better choice.

Active: <u>Paul</u> drove the <u>car</u>.

(The verb is <u>drove</u>, and <u>Paul</u>, the subject, did the driving.)

Passive: The <u>car</u> <u>was driven</u> by Paul.

(The verb is <u>was driven</u>, and <u>the car</u>, the subject, was acted upon.)

Active verbs are clearer than passive because they indicate who is doing the action and add a better sense of immediateness and vigor. In a sentence with a passive verb, the "by the" phrase where the doer of the action is indicated maybe left out or put far from the verb. Compare these sentences:

Passive: The photographs showing the tornado were snapped in a hurry by me.

Active: I hurriedly snapped the photographs of the tornado.

Because active verbs add directness and force, they are often a better choice for sentences containing action that begin with "there is" or "there are."

Original	Revised
There were six victims of the crime whose accounts of what happened agreed.	Six of the crime victims gave the same accounts of the crime.

However, there are occasions to use the passive:

- When the doer of the action is not known or not important:
  The water temperature was recorded.

- When you want to focus on the receiver of the action:
  Historical fiction is not widely read.

- When you want to focus on the action, not the doer:
  The records have been destroyed.

- When you want to avoid blaming or giving credit:
  The candidate concedes that the election is lost.

- When you want a tone of objectivity or wish to exclude yourself.
  The complete report was drafted and on the president's desk yesterday.

8 PARALLELISM

Parallel structure exists when the same grammatical form or structure is used for equal ideas in a list or in a comparison. That similar form helps your reader locate the similar or compared ideas. Often, the equal elements repeat words or sounds.

Parallel: The computer manual explained <u>how to boot</u>
(1)

<u>up the hard drive</u> and <u>how to install the software</u>.
(2)

(Phrases (1) and (2) are parallel because both start with <u>how to</u>.)

Parallel: <u>Watching Walt fumble with his headgear</u> was
(1)

as funny as <u>seeing him try to skate</u>.
(2)

*(Phrases (1) and (2) are parallel because both start with
-ing verb forms.)*

Parallel: Three keys to marketing success include the
following:

1. <u>To listen</u> to the customer's wishes
(1)

2. <u>To offer</u> several alternatives
(2)

3. <u>To motivate</u> the customer to buy
(3)

*(Phrases (1), (2), and (3) are parallel items in a list because
all begin with to + verb.)*

Parallel is also needed when you link items using the following:

both . . . as   either . . . or
not only . . . but   neither . . . nor
coordinating conjunctions: *for, and, nor, but, or, yet, so*
comparisons using *than* or *as*

Parallel: Job opportunities are not only <u>increasing</u> in
(1)

the health fields but <u>expanding</u> in many areas
(2)

of manufacturing as well.

*(1) and (2) are parallel items using -ing verbs linked with
but.)*

Faulty parallelism is not only grammatically incorrect but can
also lead to possible lack of clarity.

Dr. Willo, explained that <u>either</u> starting the
(1)

avoiding
treatment or to avoid surgery was impossible.
(2)

HINT

PARALLEL STRUCTURE

As you proofread, do the following:

- Listen to the sound when you are linking equal
  ideas or comparing two or more elements. (Parallelism can add emphasis to your writing and
  public speaking by that repetition of sound.)

- Visualize similar elements in a list and check to
  see if they are in the same grammatical structure.

Tara wondered whether it was better <u>to tell</u> her mother
(1)

to fix
that she had wrecked the car or maybe <u>fixing</u> it herself.
(2)^

9   TRANSITIONS

Transitions are words and phrases that build bridges to connect sentences, parts of sentences, and paragraphs together.
These bridges show relationships and add smoothness (or
"flow") to your writing.

There are several types of transitions you can use:

- **Repetition of a key term or phrase**

  Delegates at the conference could not agree on the
  degree of danger from <u>global warming.</u> But no one dis-
  puted the existence of <u>global warming.</u>

- **Synonyms**

  The <u>movie industry</u> is expanding to produce a variety of
  forms of entertainment, such as television films and
  music videos. But <u>Hollywood</u> will always have movies
  as its main focus.

- **Pronouns**

  <u>College tuition</u> has been increasing rapidly for several
  years. But <u>it</u> still does not finance needed improvements
  on many campuses.

- **Transition words and phrases**

  The investigators looking into the cause of the pollution
  pinpointed one farm. <u>Therefore,</u> the owner was forced
  to reduce his use of the fertilizers that were washing into
  the stream. <u>In the meantime,</u> the local chemical plant
  continued its dumping practices.

TRANSITIONS	
Adding:	and, besides, in addition, also, too, furthermore, third
Comparing:	similarly, likewise, in the same way, at the same time
Contrasting.	but, yet, on the other hand, instead, whereas, although
Emphasizing:	indeed, in fact, above all, and also, obviously, clearly
Ending:	after all, finally, in sum
Giving examples:	for example, for instance, namely, specifically, that is
Showing cause and effect:	thus, therefore, consequently, as a result, accordingly, so
Showing place or direction:	over, above, next to, beneath, to the left, in the distance
Showing time:	meanwhile, later, afterward, now, finally, in the meantime
Summarizing:	in brief, on the whole, in conclusion, in other words

WORDS THAT START SENTENCES

Although some instructors prefer that you don't use or overuse the transitions "But" or "And" as sentence starters, it is not wrong to use them. Any word can start a sentence.

10 NONSEXIST LANGUAGE

Language that favors the male noun or pronoun or excludes females is sexist. To avoid such language, do the following:

- Use alternatives to <u>man</u>:

<u>Man</u>	Alternative
man	person
mankind	people, human beings
man-made	machine-made, synthetic
to man	to operate

- Use alternatives for job titles:

<u>Man</u>	Alternative
chairman	chair, chairperson
mailman	letter carrier, postal worker
policeman	police officer
fireman	firefighter

Strategies for avoiding masculine pronouns:

- Use the plural instead.

  Sexist: Give the customer <u>his</u> receipt immediately.

  Revised: Give customers <u>their</u> receipts immediately.

- Reword and eliminate the male pronoun.

  Sexist: Give the customer <u>his</u> receipt.

  Revised: Give the customer <u>the</u> receipt.

USING EVERYONE ... HIS

For indefinite pronouns such as "everyone" and "anybody," the traditional practice is to use the masculine singular to refer back to that pronoun:

Traditional: Everyone brought his own pen and paper.

To avoid the male pronoun, which is seen by many people as sexist, you can use the strategies just listed.

Others, however, such as the National Council of Teachers of English, accept the plural as a way to avoid sexist language:

Everyone brought their own pen and paper.

- Replace the male pronoun with <u>one</u>, <u>you</u>, <u>he or she</u>, and so on.

  Sexist: The student can select <u>his</u> preferred residence hall.

  Revised: The student can select <u>his or her</u> preferred residence hall.

- Address the reader directly.

  Sexist: The applicant should mail two copies of <u>his</u> form by Monday.

  Revised: Mail two copies of <u>your</u> form by Monday.

III SENTENCE GRAMMAR

The sections in this part offer help with grammatical rules for most common problems. If you are not familiar with certain grammatical terms, such as "fragment," "comma splice," "subject-verb agreement," or "pronoun reference," and want to look up the rules for these, check the glossary of terms at the back of the book, use the index to find the term you are looking for, or use the question section here.

YOUR QUESTION	SECTION
• What is a sentence fragment, and how do I recognize one?	11
• Is "Because she wanted to pass the course" a sentence or a fragment?	11
• Is "Such as the way he explained the joke" a sentence or a fragment?	11
• How can I proofread for fragments?	11
• What are comma splices, fused sentences, and run-on sentences?	12
• When do I use a comma and when do I use a semicolon to join two clauses that could be sentences by themselves?	12
• What are the sentence patterns I can use to avoid comma splices, fused, or run-on sentences?	12
• What is subject-verb agreement?	13
• Is it correct to write "Nearly <u>every one</u> of the other students in my classes <u>are</u> complaining about the test"?	13
• How can I check for subject-verb agreement? And what are some of the different subject-verb complications?	13
• When I have two subject terms such as "Talesha" and "her friends," is the verb "is" or "are"?	13
• Do I use singular or plural verbs with collective nouns such as "family" and "group"? (The family is/are moving.")	13
• Do I use singular or plural verbs with plural subjects such as "mathematics" or "jeans" and names like General Foods?	13
• Do I use singular or plural verbs with phrases and clauses that start with "who" or "which" in the middle of sentences? "He is the person on those committees who want/wants to change the rules.")	13
• What are the regular verb endings?	13
• What is the past tense for irregular verbs such as "swim," "lie/lay," or "forbid"?	13

- Are the following verb forms correct?
  —may have <u>like</u> to come along
  —could <u>of</u>
  —<u>suppose</u> to   13
- What is active and passive voice?   13
- Should I use an apostrophe with pronouns such as "his" or "theirs"?   14
- Is the phrase "<u>them</u> boxes" wrong?   14
- Which is correct?
  The coach liked <u>his/him</u> pitching to the right.   14
- Which is correct?
  The musicians and <u>myself/I</u> took a short break.
  She gave <u>myself/me</u> and Tim some good advice.   14
- Should I write "between you and I" or "between you and me"?   14
- When do I use "who," and when do I use "whom"?   14
- What is "vague pronoun reference," and how can I avoid it?   14
- Is there a problem with using "this" or "they" (or "it") in sentences such as "Caitlin lost her umbrella, and <u>this</u> problem she has really bothers me" or "<u>They</u> say it's going to be a very cold winter"?   14
- What's wrong with the phrase "real bad"?   15
- When do I use "good," and when do I use "well"?   15
- How should I correct the following sentence? "After eating dinner, the doorbell rang."
- When I use modifying words such as 'only." "even," or "nearly," where should I place them in the sentence?   16
- How should I correct the following sentence? "For most <u>people</u>, the career <u>you</u> decide on isn't always the major <u>they</u> had in college."   17

11 FRAGMENTS

A sentence fragment is an incomplete sentence. To recognize a fragment consider the basic requirements of a sentence:

- A sentence is a group of words with at least one independent clause.
- An independent clause has a subject and complete verb plus an object or a complement if needed.
- An independent clause can stand alone as a thought, even though it may need other sentences before and after it to clarify the thoughts being expressed.

Independent clause: Jeremy's picture was in the newspaper.

Independent clause: He scored six three-point baskets during the game.

(We don't know who "he" is in this sentence, but a pronoun can be a subject, and we don't know which game is being referred to. But those bits of information, if needed, would be explained in accompanying sentences. The clause has a subject ["he"], a verb ["scored"], and an object ["baskets"].)

Not an independent clause: Because he scored six three-point baskets during the game.

(Say that clause out loud, and you will hear that it's not a complete sentence. The problem is that we don't know the result of the "because" clause.)

Not an independent clause: Luis who was one of my closest friends in third and fourth grade and is now moving with his family to another city.

(This is not a complete sentence because it has a subject, "Luis," but no main verb that tells us what Luis did. The verbs "was" and "is moving" belong to another subject [the pronoun "who"] and tell us what "who" did.)

Some fragments are unintended:

Unintended fragment: There were some complications with her phone bill. <u>Such as two calls she did not make and a long distance charge for a local call.</u>

(The second sentence is an unintended fragment because it has no subject and verb. It is a phrase that got disconnected from the independent clause that came before it.)

Unintended fragment: The doctor's recommendation that I get more sleep because I was becoming very stressed out while taking too many classes which I need for my major.

(The subject is "recommendation" but there is no main verb to complete that thought.)

Some fragments are intentional when they are used to add an effect such as emphasis or sudden change in tempo. However, intended fragments should only be used when the writing clearly indicates that the writer chose to include a fragment.

HINT

AVOIDING FRAGMENTS

HINT 1

To proofread for fragments caused by misplaced periods, read your paper backward, from the last sentence to the first. You will notice a fragment more easily when you hear it without the sentence to which it belongs. Most, but not all, fragments occur after the main clause.

HINT 2

Some fragments are caused by a marker word typically found at the beginning of a dependent clause that requires a second clause to finish the thought. Consider a marker word such as "if" and how it affects the clause:

    If A happens ⟶ ?

When you hear that, you want to know what B is, that is, what the result is if A happens.

    Watch for other, similar marker words such as the following:

after	before	since
although	even though	unless
because	if	when

Intended fragment: Never had there been such a decisive victory in the school's history of participating in the tournament, and no one stayed at home that night when the team's bus pulled into town. <u>No one</u>.

(The fragment, "No one," is used here to add emphasis.)

12 COMMA SPLICES AND FUSED SENTENCES

A comma splice and a fused sentence (also called a run-on sentence) are punctuation problems in compound sentences. (A compound sentence is one that joins two or more independent clauses-clauses that could have been sentences by themselves.)

To avoid comma splices and fused sentences, note the three patterns for commas and semicolons in compound sentences:

1. Join two independent clauses with a comma and one of the seven joining words (coordinating conjunctions) listed:

   Independent clause, for independent clause.
   and
   nor
   but
   or
   yet
   so

   No one was home, but the door was open.

2. Join two independent clauses with a semicolon and no joining words:

   Independent clause; independent clause.

   No one was home; the door was open.

3. Join two independent clauses with a semicolon and any connecting word other than one of the seven joining words for commas listed in number 1: for, and, nor, but, or, yet, so.

   Independent clause; however, independent clause.
   therefore,
   consequently,
   thus,

   No one was home; however, the door was open.

If you don't use one of these three patterns for compound sentences, the sentence will have a comma splice:

Comma splice: No one was home, the door was open.

USING COMMAS IN COMPOUND SENTENCES

HINT 1

When punctuating compound sentences, think of the comma as only half of the needed connection to tie two independent clauses together. The other half is the connecting word. You need both the comma and the connecting word.

HINT 2

Don't put commas before every "and" or "but" in your sentences. "And" and 'but" have other uses in sentences in addition to joining two independent clauses.

13 SUBJECTS AND VERBS

Subject verb agreement

Subjects and verbs should agree in number and person.

- To agree in number, the verb used with a plural subject should have a plural ending, and a verb used with a singular subject should have a singular ending.

  <u>The customer</u>   <u>orders</u>
  (singular subject)   (singular verb)

  <u>The customers</u>   <u>order</u>
  (plural subject)   (plural verb)

- To agree in person, the verb should be in the same person (first person = I/we, second person = you, or third person he/they) as the subject.

  I know   you know   she knows   they know

  (For verb tense endings, see page 794.)

- Some subjects are hard to find because they are buried among many other words. In that case, disregard the prepositional phrases, modifiers, and other surrounding words.

  Almost every <u>one</u> of the applicants for the job who

      (subject)

  came for interviews <u>is</u> highly qualified.

              (verb)

SUBJECT-VERB AGREEMENT

HINT 1

The letter *-s* is used both for subject endings (plural) and for verb endings (singular). Because a plural subject can't have a singular verb, and a singular verb can't have a plural subject, the letter *-s* should not normally be the ending for both subject and verb.

chimes ring   the chime rings

HINT 2

When you check subject-verb endings, start by finding the verb. The main verb is the word that changes when you change the time of the sentence, from past to present or present to past. Then, ask yourself "who" or "what" is doing that action, and you will find the subject.

HINT 3

To find a subject buried among other words and phrases, start by eliminating phrases starting with prepositions; "who," "that," or "which" clauses; or words such as the following:

including	along with	together with
accompanied by	in addition to	as well as
except	with	no less than

Compound subjects

Subjects joined by "and" take a plural verb (X and Y = more than one).

The stereo and the speaker are sold as a unit.

Sometimes, words joined by "and" act together as a unit and are thought of as one thing. If so, use a singular verb.

Peanut butter and jelly is his favorite sandwich spread.

Either/Or subjects

When the subject words are joined by "either . . . or," "neither . . . nor," or "not only . . . but also," the verb agrees with the closest subject word.

Either Maylene or her children are going to bed early.

Indefinites as subjects

Indefinite words with singular meanings such as "each," "every," and "any" take a singular subject when (1) they are the subject word, or (2) they precede the subject word.

Each book on the shelves is marked with a bar code.

However, when indefinite words such as "none," "some," "most," or "all" are the subject, the number of the verb depends on the meaning of the subject.

Some of the movie is difficult to understand.

Some of those movies are difficult to understand.

Collective nouns and amounts as subjects

Nouns that refer to groups or a collection (such as "family," "committee," or "group") are collective nouns. When the collective noun refers to the group acting as a whole or single unit, the verb is singular.

Our family needs a new car.

Occasionally, a collective noun refers to members of a group acting individually, not as a unit. In that case, the verb is plural.

The committee are happy with each other's decisions.

Plural words as subjects

Some words that have an *-s* ending, such as "news" or "mathematics," are thought of as a single unit and take a singular verb.

Physics is . . .   Economics is . . .   Measles is . . .

Some words, such as those in the following examples, are treated as plural and take a plural verb, even though they refer to one thing. (In many cases, though, there are two parts to these things.)

Jeans are . . .   Pants are . . .   Scissors are . . .

Titles, company names, and terms as subjects

For titles of written works, names of companies, and words used as terms, use singular verbs.

All the King's Men is the book assigned for this week.

General Foods is hiring people for its new plant.

"Cheers" is a word he often uses when leaving.

Linking verbs

Linking verbs agree with the subject rather than the word that follows (the complement).

Those poems are my favorites.

The poem is their favorite.

There is/There are/It

When a sentence subject is "there is," "there are," or "it," the verb depends on the complement that follows it.

There is a surprise ending to that story.

There are surprise endings in many of her stories.

Who, Which, That, and One of . . . Who/Which/That as subjects

When "who," "which," and "that" are used as subjects, the verb agrees with the previous word they refer to (the antecedent).

They are the students who want to change the parking rules.

He is the student who wants to change the parking rules.

In the phrase "one of those who," it is necessary to decide whether the "who," "which," or "that" refers only to the one or to the whole group. Only then can you decide whether the verb is singular or plural.

Mr. Liu is one of the salespersons who know the product.

(In this case, Mr. Liu is part of a large group, those salespersons who know the product.)

Mr. Liu is the only one of the salespersons who knows the product.

Verbs

Regular and irregular verb forms

Verbs that add *-ed* for the past tense and the past participle are regular verbs:

I talk   I talked   I have talked
(present)   *(past)*   *(past participle)*

The past participle is the form that has a helping verb such as "has" or "had."

VERB FORMS (REGULAR)			
	Present	**Past**	**Future**
Simple	I walk	I walked	I will walk
Progressive	I am walking	I was walking	I will be walking
Perfect	I have walked	I had walked	I will have walked
Perfect Progressive	I have been walking	I had been walking	I will have been walking

VERB FORMS (IRREGULAR)

Verb	Present		Past	
	Singular	**Plural**	**Singular**	**Plural**
	I am	we are	I was	we were
to be	you are	you are	you were	you were
	he, she, it is	they are	he, she, it was	they were
	I have	we have	I had	we had
to have	you have	you have	you had	you had
	he, she, it has	they have	he, she, it had	they had
	I do	we do	I did	we did
	you do	you do	you did	you did
to do	he, she, it does	they do	be, she, it did	they did

Some of the frequently used irregular verb forms include the following:

IRREGULAR VERBS

Base (present)	Past	Past participle
awake	awoke	awoken
be	was, were	been
beat	beat	beaten
became	become	become
begin	began	begun
bet	bet	bet
bite	bit	bitten (or) bit
bleed	bled	bled
blow	blew	blown
break	broke	broken
bring	brought	brought
build	built	built
burst	burst	burst
buy	bought	bought
catch	caught	caught
choose	chose	chosen
come	came	come
cost	cost	cost
cut	cut	cut
dig	dug	dug
do	did	done
draw	drew	drawn
drink	drank	drunk
drive	drove	driven
eat	ate	eaten
fall	fell	fallen
feed	fed	fed
feel	felt	felt
fight	fought	fought
find	found	found

Base (present)	Past	Past participle
fling	flung	flung
fly	flew	flown
forbid	forbade	forbidden
forget	forgot	forgotten
freeze	froze	frozen
get	got	gotten
give	gave	given
go	went	gone
grow	grew	grown
hang	hung	hung
have	had	had
hear	heard	heard
hit	hit	hit
hold	held	held
hurt	hurt	hurt
keep	kept	kept
know	knew	known
lay	laid	laid
lie	lay	lain
make	made	made
mean	meant	meant
meet	met	met
mistake	mistook	mistaken
pay	paid	paid
prove	proved	proved (or) proven
put	put	put
read	read	read
ride	rode	ridden
ring	rang	rung
rise	rose	risen
run	ran	run
say	said	said
see	saw	seen
sell	sold	sold
send	sent	sent
set	set	set
shake	shook	shaken
shine	shone	shone
shoot	shot	shot
shrink	shrank	shrunk
shut	shut	shut
sing	sang	sung
sit	sat	sit
sleep	slept	slept
slide	slid	slid
speak	spoke	spoken
spend	spent	spent
spin	spun	spun
split	split	split
spread	spread	spread
spring	sprang	sprung

Base (present)	Past	Past participle
stand	stood	stood
steal	stole	stolen
stick	stuck	stuck
stink	stank	stunk
strike	struck	struck
swear	swore	sworn
sweep	swept	swept
swim	swam	swum
swing	swung	swung
take	took	taken
teach	taught	taught
tear	tore	tom
tell	told	told
understand	understood	understood
wear	wore	worn
weep	wept	wept
win	won	won
wind	wound	wound
write	wrote	written

Lie/lay, sit/set, rise/raise

Three sets of verbs that cause problems are "lie/lay," "sit/set," and "rise/raise." Because they are related in meaning and sound, they are sometimes confused with each other. In each case, one of the set takes an object and the other doesn't, and each member of the set has a somewhat different meaning:

Lie (recline)   She <u>lies</u> in bed all day. (present)
　　　　　　　She <u>lay</u> in bed all last week. (past)

Lay (put)　　He <u>lays</u> his dishes on the table. (present)
　　　　　　　He <u>laid</u> his dishes on the table. (past)

Sit (be seated)   Please <u>sit</u> here by the window. (present)
　　　　　　　He <u>sat</u> by the window in class. (past)

Set (put)　　Please <u>set</u> the flowers on the table. (present)
　　　　　　　He <u>set</u> the flowers on the chair before
　　　　　　　　(past)
　　　　　　　he left.

Rise (get up)   They all <u>rise</u> early in the morning. (present)
　　　　　　　They all <u>rose</u> early yesterday too. (past)

Raise (lift up)   Can you <u>raise</u> that weight above your head?
　　　　　　　　(present)
　　　　　　　He <u>raised</u> the curtain for the play. (past)

Verb tense

The four verb tenses for past, present, and future are as follows:

- Simple:

  I see　　　I saw

- Progressive: "be" + *-ing* form of the verb

  I am seeing　I was seeing

- Perfect: "have," "had," or "shall" + the *-ed* form of the verb

  I have walked　　　I had walked

- Perfect progressive: "have" or "had" + "been" + *-ing* form of the verb

  I have been singing　　I had been singing

(For a guide to using the tenses, see section 30.)

Verb voice

Verb voice tells whether the verb is in the active or passive voice, In the active voice, the subject performs the action of the verb. In the passive voice, the subject receives the action.

HINT

VERB ENDINGS

Avoid the verb ending problem that omits the final "d" or uses "of" instead of "have" in such forms as the following:

might have <u>like</u> to (should be: might have <u>liked</u> to)

could <u>of</u> (should be: could <u>have</u>)

suppose to (should be: <u>supposed</u> to)

The doer of the action in the passive voice may be omitted or may appear in a "by the . . ." phrase.

Active:　　The child sang the song.

Passive:　The song was sung by the child. *Verb mood*

The mood of a verb tells whether it expresses a fact or opinion (**indicative** mood); expresses a command, request, or advice (**imperative** mood); or expresses a doubt, wish, recommendation, or something contrary to fact (**subjunctive** mood).

Indicative:　　The new software <u>runs</u> well on this computer.

Imperative:　<u>Watch</u> for falling rock.

Subjunctive:　In the subjunctive, present tense verbs stay in the simple base form and do not indicate the number and person of the subject. Use the subjunctive mood in "that" clauses following, verbs such as "ask," "insist," and "request."

　　　In the past tense, the same form as simple past is used; however, for the verb "be," "were" is used for all persons and numbers.

It is important that he <u>join</u> the committee.

He insisted that she <u>be</u> one of the leaders of the group.

If I <u>were</u> you, I wouldn't ask that question.

14　PRONOUNS

Pronoun case

Pronouns, the words that substitute for nouns, change case according to their use in a sentence.

Subject:　　<u>He</u> bought some film. It is <u>he</u>.

Object: Cherise gave <u>him</u> the film.

Possessive: No one used <u>his</u> film.

PRONOUN CASE

	Subject		Object		Possessive	
	sing.	pl.	sing.	pl.	sing.	pl.
1st person	I	we	me	us	my, mine	our, ours
2nd person	you	you	you	you	your, yours	your, yours
3rd person	he	they	him	them	his	their, theirs
	she	they	her	them	her, hers	their, theirs
	it	they	it	them	it, its	their, theirs

Pronoun case in compound constructions

To find the right case when your sentence has two pronouns or a noun and a pronoun, temporarily eliminate the noun or one of the pronouns as you read the sentence to yourself. You will hear the case that is needed.

Which is correct?   Nathan and him ordered a pizza.

                (or)

                Nathan and he ordered a pizza.

Test: Would you say "Him ordered a pizza"? The correct pronoun here is "he."

Which is correct?   I gave those tickets to Mikki and she.

                (or)

                I gave those tickets to Mikki and her.

COMMON PROBLEMS WITH PRONOUNS

HINT 1

Remember that "between," "except," and "with" are prepositions, and they take pronouns in the object case:

        between you and me   (*not* between you and I)

        except Amit and her   (*not* except Amit and she)

HINT 2

Possessive case pronouns never take apostrophes:

        his shoes (*not* his' shoes)

        its eye (*not* it's eye)

HINT 3

Don't use "them" as a pointing pronoun in place of "those" or "these." Use "them" only as the object by itself.

        those pages   (*not* them pages)

HINT 4

Use possessive case before *-ing* verb forms.

        They applauded his scoring a goal. (*not* him scoring)

HINT 5

Reflexive pronouns are those that end in *"-self"* or *"-selves"* and are used to intensify the nouns they refer back to:

        I soaked myself in suntan oil.

        Please help yourself.

Don't use the reflexive pronoun in other cases because it sounds as if it might be more correct. (It isn't.)

                    I
        Joseph and ~~myself~~ went to pick up the tickets.
                    ^

                    me
        They included ~~myself~~ in the group.
                        ^

Test: Would you say "I gave those tickets to she"? When in doubt, some writers mistakenly choose the subject case, thinking it sounds more formal. But the correct pronoun here is "her," the object case, because it is the object of the preposition "to."

Who/Whom

In informal speech, some writers do not distinguish between "who" and "whom." But for formal writing, the cases are as follows:

Subject	Object	Possessive
who	whom	whose
whoever	whomever	

Subject: <u>Who</u> is going to drive that van?

Object: To <u>whom</u> should I give this booklet?

Possessive: Everyone wondered <u>whose</u> coat that was.

USING "WHO" AND "WHOM"

If you aren't sure whether to use "who" or "whom," turn a question into a statement or rearrange the order of the phrase:

Question: (Who, Whom) are you looking for?

Statement: You are looking for <u>whom</u>.

Sentence: She is someone (who, whom) I have already met.

Rearranged order: I have already met <u>whom</u>.

Pronoun case after "than" or "as"

In comparisons using "than" or "as," choose the correct pronoun case by recalling the words that are omitted:

> He is taller than (I, me). (The omitted words are "am tall.")

> He is taller than I (am tall).

> My sister likes her cat more than (I, me). (The omitted words are "she likes.")

> My sister likes her cat more than (she likes) me.

"We" or "us" before nouns

When combining "we" or "us" with a noun, such as "we players," use the case that is appropriate for the noun. You can hear that by omitting the noun and seeing which sounds correct.

> (We, Us) players chose to pay for our own equipment.

> **Test:** Would you say "Us chose to pay for our own equipment"? The correct pronoun here is "We."

"WE" OR "US"?

Remember the famous opening words of the U.S. Declaration of Independence: "We the people. . . ."

Pronoun case with infinitives ("to" + verb)

When using pronouns after infinitives, verb forms with "to" + verb, use the object case. (You can also hear this by omitting a noun that may precede the pronoun.)

> She offered to drive Orin and (I, me) to the meeting.

> **Test:** Would you say, "She offered to drive I to the meeting? The correct pronoun here is "me."

Pronoun case before gerunds (-ing verb forms)

If a pronoun is used to modify a gerund, an -*ing* word, use the possessive case.

> She was proud of (us, our) walking in the fund-raising marathon.

> The correct form here is the possessive "our."

Pronoun antecedents

Pronouns substitute for nouns. In the sentence "Emilio washed his car," the pronoun "his" is a substitute for the noun Emilio (to avoid unnecessary repetition) and refers back to Emilio. The noun that a pronoun refers back to is its antecedent. For clarity, then, pronouns should agree in number and gander with their antecedents. (In the sentences "Emilio washed their car" or "Emilio washed her car," you would assume a different car is being referred to.)

Singular.   The student turned in her lab report.

Plural: The students turned in their lab reports.

Indefinite pronouns

Indefinite pronouns are those pronouns that don't refer to any specific person or thing such as "anyone," "no one," "someone," "something," "everybody," "none," and "each." Some of them may seem to have a plural meaning, but in formal writing, treat them as singular. Others, such as "many," are always plural, and "some" can be singular or plural depending on the meaning of the sentence.

When using indefinite pronouns that are normally treated as singular, some writers prefer to use the plural to avoid sexist language (section 10). But as an alternative, you can use "his or her" (which can be wordy) or switch to plural.

> Everyone in the class took out his notebook.

> Everyone in the class took out his or her notebook.

> The students took out their notebooks.

Collective nouns

When you use collective nouns such as "committee," "family," "group," and "audience," treat them as singular because they are acting as a group.

> The jury handed in its verdict.

Generic or general nouns

When you use generic nouns to indicate members of a group, such as "voter," "student," and "doctor," treat them as singular. To avoid sexist language, switch to plural.

> A truck driver should keep his road maps close at hand.

> Truck drivers should keep their road maps close at hand.

Pronoun reference

To avoid reader confusion, be sure your pronouns have a clear reference to their antecedents. Here are several possible problems to avoid:

Ambiguous pronoun reference

When a pronoun does not clearly indicate which of two or more possible antecedents it refers to, the reference is ambiguous. Rewrite the sentence to make sure the reference is clear.

Unclear reference: Marina told Michelle that she took her bike to the library.

her = Marina?   her = Michelle?

(Did Marina take Michelle's bike to the library, or did Marina take her own bike to the library?)

Clear reference: When Marina took Michelle's bike to the library, she told Michelle she was borrowing it.

Vague pronoun reference ("this," "that," and "which")

When you use "this," "that," and "which" to refer to something, be sure the word refers to a specific antecedent that has been named.

Vague pronoun reference: Ray worked in a national forest last summer, and this may be his career choice.

this = ?

(What does "this" refer to? Because no word or phrase in the first part of the sentence refers to the pronoun, the sentence needs to be revised so the antecedent is stated.)

Clear pronoun reference: Ray worked in a national forest last summer, and <u>working as a forest ranger</u> may be his career choice.

Indefinite use of "you," "it," and "they"

Avoid the use of "you, it," and "they" that doesn't refer to any specific group.

Vague pronoun reference: Everyone knows <u>you</u> should use sunscreen lotion when out in bright summer sun.

you = ?

Clear pronoun reference: It is well known that <u>people</u> should use sunscreen lotion when out in bright summer sun.

Vague pronoun reference: In Hollywood <u>they</u> don't know what type of movies the American public wants to see.

they = ?

Clear pronoun reference: In Hollywood, <u>screenwriters and producers</u> don't know what type of movies the American public wants to see.

15 ADJECTIVES AND ADVERBS

Adjectives and adverbs are modifiers, but they modify different kinds of words:

Adjectives
- modify nouns and pronouns
- answer the questions "which?" "how many?" and "what kind?"

<u>six</u> packages (how many? <u>six</u>)

<u>cheerful</u> smile (what kind? <u>cheerful</u>)

It is <u>cold</u>.

(An adjective after a linking verb modifies the subject and is called a "subject complement." Here, "cold" is an adjective modifying the subject pronoun "it.")

The water tastes <u>salty</u>. water = salty

(Some verbs, like "taste," "feel," "appear" and "smell," can be linking verbs.)

Adverbs
- modify verbs, verb forms, adjectives, and other adverbs
- answer the questions "how?" "when?" "where?" and "to what extent?"
- Most (but not all) adverbs end in *-ly*

danced <u>gracefully</u> (how? <u>gracefully</u>)

<u>very</u> long string (to what extent? <u>very</u>)

quickly
He ran ~~quick~~. (How did he run? <u>quickly</u>)

really
They sing ~~real~~ loud. (How loud? <u>really</u>)

Adjectives	Adverbs
sure	surely
real	really
good	well
bad	badly

"Good," "bad," "badly," and "well"

The modifiers "good," "bad," "badly," and "well" can cause problems because they are occasionally misused in speech. In addition, "well" can function as an adjective or an adverb.

well (adjective) = healthy   well (adverb) = done satisfactorily

He played <u>well</u>. (*not* good)

Despite the surgery, I feel <u>well</u>. (*not* good)

The linebacker played <u>badly</u> today. (*not* bad)

She feels <u>bad</u> about missing that meeting. (*not* badly)

He looked <u>good</u> in that suit. (*not* well)

HINT

COMPLETING COMPARISONS

When you use adverbs such as "so," "such," and "too," be sure to complete the phrase or clause.

that she laughed out loud
She is so happy.
^

to ask for help
Tran's problem is that he is too proud.
^

Comparatives and superlatives

Adjectives and adverbs are often used to show comparison, and the degree of comparison is indicated in their forms. Adjectives and adverbs with one or two syllables add *-er* and *-est* as endings, and longer adjectives and adverbs combine with the words "more" and "most" or "less" and "least."

- **Positive** (when no comparison is made):

  a <u>large</u> box   a <u>cheerful</u> smile

- **Comparative** (when two things are compared):

  the <u>larger</u> of the two boxes   a <u>more cheerful</u> smile

- **Superlative** (when three or more things are compared):

  the <u>largest</u> box   the <u>most cheerful</u> smile

HINT

REGULAR FORMS OF COMPARISON

Positive	Comparative	Superlative
(for one)	(for two)	(for three or more)
tan	taller	tallest
pretty	prettier	prettiest
selfish	more selfish	most selfish
unusual	more unusual	most unusual

IRREGULAR FORMS OF COMPARISON

Positive	Comparative	Superlative
(for one)	(for two)	(for three or more)
good	better	best
well	better	best
little	less	least
some	more	most
much	more	most
many	more	most
bad, badly	worse	worst

Absolute adjectives and adverbs

Some adjectives and adverbs such as "unique," "perfect," and "final" cannot logically be compared because there can't be degrees of being final or unique or perfect.

Terri has a ~~most~~ unique smile.

16 MODIFIERS

Dangling modifiers

A dangling modifier is a word or group of words that refers to (or modifies) a word or phrase that has not been clearly stated in the sentence. When an introductory phrase does not name the doer of the action, the phrase is assumed to refer to the subject of the independent clause that follows.

Having finished the assignment, Jeremy turned on the TV.

("Jeremy," the subject of the independent clause, is the doer of the action in the introductory phrase.)

However, when the intended subject (or doer of the action) of the introductory phrase is not stated, the result is a dangling modifier.

Having finished the assignment, the TV was turned on.

(This sentence is not logical because it implies that the TV finished the homework.)

Dangling modifiers most frequently occur at the beginning of the sentence but can also appear at the end. They often have an *-ing* verb or a "to" + verb phrase near the start of the phrase. To repair a dangling modifier, name the subject in the dangling phrase or as the subject of the sentence.

Dangling: After completing a degree in education, more experience in the classroom is also needed to prepare a good teacher.

Revised: After completing a degree in education, good teachers also need to gain more experience in the classroom.

Dangling: To work as a lifeguard, practice in CPR is required.

Revised: To work as a lifeguard, applicants are required to have practice in CPR.

Misplaced modifiers

Misplaced modifiers are words or groups of words placed so far away from what is being referred to that the reader may be confused.

Misplaced modifier. The assembly line workers were told that they had been fired by the personnel director.

(Were the workers told by the personnel director that they had been fired? Or were they told by someone else that the personnel director had fired them?)

Revised: The assembly line workers were told by the personnel director that they had been fired.

Misplaced modifiers are often the source of comedians' humor, as in the classic often used by Groucho Marx and others:

The other day I shot an elephant in my pajamas. How he got in my pajamas I'll never know.

Single-word modifiers such as "only," "even," and "hardly" should be placed immediately before the words they modify or as close to that word as possible. Note the difference in meaning in these two sentences:

I earned nearly $50. (The amount was almost $50, but not quite.)

I nearly earned $50. (1 almost had the opportunity to earn $50, but it didn't work out.)

HINT

PLACING MODIFIERS CORRECTLY

Some one-word modifiers that may get misplaced:

almost	hardly	merely	only
even	just	nearly	simply

Split infinitives

Split infinitives occur when modifiers are inserted between "to" and the verb. Some people object to split infinitives, but others consider them grammatically acceptable when other phrasing would be less natural.

to quickly reach

(Here "quickly" fits naturally between the "to" and the verb "reach.")

Some split infinitives such as "to more than double" are almost impossible to rephrase, so there is no modifier between "to" and the verb.

17 SHIFTS

To maintain consistency in writing, use the same perspective throughout a paper by maintaining the same person: first person ("I"), second person ("you"), and third person ("he," "she," "it," "one," or "they"). Maintain consistency in number, tense, and tone also.

Unnecessary shift in person:	In a <u>person's</u> life, the most important (3rd person)
	thing <u>you</u> do is to decide on a career. (2nd person)
Revised:	In a <u>person's</u> life, the most important thing <u>he or she</u> does is to decide on a career.
Unnecessary shift in number:	The working <u>woman</u> faces many (singular)
	challenges to advancement. When <u>they</u> (plural)
	marry and have children, <u>they</u> may need to take a leave of absence.
Revised:	Working <u>women</u> face many challenges to advancement. When <u>they</u> marry and have children, <u>they</u> may need to take a leave of absence.

Keep writing with verbs in the same time (past, present, or future) unless the logic of what you are writing about requires a switch.

Unnecessary shift in tense:	While we <u>were watching</u> the last game (past)
	of the World Series, the picture suddenly <u>gets</u> fuzzy. (present)
Revised:	While we <u>were watching</u> the last game of the World Series, the picture suddenly <u>got</u> fuzzy.

Once you choose a formal or informal tone for a paper, keep that tone consistent in your word choices. A sudden intrusion of a very formal word or phrase in an informal narrative or the use of slang or informal words in a formal report or essay indicate the writer's loss of control over tone.

| Unnecessary shift in tone: | The job of the welfare worker is to assist in a family's struggle to obtain funds for the <u>kids'</u> clothing and food. |

("Kids" is a very informal word choice here for a sentence that is somewhat formal in tone.)

| Revised: | The job of the welfare worker is to assist in a family's struggle to obtain funds for the <u>children's</u> clothing and food. |

IV PUNCTUATION

This section includes the rules you'll need to use punctuation correctly.

YOUR QUESTION   SECTION

- When I have a list, do I use a colon such as in the following:

  —Three parts in that watch are guaranteed: the battery; the spring, and the crystal. 23

- Which abbreviations don't need a period? 23

- Where does the question mark go in this sentence:

  —Did she say, "I don't have the answer"? 23

- When is it acceptable to use dashes? 23

- How do I indicate left-out words in a quotation? 23

18　SENTENCE PUNCTUATION PATTERNS (FOR COMMAS, SEMICOLONS, AND COLONS)

- (Independent clause).

  Everyone agreed to her suggestion.

- (Independent clause), for (independent clause)

  and
  nor
  but
  or
  yet
  so (the coordinating conjunctions)

  It took four years for the tree to produce a crop, but the fruit was abundant.

- (Independent clause); (independent clause).

  They arrived later; they offered no excuse.

- (Independent clause); *thus*, (independent clause).
  however,
  nevertheless,
  (or other independent clause markers)

  They arrived late; however, they offered no excuse.

- (Independent clause): (example, list of items, or explanation).

  They needed three items: a contract, her signature, and payment.

  He had only one fault: stupidity.

- (Independent clause): (independent clause).

  The candidate promised fewer taxes: he campaigned on a platform of eliminating property taxes.

- *If* (dependent clause), (independent clause).
  After
  Because
  Since
  When
  (or other dependent clause markers)

  If he studies more, his grades will improve.

- (Independent clause) *if* (dependent clause).
  because
  since
  when
  after

  His grades will improve if he studies more.

- Subject, (nonessential dependent clause), verb/predicate.

  Mako, who is my cousin, is going to major in economics.

- Subject (essential dependent clause) verb/predicate.

  The movie that I rented last night was a box office failure.

19　COMMAS

Commas between independent clauses

To use commas in independent clauses, you need to know the following:

> **Independent clause:** a clause that can stand alone as a sentence

> **Compound sentence:** a sentence with two or more independent clauses

When you join two or more independent clauses to make a compound sentence, use a comma and any of the seven joining words (coordinating conjunctions) listed here. Place a comma before the joining word. A compound sentence that does not have both the comma and the joining word is called a "comma splice" or "fused sentence." (See page 00.)

The seven joining words

For
And
Nor
But
Or
Yet
So

Vanilla is my favorite ice cream, but chocolate is a close second.

(Some writers remember this list as "FAN BOYS," spelled out by the first letters of each word.)

Alternative: If the two independent clauses are very short, some writers leave out the comma.

It started raining but the game continued.

Alternative: If one of the independent clauses has a comma in it, use a semicolon instead as part of the joining pair.

Jillian, not Alesha, is captain of the team; but Alesha assists the coach during practices.

Commas after introductory word groups

If you include introductory words, phrases, or clauses before the main part of your sentence, place a comma after the introductory part to indicate the break.

Word:　However, the farmer switched his crop to hay.

Phrase:　Having lived in Korea, he enjoyed eating kim-chee.

Clause:　While I was working on my car, it started to rain.

Alternative: If the introductory element is short (no more than four or five words) and not likely to cause confusion, some writers leave out the comma. In the second example here, some readers might, at first, misread the sentence as stating that Matt was eating the cat, so a comma is needed.

In most cases the statistics were reliable.

While Matt ate, the cat watched intently.

Commas before and after nonessential elements

When you include words, phrases, or clauses that are not essential to the meaning of the sentence and could be included in another sentence instead, place commas before and after the nonessential element.

> Dr. Gupta **,** who is a cardiac surgeon **,** retired after fifty years of practice.

Commas in series and lists

Use commas when you have three or more items in a series or list.

> The painting was done in blues **,** greens **,** and reds.
>
> Wherever Mr. Chaugh went in the town **,** whatever he saw **,** and whomever he met, he was reminded of his childhood days there.

ESSENTIAL AND NONESSENTIAL CLAUSES

You can decide if an element is essential by reading the sentence without it. If the meaning changes, that element is essential.

Essential:

Apples *that are green* are usually very tart.

> If you remove the phrase "that are green," the statement changes to indicate that all apples are usually very tart.

Nonessential:

Apples, *which are Bryna's favorite fruit,* are on sale this week at the market.

> Whether or not apples are Bryna's favorite fruit, they are still on sale this week at the market. Thus *which are Bryna's favorite fruit* is a nonessential element.

COMMAS WITH LISTS

Remember that there must be at least three items in a list in order to use commas. Some writers mistakenly put a comma between two items (often verbs) in a sentence:

> No one had ever been able to locate the source of the river **,** and follow all its tributaries.

Alternative: Although most writers prefer to use the comma before "and" in a list of three or more items, some writers omit it.

> The menu included omelets **,** pastas ⌄ and salads.

Commas with adjectives

When you include two or more adjectives that describe a noun equally, separate the adjectives by commas. But not all adjectives describe a noun equally. A quick test to see if the adjec-

tives are equal is to switch the order. If that still sounds correct to your ear, they're equal. Another test is to insert "and" between the adjectives.

> happy **,** healthy child

(You can switch this to "healthy, happy child." You can also write "happy and healthy child.")

> six large dogs

(You cannot switch this to "large six dogs" or "six and large dogs.")

Commas with interrupting words or phrases

Use commas to set off words and phrases that interrupt the sentence.

> Louisa Marcos **,** a math teacher **,** won the award.
>
> The committee was **,** however **,** unable to agree.
>
> The weather prediction **,** much to our surprise **,** was accurate.

Commas with dates, addresses, geographical names, and numbers

- With dates

  In a heading or list:

  February 25, 1999 (or) 25 February, 1999

  (No commas are needed to separate the day and month if the day is placed before the month.)

  In a sentence:

  The order was shipped on March 9, 1998, and not received until January 18, 1999.

- With addresses

  In a letter heading or on an envelope:

  Michael Cavanaugh, Jr.
  1404 Denton Drive
  Mineola, NM 43723

  In a sentence:

  If you need more information, write to General Investment, 132 Maple Avenue, Martinsville, IL 60122.

- With geographical names

  Put a comma after each item in a place name.

  The conference next year will be in Chicago, Illinois, and in New Orleans, Louisiana, the year after that.

- With numbers

  8,190,434   27,000   1,300 (or) 1300

  The herd included 9,200 head of cattle.

Commas with quotations

Use a comma after expressions such as "he said."

> Everyone was relieved when the chairperson said, "I will table this motion until the next meeting.

Unnecessary commas

Don't separate a subject from its verb.

Unnecessary: Increasing numbers of eighteen-year-olds who vote **/** are calling for stricter laws.

Don't put a comma between two verbs.

Unnecessary: We offered to lend her our notes⸝ and help her with the homework.

Don't put a comma before every "and" or "but."

Unnecessary: The automobile industry now designs fuel-efficient cars⸝ and is finding a large market for them.

Don't put a comma before a direct object, especially a clause that starts with "that."

Unnecessary: Shaundra explained to me⸝ that she was interested in hearing my view.

Don't put a comma before a dependent clause when it comes after an independent clause, except for extreme contrast.

Unnecessary: Deer populations are exploding⸝ because their natural enemies are disappearing.

But: He was delighted with the news⸝ although he needed some time to absorb it.

Don't put a comma after "such as" or "especially."

Unnecessary: The take-out shop sold various soups, such as⸝ minestrone, bean, and chicken noodle.

20 APOSTROPHES

Apostrophes with possessives

The apostrophe indicates a form of ownership, but this is not always obvious. To test for possession, turn the two words around into an "of the" phrase.

Manuel's skates → the skates of Manuel

day's pay → the pay of the day

- For singular nouns, use 's.

  book's cover   river's edge

- For singular nouns ending in -s, the -s after the apostrophe is optional if adding that -s makes the pronunciation difficult.

  Jame's coat   (or)   James's coat
  Mr. Martinez's coat   grass's color

- For plural nouns ending in -s, add an apostrophe.

  books's covers   river's edges

- For plural nouns that do not end in -s, use -'s.

  children's toys   oxen's tails

- For possession with two or more nouns:

When jointly owned: Jim and Sabrina's house

(The house belongs jointly to both Jim and Sabrina.)

When individually owned: Jim's and Sabrina's plans

(Jim and Sabrina each have their own plans.)

- For compound nouns:

  sister-in-law's car   secretary of state's office

- For indefinite pronouns (someone, everybody, etc.), use -'s.

  no one's fault   somebody's hat

Apostrophes with contractions

Use the apostrophe to mark the omitted letter or letters in contractions.

it's = it is   don't = do not   they're = they are
o'clock = of the clock   '89 = 1989

HINT

USING APOSTROPHES

When you aren't sure where the apostrophe goes, follow this order. Notice that everything to the left of the apostrophe is the word and its plural. Everything after the plural is the possessive marker.

1. Write the word.
2. Put in the plural if needed.
3. Put in the apostrophe for possession.

	Word	Plural	Possessive marker
girl's glove:	girl		-'s
girls' gloves:	girl	-s	-'
men's gloves	men		-'s

Apostrophes with plurals

Use the apostrophe to form the plurals of letters, abbreviations with periods, numbers, and words used as words.

She got all A's last semester.
They all have Ph.D.'s.
He picked all 5's in the lottery.
Melissa's *maybe's* were irritating.

Alternative: For some writers, the apostrophe is optional if the plural is clear.

9s   (or)   9's
1960s   (or)   1960's

But the apostrophe is needed here:

a's   A's

(Without the apostrophe, these might be the word "as.")

Unnecessary apostrophes

- Don't use an apostrophe with possessive pronouns (his, hers, its, yours, whose, etc.).

  his⸝arm   it⸝s edge   yours⸝

- Don't use an apostrophe with regular forms of plurals that do not show possession.

  The apple⸝s were ripe. They reduced the prices⸝.

21 SEMICOLONS

The semicolon is a stronger mark of punctuation than the comma, and it is used with two kinds of closely related elements:

1. between independent clauses
2. between items in a series.

It is almost like a period but does not come at the end of the sentence.

Semicolons to separate independent clauses

Use a semicolon when joining two independent clauses (see page 00) not joined by the seven connectors that require a comma: *for, and, nor, but, or, yet,* and *so*. Two patterns for using semicolons:

1. independent clause + **semicolon** + independent clause

The television shots showed extensive flood damage; houses were drifting downriver.

2. independent clause + semicolon + joining word + comma + independent clause

There was no warning before the flood; however, no lives were lost.

Some frequently used joining words:

also,	finally,	instead,
besides,	for example,	on the contrary,
consequently,	however,	still,
even so,	in addition,	therefore,

Alternative: You can use a semicolon between two independent clauses joined with a coordinating conjunction (section 19a) when one of those clauses has its own comma. The semicolon makes the break between the two clauses clearer.

The police officer, who was the first person on the scene, wrote down the information; and the newspaper relied on her account of the accident.

Semicolons in a series

Normally, commas are used between three or more items in a series (page 00), but when each item has its own comma, a semicolon can be used between items for clarity.

Luanne, Mina, and Karla

(or)

Luane, my first cousin; Mim, my best friend; and Karla, my neighbor

Semicolons with quotation marks

If a semicolon is needed, put it after the quotation marks.

Her answer to every question was, "I'll think about that"; she wasn't ready to make a decision.

Unnecessary semicolons

- Don't use a semicolon between a clause and a phrase.

Mexico is my favorite vacation place; especially the
(should be a comma)
beaches in Cancun.

- Don't use a semicolon in place of a dash, comma, or colon.

She spent the funds on a necessary piece of equipment for her home office; a computer.
(should be a colon)

22 QUOTATION MARKS

Quotation marks with direct quotations of prose, poetry, and dialogue

When you are writing the exact words you've seen in print or heard, use quotation marks when the quotation is less than four lines. For quotations that are four lines or longer, indent with no quotation marks.

Mrs. Alphonse said, "The test scores show improved reading ability."

In his poem, "Mending Wall," Robert Frost says: "Something there is that doesn't love a wall, / That sends the frozen-ground-swell under it."
(Notice the use of the slash to separate the two lines of poetry here.)

In his poem, "Mending Wall," Robert Frost questions the building of barriers and walls:

Something there is that doesn't love a wall,
That sends the frozen-ground-swell under it,
And spills the upper boulders in the sun;
And makes gaps even two can pass abreast.

- If you have a quotation within a quotation, use single quotation marks (') to set off the quotation enclosed inside the longer quotation.

The newspaper reporter explained: "When I interviewed the lawyer, he said, 'No comment.'"

- If you leave words out of a quotation, use an ellipsis mark (three periods; see page 00) to indicate the missing words.

The lawyer stated that he "would not . . . under any circumstances violate the client's desire for privacy."

- If you add material within a quotation, use brackets [] (see page 00).

No one, explained the scientist, "could duplicate [Mayniew's] experiment without having his notes."

- If you quote dialogue, write each person's speech as a separate paragraph. Closely related bits of narrative can be included with a paragraph with dialogue. If one person's speech goes on for several paragraphs, use quotation marks at the beginning of each paragraph, but not at the end of all paragraphs before the last one. To signal the end of the person's speech, put quotation marks at the end of the last paragraph.

Quotation marks for minor titles and parts of wholes

Use quotation marks for titles of parts of larger works (titles of book chapters, magazine articles, and episodes of television and radio series) and for short minor works (songs, short stories, essays, short poems, one-act plays). Do not use quotation marks when referring to the Bible or legal documents. For larger, more complete works, use italics (see page 00).

"The Star Spangled Banner" Exodus 2:1
"Think Warm Thoughts" (an episode on *ER*)

Quotation marks for words

- Use quotation marks (or italics) for words that are used as words rather than for their meaning.

It was tiresome to hear her always inserting "cool" or "like" in each sentence she spoke.

Use of other punctuation with quotation marks

- Place commas and periods inside quotation marks. However, in MLA format, when you include a page reference, put the period after the page reference.

    . . . was an advantage, " she said.

    . . . until tomorrow. "

    . . . when the eclipse occurs " (9).

- Place colons and semicolons outside the quotation marks.

    . . . until tomorrow ": moreover, this will be . . .

- Place the dash, exclamation mark, and question mark before the end set of quotation marks when the punctuation mark applies to the quotation. When the punctuation mark does not apply to the quotation, put the punctuation after the end set of quotation marks.

    He asked, "Should I return the book to her? "

    "Should I return the book to her? " he asked.

    Did Professor Sandifur really say, "No class tomorrow "?

Unnecessary quotation marks

Don't use quotation marks around titles of your essays, common nicknames, bits of humor, technical terms, and trite or well-known expressions.

"Bubba " wanted the fastest "modem " available.

23 OTHER PUNCTUATION

Hyphens

Hyphens have a variety of uses:

- For compound words

    Some compound words are one word:

    weekend   mastermind   watercooler

    Some compound words are two words:

    high school   executive director   home run

    Some compound words are joined by hyphens:

    father-in-law   president-elect   clear-cut

    Fractions and numbers from twenty-one to ninety-nine that are spelled out have hyphens.

    one-half   thirty-six   nine-tenths

    Particularly for new words or compounds that you are forming, check your dictionary. You may find an answer there, but not all hyphenated words appear there yet, especially new ones. Also, you will find that usage varies between dictionaries for some compounds.

    e-mail (or) email   witch-hunt (or) witch hunt

    wave-length (or) wavelength (or) wave length

    For hyphenated words in a series, use hyphens as follows:

    four-, five-, and six-page essays

- For two-word units

    Use a hyphen when two or more words placed before a noun work together as a single unit to describe the noun. When these words come after the noun, they are usually not hyphenated. But don't use hyphens with adverbs such as -ly modifiers.

    He needed up-to-date statistics. (or) He needed statistics that were up to date.

    They repaired the six-inch pipe. (or) They repaired the pipe that was six inches long.

    That was a widely known fact.

    (not with adverbs such as those ending in -ly)

- *For prefixes, suffixes, and letters joined to a word:*

    Use hyphens between words and prefixes self-, all-, and *ex-*.

    self-contained   all-American   ex-president

    For other prefixes, such as *anti-*, *pro-*, and *co-*, use the dictionary as a guide.

    co-author   antibacterial   pro-choice

    Use a hyphen when joining a prefix to a capitalized word or to figures and numbers.

    anti-American   non-Catholic   pre-1998

    Use a hyphen when you add the suffix *-elect*.

    president-elect

    Use a hyphen to avoid doubling vowels and tripling consonants and to avoid ambiguity.

    anti-intellectual bell-like   re-creation

- To divide words between syllables when the last part of the word appears on the next line.

    Every spring the nation's capitol is flooded with tourists snapping pictures of the cherry blossoms.

    When dividing words at the end of a line:

- Don't divide one-syllable words.
- Don't leave one or two letters at the end of a line.
- Don't put fewer than three letters on the next line.
- Don't divide the last word in a paragraph or the last word on a page.
- Divide compound words so the hyphen for the compound comes at the end of the line. Or put the whole compound word on the next line.

Colons

Use colons as follows:

- To announce elements at the end of the sentence

    The company sold only electronics they could service : computers, stereos, CD players, and television sets.

- To separate independent clauses

    Use a colon instead of a semicolon to separate two independent clauses when the second restates or amplifies the first.

    The town council voted not to pave the gravel made outside of town: they did not have the funds for road improvement.

- To announce long quotations

    Use a colon to announce a long quotation (more than one sentence) or a quotation not introduced by words such as "said," "remarked," or "stated."

The candidate for office offered only one reason to vote for her **,** "I will not raise parking meter rates.

- In salutations and between elements

  Dear Dr. Philippa: 6:12 A.M. Chicago: Howe Books
  Genesis 1:8 scale of 1:3 Maryland: My Home

- With quotation marks

  If a colon is needed, put it after the closing quotation mark

  "One sign of intelligence is not arguing with your boss" **,** that was her motto for office harmony.

- Unnecessary colons

  Do not use a colon after a verb or phrases like "such as" or "consisted of."

  The two most valuable players were **/** Timon Lasmon and Maynor Field.

  The camping equipment consisted of **/** tents, bug spray, lanterns, matches, and dehydrated food.

End punctuation

- Periods

  Use a period at the end of a sentence that is a statement, mild command, indirect question, or polite question where an answer isn't expected.

  Electric cars are growing in popularity. (statement)

  Do not use your calculator during the test. (mild command)

  Would you please let me know when you're done? (polite question)

  Use a period with abbreviations, but don't use a second period if the abbreviation is at the end of the sentence.

  R.S.V.P. U.S.A. Dr. Mr. 8 A.M.

  A period is not needed after agencies, common abbreviations, names of well-known companies, and state abbreviations used by the U.S. Postal Service.

  NATO NBA IBM DNA TX

  Put periods that follow quotations inside the quotation mark. But if there is a reference to a source, put the period after the reference.

  She said, "I'm going to Alaska next week."
  Neman notes "the claim is unfounded" (6).

- Question marks

  Use a question mark after a direct question but not after an indirect one.

  Did anyone see my laptop computer? (direct question)

  Jules wonders if he should buy a new stereo. (indirect question)

  Place a question mark inside the quotation marks if the quotation is a question. Place the question mark outside the quotation marks if the whole sentence is a question.

  Drora asked, "Is she on time?
  Did Eli really say, "I'm in love"?

Question marks may be used between parts of a series.

Would you like to see a movie? go shopping? eat at a restaurant?

Use a question mark to indicate doubt about the correctness of the preceding date, number, or other piece of information. But do not use it to indicate sarcasm.

The ship landed in Greenland about 1521 (**?**) but did not keep a record of where it was.
Matti's sense of humor (**?**) evaded me.

- Exclamation marks

  Use the exclamation mark after a strong command or a statement said with great emphasis or with strong feeling. But do not overuse the exclamation mark.

  I'm absolutely delighted!

  Unnecessary: Wow! What a great party! I enjoyed every minute of it!

  Enclose the exclamation mark within the quotation marks only if it belongs to the quotation.

  He threw open the door and exclaimed, "I've won the lottery!"

Dashes

The dash is somewhat informal but can be used to add emphasis or clarity, to mark an interruption or shift in tone, or to introduce a list. Use two hyphens to indicate the dash when you are typing, and do not leave a space before or after the hyphen.

To be a millionaire, the owner of a yacht, and a race car driver—this was his goal.

The cat looked at me so sweetly—with a dead rat in its mouth.

Slashes

Use the slash to mark the end of a line of poetry and to indicate acceptable alternatives. For poetry, leave a space before and after the slash. For alternatives, leave no space. The slash is also used in World Wide Web addresses.

pass/fail and/or

He reiterated Milton's great lines: "The mind is its own place, and in itself / Can make a Heaven of Hell, a Hell of Heaven."

http://www.whitehouse.gov

Parentheses

Use parentheses to enclose supplementary or less important material added as further explanation or example or to enclose figures or letters that enumerate a list.

The newest officers of the club (those elected in May) were installed at the ceremony.

They had three items on the agenda: (1) a revised budget, (2) the parking permits, and (3) a new election procedure.

Brackets

Use brackets to add your comments or additional explanation within a quotation and (to replace parentheses within parentheses. The Latin word *sic* in brackets means you copied the original quotation exactly as it appeared, but you think there is an error there.

We all agreed with Fellner's claim that "this great team [the Chicago Bears] is destined to go to the Super Bowl next year."

The lawyer explained, "We discussed the matter in a fiendly [*sic*] manner."

Omitted words/ellipsis

Use an ellipsis (a series of three spaced periods) to indicate that you are omitting words or part of a sentence from the source you are quoting. If you omit a whole sentence or paragraph, add a fourth period with no space after the last word preceding the ellipsis.

"modern methods . . . with no damage."

"the National Forest System . . ." (Smith 9).

"federal lands. . . . They were designated for preservation."

If you omit words immediately after a punctuation mark in the original, include that mark in your sentence.

"because of this use of the forest, . . ."

V MECHANICS

In this section are the rules for matters of mechanics, including capitalization, italics, numbers, abbreviations, and spelling.

YOUR QUESTION   SECTION

- Should I capitalize words such as "spring," "kleenex," "sister," and "history"?   24
- When I have a quotation, when do I capitalize the first word of the quotation?   24
- Should I capitalize the first word in each item in a list?   24
- Are book titles underlined, or should I use italics?   25
- When do I use italics instead of quotation marks with various kinds of titles and names?   25
- Should I write out "six" or use the numeral "6"?   26
- When do I write "May fifth," and when do I write "May 5"?   26
- Which titles of people can be abbreviated?   27
- Do I write "U.S." or "United States"?   27
- Do I write "six million" or "6,000,000"?   27
- Should I write CIA or C.I.A.?   27
- What do Latin abbreviations such as "cf." and "e.g." mean?   27
- Should I use "e.g." or the same phrase in English?   27
- How useful is a spell checker?   28
- What are some strategies for checking spelling?   28
- What's the "ie/ei" rule that tells me whether to write "receive" or "recieve"?   28
- How can I remember whether to write words such as "beginning" (with to "n's") or "beginning" (with one "n")?   28
- What's the rule for the "-e" in words such as "truly" or "likely"?   28
- Are words such as "data" and "media" singular or plural?   28

- Is it correct to include an apostrophe for plurals such as the following:
  –three crayons' in the box   28
- What's the difference in meaning and spelling for words such as these sound-alike words?
  —accept and except
  —affect and effect
  —its and it's
  —your and you're   28

24 CAPITALIZATION

Proper nouns vs. common nouns

Capitalize proper nouns, words that name one particular thing, most often a person or place rather than a general type or group of things.

Listed here are categories of words that should be capitalized. If you are not sure about a particular word, check your dictionary

Proper noun	Common noun
James Joyce	man
Thanksgiving	holiday
University of Maine	state university
Macintosh	personal computer
May	spring

HINT

HINT: CAPITALIZING ACADEMIC SUBJECTS

Remember that general names of academic subjects, such as "history" or "economics," are not capitalized. However, the name of a specific department is capitalized: the History Department. (This proper noun describes a particular department. Another history department might be called the Department of Historical Studies.)

- Persons

  Caitlin Baglia   Hannah Kaplan   Masuto Tatami

- Places, including geographical regions

  Indianapolis   Ontario   Midwest

- Peoples and their languages

  Spanish   Dutch   English

- Religions and their followers

  Buddhist   Judaism   Christianity

- Members of national, political, racial, social, civic, and athletic groups

Democrat	African American	Chicago Bears
Friends of	Danes	Olympics
the Library		Committee

- Institutions and organizations

Supreme Court	Legal Aid	Lions Club
	Society	

- Historical documents

  Magna Carta   The Declaration of Independence

- Periods and events (but not century numbers)

  Middle Ages   Boston Tea   eighteenth century
                 Party

- Days, months, and holidays (but not seasons)

  Monday   Thanksgiving   winter

- Trademarks

  Coca-Cola   Kodak   Ford

- Holy books and words denoting the Supreme Being (including pronouns)

  Talmud   wonders of   the Bible
            His creation

- Words and abbreviations derived from specific names (but not the names of things that have lost the specific association and now refer to the general type)

  Stalinism   NATO   CBS
  french fry   pasteurize   italics

- Place words ("City" or "Mountain") that are part of specific names

  New York City   Zion National   Wall Street
                   Park

- Titles that precede people's names (but not titles that follow names)

  Aunt Sylvia   President Taft   Governor Sam Parma
  Sylvia, my aunt   Sam Parma, governor

- Words that indicate family relationships when used as a substitute for a specific name

  Here is a gift for Mother.   She sent a gift to her mother.

- Titles of books, magazines, essays, movies, and other works, but not articles ("a," "an," "the"), short prepositions ("to," "by," "on"), or short joining words ("and," "or") unless they are the first or last word. With hyphenated words, capitalize the first and other important words. (For APA style, which has different rules, see section 40.)

  The Taming of the Shrew   "The Sino-Soviet Conflict"
  A Dialogue Between Body and Soul   "My Brother-in-Law"

- The Pronoun "I" and the interjection "O" (but not the word "oh")

  "Sail on, sail on, O ship of state," I said as the canoe sank.

- Words placed after a prefix that are normally capitalized

  un-American   anti-Semitic   ex-wife

Capitals in sentences, quotations, and lists

- Capitalize the first word in a sentence.
- Capitalize the first word of a sentence in parentheses but not when the parenthetical sentence is inserted within another sentence.

- Do not capitalize the first word in a series of questions in which the questions are not full sentences.

  What did the settlers want from the natives? food? horses?

- Capitalize the first word of directly quoted speech, but not for the second portion of interrupted direct quotations or quoted phrases or clauses integrated into the sentence.

  She answered, "No one will understand."
  "No one," she answered, "will understand."
  When Hemmings declined the nomination, he said that "this is not a gesture of support for the other candidate."

- The first word in a list after a colon if each item in the list is a complete sentence.

  The rule books were very clear. (1) No player could continue to play after committing two fouls. (2) Substitute players would be permitted only with the consent of the other team. (3) Every eligible player had to be designated before the game.

  (or)

  The rule books were very clear.

  1. No player could continue to play after committing two fouls.
  2. Substitute players would be permitted only with the consent of the other team.
  3. Every eligible player had to be designated before the game.

25 ITALICS

When you are typing or writing by hand, use underlining (a printer's mark to indicate words to be set in italic type font) for those types of titles and names indicated here. If your computer has an italic font, use that instead of underlining.

  italics type font = *italics*
  underlining = underlining

Titles

Use italics for titles and names of long or complete works, including the following:

Books:	*Catcher in the Rye*
Magazines:	*Time*
Newspapers:	*The New York Times*
Works of art (visual and performance):	*Swan Lake*
Pamphlets:	*Coping with Diabetes*
Television and radio series (not titles of individual episodes):	*Sixty Minutes*
Films and videos:	*Titanic*
Long plays:	*Macbeth*
Long musical works:	*Symphony in B Minor*
Long poems:	*Paradise Lost*
Software:	*PageMaker*
Recordings:	*Yellow Submarine*

- Do not italicize or use quotation marks for the Bible and other major religious works or for legal documents.

  Koran   The Constitution   Bible

- For shorter works or parts of whole works, use quotation marks (see section 22).

Other uses of italics/underlining

- Names of ships, airplanes, and trains

  Queen Mary   *Concorde*   *Orient Express*

- Foreign words and scientific names of plants and animals

  in vino veritas   *Canis lupis*

- Words used as words or letters, numbers, and symbols used as examples or terms

  Some words, such as *Kleenex,* are brand names.

  The letters *ph* and *f* often have the same sound.

- Words being emphasized

  It *never* snows here in April.

  (Use italics or underlining for emphasis only sparingly.)

  Do not use italics or underlining for the following:

- Words of foreign origin that are now part of English:

  alumni   karate   hacienda

- Titles of your own papers

26 NUMBERS

Style manuals for different fields and companies vary. The suggestions for writing numbers offered here are generally useful as a guide for academic writing.

- Spell out numbers that can be expressed in one or two words and use figures for other numbers.

Words	Figures
two pounds	126 days
six million dollars	$31.95
thirty-one years	6.381 bushels
eighty-three people	4.6 liters

- When you write several numbers, be consistent in choosing words or figures.

  He didn't know whether to buy ~~nine~~ ⁹ gallons of milk or 125 separate small containers. ^

- Use figures for the following:

  Days and years

December 12,1921	(or)	12 December 1921
in 1971–72	(or)	in 1971–1972
the 1990's	(or)	the 1990s
A.D. 1066		

  Time of day

8 p.m. (or) P.M.	(or)	eight o'clock in the evening
2:30 a.m. (or) A.M.	(or)	half past two in the morning

Addresses

15 Tenth Avenue
350 West 114 Street (or)   350 West 114th Street
Prescott, AZ 86301

Identification numbers

Room 8	Channel 18
Interstate 65	Henry VIII

Page and division of books and plays

page 00	chapter 6
act 3, scene 2 (or)	Act III, Scene ii

Decimals and percentages

2.7 average	12 and ½ percent
0.036 metric ton	

Numbers in series and statistics

two apples, six oranges, and three bananas
115 feet by 90 feet

(Be consistent whichever form you choose.)

Large round numbers

four billion dollars	(or)	$4 billion
16,500,000	(or)	16.5 million

Repeated numbers (in legal or commercial writing)

The bill will not exceed one hundred (100) dollars.

- Do not use figures for the following:

Numbers that can be expressed in one or two words

the eighties   the twentieth century

Dates when the year is omitted

June sixth   May fourteenth

Numbers beginning a sentence

Ten percent of the year's crop was harvested.

27 ABBREVIATIONS

In the fields of social science, science, and engineering, abbreviations are used frequently. But in other fields and in academic writing in the humanities, only a limited number of abbreviations are generally used.

Abbreviating titles

- "Mr.," "Mrs.," and "Ms." are abbreviated when used as titles before the name.

  Mr. Tanato   Ms. Ojebwa.

- "Dr." and "St." ("Saint") are abbreviated only when they immediately precede a name; they are written out when they appear after the name.

  Dr. Martin Klein (but) Martin Klein, doctor of pediatrics

- "Prof.," "Sen.," "Gen.," "Capt.," and similar abbreviated titles can be used when they appear in front of a name or before initials and a last name. But they are not abbreviated when they appear with the last name only.

  Gen. R.G. Fuller   (but)   General Fuller

- "Sr.," "Jr.," "Ph.D.," "M.F.A.," "C.P.A.," and other abbreviated academic titles and professional degrees can be used after the name.

  Lisle Millen, Ph.D.   Charleen Dyer, C.RA.

- "Bros.," "Co.," and similar abbreviations are used only if they are part of the exact name.

  Marshall Field & Co.   Brown Bros.

Abbreviating places

In general, spell out names of states, countries, continents, streets, rivers, and so on. But there are several exceptions:

- Use the abbreviation "D.C." in Washington, D.C.
- Use "U.S." only as an adjective, not as a noun.

  U.S. training bases   training bases in the United States

- If you include a full address in a sentence, citing the street, city, and state, you can use the postal abbreviation for the state.

  For further information, write to the company at 100 Peachtree Street, Atlanta, GA 30300 for a copy of their catalog.

  (but)

  The company's headquarters in Atlanta, Georgia, will soon be moved.

Abbreviating numbers

- Write out numbers that can be expressed in one or two words.

  twenty-seven   135

- The dollar sign abbreviation is generally acceptable when the whole phrase will be more than three words.

  $36 million   one million dollars

- For temperatures, use words if only a few temperatures are cited, but use figures if temperatures are cited frequently in a paper.

  ten degrees below zero, Fahrenheit   –10°F

Abbreviating measurements

Spell out units of measurement, such as acre, meter, foot, and percent, but use abbreviations in tables, graphs, and figures.

Abbreviating dates

Spell out months and days of the week. With dates and times, the following are acceptable:

57 A.D. (or) 57 B.C.E. (Before the Common Era)

A.D. 329 (The abbreviation A.D. is placed before the date.)

a.m., p.m. (or) A.M., P.M.

EST (or) E.S.T., est

Abbreviating names of familiar organizations and other entities

Use abbreviations for names of organizations, agencies, countries, and things usually referred to by their capitalized initials.

NASA	IBM	VCR	AFL-CIO
UNICEF	USSR	CNN	YMCA

If an abbreviation may not be familiar to your readers, spell out the term the first time you use it, with the abbreviation in parentheses. From then on, you can use the abbreviation.

The Myer-Briggs Type inventory (MBTI) is offered in the dean's Career Counseling Office. Students who take the MIM can then speak to a career counselor about the results.

Abbreviating Latin expressions and documentation terms

Some Latin expressions always appear as abbreviations:

Abbreviation	Meaning
cf.	compare
e.g.	for example
et al.	and others
etc.	and so forth
vs. (or) v.	versus
N.B.	note well

These abbreviations are appropriate for bibliographies and footnotes, as well as in informal writing, but for formal writing, use the English phrase instead.

Because the format for abbreviations in documentation may vary from one style manual to another, use the abbreviations listed in the particular style manual you are following.

Abbreviation	Meaning
abr.	abridged
anon.	anonymous
ed., eds.	editor, editors
p., pp.	page, pages
vol., vols.	volume, volumes

28 SPELLING

English spelling is difficult because it contains so many words from other languages that have different spelling conventions. In addition, unlike some other languages that have only one spelling for a sound, English has several ways to spell some sounds. But it's important to spell correctly, to be sure your reader understands your writing. Also, misspelled words can signal to the reader that the writer is careless and not very knowledgeable.

Because no writer wants to lose credibility, correct spelling is necessary. The following suggestions should help ensure that your papers are spelled correctly:

- Learn some spelling rules.

See the following pages for some useful rules.

- Learn your own misspelling patterns and troublesome words.

When you identify a word that tends to cause you problems, write it down in a list and if possible, make up your own memory aid. For example, if you can't remember whether "dessert" is that barren sandy place like the Mohave (or Sahara) or the sweet treat you eat after a meal, try making up some rule or statement that will stick in your mind. For "dessert" and "desert," you might try a reminder such as the fact that the word for the sweet treat has two s's, and you like seconds on desserts.

- Use a spell checker.

Spell checkers are helpful tools, but they can't catch all spelling errors. Although different spell checking programs have different capabilities, they are not foolproof, and they do make mistakes. Most spell checkers do not catch the following types of errors:

1. **Omitted words**

2. **Sound-alike words (homonyms)**

   Some words sound alike but are spelled differently. For example, the spell checker cannot distinguish between "there" and "their."

3. **Substitution of one word for another**

   If you meant to write "one" and typed "own" instead, the spell checker will not flag that.

4. **Proper nouns**

   Some well-known proper nouns, such as "Washington," may be in the spell checker, but many will not be.

5. **Misspelled words**

   If you have misspelled a word, the spell checker may be able to suggest the correct spelling. But for other misspellings, the spell checker will not be able to offer the correct spelling. You will need to know how to use a dictionary to look it up.

- Learn how to proofread.

  Proofreading requires slow and careful reading to catch misspellings and typographical errors. This is hard to do because we are used to reading quickly and seeing a group of words together. Some useful proofreading strategies are the following:

1. **Slow down.** For best results, slow down your reading rate so you actually see each word.

2. **Focus on each word.** One way to slow down is to point a pencil or pen at each word as you say it aloud or to yourself. Note with a check in the margin any word that doesn't look quite right, and come back to it later.

3. **Read backward.** Don't read right to left, as you usually do, or you will soon slip back into a more rapid reading rate. Instead, move backward through each line from right to left. In this way, you won't be listening for meaning or checking for grammatical errors.

4. **Cover up distractions.** To focus on each word, hold a sheet of paper or a note card under the line being read. That way you won't be distracted by other words on the page.

Some spelling guidelines

1. *ie/ei*

   Write *i* before *e*

   Except after *c*

   Or when sounded like "ay"

   As in "neighbor" or "weigh."

This rhyme reminds you to write *ie*, except under two conditions:

- When the two letters follow a *c*
- When the two letters sound like "ay" (as in "day")

Some *ie* words		Some *ei* words	
believe	niece	ceiling	eight
chief	relief	conceit	receive
field	yield	deceive	vein

Some exceptions to this rule:

conscience	foreign	neither	species
counterfeit	height	science	sufficient
either	leisure	seize	weird

2. *Doubting consonants*

One-syllable words

If the word ends in a single short vowel and then a consonant, double the last consonant when you add a suffix beginning with a vowel.

drag	dragged	dragging
star	starred	starring
tap	tapped	tapping
wet	wetted	wetting

Two-syllable words

If the word has two or more syllables and then a single vowel and a consonant, double the consonant when (1) you are adding a suffix that begins with a vowel, and (2) the last syllable of the base word is accented.

begin		beginning
occur	occurred	occurring
onlit	omitted	omitting
prefer	preferred	preferring
refer	referred	referring

3. *Final silent -e*

Drop the final silent -*e* when you add a suffix beginning with a vowel.

line	lining
smile	smiling

But keep the final -*e* when the suffix begins with a consonant.

care	careful
like	likely

Words such as "true/truly" and "argue/argument" are exceptions to this.

4. *Plurals*

Generally, most words add -*s* for plurals. But add -*es* when the word ends in -*s*, -*sh*, -*ch*, -*x*, or -*z* because another syllable is needed.

one apple	two apples
one box	two boxes
a brush	some brushes

With phrases and hyphenated words, pluralize the last word unless another word is more important.

one videocassette recorder	two videocassette recorders
one sister-in-law	two sisters-in-law

HINT

AVOIDING WRONG APOSTROPHES

Some writers mistakenly add an apostrophe for plurals.

one book	two books
a monkey	the cage of monkeys

For words ending in a consonant plus -y, change the -y to and add -es. For proper nouns, keep the -y.

one boy	two boys
one company	two companies
Mr. Henry	the Henrys

For some words. the plural is formed by changing the base word. Some other words have the same form for singular and plural. And other words, taken from other languages, form the plural in the same way as the original language.

one child	two children
one woman	two women
one deer	two deer
one datum	much data
one medium	many media
a phenomenon	some phenomena

Sound-alike words (homonyms)

accept:	to agree/receive	accept a gift
except:	other than	all except her
affect:	to influence	Insomnia affects me.
effect:	a result	What was the effect?
hear:	(verb)	Did you hear that?
here:	indicates a place	Come here.
its:	shows possession	Its log is broken.
it's:	= it is	It's raining out.
quiet:	no noise	Be quiet!
quite:	very	That's quite nice.
quit:	give up	He quit his job.
than:	used to compare	taller than I
then:	time word	Then he went home.
to:	preposition	to the house
too:	also	She is too tired to work.
were:	verb	were singing
we're:	= we are	We're going. on vacation.
where:	in what place	Where is he?
who's:	= who is	Who's going to the movies?
whose:	shows possession	Whose book is this?
your:	shows possession	What is your name?
you're:	= you are	You're right!

VI MULTILINGUAL SPEAKERS (ESL)

This section includes topics that are especially useful for students whose first language is not English.

YOUR QUESTION   SECTION

- What are some of the characteristics of American style in writing?   29
- Are conciseness and a clearly announced topic important in American writing?   29
- How important is it in American writing to cite sources? Why?   29

- What are the verb tenses in English and how are they used?   30
- How do the helping verbs ("be," "do," "have") and the modal verbs (such as "may" and "could") combine with the main verb?   30
- Is there a difference in meaning when different second (or third) words are combined with the main verb (such as "turn on" and "turn out")?   30
- Which is correct?
  –She enjoys to go/going swimming.
  –Mi Lan wants to ask/asking you for a favor.   30
- Is it correct to say "two furnitures" and "six chairs"?   31
- When do I use "the," and when do I use "a/an"?   32
- Which is the correct preposition?
  –in/on/at Friday   on/at/in the bottle
  –in/on/at 2 P.M.   six of/for the books
  –on/at/in home   a gift of/for her   33
- Why are the following sentences not correct?
  –Is going to be a test tomorrow.   34
  –My teacher very good speaker.   34
- Some phrases in English do not mean exactly what the words mean (such as "dead as a doornail"). How do I learn what these mean?   35

29 AMERICAN STYLE IN WRITING

If your first language is not English, you may have some writing style preferences and some questions about English grammar and usage. Some of these matters are addressed in this section, and if you have individual questions and are a student at an institution with a writing center, talk with a tutor in the writing center.

Your style preferences and customs will depend on what language(s) you are more familiar with, but in general, consider the following differences between the language(s) you know and academic style in American English. Academic style in American English is characterized by the following:

Conciseness

In some languages, writers strive for a type of eloquence marked by a profusion of words and phrases that elaborate on the same topic. Effective academic style in American English, however, is concise, eliminating extra or unnecessary words.

Clearly announced topic at the beginning of the paper

In some languages, the topic is delayed or not immediately announced. Instead, suggestions lead readers to formulate the main ideas for themselves. In American English, there is a decided preference for announcing the topic in the opening paragraph or somewhere near the beginning of the paper.

Tight organization

Although digressions into side topics or related matters can be interesting and are expected in writing in some languages, American academic writing stays on topic and does not digress.

Sources are clearly cited

In some languages, less attention is paid to citing sources of information, ideas, or the exact words used by others. In American academic writing, however, writers are expected to cite all sources other than what is generally known by most people. Otherwise, the writer is in danger of being viewed as plagiarizing.

When you are considering matters of grammar and usage, the following are topics that may cause difficulty as you write in English.

30 VERBS

Unlike some other languages, verbs are required in English sentences. Verbs are very important parts of English sentences because they indicate time and person (see section 13).

Verb tenses

Progressive tenses: use a form of "be" plus *-ing* form of the verb such as "going" or "running."

She is going to the concert tonight.

Perfect tenses: use a form of "have" plus the past participle, such as "walked" or "gone."

1. *Present tense*

    - Simple present

      presents action or condition

      > They <u>ride</u> their bikes.

      general or literary truth

      > States <u>defend</u> their rights.
      > Shakespeare <u>uses</u> humor effectively.

      habitual action

      > I <u>like</u> orange juice for breakfast.

      future time

      > The plane <u>arrives</u> at 10 P.M. tonight.

    - Present progressive

      activity in progress, not finished, or continued

      > I <u>am majoring</u> in engineering.

    - Present perfect

      action that began in the past and leads up to and continues into the present

      > He <u>has worked</u> here since May.

    - Present perfect progressive

      action begun in the past, continues to the present, and may continue into the future

      > I <u>have been thinking</u> about buying a car.

2. *Past tense*

    - Simple past

      completed action or condition

      > She <u>walked</u> to class.

    - Past progressive

      past action over a period of time or that was interrupted by another action

      > The engine <u>was running</u> while he waited.

    - Past perfect

      action or event completed before another event in the past

      > He <u>had</u> already <u>left</u> when I arrived.

    - Past perfect progressive

      ongoing condition in the past that has ended

      > She <u>had been speaking</u> to that group.

3. *Future tense*

    - Simple

      actions or events in the future

      > They <u>will arrive</u> tomorrow.

    - Future progressive

      future action that will continue for some time

      > I <u>will be expecting</u> you.

    - Future perfect

      actions that will be completed by or before a specified time in the future

      > By Monday, <u>I will have cleaned up</u> that cabinet.

    - Future perfect progressive

      ongoing actions or conditions until a specified time in the future

      > She <u>will have been traveling</u> for six months by the time she arrives here.

Helping verbs with main verbs

Helping (or auxiliary) verbs combine with other verbs. (See section 13.)

> **be**: "be," "am," "is," "are," "was," "were," "being," "been"
>
> **do**: "do," "does," "did"
>
> **have**: "have," "has," "had"
>
> **modals**: "can," "could," "may," "might," "must," "shall," "should," "will," "would," "ought to"

Modal verbs are helping verbs that indicate possibility, uncertainty, necessity, or advisability. Use the bass form of the verb after a modal.

> <u>May</u> I <u>ask</u> you a question?

Two-word (phrasal) verbs

Some verbs are followed by a second (and sometimes a third) word that combine to indicate the meaning. Many dictionaries indicate the meanings of these phrasal verbs.

look over (examine):	She <u>looked over</u> the contract.
look up (search):	I will <u>look up</u> his phone number.
look out for (watch for):	<u>Look out for</u> the puddle.

The second word of some of these verbs can be separated from the main verb by a noun or pronoun:

> add (it) up   put (the phone call) off

In other cases, the second word cannot be separated from the main verb:

> back out of the garage   get through the mob

Verbs with *-ing* and with "to" + verb form

Some verbs combine only with the *-ing* form of the verb; some verbs combine only with the "to" + verb form (the infinitive form); some verbs can be followed by either form.

Verbs followed only by -ing forms

admit	enjoy	recall
appreciate	finish	regret
deny	keep	stop
dislike	practice	suggest

He <u>admits spending</u> that money.

Verbs followed only by 'to" + verb

agree	have	plan
ask	mean	promise
claim	need	wait
decide	offer	want

She <u>needs to take</u> that medicine.

Verbs that can be followed by either form

begin	intend	prefer
continue	like	start
hate	love	try

They <u>begin to sing</u>. (or) They <u>begin singing</u>.

31 NOUNS (COUNT AND NONCOUNT)

Nouns are either proper nouns that name specific things and begin with capital letters (see section 24) or common nouns. There are two kinds of common nouns, those that can be counted (count nouns) and those that cannot be counted (non-count nouns).

1. **Count nouns:** name things that can be counted because those things can be divided into separate and distinct units. Count nouns have plurals and usually refer to things that can be seen, heard, touched, tasted, or smelled.

one apple	some apples
a chair	six chairs
the child	all of the children

2. **Noncount nouns:** name things that cannot be counted because they are abstractions or things that cannot be cut into parts. Noncount nouns do not have plurals, do not have "a" or "an" preceding them, and may have a collective meaning.

air	humor	oil weather
furniture	money	beauty clothing

   To indicate amounts for noncount nouns, use a count noun that quantifies:

a pound of coffee	a loaf of bread
a quart of milk	a great deal of money

HINT

NONCOUNT NOUNS

Many foods are noncount nouns:

coffee	tea	corn	water
cereal	milk	candy	flour

32 ARTICLES ("A," "AN," AND "THE")

A/An

"A" and "an" identify, nouns in a general or indefinite way and refer to any member of a group. "A" and "an" are generally used with singular count nouns.

Please hand me <u>a</u> towel

(This sentence does not specify which towel, just any towel that is handy.)

The

"The' identifies a particular or specific noun in a group or a noun already specified in a previous phrase or sentence. "The" may be used with singular or plural nouns.

Please hand me <u>the</u> towel that is on the table.

(This sentence means that not just any towel is being requested—only the one particular towel that is on the table.)

<u>A</u> new computer model is being introduced. <u>The</u> new model will probably cost more.

("A" is used first in a general way to mention the model, and then, because it has been specified, it is referred to as "the" model.)

Some uses of "the."

- Use "the" when an essential phrase or clause follows the noun.

  <u>The</u> man who addressed the group is my art teacher.

- Use "the" when the noun refers to a class as a whole.

  He explained that <u>the</u> fox is a nocturnal animal.

- Use "the" with names composed partly of common nouns and plural nouns.

<u>the</u> British Commonwealth	<u>the</u> United States
<u>the</u> Netherlands	<u>the</u> University of Illinois

- Use "the" with names that refer to rivers, oceans, seas, deserts, forests, gulfs, and peninsulas and with points of the compass used as names.

<u>the</u> Nile	<u>the</u> Persian Gulf	<u>the</u> South

- Use "the" with superlatives

<u>the</u> best reporter	<u>the</u> most expensive car

No articles

Articles are not used with names of streets, cities, states, countries, continents, lakes, parks, mountains, names of languages, sports, holidays, universities, and academic subjects.

He traveled to Africa.	She is studying Chinese.
He likes to watch tennis.	He graduated from Brandeis.

33 PREPOSITIONS

Prepositions in English show relationships between words and are difficult to master because they are idiomatic. The following lists some of the most commonly used prepositions and the relationships they indicate.

Prepositions of time

On	used with days	
	<u>on</u> Monday	
At	used with hours of the day	
	<u>at</u> 9 A.M.	
In	used with other parts of the day	
	<u>in</u> the afternoon	

Prepositions of place

On	indicates a surface on which something rests
	The car was parked <u>on</u> the street.
At	indicates a point in relation to another subject
	My sister is <u>at</u> home.
In	indicates a subject is inside the boundaries of an area or volume
	The sample is <u>in</u> the bottle.

Prepositions to show logical relationships

Of	shows relationship between a part and the whole
	Two <u>of</u> her teachers gave quizzes today.
	shows material or content
	That basket <u>of</u> fruit is a present
For	shows purpose
	He bought some plants <u>for</u> the garden.

34   OMITTED/REPEATED WORDS

Omitted words

Subjects and verbs can be omitted in some languages, but they are necessary in English sentences and must appear. The only exception in English is the command that has an understood subject: "Put that box here." (The understood subject here is "you.")

Subjects

Include a subject in the main clause and in all other clauses as well. "There" and "it" may sometimes serve as subject words.

All the children laughed while *they* were watching cartoons.

*It i*s raining today.

Verbs

Although verbs such as "is" and "are" and helping verbs can be omitted in some languages, they must appear in English.

She *is* an effective Spanish teacher.

No one *had* gone to the lecture.

Repeated verbs

In some languages, the subject can be repeated as a pronoun before the verb. However, in English, the subject is included only once.

Bones in the body they become brittle as people grow older.

In some languages, objects of verbs or prepositional phrases are repeated, but not in English.

The woman tried on the hat that I left it on the seat.

The city where I live there has two soccer fields.

35   IDIOMS

An idiom is an expression that mean something beyond the simple definition or literal translation into another language. Am idiom such as "kick the bucket" (meaning "die") is not understandable from the meanings of the individual words.

Dictionaries of American English idioms can define many of the commonly used ones. The second word in two-word verb phrases (phrasals; see section 30) is idiomatic and changes meanings.

dark horse = someone not likely to be the winner

under the table = something done illegally

to turn <u>off</u> the light = to shut the light, to stop it

to set an alarm to go <u>off</u> = to make an alarm work, to start it

VII   RESEARCH

This section covers the process of doing research, from finding and narrowing a topic to evaluating and collecting information, and it includes some useful sources for finding relevant material. It also discusses summaries, paraphrases, and quotations and suggestions for integrating them into your paper, as well as plagiarism and how to avoid it.

YOUR QUESTION	SECTION
• What are some suggestions for finding a topic fox my research paper?	36
• What should I do to come up with a thesis statement after I have a topic?	36
• What's the difference between primary and secondary sources?	36
• When I'm looking for information in the library, what are some useful catalogs, databases, and other sources?	36
• In addition to the library, where else can I search for information?	36
• What are some journals and magazines I can search through?	36
• When I'm collecting information, how should I handle note cards?	36
• What is a working bibliography?	36
• What are some criteria to think about when I evaluate whether to use a source I find?	37
• What should I consider when I'm deciding if the author is a reliable source of information?	37
• Are there other criteria to keep in mind to evaluate the bibliographic citation before I decide to spend time finding and reading it?	37

36 DOING RESEARCH

Some research papers that present your findings on a topic may be objective and discuss only the information you found but not your opinion or perspective on the topic. Other research papers may be persuasive because after presenting the information you found, you come to conclusions or argue your opinion or viewpoint. Be sure that you know which one is the appropriate goal as you write the paper.

Doing research is a process of selecting a topic, formulating the research question(s) you will address, searching for information, taking notes and keeping a list of the citations, evaluating what you have found, organizing the material, writing the paper, including adequate support for your thesis, and citing your sources.

Selecting a topic

There are four steps to selecting a topic

1. **Find** a general subject that interests you (if one has not been assigned). One way to locate an interesting subject is to browse through any book or catalog of subject headings such as the *Library of Congress Subject Headings* or the *Readers Guide to Periodical Literature*. On the Web, you can browse through subject directories (see section 40 for specific sites to try). You can also browse through journals in a field you are interested in to look at topics discussed there. (See page 818 for a resource list of journals you are likely to find in your library) For example:

- Imagery in Maya Angelou's poetry
- Careers in technical writing

2. **Narrow** that subject to a topic to fit the assignment and the length of the paper you will write. To narrow a topic, begin by listing some subtopics or smaller aspects of the larger topic, and choose one of those subtopics.

- Maya Angelou's use of fire imagery
- Technical writing jobs in the computer industry

3. **Formulate a research question** about your topic. Your research question will help you decide what information is relevant. Try formulating the question with any of the reporter's "who," "what," "where," "why," "when," and "how" questions.

- How does Maya Angelou use fire imagery in her poetry?
- What kinds of jobs are available in the computer industry for technical writing majors?

4. **Formulate a thesis statement** that answers your research question. After completing your research and reviewing your information, you will be able to formulate a tentative thesis. This statement will answer your research question, but it may need to be revised somewhat as you write your paper.

- Maya Angelou uses images of fire in her poetry to convey cleansing and rebirth.
- In the computer industry, technical writing majors are hired to write documents such as computer manuals, training materials, and in-house newsletters.

Finding information

The two categories of information to use are primary and secondary sources:

1. Primary sources

Primary sources are original or firsthand materials such as the poem or novel by the author you are writing about. Other primary materials are surveys, speeches, interviews you conduct, or firsthand accounts of events. Primary sources are not filtered through a second person. They may be more accurate because they have not been distorted or misinterpreted.

HINT

ESL HINT: STATING YOUR IDEAS

Although some cultures place more value on student writing that primarily brings together or collects the thoughts of great scholars or experts, readers of research papers in American institutions value the writer's own interpretations and thinking about the subject.

2. Secondary sources

Secondary sources are secondhand accounts, information, or reports about primary sources. Typical secondary sources include reviews, biographies about a person you are studying, documentaries, encyclopedia articles, and other materials interpreted or studied by others. Remember that secondary sources are interpretations or analyses that may be biased, inaccurate, or incomplete.

Sources of information

1. *Libraries*

Libraries have various printed guides for users and an information desk where you can talk with helpful librarians. In addition to the catalogs, some other indexes, catalogs, and databases you can browse through include the following:

- *Books in Print*
- *Reader's Guide to Periodical Literature*
- Encyclopedias such as *Collier's Encyclopedia* and more specialized ones such as the *Harvard Guide to American History* and the *Oxford Companion to English Literature*
- Abstracts and almanacs such as *Statistical Abstracts of the United States*
- Online searches of the library's holdings by author, title, keyword, and subject heading
- *Library of Congress Subject Headings* (This is helpful for subject heading searches when working

online and can suggest alternate ways of phrasing keywords for your topic.)

- Electronic databases such as the following: Business and Industry

  Business Periodicals Index

  CINAHL: Cumulative Index to Nursing and Allied Health Literature

  Contemporary Authors

  Contemporary Literary Criticism Select

  Dictionary of Literary Biography

  ERIC

  GPO Index

  Humanities Index

  Literary Resource Center

  MathSciNet

  MedLine

  Newspaper Source

  PsycINFO

  Social Science Index

- Computerized bibliographic utilities such as *First-Search* (that accesses databases such as academic journals, corporations, congressional publications, and medical journals) or *Newsbank CD News* (that indexes articles from a variety of newspapers). Nexis/Lexis is a commercial service available online and has abstracts and full texts of magazines, newspapers, publications from industry and government, wire services, and other sources.

2. *The Internet*

   See section 39 for a discussion of finding information on the Internet (page 000), searching useful sites (section 40), evaluating resources you find there (section 41), and citing Internet sources (section 42). Formats listed for MLA, APA, and others are in sections 43, 44, and 45.

3. *Community sources*

   Your community has a variety of resources to tap, including public records and other local government information in a city hall or county courthouse. Other sources are community service workers, social service agencies, schoolteachers and school administrators, community leaders, religious leaders, coordinators in nonprofit groups, the local chamber of commerce or visitors and convention bureau, local public library, museums or historical societies, the newspaper, and your campus offices and faculty.

4. *Interviews and surveys*

   You can do field research by interviewing people, sending e-mail messages, conducting surveys, and taking notes on your own observations.

Sources in various disciplines

When you are seeking information in various fields of study, the following lists of journals should help you. Also, check the Web sites listed in section 40 (Web Resources).

RESEARCH SOURCES IN VARIOUS FIELDS: JOURNALS AND MAGAZINES

Art
- *American Artist*
- *Art History*
- *Artforum*
- *Metropolis*

Biology
- *JAMA: Journal of the American Medical Association*
- *Quarterly Review of Biology*

Business/Economics/Management
- *Business Week*
- *Economist*
- *Harvard Business Review*
- *Journal of Business*
- *Sloan Management Review*

Communication
- *Communication Monographs*
- *Journalism & Mass Communication*
- *Quarterly Journal of Speech*

Composition and Rhetoric
- *College Composition and Communication*
- *College English*

Computer Science
- *Artificial Intelligence*
- *Byte*
- *Harvard Computer Review*
- *Technical Computing*

Culture
- *Common Knowledge*
- *Language, Society, and Culture*

Education
- *Education Research and Perspectives*
- *Education Week*

Engineering
- *Space Technology*
- *Automotive Engineering*
- *Industrial Engineering*

Environment
- *Amicus*
- *Atmospheric Environment*
- *Center for Health and the Global Environment Newsletter*
- *The Earth Times*
- *Environmental Health Perspectives Journals*

History
- *American Historical Review*
- *American History*
- *History Today*
- *Journal of Social History*
- *Journal of Modern History*
- *Journal of World History*

Law
- *IDEA: The Journal of Law and Technology*
- *Intellectual Property*
- *Journal of Information Law and Technology*

Literature	• *African-American Review*
	• *Journal of Modern Literature*
	• *PMLA*
Movies	• *KINEMA: A Journal for Film and Audiovisual Media*
Music	• *Journal of Musicology*
	• *Musical Quarterly*
Physics	• *Physical Review*
	• *Physical Letters*
Political Science	
	• *The Americana*
	• *Foreign Affairs*
	• *Harvard Political Review*
	• *Political Science Quarterly*
	• *Yale Political Monthly*
Psychology	• *American Journal of Psychology*
	• *Counseling Psychologist*
	• *Journal of Personality and Social Psychology*
	• *Psychological Review*
Religion	• *Cross Currents*
	• *Religion and Literature*
Sociology	• *American Sociological Review*
	• *Human Development and Family Life*
	• *Journal of Sociology and Social Welfare*
Women's Studies	
	• *Journal of Women's History*
	• *Sister*
	• *Womanist Theory and Research*
	• *Women's Review of Books*

Taking notes

Working bibliography

As you collect information, start a working bibliography that lists a the sources you will read. See section 37 on evaluating the citation before you spend time locating it. Some sources won't be as useful, and there are some questions you can ask yourself as you look at citations to see whether they belong in your working bibliography.

Because you may not use all those sources in your paper, the final list of sources is likely to be shorter than the working bibliography. Make bibliographic entries on separate 3 ¥ 5≤ cards so that you can easily insert new entries in alphabetical order. If you are using a computer, construct your list in a file that is separate from the paper. In each entry include all the information you will need in your bibliography. You may also want to include information you would need (such as the library call number) to find that source again.

Note cards

After you evaluate the source (see section 37) and decide the information may be useful, record the information on note cards (either 3 × 5″ or 4 × 6″ cards), to summarize, paraphrase, or record a quotation (see section 38). Use parentheses or brackets to record your thoughts on how you can use this source in your paper. Limit each note card to one short aspect of a topic so you can reorder the cards later as you organize the whole project.

Label the note card with the author's last name and shortened title, if needed, in the upper right-hand corner. Use a short phrase in the upper left-hand corner as a subject heading. As you write the information, include the exact page numbers and indicate whether the information is a summary; paraphrase, or quotation.

37 EVALUATING SOURCES

We live in an age of information—such vast amounts of information that we cannot know everything about a subject. All of that information that comes streaming at us in newspapers, magazines, the media, books, journals, brochures, Web sites, and so on, is also of very uneven quality. People want to convince us to depend on their data, buy their products, accept their viewpoints, vote for their candidates, agree with their opinions, and rely on them as experts.

We sift and make decisions all the time about which information we will use based on how we evaluate the information. Evaluating sources, then, is a skill we need all the time, and applying that skill to research papers is equally important. Listed here are some stages in the process of evaluating sources for those research papers.

For additional material on evaluating sources on the Internet, see section 41.

Getting started

As you begin searching for information, ask yourself what kinds of information you are looking for and where you are likely to find appropriate sources for that kind of information. You want to be sure that you are headed in the right direction as you launch into your search, and this too is part of the evaluation process—evaluating where you are most likely to get the right kind of information for your purpose.

What kind of information are you looking for?

Do you want facts? Opinions? News reports? Research studies? Analyses? Historical accounts? Personal reflections? Data? Public records? Scholarly essays reflecting on the topic? Reviews?

Where would you find such information?

Which sources are most likely to be useful? Libraries with scholarly journals, books, and government publications? Public libraries with popular magazines? The Internet? Newspapers? Community records? Someone on your campus?

If, for example, you are searching for information on some current event, a reliable newspaper such as the *New York Times* will be a useful source, and it is likely to be available in a university library; a public library, and on the Web. If you need some statistics on the U.S. population, government census documents in libraries and on the Internet will be appropriate places to search. But if you want to do research into local history, the archives of the local government offices and local newspaper are better places to start. Consider whether there are organizations designed to gather and publish the kinds of information you are seeking. And be sure to ask yourself if the organization's goal is to be objective or to gain support for its viewpoint.

Evaluating bibliographic citations

Before you spend time hunting for a source or read it, begin by looking at the following information in the citation to evaluate whether it is worth finding or reading.

1. *Author*

   Credentials

   * How reputable is the person (or organization) listed there?

     What is the author's educational background?

     What has the author written in the past about this topic?

     Why is this person considered an expert or a reliable authority?

   You can learn more about the person by checking the Library of Congress to see what else the person has written, and the *Book Review Index* and *Book Review Digest* may lead you to reviews of other books by this person. Your library may have citation indexes in the person's field that will lead you to other articles and short pieces by this person that have been cited by others.

   For biographical information you can read *Who's Who in America* or the *Biography index*. There may also be information about the person in the publication such as listing of previous writings, awards, and notes about the author. Your goal is to get some sense of who this person is and why it is worth reading what that person wrote before you plunge in and begin reading. That may be important as you write the paper and build your case. For example, if you are citing a source to show the spread of AIDS in Africa, which of these sentences strengthens your argument?

   > "Dr. John Smith notes that the incidence of AIDS in Africa has more than doubled in the last five years.

   (or)

   > "Dr. John Smith, head of the World Health Organization committee studying AIDS in African countries, notes that the incidence of AIDS in Africa has more than *doubled in* the last five years.

   References

   Did a teacher or librarian or some other person who is knowledgeable about the topic mention this person?

   Did you see the person listed in other sources that you've already determined to be trustworthy?

   When someone is an authority, you may find other references to this person. Or this person's viewpoint or perspective may be important to read.

   Institution or affiliation

   * What organization, institution, or company is the person associated with?

   * What are the goals of this group?

   * Does it monitor or review what is published under its name?

   * Might this group be biased in some way? Are they trying to sell you something or convince you to accept their views? Do they conduct disinterested research?

2. *Timeliness*

   * When was the source published? (For Web sites, look at the "last revised" date at the end of the page.)

   * Is that date current enough to be useful, or might there be outdated material?

   * Is the source a revision of an earlier edition? If so, it is likely to be more current, and a revision indicates that the source is sufficiently valuable to revise. Check a library catalog or *Books in Print* to see if you have the latest edition.

3. *Publisher/producer*

   * Who published or produced the material?

   * Is that publisher reputable? For example, a university press or a government agency is likely to be a reputable source that reviews what it publishes.

   * Is the group recognized as an authority?.

   * Is the publisher or group an appropriate one for this topic?

   * Might the publisher be likely to have a particular bias? (For example, a brochure printed by an anti-abortion, right-to-life group is not going to argue for abortion.)

   * Is there any review process or fact checking? (If a pharmaceutical company publishes data on a new drug A is developing, is there evidence of outside review of the data?)

4. *Audience*

   * Can you tell who the intended audience is? Is that audience appropriate for your purposes?

   * Is the material too specialized or too popular or brief to be useful? (A three-volume study of gene splitting is more than you need for a five-page paper on some genetically transmitted disease. But a half-page article an a visit to the Antarctica won't tell you much about research into ozone depletion going on there.)

Evaluating content

When you have the source in hand, you can evaluate the content by keeping in mind the following important criteria:

* **Accuracy** Are the facts accurate? Do they agree with other information you've read? Are there sources for the data given?

* **Comprehensiveness** Is the topic covered in adequate depth? Or is it too superficial or limited to only one aspect that, therefore, overemphasizes only one part of the topic?

* **Credibility** Is the source of the material generally considered trustworthy? Does the source have a review process or do fact checking? Is the author an expert? What are the author's credentials for writing about this topic?

* **Fairness** If the author has a particular viewpoint, are differing views presented with some sense of fairness? Or are they presented as irrational or stupid?

* **Objectivity** Is the language objective or emotional? Does the author acknowledge differing viewpoints? Are the various perspectives fairly presented? If you are reading an article in a magazine, do other articles in that source promote a particular viewpoint?

* **Relevance** How closely related is the material to your topic? Is it really relevant or merely related? Is it too general or too specific? Too technical?

* **Timeliness** Is the information current enough to be useful? How necessary is timeliness for your topic?

To help you determine the degree to which the criteria just listed are present in the source, try the following:

* Read the preface. What does the author want to accomplish?

- Browse through the table of contents and the index. Is the topic covered in enough depth to be helpful?

- Is there a list of references that look as if the author has consulted other sources and may lead you to useful related material?

- Are you the intended audience? Consider the tone, style, level of information, and assumptions the author makes about the reader. Are they appropriate to your needs?

- Is the content of the source fact, opinion, or propaganda? If the material is presented as factual, are the sources of the facts clearly indicated? Do you think there's enough evidence offered? Is the coverage comprehensive? (As you learn more about the topic, you will notice that this gets easier as you become more of an expert.) Is the language emotional or objective?

- Are there broad, sweeping generalizations that overstate or simplify the matter?

- Does the author use a mix of primary and secondary sources?

- To determine accuracy, consider whether the source is outdated. Do some cross-checking. Do you find some of the same information elsewhere?

- Are there arguments that are one-sided with no acknowledgment of other viewpoints?

38 USING SOURCES

Integrating sources

As you write your paper, you may be incorporating summaries, paraphrases, and quotations from your sources. It is important to weave these in smoothly, to distinguish for the reader whether you are summarizing, paraphrasing, or quoting, and to signal your reader by using signal words (page 000) to introduce and smoothly blend these sources into your writing.

Summaries

Characteristics of summaries:

- Are written in your own words

- Include only the main points, omitting details, facts. examples, illustrations, direct quotations, and other specifics

- Use fewer words than the source

- Do not have to be in the same order as the source

- Are objective and do not include your own interpretation.

Use summaries if the source has unnecessary detail. the writing is not particularly memorable or worth quoting, or you want to keep your writing concise.

WRITING A SUMMARY

A summary is a brief statement of the main idea in a source, using your own words. Include a citation to the source.

2. *Paraphrases*

Characteristics of paraphrases:

- Have approximately the same number of words as the source

- Include all main points and important details in the source

- Use your own words, not those of the source

- Keep the same organization as the source

- Are more detailed than a summary

- Are objective and do not include your interpretation.

WRITING A PARAPHRASE

A paraphrase restates the information fro a source, using your own words. Include a citation to the source.

3. *Quotations*

Guidelines for using quotations:

- Use quotations as evidence, support, or further explanation of what you have written. Quotations are not substitutes for stating your point in your own words.

- Before you include the quotation, indicate what point the quotation is making and why that quotation is relevant.

- Use quotations sparingly Too many quotations strung together with very little of your own writing make a paper look like a scrapbook of pasted-together sources, not a thoughtful integration of what is known about a subject.

- Use quotations that illustrate an authority's viewpoint or style or that would not be as effective if rewritten in different words.

- Introduce quotations with signal words (see next page for examples).

For guidelines on punctuation quotations, see section 22.

USING QUOTATIONS

A quotation is the record of the exact words of a written or spoken source and is set off by quotation marks. All quotations should have citation to the source.

Using signal words

As you include summaries, paraphrases, and quotations in your papers, you need to integrate them smoothly so that there is no sudden jump or break between the flow of your words and the source material. Use the following strategies to prepare your readers and to create that needed smooth transition into the inserted material.

Use signal phrases

Signal words or phrases let the reader know a quotation will follow. Choose a phrase or word that is appropriate to the quotation and indicates the relationship to the ideas being discussed.

In 1990 when the United Nations International Human Rights Commission predicted "there will be an outburst of major violations of human rights m Yugoslavia within the next few years" (14), few people in Europe or the United States paid attention to the warning.

(UNITED NATIONS INTERNATIONAL HUMAN RIGHTS COMMISSION. *The Future of Human Rights in Eastern Europe.* New York: United Nations, 1990.)

Explain the connection

Always explain the connection between a quotation you use and the point you are making. Show the logical link, or add a follow-up comment that integrates the quotation into your paragraph.

Quotation not integrated into the paragraph

Modern farming techniques are different from those used twenty years ago. John Hession, an Iowa soybean grower, says, "Without a computer program to plan my crop allotments or to record my expenses, I'd be back in the dark ages of guessing what to do." New computer software program are being developed commercially and are selling well.

(HESSION, JOHN. PERSONAL INTERVIEW. 27 JULY 1998.)

SOME COMMON SIGNAL WORDS

according to	considers	observes
acknowledges	denies	points out
admits	describes	predicts
argues	disagrees	proposes
asserts	emphasizes	rejects
comments	explains	reports
complains	finds	responds
concedes	insists	suggests
concludes	maintains	thinks
condemns	notes	warns

The quotation here is abruptly dropped into the paragraph, without an introduction and without a clear indication from the writer how Mr. Hession's statement fits into the ideas being discussed.

Quotation revised

Modern farming techniques differ from those of twenty years ago, *particularly in the use of computer programs for planning and budgeting.* John Hession, an Iowa soybean grower *who relies heavily on computers, confirms this* when he notes, "Without a computer program to plan my crop allotments or to record my expenses, I'd be back in the dark ages of guessing what to do." Commercial software programs *such as those used by Mr. Hession, for crop allotments and budgeting,* are being developed and are selling well.

The added words in italics explain how Mr. Hession's statement confirms the point being made.

PLAGIARISM

Plagiarism results when writers fail to document a source so that the words and ideas of someone else are presented as the writer's own work.

Avoiding Plagiarism

What information needs to be documented?

When we use the ideas, findings, data, and conclusions, arguments, and words of others. we need to acknowledge that we are borrowing their work and inserting in our own by documenting. Consciously or unconsciously passing off the work of others as our own results in the form of stealing known as plagiarism, an act that has serious consequences for the writer who plagiarizes.

If you summarize, paraphrase, or use the words of someone else (see pages 116–117), provide documentation for those sources.

CITING SOURCES

In some cultures, educated writers are expected to know and incorporate the thinking of great scholars. It may be considered an insult to the reader to mention the names of the scholars, implying that the reader is not educated enough to recognize the references. However, in American writing this is not the case, and writers are always expected to acknowledge their sources and give public credit to the source.

What information does not need to be documented?

Common knowledge, that body of general ideas we share with our readers, does not have to be documented. Common knowledge consists of the following:

- Standard information on a subject that your readers know
- Information that is widely shared and can be found in numerous sources.

AVOIDING PLAGIARISM

To avoid plagiarism, read over your paper and ask yourself whether your readers can properly identify which ideas and words are yours and which are from the sources you cite. If that is clear, then you are not plagiarizing.

For example, it is common knowledge among most Americans aware of current energy problems that solar power is one answer to future energy needs. But forecasts about how widely solar power may be used in twenty years or estimates of the cost effectiveness of using solar energy would be the work of some person or group studying the subject, and documentation would be needed.

Field research you conduct also does not need to be documented, although you should indicate you are reporting your own findings.

Original version

Researchers studying human aggression are discovering that, in contrast to the usual stereotypes, patterns of aggression among girls and women under circumstances may mirror or even exaggerate those seen in boys and men. And while women's weapons are often words, fists may be used too.

ABIGAL ZUGER, "A FISTFUL OF HOSTILITY IS FOUND IN WOMEN." *NEW YORK TIMES* 28 Jul. 1998, B9.

Plagiarized version

Women can be as hostile as men. According to Abigail Zuger, researchers studying human aggression are discovering that, in contrast to the usual stereotypes, patterns of aggression among girls and women under some circumstances may mirror or even exaggerate those seen in boys and men. And while women's weapons are often words, fists may be used too (B9).

Revision

An acceptable revision of the plagiarized version would have either a paraphrase or summary in the writer's own words or would indicate, with quotation marks, the exact words of the source. (See page 821.)

GLOSSARY OF USAGE

This list includes words and phrases you may be uncertain about when writing. If you have questions about words riot included here, try the index at the back of this book to see whether the word is discussed elsewhere. You can also check a recently published dictionary.

A, An Use *a* before words beginning with a consonant and before words beginning with a vowel that sounds like a consonant:

a cat   a house   a one-way street   a union   a history

Use *an* before words that begin with a vowel and before words with a silent *h*.

an egg   an ice cube   an hour   an honor

Accept, Except *Accept,* a verb, means to agree to, to believe, or to receive.

The detective **accepted** his account of the event.

Except, a verb, means to exclude or leave out, and *except,* a preposition, means leaving out.

Because he did not know the answers, he was **excepted** from the list of contestants and asked to leave.

Except for brussel sprouts, I eat most vegetables.

Advice, Advise *Advice* is a noun, and *advise* is a verb.

She always offers too much **advice.**

Would you **advise** me about choosing the right course?

Affect, Effect Most frequently, *affect,* which means to influence, is used as a verb, and *effect,* which means a result, is used as a noun.

The weather **affects** my ability to study.

What **effect** does coffee have on your concentration?

However, *effect,* meaning to cause or bring about, is also used as a verb.

The new traffic enforcement laws **effected** a change in people's driving habits.

Common phrases with *effect* include the following:

in effect   to that effect

Ain't This is a nonstandard way of saying *am not, is not, has not, have not,* and so on.

All Ready, Already *All ready* means *prepared; already* means *before* or *by this time.*

The courses for the meal are **all ready** to be served.

When I got home, she was **already** there.

All Right, Alright *All right* is two words, not one. *Alright* is an incorrect form.

All Together, Altogether *All together* means *in a group,* and *altogether* means *entirely, totally.*

We were **all together** again after having separate vacations.

He was not **altogether** happy about the outcome of the test.

Alot, A Lot *Alot* is an incorrect form of *a lot.*

a.m., p.m. (or) A.M., P.M. Use these with numbers, not as substitutes for the words *morning* or *evening.*

                    morning at 9 a.m.
We meet every ~~a.m.~~, for an exercise class.
                    ^

Among, Between Use *among* when referring to three or more things and *between* when referring to two things.

The decision was discussed **among** all the members of the committee.

I had to decide **between** the chocolate mousse pie and the almond ice cream.

Amount, Number Use *amount* for things or ideas that are general or abstract and cannot be counted. For example, furniture is a general term and cannot be counted. That is, we cannot say *one furniture* or *two furnitures.* Use number for things that can be counted (for example, *four chairs* or *three tables*).

He had a huge **amount** of work to finish before the deadline.

There were a **number** of people who saw the accident.

An See the entry for *a, an.*

And Although some people discourage the use of *and* as the first word in a sentence, it is an acceptable word with which to begin a sentence.

And Etc. Adding *and* is redundant because *et* means *and* in Latin. See the entry for **etc.**

Anybody, Any Body See the entry for **anyone, any one.**

Anyone, Any One *Anyone* means *any person at all. Any one* refers to a specific person or thing in a group. There are similar distinctions for other words ending in *-body* and *-one* (for example, *everybody, every body, anybody, any body, someone,* and *some one*).

The teacher asked if **anyone** knew the answer.

Any one of those children could have taken the ball.

Anyways, Anywheres These are nonstandard forms for *anyway* and *anywhere.*

As, As if, As Though, Like Use *as* in a comparison (not *like,* when there is an equality intended or when the meaning is *in the function of.*

Celia acted **as** [not *like*] the leader when the group was getting organized.   (Celia = leader)

Use *as if* or *as though* for the subjunctive.

He spent his money **as if** [or **as though**] he were rich.

Use *like* in a comparison (not *as*) when the meaning is *in the manner of* or *to the same degree as.*

The boy swam **like** a fish.

Don't use *like* as the opening word in a clause in formal writing:

Informal: **Like** I thought, he was unable to predict the weather.

Formal: **As** I thought, he was unable to predict the weather.

Assure, Ensure, Insure *Assure* means *to declare or promise, ensure* means *to make safe or certain,* and *insure* means *to protect with a contract of insurance.*

I **assure** you that I am trying to find your lost package.

Some people claim that eating properly **ensures** good health.

This insurance policy also **insures** my car against theft.

Awful, Awfully *Awful* is an adjective meaning *inspiring awe* or *extremely unpleasant.*

> He was involved in an **awful** accident.

Awfully is an adverb used in very informal writing to mean *very.* Avoid it in formal writing.

> **Informal:**   The dog was **awfully** dirty

Awhile, A While *Awhile* is an adverb meaning a *short time* and modifies a verb:

> He talked **awhile** and then left

A while is an article with the noun *while* and means *a period of time:*

> I'll be there in **a while.**

Bad, Badly *Bad* is an adjective and is used after linking verbs. *Badly* is an adverb. (See section 15.)

> The wheat crop looked **bad** [not *badly*] because of lack of rain.

> There was a **bad** flood last summer.

> The building was **badly** constructed and unable to withstand the strong winds.

Beside, Besides *Beside* is a preposition meaning *at the side of*, *compared with*, or *having nothing to do with*. *Besides* is a preposition meaning *in addition to* or *other than*. *Besides* as an adverb which means *also* or *moreover*. Don't confuse *beside* with *besides*.

> That is **beside** the point.

> **Besides** the radio, they had no other means of contact with the outside world.

> **Besides,** I enjoyed the concert.

Between, Among See the entry for **among, between.**

Breath, Breathe *Breath* is a noun, and *breathe* is a verb.

> She held her **breath** when she dived into the water.

> Learn to **breathe** deeply when you swim.

But Although some people discourage the use of *but* as the first word in a sentence, it is an acceptable word with which to begin a sentence.

Can, May *Can* is a verb that expresses *ability, knowledge,* or *capacity:*

> He **can** play both the violin and the cello.

May is a verb that expresses possibility or permission. Careful writers avoid using can to mean permission:

> **May** [not *can*] I sit here?

Can't Hardly This is incorrect because it is a double negative.

> She ~~can't~~ ^can^ hardly hear normal voice levels.

Choose, Chose *Choose* is the present tense of the verb, and *chose* is the past tense:

> Jennie should **choose** strawberry ice cream.

> Yesterday, she **chose** strawberry-flavored popcorn.

Cite, Site *Cite* is a verb that means *to quote an authority or source*; *site* is a noun referring to a place.

> Be sure to **cite** your sources in the paper.

> That is the **site** of the new city swimming pool.

Cloth, Clothe *Cloth* is a noun, and *clothe* is a verb.

> Here is some **cloth** for a new scarf.

> His paycheck helps to feed and **clothe** many people in his family.

Compared to, Compared with Use *compared to* when showing that two things are alike. Use *compared with* when showing similarities and differences.

> The speaker **compared** the economy **to** a roller coaster because both have sudden ups and downs.

> The detective **compared** the fingerprints **with** other sets from a previous crime.

Could of This is incorrect. Instead use *could have.*

Data This is the plural form of *datum.* In informal usage data is used as a singular noun, with a singular verb. However, because dictionaries do not accept this, use data as a plural form for academic writing.

> **Informal Usage:**   The **data** is inconclusive.
> **Formal Usage:**   The **data** are inconclusive.

Different from, Different than *Different from* is always correct, but some writers use *different than* if a clause follows this phrase.

> This program is **different from** the others.

> That is a **different** result **than** they predicted.

Done The past tense forms of the verb *do* are *did* and *done.* *Did* is the simple form that needs no additional verb as a helper. *Done* is the past form that requires the helper *have.* Some writers make the mistake of interchanging *did* and *done.*

> They ~~done~~ ^did^ it again.   (or)   They done ^have^ it again.

Effect, Affect See the entry for **affect, effect.**

Ensure See the entry for **assure, ensure, insure.**

Etc. This is an abbreviation of the Latin *et cetera,* meaning *and the rest.* Because it should be used sparingly if at all in formal academic writing, substitute other phrases such as *and so forth* or *and so on.*

Everybody, Every Body See the entry for **anyone, any one.**

Everyone, Every One See the entry for **anyone, any one.**

Except, Accept See the entry for **accept, except.**

Farther, Further Although some writers use these words interchangeably, dictionary definitions differentiate them. *Farther* is used when actual distance is involved, and *further* is used to mean *to a greater extent, more.*

> The house is **farther** from the road than I realized.

> That was **furthest** from my thoughts at the time.

Fewer, Less *Fewer* is used for things that can be counted (*fewer trees, fewer people*), and *less* is used for ideas, abstractions, things that are thought of collectively, not separately (*less trouble, less furniture*), and things that are measured by amount, not number (*less milk, less fuel*).

Fun This noun is used informally as an adjective.

> **Informal:**   They had a **fun** time.

Goes, Says *Goes* is a nonstandard form of *says.*

> Whenever I give him a book to read, he ~~goes~~ ^says^ "What's it about?"

Gone, Went Past tense forms of the verb *go*. *Went* is the simple form that needs no additional verb as a helper. *Gone* is the past form that requires the helper *have*. Some writers make the mistake of interchanging *went* and *gone*. (See section 13.)

went (or) have gone
They ~~gone~~ away yesterday.

Good, Well *Good* is an adjective and therefore describes only nouns. *Well* is an adverb and therefore describes adjectives, other adverbs, and verbs. The word *well* is used as an adjective only in the sense of *in good health*. (See section 15.)

well   well
The stereo works ~~good~~. I feel ~~good~~.

She is a **good** driver.

Got, Have *Got* is the past tense of *get* and should not be used in place of *have*. Similarly, *got to* should not be used as a substitute for *must*. *Have got to* is an informal substitute for *must*.

have
Do you ~~got~~ any pennies for the meter?

must
I ~~got to~~ go now.

Informal: You have **got to** see that movie.

Great This adjective is overworked in its formal meaning of *very enjoyable*, *good*, or *wonderful* and should be reserved for its more exact meanings such as of *remarkable ability, intense, high degree of*, and so on.

Informal: That was a **great** movie.

More exact uses of *great:*
The vaccine was a **great** discovery.
The map went into **great** detail.

Have, Got See the entry for **got, have.**

Have, Of *Have*, not *of*, should follow verbs such as *could*, *might*, *must*, and *should*.

have
They should ~~of~~ called by now.

Hisself This is a nonstandard substitute for *himself*.

Hopefully This adverb mean *in a hopeful way*. Many people consider the meaning *it is to be hoped* as unacceptable.

Acceptable: He listened **hopefully** for the knock at the door.

Often considered unacceptable: **Hopefully**, it will not rain tonight.

I Although some people discourage the use of *I* in formal essays, it is acceptable. If you wish to eliminate the use of *I*, see section 7 on passive verbs.

Imply, Infer Some writers use these interchangeably, but careful writers maintain the distinction between the two words. *Imply* means to *suggest without stating directly to hint*. *Infer* means to reach an opinion from facts or reasoning.

The tone of her voice **implied** he was stupid.
The anthropologist **inferred** this was a burial site for prehistoric people.

Insure See the entry for **assure, ensure, insure.**

Irregardless This is an incorrect form of the word *regardless*.

Is When, Is Why, Is Where, Is Because These are incorrect forms for definitions. See section 6 and the Glossary of Grammatical Terms on faulty predication.

Faulty predication: Nervousness is when my palms sweat.

Revised: When I am nervous, my palms sweat.
(or)
Nervousness is a state of being very uneasy or agitated.

Its, It's *Its* is a personal pronoun in the possessive case. *It's* is a contraction for *it is*.

The kitten licked **its** paw.
It's a good time for a vacation.

Kind, Sort These two forms are singular and should be used with *this* or *that*. Use *kinds* or *sorts* with *these* or *those*.

This **kind** of cloud indicates heavy rain.
These **sorts** of plants are regarded as weeds.

Lay, Lie *Lay* is a verb that needs an object and should not be used in place of *lie*, a verb that takes no direct object. (See section 13.)

lie
He should ~~lay~~ down and rest awhile.

lay
You can ~~lie~~ that package on the front table.

Leave, Let *Leave* means *to go away*, and *let* means to *permit*. It is incorrect to use *leave* when you mean *let:*

Let
~~Leave~~ me get that for you.

Less, Fewer See the entry for **fewer, less.**

Let, Leave See the entry for **leave, let.**

Like, As See the entry for **as, as if, like.**

Like for The phrase "I'd like *for* you to do that" is incorrect. Omit *for*.

May, Can See the entry for **can, may.**

Most It is incorrect to use *most* as a substitute for *almost*.

Nowheres This is an incorrect form of *nowhere*.

Number, Amount See the entry for **amount, number.**

Of, Have See the entry for **have, of.**

Off of It is incorrect to write *off of* for *off* in a phrase such as *off the table*.

O.K., Ok, Okay These can be used informally but should not be used in formal or academic writing.

Reason . . . Because This is redundant. Instead of *because*, use *that*.

that
The reason she dropped the course is ~~because~~ she couldn't keep up with the homework.

Less wordy revision: She dropped the course **because** she couldn't keep up with the homework.

Reason Why Using *why* is redundant. Drop the word *why*.

The reason why I called is to remind you of your promise.

Saw, Seen Past tense forms of the verb *see*. *Saw* is the simple form that needs no additional verb as a helper. *Seen* is the past form that requires the helper *have*. Some writers make the mistake of interchanging saw *and* seen. (See section 13.)

They ~~seen~~ it happen.   (or)   They seen it happen.
   saw   *have*

Set, Sit *Set* means to place and is followed by a direct object. *Sit* means to be seated. It is incorrect to substitute *set* for *sit*.

Come in and ~~set~~ down.
             sit

~~Sit~~ the flowers on the table.
Set

Should of This is incorrect. Instead use *should have*.

Sit, Set See the entry for **set, sit.**

Site, Cite See the entry for **cite, site.**

Somebody, Some Body See the entry for **anyone, any one.**

Someone, Some One See the entry for **anyone, any one.**

Sort, Kind See the entry for **kind, sort.**

Such This is an overworked word when used in place of *very* or *extremely*.

Suppose to, Use to These are nonstandard forms for *supposed to* and *used to*.

Sure The use of *sure* as an adverb is informal. Careful writers use *surely* instead.

Informal: I **sure** hope you can join us.
Revised: I **surely** hope you can join us.

Than, Then *Than* is a conjunction introducing the second element in comparison. *Then* is an adverb meaning *at that time, next, after that, also,* or *in that case.*

She is taller **than** I am.

He picked up the ticket and **then** left the house.

That There, This Here, These Here, Those There These are incorrect forms for *that, this, these, those.*

That, Which Use *that* for essential clauses and *which* for nonessential clauses. Some writers, however, also use *which* for essential clauses. (See section 19.)

Their, There, They're *Their* is a possessive pronoun; *there* means *in, at,* or *to that place;* and *they're is a contraction for* they are.

Their house has been sold.

There is the parking lot.

They're both good swimmers.

Theirself, Theirselves, Themself These are all incorrect forms for *themselves.*

Them It is incorrect to use this in place of either the pronoun *these* or *those.*

Look at ~~them~~ apples.
        those

Then, Than See the entry for **than, then.**

Thusly This is an incorrect substitute for *thus.*

To, Too, Two *To* is a preposition; *Too* is an adverb meaning *very* or *also;* and *two* is a number.

He brought his bass guitar **to** the party.

He brought his drums **too**.

He had **two** music stands.

Toward, Towards Both are accepted forms with the same meaning although *toward* is preferred in American usage.

Use to This is incorrect for the modal meaning *formerly.* Instead, use *used to.*

Use to, Suppose to See the entry for **suppose to, use to.**

Want for Omit the incorrect for in phrases such as "I want *for* you to come here."

Well, Good See the entry for **good, well.**

Went, Gone See the entry for **gone, went.**

Where It is incorrect to use *where* to mean *when* or *that.*

The Fourth of July is a holiday ~~where~~ the town council shoots off fireworks.
                                when

I see ~~where~~ there is now a ban on shooting panthers.\
       that

Where . . . at This is a redundant form. Omit *at*.

This is where the picnic is at.

Which, That See the entry for **that, which.**

While, Awhile See the entry for **awhile, a while.**

Who, Whom Use *who* for the subject case; use *whom* for the object case.

He is the person **who** signs that form.

He is the person **whom** I asked for help.

Who's, Whose *Who's* is a contraction for *who is, whose* is a possessive pronoun.

Who's included on that list?

Whose wristwatch is this?

Your, You're *Your* is a possessive pronoun; *you're* is a contraction for *you are.*

Your hands are cold.

You're a great success.

GLOSSARY OF GRAMMATICAL TERMS

Absolutes Words or phrases that modify whole sentences rather than parts of sentences or individual words. Am absolute phrase, which consists of a noun and participle, can be placed anywhere in the sentence but needs to be set off from the sentence by commas.

The snow having finally stopped, the football
(absolute phrase)
game began.

Abstract Nouns Nouns that refer to ideas, qualities, generalized concepts, and conditions and do not have plural forms. (See section 31.)

happiness, pride, furniture, trouble, sincerity

Active Voice See **Voice.**

Adjectives Words that modify nouns and pronouns. (See section 15.)

Descriptive adjectives (*red, clean, beautiful, offensive*, for example) have three forms:

Positive:	red, clean, beautiful, offensive
Comparative (for comparing two things):	cleaner, more beautiful, less offensive
Superlative (for comparing more than two things):	cleanest, most beautiful, least offensive

Adjective Clauses See **Dependent Clauses.**

Adverbs Modify verbs, verb forms, adjectives, and other adverbs. (See section 15.) Descriptive adverbs (for example, *fast, graceful, awkward*) have three forms:

Positive:	fast, graceful, awkward
Comparative (for comparing) two things):	faster, more graceful, less awkward
Superlative (for comparing more than two things):	fastest, most graceful, least awkward

Adverb Clauses See **Dependent Clauses.**

Agreement The use of the corresponding form for related words in order to have them agree in number, person, or gender. (See sections 13 and 14.)

John runs. (Both subject and verb are singular.)

It is necessary to flush the **pipes** regularly so that **they** don't freeze.

(Both subjects, *it* and *they*, are in third person; *they* agrees in number with the antecedent, *pipes*.)

Antecedents Words or groups of words to which pronouns refer.

When the **bell** was rung, **it** sounded very loudly.

(*Bell* is the antecedent of *it*.)

Antonyms Words with opposite meanings.

Word	Antonym
hot	cold
fast	slow
noisy	quiet

Appositives Nonessential phrases and clauses that follow nouns and identify or explain them. (See section 19.)

My uncle, **who lives in Wyoming,** is taking windsurfing
(appositive)
lessons in Florida.

Articles See noun determiners and section 32.

Auxiliary Verbs Verbs used with main verbs in verb phrases.

should be going   **has** taken
(auxiliary verb)   *(auxiliary verb)*

Cardinal Numbers See **Noun Determiners.**

Case The form or position of a noun or pronoun that shows its use or relationship to other words in a sentence. The three cases in English are (1) subject (or subjective or nominative), (2) object (or objective), and (3) possessive (or genitive). (See section 14.)

Clauses Groups of related words that contain both subjects and predicates and function either as sentences or as parts of sentences. Clauses are either independent (or main) or dependent (or subordinate). (See section 11.)

Collective Nouns Nouns that refer to groups of people or things, such as a *committee, team,* or *jury*. When the group includes a number of members acting as a unit and is the subject of the sentence, the verb is also singular, (See section 13.)

The **jury** has made a decision.

Comma Splices Punctuation errors in which two or more independent clauses in compound sentences are separated only by commas and no coordinating conjunctions. (See section 12.)

but (or) ;
Jessie said he could not help, that was typical of his responses to requests.

Common Nouns Nouns that refer to general rather than specific categories of people, places, and things and are not capitalized. (See section 24.)

basket, person, history, tractor

Comparative The form of adjectives and adverbs used when two things are being compared. (See section 15.)

higher, more intelligent, less friendly

Complement When linking verbs link subjects to adjectives or nouns, the adjectives or nouns are complements.

Phyllis was **tired.**
(complement)

She became a **musician.**
(complement)

Complex Sentences Sentences with at least one independent clause and at least one dependent clause arranged in any order.

Compound Nouns Words such as swimming pool, dropout, roommate, and stepmother, in which more than one word is needed.

Compound Sentences Sentences with two or more independent clauses and no dependent clauses. (See section 12.)

Compound-Complex Sentences Sentences with at least two independent clauses and at least one dependent clause arranged in any order.

Conjunctions Words that connect other words, phrases, and clauses in sentences. Coordinating conjunctions connect independent clauses; subordinating conjunctions connect dependent or subordinating clauses with independent or main clauses.

Coordinating Conjunctions:	and, but, for, or, nor, so, yet
Some Subordinating Conjunctions:	after, although, because, if, since, until, while

Conjunctive Adverbs Words that begin or join independent clauses. (See section 19.)

consequently, however, therefore, thus, moreover

Connotation The attitudes and emotional overtones beyond the direct definition of a word.

> The words *plump* and *fat* both mean fleshy, but *plump* has a more positive connotation than *fat*.

Consistency Maintaining the same voice with pronouns, the same tense with verbs, and the same tone, voice, or mode of discourse. (See section 17.)

Coordinating Conjunctions See **Conjunctions**.

Coordination Of equal importance. Two independent clauses in the same sentence are coordinate because they have equal importance and the same emphasis.

Correlative Conjunctions Words that work in pairs and give emphasis.

> both . . . and   neither . . . nor   either . . . or
> not . . . but also

Dangling Modifiers Phrases or clauses in which the doer of the action is not clearly indicated. (See section 16.)

> Tim thought
> Missing an opportunity to study, the exam seemed especially difficult.
> ^

Declarative Mood See **Mood**.

Demonstrative Pronouns Pronouns that refer to things. (See Noun Determiners.)

> this, that, these, those

Denotation The explicit dictionary definition of a word, as opposed to the connotation of a word. (See **Connotation**.)

Dependent Clauses (Subordinate Clauses) Clauses that cannot stand alone as complete sentences. (See section 11.) There are two kinds of dependent clauses: adverb clauses and adjective clauses.

> **Adverb clauses:**   Begin with subordinating conjunctions such as *after, if, because, while, when.*
>
> **Adjective clauses:**   Tell more about nouns or pronouns in sentences and begin with words such as *who, which, that, whose, whom.*

Determiner See **Noun Determiner**.

Diagrams See **Sentence Diagrams**.

Direct Discourse See **Mode of Discourse**.

Direct/Indirect Quotations Direct quotations are the exact words said by someone or the exact words in print that are being copied. Indirect quotations are not the exact words but the rephrasing or summarizing of someone else's words. (See section 22.)

Direct Objects Nouns or pronouns that follow a transitive verb and complete the meaning or receive the action of the verb. The direct object answers the question *what?* or *whom?*

Ellipsis A series of three dots to indicate that words or parts of sentences are being omitted from material being quoted. (See section 23.)

Essential and Nonessential Clauses and Phrases *Essential* (also called *restrictive*) clauses and phrases appear after nouns and are necessary or essential to complete the meaning of the sentence. *Nonessential* (also called *nonrestrictive*) clauses and phrases appear after nouns and add extra informa-

tion, but that information can be removed from the sentence without altering the meaning. (See section 19.)

> Apples **that are green** are not sweet.
> (essential clause)

> Golden Delicious apples, **which are yellow,** are sweet.
> (nonessential clause)

Excessive Coordination Occurs when too many equal clauses are strung together with coordinators into one sentence.

Excessive Subordination Occurs when too many subordinate clauses are strung together in a complex sentence.

Faulty Coordination Occurs when two clauses that are either unequal in importance or that have little or no connection to each other are combined in one sentence and written as independent clauses.

Faulty Parallelism See **Parallel Construction**.

Faulty Predication Occurs when a predicate does not appropriately fit the subject. This happens most often after forms of the *to be* verb. (See section 6.)

> He
> ~~The reason he~~ was late ~~was~~ because he had to study.
> ^

Fragments Groups of words punctuated as sentences that either do not have both a subject and a complete verb or that are dependent clauses. (See section 11.)

> Whenever we wanted to pick fresh fruit while we were
> , we would head for the orchard with buckets
> staying on my grandmother's farm.
> ^

Fused Sentences Punctuation errors (also called *run-ons*) in which there is no punctuation between independent clauses in the sentence. (See section 12.)

> Jennifer never learned how to ask politely she just took what she wanted.
> ^

Gerunds Verbal forms ending in *-ing* that function as nouns. (See **Phrases** and **Verbals**.)

> Arnon enjoys **cooking**.
> (gerund)

> **Jogging** is another of his pastimes.
> (gerund)

Homonyms Words that sound alike but are spelled differently and have different meanings. (See section 28.)

> hear/here   passed/past   buy/by

Idioms Expressions meaning something beyond the simple definition or literal translation into another language. For example, idioms such as "short and sweet" or "wearing his heart on his sleeve" are expressions in English that cannot be translated literally into another language. (See section 35.)

Imperative Mood See **Mood**.

Indefinite Pronouns Pronouns that make indefinite reference to nouns.

> anyone, everyone, nobody, something

Independent Clauses Clauses that can stand alone as complete sentences because they do not depend on other clauses to complete their meanings. (See section 11.)

Indirect Discourse See **Mode of Discourse**.

Indirect Objects Words that follow transitive verbs and come before direct objects. They indicate the one to whom or for whom something is given, said, or done and answer the questions *to what?* or *to whom?* Indirect objects can always be paraphrased by a prepositional phrase beginning with *to* or *for*.

Alice gave **me** some money.
(indirect object)

Paraphrase: Alice gave some money to me.

Infinitives Phrases made up of the present form of the verb preceded by *to*. Infinitives can have subjects, objects, complements, or modifiers. (See section 16.)

Everyone wanted **to swim** in the new pool.
(infinitive)

Intensifiers Modifying words used for emphasis.

She **most certainly** did fix that car!
(intensifiers)

Interjections Words used as exclamations.

Oh, I don't think I want to know about that.
(interjection)

Interrogative Pronouns Pronouns used in questions.

who, whose, whom, which, that

Irregular Verbs Verbs in which the past tense forms and/or the past participles are not formed by adding *-ed* or *-d*. (See section 13.)

do, did, done   begin, began, begun

Jargon Words and phrases that are either the specialized language of various fields or, in a negative sense, unnecessarily technical or inflated terms. (See section 5.)

Intransitive Verbs See **Verbs**.

Linking Verbs Verbs linking the subject to the subject complement. The most common linking verbs are *appear, seem, become, feel, look, taste, sound,* and *be*.

I **feel** sleepy.   He **became** the president.
↗ (linking verb)   ↗ (linking verb)

Misplaced Modifiers Modifiers not placed next to or close to the word(s) being modified. (See section 16.)

on television
We saw an advertisement for an excellent new stereo
^
system with dual headphones ~~on television~~.

Modal Verbs Helping verbs such as *shall, should, will, would, can, could, may, might, must, ought to,* and *used to* that express an attitude such as interest, possibility, or obligation. (See section 30.)

Mode of Discourse Direct discourse repeats the exact words that someone says, and indirect discourse reports the words but changes some of the words.

Everett said, **"I want to become a physicist."**
(direct discourse)

Everett said **that he wants to become a physicist.**
(indirect discourse)

Modifiers Words or groups of words that describe or limit other words, phrases, and clauses. The most common modifiers are adjectives and adverbs. (See section 16.)

Mood Verbs indicate whether a sentence expresses a fact (the declarative or indicative mood); expresses some doubt or something contrary to fact or states a recommendation (the subjunctive mood); or issues a command (the imperative mood).

Nonessential Clauses and Phrases See **Essential and Nonessential Clauses and Phrases**.

Nonrestrictive Clauses and Phrases See **Essential and Nonessential Clauses and Phrases**.

Nouns Words that name people, places, things, and ideas and have plural or possessive endings. Nouns function as subjects, direct objects, predicate nominatives, objects of prepositions, and indirect objects.

Noun Clauses Subordinate clauses used as nouns.

What I see here is adequate.
↗ (noun clause)

Noun Determiners Words that signal a noun is about to follow. They stand next to their nouns or can be separated by adjectives. Some noun determiners can also function as nouns. There are five types of noun determiners:

1. Articles: definite: the, indefinite: a, an
2. Demonstratives: this, that, these, those
3. Possessives: my, our, your, his, her, its, their
4. Cardinal numbers: one, two, three, and so on
5. Miscellaneous: all, another, each, every, much, and others

Noun Phrases See **Phrases**.

Number The quantity expressed by a noun or pronoun, either singular (one) or plural (more than one).

Objects See **Direct Objects** and **Object Complements**.

Object Complements The adjectives in predicates modifying the object of the verb (not the subject).

The enlargement makes the picture **clear**.
↗ (object complement)

Object of the Preposition Noun following the preposition. The preposition, its object, and any modifiers make up the prepositional phrase.

For **Daniel**
↗ (object of the preposition for)

She knocked twice **on the big wooden door.**
↗ (prepositional phrase)

Objective Case of Pronouns The case needed when the pronoun is the direct or indirect object of the verb or the object of a preposition.

Singular	**Plural**
First person: me	First person: us
Second person: you	Second person: you
Third person: him, her, it	Third person: them

Parallel Construction When two or more items are listed or compared, they must be in the same grammatical form as equal elements. When items are not in the same grammatical form, they lack parallel structure (often called *faulty parallelism*). (See section 8.)

She was sure that **being an apprentice in a photographer's studio** would be more useful than **being a student in photography classes**.

(The phrases in bold type are parallel because they have the same grammatical form.)

Parenthetical Elements Nonessential words, phrases, and clauses set off by commas, dashes, or parentheses.

Participles Verb forms that may be part of the complete verb or function as adjectives or adverbs. The present participle ends in *-ing*, and the past participle usually ends in *-ed*, *-d*, *-n*, or *-t*. (See **Phrases**.)

Present participles: running, sleeping, digging

She is **running** for mayor in this campaign.
↗ (present participle)

Past participles: walked, deleted, chosen

The **elected** candidate will take office in January.
↗ (past participle)

Parts of Speech The eight classes into which words are grouped according to their function, place, meaning, and use in a sentence: nouns, pronouns, verbs, adjectives, adverbs, propositions, conjunctions, and interjections.

Passive Voice See **Voice**.

Participle See **Participles**.

Perfect Progressive Tense See **Verb Tenses**.

Person There are three "persons" in English.

First person: the person(s) speaking
I or we

Second person: the person(s) spoken to
you

Third person: the person(s) spoken about
he, she, it, they, anyone, etc.

Personal Pronouns Refer to people or things.

	Subject	Object	Possessive
Singular			
First person	I	me	my, mine
Second person	you	you	your, yours
Third person	he, she, it	him, her, it	his, her, hers, its
	Subject	**Object**	**Possessive**
Plural			
First person	we	us	our, ours
Second person	you	you	your, yours
Third person	they	them	their, theirs

Phrases Groups of related words without subjects and predicates. Verb phrases function as verbs.

She **has been eating** too much sugar.
↗ (verb phrase)

Noun phrases function as nouns.

A **major winter storm** hit **the eastern coast of Maine**.
↗ (noun phrase)   ↗ (noun phrase)

Prepositional phrases usually function as modifiers.

That book **of hers** is overdue at the library.
↗ (prepositional phrase)

Participial phrases, gerund phrases, infinitive phrases, appositive phrases, and absolute phrases function as adjectives, adverbs, or nouns.

Participial Phrase: I saw people **staring at my peculiar-looking haircut**.

Gerund Phrase: **Making copies of videotapes** can be illegal.

Infinitive Phrase: He likes **to give expensive presents**.

Appositive Phrase: You ought to see Dr. Elman, **a dermatologist**.

Absolute Phrase: **The test done**, he sighed with relief.

Possessive Pronouns See **Personal Pronouns, Noun Determiners,** and section 14.

Predicate Adjectives See **Subject Complements**.

Predicate Nominatives See **Subject Complements**.

Predication Words or groups of words that express action or state of beginning in a sentence and consist of one or more verbs, plus any complements or modifiers.

Prefixes Word parts added to the beginning of words.

Prefix	Word
bio- (life)	biography
mis- (wrong, bad)	misspell

Prepositions Link and relate their objects (usually nouns or pronouns) to some other word or words in a sentence. Prepositions usually precede their objects but may follow the objects and appear at the end of the sentence.

The waiter gave the check **to my date** by mistake.
↗ (prepositional phrase)

I wonder **what** she is asking **for**.
↗ (object of   ↗ (preposition)
the preposition)

Prepositional Phrases See **Phrases**.

Progressive Tenses See **Verb Tenses**.

Pronouns Words that substitute for nouns. (See section 14.) Pronouns should refer to previously stated nouns, called antecedents.

When **Josh** came in, **he** brought some firewood.
↗ (antecedent) ↗ (pronoun)

Forms of pronouns: personal, possessive, reflexive, interrogative, demonstrative, indefinite, and relative.

Pronoun Case Refers to the form of the pronoun that is needed in a sentence. See **Subject, Object,** and **Possessive Cases** and section 14.

Proper Nouns Refer to specific people, places, and things. Proper nouns are always capitalized. (See section 24.)

Copenhagen   Honda   House of Representatives   Spanish

Reflexive Pronouns Pronouns that show someone or something in the sentence is acting for itself or on itself. Because a reflexive pronoun must refer to a word in a sentence, it is not the subject or direct object. If used to show emphasis, reflexive pronouns are called *intensive pronouns*. (See section 14.)

Singular	Plural
First person: myself	First person: ourselves
Second person: yourself	Second person: yourselves

Third person: himself,   Third person: themselves
herself, itself

She returned the book **herself** rather than giving it to her
↗ (reflexive pronoun)
roommate to bring back.

Relative Pronouns Pronouns that show the relationship of a
dependent clause to a noun in the sentence. Relative pronouns
substitute for nouns already mentioned in sentences and introduce adjective or noun clauses.

Relative pronouns: that, which, who, whom, whose

This was the movie **that** won the Academy Award.

Restrictive Clauses and Phrases See **Essential and
Nonessential Clauses and Phrases.**

Run-on Sentences See **fused sentences** and section 12.

Sentences Groups of words that have at least one independent clause (a complete unit of thought with a subject and
predicate). Sentences can be classified by their structure as
simple, compound, complex, and compound-complex.

Simple:	one independent clause
Compound:	two or more independent clauses
Complex:	one or more independent clauses and one or more dependent clauses
Compound-complex:	two or more independent clauses and one or more dependent clauses

Sentences can also be classified by their function as declarative, interrogative, imperative, and exclamatory.

Declarative:	makes a statement
Interrogative:	asks a question
Imperative:	issues a command
Exclamatory:	makes an exclamation

Sentence Diagrams A method of showing relationships
within a sentence.

Mamie's **cousin,** who has no taste in food, **ordered a
hamburger** with coleslaw at the Chinese restaurant.

Sentence Fragment See **Fragment.**

Simple Sentence See **Sentence.**

Simple Tenses See **Verb Tenses.**

Split Infinitives Phrases in which modifiers are inserted
between *to* and the verb. Some people object to split infinitives, but others consider them grammatically acceptable.

to quickly turn   to easily reach   to forcefully enter

Subject The word or words in a sentence that act or are acted
upon by the verb or are linked by the verb to another word or

words in the sentence. The *simple subject* includes only the
noun or other main word or words, and the complete subject
includes all the modifiers with the subject.

Harvey objected to his roommate's alarm going off at
9 A.M.
(Harvey is the subject.)

Every single one of the people in the room heard her
giggle.
(The simple subject is one; the complete subject is the
whole phrase.)

Subject Complement The noun or adjective in the predicate
(predicate noun or adjective) that refers to the same entity as
the subject in sentences with linking verbs, such as *is/are, feel,
took, small, sound, taste,* and *seem.*

She feels **happy.**   He is a **pharmacist.**
↗ (subject complement) ↗ (subject complement)

Subject Case of Pronouns See **Personal Pronouns** and section 14.

Subjunctive Mood See **Mood.**

Subordinating Conjunctions Words such as *although, if,
until,* and *when,* that join two clauses and subordinate one to
the other.

She is late. She overslept.

She is late **because** she overslept.

Subordination The act of placing one clause in a subordinate
or dependent relationship to another in a sentence because it is
less important and is dependent for its meaning on the other
clause.

Suffix Word part added to the end of a word.

Suffix	**Word**
-ful	careful
-less	nameless

Superlative Forms of Adjectives and Adverbs See **Adjectives and Adverbs** and section 15.

Synonyms Words with similar meanings.

Word	**Synonym**
damp	moist
pretty	attractive

Tense See **Verb Tense.**

Tone The attitude or level of formality reflected in the word
choices in a piece of writing. (See section 5.)

Transitions Words in sentences that show relationships
between sentences and paragraphs. (See section 9.)

Transitive Verbs See **Verbs.**

Verbals Words that are derived from verbs but do not act as
verbs in sentences. Three types of verbals are infinitives, participles, and gerunds.

Infinitives	to + **verb**
to wind	to say
Participles:	Words used as modifiers or with helping verbs. The present participle ends in *-ing,* and many past participles end in *-ed.*

The dog is **panting.**   He bought only **used** clothing.
↗ (present participle)   ↗ (past participle)

Gerunds: Present participles used as nouns.

Smiling was not a natural act for her.
    ↗ (gerund)

Verbs Words or groups of words (verb phrases) in predicates that express action, show a state of being, or act as a link between the subject and the rest of the predicate. Verbs change form to show time (tense), mood, and voice and are classified as transitive, intransitive, and linking verbs. (See section 30.)

Transitive verbs: Require objects to complete the predicate.

He **cut** the cardboard **box** with his knife.
    ↗ (transitive verb)   ↗ (object)

Intransitive verbs: Do not require objects.

My ancient cat often **lies** on the porch.
    ↗ (intransitive verb)

Linking verbs: Link the subject to the following noun or adjective.

The trees **are** bare.
    ↗ (linking verb)

Verb Conjugations The forms of verbs in various tenses. (See section 30.)

Regular:

Present

Simple present:

I walk	we walk
you walk	you walk
he, she, it walks	they walk

Present progressive:

I am walking	we are walking
you are walking	you are walking
he, she, it is walking	they are walking

Present perfect:

I have walked	we have walked
you have walked	you have walked
he, she, it has walked	they have walked

Present perfect progressive:

I have been walking	we have been walking
you have been walking	you have been walking
he, she, it has been walking	they have been walking

Past

Simple past:

I walked	we walked
you walked	you walked
he, she, it walked	they walked

Past progressive:

I was walking	we were walking
you were walking	you were walking
he, she, it was walking	they were walking

Past perfect:

I had walked	we had walked
you had walked	you had walked
he, she, it had walked	they had walked

Past perfect progressive:

I had been walking	we had been walking
you had been walking	you had been walking
he, she, it had been walking	they had been walking

Future

Simple future:

I shall walk	we shall walk
you will walk	you will walk
he, she, it will walk	they will walk

Future progressive:

I shall be walking	we shall be walking
you will be walking	you will be walking
he, she, it will be walking	they will be walking

Future perfect:

I shall have walked	we shall have walked
you will have walked	you will have walked
he, she, it will have walked	they will have walked

Future perfect progressive:

I shall have been walking	we shall have been walking
you will have been walking	you will have been walking
he, she, it will have been walking	they will have been walking

Irregular:

Present

Simple present:

I go	we go
you go	you go
he, she, it goes	they go

Present progressive:

I am going	we are going
you are going	you are going
he, she, it is going	they are going

Present perfect:

I have gone	we have gone
you have gone	you have gone
he, she, it has gone	they have gone

Present perfect progressive:

I have been going	we have been going
you have been going	you have been going
he, she, it has been going	they have been going

Past

Simple past:

I went	we went
you went	you went
he, she, it went	they went

Past progressive:

I was going	we were going
you were going	you were going
he, she, it was going	they were going

Past perfect:

I had gone	we had gone
you had gone	you had gone

he, she, it had gone · they had gone

Past perfect progressive:

I had been going · we had been going

you had been going · you had been going

he, she, it had been going · they had been going

Future

Simple:

I shall go · we shall go

you will go · you win go

he, she, it will go · they will go

Future progressive:

I shall be going · we shall be going

you will be going · you will be going

he, she, it will be going · they will be going

Future perfect:

I shall have gone · we shall have gone

you will have gone · you will have gone

he, she, it will have gone · they will have gone

Future perfect progressive:

I shall have been going · we shall have been going

you will have been going · you will have been going

he, she, it will have been going · they will have been going

Verb Phrases See **Verbs.**

Verb Tenses The times indicated by the verb forms in the past, present, or future. (For the verb forms, see **verb conjugations** and section 30.)

Present

Simple present: Describes actions or situations that exist now and are habitually or generally true.

I **walk** to class every afternoon.

Present progressive: Indicates activity in progress, something not finished, or something continuing.

He **is studying** Swedish.

Present perfect: Describes single or repeated actions that began in the past and lead up to and include the present.

She **has lived** in Alaska for two years.

Present perfect progressive: Indicates action that began in the past, continues to the present, and may continue into the future.

They **have been building** that garage for six months.

Past

Simple past: Describes completed actions or conditions in the past.

They **ate** breakfast in the cafeteria.

Past progressive: Indicates past action that took place over a period of time.

He **was swimming** when the storm began.

Past perfect: Indicates an action or event was completed before another event in the past.

No one **had heard** about the crisis when the newscast began.

Past perfect progressive: Indicates an ongoing condition in the past that has ended.

I **had been planning** my trip to Mexico when I heard about the earthquake.

Future

Simple future: Indicates actions or events in the future.

The store **will open** at 9 A.M.

Future progressive: Indicates future action that will continue for some time.

I **will be working** on that project next week.

Future perfect: Indicates action that will be completed by or before a specified time in the future.

Next summer, they **will have been** here for twenty years.

Future perfect progressive: Indicates ongoing actions or conditions until a specific time in the future.

By tomorrow, I **will have been waiting** for the delivery for one month.

Voice Verbs are either in the active or passive voice. In the active voice, the subject performs the action of the verb. In the passive, the subject receives the action. (See section 7.)

The dog **bit** the boy.

↗ (active verb)

The boy **was bitten** by the dog.

↗ (passive verb)

CORRECTION SYMBOLS

Symbol	Problem	Section
ab	abbreviation error	27
ad	adjective/adverb error	15
agr	agreement error	13
art	article	32
cap(s)	capitalization error	24
ca	case	14
cs	comma splice	12
dro.	dangling modifier	16
frag	fragment	11
fs	fused/run-on sentence	12
hyph	hyphen	23
ital	italics	25
lc	lowercase	24
mix(ed)	mixed construction	6
mm	misplaced modifier	16
num	number use error	26
om	omitted word	34
//	parallelism error	8

Symbol	Problem	Section
p	punctuation error	18–23
pl	plural needed	31
ref	reference error	14
shft	shift error	17
sp,	spelling error	28
t	verb-tense error	30
trans	transition needed	9
usage	usage error	Glossary of Usage
v	verb error	13,30
var	variety needed	4
w	wordy	3
wc/ww	word choice/wrong word	5,7,10
x	obvious error	
∧	insert	
∾tr	transpose	
℘	delete	

Answers to Selected Exercises for Part V

Chapter R Prealgebra Review

Exercise Set R.2 1. $\frac{16}{20}$ 2. $\frac{20}{25}$ 3. $\frac{1}{2}$ 4. $\frac{3}{4}$ 5. $\frac{3}{7}$ 6. $\frac{5}{9}$ 7. 1 8. 5 9. 7 10. $\frac{3}{5}$ 11. $\frac{14}{15}$ 12. $\frac{4}{5}$
13. $\frac{1}{5}$ 14. $\frac{8}{3}$ 15. $\frac{30}{61}$ 16. $\frac{18}{35}$ 17. $\frac{1}{8}$ 18. $\frac{1}{6}$ 19. $\frac{2}{147}$ 20. $\frac{3}{80}$ 21. $\frac{5}{72}$ 22. $\frac{4}{11}$ 23. $\frac{1}{3}$ 24. $\frac{1}{5}$ 25. $\frac{23}{21}$
26. $\frac{11}{12}$ 27. $\frac{65}{21}$ 28. $\frac{52}{35}$ 29. $\frac{5}{7}$ 30. $\frac{2}{5}$ 31. $\frac{5}{66}$ 32. $\frac{1}{6}$ 33. $\frac{7}{5}$ 34. $\frac{13}{8}$ 35. $\frac{17}{18}$ 36. $\frac{43}{44}$

Exercise Set R.3 1. $\frac{6}{10}$ 2. $\frac{9}{10}$ 3. $\frac{186}{100}$ 4. $\frac{723}{100}$ 5. 8.71 6. 34.0734 7. 6.5 8. 5.5 9. 15.22 10. 24.69
11. 56.431 12. 76.9931 13. 598.23 14. 823.75 15. 0.12 16. 0.63 17. 67.5 18. 891 19. 43.274 20. 44.165
21. 84.97593 22. 115.03226 23. 0.094 24. 5.85 25. 70 26. 40 27. 5.8 28. 3.6 29. 840 30. 960
31. 0.6 32. 0.6 33. 0.23 34. 0.45 35. 0.594 36. 63.452 37. 98,207.2 38. 68,936.5 39. 12.35 40. 42.988
41. $0.\overline{3} \approx 0.333$ 42. 0.4375 43. 0.625 44. $0.\overline{54} \approx 0.55$ 45. $0.\overline{16} \approx 0.17$ 46. 0.36 47. 0.031 48. 0.022
49. 1.35 50. 4.17 51. 0.8149 52. 0.61 53. 0.823; 0.816 54. 87.6% 55. 52.1% 56. 50% 57. 10%
58. 6.98 years 59. a. 14.8 pounds b. 89.2 pounds c. 211.2 pounds 60. 64%

Chapter 1 Real Numbers and Introduction to Algebra

Exercise Set 1.2 1. 9 2. 4 3. 1 4. $\frac{3}{10}$ 5. 1 6. 10 7. 11 8. 19 9. 8 10. 14 11. 45 12. 50

Exercise Set 1.3 1. 9 2. -3 3. -14 4. -20 5. 1 6. 2 7. -12 8. -5 9. -5 10. -11 11. -12
12. 2 13. -4 14. 4 15. 7 16. 2 17. -2 18. -3 19. 0 20. 0

Exercise Set 1.4 1. -10 2. -20 3. -5 4. -3 5. 19 6. 17 7. $\frac{1}{6}$ 8. $-\frac{1}{8}$ 9. 2 10. 28 11. -11
12. -12 13. 11 14. 9 15. 5 16. 12 17. 37 18. 48 19. -6.4 20. -2.9

Exercise Set 1.5 1. -24 2. -40 3. -2 4. -28 5. 50 6. 66 7. -12 8. -16 9. 42 10. 54
11. -18 12. -15 13. $\frac{3}{10}$ 14. $\frac{1}{24}$ 15. $\frac{24}{36} = \frac{2}{3}$ 16. $\frac{15}{60} = \frac{1}{4}$ 17. -7 18. -15 19. 0.14 20. 0.15

Exercise Set 1.6 1. -9 2. -2 3. 4 4. 3 5. -4 6. -12 7. 0 8. 0 9. -5 10. -3 11. undefined
12. undefined 13. 3 14. 5 15. -15 16. -7 17. $-\frac{18}{7}$ 18. $-\frac{8}{5}$ 19. $\frac{20}{27}$ 20. $\frac{11}{80}$ 21. -1

Exercise Set 1.7 1. $4x + 4y$ 2. $7a + 7b$ 3. $9x - 54$ 4. $11y - 44$ 5. $6x + 10$ 6. $35 + 40y$ 7. $28x - 21$
8. $24x - 3$ 9. $18 + 3x$ 10. $2x + 10$ 11. $-2y + 2z$ 12. $-3z + 3y$ 13. $-21y - 35$ 14. $-10r - 55$
15. $5x + 20m + 10$ 16. $24y + 8z - 48$ 17. $-4 + 8m - 4n$ 18. $-16 - 8p - 20$ 19. $-5x - 2$ 20. $-9r - 5$
21. $-r + 3 + 7p$

Exercise Set 1.8 1. approx. 7.8 million 2. approx. 1.6 million 3. 2002 4. approx. 7.4 million
5. 1994 6. 1995 7. 1986, 1987, or 1989, 1990 8. 1986 or 1987 9. approx. 52%
10. approx. 49% 11. 1994 12. answers may vary

Chapter 2 Equations, Inequalities, and Problem Solving

Exercise Set 2.1 1. -7 2. 17 3. like 4. unlike 5. $15y$ 6. $5x$ 7. $13w$ 8. $-4c$ 9. $-7b - 9$
10. $3g - 2$ 11. $-m - 6$ 12. $-3a - 2$ 13. -8 14. -11 15. $5y + 20$ 16. $7r + 21$ 17. $7d - 11$
18. $10x - 3$ 19. $4y + 11$ 20. $-4x - 9$

Exercise Set 2.2 1. $x = 3$ 2. $x = 11$ 3. $x = -2$ 4. $y = 10$ 5. $x = -14$ 6. $z = -16$ 7. $r = 0.5$ 8. $t = 2.4$
9. $f = \frac{5}{12}$ 10. $c = \frac{5}{24}$

MENTAL MATH   **1.** $a = 9$ **2.** $c = 6$ **3.** $b = 2$ **4.** $t = 2$ **5.** $x = -5$

EXERCISE SET 2.3   **1.** $x = -4$ **2.** $x = 7$ **3.** $x = 0$ **4.** $x = 0$ **5.** $x = 12$ **6.** $y = -8$ **7.** $x = -12$ **8.** $n = -20$
9. $d = 3$ **10.** $v = 2$ **11.** $x = 10$ **12.** $x = 9$ **13.** $x = -20$ **14.** $x = 28$ **15.** $a = 0$

EXERCISE SET 2.6   **1.** $h = 3$ **2.** $r = 65$ **3.** $h = 3$ **4.** $V = 336$ **5.** $h = 20$ **6.** $h = 12$ **7.** $c = 12$ **8.** $A = 27$
9. $r = 2.5$ **10.** $A = 63.585$ **11.** $T = 3$ **12.** $P = 3{,}200{,}000$ **13.** $h = 15$ **14.** $V = 113.04$ **15.** $h = \dfrac{f}{5g}$ **16.** $r = \dfrac{C}{2\pi}$
17. $W = \dfrac{V}{LH}$ **18.** $n = \dfrac{T}{mr}$ **19.** $y = 7 - 3x$ **20.** $y = 13 + x$

EXERCISE SET 2.7   **1.** 11.2   **2.** 880   **3.** 55%   **4.** 20%   **5.** 180   **6.** 360   **7.** 4.6   **8.** 120.4   **9.** 50
10. 125   **11.** 30%   **12.** 120%   **13.** $x = 4$ **14.** $x = \dfrac{16}{3}$ **15.** $x = \dfrac{50}{9}$ **16.** $x = \dfrac{3}{4}$ **17.** $x = \dfrac{21}{4}$ **18.** $a = \dfrac{15}{2}$
19. $x = 7$ **20.** $y = 40$

MENTAL MATH   **1.** $\{x \mid x < 6\}$ **2.** $\{x \mid x > 7\}$ **3.** $\{x \mid x \geq 10\}$ **4.** $\{x \mid x \leq 7\}$ **5.** $\{x \mid x > 4\}$ **6.** $\{x \mid x < 4\}$
7. $\{x \mid x \leq 2\}$ **8.** $\{x \mid x \geq 8\}$

EXERCISE SET 2.8   **1.** $(-\infty, -1]$ **2.** $(-3, \infty]$ **3.** $(-\infty, 11]$ **4.** $(4, \infty)$ **5.** $(-13, \infty)$ **6.** $\left[\dfrac{8}{3}, \infty\right)$ **7.** $(-\infty, 7]$
8. $\left(\dfrac{1}{2}, \infty\right)$ **9.** $[0, \infty)$ **10.** $\left(\dfrac{14}{3}, \infty\right)$ **11.** $(-\infty, -5]$ **12.** $[-1, \infty)$

Chapter 3 GRAPHING EQUATIONS AND INEQUALITIES

MENTAL MATH   **1.** answers may vary; Ex. $(5, 5), (7, 3)$ **2.** answers may vary; Ex. $(0, 6), (6, 0)$

EXERCISE SET 3.1

1. $(1, 5)$ is in quadrant I, $\left(-1, 4\dfrac{1}{2}\right)$ is in quadrant II, $(-5, -2)$ is in quadrant III, $(2, -4)$ is in quadrant IV, $(-3, 0)$ lies on the x-axis, $(0, 1)$ lies on the y-axis

2. $(2, 4)$ is in quadrant I, $(-2, 1)$ is in quadrant II, $(-3, -3)$ is in quadrant III, $(5, -4)$ is in quadrant IV, $\left(3\dfrac{3}{4}, 0\right)$ lies on the x-axis, $(0, 2)$ lies on the y-axis

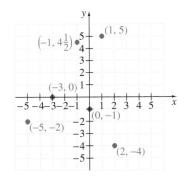 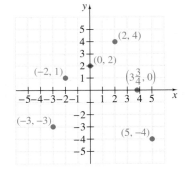

3. $a = b$ **4.** answers may vary   **5.** $(0, 0)$ **6.** $\left(3\dfrac{1}{2}, 0\right)$ **7.** $(3, 2)$ **8.** $(-1, 3)$ **9.** $(-2, -2)$ **10.** $(0, -1)$ **11.** $(2, -1)$
12. $(2, 0)$ **13.** $(0, -3)$ **14.** $(-2, 3)$ **15.** $(1, 3)$ **16.** $(1, -1)$ **17.** $(-3, -1)$ **18.** $(-2, 0)$

EXERCISE SET 3.2

1. $(6, 0); (4, -2); (5, -1)$ **2.** $(0, -4); (6, 2); (-1, -5)$ **3.** $(1, -4); (0, 0); (-1, 4)$

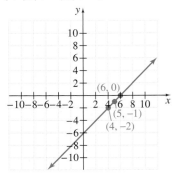 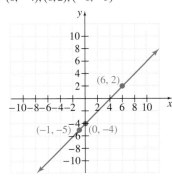 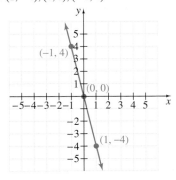

A2

4. $(1, -5); (0, 0); (-1, 5)$

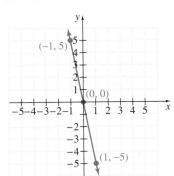

5. $(0, 0); (6, 2); (-3, -1)$

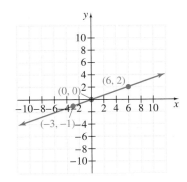

6. $(0, 0); (-4, -2); (2, 1)$

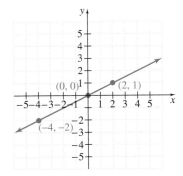

7. $(0, 3); (1, -1); (2, -5)$

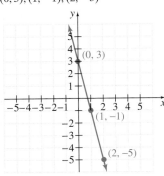

8. $(0, 2); (1, -3); (2, 8)$

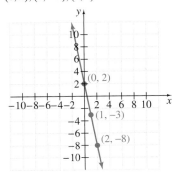

9.

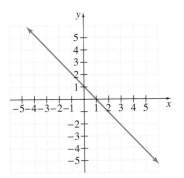

10.

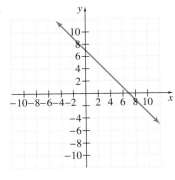

11.

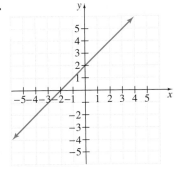

12.

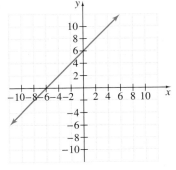

13.

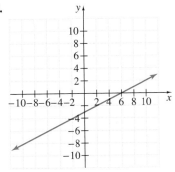

14.

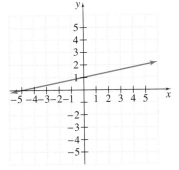

15.

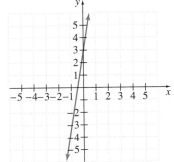

16.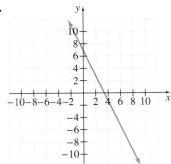

17.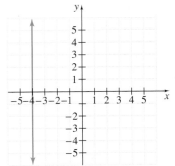

18.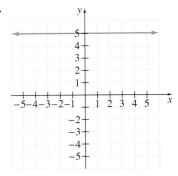

19.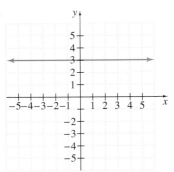

20.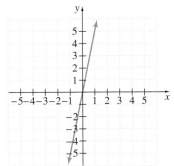

21.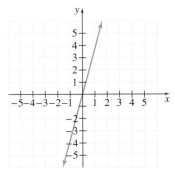

22.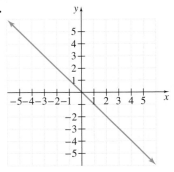

23.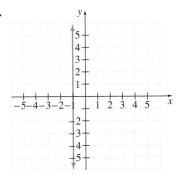

24.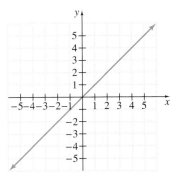

25.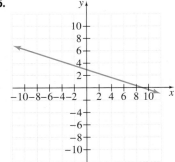

1.

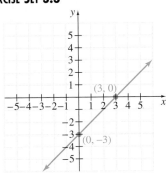

2.

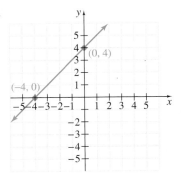

3.

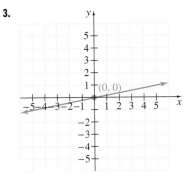

4.

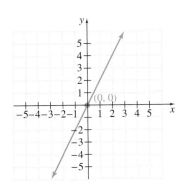

5.

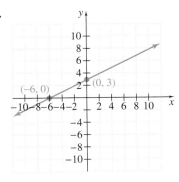

6.

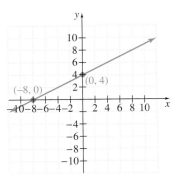

7.

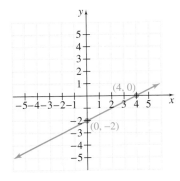

8.

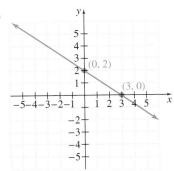

9.

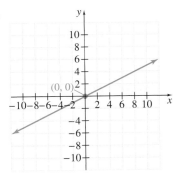

10.

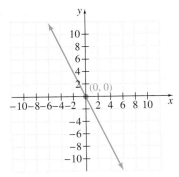

11.

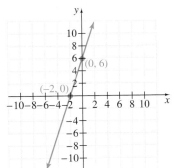

12.

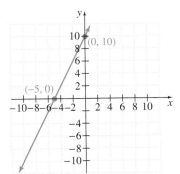

13.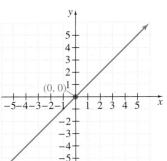

14.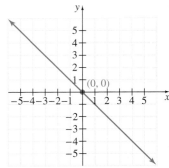

15.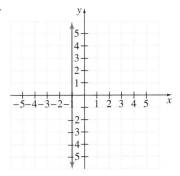

16.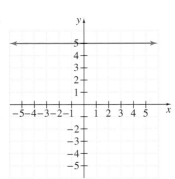

17.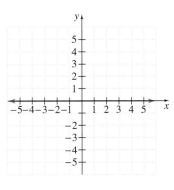

18.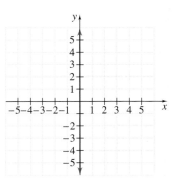

MENTAL MATH　　**1.** upward　　**2.** downward　　**3.** horizontal　　**4.** vertical

EXERCISE SET 3.4　　**1.** $m = -\frac{4}{3}$　　**2.** $m = \frac{5}{2}$　　**3.** $m = \frac{5}{2}$　　**4.** $m = -\frac{7}{4}$　　**5.** $m = \frac{8}{7}$　　**6.** $m = -5$　　**7.** $m = -1$　　**8.** $m = \frac{5}{2}$

9. $m = -\frac{1}{4}$　　**10.** $m = -5$　　**11.** $m = -\frac{2}{3}$　　**12.** $m = 5$　　**13.** $m = -2$　　**14.** $m = -2$　　**15.** $m = 5$　　**16.** $m = \frac{2}{3}$

Chapter 4　EXPONENTS AND POLYNOMIALS

MENTAL MATH　　**1.** base: 3; exponent: 2　　**2.** base: 5; exponent: 4　　**3.** base: -3; exponent: 6　　**4.** base: 3; exponent: 7
5. base: 4; exponent: 2

EXERCISE SET 4.1　　**1.** 49　　**2.** -9　　**3.** -5　　**4.** 9　　**5.** -16　　**6.** 4　　**7.** -8　　**8.** 135　　**9.** 4　　**10.** 150
11. x^7　　**12.** y^3　　**13.** x^{36}　　**14.** y^{35}　　**15.** $p^7 q^7$　　**16.** x^2　　**17.** y　　**18.** 4　　**19.** -125　　**20.** $p^6 q^5$

MENTAL MATH　　**1.** $\frac{5}{x^2}$　　**2.** $\frac{3}{x^3}$　　**3.** y^6　　**4.** x^3　　**5.** $4y^3$　　**6.** $16y^7$

EXERCISE SET 4.2　　**1.** $\frac{1}{64}$　　**2.** $\frac{1}{36}$　　**3.** $\frac{7}{x^3}$　　**4.** $\frac{1}{343x^3}$　　**5.** x^4　　**6.** y^3　　**7.** p^4　　**8.** y^6　　**9.** 7.8×10^4　　**10.** 9.3×10^9
11. 1.67×10^{-6}　　**12.** 1.7×10^{-7}　　**13.** 6.35×10^{-3}　　**14.** 1.94×10^{-3}

EXERCISE SET 4.3　　**1.** $1; -3x; 5$　　**2.** $2x^3; -1; 4$　　**3.** $-5; 3.2; 1; -5$　　**4.** $9.7; -3; 1; -\frac{1}{4}$　　**5.** (a) 6; (b) 5　　**6.** (a) -10; (b) -12

7. (a) -2; (b) 4　　**8.** (a) -4; (b) -3　　**9.** (a) -15; (b) -16　　**10.** (a) -6; (b) -1　　**11.** $23x^2$　　**12.** $14x^3$　　**13.** $12x^2 - y$
14. $3k^3 + 11$　　**15.** $7s$　　**16.** $9ab - 11a$　　**17.** $-10xy + y$　　**18.** $4x^2 - 7xy + 3y^2$　　**19.** $3a^2 - 16ab + 4b^2$　　**20.** $-3xy^2 + 4$

EXERCISE SET 4.4　　**1.** $12x + 12$　　**2.** $6x^2 + 16$　　**3.** $-3x^2 + 10$　　**4.** $4x^2 - 5$　　**5.** $-3x^2 + 4$　　**6.** $-x + 14$　　**7.** $5x^2 + 2y^2$
8. $-2x + 9$　　**9.** $y + 8$　　**10.** $2x^2 + 7x - 16$　　**11.** $5x - 9$　　**12.** $4x - 3$　　**13.** $6y + 13$　　**14.** $11y + 7$　　**15.** $-2x^2 + 8x - 1$
16. $-2a - b + 1$　　**17.** $10x - 4 + 5y$　　**18.** $3x^2 + 5$　　**19.** $9a^2 - 4b^2 + 22$　　**20.** $6x^2 - 2xy + 19y^2$

MENTAL MATH　　**1.** x^8　　**2.** x^8　　**3.** y^5　　**4.** y^{10}　　**5.** x^{14}

EXERCISE SET 4.5　　**1.** $24x^3$　　**2.** $18x^3$　　**3.** $-12.4x^{12}$　　**4.** $-15.6x^8$　　**5.** x^4　　**6.** $x^2 + 7x + 12$　　**7.** $x^2 + 11x + 18$
8. $x^3 - 5x^2 + 13x - 14$　　**9.** $x^3 + 8x^2 + 7x - 24$　　**10.** $x^4 + 5x^3 - 3x^2 - 11x + 20$　　**11.** $a^4 - a^3 - 6a^2 + 7a + 14$
12. $10a^3 - 27a^2 + 26a - 12$　　**13.** $-3b^3 - 14b^2 - 13b + 6$　　**14.** $49x^2y^2 - 14xy^2 + y^2$　　**15.** $x^4 - 8x^2 + 16$

EXERCISE SET 4.6　　**1.** $x^2 + 7x + 12$　　**2.** $x^2 + 6x + 5$　　**3.** $x^2 + 5x - 50$　　**4.** $y^2 - 8y - 48$　　**5.** $5x^2 + 4x - 12$
6. $6y^2 - 31y + 35$　　**7.** $4y^2 - 25y + 6$　　**8.** $2x^2 - 31x + 99$　　**9.** $6x^2 + 13x - 5$　　**10.** $6x^2 - 10x - 4$　　**11.** $a^2 - 49$
12. $b^2 - 9$　　**13.** $x^2 - 36$　　**14.** $x^2 - 64$　　**15.** $9x^2 - 1$　　**16.** $16x^2 - 25$　　**17.** $x^4 - 25$　　**18.** $a^4 - 36$　　**19.** $4y^4 - 1$　　**20.** $9x^4 - 1$

Exercise Set 4.7   1. $4x^3 - 2x^2 + 3x + 1$ 2. $12x^3 + 3x$ 3. $15x - 9x^4$ 4. $5p^2 + 6p$ 5. $2m - \dfrac{27m^2}{7}$ 6. $2x + 1$
7. $ab - b^2$ 8. $m^2n - n^3$

Chapter 5 Factoring Polynomials

Exercise Set 5.1   1. $3(a + 2)$ 2. $6(3a + 2)$ 3. $15(2x - 1)$ 4. $7(6x - 1)$ 5. $x^2(x + 5)$ 6. $y^4(y - 6)$
7. $2y(3y^3 - 1)$ 8. $5x^2(1 + 2x^4)$ 9. $2x(16y - 9x)$ 10. $5x(2y - 3x)$ 11. $a^2b^2(a^5b^4 - a + b^3 - 1)$
12. $x^3y^3(x^6y^3 + y^2 - x + 1)$ 13. $5xy(x^2 - 3x + 2)$ 14. $7xy(2x^2 + x - 1)$ 15. $4(2x^5 + 4x^4 - 5x^3 + 3)$
16. $3(3y^6 - 9y^4 + 6y^2 + 2)$ 17. $\dfrac{1}{3}x(x^3 + 2x^2 - 4x + 1)$ 18. $\dfrac{1}{5}y(2y^6 - 4y^4 + 3y - 2)$ 19. $(x + 2)(y + 3)$
20. $(y + 4)(z + 3)$

Exercise Set 5.2   1. $(x + 6)(x + 1)$ 2. $(x + 4)(x + 2)$ 3. $(x - 9)(x - 1)$ 4. $(x - 3)(x - 3)$ 5. $(x - 6)(x + 3)$
6. $(x - 6)(x + 5)$ 7. $(x + 10)(x - 7)$ 8. $(x + 8)(x - 4)$ 9. prime   10. prime   11. $2(z + 8)(z + 2)$
12. $3(x + 7)(x + 3)$ 13. $2x(x - 5)(x - 4)$ 14. $x(x - 8)(x + 7)$ 15. $(x - 4y)(x + y)$ 16. $(x - 11y)(x + 7y)$
17. $(x + 12)(x + 3)$ 18. $(x + 4)(x + 15)$ 19. $(x - 2)(x + 1)$ 20. $(x - 7)(x + 2)$

Exercise Set 5.3   1. $x + 4$ 2. $y + 5$ 3. $10x - 1$ 4. $3y - 2$ 5. $4x - 3$ 6. $(2x + 3)(x + 5)$ 7. $(3x + 2)(x + 2)$
8. $(y - 1)(8y - 9)$ 9. $(7x - 2)(3x - 5)$ 10. $(2x + 1)(x - 5)$ 11. $(9r - 8)(4r + 3)$ 12. $(4r - 1)(5r + 8)$
13. $(3x - 7)(x + 9)$ 14. $(5x + 1)(2x + 3)$ 15. $(2x + 5)(x + 1)$ 16. $x(3x + 2)(4x + 1)$ 17. $a(4a + 1)(2a + 3)$
18. $3(7x + 5)(x - 3)$ 19. $2(3x - 5)(2x + 1)$ 20. $(3x + 4)(4x - 3)$

Exercise Set 5.4   1. $(x + 3)(x + 2)$ 2. $(x + 3)(x + 5)$ 3. $(x - 4)(x + 7)$ 4. $(x - 6)(x + 2)$ 5. $(y + 8)(y - 2)$
6. $(z + 10)(z - 7)$ 7. $(3x + 4)(x + 4)$ 8. $(2x + 5)(x + 7)$ 9. $(8x - 5)(x - 3)$ 10. $(4x - 9)(x - 8)$

Mental Math   1. 1^2 2. 5^2 3. $(3x)^2$ 4. $(4y)^2$

Exercise Set 5.5   1. yes   2. yes   3. no   4. no   5. $(x + 11)^2$ 6. $(x + 9)^2$ 7. $(x - 8)^2$ 8. $(x - 6)^2$
9. $(4a - 3)^2$ 10. $(5x + 2)^2$ 11. $(x^2 + 2)^2$ 12. $(m^2 + 5)^2$ 13. $(x - 2)(x + 2)$ 14. $(x + 6)(x - 6)$ 15. $(9 - p)(9 + p)$
16. $(10 - t)(10 + t)$ 17. $(2r - 1)(2r + 1)$ 18. $(3t - 1)(3t + 1)$ 19. $(n^2 + 4)(n + 2)(n - 2)$ 20. $xy(x - 2y)(x + 2y)$
21. $(4 - ab)(4 + ab)$ 22. $\left(x - \dfrac{1}{2}\right)\left(x + \dfrac{1}{2}\right)$ 23. $\left(y - \dfrac{1}{4}\right)\left(y + \dfrac{1}{4}\right)$

Exercise Set 5.6   1. $x = 2, x = -1$ 2. $x = -3, x = -2$ 3. $x = 6, x = 7$ 4. $x = -4, x = 10$ 5. $x = -9, x = -17$
6. $x = 9, x = 4$ 7. $x = -9, x = 7$ 8. $x = -4, x = 2$ 9. $x = 3, x = 2$ 10. $x = 0, x = 7$ 11. $x = 0, x = 3$
12. $x = 0, x = -20$ 13. $x = 0, x = -15$ 14. $x = 4, x = -4$ 15. $x = 3, x = -3$ 16. $x = 8, x = -4$ 17. $x = 8, x = -3$
18. $x = \dfrac{7}{3}, x = -2$ 19. $x = -\dfrac{1}{4}, x = 3$ 20. $x = \dfrac{8}{3}, x = -9$

Exercise Set 5.7   1. width $= x$; length $= x + 4$ 2. width: x; length: $2x$ 3. x and $x + 2$ if x is an odd integer
4. x and $x + 2$ if x is an even integer   5. base $= x$; height $= 4x + 1$ 6. height $= x$; base $= 5x - 3$
7. 11 units   8. length $= 12$ in.; width $= 7$ in.   9. 15 cm, 13 cm, 70 cm, 22 cm   10. 14 ft, 19 ft, 52 ft

Chapter 6 Rational Expressions

Exercise Set 6.1   1. $\dfrac{1}{4(x + 2)}$ 2. $\dfrac{1}{3x + 2}$ 3. $\dfrac{1}{x + 2}$ 4. $\dfrac{1}{x - 5}$ 5. can't simplify   6. $\dfrac{3}{4}$ 7. 1   8. 1
9. -1 10. -1 11. -5 12. $\dfrac{7}{x}$ 13. $\dfrac{1}{x - 9}$ 14. $\dfrac{1}{x - 3}$ 15. $5x + 1$ 16. $6x - 1$ 17. $\dfrac{1}{x - 2}$ 18. $\dfrac{1}{x - 7}$
19. $x + 2$ 20. $4x$

Mental Math   1. $\dfrac{2x}{3y}$ 2. $\dfrac{3x}{4y}$ 3. $\dfrac{5y^2}{7x^2}$ 4. $\dfrac{4x^5}{11z^3}$ 5. $\dfrac{9}{5}$

Exercise Set 6.2   1. $\dfrac{21}{4y}$ 2. 12   3. x^4 4. $\dfrac{1}{4}$ 5. $-\dfrac{b^2}{6}$ 6. $\dfrac{x^2}{10}$ 7. $\dfrac{1}{x}$ 8. $\dfrac{1}{3}$ 9. $2x$ 10. 1   11. x^4 12. $\dfrac{9y}{2}$
13. $\dfrac{12}{y^6}$ 14. $\dfrac{14}{9b^2}$ 15. $x(x + 4)$ 16. $\dfrac{5}{6}$ 17. 4   18. $\dfrac{3x}{8}$ 19. $\dfrac{9x}{4}$ 20. $\dfrac{3}{2}$

Mental Math   1. 1   2. $\dfrac{6}{11}$ 3. $\dfrac{7x}{9}$ 4. $\dfrac{5y}{8}$ 5. $\dfrac{1}{9}$ 6. $\dfrac{17y}{5}$

Exercise Set 6.3   1. $\dfrac{a + 9}{13}$ 2. $\dfrac{x + 7}{7}$ 3. $\dfrac{3m}{n}$ 4. $7p$ 5. 4   6. 8   7. $\dfrac{y + 10}{3 + y}$ 8. 1   9. 3   10. $x - 2$
11. $\dfrac{1}{a + 5}$ 12. $\dfrac{3}{y + 5}$ 13. $\dfrac{1}{x - 6}$ 14. $\dfrac{1}{x - 1}$

Exercise Set 6.4   1. $\dfrac{5}{x}$ 2. $\dfrac{73}{21a}$ 3. $\dfrac{75a + 6b^2}{5b}$ 4. $\dfrac{20c - 8dx}{5d}$ 5. $\dfrac{6x + 5}{2x^2}$ 6. $\dfrac{3x - 7}{(x - 2)(x + 2)}$ 7. $\dfrac{35x - 6}{4x(x - 2)}$

A7

8. $\dfrac{5 + 10y - y^3}{y^2(2y + 1)}$ **9.** $-\dfrac{2}{x - 3}$ **10.** 0 **11.** 2 **12.** 3 **13.** $3x^3 - 4$ **14.** $\dfrac{5x + 45x^2}{6}$ **15.** $\dfrac{x + 2}{(x + 3)^2}$ **16.** $\dfrac{2(x + 3)}{(x - 2)^2}$

17. $\dfrac{9b - 4}{5b(b - 1)}$ **18.** $\dfrac{5(y + 2)}{3y(y + 5)}$ **19.** $\dfrac{2 + m}{m}$ **20.** $\dfrac{6 - x}{x}$

MENTAL MATH **1.** $x = 10$ **2.** $x = 32$ **3.** $z = 36$ **4.** $y = 56$

EXERCISE SET 6.5 **1.** $x = 30$ **2.** $x = 55$ **3.** $x = 0$ **4.** $x = 0$ **5.** $x = -2$ **6.** $x = -1$ **7.** $x = -5, x = 2$
8. $y = -1, y = 7$ **9.** $a = 5$ **10.** $b = 10$ **11.** $x = 3$ **12.** $y = 1$ **13.** $y = 5$ **14.** $x = -3$ **15.** $x = 9$ **16.** no solution

EXERCISE SET 6.6 **1.** 2 **2.** 4 **3.** -3 **4.** -3 or -2 **5.** 5 **6.** 3 **7.** 2 **8.** 4 **9.** $2\frac{2}{9}$ hr **10.** 2 hr **11.** $1\frac{1}{2}$ min
12. 6 min **13.** $\$108.00$ **14.** $2\frac{2}{9}$ days **15.** 3 hr **16.** $6\frac{2}{3}$ hr **17.** 20 hr **18.** first pump: 28 min; second pump: 84 min
19. 6 mph **20.** time traveled by jet: 3 hr; time traveled by car: 4 hr

EXERCISE SET 6.7 **1.** $\dfrac{2}{3}$ **2.** $-\dfrac{3}{10}$ **3.** $\dfrac{2}{3}$ **4.** $-\dfrac{27}{22}$ **5.** $-\dfrac{4x}{15}$ **6.** $-\dfrac{1}{6}$ **7.** $\dfrac{4}{3}$ **8.** $\dfrac{1}{5}$ **9.** $\dfrac{27}{16}$ **10.** $\dfrac{37}{63}$ **11.** $\dfrac{m - n}{m + n}$

12. $\dfrac{x + 4}{x - 4}$ **13.** $\dfrac{2x(x - 5)}{7x^2 + 10}$ **14.** $\dfrac{6 + 4y^2}{6y - 5y^2}$ **15.** $\dfrac{1}{y - 1}$ **16.** $2x - 1$ **17.** $\dfrac{1}{6}$ **18.** $\dfrac{56}{9}$ **19.** $\dfrac{x + y}{x - y}$ **20.** $\dfrac{3 + 40y}{3 - 40y}$

Chapter 7 GRAPHS AND FUNCTIONS

MENTAL MATH **1.** $m = -4, b = 12$ **2.** $m = \dfrac{2}{3}, b = -\dfrac{7}{2}$ **3.** $m = 5, b = 0$ **4.** $m = -1, b = 0$ **5.** $m = \dfrac{1}{2}, b = 6$

EXERCISE SET 7.1

1.
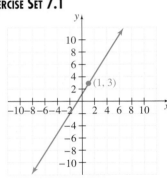

2.
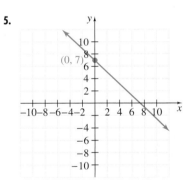

3.
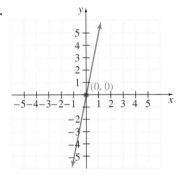

4.

5.
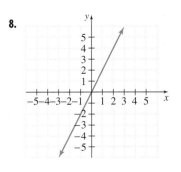

6.
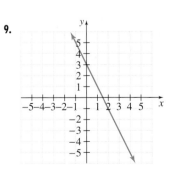

7.

8.

9.

10.

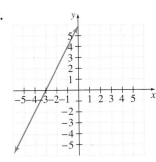

11.

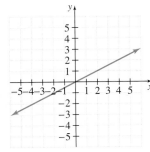

12. $y = -x + 1$ **13.** $y = \frac{1}{2}x - 6$ **14.** $y = 2x + \frac{3}{4}$ **15.** $y = -3x - \frac{1}{5}$

MENTAL MATH   **1.** $m = -2; (1, 4)$ **2.** $m = -3; (4, 6)$ **3.** $m = \frac{1}{4}; (2, 0)$ **4.** $m = -\frac{2}{3}; (0, 1)$ **5.** $m = 5; (3, -2)$

EXERCISE SET 7.2   **1.** $y = 3x - 1$ **2.** $y = 4x - 19$ **3.** $y = -2x - 1$ **4.** $y = -4x + 4$ **5.** $y = \frac{1}{2}x + 5$

6. $y = 3x - 6$ **7.** $y = 2x - 6$ **8.** $y = -2x + 1$ **9.** $y = -2x + 10$ **10.** $y = -\frac{1}{2}x - 5$ **11.** $x = 2$ **12.** $y = -4$

13. $y = 1$ **14.** $x = 4$ **15.** $x = 0$ **16.** $y = 4x - 4$ **17.** $y = 3x + 2$ **18.** $y = \frac{1}{2}x - 6$ **19.** $y = \frac{1}{4}x + 9$

20. $y = -\frac{3}{2}x - 6$

EXERCISE SET 7.4   **1.** 57   **2.** 79   **3.** 499   **4.** 13   **5.** 1   **6.** -1 **7.** 9   **8.** 47   **9.** -16 **10.** -6 **11.** $\frac{10}{3} = 3\frac{1}{3}$

12. 1.6   **13.** 0.4   **14.** -8 **15.** $-\frac{7}{3} = -2\frac{1}{3}$ **16.** 0.7   **17.** 6   **18.** 2.5   **19.** -8 **20.** -4.5

EXERCISE SET 7.5   **1. a.** domain, $(-\infty, \infty)$; range, $(-\infty, 5]$ **b.** x-intercept points, $(-2, 0), (6, 0)$; y-intercept point, $(0, 5)$ **c.** $(0, 5)$
d. There is no such point.   **2. a.** domain, $(-\infty, \infty)$; range, $[1, \infty)$ **b.** no x-intercept points; y-intercept point, $(0, 4)$
c. There is no such point.   **d.** $(-3, 1)$ **3. a.** domain, $(-\infty, \infty)$; range, $[-4, \infty)$ **b.** x-intercept points, $(-3, 0), (1, 0)$; y-intercept
point, $(0, -3)$ **c.** There is no such point.   **d.** $(-1, -4)$ **4. a.** domain, $(-\infty, \infty)$; range, $[1, \infty)$ **b.** no x-intercept points;
y-intercept point, $(0, 13)$ **c.** There is no such point.   **d.** $(2, 1)$ **5. a.** domain, $(-\infty, \infty)$; range, $(-\infty, \infty)$ **b.** x-intercept points,
$(-2, 0), (0, 0), (2, 0)$; y-intercept point, $(0, 0)$ **c.** There is no such point.   **d.** There is no such point.   **6. a.** domain, $(-\infty, \infty)$; range,
$(-\infty, \infty)$ **b.** x-intercept point, $(0, 0)$; y-intercept point, $(0, 0)$ **c.** There is no such point.   **d.** There is no such point.

7.

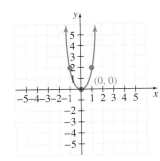

8.

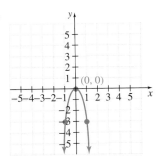

9.

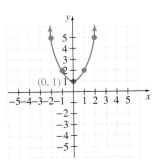

10.

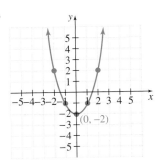

11.

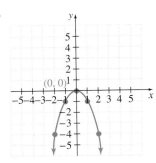

12.

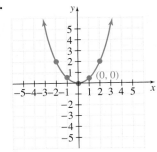

13.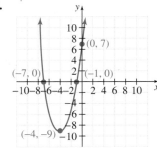

14.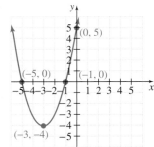

15.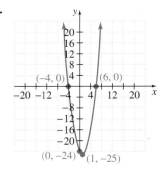

16.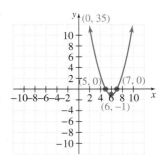

17.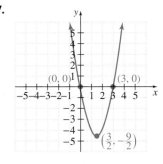

18.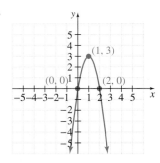

19.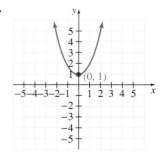

20.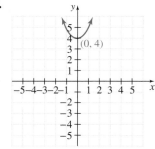

Chapter 8 Systems of Equations and Inequalities

Exercise Set 8.2   **1.** $(2,1)$ **2.** $(15,5)$ **3.** $(-3,9)$ **4.** $(-2,8)$ **5.** $(4,2)$ **6.** $(3,4)$ **7.** $(10,5)$ **8.** $(-2,-4)$

9. $(2,7)$ **10.** $(1,5)$ **11.** $\left(-\dfrac{1}{5},\dfrac{43}{5}\right)$ **12.** $\left(-\dfrac{7}{3},-\dfrac{44}{3}\right)$ **13.** $(-2,4)$ **14.** $(7,-4)$ **15.** $(-2,-1)$ **16.** no solution

17. no solution   **18.** $(-3,1)$ **19.** $(3,-1)$ **20.** $(-4,-2)$ **21.** $(3,5)$ **22.** $(1,4)$ **23.** $\left(\dfrac{2}{3},-\dfrac{1}{3}\right)$ **24.** $\left(-\dfrac{9}{5},\dfrac{3}{5}\right)$

25. $(-1,-4)$

Chapter 9 Rational Exponents, Radicals, and Complex Numbers

Exercise Set 9.1   **1.** $2,-2$ **2.** $3,-3$ **3.** 10 **4.** 20 **5.** $\dfrac{1}{2}$ **6.** $\dfrac{3}{5}$ **7.** 0.01 **8.** 0.2 **9.** 2.646 **10.** 3.317

11. 4 **12.** 3 **13.** $\dfrac{1}{2}$ **14.** $\dfrac{3}{4}$ **15.** -1 **16.** -5 **17.** -2 **18.** -3 **19.** not a real number   **20.** not a real number

21. -2 **22.** $\sqrt{3}$ **23.** -2 **24.** -1 **25.** 1 **26.** -3

Exercise Set 9.3   **1.** $\sqrt{14}$ **2.** $\sqrt{110}$ **3.** 2 **4.** 3 **5.** $\sqrt[3]{36}$ **6.** $\sqrt[3]{50}$ **7.** $\sqrt{6x}$ **8.** $\sqrt{15xy}$ **9.** $\sqrt{\dfrac{14}{xy}}$

10. $\sqrt{\dfrac{6n}{5m}}$ **11.** $\sqrt[4]{20x^3}$ **12.** $\sqrt[4]{27a^2b^3}$ **13.** $\dfrac{\sqrt{6}}{7}$ **14.** $\dfrac{2\sqrt{2}}{9}$ **15.** $\dfrac{\sqrt{2}}{7}$ **16.** $\dfrac{\sqrt{5}}{11}$ **17.** $\dfrac{\sqrt[4]{x^3}}{2}$ **18.** $\dfrac{\sqrt[4]{y}}{3x}$ **19.** $\dfrac{\sqrt[3]{4}}{3}$

20. $\dfrac{\sqrt[3]{3}}{4}$ **21.** $\dfrac{\sqrt[4]{8}}{x^2}$ **22.** $\dfrac{\sqrt[4]{a^3}}{3}$ **23.** $\dfrac{\sqrt[3]{2x}}{3y^4\sqrt[3]{3}}$ **24.** $\dfrac{\sqrt[3]{3}}{2x^2}$ **25.** $\dfrac{x\sqrt{y}}{10}$ **26.** $\dfrac{y\sqrt{z}}{6}$ **27.** $\dfrac{\sqrt{5x}}{2y}$

MENTAL MATH   **1.** $6\sqrt{3}$ **2.** $8\sqrt{7}$ **3.** $3\sqrt{x}$ **4.** $13\sqrt{y}$ **5.** $12\sqrt[3]{x}$ **6.** $6\sqrt[3]{z}$ **7.** 3 **8.** $4x+1$

EXERCISE SET 9.4   **1.** $-2\sqrt{2}$ **2.** $-2\sqrt{3}$ **3.** $10x\sqrt{2x}$ **4.** $10x\sqrt{5x}$ **5.** $17\sqrt{2}-15\sqrt{5}$ **6.** $29\sqrt{2}$ **7.** $-\sqrt[3]{2x}$

8. $-7a\sqrt[3]{3a}$ **9.** $5b\sqrt{b}$ **10.** $6x^3\sqrt{x}$ **11.** $\dfrac{31\sqrt{2}}{15}$ **12.** $\dfrac{11\sqrt{3}}{6}$ **13.** $\dfrac{\sqrt[3]{11}}{3}$ **14.** $\dfrac{3\sqrt[3]{4}}{14}$ **15.** $\dfrac{5\sqrt{5x}}{9}$

16. $\dfrac{7x\sqrt{7}}{10}$ **17.** $14+\sqrt{3}$

Chapter 10 QUADRATIC EQUATIONS AND FUNCTIONS

EXERCISE SET 10.1   **1.** $\{-4,4\}$ **2.** $\{-7,7\}$ **3.** $\{-\sqrt{7},\sqrt{7}\}$ **4.** $\{-\sqrt{11},\sqrt{11}\}$ **5.** $\{-3\sqrt{2},3\sqrt{2}\}$ **6.** $\{-2\sqrt{5},2\sqrt{5}\}$
7. $\{-\sqrt{10},\sqrt{10}\}$ **8.** $\{-\sqrt{2},\sqrt{2}\}$ **9.** $\{-8,-2\}$ **10.** $\{1,5\}$ **11.** $\{6-3\sqrt{2},6+3\sqrt{2}\}$ **12.** $\{-4-3\sqrt{3},-4+3\sqrt{3}\}$
13. $\left\{\dfrac{3-2\sqrt{2}}{2},\dfrac{3+2\sqrt{2}}{2}\right\}$ **14.** $\left\{\dfrac{-9-\sqrt{6}}{4},\dfrac{-9+\sqrt{6}}{4}\right\}$ **15.** $\{-3i,3i\}$ **16.** $\{-2i,2i\}$ **17.** $\{-\sqrt{6},\sqrt{6}\}$
18. $\{-\sqrt{10},\sqrt{10}\}$ **19.** $\{-2i\sqrt{2},2i\sqrt{2}\}$ **20.** $\{-2i\sqrt{3},2i\sqrt{3}\}$ **21.** $\{1-4i,1+4i\}$ **22.** $\{-1-5i,-2+5i\}$
23. $\{-7-\sqrt{5},-7+\sqrt{5}\}$ **24.** $\{-10-\sqrt{11},-10+\sqrt{11}\}$ **25.** $\{-3-2i\sqrt{2},-3+2i\sqrt{2}\}$

MENTAL MATH   **1.** $a=1,b=3,c=1$ **2.** $a=2,b=-5,c=-7$ **3.** $a=7,b=0,c=-4$ **4.** $a=1,b=0,c=9$

5. $a=6,b=-1,c=0$

EXERCISE SET 10.2   **1.** $\{-6,1\}$ **2.** $\{-12,1\}$ **3.** $\left\{-\dfrac{3}{5},1\right\}$ **4.** $\left\{-\dfrac{1}{5},3\right\}$ **5.** $\{3\}$ **6.** $\{-5\}$

7. $\left\{\dfrac{7-\sqrt{33}}{2},\dfrac{-7+\sqrt{33}}{2}\right\}$ **8.** $\left\{\dfrac{-5-\sqrt{13}}{2},\dfrac{-5+\sqrt{13}}{2}\right\}$ **9.** $\left\{\dfrac{1-\sqrt{57}}{8},\dfrac{1+\sqrt{57}}{8}\right\}$ **10.** $\left\{\dfrac{9-5\sqrt{5}}{22},\dfrac{9+5\sqrt{5}}{22}\right\}$

11. $\left\{\dfrac{7-\sqrt{85}}{6},\dfrac{7+\sqrt{85}}{6}\right\}$ **12.** $\left\{\dfrac{5-\sqrt{77}}{2},\dfrac{5+\sqrt{77}}{2}\right\}$ **13.** $\{1-\sqrt{3},1+\sqrt{3}\}$ **14.** $\{-3-\sqrt{7},-3+\sqrt{7}\}$

15. $\left\{-\dfrac{3}{2},1\right\}$ **16.** $\{-10,2\}$ **17.** $\left\{\dfrac{3-\sqrt{11}}{2},\dfrac{3+\sqrt{11}}{2}\right\}$ **18.** $\{1-\sqrt{2},1+\sqrt{2}\}$ **19.** $\left\{\dfrac{-5-i\sqrt{5}}{10},\dfrac{-5+i\sqrt{5}}{10}\right\}$

20. $\left\{\dfrac{-3-i\sqrt{6}}{3},\dfrac{-3+i\sqrt{6}}{3}\right\}$

Chapter 11 EXPONENTIAL AND LOGARITHMIC FUNCTIONS

EXERCISE SET 11.3

1.
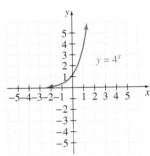

2.
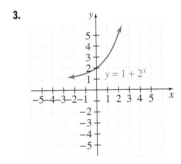

3.
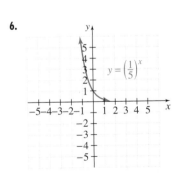

4.
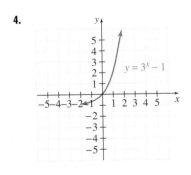

5.
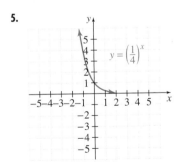

6.

7.

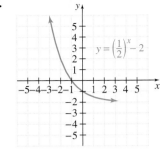

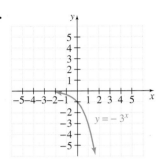

8.

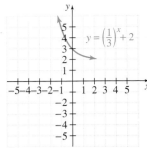

9.

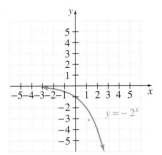

10.

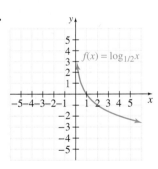

EXERCISE SET 11.4   **1.** $6^2 = 36$ **2.** $2^5 = 32$ **3.** $3^{-3} = \dfrac{1}{27}$ **4.** $5^{-2} = \dfrac{1}{25}$ **5.** $10^3 = 1000$ **6.** $\log_2 16 = 4$

7. $\log_5 125 = 3$ **8.** $\log_{10} 100 = 2$ **9.** $\log_{10} 1000 = 4$ **10.** $\log_e x = 3$ **11.** 3   **12.** 2   **13.** -2 **14.** -5 **15.** $\dfrac{1}{2}$

16. {2}   **17.** {3}   **18.** {81}   **19.** {8}   **20.** {7}

21.

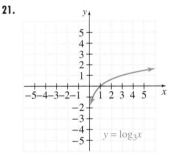

22.

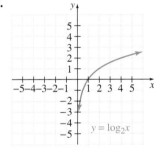

23.

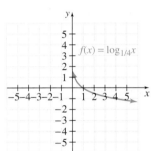

24.

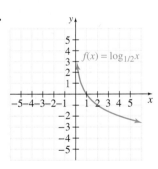

25.

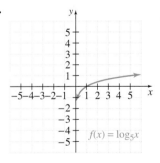

EXERCISE SET 11.6   **1.** 0.9031   **2.** 0.7782   **3.** 0.3636   **4.** 0.6866   **5.** 0.6931   **6.** 1.0986   **7.** -2.6367 **8.** -5.7446

9. 1.1004   **10.** 1.4133   **11.** 2   **12.** 4   **13.** -3 **14.** -1 **15.** 2   **16.** 4   **17.** $\dfrac{1}{4}$ **18.** $\dfrac{1}{5}$ **19.** 3   **20.** 5

Appendix A GEOMETRY

EXERCISE SET A.3   1. 64 ft   2. 28 m   3. 36 cm   4. 184 mi   5. 21 in.   6. 22 units   7. 120 cm   8. 10 yd   9. 48 ft
10. 66 m   11. 96 m   12. 86 in.   13. 66 ft   14. 58 cm   15. 128 mi   16. 90 km   17. 17π cm; 53.38 cm
18. 12π in. 37.68 in   19. 16π mi; 50.24 mi   20. 50π ft; 157 ft

EXERCISE SET A.4   1. 7 sq. m   2. 19.25 sq. ft   3. $9\frac{3}{4}$ sq. yd   4. $11\frac{1}{4}$ sq. ft   5. 15 sq. yd   6. 17.5 sq. ft

7. 2.25π sq. in. $\approx$ 7.065 sq. in.   8. 4π sq. cm $\approx 12\frac{4}{7}$ sq. cm   9. 36.75 sq. ft   10. 12.75 sq. cm   11. 28 sq. m   12. $40\frac{1}{2}$ sq. in.

13. 22 sq. yd   14. 22.5 sq. ft   15. $36\frac{3}{4}$ sq. ft   16. $12\frac{3}{4}$ sq. cm   17. $22\frac{1}{2}$ sq. in.   18. 24 sq. m   19. 25 sq. cm
20. 82 sq. km   21. 86 sq. mi   22. 360 sq. cm   23. 24 sq. cm   24. 45 sq. in.   25. 36π sq. in. $\approx$ 113.1 sq. in.

EXERCISE SET A.5   1. 72 cu. in.   2. $113\frac{1}{7}$ cu. mi   3. 512 cu. cm   4. 128 cu. cm   5. $12\frac{4}{7}$ cu. yd   6. $471\frac{3}{7}$ cu. ft

7. $523\frac{17}{21}$ cu. in.   8. $28\frac{7}{8}$ cu. in.   9. 75 cu. cm   10. 18 cu. ft   11. $2\frac{10}{27}$ cu. in.   12. 3080 cu. ft   13. 8.4 cu. ft

14. 125 cu. ft   15. $10\frac{5}{6}$ cu. in.

EXERCISE SET A.6   1. 13 in.   2. 34.655 ft   3. 6.633 cm   4. 8.485 yd   5. 5   6. 15   7. 8   8. 22.472
9. 17.205   10. 37.216   11. 16.125   12. 45   13. 12   14. 108   15. 44.822   16. 33.541   17. 42.426   18. 171.825
19. 1.732   20. 3.606

Index

- Browse through the table of contents and the index. Is the topic covered in enough depth to be helpful?

- Is there a list of references that look as if the author has consulted other sources and may lead you to useful related material?

- Are you the intended audience? Consider the tone, style, level of information, and assumptions the author makes about the reader. Are they appropriate to your needs?

- Is the content of the source fact, opinion, or propaganda? If the material is presented as factual, are the sources of the facts clearly indicated? Do you think there's enough evidence offered? Is the coverage comprehensive? (As you learn more about the topic, you will notice that this gets easier as you become more of an expert.) Is the language emotional or objective?

- Are there broad, sweeping generalizations that overstate or simplify the matter?

- Does the author use a mix of primary and secondary sources?

- To determine accuracy, consider whether the source is outdated. Do some cross-checking. Do you find some of the same information elsewhere?

- Are there arguments that are one-sided with no acknowledgment of other viewpoints?

38   USING SOURCES

Integrating sources

As you write your paper, you may be incorporating summaries, paraphrases, and quotations from your sources. It is important to weave these in smoothly, to distinguish for the reader whether you are summarizing, paraphrasing, or quoting, and to signal your reader by using signal words (page 000) to introduce and smoothly blend these sources into your writing.

Summaries

Characteristics of summaries:

- Are written in your own words
- Include only the main points, omitting details, facts. examples, illustrations, direct quotations, and other specifics
- Use fewer words than the source
- Do not have to be in the same order as the source
- Are objective and do not include your own interpretation.

Use summaries if the source has unnecessary detail. the writing is not particularly memorable or worth quoting, or you want to keep your writing concise.

WRITING A SUMMARY

A summary is a brief statement of the main idea in a source, using your own words. Include a citation to the source.

2.   *Paraphrases*

Characteristics of paraphrases:

- Have approximately the same number of words as the source

- Include all main points and important details in the source
- Use your own words, not those of the source
- Keep the same organization as the source
- Are more detailed than a summary
- Are objective and do not include your interpretation.

WRITING A PARAPHRASE

A paraphrase restates the information fro a source, using your own words. Include a citation to the source.

3.   *Quotations*

Guidelines for using quotations:

- Use quotations as evidence, support, or further explanation of what you have written. Quotations are not substitutes for stating your point in your own words.

- Before you include the quotation, indicate what point the quotation is making and why that quotation is relevant.

- Use quotations sparingly Too many quotations strung together with very little of your own writing make a paper look like a scrapbook of pasted-together sources, not a thoughtful integration of what is known about a subject.

- Use quotations that illustrate an authority's viewpoint or style or that would not be as effective if rewritten in different words.

- Introduce quotations with signal words (see next page for examples).

For guidelines on punctuation quotations, see section 22.

USING QUOTATIONS

A quotation is the record of the exact words of a written or spoken source and is set off by quotation marks. All quotations should have citation to the source.

Using signal words

As you include summaries, paraphrases, and quotations in your papers, you need to integrate them smoothly so that there is no sudden jump or break between the flow of your words and the source material. Use the following strategies to prepare your readers and to create that needed smooth transition into the inserted material.

Use signal phrases

Signal words or phrases let the reader know a quotation will follow. Choose a phrase or word that is appropriate to the quotation and indicates the relationship to the ideas being discussed.

In 1990 when the United Nations International Human Rights Commission predicted "there will be an outburst of major violations of human rights m Yugoslavia within the next few years" (14), few people in Europe or the United States paid attention to the warning.

(UNITED NATIONS INTERNATIONAL HUMAN RIGHTS COMMISSION. *The Future of Human Rights in Eastern Europe.* New York: United Nations, 1990.)

Explain the connection

Always explain the connection between a quotation you use and the point you are making. Show the logical link, or add a follow-up comment that integrates the quotation into your paragraph.

Quotation not integrated into the paragraph

Modern farming techniques are different from those used twenty years ago. John Hession, an Iowa soybean grower, says, "Without a computer program to plan my crop allotments or to record my expenses, I'd be back in the dark ages of guessing what to do." New computer software program are being developed commercially and are selling well.

(HESSION, JOHN. PERSONAL INTERVIEW. 27 JULY 1998.)

SOME COMMON SIGNAL WORDS

according to	considers	observes
acknowledges	denies	points out
admits	describes	predicts
argues	disagrees	proposes
asserts	emphasizes	rejects
comments	explains	reports
complains	finds	responds
concedes	insists	suggests
concludes	maintains	thinks
condemns	notes	warns

The quotation here is abruptly dropped into the paragraph, without an introduction and without a clear indication from the writer how Mr. Hession's statement fits into the ideas being discussed.

Quotation revised

Modern farming techniques differ from those of twenty years ago, *particularly in the use of computer programs for planning and budgeting.* John Hession, an Iowa soybean grower *who relies heavily on computers, confirms this* when he notes, "Without a computer program to plan my crop allotments or to record my expenses, I'd be back in the dark ages of guessing what to do." Commercial software programs *such as those used by Mr. Hession, for crop allotments and budgeting,* are being developed and are selling well.

The added words in italics explain how Mr. Hession's statement confirms the point being made.

PLAGIARISM

Plagiarism results when writers fail to document a source so that the words and ideas of someone else are presented as the writer's own work.

Avoiding Plagiarism

What information needs to be documented?

When we use the ideas, findings, data, and conclusions, arguments, and words of others. we need to acknowledge that we are borrowing their work and inserting in our own by documenting. Consciously or unconsciously passing off the work of others as our own results in the form of stealing known as plagiarism, an act that has serious consequences for the writer who plagiarizes.

If you summarize, paraphrase, or use the words of someone else (see pages 116–117), provide documentation for those sources.

CITING SOURCES

In some cultures, educated writers are expected to know and incorporate the thinking of great scholars. It may be considered an insult to the reader to mention the names of the scholars, implying that the reader is not educated enough to recognize the references. However, in American writing this is not the case, and writers are always expected to acknowledge their sources and give public credit to the source.

What information does not need to be documented?

Common knowledge, that body of general ideas we share with our readers, does not have to be documented. Common knowledge consists of the following:

- Standard information on a subject that your readers know
- Information that is widely shared and can be found in numerous sources.

AVOIDING PLAGIARISM

To avoid plagiarism, read over your paper and ask yourself whether your readers can properly identify which ideas and words are yours and which are from the sources you cite. If that is clear, then you are not plagiarizing.

For example, it is common knowledge among most Americans aware of current energy problems that solar power is one answer to future energy needs. But forecasts about how widely solar power may be used in twenty years or estimates of the cost effectiveness of using solar energy would be the work of some person or group studying the subject, and documentation would be needed.

Field research you conduct also does not need to be documented, although you should indicate you are reporting your own findings.

Original version

Researchers studying human aggression are discovering that, in contrast to the usual stereotypes, patterns of aggression among girls and women under circumstances may mirror or even exaggerate those seen in boys and men. And while women's weapons are often words, fists may be used too.

ABIGAL ZUGER, "A FISTFUL OF HOSTILITY IS FOUND IN WOMEN." *NEW YORK TIMES* 28 Jul. 1998, B9.

Plagiarized version

Women can be as hostile as men. According to Abigail Zuger, researchers studying human aggression are discovering that, in contrast to the usual stereotypes, patterns of aggression among girls and women under some circumstances may mirror or even exaggerate those seen in boys and men. And while women's weapons are often words, fists may be used too (B9).

Revision

An acceptable revision of the plagiarized version would have either a paraphrase or summary in the writer's own words or would indicate, with quotation marks, the exact words of the source. (See page 821.)

GLOSSARY OF USAGE

This list includes words and phrases you may be uncertain about when writing. If you have questions about words not included here, try the index at the back of this book to see whether the word is discussed elsewhere. You can also check a recently published dictionary.

A, An Use *a* before words beginning with a consonant and before words beginning with a vowel that sounds like a consonant:

a cat   a house   a one-way street   a union   a history

Use *an* before words that begin with a vowel and before words with a silent *h*.

an egg   an ice cube   an hour   an honor

Accept, Except *Accept*, a verb, means to agree to, to believe, or to receive.

The detective **accepted** his account of the event.

Except, a verb, means to exclude or leave out, and *except*, a preposition, means leaving out.

Because he did not know the answers, he was **excepted** from the list of contestants and asked to leave.

Except for brussel sprouts, I eat most vegetables.

Advice, Advise *Advice* is a noun, and *advise* is a verb.

She always offers too much **advice**.

Would you **advise** me about choosing the right course?

Affect, Effect Most frequently, *affect*, which means to influence, is used as a verb, and *effect*, which means a result, is used as a noun.

The weather **affects** my ability to study.

What **effect** does coffee have on your concentration?

However, *effect*, meaning to cause or bring about, is also used as a verb.

The new traffic enforcement laws **effected** a change in people's driving habits.

Common phrases with *effect* include the following:

in effect   to that effect

Ain't This is a nonstandard way of saying *am not, is not, has not, have not*, and so on.

All Ready, Already *All ready* means *prepared; already* means *before* or *by this time*.

The courses for the meal are **all ready** to be served.

When I got home, she was **already** there.

All Right, Alright *All right* is two words, not one. *Alright* is an incorrect form.

All Together, Altogether *All together* means *in a group*, and *altogether* means *entirely, totally*.

We were **all together** again after having separate vacations.

He was not **altogether** happy about the outcome of the test.

Alot, A Lot *Alot* is an incorrect form of *a lot*.

a.m., p.m. (or) A.M., P.M. Use these with numbers, not as substitutes for the words *morning* or *evening*.

       morning at 9 a.m.
We meet every ~~a.m.~~, for an exercise class.

Among, Between Use *among* when referring to three or more things and *between* when referring to two things.

The decision was discussed **among** all the members of the committee.

I had to decide **between** the chocolate mousse pie and the almond ice cream.

Amount, Number Use *amount* for things or ideas that are general or abstract and cannot be counted. For example, furniture is a general term and cannot be counted. That is, we cannot say *one furniture* or *two furnitures*. Use number for things that can be counted (for example, *four chairs* or *three tables*).

He had a huge **amount** of work to finish before the deadline.

There were a **number** of people who saw the accident.

An See the entry for *a, an*.

And Although some people discourage the use of *and* as the first word in a sentence, it is an acceptable word with which to begin a sentence.

And Etc. Adding *and* is redundant because *et* means *and* in Latin. See the entry for **etc.**

Anybody, Any Body See the entry for **anyone, any one**.

Anyone, Any One *Anyone* means *any person at all*. *Any one* refers to a specific person or thing in a group. There are similar distinctions for other words ending in *-body* and *-one* (for example, *everybody, every body, anybody, any body, someone*, and *some one*).

The teacher asked if **anyone** knew the answer.

Any one of those children could have taken the ball.

Anyways, Anywheres These are nonstandard forms for *anyway* and *anywhere*.

As, As if, As Though, Like Use as in a comparison (not *like*, when there is an equality intended or when the meaning is *in the function of*.

Celia acted **as** [not *like*] the leader when the group was getting organized.   (Celia = leader)

Use *as if* or *as though* for the subjunctive.

He spent his money **as if** [or **as though**] he were rich.

Use *like* in a comparison (not *as*) when the meaning is *in the manner of* or *to the same degree as*.

The boy swam **like** a fish.

Don't use *like* as the opening word in a clause in formal writing:

Informal: **Like** I thought, he was unable to predict the weather.

Formal: **As** I thought, he was unable to predict the weather.

Assure, Ensure, Insure *Assure* means *to declare or promise, ensure* means *to make safe or certain*, and *insure* means *to protect with a contract of insurance*.

I **assure** you that I am trying to find your lost package.

Some people claim that eating properly **ensures** good health.

This insurance policy also **insures** my car against theft.

Awful, Awfully *Awful* is an adjective meaning *inspiring awe* or *extremely unpleasant.*

He was involved in an **awful** accident.

Awfully is an adverb used in very informal writing to mean *very.* Avoid it in formal writing.

Informal: The dog was **awfully** dirty

Awhile, A While *Awhile* is an adverb meaning a *short time* and modifies a verb:

He talked **awhile** and then left

A while is an article with the noun *while* and means *a period of time:*

I'll be there in **a while.**

Bad, Badly *Bad* is an adjective and is used after linking verbs. *Badly* is an adverb. (See section 15.)

The wheat crop looked **bad** [not *badly*] because of lack of rain.

There was a **bad** flood last summer.

The building was **badly** constructed and unable to with-stand the strong winds.

Beside, Besides *Beside* is a preposition meaning *at the side of, compared with,* or *having nothing to do with. Besides* is a preposition meaning *in addition to* or *other than. Besides* as an adverb which means *also* or *moreover.* Don't confuse *beside* with *besides.*

That is **beside** the point.

Besides the radio, they had no other means of contact with the outside world.

Besides, I enjoyed the concert.

Between, Among See the entry for **among, between.**

Breath, Breathe *Breath is* a noun, and *breathe* is a verb.

She held her **breath** when she dived into the water.

Learn to **breathe** deeply when you swim.

But Although some people discourage the use of *but* as the first word in a sentence, it is an acceptable word with which to begin a sentence.

Can, May *Can* is a verb that expresses *ability, knowledge,* or *capacity:*

He **can** play both the violin and the cello.

May is a verb that expresses possibility or permission. Careful writers avoid using can to mean permission:

May [not *can*] I sit here?

Can't Hardly This is incorrect because it is a double negative.

                can
She ~~can't~~ hardly hear normal voice levels.
      ^

Choose, Chose *Choose* is the present tense of the verb, and *chose* is the past tense:

Jennie should **choose** strawberry ice cream.

Yesterday, she **chose** strawberry-flavored popcorn.

Cite, Site *Cite* is a verb that means *to quote an authority or source; site* is a noun referring to a place.

Be sure to **cite** your sources in the paper.

That is the **site** of the new city swimming pool.

Cloth, Clothe *Cloth* is a noun, and *clothe* is a verb.

Here is some **cloth** for a new scarf.

His paycheck helps to feed and **clothe** many people in his family.

Compared to, Compared with Use *compared to* when showing that two things are alike. Use *compared with* when showing similarities and differences.

The speaker **compared** the economy **to** a roller coaster because both have sudden ups and downs.

The detective **compared** the fingerprints **with** other sets from a previous crime.

Could of This is incorrect. Instead use *could have.*

Data This is the plural form of *datum.* In informal usage data is used as a singular noun, with a singular verb. However, because dictionaries do not accept this, use data as a plural form for academic writing.

Informal Usage: The **data** is inconclusive.

Formal Usage: The **data** are inconclusive.

Different from, Different than *Different from* is always cor-rect, but some writers use *different than* if a clause follows this phrase.

This program is **different from** the others.

That is a **different** result **than** they predicted.

Done The past tense forms of the verb *do* are *did* and *done. Did* is the simple form that needs no additional verb as a helper. *Done* is the past form that requires the helper *have.* Some writers make the mistake of interchanging *did* and *done.*

                did   *have*
They ~~done~~ it again.   (or)   They done it again.
         ^   ^

Effect, Affect See the entry for **affect, effect.**

Ensure See the entry for **assure, ensure, insure.**

Etc. This is an abbreviation of the Latin *et cetera,* meaning *and the rest.* Because it should be used sparingly if at all in for-mal academic writing, substitute other phrases such as *and so forth* or *and so on.*

Everybody, Every Body See the entry for **anyone, any one.**

Everyone, Every One See the entry for **anyone, any one.**

Except, Accept See the entry for **accept, except.**

Farther, Further Although some writers use these words interchangeably, dictionary definitions differentiate them. *Far-ther* is used when actual distance is involved, and *further* is used to mean *to a greater extent, more.*

The house is **farther** from the road than I realized.

That was **furthest** from my thoughts at the time.

Fewer, Less *Fewer* is used for things that can be counted *(fewer trees, fewer people),* and *less* is used for ideas, abstrac-tions, things that are thought of collectively, not separately *(less trouble, less furniture),* and things that are measured by amount, not number *(less milk, less fuel).*

Fun This noun is used informally as an adjective.

Informal: They had a **fun** time.

Goes, Says *Goes* is a nonstandard form of *says.*

                                                    says
Whenever I give him a book to read, he ~~goes~~ "What's it about?"
                                                       ^

Gone, Went Past tense forms of the verb *go*. *Went* is the simple form that needs no additional verb as a helper. *Gone* is the past form that requires the helper *have*. Some writers make the mistake of interchanging *went* and *gone*. (See section 13.)

> went (or) have gone
> They ~~gone~~ away yesterday.

Good, Well *Good* is an adjective and therefore describes only nouns. *Well* is an adverb and therefore describes adjectives, other adverbs, and verbs. The word *well* is used as an adjective only in the sense of *in good health*. (See section 15.)

> well   well
> The stereo works ~~good~~. I feel ~~good~~.

> She is a **good** driver.

Got, Have *Got* is the past tense of *get* and should not be used in place of *have*. Similarly, *got to* should not be used as a substitute for *must*. *Have got to* is an informal substitute for *must*.

> have
> Do you ~~got~~ any pennies for the meter?

> must
> I ~~got to~~ go now.

> **Informal:**   You have **got to** see that movie.

Great This adjective is overworked in its formal meaning of *very enjoyable, good,* or *wonderful* and should be reserved for its more exact meanings such as of *remarkable ability, intense, high degree of,* and so on.

> **Informal:**   That was a **great** movie.

> **More exact uses of** *great:*

> > The vaccine was a **great** discovery.

> > The map went into **great** detail.

Have, Got See the entry for **got, have.**

Have, Of *Have*, not *of*, should follow verbs such as *could, might, must,* and *should*.

> have
> They should ~~of~~ called by now.

Hisself This is a nonstandard substitute for *himself*.

Hopefully This adverb mean *in a hopeful way*. Many people consider the meaning *it is to be hoped* as unacceptable.

> **Acceptable:**   He listened **hopefully** for the knock at the door.

> **Often considered**   **Hopefully,** it will not rain tonight.
> **unacceptable:**

I Although some people discourage the use of *I* in formal essays, it is acceptable. If you wish to eliminate the use of *I*, see section 7 on passive verbs.

Imply, Infer Some writers use these interchangeably, but careful writers maintain the distinction between the two words. *Imply* means to *suggest without stating directly to hint*. *Infer* means to reach an opinion from facts or reasoning.

> The tone of her voice **implied** he was stupid.

> The anthropologist **inferred** this was a burial site for prehistoric people.

Insure See the entry for **assure, ensure, insure.**

Irregardless This is an incorrect form of the word *regardless*.

Is When, Is Why, Is Where, Is Because These are incorrect forms for definitions. See section 6 and the Glossary of Grammatical Terms on faulty predication.

> **Faulty predication:**   Nervousness is when my palms sweat.

> **Revised:**   When I am nervous, my palms sweat.

> > (or)

> > Nervousness is a state of being very uneasy or agitated.

Its, It's *Its* is a personal pronoun in the possessive case. *It's* is a contraction for *it is*.

> The kitten licked **its** paw.

> **It's** a good time for a vacation.

Kind, Sort These two forms are singular and should be used with *this* or *that*. Use *kinds* or *sorts* with *these* or *those*.

> This **kind** of cloud indicates heavy rain.

> These **sorts** of plants are regarded as weeds.

Lay, Lie *Lay* is a verb that needs an object and should not be used in place of *lie*, a verb that takes no direct object. (See section 13.)

> lie
> He should ~~lay~~ down and rest awhile.

> lay
> You can ~~lie~~ that package on the front table.

Leave, Let *Leave* means *to go away*, and *let* means to *permit*. It is incorrect to use *leave* when you mean *let*:

> Let
> ~~Leave~~ me get that for you.

Less, Fewer See the entry for **fewer, less.**

Let, Leave See the entry for **leave, let.**

Like, As See the entry for **as, as if, like.**

Like for The phrase "I'd like *for* you to do that" is incorrect. Omit *for*.

May, Can See the entry for **can, may.**

Most It is incorrect to use most as a substitute for *almost*.

Nowheres This is an incorrect form of *nowhere*.

Number, Amount See the entry for **amount, number.**

Of, Have See the entry for **have, of.**

Off of It is incorrect to write *off of* for *off* in a phrase such as *off the table*.

O.K., Ok, Okay These can be used informally but should not be used in formal or academic writing.

Reason . . . Because This is redundant. Instead of *because*, use *that*.

> that
> The reason she dropped the course is ~~because~~ she couldn't keep up with the homework.

Less wordy revision: She dropped the course **because** she couldn't keep up with the homework.

Reason Why Using *why* is redundant. Drop the word *why*.

> The reason why I called is to remind you of your promise.

Saw, Seen Past tense forms of the verb *see*. *Saw* is the simple form that needs no additional verb as a helper. *Seen* is the past form that requires the helper *have*. Some writers make the mistake of interchanging *saw* and *seen*. (See section 13.)

> They ~~seen~~ *saw* it happen.   (or)   They ~~seen~~ *have* it happen.

Set, Sit *Set* means to place and is followed by a direct object. *Sit* means to be seated. It is incorrect to substitute *set* for *sit*.

> Come in and ~~set~~ *sit* down.

> ~~Sit~~ *Set* the flowers on the table.

Should of This is incorrect. Instead use *should have*.

Sit, Set See the entry for **set, sit.**

Site, Cite See the entry for **cite, site.**

Somebody, Some Body See the entry for **anyone, any one.**

Someone, Some One See the entry for **anyone, any one.**

Sort, Kind See the entry for **kind, sort.**

Such This is an overworked word when used in place of *very* or *extremely*.

Suppose to, Use to These are nonstandard forms for *supposed to* and *used to*.

Sure The use of *sure* as an adverb is informal. Careful writers use *surely* instead.

> **Informal:**   I **sure** hope you can join us.
> **Revised:**   I **surely** hope you can join us.

Than, Then *Than* is a conjunction introducing the second element in comparison. *Then* is an adverb meaning *at that time, next, after that, also,* or *in that case.*

> She is taller **than** I am.

> He picked up the ticket and **then** left the house.

That There, This Here, These Here, Those There These are incorrect forms for *that, this, these, those.*

That, Which Use *that* for essential clauses and *which* for nonessential clauses. Some writers, however, also use *which* for essential clauses. (See section 19.)

Their, There, They're *Their* is a possessive pronoun; *there* means *in, at,* or *to that place;* and *they're* is a contraction for they are.

> **Their** house has been sold.

> **There** is the parking lot.

> **They're** both good swimmers.

Theirself, Theirselves, Themself These are all incorrect forms for *themselves.*

Them It is incorrect to use this in place of either the pronoun *these* or *those.*

> Look at ~~them~~ *those* apples.

Then, Than See the entry for **than, then.**

Thusly This is an incorrect substitute for *thus.*

To, Too, Two *To* is a preposition; *Too* is an adverb meaning *very* or *also;* and *two* is a number.

> He brought his bass guitar **to** the party.
> He brought his drums **too**.
> He had **two** music stands.

Toward, Towards Both are accepted forms with the same meaning although *toward* is preferred in American usage.

Use to This is incorrect for the modal meaning *formerly.* Instead, use *used to.*

Use to, Suppose to See the entry for **suppose to, use to.**

Want for Omit the incorrect *for* in phrases such as "I want *for* you to come here."

Well, Good See the entry for **good, well.**

Went, Gone See the entry for **gone, went.**

Where It is incorrect to use *where* to mean *when* or *that.*

> The Fourth of July is a holiday ~~where~~ *when* the town council shoots off fireworks.

> I see ~~where~~ *that* there is now a ban on shooting panthers.\

Where . . . at This is a redundant form. Omit *at.*

> This is where the picnic is at.

Which, That See the entry for **that, which.**

While, Awhile See the entry for **awhile, a while.**

Who, Whom Use *who* for the subject case; use *whom* for the object case.

> He is the person **who** signs that form.
> He is the person **whom** I asked for help.

Who's, Whose *Who's* is a contraction for *who is, whose* is a possessive pronoun.

> **Who's** included on that list?
> **Whose** wristwatch is this?

Your, You're *Your* is a possessive pronoun; *you're* is a contraction for *you are.*

> **Your** hands are cold.
> **You're** a great success.

GLOSSARY OF GRAMMATICAL TERMS

Absolutes Words or phrases that modify whole sentences rather than parts of sentences or individual words. Am absolute phrase, which consists of a noun and participle, can be placed anywhere in the sentence but needs to be set off from the sentence by commas.

The snow having finally stopped, the football
(*absolute phrase*)
game began.

Abstract Nouns Nouns that refer to ideas, qualities, generalized concepts, and conditions and do not have plural forms. (See section 31.)

happiness, pride, furniture, trouble, sincerity

Active Voice See **Voice.**

Adjectives Words that modify nouns and pronouns. (See section 15.)

Descriptive adjectives (*red, clean, beautiful, offensive,* for example) have three forms:

Positive:	red, clean, beautiful, offensive
Comparative (for comparing two things):	cleaner, more beautiful, less offensive
Superlative (for comparing more than two things):	cleanest, most beautiful, least offensive

Adjective Clauses See **Dependent Clauses.**

Adverbs Modify verbs, verb forms, adjectives, and other adverbs. (See section 15.) Descriptive adverbs (for example, *fast, graceful, awkward*) have three forms:

Positive:	fast, graceful, awkward
Comparative (for comparing) two things):	faster, more graceful, less awkward
Superlative (for comparing more than two things):	fastest, most graceful, least awkward

Adverb Clauses See **Dependent Clauses.**

Agreement The use of the corresponding form for related words in order to have them agree in number, person, or gender. (See sections 13 and 14.)

John runs. (Both subject and verb are singular.)

It is necessary to flush the **pipes** regularly so that they don't freeze.

(Both subjects, *it* and *they,* are in third person; *they* agrees in number with the antecedent, *pipes.*)

Antecedents Words or groups of words to which pronouns refer.

When the **bell** was rung, **it** sounded very loudly.

(*Bell* is the antecedent of *it.*)

Antonyms Words with opposite meanings.

Word	Antonym
hot	cold
fast	slow
noisy	quiet

Appositives Nonessential phrases and clauses that follow nouns and identify or explain them. (See section 19.)

My uncle, **who lives in Wyoming,** is taking windsurfing
(*appositive*)
lessons in Florida.

Articles See noun determiners and section 32.

Auxiliary Verbs Verbs used with main verbs in verb phrases.

should be going	**has** taken
(*auxiliary verb*)	(*auxiliary verb*)

Cardinal Numbers See **Noun Determiners.**

Case The form or position of a noun or pronoun that shows its use or relationship to other words in a sentence. The three cases in English are (1) subject (or subjective or nominative), (2) object (or objective), and (3) possessive (or genitive). (See section 14.)

Clauses Groups of related words that contain both subjects and predicates and function either as sentences or as parts of sentences. Clauses are either independent (or main) or dependent (or subordinate). (See section 11.)

Collective Nouns Nouns that refer to groups of people or things, such as a *committee, team,* or *jury.* When the group includes a number of members acting as a unit and is the subject of the sentence, the verb is also singular, (See section 13.)

The **jury** has made a decision.

Comma Splices Punctuation errors in which two or more independent clauses in compound sentences are separated only by commas and no coordinating conjunctions. (See section 12.)

but (or) ;
Jessie said he could not help, that was typical of his responses to requests.

Common Nouns Nouns that refer to general rather than specific categories of people, places, and things and are not capitalized. (See section 24.)

basket, person, history, tractor

Comparative The form of adjectives and adverbs used when two things are being compared. (See section 15.)

higher, more intelligent, less friendly

Complement When linking verbs link subjects to adjectives or nouns, the adjectives or nouns are complements.

Phyllis was **tired.**
(*complement*)

She became a **musician.**
(*complement*)

Complex Sentences Sentences with at least one independent clause and at least one dependent clause arranged in any order.

Compound Nouns Words such as swimming pool, dropout, roommate, and stepmother, in which more than one word is needed.

Compound Sentences Sentences with two or more independent clauses and no dependent clauses. (See section 12.)

Compound-Complex Sentences Sentences with at least two independent clauses and at least one dependent clause arranged in any order.

Conjunctions Words that connect other words, phrases, and clauses in sentences. Coordinating conjunctions connect independent clauses; subordinating conjunctions connect dependent or subordinating clauses with independent or main clauses.

Coordinating Conjunctions:	and, but, for, or, nor, so, yet
Some Subordinating Conjunctions:	after, although, because, if, since, until, while

Conjunctive Adverbs Words that begin or join independent clauses. (See section 19.)

consequently, however, therefore, thus, moreover

Connotation The attitudes and emotional overtones beyond the direct definition of a word.

> The words *plump* and *fat* both mean fleshy, but *plump* has a more positive connotation than *fat*.

Consistency Maintaining the same voice with pronouns, the same tense with verbs, and the same tone, voice, or mode of discourse. (See section 17.)

Coordinating Conjunctions See **Conjunctions**.

Coordination Of equal importance. Two independent clauses in the same sentence are coordinate because they have equal importance and the same emphasis.

Correlative Conjunctions Words that work in pairs and give emphasis.

> both . . . and   neither . . . nor   either . . . or
> not . . . but also

Dangling Modifiers Phrases or clauses in which the doer of the action is not clearly indicated. (See section 16.)

> *Tim thought*
> Missing an opportunity to study, the exam seemed especially difficult.
>   ^

Declarative Mood See **Mood**.

Demonstrative Pronouns Pronouns that refer to things. (See Noun Determiners.)

> this, that, these, those

Denotation The explicit dictionary definition of a word, as opposed to the connotation of a word. (See **Connotation**.)

Dependent Clauses (Subordinate Clauses) Clauses that cannot stand alone as complete sentences. (See section 11.) There are two kinds of dependent clauses: adverb clauses and adjective clauses.

> **Adverb clauses:**   Begin with subordinating conjunctions such as *after, if, because, while, when.*
>
> **Adjective clauses:** Tell more about nouns or pronouns in sentences and begin with words such as *who, which, that, whose, whom.*

Determiner See **Noun Determiner**.

Diagrams See **Sentence Diagrams**.

Direct Discourse See **Mode of Discourse**.

Direct/Indirect Quotations Direct quotations are the exact words said by someone or the exact words in print that are being copied. Indirect quotations are not the exact words but the rephrasing or summarizing of someone else's words. (See section 22.)

Direct Objects Nouns or pronouns that follow a transitive verb and complete the meaning or receive the action of the verb. The direct object answers the question *what?* or *whom?*

Ellipsis A series of three dots to indicate that words or parts of sentences are being omitted from material being quoted. (See section 23.)

Essential and Nonessential Clauses and Phrases *Essential* (also called *restrictive*) clauses and phrases appear after nouns and are necessary or essential to complete the meaning of the sentence. *Nonessential* (also called *nonrestrictive*) clauses and phrases appear after nouns and add extra informa-

tion, but that information can be removed from the sentence without altering the meaning. (See section 19.)

> Apples **that are green** are not sweet.
>   (essential clause)
>
> Golden Delicious apples, **which are yellow,** are sweet.
>   (nonessential clause)

Excessive Coordination Occurs when too many equal clauses are strung together with coordinators into one sentence.

Excessive Subordination Occurs when too many subordinate clauses are strung together in a complex sentence.

Faulty Coordination Occurs when two clauses that are either unequal in importance or that have little or no connection to each other are combined in one sentence and written as independent clauses.

Faulty Parallelism See **Parallel Construction**.

Faulty Predication Occurs when a predicate does not appropriately fit the subject. This happens most often after forms of the *to be* verb. (See section 6.)

> *He*
> ~~The reason he~~ was late ~~was~~ because he had to study.
>   ^

Fragments Groups of words punctuated as sentences that either do not have both a subject and a complete verb or that are dependent clauses. (See section 11.)

> Whenever we wanted to pick fresh fruit while we were
>   *, we would head for the orchard with buckets*
> staying on my grandmother's farm.
>   ^

Fused Sentences Punctuation errors (also called *run-ons*) in which there is no punctuation between independent clauses in the sentence. (See section 12.)

> Jennifer never learned how to ask politely she just took
> what she wanted.   ^

Gerunds Verbal forms ending in *-ing* that function as nouns. (See **Phrases** and **Verbals**.)

> Arnon enjoys **cooking.**
>   (gerund)
>
> **Jogging** is another of his pastimes.
> (gerund)

Homonyms Words that sound alike but are spelled differently and have different meanings. (See section 28.)

> hear/here   passed/past   buy/by

Idioms Expressions meaning something beyond the simple definition or literal translation into another language. For example, idioms such as "short and sweet" or "wearing his heart on his sleeve" are expressions in English that cannot be translated literally into another language. (See section 35.)

Imperative Mood See **Mood**.

Indefinite Pronouns Pronouns that make indefinite reference to nouns.

> anyone, everyone, nobody, something

Independent Clauses Clauses that can stand alone as complete sentences because they do not depend on other clauses to complete their meanings. (See section 11.)

Indirect Discourse See **Mode of Discourse.**

Indirect Objects Words that follow transitive verbs and come before direct objects. They indicate the one to whom or for whom something is given, said, or done and answer the questions *to what?* or *to whom?* Indirect objects can always be paraphrased by a prepositional phrase beginning with *to* or *for.*

> Alice gave **me** some money.
> (indirect object)

> **Paraphrase:**   Alice gave some money to me.

Infinitives Phrases made up of the present form of the verb preceded by *to.* Infinitives can have subjects, objects, complements, or modifiers. (See section 16.)

> Everyone wanted **to swim** in the new pool.
> (infinitive)

Intensifiers Modifying words used for emphasis.

> She **most certainly** did fix that car!
> (intensifiers)

Interjections Words used as exclamations.

> **Oh,** I don't think I want to know about that.
> (interjection)

Interrogative Pronouns Pronouns used in questions.

> who, whose, whom, which, that

Irregular Verbs Verbs in which the past tense forms and/or the past participles are not formed by adding *-ed* or *-d.* (See section 13.)

> do, did, done   begin, began, begun

Jargon Words and phrases that are either the specialized language of various fields or, in a negative sense, unnecessarily technical or inflated terms. (See section 5.)

Intransitive Verbs See **Verbs.**

Linking Verbs Verbs linking the subject to the subject complement. The most common linking verbs are *appear, seem, become, feel, look, taste, sound,* and *be.*

> I **feel** sleepy.   He **became** the president.
> (linking verb)   (linking verb)

Misplaced Modifiers Modifiers not placed next to or close to the word(s) being modified. (See section 16.)

> on television
> We saw an advertisement for an excellent new stereo
> ^
> system with dual headphones ~~on television~~.

Modal Verbs Helping verbs such as *shall, should, will, would, can, could, may, might, must, ought to,* and *used to* that express an attitude such as interest, possibility, or obligation. (See section 30.)

Mode of Discourse Direct discourse repeats the exact words that someone says, and indirect discourse reports the words but changes some of the words.

> Everett said, **"I want to become a physicist."**
> (direct discourse)

> Everett said **that he wants to become a physicist.**
> (indirect discourse)

Modifiers Words or groups of words that describe or limit other words, phrases, and clauses. The most common modifiers are adjectives and adverbs. (See section 16.)

Mood Verbs indicate whether a sentence expresses a fact (the declarative or indicative mood); expresses some doubt or something contrary to fact or states a recommendation (the subjunctive mood); or issues a command (the imperative mood).

Nonessential Clauses and Phrases See **Essential and Nonessential Clauses and Phrases.**

Nonrestrictive Clauses and Phrases See **Essential and Nonessential Clauses and Phrases.**

Nouns Words that name people, places, things, and ideas and have plural or possessive endings. Nouns function as subjects, direct objects, predicate nominatives, objects of prepositions, and indirect objects.

Noun Clauses Subordinate clauses used as nouns.

> **What I see here** is adequate.
> (noun clause)

Noun Determiners Words that signal a noun is about to follow. They stand next to their nouns or can be separated by adjectives. Some noun determiners can also function as nouns. There are five types of noun determiners:

1. Articles: definite: the, indefinite: a, an
2. Demonstratives: this, that, these, those
3. Possessives: my, our, your, his, her, its, their
4. Cardinal numbers: one, two, three, and so on
5. Miscellaneous: all, another, each, every, much, and others

Noun Phrases See **Phrases.**

Number The quantity expressed by a noun or pronoun, either singular (one) or plural (more than one).

Objects See **Direct Objects** and **Object Complements.**

Object Complements The adjectives in predicates modifying the object of the verb (not the subject).

> The enlargement makes the picture **clear.**
> (object complement)

Object of the Preposition Noun following the preposition. The preposition, its object, and any modifiers make up the prepositional phrase.

> For **Daniel**
> (object of the preposition for)

> She knocked twice **on the big wooden door.**
> (prepositional phrase)

Objective Case of Pronouns The case needed when the pronoun is the direct or indirect object of the verb or the object of a preposition.

Singular	Plural
First person: me	First person: us
Second person: you	Second person: you
Third person: him, her, it	Third person: them

Parallel Construction When two or more items are listed or compared, they must be in the same grammatical form as equal elements. When items are not in the same grammatical form, they lack parallel structure (often called *faulty parallelism*). (See section 8.)

She was sure that **being an apprentice in a photographer's studio** would be more useful than **being a student in photography classes.**

(The phrases in bold type are parallel because they have the same grammatical form.)

Parenthetical Elements Nonessential words, phrases, and clauses set off by commas, dashes, or parentheses.

Participles Verb forms that may be part of the complete verb or function as adjectives or adverbs. The present participle ends in *-ing*, and the past participle usually ends in *-ed, -d, -n,* or *-t.* (See **Phrases.**)

> **Present participles:** running, sleeping, digging
>
> She is **running** for mayor in this campaign.
> ↗ (present participle)
>
> **Past participles:** walked, deleted, chosen
>
> The **elected** candidate will take office in January.
> ↗ (past participle)

Parts of Speech The eight classes into which words are grouped according to their function, place, meaning, and use in a sentence: nouns, pronouns, verbs, adjectives, adverbs, propositions, conjunctions, and interjections.

Passive Voice See **Voice.**

Participle See **Participles.**

Perfect Progressive Tense See **Verb Tenses.**

Person There are three "persons" in English.

First person:	the person(s) speaking I or we
Second person:	the person(s) spoken to you
Third person:	the person(s) spoken about he, she, it, they, anyone, etc.

Personal Pronouns Refer to people or things.

	Subject	Object	Possessive
Singular			
First person	I	me	my, mine
Second person	you	you	your, yours
Third person	he, she, it	him, her, it	his, her, hers, its

	Subject	Object	Possessive
Plural			
First person	we	us	our, ours
Second person	you	you	your, yours
Third person	they	them	their, theirs

Phrases Groups of related words without subjects and predicates. Verb phrases function as verbs.

> She **has been eating** too much sugar.
> ↗ (verb phrase)

Noun phrases function as nouns.

> A **major winter storm** hit **the eastern coast of Maine.**
> ↗ (noun phrase)   ↗ (noun phrase)

Prepositional phrases usually function as modifiers.

> That book **of hers** is overdue at the library.
> ↗ (prepositional phrase)

Participial phrases, gerund phrases, infinitive phrases, appositive phrases, and absolute phrases function as adjectives, adverbs, or nouns.

Participial Phrase:	I saw people **staring at my peculiar-looking haircut.**
Gerund Phrase:	**Making copies of videotapes** can be illegal.
Infinitive Phrase:	He likes **to give expensive presents.**
Appositive Phrase:	You ought to see Dr. Elman, **a dermatologist.**
Absolute Phrase:	**The test done,** he sighed with relief.

Possessive Pronouns See **Personal Pronouns, Noun Determiners,** and section 14.

Predicate Adjectives See **Subject Complements.**

Predicate Nominatives See **Subject Complements.**

Predication Words or groups of words that express action or state of beginning in a sentence and consist of one or more verbs, plus any complements or modifiers.

Prefixes Word parts added to the beginning of words.

Prefix	Word
bio- (life)	biography
mis- (wrong, bad)	misspell

Prepositions Link and relate their objects (usually nouns or pronouns) to some other word or words in a sentence. Prepositions usually precede their objects but may follow the objects and appear at the end of the sentence.

> The waiter gave the check **to my date** by mistake.
> ↗ (prepositional phrase)
>
> I wonder **what** she is asking **for.**
> ↗ (object of   ↗ (preposition)
> the preposition)

Prepositional Phrases See **Phrases.**

Progressive Tenses See **Verb Tenses.**

Pronouns Words that substitute for nouns. (See section 14.) Pronouns should refer to previously stated nouns, called antecedents.

> When **Josh** came in, **he** brought some firewood.
> ↗ (antecedent) ↗ (pronoun)

Forms of pronouns: personal, possessive, reflexive, interrogative, demonstrative, indefinite, and relative.

Pronoun Case Refers to the form of the pronoun that is needed in a sentence. See **Subject, Object,** and **Possessive Cases** and section 14.

Proper Nouns Refer to specific people, places, and things. Proper nouns are always capitalized. (See section 24.)

> Copenhagen   Honda   House of Representatives
> Spanish

Reflexive Pronouns Pronouns that show someone or something in the sentence is acting for itself or on itself. Because a reflexive pronoun must refer to a word in a sentence, it is not the subject or direct object. If used to show emphasis, reflexive pronouns are called *intensive pronouns.* (See section 14.)

Singular	Plural
First person: myself	First person: ourselves
Second person: yourself	Second person: yourselves

Third person: himself,   Third person: themselves
herself, itself

She returned the book **herself** rather than giving it to her
           ↗ (reflexive pronoun)
roommate to bring back.

Relative Pronouns Pronouns that show the relationship of a
dependent clause to a noun in the sentence. Relative pronouns
substitute for nouns already mentioned in sentences and introduce adjective or noun clauses.

Relative pronouns: that, which, who, whom, whose

This was the movie **that** won the Academy Award.

Restrictive Clauses and Phrases See **Essential and
Nonessential Clauses and Phrases.**

Run-on Sentences See **fused sentences** and section 12.

Sentences Groups of words that have at least one independent clause (a complete unit of thought with a subject and
predicate). Sentences can be classified by their structure as
simple, compound, complex, and compound-complex.

Simple:	one independent clause
Compound:	two or more independent clauses
Complex:	one or more independent clauses and one or more dependent clauses
Compound-complex:	two or more independent clauses and one or more dependent clauses

Sentences can also be classified by their function as declarative, interrogative, imperative, and exclamatory.

Declarative:	makes a statement
Interrogative:	asks a question
Imperative:	issues a command
Exclamatory:	makes an exclamation

Sentence Diagrams A method of showing relationships
within a sentence.

Mamie's **cousin**, who has no taste in food, **ordered a
hamburger** with coleslaw at the Chinese restaurant.

Sentence Fragment See **Fragment.**

Simple Sentence See **Sentence.**

Simple Tenses See **Verb Tenses.**

Split Infinitives Phrases in which modifiers are inserted
between *to* and the verb. Some people object to split infinitives, but others consider them grammatically acceptable.

to quickly turn   to easily reach   to forcefully enter

Subject The word or words in a sentence that act or are acted
upon by the verb or are linked by the verb to another word or

words in the sentence. The *simple subject* includes only the
noun or other main word or words, and the complete subject
includes all the modifiers with the subject.

Harvey objected to his roommate's alarm going off at
9 A.M.
(Harvey is the subject.)

Every single one of the people in the room heard her
giggle.
(The simple subject is one; the complete subject is the
whole phrase.)

Subject Complement The noun or adjective in the predicate
(predicate noun or adjective) that refers to the same entity as
the subject in sentences with linking verbs, such as *is/are, feel,
took, small, sound, taste,* and *seem.*

She feels **happy**.   He is a **pharmacist**.
    ↗ (subject complement) ↗ (subject complement)

Subject Case of Pronouns See **Personal Pronouns** and section 14.

Subjunctive Mood See **Mood.**

Subordinating Conjunctions Words such as *although, if,
until,* and *when,* that join two clauses and subordinate one to
the other.

She is late. She overslept.

She is late **because** she overslept.

Subordination The act of placing one clause in a subordinate
or dependent relationship to another in a sentence because it is
less important and is dependent for its meaning on the other
clause.

Suffix Word part added to the end of a word.

Suffix	Word
-ful	careful
-less	nameless

Superlative Forms of Adjectives and Adverbs See **Adjectives and Adverbs** and section 15.

Synonyms Words with similar meanings.

Word	Synonym
damp	moist
pretty	attractive

Tense See **Verb Tense.**

Tone The attitude or level of formality reflected in the word
choices in a piece of writing. (See section 5.)

Transitions Words in sentences that show relationships
between sentences and paragraphs. (See section 9.)

Transitive Verbs See **Verbs.**

Verbals Words that are derived from verbs but do not act as
verbs in sentences. Three types of verbals are infinitives, participles, and gerunds.

Infinitives	to + **verb**
to wind	to say
Participles:	Words used as modifiers or with helping verbs. The present participle ends in -*ing*, and many past participles end in -*ed*.

The dog is **panting**.   He bought only **used** clothing.
   ↗ (present participle)   ↗ (past participle)

Gerunds: Present participles used as nouns.

Smiling was not a natural act for her.
↗ (gerund)

Verbs Words or groups of words (verb phrases) in predicates that express action, show a state of being, or act as a link between the subject and the rest of the predicate. Verbs change form to show time (tense), mood, and voice and are classified as transitive, intransitive, and linking verbs. (See section 30.)

Transitive verbs: Require objects to complete the predicate.

He **cut** the cardboard **box** with his knife.
↗ (transitive verb) ↗ (object)

Intransitive verbs: Do not require objects.

My ancient cat often **lies** on the porch.
↗ (intransitive verb)

Linking verbs: Link the subject to the following noun or adjective.

The trees **are** bare.
↗ (linking verb)

Verb Conjugations The forms of verbs in various tenses. (See section 30.)

Regular:

Present

Simple present:

I walk	we walk
you walk	you walk
he, she, it walks	they walk

Present progressive:

I am walking	we are walking
you are walking	you are walking
he, she, it is walking	they are walking

Present perfect:

I have walked	we have walked
you have walked	you have walked
he, she, it has walked	they have walked

Present perfect progressive:

I have been walking	we have been walking
you have been walking	you have been walking
he, she, it has been walking	they have been walking

Past

Simple past:

I walked	we walked
you walked	you walked
he, she, it walked	they walked

Past progressive:

I was walking	we were walking
you were walking	you were walking
he, she, it was walking	they were walking

Past perfect:

I had walked	we had walked
you had walked	you had walked
he, she, it had walked	they had walked

Past perfect progressive:

I had been walking	we had been walking
you had been walking	you had been walking
he, she, it had been walking	they had been walking

Future

Simple future:

I shall walk	we shall walk
you will walk	you will walk
he, she, it will walk	they will walk

Future progressive:

I shall be walking	we shall be walking
you will be walking	you will be walking
he, she, it will be walking	they will be walking

Future perfect:

I shall have walked	we shall have walked
you will have walked	you will have walked
he, she, it will have walked	they will have walked

Future perfect progressive:

I shall have been walking	we shall have been walking
you will have been walking	you will have been walking
he, she, it will have been walking	they will have been walking

Irregular:

Present

Simple present:

I go	we go
you go	you go
he, she, it goes	they go

Present progressive:

I am going	we are going
you are going	you are going
he, she, it is going	they are going

Present perfect:

I have gone	we have gone
you have gone	you have gone
he, she, it has gone	they have gone

Present perfect progressive:

I have been going	we have been going
you have been going	you have been going
he, she, it has been going	they have been going

Past

Simple past:

I went	we went
you went	you went
he, she, it went	they went

Past progressive:

I was going	we were going
you were going	you were going
he, she, it was going	they were going

Past perfect:

I had gone	we had gone
you had gone	you had gone

he, she, it had gone   they had gone

Past perfect progressive:

I had been going   we had been going

you had been going   you had been going

he, she, it had been going   they had been going

Future

Simple:

I shall go   we shall go

you will go   you win go

he, she, it will go   they will go

Future progressive:

I shall be going   we shall be going

you will be going   you will be going

he, she, it will be going   they will be going

Future perfect:

I shall have gone   we shall have gone

you will have gone   you will have gone

he, she, it will have gone   they will have gone

Future perfect progressive:

I shall have been going   we shall have been going

you will have been going   you will have been going

he, she, it will have been going   they will have been going

Verb Phrases See **Verbs.**

Verb Tenses The times indicated by the verb forms in the past, present, or future. (For the verb forms, see **verb conjugations** and section 30.)

Present

Simple present: Describes actions or situations that exist now and are habitually or generally true.

I **walk** to class every afternoon.

Present progressive: Indicates activity in progress, something not finished, or something continuing.

He **is studying** Swedish.

Present perfect: Describes single or repeated actions that began in the past and lead up to and include the present.

She **has lived** in Alaska for two years.

Present perfect progressive: Indicates action that began in the past, continues to the present, and may continue into the future.

They **have been building** that garage for six months.

Past

Simple past: Describes completed actions or conditions in the past.

They **ate** breakfast in the cafeteria.

Past progressive: Indicates past action that took place over a period of time.

He **was swimming** when the storm began.

Past perfect: Indicates an action or event was completed before another event in the past.

No one **had heard** about the crisis when the newscast began.

Past perfect progressive: Indicates an ongoing condition in the past that has ended.

I **had been planning** my trip to Mexico when I heard about the earthquake.

Future

Simple future: Indicates actions or events in the future.

The store **will open** at 9 A.M.

Future progressive: Indicates future action that will continue for some time.

I **will be working** on that project next week.

Future perfect: Indicates action that will be completed by or before a specified time in the future.

Next summer, they **will have been** here for twenty years.

Future perfect progressive: Indicates ongoing actions or conditions until a specific time in the future.

By tomorrow, I **will have been waiting** for the delivery for one month.

Voice Verbs are either in the active or passive voice. In the active voice, the subject performs the action of the verb. In the passive, the subject receives the action. (See section 7.)

The dog **bit** the boy.
    ↗ (active verb)

The boy **was bitten** by the dog.
    ↗ (passive verb)